Environmental GEOLOGY

James S. Reichard

Georgia Southern University

McGraw Hill

Connect
Learn
Succeed™

ENVIRONMENTAL GEOLOGY

Published by McGraw-Hill, a business unit of The McGraw-Hill Companies, Inc., 1221 Avenue of the Americas, New York, NY 10020. Copyright © 2011 by The McGraw-Hill Companies, Inc. All rights reserved. No part of this publication may be reproduced or distributed in any form or by any means, or stored in a database or retrieval system, without the prior written consent of The McGraw-Hill Companies, Inc., including, but not limited to, in any network or other electronic storage or transmission, or broadcast for distance learning.

Some ancillaries, including electronic and print components, may not be available to customers outside the United States.

This book is printed on acid-free paper.

1 2 3 4 5 6 7 8 9 0 RJE/RJE 1 0 9 8 7 6 5 4 3 2 1 0

ISBN 978–0–07–304680–8
MHID 0–07–304680–9

Vice President, Editor-in-Chief: *Marty Lange*
Vice President, EDP: *Kimberly Meriwether David*
Director of Development: *Kristine Tibbetts*
Publisher: *Ryan Blankenship*
Executive Editor: *Margaret J. Kemp*
Senior Developmental Editor: *Joan M. Weber*
Senior Marketing Manager: *Lisa Nicks*
Senior Project Manager: *Gloria G. Schiesl*
Senior Production Supervisor: *Sherry L. Kane*
Lead Media Project Manager: *Judi David*
Senior Designer: *Laurie B. Janssen*
(USE) Cover Image: © *Steve Allen Gettyimages*
Senior Photo Research Coordinator: *Lori Hancock*
Compositor and Art Studio: *Electronic Publishing Services Inc., NYC*
Typeface: *10/12 ITC Slimbach Std.*
Printer: *R. R. Donnelley*

All credits appearing on page or at the end of the book are considered to be an extension of the copyright page.

Library of Congress Cataloging-in-Publication Data

Reichard, James S.
  Environmental Geology / James S. Reichard. -- 1st ed.
      p. cm.
  Includes Index.
  ISBN 978–0–07–304680–8 — ISBN 0–07–304680–9 (hard copy : alk. paper) 1. Environmental geology--Textbooks. I. Title.
  QE38.R45 2011
  550--dc22
                    2009034025

www.mhhe.com

This book is dedicated to my wife Linda and children, Brett and Kristen. Linda has worked tirelessly and selflessly in support of this major undertaking, and Brett and Kristen have lost precious time with their father, who spent countless hours away from home. Words cannot express my love and gratitude.

—James Reichard

Jim Reichard and his family on the Highline Trail in Glacier National Park, Montana. From left to right: Jim, son Brett, daughter Kristen, and wife Linda.

Brief Contents

Contents

Chapter 8

Streams and Flooding 223

Chapter 9

Coastal Hazards 257

PART THREE
Earth Resources

Chapter 10

Soil Resources 292

Chapter 11

Water Resources 325

Chapter 12

Mineral and Rock Resources   359

Chapter 13

Conventional Fossil Fuel Resources   397

Chapter 14

Alternative Energy Resources   435

PART FOUR
The Health of Our Environment

Chapter 15

Pollution and Waste Disposal   473

Chapter 16

Global Climate Change   509

Preface

Environmental Geology is a new textbook that focuses on the fascinating interaction between humans and the geologic processes that shape Earth's environment. Because this text emphasizes how human survival is highly dependent on the natural environment, students should find the topics to be quite relevant to their own lives, and therefore, more interesting. One of the key themes of this textbook centers on a serious challenge facing modern society: the need to continue obtaining large quantities of energy, and at the same time, move from fossil fuels to clean sources of energy that do not impact the climate system. *Environmental Geology* provides extensive coverage of the problems associated with our conventional fossil fuel supplies (Chapter 13), and an equally in-depth discussion on alternative energy sources (Chapter 14). The two chapters on energy are then intimately linked to a comprehensive overview of global climate change (Chapter 16), which is arguably civilization's most critical environmental challenge.

Another major theme in *Environmental Geology* is that humans are an integral part of a complex and interactive system scientists call the Earth system. Throughout the text the author explains how the Earth system responds to human activity, which then affects the very environment in which we live. A key point is that our activity often produces unintended and undesirable consequences. An example from the text is how engineers have built dams and artificial levees to control flooding on the Mississippi River. This has caused unintended changes in the geologic environment. For thousands of years, the rate at which the river deposited sediment in the Mississippi Delta was approximately equal to the rate that the sediment compacted under its own weight. The land surface remained above sea level because the two rates had been similar. However, by using dams and artificial levees to confine the Mississippi River to its channel, humans had disrupted the delicate balance between sediment deposition and compaction. Today large sections of the Louisiana coast, including New Orleans, are sinking below sea level. This has not only caused severe coastal erosion, but has greatly increased the chance that New Orleans will be inundated during a major hurricane.

"My overall impression after reading Chapter 16 – Global Climate Change was that of an excellent coverage of a still very controversial topic. Reichard has managed to cover the most fundamental societal and scientific issues related to global climate change in a format accessible to undergraduate students with or without strong science background. Reichard provides an unbiased representation of facts and does not shy away from a critical discussion of opposing arguments resulting from the interpretation of the facts."
—Thomas Boving, *University of Rhode Island*

Environmental Geology includes a sufficient amount of background material on physical geology for students who have never taken a geology course. The author believes this additional coverage is critical. Without a basic understanding of physical science, students would be unable to fully appreciate the interrelationships between humans and the geologic environment. To meet the needs of courses with a physical geology prerequisite, the book was organized so instructors could easily omit the few chapters that contain mostly background material. In addition, *Environmental Geology* does more than provide a physical description of water, mineral, and energy resources; it explores the difficult problems associated with extracting the enormous quantities of resources needed to sustain modern societies. With respect to geologic hazards (e.g., earthquakes, volcanic eruptions, and floods), the textbook goes beyond the physical science and examines the societal impacts along with the ways humans can minimize the risks. The author also highlights the fact that as population continues to grow, the problems related to resource depletion and hazards will become more severe.

Most importantly, this textbook includes distinct learning tools that will maximize use of this text. For example, it is unreasonable to expect students to remember everything they read. For this reason, the text utilizes numerous cross-references between chapters as a reminder that information on certain topics can be found elsewhere. It is hoped that cross-referencing will encourage students to make better use of the index for locating additional information.

> *"... I give the author credit for excelling in a very up-to-date assessment of alternative technologies, with some delightful examples of innovative systems that should interest the student reader. The author recognizes the importance of portraying the subject within the modern world that the student lives in—for example, the repeated references to the iPod [MP3 player] are a nice touch that will surely resonate with the student reader."*
> —Lee Slater, *Rutgers University—Newark*

Key Features

As with all college textbooks, there are differences among the various environmental geology books currently being offered. Some of the more significant and noticeable differences you will find in *Environmental Geology* are:

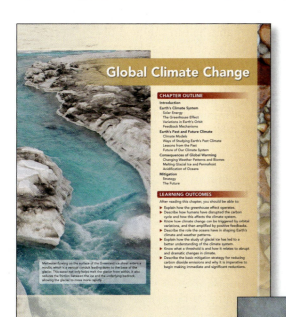

- **Learning Outcomes:** Each chapter is introduced with a list that provides valuable student guidance by stating key chapter concepts. This encourages students to be "active" learners as they complete the tasks and activities that require them to use critical thinking skills.
- **Chapter 2 Is Unique.** "Earth from a Larger Perspective" describes Earth's relationship to the solar system and universe, which helps give students the broadest possible perspective of our environment. Here students learn how the Earth system is part of even larger systems before moving on to the remaining chapters that focus on our planet. Chapter 2 also gives instructors the opportunity to discuss some of the external forces that influence Earth's environment, such as solar radiation, asteroid impacts, and the effect of the Moon on our tides and climate. In addition, this chapter helps explain why Earth supports a diverse array of complex life, and why humans are so dependent on its unique and fragile environment. This sets the stage for a theme that is woven throughout the entire text, namely that human survival is intimately linked to the environment. Students can then see how being better stewards of the Earth is in our own best interest.

- **Case Studies.** Nearly every chapter highlights a case study that is designed to give students a more in-depth look at an environmental issue. A good example is Chapter 7, where the case study examines the recurring mass wasting problems at La Conchita, California. Here students are asked to consider why some people willingly live in a hazardous area, even when the risk is well understood. In Chapter 13, the case study explores the controversy over expanding domestic oil production by drilling in the Arctic National Wildlife Refuge (ANWR). Students are given an objective overview of both sides of the issue, and are then expected to draw their own conclusion as to which side of the policy debate they would support.

- **Photos and Illustrations.** It is well established in the field of education that most people are predominantly visual learners. Therefore, the author integrated very specific photos and illustrations within the narrative so that abstract and complex concepts are easier to understand. The integrated use of visual examples within a narrative writing style should not only help increase student comprehension, but it should also encourage students to read more of the text.

FIGURE 9.26 Rip currents (A) form when backwash from the surf zone funnels through a break in underwater sand bars. Photo (B) showing a rip current flowing back out to sea through the surf zone in the Monterey Bay area of California. Note that the rip current can be recognized by how it disrupts breaking waves within the surf zone.

FIGURE 14.3 Synthetic crude oil is currently being produced from tar sand deposits (A) in Alberta, Canada. The hydrocarbons are in the form of bitumen, a highly viscous substance that is separated from the sand using steam. Photo (B) shows a bitumen sample whose viscosity has been lowered by heating. Approximately 60% of Canadian production involves strip-mining (C); the remainder is produced by steam injection and pumping wells.

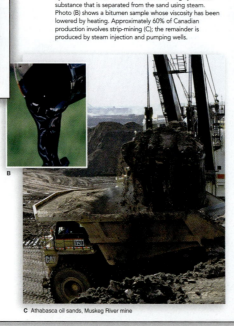

C Athabasca oil sands, Muskeg River mine

FIGURE 9.17 Storm surge (A) not only inundates areas normally above high tide, but also brings breaking waves that demolish structures. Photo (B) of Bolivar peninsula near Galveston Bay, Texas, showing the effects of the storm surge after Hurricane Ike made landfall in 2008. Arrows mark features that appear in both images. Notice how the majority of homes were destroyed.

- **Summary Points.** Each chapter concludes with a list of Summary Points to provide students with a list of important concepts that should be reviewed in preparation for exams.

- **Key Words.** The study of geologic processes can be daunting owing to the proliferation of unfamiliar terms. Each chapter includes a list of important terms and their page references, so that terms can be viewed within the context of their use. Complete definitions are also located in the Glossary at the back of the text.

- **Applications.** At the end of each chapter, there are sections called **"Student Activity,"** and **"Critical Thinking Questions,"** and **"Your Environment: YOU Decide,"** designed to encourage students to think about how their own lifestyles may be playing a role in environmental issues. For example, in Chapter 12 (Mineral and Rock Resources) they are asked to think about the social implications of buying a diamond that comes from a part of the world where illegal proceeds support violent uprisings and civil war. In Chapter 15 (Pollution and Waste Disposal), students are asked to contact their local government to determine the location of the landfill where their trash is being sent. They are then asked to investigate whether the landfill has any reported pollution problems, and if so, to describe what impacts the landfill might be having on local residents.

- **Laboratory Manual.** Twelve comprehensive laboratory exercises are available on the text website and include a list of materials needed, questions for students to complete, and corresponding answer keys on the instructor resource website.

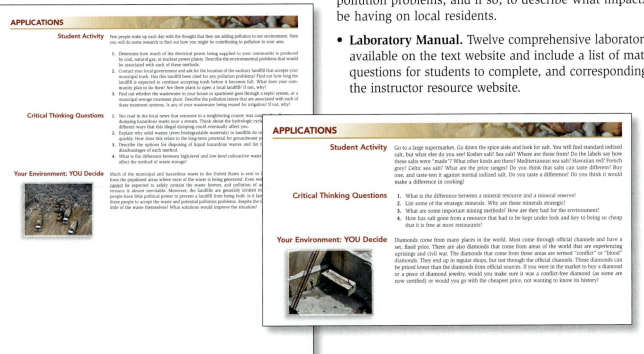

Organization

In most environmental geology courses the list of topics includes some combination of geologic hazards and resources along with waste disposal and pollution. Consequently, this book is conveniently organized so instructors can pick and choose the chapters that coincide with their particular course objectives. The chapters are organized as follows:

Part One   Fundamentals of Environmental Geology
Chapter 1 Humans and the Geologic Environment
Chapter 2 Earth from a Larger Perspective
Chapter 3 Earth Materials
Chapter 4 Earth's Structure and Plate Tectonics

Part Two   Hazardous Earth Processes
Chapter 5 Earthquakes and Related Hazards
Chapter 6 Volcanoes and Related Hazards
Chapter 7 Mass Wasting and Related Hazards
Chapter 8 Streams and Flooding
Chapter 9 Coastal Hazards

Part Three   Earth Resources
Chapter 10 Soil Resources
Chapter 11 Water Resources
Chapter 12 Mineral and Rock Resources
Chapter 13 Conventional Fossil Fuel Resources
Chapter 14 Alternative Energy Resources

Part Four   The Health of Our Environment
Chapter 15 Pollution and Waste Disposal
Chapter 16 Global Climate Change

"I found the chapter [16] to overall be very well written, very interesting, and logically organized. I am especially impressed by the thorough summary the author provides on the Earth's climate system."
—John C. White, *Eastern Kentucky University*

Supplements

McGraw-Hill offers various tools and technology products to support this text. Students can order supplemental study materials by contacting their local bookstore or by calling 800-262-4729. Instructors can obtain teaching aids by calling the Customer Service Department at 800-338-3987, visiting the McGraw-Hill website at www.mhhe.com, or contacting their McGraw-Hill sales representative.

Instructor Resources

Website: **www.mhhe.com/reichard1e**

Instructor's Manual

Answer Key to Laboratory Exercises

Presentation Center

Digital image files are available by chapter as either PowerPoint or jpg files. All line drawings, photos, maps, and tables are included here. The Presentation Center is a powerful tool that allows you to search and download illustrations from numerous McGraw-Hill science titles.

Computerized Test Bank Online

Test questions were prepared to assure that assessment directly correlates to *Environmental Geology*. The comprehensive bank of test questions is provided within a computerized test bank powered by McGraw-Hill's flexible electronic testing program, EZ Test Online (www.eztestonline.com). EZ Test Online allows you to create paper and online tests or quizzes in this easy-to-use program.

Imagine being able to create and access your test or quiz anywhere, at any time, without installing the testing software. Now, with EZ Test Online, instructors can select questions from multiple McGraw-Hill test banks or author their own, and then either print the test for paper distribution or give it online.

Test Creation
- Author/edit questions online using the 14 different question type templates.
- Create printed tests or deliver online to get instant scoring and feedback.
- Create question pools to offer multiple versions online—great for practice.
- Export your tests for use in WebCT, Blackboard, PageOut, and Apple's iQuiz.
- Compatible with EZ Test Desktop tests you've already created.
- Share tests easily with colleagues, adjuncts, and TAs.

Online Test Management
- Set availability dates and time limits for your quiz or test.
- Control how your test will be presented.
- Assign points by question or question type with a drop-down menu.
- Provide immediate feedback to students or delay until all finish the test.
- Create practice tests online to enable student mastery.
- Upload your roster to enable student self-registration.

Online Scoring and Reporting
- Automated scoring for most of EZ Test's numerous question types.
- Allows manual scoring for essay and other open-response questions.
- Manual rescoring and feedback is also available.
- EZ Test's grade book is designed to export easily to your grade book.
- View basic statistical reports.

Support and Help
- User's Guide and built-in page specific help
- Flash tutorials for getting started on the support site
- Support website: www.mhhe.com/eztest
- Product specialist: 1-800-331-5094
- Online training: auth.mhhe.com/mpss/workshops/

Course Delivery Systems

With help from our partners WebCT, Blackboard, Top-Class, eCollege, and other course management systems, professors can take complete control of their course content. Course cartridges containing website content, online testing, and powerful student tracking features are readily available for use within these platforms.

McGraw-Hill Tegrity Campus™

 Tegrity is a service that makes class time always available by automatically capturing every lecture in a searchable format for students to review when they study and complete assignments. With a simple one-click start and stop process, you capture all computer screens and corresponding audio. Students can replay any part of any class with easy-to-use browser-based viewing on a PC or Mac.

Educators know that the more students can see, hear, and experience class resources, the better they learn. With Tegrity, students quickly recall key moments by using Tegrity's unique search feature. This search helps students efficiently find what they need, when they need it across an entire semester of class recordings. Help turn all your students' study time into learning moments immediately supported by your lecture.

To learn more about Tegrity, watch a two-minute Flash demo at tegritycampus.mhhe.com.

Classroom Performance System and Questions

 The Classroom Performance System (CPS) brings interactivity into the classroom or lecture hall. CPS is a wireless response system that gives the instructor and students immediate feedback from the entire class. The wireless response pads are essentially remotes that are easy to use and engage students. CPS allows you to motivate student preparation, interactivity, and active learning so you can receive immediate feedback and know what students understand. A text-specific set of questions is available via download from the Instructor area of the *Environmental Geology* website.

Custom Publishing

Did you know that you can design your own text or lab manual, using any McGraw-Hill text and your personal materials to create a custom product that correlates specifically to your syllabus and course goals? Contact your McGraw-Hill sales representative to learn more about this option.

Student Resources

Text Website: www.mhhe.com/reichard1e

Practice Quizzes
Laboratory Manual Online

eBook

If you or your students are ready for an alternative version of the traditional textbook, McGraw-Hill has partnered with CourseSmart to bring you innovative and inexpensive electronic textbooks. Students can save up to 50% off the cost of a printed book, reduce their impact on the environment, and gain access to powerful Web tools for learning, including full text search, notes and highlighting, and email tools for sharing notes between classmates.

eBooks from McGraw-Hill are smart, interactive, searchable, and portable. To review complimentary copies or to purchase an eBook, go to www.CourseSmart.com.

Related Title of Interest

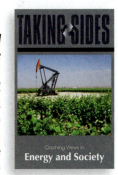

Taking Sides: Clashing Views in Energy and Society   By Thomas A. Easton, *Thomas College* (ISBN: 978-0-07-812755-7 / MHID: 0-07-812755-6) *Taking Sides: Clashing Views in Energy and Society* presents current controversial issues in a debate-style format designed to stimulate student interest and develop critical thinking skills. Each issue is thoughtfully framed with an issue summary, an issue introduction, and a postscript. An instructor's manual with testing material is available online for each volume. *Taking Sides: Clashing Views in Energy and Society* is also an excellent instructor resource with practical suggestions on incorporating this effective approach in the classroom. Each *Taking Sides* reader features an annotated listing of selected World Wide Web sites. Visit www.mhcls.com for more information. To purchase an electronic eBook version of this title, visit www.CourseSmart.com.

Acknowledgments

Developing a new environmental geology text has been an enormous undertaking that naturally involved the help of many people. First and foremost, I would like to thank my wife Linda for her unyielding support. Linda not only helped with the editing and served as a sounding board for new ideas, but she provided much needed encouragement during the more stressful and difficult periods. Most of all, Linda never complained about the endless hours her husband had to spend away from our family. I could not imagine a better partner.

Writing a new textbook is one thing, but producing a beautifully designed product requires a publisher with many dedicated team members. I would first like to thank Beva Lincoln, the McGraw-Hill sales representative who took it upon herself to forward a custom lab manual, which I had co-authored, to the executive editor, Marge Kemp. It was Marge's vision and confidence in a new author that made this new text possible. I would also like to thank the rest of the McGraw-Hill team who worked on this project, including Margaret Horn and Joan Weber (developmental editors), Gloria Schiesl (project manager), Lori Hancock (photo researcher), Laurie Janssen (designer), and Lisa Nicks (marketing manager). Cathy Conroy and Veronica Jurgena also deserve credit for doing the initial editing that resulted in a much improved final manuscript. Lastly, Ed Spencer should be acknowledged for his complete review of the manuscript.

A great deal of attention was given to the photographs and illustrations in *Environmental Geology* because they are critical in reinforcing the key concepts that I cover. Many thanks go to the art team at Electronic Publishing Services for accurately transforming my sketches into the striking illustrations found in the book. With respect to photographs, I would like to thank the many individuals, companies, and government agencies who generously supplied the noncommercial photos. In many cases, this involved people taking time from their busy schedules to search photo archives and retrieve the high-resolution photos that I wanted. Because far too many people contributed to this effort for me to acknowledge individually, their contributions are listed in the photo credits located at the back of the book. However, I do want to recognize the following individuals for producing custom photos and graphics for this textbook:

James Bunn   *National Oceanic and Atmospheric Administration*
Chris Daly   *Oregon State University*
Lundy Gammon   *IntraSearch*
Robert Gilliom   *U.S. Geological Survey*
Bob Larson   *University of Illinois*
David Levinson   *National Oceanic and Atmospheric Administration*
Naomi Nakagaki   *U.S. Geological Survey*
Matt Sares   *Colorado Geological Survey*
Jeremy Weiss   *University of Arizona*

I would also like to acknowledge different government agencies for supporting programs that address important environmental issues around the globe. This textbook made use of publically available reports, data, and photographs from the following agencies:

Australian Commonwealth Scientific and Industrial Research
    Organization
Environment Canada
Geological Survey of Canada, Natural Resources Canada
National Aeronautics and Space Administration (NASA),
    U.S. Government
National Oceanic and Atmospheric Administration (NOAA),
    U.S. Department of Commerce
Natural Resources Conservation Service, U.S. Department
    of Agriculture
United States Department of Energy
United States Environmental Protection Agency
United States Fish and Wildlife Service, Department of the Interior
United States Geological Survey, Department of the Interior

Finally, I would like to thank all those who reviewed various parts of the manuscript during the course of this project. Their insightful comments, suggestions, and criticisms were immensely valuable, particularly for a first-edition text such as this.

Lewis Abrams   *University of North Carolina–Wilmington*
Christine N. Aide   *Southeast Missouri State University*
Michael T. Aide   *Southeast Missouri State University*
Erin P. Argyilan   *Indiana University Northwest*
Richard W. Aurisano   *Wharton County Junior College*
Dirk Baron   *California State University—Bakersfield*
Jessica Barone   *Monroe Community College*
Mark Baskaran   *Wayne State University*
Robert E. Behling   *West Virginia University*
Prajukti Bhattacharyya   *University of Wisconsin-Whitewater*
Thomas Boving   *University of Rhode Island*
David A. Braaten   *University of Kansas*
Eric C. Brevik   *Dickinson State University*
Patrick Burkhart   *Slippery Rock University of Pennsylvania*
Ernest H. Carlson   *Kent State University*
James R. Carr   *University of Nevada–Reno*
Patricia H. Cashman   *University of Nevada–Reno*
Elizabeth Catlos   *Oklahoma State University*
Robert Cicerone   *Bridgewater State College*
Katherine Folk Clancy   *University of Maryland*
Jim Constantopoulos   *Eastern New Mexico University*

Geoffrey W. Cook   *University of Rhode Island*
Heather M. Cook   *University of Rhode Island*
Raymond Coveney   *University of Missouri–Kansas City*
Ellen A. Cowan   *Appalachian State University*
Anna M. Cruse   *Oklahoma State University*
Gary J. Cwick   *Southeast Missouri State University*
George E. Davis   *California State University–Northridge*
Hailang Dong   *Miami University of Ohio*
Yoram Eckstein   *Kent State University*
Dori J. Farthing   *SUNY–Geneseo*
Larry A. Fegel   *Grand Valley State University*
James R. Fleming   *Colby College*
Christine A. M. France   *University of Maryland–College Park*
Alan Fryar   *University of Kentucky*
Heather L. Gallacher   *Cleveland State University*
Alexander E. Gates   *Rutgers University–Newark*
John R. Griffin   *University of Nebraska–Lincoln*
Syed E. Hasan   *University of Missouri–Kansas City*
Chad Heinzel   *Minot State University*
Neil E. Johnson   *Appalachian State University*
Steven Kadel   *Glendale Community College*
Chris R. Kelson   *University of Georgia*
John Keyantash   *California State University–Dominguez Hills*
Gerald H. Krockover   *Purdue University*
Glenn C. Kroeger   *Trinity University*
Michael A. Krol   *Bridgewater State College*
Jennifer Latimer   *Indiana State University*
Liliana Lefticariu   *Southern Illinois University*
Adrianne A. Leinbach   *Wake Technical Community College*
Gene W. Lené   *St. Mary's University of San Antonio*
Nathaniel Lorentz   *California State University–Northridge*
James B. Maynard   *University of Cincinnati*
Richard V. McGehee   *Austin Community College*
Gretchen L. Miller   *Wake Technical Community College*
Klaus Neumann   *Ball State University*
Duke Ophori   *Montclair State University*
David L. Ozsvath   *University of Wisconsin-Stevens Point*
Evangelos K. Paleologos   *University of South Carolina–Columbia*
Alyson Ponomarenko   *San Diego City College*
Libby Prueher   *University of Northern Colorado*
Fredrick J. Rich   *Georgia Southern University*
Paul Robbins   *University of Arizona*
Michael Roden   *University of Georgia*
Lee D. Slater   *Rutgers University–Newark*
Edgar W. Spencer   *Washington and Lee University*
Michelle Stoklosa   *Boise State University*
Eric C. Straffin   *Edinboro University of Pennsylvania*
Benjamin Surpless   *Trinity University*
Sam Swanson   *University of Georgia*
Gina Seegers Szablewski   *University of Wisconsin–Milwaukee*
James V. Taranik   *University of Nevada–Reno*
J. Robert Thompson   *Glendale Community College*
Jody Tinsley   *Clemson University*
Daniel L. Vaughn   *Southern Illinois University*
Adil M. Wadia   *University of Akron, Wayne College*
Miriam Weber   *California State University–Monterey Bay*
David B. Wenner   *University of Georgia*
John C. White   *Eastern Kentucky University*
David Wilkins   *Boise State University*
Ken Windom   *Iowa State University*

Meet the Author

James Reichard James Reichard is an Associate Professor in the Department of Geology and Geography at Georgia Southern University. He obtained his Ph.D. in Geology (1995) from Purdue University, specializing in hydrogeology, and his M.S. (1984) and B.S. (1981) degrees from the University of Toledo, where he focused on structural and petroleum geology. Prior to his Ph.D., he worked as an environmental consultant in Cleveland, Ohio, and as a photogeologist in Denver, Colorado.

James (Jim) grew up in the flat glacial terrain of northwestern Ohio. Each summer, he went on an extended road trip with his family and traveled the American West. It was during this time that Jim was exposed to a wide variety of beautiful landscapes. Although he had no idea how these landscapes formed, he knew that one day he wanted to live in an area with more scenic terrain. It was not until college, when Jim had to satisfy a science requirement, that he finally came across the field of geology. Here, he discovered a science that could explain how beautiful landscapes actually form. From that moment on, he was hooked on geology. This eventually led Jim to a graduate degree in geology, after which he was able to fulfill his dream of living and working in Colorado. Then, due to one of life's many unexpected opportunities, he took an environmental job back in Ohio. This ultimately led to a Ph.D. from Purdue and a faculty position at Georgia Southern University, where he currently enjoys teaching and doing research in environmental geology and hydrogeology. His personal interests include hiking, camping, and biking.

It is through this textbook that Professor Reichard hopes to excite students as to how geology shapes the environment in which we live, similar to the way he became excited about geology nearly thirty years ago. To help meet this goal, he has tried to write this book with the student's perspective in mind to keep it more interesting and relevant. Hopefully, students who read the text will begin to share some of Professor Reichard's fascination of how geology plays an integral role in our everyday lives.

Grinnell Glacier overlook, Glacier National Park, Montana.

Environmental
GEOLOGY

Humans and the Geologic Environment

LEARNING OUTCOMES

After reading this chapter, you should be able to:

▶ Explain what the subdiscipline of environmental geology addresses.
▶ Describe how scientists develop hypotheses and theories to understand the natural world.
▶ Understand the concept of geologic time and how the geologic time scale was constructed.
▶ Explain why the time scale and the manner in which natural processes operate affect how humans respond to environmental issues.
▶ Describe how Earth operates as a system and why humans are an integral part of the system.
▶ Understand the concept of exponential population growth and how it relates to geologic hazards and resource depletion.
▶ Describe the concept of sustainability in terms of the living standard of developed nations and also in terms of the human impact on the biosphere.

Although Earth has limited soil and water resources, agricultural food production has been able to keep ahead of human population growth because of fertilizers, pesticides, and irrigation techniques made possible by fossil fuels. As illustrated by these farm fields in the Jordanian desert, much of the irrigation water however is coming from water stored in the subsurface that is not being replenished. Long-term food production from such fields will not be sustainable as water and energy resources continue to dwindle.

Introduction

Earth is unique among the other planets in the solar system in that it has an environment where life has been able to thrive, evolving over billions of millions of years from single-cell bacteria to complex plants and animals. Ultimately there are two critical factors that led to the development of the diverse biosphere we see today. One is that Earth's distance from the Sun generates surface temperatures in the range where water can exist in both the liquid and vapor states. The other is that our planet was able to retain its atmosphere, which in turn allows the water to move between the liquid and vapor states in a cyclic manner. With respect to humans, our most direct ancestors, *Homo sapiens,* have been part of the biosphere for just the past 200,000 years, whereas other hominid species go back as much as 6 to 7 million years. Compared to Earth's 4.6-billion-year history, humans have occupied the planet for a rather brief period of time. However, rapid population growth combined with the Industrial Revolution has resulted in humans having a tremendous impact on the Earth's surface environment. The focus of this textbook will be on the interaction of humans and Earth's geologic environment. We will pay particular attention to how humans make use of resources such as soils, minerals, and fossil fuels as well as how we interact with natural processes, including floods, earthquakes, landslides, and so forth.

One of the key reasons humans have been able to thrive is our ability to understand and modify the environment in which we live. For example, consider that for most of history humans lived directly off the land. To survive they had to be keenly aware of the environment in order to find food, water, and shelter. This forced some people to travel with migrating herds of wild animals, which in turn were following seasonal changes in their own food and water supplies. Eventually people learned to clear the land and grow crops in organized settlements. From this development, humans became skilled at recognizing those parts of the landscape with the most productive soils. The best soils, however, were commonly found in low-lying areas along rivers and periodically inundated by floodwaters. To reduce the flood hazard, people learned to seek out farmland on higher ground and to place their homes even higher, thereby avoiding all but the most extreme floods. In addition to reducing the risk of floods and other natural hazards, people learned how to take advantage of Earth's mineral and energy resources. The ability to use these resources led directly to the Industrial Revolution and the modern consumer societies of today.

Although humans have benefited greatly by modifying the environment and making use of Earth's resources, these actions have resulted in unintended and undesirable consequences. For example, in order to grow crops and build cities it was necessary to remove forests and grasslands that once covered the natural landscape. This resulted in a reduction of the land's ability to absorb water, thereby increasing the frequency and severity of floods and decreasing both the quality and quantity of our water supplies. Also, the use of mineral and energy resources by modern societies creates waste by-products that can poison our streams and foul the air we breathe. In recent years scientists have shown that the prolific use of fossil fuels is altering the planet's climate system and contributing to the problem of global warming. It has become quite clear that the human race is an integral part of the Earth system and, most importantly, that our actions affect the very environment we depend upon.

While the link between environmental degradation and human activity may be clear to scientists, it is not always so obvious to large segments of the population. A well-established concept with respect to environmental degradation is known as the **tragedy of the commons,** which is where the self-interest of individuals results in the destruction of a common or shared resource. A common resource would include such things as a river used for water supply, wood in a forest, grassland for grazing animals, and fish in the sea. Consider a fishing village whose primary source of food is the local fishing grounds offshore. As long as the fish are not harvested at a faster rate than they can reproduce, everyone in the village benefits since the resource is renewable. However, if the village grows too large, the increased demand can make the fishing become unsustainable. As the fish become scarcer, the competition between individual fishermen gets more intense as they chase after the remaining fish in order to feed their families. The fisherman's self-interest creates a downward spiral where all members of society ultimately suffer as the fishery becomes so depleted that it collapses and is unable to recover.

Another phenomenon that can contribute to environmental degradation is when citizens in modern consumer societies become disconnected from the natural environment. An example is the United States, where many people now live and work in climate-controlled buildings and get their food from grocery stores as opposed to growing their own (Figure 1.1). This has helped people lose their sense of connection with the natural world, despite the fact they remain dependent upon Earth's environment just as our ancient ancestors. As with the tragedy of the commons, a lack of environmental awareness can lead to serious problems and hardships since society invariably has to interact with Earth's environment.

FIGURE 1.1 In modern consumer societies few people live directly off the land, but instead buy most of their food in stores. This trend has led to people becoming more disconnected from the natural environment upon which they still depend.

A   B   C

FIGURE 1.2   Rock and mineral deposits (A) provide the raw materials used for building (B) and operating our modern societies. The geologic resources known as fossil fuels provide the bulk of the energy used for powering (C) the industrial, transportation, and residential sectors of society.

What Is Geology?

The science of **geology** is the study of the solid earth, which includes the materials it is composed of and the various processes that shape the planet. Many students who are unfamiliar with geology tend to think it is just a study of rocks, and therefore must not be very interesting. However, this perception commonly changes once students begin to realize how intertwined their own lives are with the geologic environment. For example, the success of our high-tech society is directly tied to certain minerals whose physical properties are used to perform vital tasks. Perhaps the most important are those minerals containing the element copper, a metal whose ability to conduct electricity is absolutely essential to our modern way of life. Imagine doing without electric lights, refrigerators, televisions, cell phones, and the like. Because geologists study how minerals form and the processes that concentrate them, mining companies hire geologists to look for valuable mineral deposits (Figure 1.2). Equally important is the ability of geologists to locate deposits of oil, gas, and coal as these serve as the primary source of energy for society. Geologists also provide valuable knowledge as to how society can minimize the risk from hazardous Earth processes such as floods, landslides, earthquakes, and volcanic eruptions (Figure 1.3).

Geology has traditionally been divided into two main subdisciplines: physical geology and historical geology. **Physical geology** involves the study of the solid earth and the processes that shape and modify the planet, whereas **historical geology** interprets Earth's past by unraveling the information held in rocks. The most important geologic tool in both disciplines is Earth's 4-billion-year-old collection of rocks known as the *geologic rock record*. This vast record contains an abundance of information on topics ranging from the evolution of life-forms to the rise and fall of mountain ranges and changes in climate and sea level. Over the past 30 years or so a new subdiscipline has emerged called **environmental geology,** whereby geologic principles and knowledge are used to address problems arising from the interaction between humans and the geologic environment. Environmental geology is becoming increasingly important as population continues to expand, which in turn is leading to widespread pollution and shortages of certain resources, particularly water and energy. Population growth has also resulted in greater numbers of people living in areas where geologic processes (floods, earthquakes, volcanic eruptions, landslides, etc.) pose a serious risk to life and property.

A

B

FIGURE 1.3   In addition to locating resources, geologists study hazardous earth processes and use this knowledge to help society avoid or minimize the loss of life and property damage. Photo (A) shows a bridge that was destroyed during the 2008 earthquake in Sichuan, China, and (B) shows the results of an earthquake-induced landslide in Las Colinas, El Salvador, 2001.

The first step in solving our environmental problems is to understand the way in which various Earth processes operate and how they respond to human actions. The most effective way of doing this is through *science*, which is the methodical approach developed by humans for learning about the natural world. Once this interaction is understood, appropriate action can be taken that can reduce or minimize the problems. Because science is critical to addressing our environmental problems, we will begin by taking a brief look at how science operates.

Scientific Inquiry

By our very nature, humans are curious about our surroundings. This natural curiosity has ultimately led to the development of a systematic and logical process that tries to explain how the physical world operates. We call this process *science*, which comes from the Latin word *scientia*, meaning knowledge. Over the past several thousand years the human race has accumulated a staggering amount of scientific knowledge. Although we now understand certain aspects of nature in great detail, there is still a lot we do not understand. Throughout this period of discovery the general public has remained generally fascinated with what scientists have learned about the natural world. Evidence for this is the continued popularity of science programs currently available on cable television. It seems rather odd then that one of the common complaints in science courses is that nonscience majors find the subject boring. This raises the obvious question of what is it about science courses that tends to cause students to lose their natural interest in science?

One reason, perhaps, for the loss of interest in science is because students are often required to memorize trivial facts and terminology. The problem is compounded when it is not made clear how or why these facts are relevant to people's own lives. Focusing on just the facts is unfortunate because it is the *explanation* of the facts that makes science interesting, not necessarily the facts themselves. Take, for example, the fact that coal is found on the continent of Antarctica, which sits directly over the South Pole (Figure 1.4). This, combined with the fact that coal forms only in swamps, where vegetation and liquid water are abundant, allows us to logically conclude that Antarctica at one time must have been relatively ice-free and located closer to the equator. This leads us to ask the obvious: How did this giant landmass actually move to its present position? How long did it take to get there? Are other continents also moving? If so, how does this affect the people who are living on these continents? To answer these questions scientists must gather additional data (i.e., facts), and in the process, will likely result in even more questions that need to be answered.

Science therefore can be thought of as a method by which people use data to discover how the natural world operates. Unlocking the secrets of nature is truly exciting and is a primary reason most scientists love what they do. Anyone who has found a fossil or an old coin, for example, can relate to the thrill of discovery. A key point here is that nearly everyone, not just scientists, practices science each and every day. When we observe dark clouds moving toward us we process this information (i.e., data) along with past observations, and logically conclude that a storm is approaching and that it is wise to seek shelter. A fisherman who keeps changing lures until he or she finds one that attracts a certain type of fish is also practicing science. Because science is fundamental to the topics discussed throughout this textbook, we need to explore the actual process in somewhat more detail.

FIGURE 1.4 The basic goal of science is to use facts or data to explain different aspects of our natural world. For example, the coal beds shown here in Antarctica are a scientific fact. It's also a fact that coal forms only in lush swamps. The best explanation for these facts or data is that Antarctica was much closer to the equator and relatively ice-free at some point in the geologic past.

How Science Operates

Modern scientific studies of the physical world are based on the premise that the entire universe, not just planet Earth, behaves in a consistent and often predictable manner. When an event or phenomenon is observed repeatedly and consistently, it can be described as having a pattern. Finding patterns in nature is important because it enables future events to be predicted or anticipated. Consider how humans long ago observed that the ocean rises and falls along coastlines on a regular basis. This pattern, known as the tides, is so regular that humans can accurately predict when the sea will reach its maximum and minimum heights each day. In contrast, events such as floods and volcanic eruptions occur repeatedly, but on a more irregular basis. The random nature of irregular events means that scientists can only predict future occurrences in terms of statistical probabilities. Although recognizing natural patterns is a key component of science, the ability of scientists to explain *how* and *why* things happen in the first place is equally important.

The process by which the physical world is examined in a logical manner is commonly referred to as the **scientific method.** The basic approach is to first gather data (i.e., facts) about the world through observations or by conducting experiments. Examples of data include such things as the frequency of floods, fossils contained within rocks, velocity of earthquake waves, and animal behavior. Note that all scientific data can be observed and/or physically measured. Also, these data are considered to be facts provided that scientists working independently of each other are able to repeat the work and obtain similar results. Once data are collected scientists then seek to develop an explanation for the data itself, or any patterns it may contain. For example, suppose a researcher collects fossils, shown to be of marine origin, in rock layers that are 10,000 feet above sea level. The next step would be to develop a scientific explanation for the fossils that is consistent with other known data. In this case, all possible explanations would have to be consistent with the fact that the planet does not contain enough water for sea level to ever have been 10,000 feet above its present level. Logic then dictates that any plausible explanation must include some mechanism for uplifting the fossil-bearing rocks from sea level to their present position.

The term **hypothesis** refers to a scientific explanation of data that can be tested in such a way that shows it to be false or incorrect; something scientists refer to as being *falsifiable.* Supernatural explanations are not considered scientific simply because they are not testable and cannot be shown to be false. This concept of a hypothesis being falsifiable may seem odd since people generally think in terms of trying to prove ideas to be correct rather than false. Nevertheless, this concept is important in science because a hypothesis is considered valid provided that additional testing does not show it to be false. Take for example how fossil evidence shows that dinosaurs went extinct 65 million years ago, whereas the first fossils of primitive humans (hominids) do not show up in the rock record until around 7 million years ago. Scientists have logically concluded, or hypothesized, that humans never coexisted with dinosaurs. This hypothesis could be proven to be false if hominid fossils are ever found in rocks that correspond with the age of dinosaurs (i.e., at least 65 million years old). Because extensive searches have never yielded such hominid fossils, the hypothesis that people and dinosaurs did not coexist remains valid.

Another key aspect of the methodology we call science is that in the early stages of an investigation researchers commonly come up with more than one plausible hypothesis for a given set of data. As shown in Figure 1.5, scientists refer to these different explanations as **multiple working hypotheses,** all of which are considered valid as long as they are consistent with the existing data (i.e., facts). Because the goal of science is to seek out the best possible explanation, researchers continue to collect new data as they try to prove that one or more of the hypotheses are false. If an individual hypothesis is shown to be false, then it must either be modified or removed from consideration. Over time, this process of eliminating and refining hypotheses by gathering new data gives scientists greater and greater confidence in the validity of the remaining hypotheses. Note in Figure 1.5 that hypotheses are validated by their ability to predict future observations or experimental data. It should be emphasized that geology is more of an observational science than experimental, such as chemistry. This means that geologic hypotheses are typically tested or validated by making predictions that are confirmed through additional observations as opposed to controlled lab experiments. A good example is the hypothesis that humans and dinosaurs did not coexist, something which cannot be tested in a lab experiment, but rather only by observing more of the fossil record.

The terms *theory* and *hypothesis* are sometimes used interchangeably, but are actually different. As indicated in Figure 1.6, a **theory** describes the relationship between several different and well-accepted hypotheses, providing a more comprehensive or unified explanation of how the world operates. In other words, a theory ties together seemingly unrelated hypotheses and allows us to see the "big picture." For example, the theories of atomic matter, relativity, and evolution all tie together or unify individual hypotheses within their respective disciplines of chemistry, physics, and biology. In geology the central unifying theory is known as the theory of plate tectonics (Chapter 4). This important theory explains how Earth's rigid crust is broken up into separate plates, which are in constant motion due to forces associated with the planet's internal heat. The movement of tectonic plates influences the location of continents and circulation of ocean currents, and consequently has a strong effect on the global climate system and biosphere on which we humans depend. As with all scientific theories, the theory of plate tectonics provides scientists with a larger context for understanding an array of different hypotheses. It should be noted here that the term *theory* has an entirely different meaning to scientists compared to what it means to the general public. In common everyday language, the term "theory" is used to describe some educated guess or speculation. In science, however, a theory is a widely accepted and logical explanation of natural phenomena that has survived rigorous testing. Later we will examine how these different meanings can impact public debate and policy considerations of environmental issues.

There are some phenomena in nature where the relationship between different data occurs so regularly and with so little deviation that scientists refer to the relationship as a **law.** In some cases a law can be expressed mathematically, as in the case of Newton's three laws of motion and gravitational law. An example of a law in geology is the *law of superposition,* which states that in a sequence of layered rocks derived from weathering and erosion (i.e., sedimentary rocks, Chapter 3), the layer at the top is the youngest and the one on the bottom is the oldest. This simple and intuitive idea that sedimentary layers become progressively older with depth has been a valuable tool in using the geologic rock record to unravel Earth's history. Scientific laws, therefore, are quite useful despite the fact they do not necessarily unify different hypotheses and provide grand explanations as do theories.

A good example of how knowledge is advanced through the use of scientific theories and laws is the discovery of the planet Neptune. Early

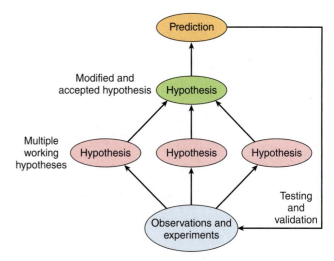

FIGURE 1.5   A scientific hypothesis is an explanation of known observations and experimental data. Multiple hypotheses are commonly developed, with most being discarded or modified as new data are gathered during testing and validation. Over time a refined hypothesis normally emerges from the process and becomes generally accepted by the scientific community. Validation involves the ability of a hypothesis to predict future events.

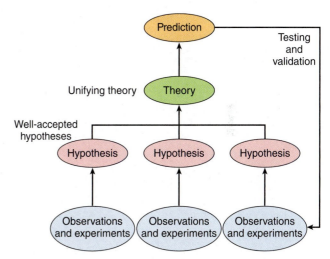

FIGURE 1.6   Scientific theories describe the relationship among different hypotheses and provide a more comprehensive or unified explanation of how the natural world operates. As with all scientific explanations, theories undergo repeated testing and validation.

astronomers noted strange wobbles in the elliptical orbits of the planets around the Sun, but could not explain the wobbling with the existing knowledge. It was not until after Isaac Newton published his theory of gravitation in 1687 that astronomers were able to explain that the wobbling was caused by the gravitational effects of planets in adjoining orbits. Then, in the 1800s, scientists were unable to explain the wobble in the orbit of Uranus since it was the outermost known planet at the time. This led some astronomers to predict that an unknown planet existed beyond Uranus' orbit and was causing the wobble. The planet Neptune was then discovered in 1846 when astronomers pointed a telescope at the exact position in the sky where Newton's laws predicted a planet would be. Because of this and other successful predictions, scientists soon accepted the validity of Newton's theories without question.

In the early 1900s, Albert Einstein stunned the scientific community when his special theory of relativity (1905) and general theory of relativity (1915) proved that Newton's gravitational law produced significant errors in situations of unusually strong gravity or high velocities. However, Einstein's work did not invalidate Newton's law, but rather represented a modified or improved version that was accurate under more extreme conditions. This example helps highlight the fact that scientific theories and hypotheses can be improved and modified through continued testing. Although rare, well-established theories are sometimes rejected should they fail to explain new data, or when a better explanation is presented. Perhaps the most well-known example is how the scientific community completely abandoned the Earth-centered theory of the solar system during the 1500s. A Sun-centered theory, based on the earlier work of Nicolaus Copernicus, eventually gained acceptance because it provided a much simpler explanation for the known movements of the planets.

Science and Society

Over the centuries scientists have accumulated a vast amount of knowledge regarding the natural world. This knowledge has allowed society to understand parts of nature in great detail, and at the same time, reap great rewards. Highly trained scientists in specialized fields are, of course, the ones who best understand the minute details, whereas all of society benefits from scientific knowledge. Consider how advances in medical research have led to an increase in doctors specializing in fields such as cardiology, neurology, gynecology, and dermatology. Today if a person has a medical problem they can go to a specialist for a more accurate diagnosis and more sophisticated treatment. This, in turn, greatly increases the chance of being cured of a serious illness. Also consider the scientists with specialized knowledge on the complex interactions between hurricanes and the atmosphere and oceans. This area of science has progressed to the point where sophisticated computer models now routinely make projections of where hurricanes will make landfall, thereby saving large numbers of lives.

Although scientists are able to make useful predictions based on well-established theories, it is important to realize that there is almost always some degree of uncertainty associated with any prediction. The amount or degree of uncertainty usually depends on the nature of the process and the amount of error involved in making measurements. Take, for example, how scientists use Newton's laws of motion and gravity to predict, with great confidence and accuracy, the future position of the planets as they orbit the Sun. These laws were used to land a spacecraft on Saturn's moon Titan in 2005, which was an impressive feat considering the craft traveled seven years through space before meeting up with Saturn and Titan in their orbit around the Sun. Also consider how hydrologists can predict with great certainty that deforestation will lead to increased flooding. Due to the sporadic nature of flood events, however, predicting an actual flood can only be done in a statistical manner using probabilities.

Nearly everyone in a modern society benefits from science, but problems can arise when people have a poor understanding of science and then tend to take its benefits for granted. For example, when you enter a friend's number into your cell phone you probably never think about all the science behind the wireless communication and satellite technology that makes your call possible. Likewise, when we put gas in our vehicles or enjoy a warm house few of us think about how the study of geology enables oil companies to locate petroleum deposits hidden deep within the Earth. Or, we may not consider how it took years of medical research before surgeons could perform open-heart surgeries that extend the lives of our loved ones. Although society reaps great benefits from science, history is full of examples where new knowledge met with considerable resistance. Much of this resistance can be attributed to the tendency of knowledge to create change. In some instances scientific advances present society with new moral and ethical questions that it must address, as is the case for stem-cell research and its potential to develop cures for different diseases. Other times people feel their religious views are being threatened by science, such as the theory of evolution and how it explains the development of the biosphere. Perhaps the most common reason scientific knowledge meets resistance is when it initiates changes that threaten certain economic interests within a society. Pollution controls on coal-burning power plants, elimination of lead in gasoline, and restriction of ozone-depleting refrigerant gases are but a few examples where science identified a human health threat, but corrective measures were greatly resisted by business interests.

It is not surprising then to find science thrust into public debates whenever new information runs counter to the interests of economic, political, or religious groups. A common reaction from interest groups is to try and discredit scientific information by referring to it as "only a theory." This, of course, takes advantage of the common misperception that a scientific theory is just speculation or an educated guess. Another tactic is to create doubt about the science by saying "scientists are not certain." The implication here is that the science is untrustworthy, conveniently ignoring that all scientific work by its very nature has some level of uncertainty. Perhaps the best example of this is how the tobacco industry, for over 30 years, used scientific uncertainty to effectively sow doubt as to the link between smoking and lung cancer. Today we see the terms "theory" and "uncertainty" again being misused in an effort to convince the public that the threat of global warming is a hoax (Chapter 16).

Another common tactic by interest groups is to try to create doubt by putting forth nonscientific work that runs counter to the legitimate work of scientists. Here the nonscientific work is simply labeled "scientific," which is usually effective because most people find it difficult to distinguish between good, solid science and so called *junk science*. The key difference is that good scientific explanations are always capable of being proved false and are consistent with *all* the data, not just selective data that fits a particular viewpoint. Another measure of good scientific work is whether it has passed the so-called *peer-review process* before being published. Here scientists submit their work to scientific organizations so it can undergo rigorous scrutiny, not only by true scientists, but by leading experts within a particular field. In this process, papers are published and presented to the public only if their conclusions are supported by physical data that have been properly collected.

FIGURE 1.7   This house erupted into flames when a volcanic lava flow moved through a residential neighborhood on the big island of Hawaii. Lava flows have been taking place on the big island for the past 700,000 years and have resulted in the formation of the island itself. The other Hawaiian islands are much older. Although lava flows are a natural geologic process that creates habitable living space for people, they also pose a serious risk to human life and property.

Finally, because scientific knowledge is so vital to society, scientists often find themselves trying to educate policymakers on important environmental issues, many of which are politically charged. Although science itself is objective and nonpolitical, scientists commonly speak out publicly so that policymakers can base decisions on the best information available. Since today's students will be the decision makers of tomorrow, it is especially important for you not only to learn the science behind environmental issues, but also how science itself operates.

Environmental Geology

Environmental problems related to geology generally fall into one of two categories: *hazards* and *resources.* We will define a **geologic hazard** as any geologic condition, natural or artificial, that creates a potential risk to human life or property. Examples include earthquakes, volcanic eruptions, floods, and pollution. There are some geologic processes, such as the eruption of volcanic lava (Figure 1.7), that present a clear hazard to society. Ironically, what we consider as geologic hazards are commonly processes that play an important role in maintaining our habitable environment, and have been operating throughout Earth's history. Volcanic eruptions, for example, are as old as the Earth itself and have been instrumental in the development of the atmosphere and oceans. Also interesting is the fact that human activity can affect certain types of geologic processes themselves, increasing the severity of an existing hazard and making it more costly in terms of the loss of life and property. A good example is the use of engineering controls to minimize flooding in one area, but which oftentimes ends up causing increased flooding somewhere else. Moreover, human interference in natural processes commonly produces unintended consequences, as in the loss of wetlands resulting from our efforts to control flooding. By destroying wetlands, humans inadvertently disrupt the food web within critical ecosystems, which ultimately results in the loss of sport and commercial fisheries.

Pollution is a type of hazard because it directly impacts human health and the ecosystems on which we depend. Although pollution can occur naturally, human activity is by far the most common cause (Figure 1.8). An example is metallic mercury, which cycles through the biosphere after being released during the natural breakdown of certain types of minerals. Mercury tends to accumulate in wetlands due to the acidic and low oxygen conditions, but is then periodically released into the atmosphere when droughts allow fires to sweep through the dried-out wetlands. Since the Industrial Revolution, humans have been releasing mercury into the environment in a similar manner by burning vast quantities of coal, which are ancient swamp deposits that naturally contain mercury-bearing minerals (Chapters 13 and 15). Consequently, the amount of mercury in the biosphere is now significantly higher than natural background levels. The problem is that mercury forms bonds with carbon atoms, creating highly toxic compounds capable of moving through the aquatic food chain and into humans. Of particular concern are pregnant or nursing women who eat mercury-contaminated fish and unknowingly pass the toxic compounds on to their small children. The result is

FIGURE 1.8   Pollution of air and water resources is a serious issue that affects both natural ecosystems and our quality of life, particularly human health.

that children with elevated levels of mercury run a higher risk of developing severe and irreversible brain damage. A somewhat different form of pollution is the emission of greenhouse gases (e.g., carbon dioxide) from the burning of fossil fuels, which is contributing to the problem of global warming (Chapter 16). Although greenhouse gases are natural and have helped regulate Earth's climate system for millions of years, the volume of gases being released into the atmosphere by humans is large, threatening to disrupt the planet's entire ecosystem.

The other major area of environmental geology relates to **earth resources,** which include water, soil, mineral, and energy resources. Resource issues in society generally involve trying to maintain adequate supplies and minimizing the pollution that results from the extraction of a particular resource and disposal of waste by-products. Freshwater and soil resources are the most critical to society since they make land plants possible, which are the basis of our food supply (Figure 1.9). As soils continue to be lost through erosion and irrigation water supplies become stretched to their limit, rapid population growth may soon outstrip world food production. Depletion of soil and water resources therefore is potentially one of humanity's greatest challenges. Moreover, by removing the natural vegetation from the landscape in order to grow food and develop cities, we inadvertently create the problem known as *sediment pollution.* When soils are left exposed, we allow excessive amounts of sediment to wash off the landscape and into our natural waterways. The additional sediment destroys the natural ecology of streams and fills the channels, leaving them more prone to flooding.

In addition to soil and water resources, modern society is also highly dependent on nonrenewable supplies of energy and minerals. Mineral resources (Chapter 12) are critical because they provide most of the raw materials used in building our modern infrastructure. Of particular importance is the iron for making steel, copper for anything involving electricity, and limestone for making concrete. Despite being nonrenewable, some mineral resources such as limestone, sand, and gravel are so abundant that their supplies are essentially inexhaustible. In contrast, there are other minerals that have very specific and critical applications, but whose supply is so limited that they are considered to be of strategic importance. Examples include chromium and cobalt minerals that are required to produce high-performance jet engines for military aircraft. Equally important to modern society are the energy resources that power the industrial, transportation, commercial, and residential sectors of an economy. Crude oil is especially important because it is the primary source of our transportation fuels, and it serves as the raw material for making plastics and agricultural chemicals. Because oil is truly the lifeblood of modern societies, one of our major challenges is that of replacing our dwindling oil supplies with alternative sources of energy (Chapters 13 and 14).

FIGURE 1.9   The ability of people to use water and soil resources in conjunction with fossil fuel–based fertilizers and pesticides has resulted in the present high rates of food production. Depletion of these resources threatens society's long-term ability to feed our growing population.

Environmental Problems and Time Scales

One of the key factors that influences the way in which humans respond to environmental problems is the nature of the geologic processes that are involved and the time scales over which they operate. Some geologic hazards such as earthquakes, for example, happen suddenly and have unmistakable consequences that we easily recognize. In areas where earthquakes are common, society will typically takes steps to minimize the future loss of life and property damage. This usually involves constructing buildings that resist ground shaking and having emergency personnel better equipped

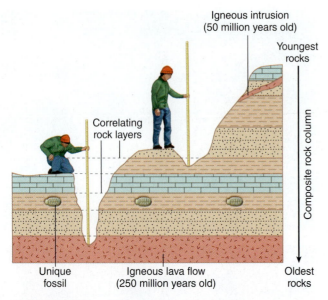

FIGURE 1.10 In a sequence of undisturbed sedimentary rocks, the oldest rocks always lie on the bottom. By correlating specific layers at different exposures across a wide area, geologists can establish the relative age of an entire rock column. Should igneous rocks which formed from magma be present, radiometric dating can be used to determine an age range for the sedimentary column.

and trained for earthquake disasters. Pollution, in contrast, is a problem that may occur gradually over many years and where its effects on human health are not so readily apparent. This can cause society to delay taking corrective measures until the consequences become more severe and noticeable.

With respect to geologic hazards that occur suddenly and in random or sporadic manner, society's response is governed in part by the frequency at which the events recur relative to the life span of people. Generally speaking, once those with a living memory of a hazardous event begin to die off, the generations that follow tend to forget or become complacent about the hazard and its consequences. On the other hand, if a hazard happens frequently enough, we are less complacent and tend to take steps to reduce the risk. A good analogy is how people commonly carry an umbrella or wear rain gear in areas where it rains frequently. In desert climates hardly anyone is going to bother taking similar precautions since the risk of getting caught in the rain is quite low. In much the same way, humans tend to become complacent about geologic hazards that recur on the order of decades or centuries. Because of the importance of time and the fact that geologic processes often operate on unfamiliar time scales, we will briefly explore the concept of geologic time.

Geologic Time

During the late 1800s and early 1900s, geologists began a systematic effort at studying sections of sedimentary rocks that are exposed at the surface. Sedimentary rocks (Chapter 3) are unique in that they are derived from the erosion and weathering of older rocks and are deposited in layers. Such rocks are important because they hold clues to the environmental conditions and life-forms that were present at the time the rocks were deposited. As indicated in Figure 1.10, when geologists study the exposed parts of a sequence of sedimentary rocks they typically record the composition of each layer and describe any fossils it may contain. They also determine the **relative age** of each layer using the *law of superposition,* which is based on the principle that the bottom layers were deposited first, and thus are the oldest. Also note in Figure 1.10 how individual layers can sometimes be correlated from one exposure to another. By finding exposed sections that correlate and overlap with one another, geologists have been able to construct large, composite sections called *rock columns* (Figure 1.10). Sedimentary rock columns from different locations around the world have provided scientists with a reliable record of Earth's environmental changes and evolutionary history of its plants and animals. This worldwide rock record has also led to the development of the **geologic time scale,** which classifies all rocks according to their relative or chronological age. From Figure 1.11 one can see that the geologic time scale uses various names to subdivide Earth's rock record into progressively smaller time intervals. Perhaps the most familiar of these divisions is the Jurassic period of the Mesozoic era, whose rocks contain dinosaur fossils.

Although early geologists understood that the rocks that comprise the geologic time scale represent an enormous amount of time, they had no way of quantifying the actual or **absolute age** of the rocks in terms of years. What they needed was some characteristic or feature within rocks that forms at a steady and reliable rate. By knowing the rate one could then measure time, similar to a stopwatch. In the early 1900s, scientists discovered that as uranium atoms undergo radioactive decay they are transformed into lead atoms at a dependable rate; the term *half-life* describes the time required for half of the radioactive atoms in a sample to decay into stable atoms (Chapter 15). Because nearly all igneous rocks contain uranium-

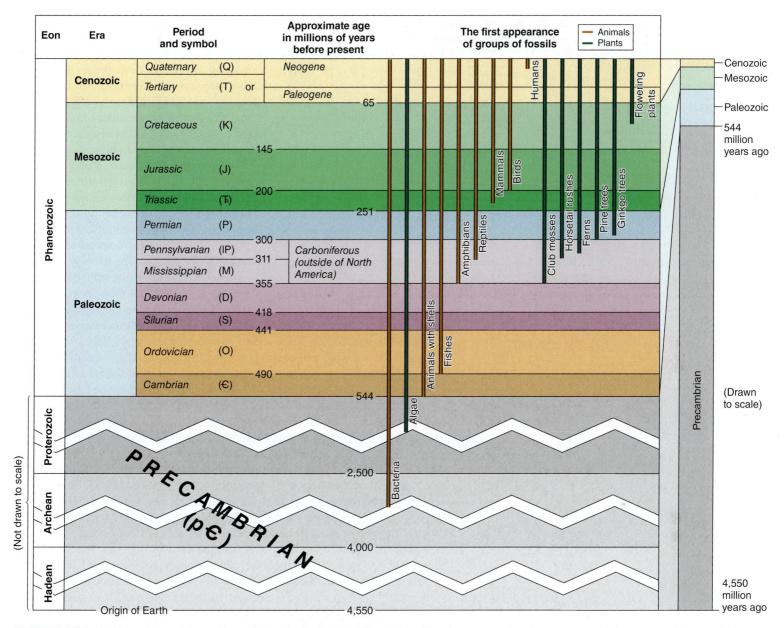

FIGURE 1.11   The geologic time scale was first developed by determining the relative age of sedimentary rocks from around the world. Radiometric dating was later used to establish the absolute dates of the various divisions within the scale. This 4.6-billion-year-old rock record holds the key to understanding the evolution of plants and animals as well as the changes in Earth's environment over time.

bearing minerals, scientists could now use uranium's known decay rate (i.e., half-life) to calculate the number of years since the original magma (molten rock) cooled and solidified into igneous rock. Older rocks contain progressively more lead atoms. By the 1950s, instruments called *mass spectrometers,* which measure the ratio of different atoms, were refined to the point that lead-uranium dating became quite precise. This was highly significant because igneous rocks often come into contact and cut across sedimentary sequences (see Figure 1.10). Therefore, sedimentary rocks within the geologic time scale could now be assigned absolute dates in terms of years. Note that lead-uranium dating is a specific type of **radiometric dating,** which is the general term applied to absolute dating techniques involving any type of radioactive element and its decay product.

TABLE 1.1   Milestones in Earth's 4.6-billion-year history as represented on a compressed calendar year consisting of 365, 24-hour days.

Beginning of Earth history	January 1
Oldest surviving rocks	Middle February
Oldest fossils—single-cell cyanobacteria	Early March
First fossils of animals with hard body parts	Middle October
First dinosaur fossils	December 11
Last dinosaur fossils	December 26
First modern human fossils	23 minutes before midnight, December 31
Egyptian civilization	35–14 seconds before midnight
Roman civilization	18–11 seconds before midnight
Columbus arrives in North America	3.5 seconds before midnight
Past 20 years	0.14 seconds before midnight

Moreover, different radioactive elements decay at different rates, which allows scientists to obtain reliable dates for events ranging anywhere from thousands to billions of years old. Here elements that decay rapidly are used to date younger events, and those with slower decay rates are better suited for older events (see Chapters 14 and 15 for details on radioactive decay).

Another important outcome of radiometric dating is the determination by scientists that the Earth solidified approximately 4.6 billion years ago. Because humans personally experience time in intervals ranging from seconds to decades, time can be difficult for us to grasp when measured in millions or billions of years, something geologists refer to as **geologic time.** To better understand geologic time, imagine someone handing you a dollar bill every second, 24 hours a day. In order to reach a million dollars you would have to stay awake for 11.6 days straight. To reach a billion dollars would take 32 years of continuous counting. Another useful analogy is to compress all 4.6 billion years of Earth's history into a single calendar year consisting of 365 days. This means that one 24-hour day would equal 12.6 million years of actual time. Table 1.1 lists some important historical milestones in terms of this compressed calendar, with January 1 representing the beginning of Earth's history and December 31 being the most recent. On this scale the first rudimentary life-forms show up in the rock record in early March, whereas the first dinosaurs do not come into existence until December 11. Note that the dinosaurs ruled for nearly 200 million years, which represents only 15 days on the compressed calendar. In contrast, the entire 200,000 years of modern human history occupies the most recent 23 minutes of time. Also notice how Columbus reaches North America a mere 3.5 seconds before the clock strikes midnight on New Year's Eve. Even more striking is how the life span of a 20-year-old student represents just fourteen hundredths (0.14) of a second! It should be quite clear from this analogy that Earth's 4.6-billion-year history represents an immense amount of time, and that our human presence on the planet has been exceedingly brief.

Determining the age of the Earth through radiometric dating also confirmed something geologists had long suspected, namely that our planet's physical features formed by both sudden and slow processes. Prior to radiometric dating, a popular concept called *catastrophism* held that Earth's features were the result of sudden, catastrophic events, such as earthquakes, floods, and volcanic eruptions. Once the true age of the Earth was established, a concept known as *uniformitarianism* was used to explain how most of the planet's features are formed by slow processes acting over long periods of time. While geologists recognized that some features do indeed form by catastrophic processes, features such as deep canyons and thick sedimentary sequences could be more easily explained by very slow processes. For example, suppose that sediment is being deposited at a rate of only one millimeter per year at the mouth of a river. If this were allowed to continue for just a million years, the result would be a sediment sequence 1,000 meters or 3,280 feet thick. Likewise, given a sufficient amount of geologic time, slow rates of erosion can carve deep canyons and wear down entire mountain ranges. The Grand Canyon shown in Figure 1.12 is an excellent example of how slow processes can do tremendous work over time. Based on the geologic time scale and absolute dating techniques, geologists now know that the sedimentary sequence within the Grand Canyon accumulated over hundreds of millions of years. The area was later uplifted along with the Rocky Mountains by forces within the Earth, causing the Colorado River to begin cutting downward into the sequence of sedimentary rocks. The canyon we see today formed over the past 5 to 17 million years by the slow down-cutting of the river combined with rock falls and other mass-wasting processes (Chapter 7) that have acted to widen the canyon.

Environmental Risk and Human Reaction

We will define an **environmental risk** as the chance that some natural process or event will produce negative consequences for an individual or society as a whole. There are actually two separate components of risk: probability and consequences. They can perhaps be better understood through the following relationship:

risk = (probability of an event) × (expected consequences)

An example of an environmental risk is that posed by asteroids hitting the Earth. Although small fist-sized asteroids routinely strike our planet (i.e., high probability), the actual risk is low since the potential damage (i.e., consequence) is very minor. Although the consequences from the impact of an asteroid a mile in diameter would be globally catastrophic, the risk can still be considered low since the probability of such a strike in the near future is extremely low. Here the differences in probability are due to the fact that there are very few mile-sized asteroids compared to fist-sized. Therefore, when evaluating risk, one should not simply focus on the severity of the consequences without considering the probability of the event itself.

This leads us to the concept of *risk management,* which involves taking steps to reduce a specific risk. Before any action can be taken, it is of course necessary to identify the threat itself. This requires that the scientific method be used to properly identify, and understand, those aspects of the physical world that pose a risk to society. Once a threat is identified, the risk can be lowered by applying science and engineering to: (a) reduce the probability of an event taking place; and (b) minimize the impact or consequences. For example, to lower the risk of flooding, the probability of flooding can be reduced through the construction of dams and levees, and the consequences can be minimized via zoning laws that limit development on floodplains. Throughout this textbook we will focus on how the field of geology can be used to identify environmental risks, and then develop solutions designed to lower the probability of an event and to minimize its consequences.

One of the key factors affecting how people respond to environmental risk is the nature of different geologic processes and the time scale over which they operate. Natural processes can be classified as being either incremental or sporadic. An *incremental process* is one that generates small changes with time, such as the forces that are uplifting and eroding the sedimentary rocks making up the Grand Canyon. Although these incremental forces are operating continuously, the rates of change are so small that their impact on the day-to-day lives of people is inconsequential. On the other hand, the incremental loss of topsoil due to erosion is

FIGURE 1.12   The Grand Canyon is an example of how slow processes can create dramatic features when given a sufficient amount of time. The thick sequence of sedimentary rocks in the canyon required hundreds of millions of years to accumulate. Later, as the area was uplifted by forces within the Earth, the Colorado River began cutting downward into the sedimentary section. The canyon we see today is the result of between 5 and 17 million years of down-cutting by the river and mass wasting processes, which have widened the canyon.

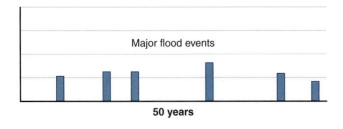

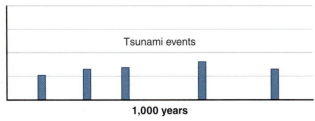

A Sporadic events

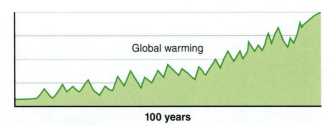

B Incremental changes

FIGURE 1.13   Natural processes generally operate in either a sporadic (A) or incremental manner (B). When sporadic events occur infrequently, it becomes more likely that a human generation will pass between events, causing people to be complacent about the risk. With incremental processes, people are typically slow to recognize environmental problems because the changes from year to year are small. Most difficult to recognize are incremental problems where upward and downward swings mask the overall change.

a serious environmental issue since it reduces society's ability to grow food (Chapter 10). In this case the rate of soil erosion in agricultural areas is great enough to result in significant soil loss in a matter of a few decades. Therefore, whenever the rate of an incremental process is relatively high, there is the potential for undesirable changes to occur within the life span of humans.

In contrast, *sporadic processes* are those that take place somewhat randomly as discrete events. Examples include floods, volcanic eruptions, earthquakes, and landslides. Sporadic processes are commonly referred to as *hazards* when they produce dramatic and sudden changes that have undesirable consequences for humans. Events with particularly severe consequences are typically called *disasters* or *catastrophes*. Although sporadic processes occur randomly, they generally repeat in somewhat regular intervals, with the frequency depending on the process itself. For example, floods occur with greater frequency than do volcanic eruptions. Also important is the fact that small-magnitude events are more common than large, catastrophic ones. Consider how minor floods occur more frequently than do major floods—similar to how the odds of winning a small lottery is much higher than those of winning a multimillion-dollar jackpot.

The fact that geologic processes operate differently is important because of the way these differences influence how humans respond to environmental risks. To illustrate this point consider the sporadic nature of major floods and tsunamis shown by the graphs in Figure 1.13A. Notice that the floods occur frequently enough that an individual person is likely to experience several major floods within his or her lifetime. The relatively high risk of flooding helps explain why people living along rivers have historically reduced their risk by building settlements on the highest ground possible. On the other hand, major tsunamis (Chapters 5 and 9) occur much less frequently, which means several generations may pass between events. Humans that have no living memory of the previous tsunami are inhabiting vulnerable coastal areas. Thus, some people may be completely unaware of the danger, whereas others are aware but feel that the benefits of living on flat ground next to the sea outweigh the risk. Unfortunately, humans often put themselves in harm's way because they are ignorant of the hazard or simply because they decide to accept the risk. A tragic example of this phenomenon is the massive tsunami that swept into coastal villages around the Indian Ocean in 2004. Nearly 250,000 people lost their lives when they suddenly found themselves in harm's way with no means of escape (Figure 1.14).

Finally, environmental problems associated with incremental processes are particularly challenging for society due to the fact they can be hard to recognize. Take for example the process of deforestation (Figure 1.13B), where the amount of forest lost each year through logging can be so small that many people do not notice the overall change taking place. Because the forest looks pretty much the same as it did the year before, people tend to believe that things are "normal"—a phenomenon some refer to as *creeping normalcy*. By the time the forest is gone and erosion has washed the topsoil into nearby stream channels, there may be no one with a living memory of the environment that once existed. Even more difficult to recognize are slow incremental processes that vary naturally from year to year, such as the climate change example in Figure 1.13B. Here natural upward and downward swings in temperature from year to year can easily mask the incremental change taking place. It was because of such natural swings in temperature that scientists were slow to recognize that Earth is currently in a warming trend.

A

B

FIGURE 1.14 Before and after photos of a coastal city in Indonesia that was completely obliterated by the 2004 tsunami. The low frequency of tsunamis means that several generations may pass between events, thereby lulling people to live in harm's way. People either become ignorant of the hazard or choose to accept the risk as they understand that the probability of such an event is low.

Earth as a System

It is clearly in society's best interest to avoid environmental problems and minimize their impacts, regardless of whether a particular problem is one we create ourselves (e.g., pollution) or cannot prevent (e.g., earthquakes). Having an appreciation of the time scale over which natural processes operate not only helps us identify problems, but allows us to make better risk management decisions. From the field of geology we also learn that Earth's history is one of constant change, including the evolution of its atmosphere, oceans, continents, and of course, plants and animals. In addition to geologic time, another important concept is that humans are part of a complex natural system, and that our actions impact the very environment in which we live and on which we depend. For example, modern societies have grown and prospered because of our ability to modify the landscape for growing food, extracting natural resources, and constructing cities and transportation networks. These activities, however, also have unintended consequences that are highly undesirable, such as increased flooding, pollution, and destruction of wildlife habits. By using science to understand how the entire Earth operates as a large system, we can learn to minimize our existing environmental problems and avoid creating new ones.

Recall from our previous discussion that science strives to understand how the natural world operates, which in turn has led to many remarkable achievements. As our knowledge of the world has grown more detailed over the course of history, we have seen the emergence of specialized fields within science, namely mathematics, physics, chemistry, biology, and geology. Today there are subdisciplines within each of these fields where specialists study rather narrow aspects of the physical world in great detail. In geology there are many specialized fields, including the study of minerals (mineralogy), rocks (petrology), ancient life (paleontology), chemical reactions within the earth (geochemistry), movement of groundwater (hydrogeology), and earth's internal structure (geophysics). Note that many of

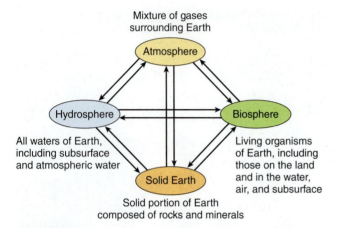

Mixture of gases
surrounding Earth

Atmosphere

Hydrosphere   Biosphere

All waters of Earth,   Living organisms
including subsurface   of Earth, including
and atmospheric water   those on the land
and in the water,
air, and subsurface

Solid Earth

Solid portion of Earth
composed of rocks and minerals

FIGURE 1.15   The Earth system is composed of four major subsystems that interact in a highly complex and integrated manner, where changes in one subsystem affect the others.

these subdisciplines are *interdisciplinary* in nature, which means they are a combination of two or more major scientific fields. For example, geochemistry is the study of both geology and chemistry.

The downside of specialization has been that scientists sometimes lose sight of nature as a whole. In recent years, however, there has been a growing interest in studying the interconnections that exist between major scientific fields. For instance, fisheries biologists have learned that individual fish species breed or spawn in very specific locations or habitats within river systems. These unique habitats are often found to be related to the type of sediment in the riverbed and to zones of groundwater discharge (springs), both of which are controlled by geologic processes. Scientists are learning that the natural world is highly interrelated, where individual processes depend on one or more different processes. Moreover, the entire Earth is now seen as operating as a single system where all natural processes are interconnected in one way or another. Scientists also now have a clear understanding that Earth's entire history is one of constant change, as evident by the evolution of the atmosphere, rise and fall of mountains, and evolution and extinction of different species.

Today this concept that all natural processes are interrelated in a constant state of change has led to a new, comprehensive field of study called **Earth systems science.** In this field the Earth is viewed as operating as a dynamic system made up of four major subsystems or components: the *atmosphere, hydrosphere, biosphere,* and *solid earth.* As illustrated in Figure 1.15, these subsystems are an integral part of the overall larger system, and they are continually interacting with one another. The Earth system is said to be *dynamic* because when one component undergoes a change it almost invariably affects one or more of the other subsystems. A good example is the damming of the Columbia River and its tributaries in North America (Figure 1.16). Groups that originally promoted the dams emphasized that the project would bring cheap electrical power and jobs to the Pacific Northwest. Plus, the vast amounts of water stored behind the dams would make large-scale agricultural production possible throughout the region. Unfortunately the dams also disrupted the region's natural hydrology, resulting in the collapse of the salmon fisheries within the biosphere. The once thriving fisheries that had sustained indigenous populations for thousands of years are now all but gone. This has resulted in a loss of fishing-related jobs and a host of changes for the small communities throughout the region, both in terms of their local economies and traditional way of life. By modifying the regional hydrology to achieve certain benefits, humans had set in motion changes within the Earth system that ended up producing some very undesirable consequences.

In some cases human activity produces changes that ripple through the entire Earth system. A good example is clearing the land through deforestation. The sequence of satellite images in Figure 1.17 shows the tremendous loss of forest in parts of South America in just a 25-year time span. From the example photo in Figure 1.18, taken of Madagascar, we can get a better sense of the devastating environmental impact of deforestation, particularly in areas of more rugged terrain. What

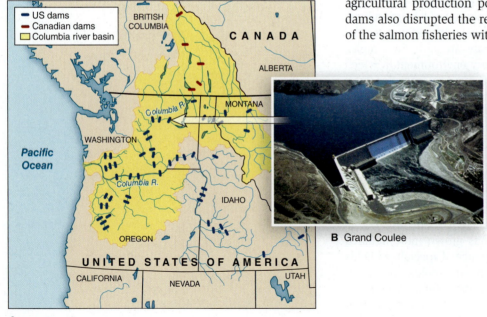

B Grand Coulee

A

FIGURE 1.16   The construction of dams along the Columbia River system, such as the Grand Coulee Dam shown here, provides electrical power and water to large areas of the Pacific Northwest. The dams also prevent the migration of salmon, which has led to the collapse of the region's salmon fishing industry and traditional way of life.

1975   **1992**   **2000**

FIGURE 1.17   Sequence of false-color satellite images showing deforestation near Santa Cruz de la Sierra, Bolivia, from 1975 to 2000. Areas of forests and dense vegetation are shown in deeper shades of red, whereas areas cleared of trees are in lighter shades.

was once a tropical forest, incredibly rich in plants and animals, is now relatively devoid of life. A closer examination of this photo reveals not only a devastated biosphere, but a landscape heavily scarred by landslides and where the soils have been lost due to the lack of forest cover. This illustrates how deforestation of the biosphere can have a major impact on the solid Earth component of the Earth system (see Figure 1.15). The example photo from Madagascar also reveals a river choked with sediment that has been washed off the landscape. This presents a serious problem in that the excess sediment has destroyed the habitat of aquatic species within the river, which in turn disrupts the entire stream ecosystem. The additional sediment also impacts the hydrologic component of the Earth system by filling the stream channel, leaving less room for water and increasing the chance of a flood during heavy rains. Finally, we need to consider what happens to the biosphere itself. When the land undergoes deforestation, much of the wood is converted into lumber, but a significant portion is simply burned, which releases carbon dioxide gas into the atmosphere. Deforestation then contributes to the problem of global warming since carbon dioxide is known to help control the heat balance within the atmosphere (Chapter 16). The effect on global warming is compounded by the fact that the trees had at one time been an important means of removing carbon dioxide from the atmosphere.

Throughout this text we will examine numerous environmental issues in which scientists use the concept of Earth as a system to first identify the underlying cause of a problem, and then develop effective solutions. Remember that a problem must first be identified before it can be addressed, and effective solutions are possible only if we use the scientific method to understand the nature of the problem.

FIGURE 1.18   Madagascar at one time had a rainforest ecosystem that was rich in natural resources, but deforestation rendered the land practically useless as shown here. In addition to losing the plants and animals, removing the tree cover triggered landslides and massive soil erosion, causing the streams to become choked with sediment. This led to the collapse of the aquatic ecosystem and increased flooding as the channels became clogged with sediment.

The Earth and Human Population

When humans modify the natural environment (e.g., damming rivers, deforestation, etc.), it is usually done with good intentions so as to achieve some positive benefit. However, our activity almost always impacts one or more natural processes, generating secondary effects that ripple through the various components of the Earth system. The result is a series of unintended consequences that ultimately affect the lives of people and other living organisms that inhabit the planet. There are certain types of geologic processes though, such as earthquakes and volcanic eruptions, in which humans have no ability to control or influence the process itself.

Regardless of whether an environmental problem is a result of our own actions or is a natural hazard we cannot control, our focus in this text will be on how humans interact with the geologic environment. It should be obvious that as human population continues to increase there will be greater interaction between people and the environment. For example, as population grows, more people will be living in areas at risk of floods, earthquakes, volcanic eruptions, landslides, and hurricanes. More people also mean greater resource depletion, pollution, modifications to the landscape, and habitat destruction. Therefore, with population growth our ability to impact the Earth system becomes greater. Because human population plays such a key role in environmental issues, we will begin by taking a closer look at population growth.

Population Growth

We are all familiar with things that increase over time, such as gas prices, traffic congestion, or college tuition. Our interest here is about *how* things increase. Scientists classify growth rates as being either linear or nonlinear, meaning their graphs will plot either as a straight line or a curve as shown in Figure 1.19. **Linear growth** can be defined as when the amount added over successive time periods remains the same. In other words, if 10 is added to the total one month, then 10 more is added the next month, and so on. When you plot the growing total against time, the result is a straight line with a constant slope. Nonlinear or **exponential growth** occurs when the amount added over successive time increments keeps increasing. Adding 10 to the total one month, 15 the next, and then 25, would be considered exponential growth, generating a graph where the slope increases with time.

Because the slope keeps getting steeper, exponential growth leads to much greater increases over time compared to linear growth. For example, the data used in the plots shown in Figure 1.19 are listed in Table 1.2. Notice how the initial sum of $1,000 grows linearly at $100 per year. After 10 years this $1,000 would grow to $2,000. If the money grew exponentially at a fixed percentage of 10% per year, then after 10 years the total would be $2,593 rather than $2,000. Why does this happen? The answer is that with linear growth the amount added each year stays the same since it *does not* depend on the total. However, with exponential growth the yearly increase *does* depend on the total because the increase itself is a percentage of the total. This is why the amount added each year in Table 1.2 keeps getting larger under exponential growth, but remains constant under linear growth.

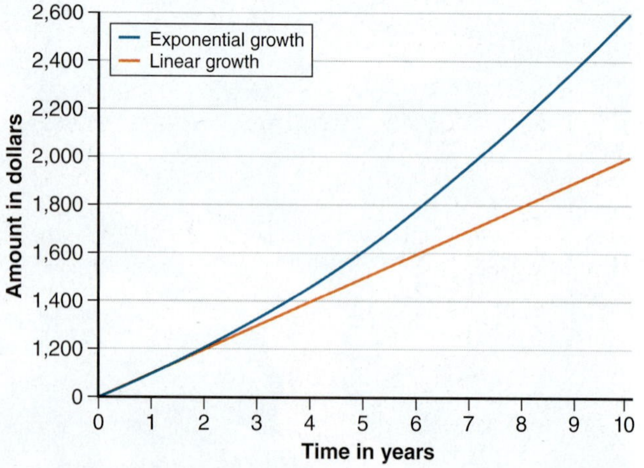

There are many natural processes that expand exponentially, but the most important in terms of our discussion is human population growth.

FIGURE 1.19 Plots showing how an initial sum of $1,000 responds to a linear growth rate of $100 per year versus an exponential (nonlinear) rate of 10% per year.

From Figure 1.20 one can see that throughout most of history the population of the modern human species, known as *Homo sapiens,* grew quite slowly. Then, around the 1700s, population growth accelerated rapidly in response to increased industrialization and advances in medicine. This brought about a decline in death rates while birth rates remained steady. Also note in the graph that it took our species until 1830 to reach a population of 1 billion, which represented a time span of nearly 200,000 years. Incredibly, the population then doubled and reached 2 billion in just 100 years. It doubled again to 4 billion in a mere 45 years and is now projected to reach 9.5 billion by 2050. This leads to an obvious question: can our population continue to expand exponentially or is there a limit to how many people Earth can hold?

TABLE 1.2   Calculations showing how an initial sum of $1,000 responds to linear and exponential growth rates. Each interval lists the beginning and ending totals as well as the yearly increase. Note how the yearly increase is constant under linear growth but keeps expanding under exponential growth.

	Linear Growth	Exponential Growth
End of year 1	$1,000 + $100 = $1,100	$1,000 + $100 = $1,100
End of year 2	$1,100 + $100 = $1,200	$1,100 + $110 = $1,210
End of year 3	$1,200 + $100 = $1,300	$1,210 + $121 = $1,331
End of year 4	$1,300 + $100 = $1,400	$1,331 + $133 = $1,464
End of year 5	$1,400 + $100 = $1,500	$1,464 + $146 = $1,610
End of year 6	$1,500 + $100 = $1,600	$1,610 + $161 = $1,771
End of year 7	$1,600 + $100 = $1,700	$1,771 + $177 = $1,948
End of year 8	$1,700 + $100 = $1,800	$1,948 + $195 = $2,143
End of year 9	$1,800 + $100 = $1,900	$2,143 + $214 = $2,357
End of year 10	$1,900 + $100 = $2,000	$2,357 + $236 = $2,593

Limits to Growth

The idea of there being a limit to the number of humans that Earth can support was first proposed around 1800 by a British political economist named Thomas Malthus. Malthus attributed the deteriorating living conditions of England's poor at the time to food production not being able to keep pace with the rapidly expanding population. From his mathematics studies, he recognized that human population growth was exponential, whereas increases in food production were linear. Malthus concluded that unless population was brought under control, it would outstrip food supply and lead to a future of poverty and recurring famine. Famine, after all, was nature's way of keeping the population of animals in line with their available food supply. Take deer for example, whose population if left unchecked will quickly grow to the point where they outstrip their food supply. Starvation and disease then follow, resulting in a population crash. Eventually the deer population rebounds, marking the beginning of a new cycle.

Malthus's ideas on population limits have been highly influential over the years, leading to various predictions that human population would collapse in a catastrophic manner. What his population model failed to take into account, however, was the ability of humans to increase food production at an exponential rate through technology and innovation. The driving force behind this modern production increase has been the use of fossil fuels, particularly crude oil and natural gas (Chapter 13). Oil has made the use of mechanized farm equipment possible along with irrigation techniques that depend on diesel-powered pumps. Equally important to increased agricultural yields has been the use of synthetic fertilizers and pesticides that are produced from oil and gas. All this has allowed world population to continue to expand, rather than collapse as Malthus's model predicted. The result is that cities around the world grow ever larger and spread across the landscape in a process known as *urbanization* or *urban sprawl* (Figure 1.21).

Despite the fact that technology and innovation have allowed food supply to keep pace with population, environmentalists continue to sound the alarm that population growth will eventually outstrip the planet's ability to support all of its human inhabitants. One of the problems facing the

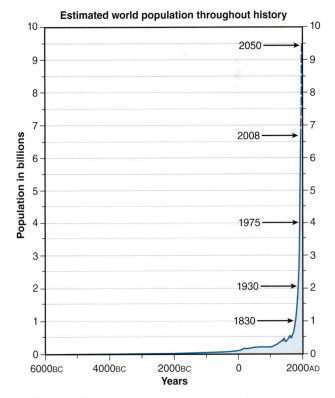

FIGURE 1.20   Graph showing exponential growth of world population throughout history. Population is projected to reach 9.5 billion by the year 2050.

FIGURE 1.21   Sequence of false-color images (A) showing urban growth over a 29-year period in the Dallas/Fort Worth metro area (urban area is shown in shades of gray, vegetation cover is in red). Note the water-supply reservoir that was added between 1974 and 1989. Close-up view (B) showing high-density development in the Cuautepec area of Mexico City.

world in terms of future food production is that topsoils worldwide are slowly being lost. The culprit here is agricultural practices that leave soils exposed to erosion (Chapter 10). Another critical issue is that water supplies are being stretched to the limit in many regions, leaving little room for expanding our current irrigation practices (Chapter 11). Equally troublesome is the fact that the production of conventional supplies of oil and gas is expected to decline soon, causing shortages and price increases (Chapter 13). This in turn will result in higher food prices, leaving many people unable to afford the higher costs.

It seems reasonable to expect that food production will ultimately reach a maximum, particularly since humanity is already approaching the limits of what Earth can supply in terms of soil, water, and fossil fuel resources. Moreover, the modern conveniences and consumer goods that people are accustomed to in developed countries are also highly dependent on Earth resources. The industrial, commercial, and transportation sectors of these nations' economies require significant quantities of mineral, energy, and water resources in order to function. As population growth causes these resources to become scarcer, developed nations will find it increasingly more difficult to maintain the current state of their economies.

Sustainability

The limiting factor to future economic and population growth basically comes down to Earth's ability to provide natural resources. This leads us to the concept of **sustainability,** which simply means being able to maintain a system or process for an indefinite period of time. With respect to humans living within the Earth system, the term *sustainable society* is used to describe a society that lives within Earth's capacity to provide resources such that resources remain available for future generations. The concept of sustainability is something that is readily observed in both nature and in our daily lives. Consider the earlier example where a deer population is allowed to grow to the point it outstrips its natural food supply, resulting in starvation and a population crash. Another example is a person's finances. Suppose you make $50,000 a year, but spend $60,000. This deficit spending is not sustainable as it will eventually lead to a level of debt where the interest makes it impossible to pay off. To avoid bankruptcy you would have to either earn more money or spend less by changing your lifestyle. There are no other choices. The relationship exists between the human demand for resources and Earth's limited ability to provide resources. Therefore, the only options for humans are to maintain a steady population, change their lifestyles, or suffer the consequences of living beyond Earth's means.

Suppose that the human race is unable to control its exploding population and ends up overwhelming the planet's ability to provide resources. The consequences would likely vary among nations due to differences in population growth rates, level of economic development, and types of resources being consumed. From Figure 1.22 one can see that there are considerable differences in population growth rates between developing nations and those that are more developed with higher living standards and more consumer goods. Note how the population in developing countries is presently growing very rapidly compared to developed countries (e.g., the United States, members of the European Union, and Japan). The difference is partly due to the high birth and death rates in developing nations, which creates a pyramid-shaped population distribution as illustrated in Figure 1.23. Here large numbers of people are in age groups with the potential to bear children, generating exponential population growth. On the other hand, developed nations typically have much lower birth and death rates, which creates a population distribution with relatively few people who have the potential to bear children, hence the low growth rates.

Because of the vastly different growth rates, the population of less-developed countries is projected to rise dramatically, whereas the population in developed nations will decline (see Figure 1.22). In fact, the population of some European countries has already started to decline, making further economic expansion more difficult due to a surplus of retirees and a shortage of workers. The rapid population growth of less-developed countries, on the other hand, will make it increasingly more difficult for them to obtain adequate supplies of food and water. Even if these countries had enough money to import food, world food production is limited by the availability of Earth's soil and water resources. To help illustrate this point, notice the uneven distribution of the world's

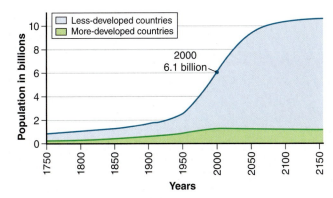

FIGURE 1.22   Graph showing world population growth and projected trends in both developed and developing countries.

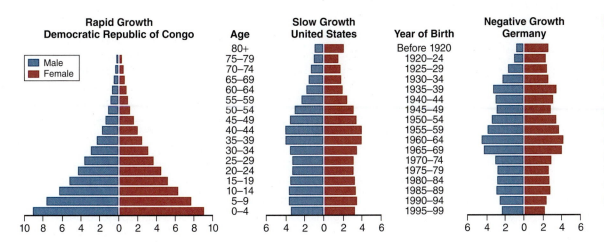

FIGURE 1.23   Developing nations generally have rapid population growth because of high birth and death rates that result in age groups with large numbers of people who potentially can have children. Developed nations typically experience slow or negative growth because of low birth and death rates that lead to relatively few people who may potentially have children.

population shown in Figure 1.24. Here one can see that the areas with low population density closely correspond to those areas with desert and polar climates. People historically have tended to avoid living in these climatic zones due to the lack of liquid water and difficulty of finding or growing food.

Some people believe that developing nations will be able to increase their food production in a similar manner as have developed nations, namely through mechanization, irrigation, and synthetic fertilizers and pesticides. They also point out that with increased economic prosperity the high birth rates in these countries will fall, at which point global population would stabilize and become sustainable. The problem is that as developing nations begin to modernize they naturally increase their per capita (per person) consumption of all resources. This places additional demands on Earth's finite mineral and energy resources, and thereby threatens the living standards of the developed countries. For example, China's rapid industrialization is currently placing such an additional demand on the world's dwindling supplies of crude oil (Chapter 13) that the market is struggling to meet world demand. This in turn is driving up oil prices and could soon strain the global economy to the point where living standards start to decline.

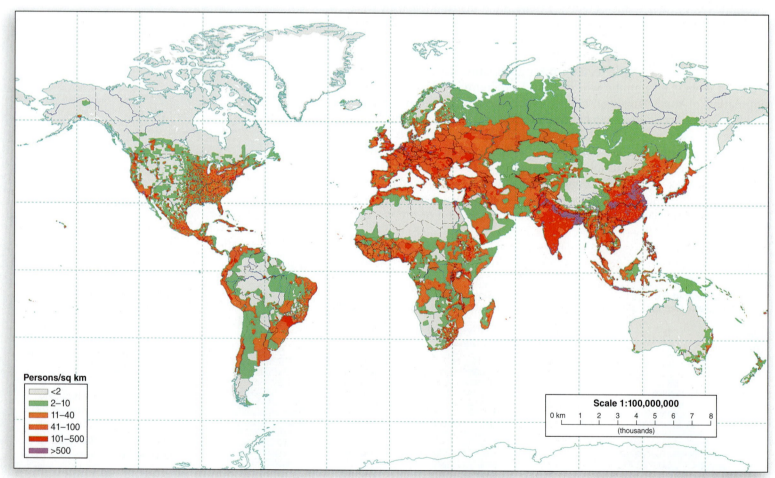

Persons/sq km
- <2
- 2–10
- 11–40
- 41–100
- 101–500
- >500

Scale 1:100,000,000
0 km 1 2 3 4 5 6 7 8
(thousands)

FIGURE 1.24   Map showing global population density in 1994. Note how vast regions that are sparsely populated correspond to desert and polar climates where food production is limited or nonexistent.

The key to sustainability, therefore, is not just Earth's total population, but also humanity's per capita consumption rate of resources. More people naturally means greater demand on Earth's resources, but when per capita consumption rates increase along with population, the depletion of resources will follow a nonlinear downward path, as shown in Figure 1.25. Consider for a moment that if China's entire population were to achieve the present living standards of developed nations, then humanity's impact on the planet would double. If all of Earth's inhabitants were to attain these higher living standards, then demand on resources would increase 14-fold. Achieving a sustainable society at this level of consumption would be virtually impossible.

Ecological Footprint

It should be apparent that one way to view sustainability is from the perspective of how many people Earth can feed with its existing water and soil resources. Another is the number of people living at developed world standards that can be supported by the planet's mineral and energy resources. We can also look at sustainability in terms of maintaining the present-day biosphere of the Earth system. A particularly useful concept here is the idea of an **ecological footprint,** which is simply the amount of *biologically productive* land/sea area needed to support the lifestyle of humans. The idea behind an ecological footprint is that every human requires a certain portion of the biosphere for extracting the resources they need and absorbing the waste they generate. Remember, we depend on the biosphere for its ability to purify water, provide forest resources, and regulate oxygen and carbon dioxide levels in the atmosphere. Simply put, we humans could not survive without the ecosystems that make up the biosphere.

Biologists have estimated that the ecological footprint for all of humanity is currently six acres per person. Due to their higher rates of per capita consumption, citizens living in developed countries have a much larger footprint than the global average. For example, the Swiss average is 10 acres per person and British 13 acres, whereas Americans require a staggering 24 acres per person. The Chinese average is presently about four acres per person, but is expected to rise as China continues its rapid industrialization and its citizens purchase more and more consumer goods. If all the people in developing counties were to achieve the living standards of developed nations, then humanity's ecological footprint would be far greater than the current average of six acres per person. According to the Global Footprint Network, a nonprofit organization, humanity's current ecological footprint is already estimated to be over 20% larger than what the planet can support. This means humans are consuming Earth's renewable resources faster than they can be replenished by the biosphere's ecosystems. Based on this footprint analysis, many people have concluded that humans have gone beyond the ecological limits of the planet and that the present state of humanity is not sustainable.

Many environmentalists believe the solution is for humanity to stabilize its population and to reduce its per capita consumption of resources through conservation. Otherwise we will have to suffer the consequences of living beyond the ability of Earth to support us. The collapse of the society on Easter Island (Case Study 1.1) is perhaps the best example of the consequences of people living in an unsustainable manner, ultimately destroying the very ecosystem on which they depended.

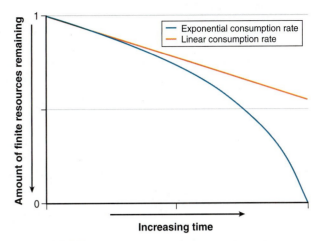

FIGURE 1.25 When an expanding population begins to increase its per capita consumption rate of a finite resource such as crude oil, the depletion of resources will accelerate and follow a nonlinear downward path.

Collapse of a Society Living Unsustainably

While exploring the remote reaches of the Pacific in 1722, Dutch sailors came upon a place they named Easter Island, which was very different from all the others they had seen in the region. Here sailors found an island inhabited by approximately 2,000 people who looked very similar to the Polynesians they had encountered throughout the Pacific. However, unlike the seafaring Polynesians with their large canoes crafted from solid tree trunks, Easter Islanders came out to greet them in small, leaky crafts made from a patchwork of small planks and timbers. Even more odd was Easter Island itself. Most other islands in this subtropical climate had rich volcanic soils and lush forests teeming with birds. Instead of a rich paradise, the Dutch found a windswept and grass-covered wasteland, completely devoid of large trees and any native animals larger than insects. Equally strange were the more than 200 stone statues (Figure B1.1) lining the coastline and another 700 in various stages of development. Some statues weighed over 80 tons and were somehow transported as much as 6 miles (9.7 km) from a single quarry in which they were carved. Uncompleted statues weighed as much as 270 tons!

Several intriguing questions have been raised in the years since Easter Island was discovered. First, why was this grass-covered wasteland so unlike the surrounding islands where lush forests, abundant birds, and flowing streams are common? How could a population of only 2,000 manage to transport and erect 200 huge statues, particularly since they had no suitable trees for making heavy timbers and ropes? Moreover, why would people who were living in caves and struggling to survive by raising chickens and growing crops in thin soils expend such great effort erecting statutes? Also puzzling was the islander's oral history, where they told of their ancestors routinely visiting a well-known reef located 260 miles (420 km) away. How could they have made such a journey in leaky canoes barely capable of sailing offshore their own island? Finally, how in the world did these people ever come to colonize Easter Island in the first place?

The answers to these questions didn't come about until modern times when scientists from various disciplines began collecting data. Here radiocarbon dates from archaeological excavations first showed that human activity began on Easter Island somewhere between 400 and 700 AD; followed by peak statue construction from 1200 to 1500. Moreover, the density of archaeological sites indicated a population of around 7,000 during the peak period—some estimates go as high as 20,000. A serious population crash obviously must have occurred since only 2,000 people were present when the Dutch arrived in 1722. Also quite revealing was the analysis of pollen spores that had fallen into wetlands and incorporated into the sediment record. By comparing the pollen grains found in the various sediment layers to known plant species, scientists could determine the abundance of different plants over time. This analysis proved that large palm trees (up to 6 feet in diameter and 80 feet tall) along with numerous species of shorter trees, ferns, and shrubs had blanketed Easter Island for more than 30,000 years prior to the first human inhabitants. These palm trees would have been ideally

suited for constructing the large ocean-going canoes and the timbers needed to transport and erect statutes. Similar palm trees today are used in other cultures for their edible nuts and sap.

By examining the bones found in old garbage dumps, scientists were able to determine that the islanders' diet consisted of large native birds and dolphins in addition to palm nuts. The fact that people feasted on dolphins rather than fish was explained by the relatively deep water and corresponding lack of coral reefs around Easter Island. This forced the people to sail offshore in order to harvest the only sea animal available in any abundance, namely dolphins. For this, of course, they needed large, seaworthy canoes. Easter Island's extensive forest then provided a direct source of food (birds, nuts, and sap) and the large trees allowed them to build canoes for hunting dolphins

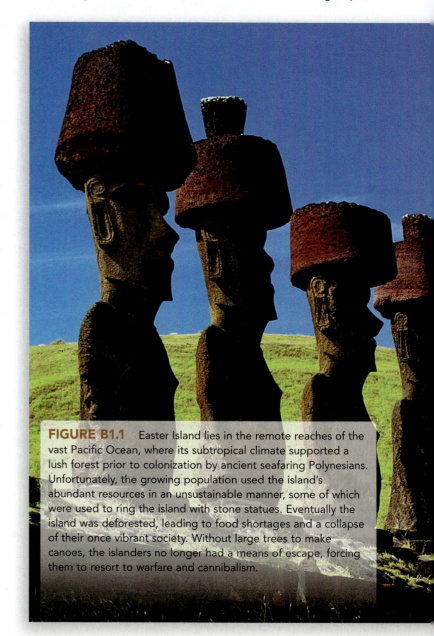

FIGURE B1.1 Easter Island lies in the remote reaches of the vast Pacific Ocean, where its subtropical climate supported a lush forest prior to colonization by ancient seafaring Polynesians. Unfortunately, the growing population used the island's abundant resources in an unsustainable manner, some of which were used to ring the island with stone statues. Eventually the island was deforested, leading to food shortages and a collapse of their once vibrant society. Without large trees to make canoes, the islanders no longer had a means of escape, forcing them to resort to warfare and cannibalism.

and traveling to distant islands. It was this abundance of natural resources that enabled the population to expand and develop into a highly organized society. Clearly, erecting monuments around the entire island, all from a single quarry, must have required the organization and cooperation of a large number of people.

The obvious question now is why did this complex society collapse? Here again the pollen grains and bones provide the answer. By 800 AD the sediment record includes large amounts of charcoal from wood fires, but far fewer pollen grains from the palm trees. After inhabiting the island for only a few centuries, the people had begun the process of clearing the forest to grow crops and to provide wood for building canoes, fueling fires, and transporting statues. Shortly after 1400 AD the palm tree finally became extinct. During this period of deforestation the analysis of bones in the islanders' garbage dumps showed that the number of birds in their diet had decreased dramatically. By around 1500 AD the bones of native birds as well as dolphins were completely lacking. As the forest was being cut, the islanders were inadvertently eliminating birds and palm nuts from their food supply. Dolphins were soon removed from their diet once there were no more large trees for making seaworthy canoes. Deforestation also led to severe soil loss and decreased the ability of rainwater to infiltrate the soils. This in turn resulted in thinner and dryer soils, reducing crop production and altering the island's hydrology such that many of its streams stopped flowing. Not surprisingly the islanders slowly began to experience both water and food shortages.

The islanders obviously tried to develop new food supplies, but unfortunately their only choices came down to raising more chickens and eating each other. Soon their highly ordered society began to unravel into smaller groups of rival tribes, forced to resort to cannibalism and warfare as they all competed for the few remaining resources. By the time the Dutch arrived in 1722, the once vibrant society had completely collapsed, leaving a struggling population that was a mere fraction of its former size. Without the large palm trees for making seaworthy canoes, the islanders had no means of escape. They were stuck.

In the recent book *Collapse* by Jared Diamond, the author asks the obvious questions regarding the demise of Easter Island's society: "Why didn't they look around, realize what they were doing, and stop before it was too late? What were they thinking when they cut down the last palm tree?" Diamond concludes that the islanders didn't see the problem because of what he refers to as *creeping normalcy*. As the years slowly passed, generation after generation of islanders saw only small changes in the amount of forest; hence it looked "normal." Islanders who understood and warned of the dangers of deforestation likely would have been drowned out by those in society whose jobs depended upon harvesting the trees. Diamond argues that by time the last palm tree was cut, the large old-growth trees were a distant memory. The only remaining trees were small and of little economic significance, thus no one would have noticed the last tree being cut.

Many environmentalists share Diamond's view that what happened on Easter Island is a small-scale example of what is currently happening on the entire globe. As Earth's rising population continues to consume key resources such as petroleum, water, and soil at an unsustainable rate, we are putting our global society at risk of collapse. Moreover, the burning of vast amounts of fossil fuels is having the unintended consequence of accelerated global warming, which may in turn threaten the very survival of humans. Similar to Easter Island in the middle of the vast Pacific, Earth is a blue speck in the vastness of space whose people have nowhere to go should they overexploit their natural resources. Unlike Easter Islanders, however, we have history from which to learn the mistakes of others. The question is, will enough of us learn the lessons from history before it is too late?

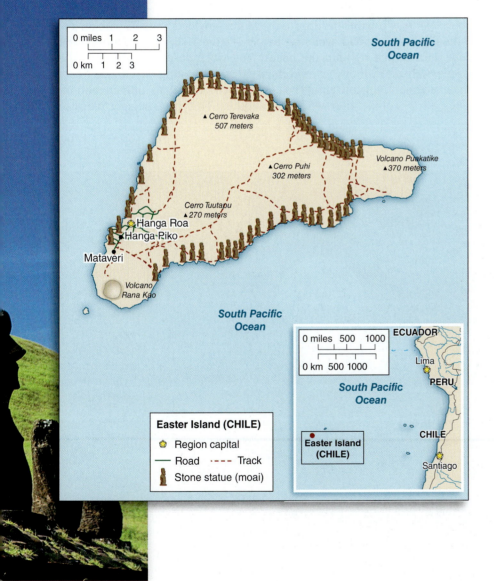

0 miles 1 2 3

0 km 1 2 3

South Pacific Ocean

▲ Cerro Terevaka
507 meters

Volcano Puakatike ▲370 meters

▲ Cerro Puhi
302 meters

Cerro Tuutapu
▲270 meters

Hanga Roa
Hanga Piko

Mataveri

Volcano Rana Kao

South Pacific Ocean

0 miles 500 1000

0 km 500 1000

ECUADOR

Lima

PERU

South Pacific Ocean

Easter Island (CHILE)

CHILE

Santiago

Easter Island (CHILE)

⚙ Region capital
— Road   --- Track
🗿 Stone statue (moai)

Environmentalism

Environmental awareness in the United States began in the 1960s and 1970s as a grassroots movement, driven in large part by widespread water and air pollution. One of the sparks for this environmental movement was Rachel Carson's classic 1962 book, *Silent Spring.* Her book helped awaken both the public and other scientists to the fact that Earth's complex web of life is sensitive to environmental change, particularly pollution. The basic problem in the United States was that industries historically had been free to discharge their waste by-products into the atmosphere and nearby water bodies. However, people soon began to demand change as they recognized that pollution was fouling their air and water as well as their beaches, fishing holes, and other recreational sites. Eventually federal laws, such as the Clean Air Act (1970) and Clean Water Act (1972), were enacted to force industries to properly dispose of their waste (Chapter 15). Businesses could no longer freely dump waste into the environment and pass the cleanup and health costs on to society. Proper waste disposal now had to be treated as a business expense.

Another defining event in the environmental movement came in 1968 when Apollo astronauts heading to the Moon provided humans with a view of the Earth no one had ever seen before (Figure 1.26). People around the globe were struck by both the beauty and isolation of our planet in the darkness of space. It also caused many to start thinking of humanity as a single race, surviving on a fragile oasis in space rather than different nationalities all in competition with one another. This new perspective also

FIGURE 1.26 Photographs of planet Earth taken by Apollo astronauts helped humans understand that Earth behaves as a system and that we depend on the system for survival.

helped us see how Earth operates as a system, providing the necessary resources that make our lives possible. Moreover, it helped us understand that we could damage this system and overuse its limited resources such that the planet becomes less hospitable for us humans. Should this occur, we would be stuck on this island in space with nowhere else to go, similar to the Easter Islanders in the vast Pacific (Case Study 1.1).

The environmental regulations passed in the United States since the 1970s succeeded in eliminating the most visible and obvious forms of pollution, leading many people to think pollution is no longer a problem. Subtle forms of pollution, however, still exist and pose a threat to the health of both humans and ecosystems in the biosphere. Another consequence of the environmental regulations was that the federal government began exercising its authority over individual and state property rights in order to protect the health of all citizens. This issue of personal property rights, combined with the perception that pollution is no longer a problem, has contributed to a backlash against the environmental movement in recent years. Today various groups and individuals portray environmentalists as "wackos," "tree huggers," or "ecoterrorists" who feel that plants and animals are more important than people. This is simply not true. Environmentalism is not just about saving owls in a forest or fish in a river; it is about saving humans from themselves. Forests and wetlands, for example, are important not simply because they contain interesting plants and animals, but because they provide people with clean water and help regulate our climate. If we end up destroying Earth's ecosystems, then humans would find it difficult to survive.

The biggest environmental issue facing the human race is sustainability. Will we be able to make use of the Earth's limited resources in a sustainable manner, or will our population outstrip the planet's ability to support us? The answer to this question will depend on whether we can control population growth and reduce per capita consumption of resources through conservation. This will also require many societies to change from one of constant growth to one of nongrowth, similar to what is taking place today in some developed countries with stable populations. The problem is that many nations define economic success in terms of increased numbers of new homes and jobs, and expanded factory production and trade. These economic indicators all depend on greater numbers of people consuming greater amounts of earth resources. Both science and common sense tell us that it is not possible to have permanent economic growth on a planet whose resources are finite. David Brower, a leading environmentalist for over 50 years, often spoke on the subject of sustainability and the rapid pace at which we have been consuming Earth's resources since the Industrial Revolution. John McPhee described Brower's thoughts on this subject in the 1971 book *Encounters with the Archdruid:*

> Sooner or later in every talk, Brower describes the creation of the world. He invites his listeners to consider the six days of Genesis as a figure of speech for what has in fact been four billion years. On this scale, a day equals something like six hundred and sixty-six million years, and thus "all day Monday and until Tuesday noon, creation was busy getting the earth going". Life began Tuesday noon, and "the beautiful organic wholeness of it" developed over the next four days. "At 4 p.m. Saturday, the big reptiles came on. Five hours later, when the redwoods appeared, there were no more big reptiles. At three minutes before midnight, man appeared. At one-fourth of a second before midnight, Christ arrived. At one-fortieth of a second before midnight, the Industrial

Revolution began. We are surrounded with people who think that what we have been doing for that one-fortieth of a second can go on indefinitely. They are considered normal, but they are stark, raving mad…. We've got to kick this addiction. It won't work on a finite planet. When rapid growth happens in an individual, we call it cancer."

It is certainly not pleasant to think of the human race as being a detriment to the planet, but we are indeed having an enormous impact on the environment. We have eliminated, and continue to eliminate, large numbers of species by destroying their habitat. Some of these species have been around for over 250 million years, a time when dinosaurs began roaming the Earth. Our impact may become so great that the planet will simply no longer be able to provide sufficient resources, causing the human population to decline to more sustainable levels. Antienvironmentalists often state that we do not need to worry because the Earth is simply too large for us to destroy, and in a sense, that is true. However, it is an undeniable fact that Earth is an interactive system that is responding to our actions. A very real concern is that we could disrupt this system to the point where the climate is no longer hospitable for humans. While the task of creating a sustainable society will certainly not be easy, it is possible because we are a species that has been given the gift of being able to make intelligent choices. Our ability to make choices that will impact our future relationship with Earth's environment is perhaps best summed up in the following excerpt from the August 18, 2002, issue of *Time* magazine:

> For starters, let's be clear about what we mean by "saving the earth." The globe doesn't need to be saved by us, and we couldn't kill it if we tried. What we do need to save—and what we have done a fair job of bollixing up so far—is the earth as we like it, with its climate, air, water and biomass all in that destructible balance that best supports life as we have come to know it. Muck that up, and the planet will simply shake us off, as it's shaken off countless species before us. In the end, then, it's us we're trying to save—and while the job is doable, it won't be easy.

SUMMARY POINTS

1. Geology is the study of the solid earth. Environmental geology is the study of how humans interact with the geologic environment, particularly with regard to geologic resources and hazards.

2. Scientists develop hypotheses in order to *explain* phenomena in the natural world that can be observed or measured. All hypotheses must be falsifiable and are considered valid as long as they remain consistent with all existing data. Supernatural explanations are not scientific because they're not falsifiable.

3. A theory describes the relationship between several different hypotheses, and thus provides a more comprehensive explanation of the natural world. Scientific laws describe natural phenomena in which the relationship between different data occurs regularly and with little deviation.

4. Environmental problems related to geology generally fall into one of two categories: hazards and the use of resources. Humans sometimes make these problems worse due to a lack of scientific understanding or appreciation for the time scale in which natural processes operate.

5. Some geologic processes operate in a sporadic manner, producing dramatic and sudden changes that can be disastrous for humans living nearby. The more frequent these hazards, the more likely people will take steps to minimize their risk. Environmental problems associated with incremental processes are challenging for society because they can be hard to recognize.

6. Earth is a complex system made up of several subsystems: atmosphere, biosphere, hydrosphere, and solid earth. A change or disruption within one of the subsystems invariably leads to changes in one or more of the others. Scientific knowledge of how the Earth system operates can help society solve existing environmental problems and avoid creating new ones.

7. Humans are part of the biosphere and are therefore an integral part of the Earth system. The way in which we interact with the Earth system can have a profound impact on the very environment upon which we depend.

8. Geologic hazards and resource issues become more pronounced as human population continues to grow. Exponential population growth exacerbates both these problems.

9. Developing countries have high population growth rates due to high birth and death rates that generate large numbers of people of potential child-bearing age. Developed nations have much lower growth rates because lower birth and death rates result in relatively few people of child-bearing age. Developed nations also have much higher per capita consumption rates of resources.

10. Modern agricultural practices have allowed world food production to keep pace with population growth. Food production and population will reach a maximum due to the worldwide loss of topsoils and limited water supplies. Other limiting factors include finite mineral and energy resources needed for fertilizers, pesticides, and mechanized farm equipment.

11. As developing nations modernize and increase their per capita consumption rates, they place exponentially greater demands on Earth's limited resources. Earth could not sustain all of humanity living at developed-country standards.

12. Sustainability can also be viewed in terms of the amount of biosphere each person requires for resources and waste disposal. Biologists estimate humanity's current ecological footprint is 20% larger than what the planet can support, thus the present state of humanity is not sustainable.

13. In order for humans to live in a sustainable manner, many environmentalists believe humanity needs to stabilize its population and reduce per capita consumption of resources through conservation.

KEY WORDS

absolute age   14
earth resources   13
Earth systems science   20
ecological footprint   27
environmental geology   6
environmental risk   17
exponential growth   22
geologic hazard   12

geologic time   16
geologic time scale   14
geology   6
historical geology   6
hypothesis   8
law   9
linear growth   22
multiple working hypotheses   9

physical geology   6
radiometric dating   15
relative age   14
scientific method   8
sustainability   24
theory   9
tragedy of the commons   4

APPLICATIONS

Student Activities   Look around your house/apartment/dorm room. Can you find any materials that relate to geology? Do you have a granite counter top? Slate floor? Salt in your kitchen? Dry wall (made from gypsum)? Do you have any decorative rocks in your living space?

Critical Thinking Questions
1. Do you think in a scientific way?
2. How do humans interact with the Earth?
3. Is the human time scale similar or different to the geologic time scale?
4. What is sustainability?

Your Environment: YOU Decide   What are the similarities and differences in sustainability in the developed world versus the developing world?

Chapter 2

Earth from a Larger Perspective

LEARNING OUTCOMES

After reading this chapter, you should be able to:

▶ Understand how the nebular hypothesis explains the formation of the solar system and how it accounts for the orbital characteristics of the planets and moons.
▶ Describe our solar system and the size of the Earth relative to the size of the solar system as well as to the size of our galaxy and the universe.
▶ Explain how extremophile bacteria are related to the origin of life on Earth and how they relate to the extraterrestrial search for life.
▶ Understand the concept of habitable zones and why complex animal life that may exist elsewhere will likely be restricted to such zones.
▶ Know what mass extinctions are and be able to name some of their possible triggering mechanisms.
▶ Understand how scientists came to appreciate the serious nature of comet and asteroid impacts and the steps being taken to reduce the risk.

In recent years humans have sent machines into space in order to learn more about the solar system and universe. This knowledge in turn helps us to better understand the Earth system and gives us a larger perspective from which to view environmental problems on our own planet. Shown here is the space shuttle *Columbia* rocketing toward space in June 1992.

Introduction

At first glance one may question why a textbook on environmental geology includes a chapter on what is beyond planet Earth. This chapter was included in part because our planet operates within an astronomical environment that has a major influence on the Earth system and the environment in which we live. Consider how the Sun generates wave energy (e.g., visible light, infrared, ultraviolet) that warms our planet and drives not only the climate system, but also the biosphere and hydrosphere. Although the amount of energy produced by the Sun has been fairly steady over much of Earth's history, subtle variations are known to produce significant changes within the Earth system. Another important astronomical or external force is the Moon's gravitational field. As the Moon orbits the Earth, its gravity is the dominant force responsible for creating the tides, where water and nutrients move within the coastal environment in a cyclic manner. Because the coastal environment serves as the nursery grounds for a large number of marine species, the Moon-induced tides therefore are critical to the ecosystem of the oceans. Finally, it is important to note that other planets in the solar system can also influence the Earth system. Here the gravitational fields of the planets will occasionally alter the trajectory of asteroids and comets such that they begin to cross Earth's orbit. This creates the potential for large impacts, whose consequences could be catastrophic for the present-day biosphere.

In addition to understanding the external forces that affect the Earth, Chapter 2 is intended to provide students with a better sense of how humanity fits into the larger scheme of things, namely the universe. Recall that one of the key themes in Chapter 1 was that humans are a small, but very important and influential part of the Earth system. Another key concept was that humans have been present on the Earth for only a small fraction of the planet's 4.6-billion-year history. Having an appreciation for the Earth as a global system, and the vastness of geologic time, is important if we are to effectively address the environmental problems facing humanity. In Chapter 2 we will go a step further and view the Earth from an even larger perspective, namely our astronomical environment, consisting not only of the solar system but the entire universe.

Humans have long sought this larger perspective by studying the stars and planets in the night sky, pondering the nature of our very existence. However, for most of history the only tools we had for learning what was beyond our planet were our naked eyes and the ability to reason. The telescope was a major advancement that provided many answers, but as is typical of the process we call science, this new knowledge led to new questions. The development of powerful rocket engines eventually allowed humans to escape Earth's gravity and send spacecraft to distant parts of our solar system, including the landing of probes on several planets and moons. Many consider the human exploration of the Moon (Figure 2.1) to be our single greatest achievement. In addition to probes and landing craft, scientists now have the orbiting Hubble space telescope along with an array of sophisticated ground-based telescopes (Figure 2.2). Together these telescopes enable us to peer into the farthest reaches of the universe with clarity that

FIGURE 2.1   Humans have long sought answers to what exists beyond planet Earth. Shown here is *Apollo 12* astronaut Alan Bean exploring the Moon in 1969.

A

B

FIGURE 2.2   Sophisticated space and ground-based instruments have allowed scientists to peer into the deepest reaches of the universe and gather data in ever-greater detail. Photo (A) is of the Hubble space telescope taken from space shuttle *Discovery;* photo (B) shows an array of 27 radio telescopes in Socorro, New Mexico, used to study everything from black holes to planetary nebula.

was unimaginable a mere 50 years ago. These instruments collect data from more than just visible light, revealing strange and incredible phenomena within the universe (Figure 2.3).

In Chapter 2 we will briefly explore some of the answers science has provided regarding Earth's place in the universe and the extraterrestrial forces that help shape our planet. We will begin by examining how our solar system formed and why Earth is the only place where life is known to exist. From there we will take a look at the relationship of our Sun and its system of planets to the other stars in the universe. We will end by examining some of the solar system hazards facing the Earth. Hopefully this larger perspective will give you a better understanding of the environmental challenges facing humanity, and a greater appreciation for the fact that Earth's environment is both rare and highly susceptible to change.

FIGURE 2.3   Modern telescopes collect data on many different types of phenomena. Shown here is a composite image from X-ray, radio, and optical telescopes showing the collision of a spiral galaxy and a black hole. A band of dust and gas is bisected by opposing jets of high-energy particles ejected away from the supermassive black hole in the nucleus. X-ray data is shown in blue, optical data in orange and yellow, and radio data in green and pink.

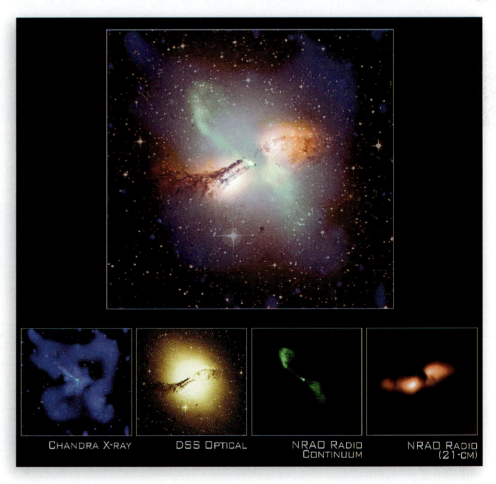

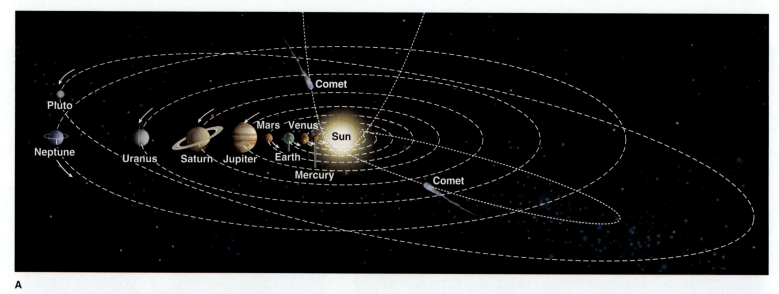

A

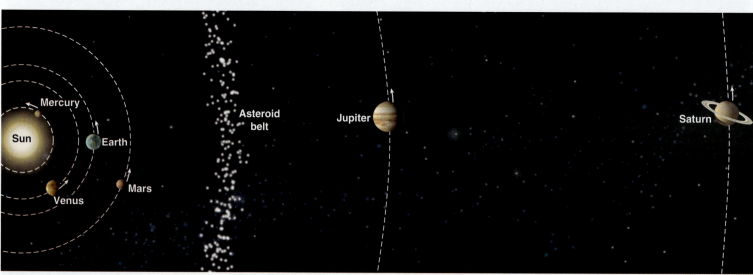

B

FIGURE 2.4 (A) Planets within the solar system and their orbital paths around the Sun—not drawn to scale (asteroids and asteroid belt not shown). (B) When the orbital paths are drawn to scale, one can begin to understand the vast distance between those planets out beyond Mars.

FIGURE 2.5 By compressing the distance between each of the planets, it becomes possible to view the relative size of the Sun and planets at the proper scale. Note how much larger the outer (gas) planets are compared to the rocky (terrestrial) planets of the inner solar system. The dwarf planet, Pluto, is shown on the far right. Images are actual photos.

38

Our Solar System

Before discussing the universe, we need to learn something about Earth's neighbors in the solar system. Our solar system consists of eight planets (Pluto has been reclassified), more than 100 named moons, a belt of asteroids, and millions of comets, all of which orbit the Sun (Figure 2.4). It is important to realize that Figure 2.4A was not drawn to scale. This was done because the distance between each of the planets is so great that it is impossible to show their correct size and orbits all in the same illustration. However, if only a portion of the orbits are drawn to scale, as shown in Figure 2.4B, then we can get a better sense for the vast distances between the planets, particularly those past Mars. In order to get an accurate view of the relative size of the planets, it is necessary to compress their distances as seen in Figure 2.5. Here the vast differences in size between the inner and outer planets becomes quite striking (Pluto again being an exception). Perhaps even more impressive is Figure 2.6 showing the Sun and all the planets all at the same scale. From this perspective we can see that the Earth is very small compared to the Sun.

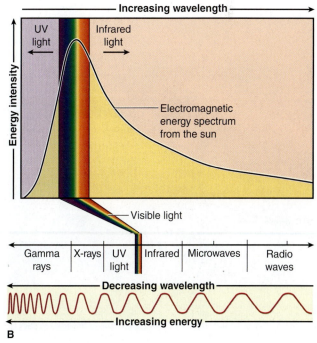

FIGURE 2.6   The Sun is enormous compared to the planets, particularly Earth and the other rocky planets of the inner solar system.

Earth

The Sun

The Sun (Figure 2.7) is an immense sphere composed mainly of hydrogen and helium atoms and relatively small amounts of the other elements. Similar to all stars, the Sun has an extremely hot and dense center surrounded by a less-dense outer region referred to as its *atmosphere*. The difference in temperature between the Sun's surface and interior is quite

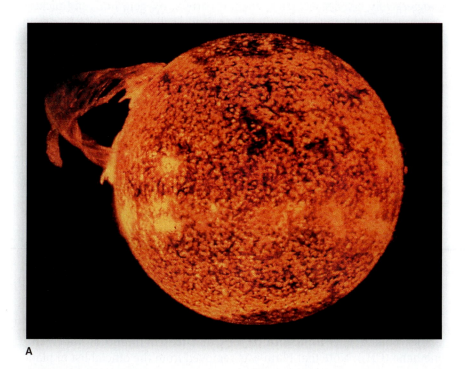

A

B

FIGURE 2.7   The Sun's immense gravity causes hydrogen atoms to undergo nuclear fusion (A) and form helium atoms. This nuclear reaction also releases a continuous spectrum of wave energy (B), known as the electromagnetic spectrum. Note that the peak energy output from the Sun lies in the visible part of the spectrum. Electromagnetic energy travels outward from the Sun and provides much of the energy that drives the Earth system.

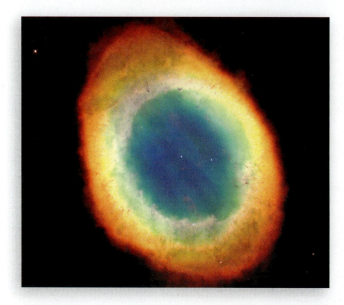

FIGURE 2.8 The Ring Nebula is a dying star that has exploded, ejecting its outer layers of material into space. Some of this material may someday combine with matter from other dying stars to form a new star with an orbiting system of planets. Note the tiny speck in the middle of the image was at one time a star larger than our own Sun.

large, causing heat and matter to rise toward the surface and then sink back into the interior in a cyclic process called *convection*. Due to the Sun's immense gravity, hydrogen atoms within the interior are packed so tightly that they collide and undergo nuclear fusion. During nuclear fusion the hydrogen atoms combine and convert into heavier helium atoms and in the process release tremendous amounts of heat energy. Note that the temperature within the Sun is estimated to be about 30,000,000°F (17,000,000°C).

In addition to heat, nuclear reactions within the Sun release what is known as **electromagnetic radiation,** which is a type of energy that travels in the form of waves. As illustrated in Figure 2.7, electromagnetic radiation is actually a continuous series of waves, called the *electromagnetic spectrum,* where individual waves vary in terms of their wavelength (distance between crests) and amount of energy they contain. Electromagnetic energy, often called solar energy or sunlight, travels outward from the Sun in all directions until it strikes a solid object, at which point it is converted into thermal or heat energy (Chapter 13). This process raises Earth's surface temperature and is a key reason life became so abundant on our planet. Also notice in Figure 2.7 how energy output from the Sun peaks in the visible portion of the spectrum. Evolution, in turn, has resulted in humans and other life-forms with eyes that can see wavelengths from this portion of the spectrum. Likewise, plants have evolved that are able to convert visible light into chemical energy (Chapter 13) via the process known as *photosynthesis*. Solar radiation also provides the energy that moves heat and water through the hydrosphere and atmosphere, making the Sun the primary driver of Earth's climate system. Therefore, without the Sun, the Earth would be a lifeless body moving through space.

Like all stars, the Sun will eventually use up its supply of hydrogen, at which point the helium atoms that had accumulated will themselves undergo nuclear fusion. At this point, nuclear fusion starts to produce progressively heavier elements, such as carbon, oxygen, iron, and nickel. A star at the end of its life cycle will typically collapse in on itself and, if it has sufficient mass, will explode violently in what is called a *supernova*. During these explosive events the various elements that had formed within the nuclear furnace are ejected outward into space (Figure 2.8). In fact, nearly all of the elements in the periodic table (see inside back cover) had at some point formed in the interior of stars that no longer exist. This means that all the atoms that make up the Earth, including those in our own bodies, are literally stardust.

The Planets

The planets closest to the Sun—Mercury, Venus, Earth, and Mars—have outer shells composed of rocky earthlike materials and are commonly referred to as the **terrestrial planets,** which comes from the Latin *terra,* meaning earth. The outer planets, Jupiter, Saturn, Uranus, and Neptune, are largely composed of hydrogen and helium gas and have surfaces that are marked by clouds of swirling gases (Figure 2.9). The outer four planets are commonly called the **gas giants** because they are composed mostly of gas and are quite large compared to the terrestrial planets (Figure 2.5). Although far less is known about Pluto because of its great distance, scientists believe it is a mixture of rock and ice composed of water and methane (CH_4). Note that Pluto was removed from the official list of planets in a controversial decision by astronomers in 2006. With modern instruments astronomers have discovered many new bodies that were both larger and closer to the Sun than Pluto. Therefore, either Pluto needed to be demoted or these additional bodies would have to be clas-

FIGURE 2.9 Photo of Jupiter taken by the *Voyager 2* spacecraft in 1979. Note the Great Red Spot, which is a giant storm, and the swirling cloud system of Jupiter's atmosphere.

sified as planets. The decision was made to define a planet as being large enough that its gravity can dominate its orbital path around the Sun, which resulted in Pluto's demotion. Ironically, just prior to this demotion NASA launched a spacecraft to Pluto called *New Horizons*. If all goes well, the craft will go into orbit around Pluto in 2015, giving us our first detailed look at the former planet.

Comets and Asteroids

Comets are relatively small bodies, 0.6 to 6 miles (1–10 km) in diameter, composed of small rocky fragments embedded in a mass of ice and frozen gases. Most have highly elliptical orbits around the Sun (Figure 2.4), with some orbits being no larger than Jupiter's. These comets can complete their journey around the Sun on the order of several years. Other comets have orbits beyond Pluto, resulting in return trips as long as 3 million years. When a comet approaches the Sun, electromagnetic (solar) radiation will cause the comet's ice to begin evaporating. This releases molecules that stream away from the comet, forming its familiar tail as shown in Figure 2.10. Astronomers believe most comets reside in a region called the Oort Cloud, which is several thousand times farther from the Sun than Pluto. It is also believed that gravitational disturbances will occasionally force a comet out of the cloud, placing it on a trajectory that takes it around the Sun.

Similar to comets, **asteroids** are small bodies that orbit the Sun, but are different in that they are composed primarily of rocky and metallic materials. Most asteroids lie in what is known as the main asteroid belt

FIGURE 2.10 A long tail develops from comet Hale-Bopp as it orbits around the Sun in 1997.

between Mars and Jupiter (Figure 2.4B). Scientists believe that during the formation of the solar system the strong gravitational influence of Jupiter prevented material in the asteroid belt from developing into a planet. It is also believed that gravitational disturbances and/or collisions with other asteroids cause some asteroids to leave their normal orbits, placing them on a collision course with planets and moons within the solar system. Note that the term *meteoroid* is used to describe a body of rock and metal that is smaller than a planet or asteroid. Such a body is called a *meteorite* if it passes through Earth's atmosphere and strikes the ground. Interestingly, when scientists date meteorites using radiometric dating techniques (Chapter 1), they find that all meteorites fall between 4.5 to 4.7 billion years old. The 4.6-billion-year age of the Earth was determined based on the radiometric dates of both asteroids and the oldest surviving rocks on our planet. In fact, scientists now believe that all the planets and moons within the solar system formed during this same 4.5- to 4.7-billion-year interval.

The Moon

As indicated earlier, Earth's Moon plays a very important special role in the Earth system. In particular, the Moon's gravitational field has a very strong effect on ocean tides, which, in turn, influence important processes that take place where the marine and terrestrial (land) environments meet (Chapter 9). These coastal processes are vital to the food web that supports the ecosystem of the oceans. Also important is how the Moon's gravity acts to minimize the amount of movement (i.e., wobble) in Earth's axis as our planet rotates. Minimizing this wobble in the axis has helped to reduce seasonal extremes between summer and winter. This has produced a more stable climate system where complex life-forms have had more time to evolve in a relatively stable environment. Were it not for the Moon, life on planet Earth would likely have evolved much differently, and possibly, not at all.

Scientists have learned a great deal about the Moon in modern times from various space and ground-based studies. For instance, the false-color image shown in Figure 2.11 was created from data collected by the *Galileo* spacecraft as it headed toward Jupiter in 1992. The different colors in this image represent rocks of different chemical composition, which, in turn, reflect the different terrain, namely the rugged lunar highlands and low-lying lunar seas, or maria. Most significant is the fact that when rocks from the *Apollo* moon landings were brought back to Earth, radiometric dating techniques proved that the age of the Moon was similar to the Earth, around 4.5 billion years old. As we will examine in the next section, this was an important piece of information that helped scientists refine their hypotheses as to the origin of the Moon.

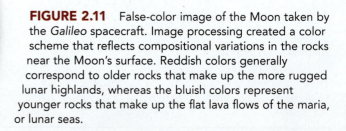

FIGURE 2.11   False-color image of the Moon taken by the *Galileo* spacecraft. Image processing created a color scheme that reflects compositional variations in the rocks near the Moon's surface. Reddish colors generally correspond to older rocks that make up the more rugged lunar highlands, whereas the bluish colors represent younger rocks that make up the flat lava flows of the maria, or lunar seas.

Origin of the Solar System

Recall from Chapter 1 that science operates by developing multiple hypotheses, each of which is capable of explaining the data (facts) that have been collected from various observations and experiments. The testing of various hypotheses through additional observations and experiments will cause some hypotheses to be refined and others thrown out. Ultimately this rigorous testing process leads to scientists having greater confidence in the hypotheses that survive. In this section we want to examine the hypothesis scientists have developed for explaining how the solar system formed.

The Nebular Hypothesis

In the previous section we learned that the Earth, Moon, and asteroids are all roughly the same age (around 4.6 billion years old), thus indicating a common origin. In addition to age data, there are several important observations that must be addressed in any explanation of how the solar system formed. First, all of the planets revolve around the Sun in the same counterclockwise direction (as viewed from above) and have regular orbits that are nearly circular (Figure 2.4). Second, the Sun and most of the planets rotate (spin) about their axes in the same counterclockwise direction—Venus is the exception as it rotates clockwise. Even the moons in the solar system spin about their axes and orbit their respective planets in a counterclockwise manner. Also interesting is the fact that all the planets and their moons lie within a plane that coincides with the Sun's equator.

Most astronomers agree that these data and observations are best explained by the **nebular hypothesis,** in which all solar system objects formed from a rotating cloud of dust and gas called a *nebula*. Note that various nebular hypotheses were first proposed in the 1600s and have been refined over the years as new data were collected. The basic idea is that the solar system formed when an exploding star (supernova) disturbed a cloud of dust and gas composed primarily of hydrogen and helium, along with smaller amounts of other elements. As illustrated in Figure 2.12, this disturbance caused the nebula to begin collapsing in on itself due to the gravitational attraction between the dust and gas particles. As the nebula continued to contract it also began to spin, causing it to flatten into the shape of a disc (Figure 2.12B). Although this contraction resulted in higher temperatures and pressures within the center of the nebula, the spinning motion caused the outer portions of the disc to thin. Because the thinner portions of the disc were farther away from the hot central region, they were allowed to cool. Eventually the temperatures became low enough for liquids to condense and solids to crystallize. In the outer reaches of the disc, it became cold enough for liquid water, ammonia (NH_3), and methane (CH_4) to turn into ice.

Once the solid materials formed within the disc, gravitational attraction caused the individual particles to clump together into larger masses, a process called **accretion.** Eventually accretion created larger bodies called *planetesimals,* which tended to form

FIGURE 2.12   Illustration showing the evolution of the solar system from a nebula. Collapsing cloud (A) increases in density and begins to rotate. Continued collapse (B) results in nuclear fusion and the formation of a star, whereas the rotation forces the nebula to take on the shape of a disc. Planetesimals develop by accretion of particles (C), while solar radiation drives off remaining parts of the nebula. When the debris is cleared, what is left are planets (D) that revolve and rotate in the same counterclockwise manner and in the same plane around the Sun.

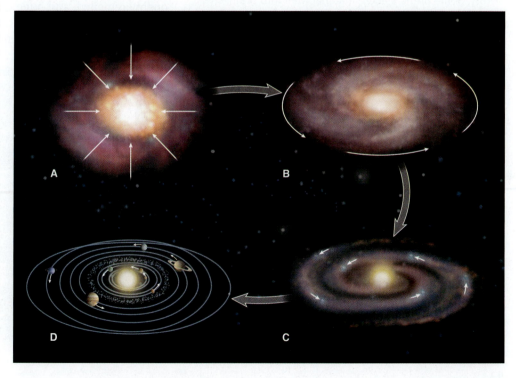

FIGURE 2.13 Artist's illustration of what a young planetary system might look like as planets clear their orbits of debris. Electromagnetic radiation and streams of charged particles emitted from the newly formed star act to clear the remaining dust and gas from the solar system.

FIGURE 2.14 The leading hypothesis for the origin of the Moon involves a Mars-sized impact (A) early in Earth's history. Some of the ejected material went into orbit around the Earth where it underwent accretion (B), forming the Moon. Such a giant impact is also believed to have caused the young Earth to melt.

in a plane within the swirling nebula (Figure 2.12C). During this period the temperature and pressure within the center of the nebula increased to the point where hydrogen atoms began to undergo nuclear fusion, and a new star was born. As electromagnetic radiation and gases began flowing outward from the Sun, the ice and lighter elements that had collected on solid objects within the nebula began to either evaporate or melt. Gradually the innermost planetesimals lost their lighter and more volatile constituents, which then ended up recondensing farther out in the new solar system.

By the time the Sun was born, the planetesimals had become large enough that their gravitational fields started attracting smaller bodies from nearby orbits. During this early stage the planetesimals grew very rapidly, with some eventually becoming **planets,** whose gravity dominates their individual orbital zones (Figure 2.12D). During this early period, when the planets were clearing their orbits of debris, it is believed that the frequency of impacts was high. Moreover, some of the bodies that the young planets were attracting must have been quite large, resulting in tremendous impacts. As time progressed the solar system was largely swept of debris, causing a dramatic decrease in the frequency and size of impacts over time. Figure 2.13 is an artist's illustration showing what the early solar system probably looked like during the heavy bombardment period when the planets were clearing out their orbits. Note how the remaining dust and gas is beginning to be cleared from the solar system by electromagnetic radiation and streams of charged particles being emitted from the newly formed Sun.

Finally, a key aspect of the nebular hypothesis is that it may also explain the origin of Earth's Moon. The current leading hypothesis is that the Moon formed early during the heavy bombardment period when the Earth experienced a Mars-sized impact as it was clearing its orbit. This proposed impact, illustrated in Figure 2.14, was so large that Earth thoroughly remelted, and huge amounts of debris were ejected out into space. Some of this debris fell back onto the Earth, but much of it went into orbit around the Earth where it underwent accretion and eventually formed the

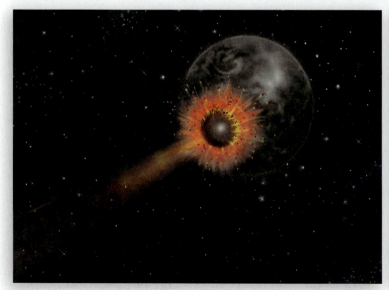

A

B

Moon. This hypothesis not only explains the nearly identical age of the Moon and Earth, but also why the Earth has such a disproportionally large moon compared to other planets in the solar system. Most significant is the fact that this single, massive impact created a moon that was large enough to have affected Earth's climate and ocean tide. This helped create an ideal environment for the evolution of life on Earth.

How Reliable Is the Nebular Hypothesis?

At this point we should take a look at how well the nebular hypothesis explains the known data regarding the solar system. As noted earlier, this hypothesis explains why nearly all the planets and moons revolve (orbit) and rotate (spin) around the Sun in the same counterclockwise manner. These bodies also all lie within a plane that coincides with the Sun's equator. For those planets and moons that do not fit this general pattern, their deviation can be accounted for by giant impacts early in the solar system history—similar to that proposed for the origin of the Moon. For example, Venus rotates about its axis very slowly and in the opposite (clockwise) direction from the other planets. One explanation is that Venus experienced a large, glancing impact such that it actually reversed its direction of spin. Similarly, a massive impact could explain why Uranus rotates on its side rather than in an upright position similar to the other planets and moons.

Direct evidence for the nebular hypothesis and the intense bombardment associated with accretion is the heavily cratered surfaces we see today on some of the planets and moons. Examples include Mercury and Earth's Moon (Figure 2.15). By analyzing the landforms and density of the craters, scientists have been able to show that the majority of the impacts occurred early in the solar system's history. This of course supports the accretion concept as proposed in the nebular hypothesis. Further evidence is that Venus and Earth have relatively few craters, which is consistent with the fact that weathering and erosion processes on these planets would have long ago erased most of the original impact record. Finally, radiometric dating has shown that the Earth, Moon, and asteroids all solidified around the same time. This too supports the nebular hypothesis.

Perhaps the most dramatic evidence for the nebular hypothesis comes from recent data astronomers have gathered using powerful space (Hubble) and ground-based telescopes. For example, about 180 planetary systems have now been detected around other stars. In some cases orbiting discs of dust have been found around stars, which likely represent the early stage of planetary development when planetesimals begin to form. In addition to proving the existence of other planets, the Hubble telescope has provided us with vivid images of new stars being born in towering clouds of dust and gas (Figure 2.16). This seems to indicate that most stars probably form in a similar manner as our Sun and that planetary systems may be a somewhat common occurrence. It is also quite possible that there are planets similar to Earth, where conditions are

A   B

FIGURE 2.15 Heavily cratered surfaces of Mercury (A) and the Moon (B). A lack of an atmosphere along with weathering and erosion processes on these bodies has preserved the record of the heavy bombardment that took place early in the solar system history.

FIGURE 2.16 Photo from the Hubble space telescope showing new stars forming in the gas and dust clouds of the Eagle Nebula.

Gas and dust: The 6-trillion-mile-high fingers, made of dust and hydrogen gas, turn blue when they are hit by ultraviolet radiation

Star eggs: Clumps of hydrogen inside the pillars called Evaporating Gaseous Globules, or EGGs, eventually hatch into stars

FIGURE 2.17 View looking down toward the central core of a clockwise-rotating galaxy. Our Sun lies on the outer band of the Milky Way Galaxy, which is similar to the spiral galaxy shown here. Such galaxies are estimated to contain hundreds of billions of stars.

favorable for life, ranging from bacteria to highly evolved and intelligent life-forms. If this is indeed the case, then it brings up the interesting question of how many Earth-like planets might there be in the universe. In the next section we will examine this issue by getting a better feel for the number of stars believed to exist in the universe.

Other Stars in the Universe

For anyone who has gazed into the night sky, it is obvious that there are many stars in the universe. Less obvious is the fact that the stars are not distributed uniformly throughout the universe, but rather are found in large groupings called **galaxies.** Figure 2.17 shows an example of what astronomers call a *spiral galaxy.* Our Sun lies within a spiral galaxy known as the Milky Way Galaxy, which is estimated to contain 200 to 400 billion stars. Because numbers this large are difficult to comprehend, it is helpful if we use an analogy. If you counted stars at a rate of one per second, it would take 32 years to count to just 1 billion, and 3,200 years to reach 100 billion. Also interesting is that parts of the Milky Way can be seen with the naked eye on dark and clear nights. As shown in Figure 2.18, this spinning disc of stars appears as a cloudy band that arcs across the night sky. Because our solar system is located out on a spiral arm of the Milky Way, what we are seeing is an edge-view looking in toward the center of the galaxy. Another interesting aspect of the Milky Way is that it takes light approximately 100,000 years to travel across the galaxy, where a *single* light-year equals 5,870 billion miles (9,450 billion km). To help grasp what this means in terms of size, consider that the average distance between the Earth and Sun is a mere 93 million miles (150 million km). Equally amazing is the fact that it takes our Sun about 250 million years to make one full revolution around this spinning disc. The Milky Way, therefore, dwarfs the Earth and our solar system on a scale that is almost beyond comprehension.

Although it may be difficult for us to comprehend the scale of our galaxy, it is even harder to grasp the size of the universe. For example, in Figure 2.18 you can see what looks like stars in the parts of the sky away from the Milky Way. The insert image shows that when astronomers focus the powerful Hubble telescope on this background, what had appeared as stars with the naked eye turn out to be galaxies. Because this deep view into space represents such a tiny fraction of the night sky, it tells us that there is a very large number of galaxies in the universe. Another important point is that the galaxies, as seen by the Hubble telescope, are so far away that the light we see has been traveling for more than 10 billion years. In fact, the light is so old that many of the stars in these galaxies no longer exist. This means that we are actually looking backward in time and seeing the galaxies as they existed 10 billion years ago. Astronomers believe that what we are seeing is the state of the galaxies shortly after the origin of the universe itself.

The scientific explanation astronomers and physicists have developed to explain the origin of the universe is called the **big bang theory.** This theory, first proposed in 1927 by a Belgian priest, states that all matter in the universe had at one time existed at a single point. Approximately 14 billion years ago, this matter then began to expand outward in all direc-

FIGURE 2.18 As seen from Earth, the Milky Way Galaxy appears as a cloudy band of stars across the night sky. The inset shows a deep view of space taken by the Hubble space telescope, where what appeared as stars with the naked eye are actually distant galaxies of various shapes and colors. This view represents a very small portion of the night sky, about the width of a dime located 75 feet away.

tions, and has been expanding ever since. Supporting evidence for the big bang theory came in 1929 when the astronomer Edwin Hubble, for whom the Hubble telescope was named, proved that all the galaxies are moving away from one another. Logic tells us that if all the galaxies are continually moving outward, similar to the surface of an expanding balloon, then if we go backward in time they must have been closer together. Figure 2.19 illustrates how galaxies within the universe have been expanding outward from a central point since the big bang. Note that the Hubble and other telescopes view distant objects as they once existed, thus we are able to look backward in time.

Additional evidence for the big bang theory came in 1964 when two astronomers inadvertently discovered microwave (electromagnetic) radiation coming from deep space. They found it odd that this radiation was not coming from a single source, such as a star, but rather they found it everywhere they pointed their instrument into the farthest reaches of space. What these scientists had discovered was the electromagnetic radiation that formed during the initial expansion of the universe (Figure 2.19). In simple terms, they found direct evidence for the big bang itself. The discovery of this background radiation not only led to a Nobel Prize, but caused the big bang theory to gain widespread acceptance within the scientific community.

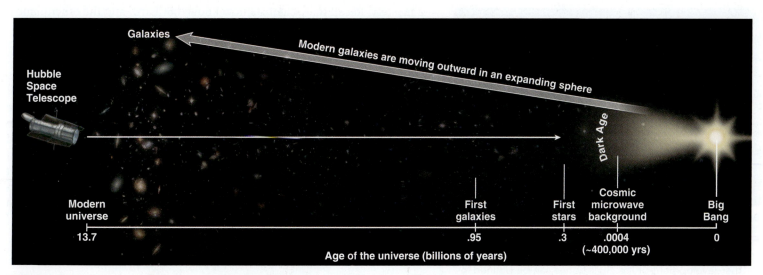

FIGURE 2.19 Conceptual diagram illustrating the big bang and the origin of the universe. Galaxies we see today have been rushing outward as an expanding sphere. The Hubble and other powerful telescopes are able to observe distant features as they appeared nearly 13 billion years ago.

Does Life Exist Beyond Earth?

Perhaps the greatest mystery in all of human history is the question of how life began here on Earth. Also fascinating is whether life exists elsewhere (i.e., extraterrestrial life), and if it does, how common is life in our galaxy and the universe. Although this subject is far beyond the scope of this book, we will take a brief look at the topic in order to give the reader a better appreciation of why Earth's environment is so special. From this discussion we can also learn how Earth's geology has played an important role in creating this environment and its incredible array of life. This topic therefore is pertinent to the study of environmental geology. In this section we will briefly examine what scientists have learned about the origin of life on Earth and what may exist beyond our planet.

Life on Earth

While studying some of Earth's oldest surviving rocks, paleontologists have discovered evidence of bacteria in rocks as old as 3.6 billion years. Such ancient bacteria prove that life began very early in Earth's 4.6-billion-year history. Moreover, life began at a time when the planet's atmosphere and climate were considerably different from today. Earth's environment, in fact, was so different that it would have been inhospitable for most of today's life-forms. Interestingly, biologists today are finding what are called **extremophile bacteria,** which thrive under extreme conditions that would be lethal to nearly all of Earth's other current life-forms. Extremophile bacteria are commonly found in places such as ancient Antarctic ice, superhot vents on the seafloor, and rocks located deep underground. Based on evolutionary changes recorded in the geologic rock record (Chapter 1), most biologists now believe that the complex plant and animal life we see today ultimately had evolved from extremophile bacteria. One possible source could have been extremophile bacteria thriving in the oceans along hot water vents associated with volcanic activity. Long periods of geologic time would have led to progressively more complex organisms and photosynthesis, ultimately creating an oxygen-rich atmosphere and the diverse animal and plant life we see today.

The question of just how nonliving material developed into the first primitive forms of bacteria (i.e., the origin of life itself) will likely remain a mystery. Recall that scientists now know that Earth and the other planets and their moons all underwent an intense bombardment of asteroids and comets early in our solar system's history. Of particular interest are comets because they are known to be composed mostly of water and organic (i.e., carbon-based) compounds, both of which are critical ingredients for life as we know it. It is quite possible then that comets acted as "seed" material, distributing the ingredients for life throughout the entire solar system. Life therefore could have developed in those places where conditions were favorable. An analogy here would be if one were to take grass seed and spread it over a wide area. While the seeds might be able to germinate in different places, they would truly flourish only in those areas with the right combination of soil, water, and nutrients. In terms of life in our solar system, Earth clearly presented the most favorable conditions, allowing life to thrive and evolve into complex plants and animals.

Habitable Zones

The only life-forms we know of today are those found on Earth, and all of them require the presence of liquid water (H_2O). This is why many scientists believe the key to finding life beyond planet Earth is to first look for liquid water. Consequently, the term **habitable zone** has been defined as

A

B

FIGURE 2.20   Habitable zones are those regions of space where conditions are believed to be most favorable for the development of life. Such zones can be defined in terms of areas where liquid water can exist around individual stars (A), and also within a galaxy (B) where there is an abundance of heavier elements, but yet fewer cosmic hazards.

that relatively narrow zone around a star where the surface temperature of orbiting planets would be such that liquid water could exist. Figure 2.20 illustrates the concept of a habitable zone around a star. For planets located too close to their star, water near the surface would vaporize and be lost, which is what scientists think happened on Mercury and Venus. On the other hand, water will remain frozen on planets located too far from their star. Earth, of course, is ideally located within the Sun's habitable zone. Note however that the position of the habitable zone will vary depending on the size and energy output of a given star. This means that when the energy output and size of a star change as it goes through its natural life cycle, the position of the habitable zone will shift accordingly. As our Sun evolves and slowly grows in size over the next several billion years, the habitable zone will move out beyond Earth's orbit. Then life here on Earth will eventually cease to exist. Finally, note in Figure 2.20 that a habitable zone can also be defined for an entire galaxy. The center galaxies are considered less favorable locations for life because they contain more hazards (exploding stars, intense radiation, black holes, etc.). The extreme outer parts of galaxies are also thought to be less favorable since there are fewer stars capable of producing the heavier elements necessary for life.

One problem with the concept of habitable zones is that it assumes that all life-forms require liquid water as a solvent (i.e., capable of dissolving different substances). Although we have no way of knowing, it is certainly possible that life-forms exist in the universe which are based on some solvent other than water, such as ammonia (NH_3), sulfuric acid (H_2SO_4), or an organic solvent like formamide (CH_3NO). Another problem is that life, particularly extremophile bacteria, could exist beyond a habitable zone as long as a planet or moon has an internal heat source like the Earth (Chapter 4). Remember, comets probably seeded the entire solar system with water and organic compounds that many believe are necessary for life to develop. Thus, an internal heat source could provide a habitable niche with liquid water far beyond the solar system's main habitable zone. After all, life on our planet may have originated on the seafloor, where superheated vents associated with volcanic activity are found. Such volcanic activity is of course powered not by the Sun, but by Earth's internal heat. The geology of a planet or moon may therefore play an important role in the development of life.

Many scientists now believe that evidence of extraterrestrial life is most likely to be found in the form of bacteria that are associated with liquid water. One promising place is Jupiter's moon Europa since there is strong evidence that it is covered with water ice (Figure 2.21). In addition, volcanic activity has been observed on a neighboring moon, called Io, whose internal heat source is thought to be caused by Jupiter's strong tidal forces. Scientists hypothesize that these same tidal forces are generating heat within Europa, thereby creating a liquid ocean beneath the moon's frozen surface. This ocean, in turn, could contain extremophile bacteria and other life-forms. Europa therefore may be harboring life despite the fact that it lies outside of the Sun's habitable zone. Another possible place we might find extraterrestrial life is on Mars. In 2004, two NASA robots landed on Mars and were able to identify rocks that were once saturated with water (Case Study 2.1). Because these rocks may contain evidence of ancient bacteria, scientists hope to someday bring samples back to Earth for detailed analysis.

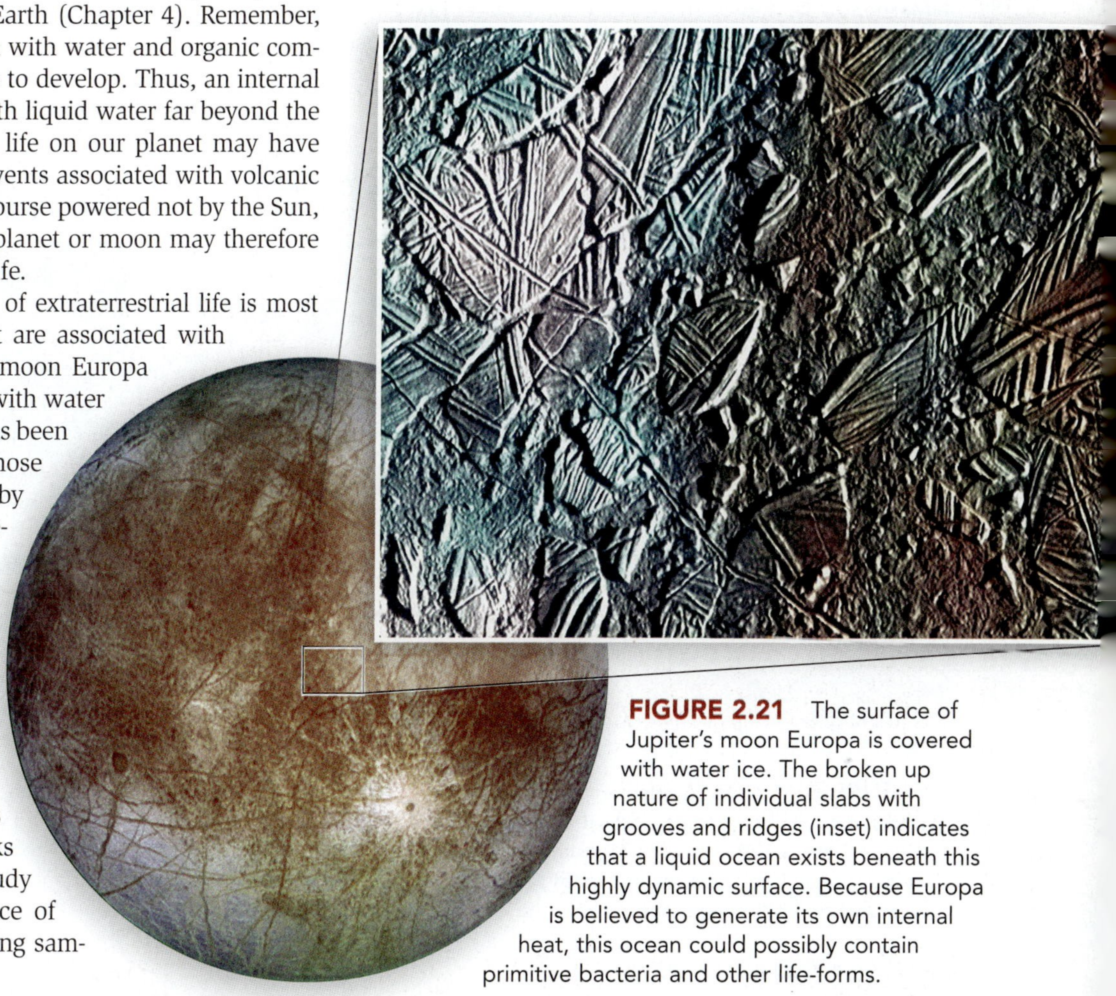

FIGURE 2.21 The surface of Jupiter's moon Europa is covered with water ice. The broken up nature of individual slabs with grooves and ridges (inset) indicates that a liquid ocean exists beneath this highly dynamic surface. Because Europa is believed to generate its own internal heat, this ocean could possibly contain primitive bacteria and other life-forms.

Search for Life on Mars

One of humanity's most intriguing questions is how life developed on Earth. Another is whether or not life is unique to Earth. Answering these basic questions is important because it can help us better understand the Earth system, and perhaps help humanity to better appreciate our planet's life-sustaining environment. Scientists therefore are searching other bodies within the solar system for evidence of primitive forms of life. Since life on Earth is always associated with water, the basic strategy is to find signs of life by first locating places where water exists, either currently or in the past. A good place to start is Mars, one of Earth's closest neighbors.

Based on relatively recent photos of Mars showing stream channels and their relationship to ancient craters, scientists now believe that liquid water was abundant on the Martian surface around 3.8 billion years ago. Although much of the original water is thought to have escaped into space as Mars' atmosphere began to thin, new data indicates that liquid water remains stored in rocks below the surface. Because liquid water is critical to life as we know it, Martian rocks could contain evidence of primitive bacteria. If such evidence could be found, then it would prove that Earth is not the only place where life developed. Consequently, the United States and other nations have sent spacecraft to study Mars more carefully, with the goal of verifying that at least some Martian rocks had been deposited in a water-rich environment. If successful, future missions could then bring samples back to Earth where they could be examined more carefully for possible evidence of life.

As part of an effort to better understand Mars, NASA put a satellite into orbit around the planet in 1997 to create more detailed surface maps. The mapping revealed tantalizing evidence of layered rocks that could have been deposited in liquid water. However, it was also possible that these rocks had accumulated from the fallout of volcanic ash. To find the answer, NASA decided to land two spacecraft and examine the rocks directly on the ground. One of the landing sites was a large impact crater (Figure B2.1A) with a clearly defined stream channel leading away from the crater, which implied it once held considerable amounts of water. At the other landing site (Figure B2.1B) the rocks were believed to contain large amounts of an iron oxide mineral known as *hematite* (Fe_2O_3). On Earth this particular form of hematite is typically associated with significant quantities of water. Finally, in 2003 NASA successfully landed robotic craft at each of the sites. The robots, called rovers, were highly mobile and equipped with instruments for taking photos and determining the different types of minerals making up the surrounding rocks and sediment.

Shortly after landing, the rovers began collecting data that finally proved liquid water had once been abundant on the Martian surface. One of the key lines of evidence at the crater site (Figure B2.1A) was the discovery of magnesium sulfate ($MgSO_4$) minerals found in significant quantities throughout the sediment. This mineral is actually a salt composed of magnesium (Mg^{2+}) and sulfate (SO_4^{2-}) ions, which often forms on Earth when salty water under-

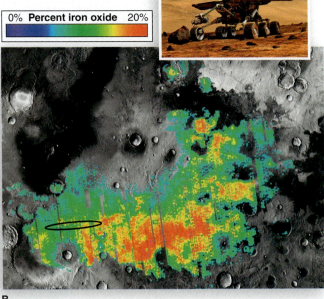

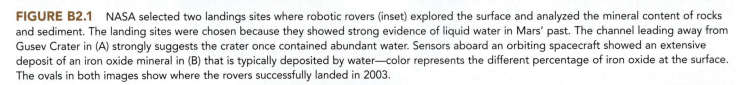

FIGURE B2.1 NASA selected two landings sites where robotic rovers (inset) explored the surface and analyzed the mineral content of rocks and sediment. The landing sites were chosen because they showed strong evidence of liquid water in Mars' past. The channel leading away from Gusev Crater in (A) strongly suggests the crater once contained abundant water. Sensors aboard an orbiting spacecraft showed an extensive deposit of an iron oxide mineral in (B) that is typically deposited by water—color represents the different percentage of iron oxide at the surface. The ovals in both images show where the rovers successfully landed in 2003.

goes evaporation, forcing the ions to bond chemically (i.e., precipitate) and form a solid. Because large quantities of this mineral were found, scientists believe the magnesium and sulfate ions originated from distant rocks that had been broken down by chemical weathering (Chapter 3). Here the ions became dissolved in water and were then carried away, eventually bonding together to form new minerals. In order for all this to occur there must have been flowing water at or very near the surface.

The other landing site (Figure B2.1B) was also found to contain mineral salts that are typically left behind when water undergoes evaporation, forcing dissolved ions to bond chemically. In addition, the rocks here contained numerous nodules the size of BBs that were composed of the iron oxide mineral hematite (Fe_2O_3). Scientists concluded that these nodules formed similar to ones on Earth, namely by growing slowly in concentric layers as the iron precipitates out from mineral-rich water. Most significant however was the discovery of what geologists call ripple marks and cross-bedding in the sediment layers (Figure B2.2). Cross-bedding forms when either wind or moving water causes sediment layers to become oriented at an angle to the main layers. The ripple marks found on Mars have wavelike peaks and troughs that only form when water moves back and forth in a shallow body of water. Scientists therefore concluded that the rocks at this site were deposited in a salty sea that may have extended for miles in every direction.

The two rover missions have proved that Mars once had water at its surface for a considerable length of time. Whether life itself was ever present in this ancient environment has yet to be determined. NASA is currently planning an international mission to Mars in which rock samples would be collected and returned to Earth for detailed analysis (due to the possibility that the rocks could contain live bacteria, the samples will be quarantined on the International Space Station). Should evidence of life be found, then one could conclude that life is relatively common throughout the universe. This would also mean a greater chance that intelligent life exists elsewhere. Perhaps most important, such a discovery might help humans better appreciate that the conditions that led to intelligent life on Earth are not only exceedingly rare, but quite fragile. Hopefully this would encourage better stewardship of our own environment.

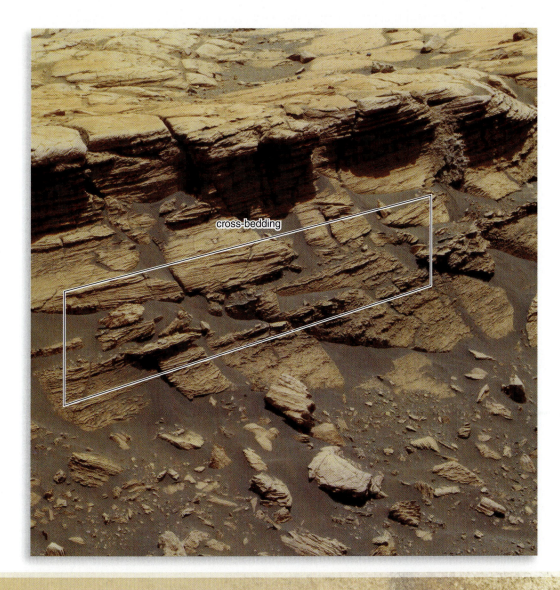

FIGURE B2.2 Photo of sedimentary rocks on Mars that are billions of years old. These rocks contain abundant sulfate minerals, verifying that the original sediment was once soaked in salt water. These fine-grained rocks also show cross-bedding and ripple marks (not visible); the ripple marks prove that the sediment was deposited in an open-body of water.

Possible Intelligent Life

Discovering evidence of life beyond the Earth would be a historic achievement, but an even greater prize would be finding evidence of intelligent life. It is commonly assumed that intelligent life would be restricted to the main habitable zone around a star. This is based on the assumption that intelligent life not only requires liquid water, but also landmasses on which to live. The idea is that intelligent life would evolve as it did on Earth, namely from marine organisms that eventually adapted to conditions on land. Such land-based or terrestrial life would require liquid water on the surface, which, in turn, would require the planet to lie within the main habitable zone. Thus, most scientists believe that any search for intelligent life should focus on stars with planets orbiting within their respective habitable zones.

Although we do not know if other intelligent life actually exists, scientists can get some idea of its likelihood by estimating the possible number of habitable planets there might be in the universe. Recall that our solar system is just a tiny speck among the 200 to 400 billion other stars making up the Milky Way Galaxy. Based on simple statistical models and the nearly 180 planetary systems already detected in our galaxy, NASA scientists estimate there are 10 billion Earth-like planets lying within the habitable zone of their star. Even if these estimates are too high by several orders of magnitude, there would still be a very large number of planets capable of supporting intelligent life. This estimate is just for the Milky Way, and astronomers estimate there are 200 to 500 billion galaxies in the universe! If one assumes each galaxy averages several hundred billion stars, then there should be something on the order of 100,000 billion billion or 10^{23} stars in the universe—a number so large it is nearly impossible to comprehend. If we consider the entire universe, then the number of planets that could potentially contain intelligent life is also beyond comprehension.

In recent years an idea called the **rare earth hypothesis** has been proposed, which contends that life is probably common throughout the universe, but complex animal life similar to Earth's is likely to be exceedingly rare. The basic premise, as described in the book *Rare Earth* by paleontologist Peter Ward and astrobiologist Donald Brownlee, is that an Earth-like planet needs a stable environment over a long period of geologic time so that complex life can evolve from more primitive forms. Below are some of the factors they describe which are believed to have been critical in helping Earth maintain a stable environment so that animal life, including humans, could evolve.

1. Energy output from the Sun has remained fairly steady, allowing Earth to have a more stable climate.
2. Earth's internal heat and plate tectonics (Chapter 4) helps regulate the amount of carbon dioxide (CO_2), which traps heat and warms the atmosphere. By controlling CO_2 levels, these processes act as a thermostat to regulate atmospheric temperature (Chapter 16).
3. Jupiter's large size has helped clear asteroids and comets from Earth's orbit, reducing the number of large, catastrophic impacts that would alter the global climate.
4. The Moon has reduced the wobble in Earth's axis, thereby helping to stabilize Earth's climate.

Obviously, no one knows how many Earth-like planets actually exist in the universe where complex animal life has evolved. It is possible we are indeed alone. Although the distances are far too great and the potential

FIGURE 2.22 Earth in the vastness of space. Image taken by *Apollo 8* astronauts on their way to the Moon.

number of planets far too many for us to explore directly, we can scan the stars of our own galaxy, listening for possible radio signals from highly advanced civilizations. The assumption is that other advanced civilizations would be broadcasting radio waves into space in the same manner humans have been doing since the invention of the radio. Since the 1960s various groups have listened for such extraterrestrial signals using radio telescopes. Although NASA started a program in 1993, it ended less than a year later when Congress eliminated its funding. Today a privately funded organization, known the SETI (Search for Extra-Terrestrial Intelligence) Institute is actively searching within our galaxy.

In the end, it seems reasonable to conclude that Earth is indeed rare and that we may never find out if intelligent life exists elsewhere in the universe. While Earth may be just a small planet orbiting an ordinary star in the vastness of space, for us it is our one and only home. When we view Earth from the perspective of space as shown in Figure 2.22, one can get a better sense that Earth's environment is not only very special, but quite fragile. In the end, despite all our efforts at trying to understand the universe and our place in it, what really counts is how well we take care of our home and its environment that makes our lives possible.

Solar System Hazards

The Earth is often thought of as being a self-contained system composed of the atmosphere, hydrosphere, biosphere, and solid earth (Chapter 1). However, as noted earlier, our planet operates within an astronomical environment that has a major influence on the Earth system, and the environment in which we live. For example, recall how the Earth system depends on solar radiation and gravitational forces within the solar system. We can think of the Earth then as a system that interacts with even larger systems, namely the solar system and our own galaxy. In this section we will briefly examine some natural hazards that originate outside of the Earth system. Our focus here will be on electromagnetic radiation and the impact of comets and asteroids. Interestingly, these extraterrestrial hazards appear to have played an important role in the evolution of life on Earth.

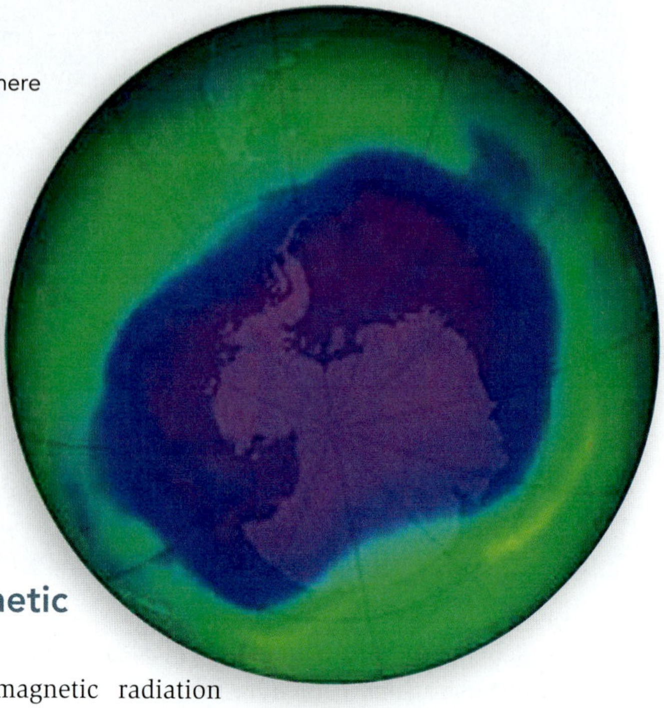

FIGURE 2.23   Map showing the concentration of ozone (O_3) in the upper atmosphere as measured by satellite instruments. Shown here is the hole in the ozone layer that develops over the Antarctic each winter—blue and purple show where ozone concentrations are the lowest and green shows the highest. Phasing out the use of ozone-destroying gases will eventually stabilize ozone levels and keep the hole from getting larger.

FIGURE 2.24   Illustration showing how a burst of gamma rays from a nearby exploding star could destroy much of Earth's ozone layer. This would allow the biosphere to be exposed to intense ultraviolet (UV) radiation, causing many species to die. Ultimately the food chain could begin to collapse and lead to a mass extinction.

Electromagnetic Radiation

Although electromagnetic radiation streaming outward from the Sun is the critical energy source that drives Earth's biosphere and climate system, it can also pose a hazard to living organisms. Of particular concern is radiation in the ultraviolet (UV) portion of the spectrum (see Figure 2.7) as these higher-energy wavelengths damage the cell tissue of carbon-based life-forms. Fortunately, Earth's upper atmosphere contains a thin layer of oxygen molecules called *ozone* (O_3), which naturally absorb much of the incoming UV radiation—the oxygen molecules we breathe are composed of two oxygen atoms (O_2). This thin layer of ozone molecules, called the *ozone layer*, acts as a protective shield for Earth's biosphere. In fact, the ozone layer has been shielding Earth's biosphere from UV rays for eons of time, making it possible for carbon-based organisms to evolve into those we see today.

In the 1930s humans began using chlorine and fluorine-based gases, called *chlorofluorocarbons,* as a coolant for refrigeration and air conditioning systems. Chlorofluorocarbons, commonly called CFCs, also became quite popular in various commercial applications, such as a propellant for cans of spray paint and hair spray. Then, in 1974, scientists discovered that the combination of CFCs and UV radiation causes ozone molecules (O_3) to chemically break down into free oxygen (O_2). They warned that CFCs released from human activity would slowly make their way to the upper atmosphere and cause the ozone layer to become dangerously thin, a problem referred to as **ozone depletion.** It was not until 1985 that researchers from the British Antarctic Survey prove that ozone levels were actually declining. Additional research showed that the combination of cold temperatures and air currents resulted in ozone depletion being most severe over the Antarctic, falling as much as 60% during the spring. This annual thinning has become known as the *ozone hole* (Figure 2.23). It was also discovered that ozone depletion is not restricted to the polar regions, but rather is a global problem. Over the United States, for example, ozone levels have fallen as much as 5–10%.

Ozone depletion as a result of human activity was soon recognized as a serious health threat to people and the biosphere as a whole. Excess exposure to UV radiation in humans, for example, is known to cause skin cancer and eye cataracts. Left unchecked, ozone depletion could impact the entire food web within the biosphere. Because this threat was serious and global in nature, an international agreement was reached in 1987, called the *Montreal Protocol,* in which nations agreed to phase out the production and use of CFCs. As a result of this agreement, the concentration of CFCs in the atmosphere appears to have peaked, and is now slowly starting to decline. NASA scientists have projected that the ozone hole will start shrinking significantly by 2018, and fully recover around 2068.

Another radiation hazard that originates in space is known as a **gamma-ray burst,** which is a short-lived burst of very high energy waves from the gamma-ray portion of the electromagnetic spectrum (see Figure 2.7). Gamma-ray bursts were first detected in the 1960s by orbiting U.S. spy satellites. More detailed studies have found that bursts average about one per day, each lasting on the order of fractions of a second to several minutes. Astronomers now know that at least some gamma-ray bursts originate in distant galaxies from stars that explode violently as they reach the end of their life cycle. Because gamma rays possess so much energy, they are even more hazardous to living organisms than ultraviolet (UV) rays. The good news is that Earth's ozone layer absorbs both gamma and UV rays. Unlike UV rays, however, gamma rays by themselves are able to destroy ozone molecules. Fortunately, gamma rays lose considerable amounts of energy during their long journey from distant galaxies. This, in turn, has allowed the production of ozone by natural earth processes to keep pace with the steady rate of ozone loss. The result is that Earth has been able to maintain a protective ozone layer for eons of geologic time.

In recent years scientists have considered the frightening possibility of a gamma-ray burst originating not in some distant galaxy, but from a star nearby in our own galaxy. Such a scenario is illustrated in Figure 2.24. Based on modeling results, astronomers estimate that if a gamma-ray burst were to originate from a star 6,000 light-years away and last just 10 seconds, half of Earth's ozone would be destroyed. The greatly thinned ozone layer would then immediately allow a steady stream of UV rays to pass through the atmosphere and begin striking the Earth. The radiation would be expected to kill off large numbers of terrestrial species as well as many species living in shallow waters in lakes and oceans. The loss could become so great that the food web reaches a tipping point and collapses, leading to a **mass extinction** where large numbers of species go extinct in a relatively short period of time. Mass extinctions are known to have occurred in the geologic past based on graphs, such as in Figure 2.25, which show abrupt decreases in the number of species in Earth's fossil record over time (Chapter 1). Scientists generally agree that these die-offs are related to significant changes in Earth's environment that are global in nature. Although gamma-ray bursts associated with exploding stars are relatively rare in our galaxy, most astronomers believe the probability is high that one or more has occurred during Earth's 4.6-billion-year history. This does not mean that all mass extinctions are related to gamma-ray bursts. In fact, there are other mechanisms that are quite capable of triggering a large global change in Earth's environment. One such mechanism is a large asteroid or comet impact, a topic we will explore next.

Asteroid and Comet Impacts

Earlier you learned that the heavily cratered surfaces of Mercury and the Moon are evidence that a heavy bombardment of comets and asteroids took place early in the solar system's history. Also interesting is the fact that the surfaces of some planets and moons have areas where the crater density is high, yet in other areas craters are relatively sparse. Because the initial cratering should have been fairly well distributed, scientists conclude that the areas we see today with few craters must have been resurfaced with younger rocks. Examples include the lava-filled basins on the Moon (Figure 2.26), which

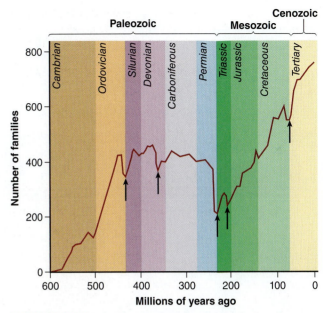

FIGURE 2.25 Graph showing the number of species recorded in the fossil record over the last 600 million years of geologic time. Mass extinctions are marked by periodic and sudden decreases in entire families of species. Scientists believe these die-offs are related to environmental changes that are global in nature.

FIGURE 2.26 The dark areas on the Moon consist of flat-lying lava deposits where the cratering density is significantly lower than the surrounding highlands. The difference in crater density combined with radiometric dating of Moon rocks proves that the heavy bombardment period ended when the solar system was about 3.6 billion years old, which is also when primitive life-forms first show up in Earth's rock record.

FIGURE 2.27   Meteor Crater in the Arizona desert formed approximately 50,000 years ago when a 150-foot (45 m) asteroid struck the Earth. The crater is 0.75 mile (1.2 km) in diameter.

show relatively light cratering compared to the heavily impacted upland areas. Based on radiometric dating (Chapter 1) of different rock samples returned from the Moon, scientists have been able to prove that the lightly cratered lava basins are considerably younger than the heavily impacted areas. From this evidence it is now known that the intense bombardment period ended about 3.6 billion years ago, a time when the solar system was approximately 1 billion years old.

Of considerable interest is the fact that the end of the heavy bombardment, 3.6 billion years ago, roughly coincides with when the first evidence of microbial life shows up in the geologic record. Many astronomers and astrobiologists interpret this to mean that the conditions on Earth during the intense bombardment period were too harsh for life to develop. Note that during this period comets are believed to have been impacting planets and moons throughout the solar system. Since comets contain carbon-based compounds and water, this means that they would have seeded the entire solar system with the essential ingredients for life. Once the heavy bombardment phase ended, life could have developed any place in the solar system where conditions became suitable. Earth just happened to have the most ideal conditions because of its location within the Sun's habitable zone. It is ironic that the early bombardment may have led to the development of life on Earth, but was then followed by occasional impacts of global consequences, triggering one or more of the mass extinctions in the geologic record (Figure 2.25). In this section we will explore how scientists came to realize that impacts are an important geologic process and how they still present a serious hazard to the Earth.

Discovering the Impact Threat

Prior to the Moon landings there was considerable debate among geologists whether the Moon's numerous craters were due to impacts or volcanic activity. The scientific debate was reasonable at the time because volcanic craters are so common on Earth. This even led some geologists to question whether the 50,000-year-old crater in the Arizona desert (Figure 2.27) was formed by an impact, especially since a large meteorite had never been found there. Geologists of course knew that asteroids continue to strike the Earth, but hardly anyone could imagine an asteroid large enough to make a crater of this size. Moreover, it was thought that a large asteroid would break up into relatively small pieces due to the stresses it would encounter as it hits Earth's atmosphere.

The mystery of the Arizona crater, called Meteor or Barringer Crater, was solved in the 1960s when a geologist named Gene Shoemaker mapped the rock layers at the site. He found that the orientation or structure of the rocks, which were once flat-lying, was identical to that of craters formed by nuclear explosions at the U.S. government's test site in Nevada. Soon after the rock structure of Meteor Crater was understood, Dr. Shoemaker and two other scientists found a rare, high-pressure mineral called *coesite* within the crater. This was significant because the only other place coesite had ever been found was in nuclear craters. Because there is no natural

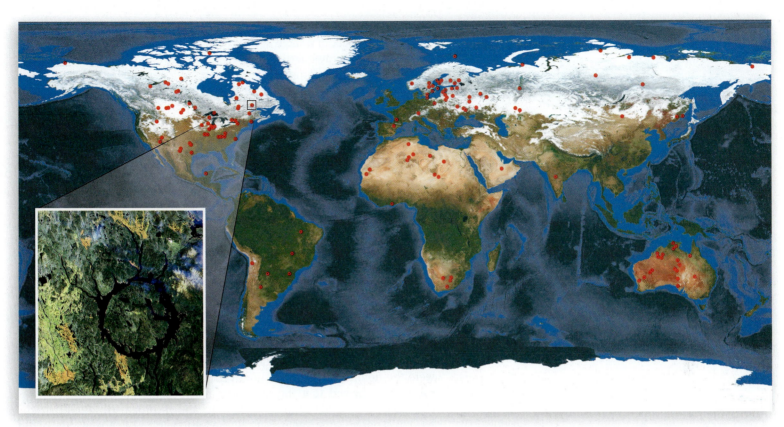

FIGURE 2.28   Approximately 160 impact sites have now been identified on Earth. Note that the number of impacts varies across the globe, in part because of accessibility and the age of the rocks exposed at the surface. Also note that very few impacts have been found in offshore areas. A 214-million-year-old impact structure in Canada (inset) is believed to have had an original crater 50 to 60 miles (75–100 km) in diameter, and formed by an asteroid over 3 miles (5 km) in diameter. Some scientists believe this event may be linked to a mass extinction where nearly 60% of all species on the planet were lost.

Earth process capable of generating the pressure necessary to create coesite, the only possible explanation was that Meteor Crater had formed from a large impact. The reason a large meteorite had never been found there could be explained by the meteor's tremendous speed, which would have caused it to completely disintegrate upon impact—similar to a high-velocity bullet hitting a solid object. What remains then is a relatively large crater and small fragments of the asteroid scattered about the site.

Although Dr. Shoemaker's work and the Apollo Moon landings ultimately proved that impacts occurred on both the Moon and Earth, it took years for the scientific community to recognize that large impacts still took place. The generally accepted idea was that after the heavy bombardment period, Earth's orbit had been swept clear of large debris. Large impacts were highly unlikely because objects the size of small mountains were simply no longer out there. This view eventually changed as geologists began finding impact craters in the more recent parts of the geologic rock record (Chapter 1). By looking for coesite and the characteristic rock structure associated with impacts, geologists have currently identified approximately 160 impact craters on Earth (Figure 2.28). The actual number is undoubtedly much higher than this, partly because most of the asteroids or comets would have struck in the oceans, making the craters extremely difficult to detect. Weathering and erosion processes (Chapter 3) would also have erased or buried the evidence for many of those that struck on land.

Despite the fact that Earth's crater record is very incomplete, it still contains evidence that very large impacts have occurred. Of particular interest are those craters known to have formed within the last 500 million years of Earth's history. This is the period when complex life-forms began to flourish, and a large impact could have led to a mass extinction, thereby affecting the evolution of life. Aside from the obvious blast and possible tsunamis, the most critical effect of a large asteroid or comet impact would be the debris ejected from the crater. Recent studies have shown that massive amounts of debris could be ejected into space, which would then generate frictional heat as the material reenters the atmosphere. This would create a global heat pulse hot enough to quickly kill those land animals not capable of burrowing into the ground. It would also cause wildfires on a global scale, sending large volumes of soot into the atmosphere. This soot, combined with fine debris from the impact, would stay suspended in the atmosphere for a considerable period of time, reducing the amount of sunlight reaching the surface. As global temperatures drop, the environment would become inhospitable for many of the species that had survived the initial heating. The end result would be mass extinction.

The Mesozoic/Cenozoic Extinction Event

Perhaps the most famous of all mass extinctions was the one that ended the reign of the dinosaurs about 65 million years ago (see Figure 2.25). This event was first recognized over 100 years ago as geologists noted that worldwide, rocks of the same relative age showed a dramatic change in their fossil content. Most fascinating was that the underlying (older) rocks showed that dinosaurs had once been the dominant type of animal, whereas mammals were the dominant species in the overlying (younger) rocks. The younger rocks also showed that the dinosaurs had become extinct. Because this change in animal life was so significant, and recorded worldwide, it was used to separate what geologists call the Mesozoic and Cenozoic eras of the geologic time scale (Chapter 1). Altogether there are three eras (Paleozoic, Mesozoic, and Cenozoic), which represent the entire 550-million-year history of complex life on our planet. Geologists have always been keenly interested in discovering what took place at the Mesozoic/Cenozoic boundary that would have caused dinosaurs to go extinct, leaving mammals to dominate the planet. This important boundary is commonly called the *Cretaceous/Tertiary boundary,* which reflects the names of the finer subdivisions within the Mesozoic and Cenozoic eras as shown in Figure 2.25.

After many years of extensive study, geologists have learned a great deal about the demise of the dinosaurs and rise of mammals. It is now known that for about 20 million years prior to the Mesozoic/Cenozoic (i.e., Cretaceous/Tertiary) boundary, dinosaurs were becoming less diversified as individual species became more specialized. Geologists generally believe that some type of global change in the environment then occurred, making it impossible for these highly specialized species to survive. Interestingly, the rock record shows that nearly 75% of *all* species on the planet became extinct—terrestrial and marine combined. In the oceans approximately 90% of the plankton were lost, which almost certainly led to a major collapse within the marine food web. The question, of course, is what triggered such a massive die-off. Most ideas center on some type of global change in Earth's climate, triggered perhaps by a large impact or a series of large volcanic eruptions, or even the movement of continental landmasses over time. Another possible mechanism would be the intense UV radiation resulting from a nearby gamma-ray burst depleting Earth's ozone layer, as described earlier.

A significant breakthrough in the mystery of the Mesozoic/Tertiary boundary came in the 1980s when physicist Luis Alvarez and his son Walter, a geologist, were studying rocks straddling the boundary in Italy. Here they discovered a thin layer of clay within the sequence of limestone rocks. Moreover, this clay layer marked a distinct change in the fossil content of the rocks, suggesting it represented the top of the Mesozoic. Laboratory analysis revealed the clay layer was 65 million years old and contained unusually large amounts of the element iridium. This was highly significant because iridium is an extremely rare element in Earth rocks, but fairly common in most meteorites. The iridium-rich clay therefore represented the first direct evidence that an asteroid or comet impact coincided with the Mesozoic/Cenozoic extinction. The Alvarezes concluded that a large asteroid or comet hit Earth, creating a huge dust cloud that led to a mass extinction.

After the discovery of the iridium layer, scientists began taking a closer look at rocks around the world where the Mesozoic and Cenozoic boundary is exposed. Astonishingly, not only was a layer of iridium-rich clay found at many sites, but it consistently yielded radiometric dates around 65 million years old. This left little doubt that the clay layer represented the atmospheric fallout of a global dust cloud. Moreover, at some of the sites the clay also contained coesite, providing conclusive proof that the dust cloud resulted from a large impact 65 million years ago. By mapping the location of coesite and other geologic data, scientists were able to determine the general area where the asteroid made impact.

The chance of finding the actual crater, though, was rather small because so much of Earth's impact record is missing. But then, in 1990, a large, 65-million-year-old crater was found whose location was consistent with the other data. This crater, called *Chicxulub,* is 112 miles (180 km) wide and located along Mexico's Yucatán peninsula (Figure 2.29). With such overwhelming evidence, most scientists became convinced that the Chicxulub crater represented the impact that created the 65-million-year-old iridium layer. Today however, some scientists are debating whether new data indicates that the Chicxulub crater formed just prior to the iridium layer, making it the wrong crater. Regardless of how the debate unfolds, there remains a great deal of evidence that a large asteroid impact took place 65 million years ago. Moreover, this event was so large that it altered the global environment such that it helped trigger a mass extinction. The large and highly specialized dinosaurs would have been very susceptible to the intense heat pulse and subsequent global cooling described earlier. In contrast, mammals at the time were mostly small, burrowing creatures that would have had a much easier time surviving the dramatic change in the environment. We humans then likely owe our current position at the top of

FIGURE 2.29   The Chicxulub impact crater on the Yucatán peninsula in Mexico is one of the largest found on Earth and is believed to be a major factor in the extinction of the dinosaurs. The crater measures 112 miles (180 km) across and was formed by an asteroid estimated to be 6.2 miles (10 km) in diameter. A recent computer-generated gravity map (inset) indicates that the crater may possibly be much larger, nearly 185 miles (300 km) across.

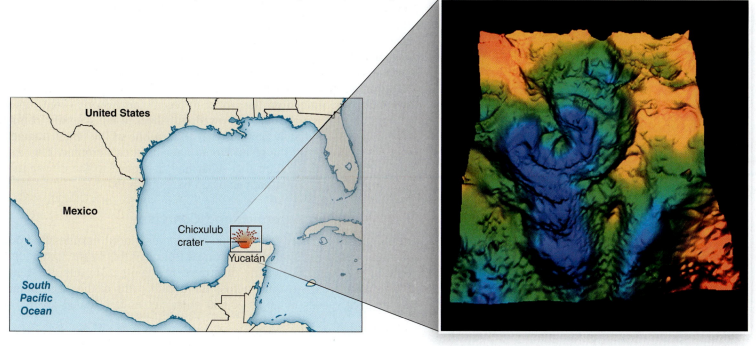

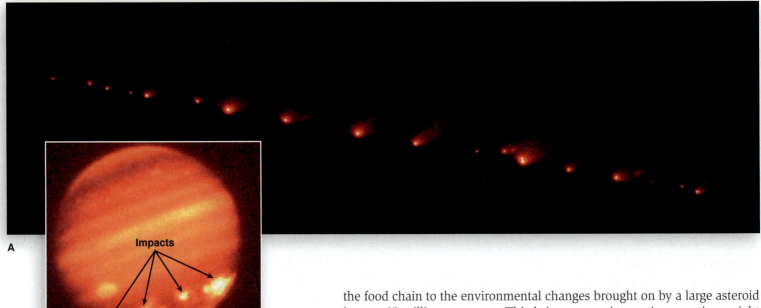

FIGURE 2.30   Image (A) taken by the Hubble space telescope of the broken fragments of comet Shoemaker-Levy 9 prior to the 1994 impact with Jupiter. An infrared image (B) showing the impact plumes made by several of the comet fragments.

the food chain to the environmental changes brought on by a large asteroid impact 65 million years ago. This brings up an interesting question: might we humans eventually succumb to the same fate as the dinosaurs, undergoing a mass extinction brought on by a global change in the environment?

Recent Impact on Jupiter

Shortly after the Chicxulub crater was discovered, scientists were fortunate enough to actually witness a major impact on Jupiter in 1994. The only reason this extremely rare event was known in advance was because of the continued research of Gene Shoemaker. Based on his earlier work, Shoemaker hypothesized that the solar system still contained large objects, perhaps a mile or more in diameter, whose orbits sent them streaking across Earth's orbit. If such Earth-crossing comets and asteroids truly existed, then they would pose a serious threat to life here on Earth. In the early 1980s Dr. Shoemaker, his wife Carolyn, and colleague David Levy began using a small telescope with a wide viewing area to map stray objects orbiting the inner solar system. In 1993 they discovered a comet (Figure 2.30), which during a previous pass by Jupiter, had broken up into more than 20 major pieces by the planets strong gravitational field. Shortly after this discovery it was determined that the comet, named Shoemaker-Levy 9, was on a collision course with Jupiter. When the broken-up comet began impacting Jupiter in July of 1994, nearly every major telescope in the world was trained on the planet to record this historic event. Images of the impact (Figure 2.30) both awed and shocked the scientific community. It was immediately obvious that if even one of the larger fragments had struck the Earth, the result would have been catastrophic. This event was a dramatic illustration that what had happened to the dinosaurs 65 million years ago, could happen again here on Earth.

Earth-Crossing Asteroids and Comets

While the vast majority of comets and asteroids orbit harmlessly around the Sun, one occasionally gets bumped from its orbit by the gravitational field of a passing planet. As illustrated in Figure 2.31, this can send a comet or asteroid streaking across the orbital paths of the inner planets, traveling at speeds of up to 70,000 miles per hour (110,000 km/hr). Of particular concern are Earth-crossing objects that are large enough that an impact would have global consequences. This of course could be catastrophic for humanity. Scientists estimate that objects with a diameter of 0.6 mile (1 km) or greater would have serious global consequences should they strike the Earth.

In response to this impact threat, NASA created the Near Earth Object (NEO) program in 1994. The mission of this program is to locate stray asteroids and comets within the solar system, then monitor those whose orbits could potentially put them on a collision course with the Earth. When the NEO program first began, only 174 large-diameter (>1 km) objects had been identified out of an estimated total of 1,000. By 2007 the number of discoveries had reached 740. Fortunately, none of the larger objects located so far are projected to make impact with the Earth in the foreseeable future. However, the paths of Earth-crossing objects can change as they orbit the solar system and encounter the gravitational fields of the planets (Figure 2.31). NASA therefore continues to monitor the orbits of objects that cross Earth's orbit.

If a large comet or asteroid is found to be on a collision course with Earth, the next step would be to have a spacecraft intercept the object in an attempt to alter its trajectory and avoid a devastating impact. All of this is not science fiction. In fact, several recent international missions have rendezvoused with comets and asteroids, gathering useful data for possible future attempts at deflecting an object. Although various methods have been proposed for altering the trajectory of a comet or asteroid, most involve using technology that already exists. One simple approach is to crash a probe directly into the object, so that the transfer of energy would change the object's trajectory. Others involve different means of exerting a small force over a long period of time, thereby slowly altering its path. Note that the preferred method in Hollywood movies is to vaporize an asteroid or comet using a nuclear weapon. In reality this would be a poor choice as even our most powerful nuclear weapons would likely result in the object being broken up into several fragments. As was the case with Shoemaker-Levy 9 (Figure 2.30), the broken fragments would likely still be quite large and cause considerable damage upon impact.

Finally, a critical aspect in any mission to deflect an asteroid or comet is to have enough advanced warning so that spacecraft can rendezvous with the object before it gets too close to the Earth. As illustrated in Figure 2.32, if the deflection attempt is made when the object is far away, then just a slight change in its trajectory would be enough to make it miss the Earth. However, if it gets too close, then the degree to which its trajectory would have to be changed would likely be beyond our capability. In this case a catastrophic collision could not be avoided. One of the reasons NASA began mapping stray asteroids and comets in the first place was to have enough time to mount a mission before the object gets too close, making the impact all but inevitable.

Impact Risk

Recall from Chapter 1 that in managing environmental risk, one needs to consider both the probability of a hazardous event taking place and its potential consequences. When it comes to the risk of impacts, tons of material strike Earth's atmosphere everyday, but the particles are so small that most burn up harmlessly in the atmosphere. The concern, of course, is a large impact whose consequences would be regional or global in extent. Fortunately, the probability of such a strike is low because there

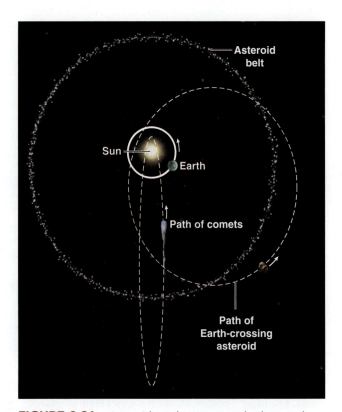

FIGURE 2.31   Asteroids and comets can be bumped from their normal orbits around the Sun by the gravitational effect of passing planets, sending them across the orbital paths of the planets. Of particular concern are comets and asteroids that cut across Earth's orbit.

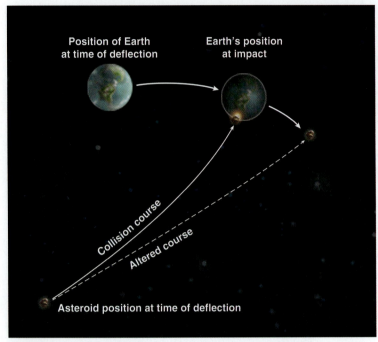

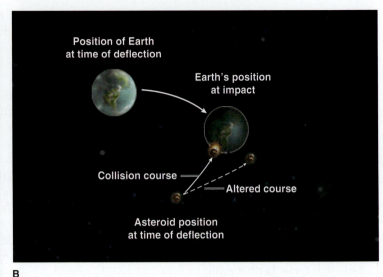

A

FIGURE 2.32 Technology already exists that would allow humans to alter the trajectory of an Earth-crossing object so that it would miss the Earth. The ideal situation (A) is for a spacecraft to intercept the object while it is still far from the Earth. In this situation only a slight change in its trajectory would be required to make it miss the Earth. If the object is intercepted too close to the Earth (B), then the degree to which it would have to be deflected would be beyond the capacity of humans.

are comparatively few large asteroids and comets that cross Earth's orbit. For example, NASA scientists estimate that an impact large enough to cause localized destruction on land, or a tsunami from a ocean strike, can be expected to occur on average once every 50 to 1,000 years. An object large enough to cause massive destruction on a regional basis, or cause a giant tsunami, could be expected once every 10,000 to 100,000 years on average. One large enough to have consequences that are catastrophic and global in nature, similar to what formed Chicxulub crater 65 million years ago, could occur once every 100,000 years or longer.

While a large asteroid or comet impact could have dire consequences for humanity, we need to keep the hazard in perspective. First of all, the probability of such an event occurring in any given year, or even your own lifetime, is exceedingly small. Moreover, society routinely faces a variety of natural Earth hazards (e.g., floods, earthquakes, volcanic hazards, hurricanes). These hazards of course do not threaten the entire human race, but cumulatively they result in a tremendous loss of life and property damage. Consider how the December 2004 Indonesian earthquake, and resulting tsunami, took the lives of nearly 250,000 people. This disaster was followed a mere 10 months later by an earthquake in Pakistan that killed an additional 100,000 people. Therefore, while it is important that society takes steps to reduce the risk of a major impact, we must also reduce the risk of those hazards that pose serious problems on a routine basis. In the end, it is ironic that humans are powerless to prevent natural disasters associated with earthquakes and volcanoes, but could prevent a large impact, which is a hazard capable of ending human civilization.

SUMMARY POINTS

1. Knowledge of the solar system and universe can help us to better understand the Earth system and provide a larger perspective from which to view our environmental problems.

2. Nuclear fusion of hydrogen atoms in the Sun produces electromagnetic radiation that travels outward, eventually striking the planets and moons within the solar system. This energy is what drives critical processes in Earth's biosphere, hydrosphere, and atmosphere.

3. The nebular hypothesis describes how the solar system formed and accounts for the orbital characteristics of the planets and moons. Here a cloud of dust and gas condensed and slowly began to rotate, flattening into a disc in which material accreted into planets. The immense gravity at the center caused nuclear fusion of hydrogen, creating the Sun.

4. Our solar system lies in the Milky Way Galaxy along with hundreds of billions of other stars. Powerful telescopes have shown there to be hundreds of billions of other galaxies in the universe. The big bang theory accounts for the fact that the universe is continuing to expand outward.

5. The oldest forms of life recorded in Earth's rocks are traces of extremophile bacteria. Scientists believe that such ancient bacteria could have formed elsewhere in the solar system wherever the temperature would allow liquid water to exist.

6. Complex animal life in other solar systems is likely to be restricted to the habitable zone around stars where liquid water can exist on planets and where the climate is stable for long periods of geologic time.

7. Mass extinctions recorded in Earth's geologic record are believed to be related to changes in the global environment. Possible triggers for such global change include large volcanic eruptions, comet or asteroid impacts, and nearby gamma-ray bursts.

8. Geologic studies have shown that large asteroids and comets have repeatedly struck the Earth and the other planets. A worldwide iridium-rich clay layer and a large crater in Mexico are strong evidence that a large impact triggered a mass extinction that ended the age of the dinosaurs 65 million years ago.

9. After witnessing a large impact on Jupiter in 1994, scientists realized that the Earth is still at risk from being hit by a stray asteroid or comet. NASA then began a program to map Earth-crossing objects to detect those on a potential collision course with Earth. With advanced warning, humans may be able to deflect the object, thus avoiding a collision.

10. A large impact with Earth could be devastating for humanity, but the probability of it occurring in any given year is exceedingly small. On the other hand, natural Earth hazards such as earthquakes and volcanoes happen relatively frequently, and their cumulative effect in terms of deaths and property loss is enormous.

KEY WORDS

accretion   43
asteroids   41
big bang theory   46
comets   41
electromagnetic radiation   40
extremophile bacteria   48

galaxies   46
gamma-ray burst   55
gas giants   40
habitable zone   48
mass extinction   55
nebular hypothesis   43

ozone depletion   54
planets   44
rare earth hypothesis   52
terrestrial planets   40

APPLICATIONS

Student Activities   Look up at the sky on a dark, clear night. (If you live in a very urban area, take a drive into the country on a clear night). Allow your eyes to adjust to the dark and look up into the sky. You should be able to see the Milky Way. It goes from east to west not quite directly above your head. It looks like a bright grey area. This is an arm of our spiral galaxy!

Critical Thinking Questions

1. How did the solar system form?
2. What are extremophile bacteria and how are they important to life on Earth?
3. What is mass extinction?
4. What is the difference between an asteroid and a comet?

Your Environment: YOU Decide   Do you think that humans should explore worlds other than Earth?

Chapter 3

Earth Materials

LEARNING OUTCOMES

After reading this chapter, you should be able to:

▶ Understand how different types of atoms are created and how they are assembled into minerals and rocks.
▶ Know why the rock-forming minerals are so important to the study of geology.
▶ Be able to explain the basic way in which each of the major rock types form.
▶ Understand the relationship between weathering, erosion, transportation, and deposition and the different types of sedimentary rocks.
▶ Know why some rocks are more susceptible to chemical weathering than others.
▶ Describe the basic paths rocks can take through the rock cycle in response to different geologic processes.
▶ Understand how rocks can be used to interpret ancient environments.

The Earth is composed of different types of rocks and minerals, which are commonly sculpted into striking landscapes by weathering and erosion processes. Shown here is an iron stained sandstone in Arizona whose layering indicates that it formed from an ancient deposit of windblown sand. Rock and mineral deposits not only let geologists interpret Earth's history, but also serve as important raw materials used in modern societies.

Introduction

In Chapter 3 we will take a closer look at minerals and how they form different types of rocks, which make up the solid earth. It is important that you have a working knowledge of rocks and minerals because nearly every topic in environmental geology in some way or another involves these materials. For example, most of the material used to build the infrastructure and produce the consumer goods in modern societies comes from rock and mineral resources (Chapter 12). To meet the high demand, enormous quantities of iron, copper, aluminum, and other metallic-bearing minerals must be removed from the solid earth and processed into pure metal. Large amounts of nonmetallic minerals are also extracted and used to produce key materials such as concrete, glass, ceramics, and drywall (i.e., sheetrock). In addition to raw materials, subsurface rocks provide critical supplies of coal, oil, and gas (Chapter 13). Without these energy resources, modern society as it exists today would simply grind to a halt.

Knowledge of rocks and minerals is also important when it comes to our food and water supplies. The crops farmers grow depend on fertile soils that are derived from the breakdown of rocks and minerals, a process that requires water percolating into the subsurface over long periods of time. In addition to producing soil, infiltrating water accumulates in porous layers of rock and sediment, which humans extract and use for a multitude of purposes. One key use of this groundwater has been to irrigate crops so as to increase food production in areas where rainfall is limited. Moreover, chemical reactions between minerals and water affect the very quality of the water we drink. In some areas water supplies are naturally contaminated with heavy metals, or radioactive gases, which have a negative impact on human health.

The chemical and physical properties of rocks and minerals also have a profound influence on many types of geologic hazards, as well as the landscape that we inhabit. For example, the physical strength of rocks and minerals helps determine the amount of energy a rock body can accumulate before it fails, at which point the stored energy is released in the form of vibrating waves. These waves produce the ground shaking we refer to as an *earthquake;* therefore, the more energy rocks can store, the stronger the earthquake (Chapter 5). The inherent strength of rocks and minerals

A   B

FIGURE 3.1 Earth's different landscapes reflect the underlying geology and erosion history of an area. A key control is the way different rocks respond to weathering and erosion processes explained in this chapter. Igneous rocks of Yosemite Valley (A) have fairly uniform chemical and physical properties, and thus weather in more rounded shapes. The properties of layered sedimentary rocks in the Grand Canyon (B) vary significantly, generating a landscape with a distinctive stair-step pattern.

also helps determine the stability of steep slopes, and therefore is important in mass wasting hazards such landslides and rock falls (Chapter 7). Last, we can see from Figure 3.1 that landscapes develop differently over time depending in part on the properties of the underlying rocks. Geologists often like to point out that our scenic landscapes are the result of a combination of interesting rocks and geologic history.

Rocks and minerals are clearly fundamental to nearly every interaction between people and the geologic environment. Consequently, the purpose of Chapter 3 is to provide readers with a basic knowledge of these materials so that they will have a better understanding of the topics discussed in this text.

Basic Building Blocks

In this section we will explore the different building blocks that make up the solid Earth. In particular we want to examine how rocks are composed of individual mineral grains, which, in turn, are made up of even smaller particles called atoms. We will begin at an even smaller scale, namely the particles that make up individual atoms.

Atoms and Elements

From chemistry we know that all matter is composed of individual *atoms,* which consist of tiny *electrons* orbiting around a nucleus of much larger particles called *protons* and *neutrons*. Atoms containing the same number of protons are referred to as *elements*. The simplest types of atoms are those known as the element hydrogen, in which a single electron orbits around a proton as illustrated in Figure 3.2. Progressively heavier elements in the periodic table (see inside back cover) have a nucleus with additional protons and neutrons along with a corresponding increase in the number of electrons. For example, all carbon atoms contain six protons and between six and eight neutrons. Note that electrons have a negative electrical charge and protons are positively charged, whereas neutrons have no charge. Thus when atoms have an equal number of electrons and protons, they are considered to be electrically neutral. However, some elements are configured in such a way that they are more stable when they can acquire an extra electron; others are more stable when they give off an electron. When this occurs, atoms will have either a positive or negative charge and are then referred to as **ions**—the actual charge depends on the number of electrons gained or lost. Examples of common ions include sodium (Na^+), iron (Fe^{3+}), chlorine (Cl^-), and oxygen (O^{2-}).

An obvious question is how do electrons, protons, and neutrons manage to get assembled into the various configurations that make up the elements in the periodic table? In other words, how do the different elements form? Recall from Chapter 2 that hydrogen atoms, which are the simplest of all, undergo nuclear fusion in stars to produce helium atoms with two protons and two neutrons in the nucleus. Eventually the hydrogen fuel in a star diminishes to the point were fusion reactions start to convert helium into progressively heavier atoms such as carbon, oxygen, iron, and nickel. As some stars approach the end of their life cycle, they will eject this material out into space. This mixture of different types of atoms (i.e., elements) then becomes part of a swirling cloud of dust and gas around a newly formed star, coalescing into planets in a process called *accretion*. Therefore, the different elements we see on Earth, making up everything from rocks to gases to our own human bodies, were created in the nuclear furnace of stars.

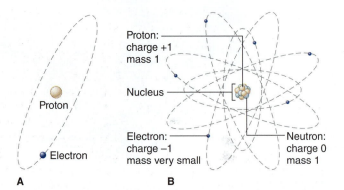

FIGURE 3.2 In a simplified view of atoms, small, negatively charged electrons orbit a nucleus composed of protons (positive charge) and neutrons (neutral). The simplest types of atoms are of the element hydrogen (A), with a single electron orbiting a proton. Each succeeding element in the periodic table contains an additional proton and varying numbers of neutrons, thereby making them heavier. A carbon atom (B) contains roughly the same number of neutrons and electrons as it does protons.

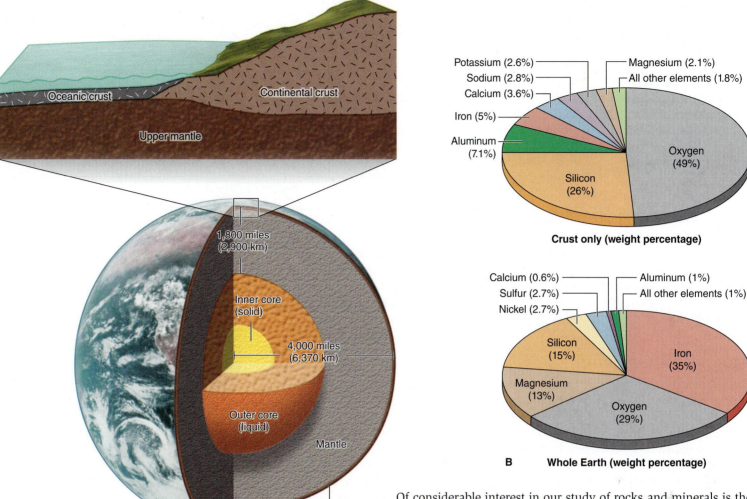

FIGURE 3.3 Earth has a layered structure (A) consisting of the core, mantle, and crust. Geologic processes have caused the heaviest elements to become concentrated in the core over time, whereas the lighter elements have tended to accumulate in the crust. The diagrams (B) show that the crust is largely composed of oxygen and silicon atoms, but when considering the planet as a whole, iron atoms are the most abundant.

Of considerable interest in our study of rocks and minerals is the relative abundance of the elements found on Earth. Although there are around 90 naturally occurring elements in the periodic table, from Figure 3.3 one can see that just a few elements make up most of the Earth. If we consider the Earth as a whole, we see that iron atoms account for 35% of the planet by weight, followed by oxygen at 29%. On the other hand, if we look at just the outermost portion of Earth called the *crust*, oxygen atoms account for an incredible 49% of the crust by weight, whereas the iron content decreases to around 5%. This compositional difference between the whole Earth and the crust is due to internal processes within the Earth that have caused the planet to become layered. These internal processes tend to cause heavier elements to migrate toward the center of the planet, called the *core*, whereas lighter elements tend to accumulate in the crust; the intermediate layer is called the *mantle*. Note that Earth's internal processes and layered structure will be described in more detail in Chapter 4.

Minerals

Atoms combine to form solids, liquids, and gases, but our interest here is in a particular type of solid known as a mineral. A **mineral** is a naturally occurring inorganic (non-carbon-based) solid composed of one or more elements, where the individual atoms have an orderly arrangement called a *crystalline structure*. In this structure the atoms are in a fixed pattern that repeats itself in a three-dimensional manner as illustrated in Figure 3.4. Solids such as glass and plastic, for example, are not minerals because their atoms are not arranged in a crystalline structure; they are referred to as

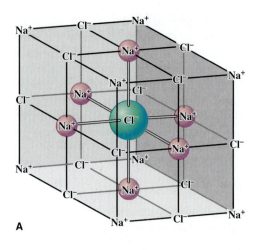

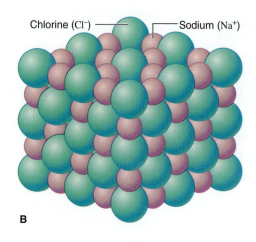

Chlorine (Cl⁻) — Sodium (Na⁺)

A   **B**

FIGURE 3.4 All minerals have an internal structure and a definite chemical composition where the atoms are arranged in a set pattern that repeats itself in a three-dimensional manner. In view (A) the distance between atoms has been exaggerated in order to illustrate the fact that the angles and distances within the crystal structure are fixed. Also note that the surface of each atom represents the outermost shell or cloud of the orbiting electrons.

noncrystalline or *amorphous*. Note that different minerals can have different crystalline structures which vary in terms of the angles and distances between atoms. In any given mineral, however, the structure is fixed and always the same no matter where it forms. This holds true for minerals that form on bodies other than the Earth, such as those found in rock samples from Mars and the Moon.

In addition to having a crystalline structure, all minerals have a definite chemical composition. What this means is that only certain elements, and in certain proportions, are allowed into the crystalline structure. For example, the chemical formula for the mineral pyrite (fool's gold) is FeS_2; thus, only iron (Fe) and sulfur (S) atoms are allowed in the structure. This structure contains exactly two sulfur atoms for every iron atom. However, some minerals form what geologists call a *continuous series,* which is where two or more types of atoms freely substitute for one another; thus, they are found in varying proportions. A good example is the mineral series called *plagioclase feldspar*—$(Ca,Na)(Al,Si)AlSi_2O_8$—where calcium (Ca) and sodium (Na) substitute for each other. Although minerals are very specific as to the type and number of atoms allowed into their structure, atoms of similar size and electrical charge will occasionally be allowed to substitute. Minerals therefore may have minor impurities due to such atomic substitutions.

Interestingly, mineralogists have identified well over 4,000 different minerals on Earth. Each mineral has a unique combination of crystalline structure and proportion of specific elements. This means that if two minerals happen to have the exact same internal structure, then their chemical composition must be different—otherwise they would be the same mineral. For example, the minerals calcite ($CaCO_3$) and siderite ($FeCO_3$) have the same crystalline structure, but siderite has iron atoms in place of calcium atoms within the structure. In other cases minerals may share the exact same chemistry, but differ in their crystalline structure—as with calcite and aragonite ($CaCO_3$).

A key point here is that physical properties of individual minerals, such as melting point, hardness, and density, are all controlled by their internal structure and chemical composition. This means that because each mineral has a unique combination of structure and chemistry, each has a corresponding unique set of physical properties. For example, graphite (common pencil "lead") and diamond are two minerals composed entirely of carbon (C) atoms, but have different crystalline structures (Figure 3.5). Despite being composed of the same type of atoms, the structural difference results in these two minerals having vastly different physical properties. Diamond is the hardest known substance, whereas graphite is gray and one of the

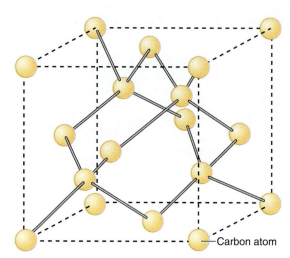

Carbon atom

Structure of diamond

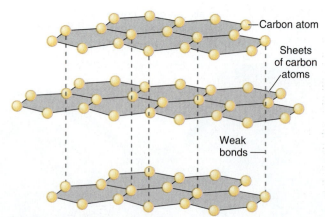

Carbon atom

Sheets of carbon atoms

Weak bonds

Structure of graphite

FIGURE 3.5 Although both diamond and graphite are composed entirely of carbon atoms, they have different crystalline structures. Diamond's structure helps make it the hardest known substance, whereas the weak bond between the sheets of carbon atoms in graphite make it one of the softest minerals.

1 cm

A

B 1 cm

FIGURE 3.6 Rocks are commonly composed of multiple types of mineral grains like the granite in (A), but some rocks contain only a single type of mineral grain, like the quartzite shown in (B).

FIGURE 3.7 Minerals can grow crystal faces, as in these quartz crystals, if conditions are favorable and there is sufficient space for the faces to develop.

softest. Humans have taken diamond's extreme hardness and put it to practical use in making tools that will cut through *any* material. With respect to graphite, the softness of this mineral has been put to use as a dry lubricant and in writing tools. Interestingly, humans have been finding uses for the physical properties of minerals for thousands of years, beginning perhaps with the extremely sharp edges of broken pieces of flint. In Chapter 12 we will explore this link between the physical properties of minerals and how people have learned to put these properties to work in various applications.

Rocks

A **rock** is defined as an aggregate or assemblage of one or more types of minerals. Most rocks are composed of several different types of mineral grains bound together, similar to the granite shown in Figure 3.6A. There are some rocks that consist almost entirely of grains from a single type of mineral. A good example is the rock known as quartzite (Figure 3.6B), where grains of the mineral called quartz are bound together. In addition to mineral composition, another important property of rocks is *texture*, which refers to the way the mineral grains themselves are arranged. The texture of rocks varies depending on such things as the size (coarse, fine, or mixed) and shape (round, angular, tabular, and elongate) of individual mineral grains. In some cases the grains are oriented in a specific direction, giving a rock what geologists refer to as a fabric, similar to how fibers within cloth are oriented parallel to one another. Later you will see that the texture of a rock provides geologists with important clues as to the rock's origin and history.

One of the ways in which mineral grains assemble themselves into rocks is when molten rock, called **magma,** begins to cool. Magma itself forms deep within the Earth when the temperature increases to a point where the chemical bonds within minerals begin to break down. As the bonds break, the individual atoms making minerals are released, forming a complex mass of positively and negatively charged ions (e.g., Na^+, Fe^{2+}, Cl^-, SO_4^{2-}). When magma encounters conditions that allow it to cool, the ions will begin to arrange themselves back into the crystalline structure of minerals. Over time the individual mineral grains will grow as ions are selectively removed from the magma. Minerals also form when ions crystallize out of waters that are rich in dissolved ions, in which case the crystallization process is referred to as *precipitation*. These solutions are often present around magma bodies where the minerals precipitate at relatively high temperatures. Other types of minerals precipitate from cooler waters associated with shallow groundwater systems. In either case, if the precipitating mineral grains have sufficient room to grow, as in a cave or small void space, then they can develop crystal faces similar to the example in Figure 3.7. Note that when precipitation occurs in open bodies of water, mineral grains will settle out over time, creating layers of sediment. The mineral halite (NaCl), or common table salt, forms in this manner.

Finally, new minerals also form when preexisting minerals undergo chemical and physical transformations while in the solid state. This process is quite common when changes in temperature and pressure deep within Earth's crust force minerals to recrystallize—here ions slowly move between individual grains within the rock body. Some minerals are transformed into new minerals during chemical reactions with the water and oxygen molecules that are associated with Earth's surface environment. This process, referred to as *weathering,* is one of the primary ways that clay minerals form. Humans, of course, have been using clay to make pottery for thousands of years.

Rock-Forming Minerals

Geologists classify minerals based primarily on the type of negatively charged ion they contain—this electrical charge is balanced by positively charged ions. For example, *sulfide* minerals all contain sulfur ions bonded to positively charged ions like lead (PbS), zinc (ZnS), and iron (FeS_2). *Carbonate* minerals all have different positive ions bonded to the carbonate ion (CO_3^{2-}). Likewise, *oxide* minerals all contain the oxygen ion (O^{2-}) and *sulfates* contain the sulfate ion (SO_4^{2-}). The *silicate* class by far contains the greatest number of minerals, where the basic building block is the silicate ion (SiO_4^{4-}). Figure 3.8 illustrates how the silicate ion itself creates a variety of complex chain and sheetlike structures, all which can be bonded together by positive ions.

Although there are over 4,000 known minerals on Earth, only a dozen or so make up most of the rocks in Earth's outermost layer known as the *crust* (see Figure 3.3). Geologists generally refer to the common minerals listed in Table 3.1 as the **rock-forming minerals.** Note that the table contains several mineral groups, each of which includes different minerals that are closely related. Also note how a majority of the rock-forming minerals are silicates (i.e., contain the SiO_4^{4-} ion), accounting for the fact that silicon and oxygen atoms make up 75% of Earth's crust by weight. While many of the remaining 4,000 or so minerals are relatively rare and insignificant in terms of the overall volume of rocks, they are oftentimes concentrated by geologic processes. Because many of the less common minerals have important applications in modern society, such localized concentrations can be of considerable economic value—a topic covered in detail in Chapter 12.

TABLE 3.1 Some of the more common minerals that make up most of the rocks in Earth's crust. For a listing of chemical symbols see inside back cover.

Mineral Name or Group	Class—Important Negative Ion	Important Positive Ions
Olivine ($[Mg,Fe]_2SiO_4$)	Silicate (SiO_4^{4-})	Mg, Fe
Pyroxene Group (e.g., augite)	Silicate (SiO_4^{4-})	Fe, Mg, Ca
Amphibole Group (e.g., hornblende)	Silicate (SiO_4^{4-})	Ca, Na, Mg, Fe, Al
Plagioclase Feldspar Group	Silicate (SiO_4^{4-})	Ca, Na, Al
Potassium Feldspar Group (e.g., orthoclase)	Silicate (SiO_4^{4-})	K, Al
Mica Group (e.g.,muscovite and biotite)	Silicate (SiO_4^{4-})	K, Fe, Mg, Al
Quartz (SiO_2)	Silicate (SiO_4^{4-})	n/a
Clay Minerals Group (e.g., kaolinite and illite)	Silicate (SiO_4^{4-})	K, Ca, Na, Mg, Fe, Al
Calcite ($CaCO_3$)	Carbonate (CO_3^{2-})	Ca
Dolomite ($CaMg[CO_3]_2$)	Carbonate (CO_3^{2-})	Ca, Mg
Gypsum ($CaSO_4 \bullet 2H_2O$)	Sulfate (SO_4^{2-})	Ca

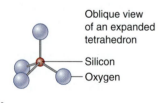

Tetrahedron viewed from above

Oblique view of an expanded tetrahedron

— Silicon
— Oxygen

A

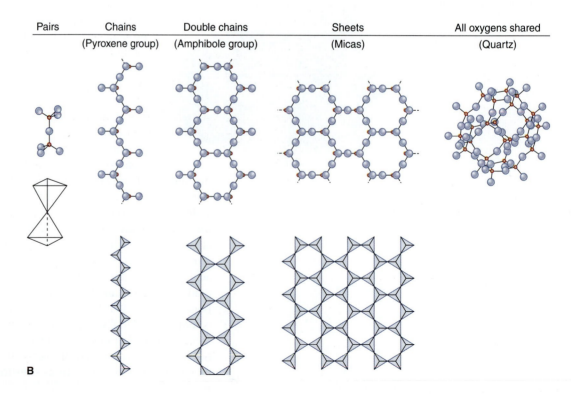

Pairs	Chains	Double chains	Sheets	All oxygens shared
	(Pyroxene group)	(Amphibole group)	(Micas)	(Quartz)

FIGURE 3.8 Minerals in the silicate class all have the silicon-oxygen tetrahedral as their basic building block, which can be linked together in the various ways shown here.

B

FIGURE 3.9 Some of the more important rock-forming minerals: (A) olivine, an iron and magnesium-rich silicate believed to be compositionally similar to the minerals making up the mantle; (B) feldspars, a mineral group that makes up the largest percentage of crustal rocks; (C) quartz, a very abundant mineral in continental rocks and sediment; and (D) micas, a group of common platy minerals.

At this point we need to take a brief look at the rock-forming minerals listed in Table 3.1. Here we can see that *olivine* and the mineral groups called *pyroxene* and *amphibole* are all silicates containing iron (Fe) and magnesium (Mg) ions. Collectively these minerals are referred to as **ferromagnesian minerals,** whose high iron content makes them relatively dense. Later in Chapter 4 you will see how the ferromagnesian content of rocks helps explain why some rocks are denser than others. Rock density is an important factor in many physical processes, including the movement of Earth's crustal plates (Chapter 4). Note that the ferromagnesian mineral called *olivine* (Figure 3.9A) is compositionally similar to minerals believed to be making up the bulk of Earth's rocky interior. Interestingly, coarse grains of olivine are sometimes found in the black volcanic rocks that make up most of the Hawaiian islands. Olivine's green color is what gives Hawaii's so-called green beaches their color.

Another important group of rock-forming minerals listed in Table 3.1 includes aluminum (Al)-rich silicate minerals called plagioclase and potassium feldspar, often simply referred to as **feldspars** (Figure 3.9B). Together, feldspar and ferromagnesian minerals make up the bulk of Earth's crust. Of particular importance is the fact that when these minerals are exposed to acidic solutions (low pH), such as natural rainwater, they readily break down and are transformed into **clay minerals,** a group of silicate minerals enriched with aluminum. Because feldspar and ferromagnesian minerals are so common, clays end up being a major component of soils and sediment found blanketing the surface. The formation of clays has been critical to the development of complex life on Earth because our vegetation and food supply depends on the clay content in soils (Chapter 10)—clays also serve as the raw material for ceramics. Another common group of silicates that are transformed into clays includes the platy minerals known collectively as *micas* (Figure 3.9D). Last there is **quartz** (SiO_2), a silicate mineral composed entirely of silicate ions (SiO_4^{4-}). Quartz (Figure 3.9C) is commonly found in significant quantities along with feldspars in the crustal rocks that underlie much of Earth's continental landmasses. Unlike feldspars though, quartz is chemically resistant to acidic solutions, and is therefore found in great abundance along with clays in soil and sediment. In fact, much of the sand-sized sediment we find near Earth's surface is composed of small, rounded grains of quartz. Note that pure deposits of quartz are used as the raw material for making glass.

Table 3.1 also lists several rock-forming minerals that are based on carbonate and sulfate ions rather than the silicate ion. The carbonate mineral called **calcite** ($CaCO_3$) is the dominant mineral in an important group of rocks called *limestone,* many of which contain shell fragments composed of calcite that are made by marine organisms. Calcite-bearing rocks are used as the raw material for making cement and concrete. A key property of calcite, and any rock made of calcite, is its tendency to dissolve in acidic water. For example, when groundwater travels through fractures in limestone it can slowly dissolve the rock to form passageways and caverns. It turns out that calcite's ability to dissolve creates some unique environmental problems. One problem is when underground caverns collapse and form sinkholes (Chapter 7), and another is when acid rain causes the rapid deterioration of monuments and concrete structures made of calcite (Chapter 15). Note that *dolomite* is a magnesium-bearing carbonate mineral that makes up a fairly common type of rock called dolomite, or dolostone. Finally, the last rock-forming mineral listed in Table 3.1 is the sulfate mineral known as *gypsum* ($CaSO_4 \cdot 2H_2O$). Gypsum is important because rocks composed primarily of gypsum provide the raw material for making drywall, also known as sheetrock.

In the following sections we will explore how rock-forming minerals are assembled into one of the three major types of rocks: *igneous*, *sedimentary*, or *metamorphic*. Because a thorough review of the classification of rocks is beyond the scope of this text, we will focus on only the most common types of rocks.

Igneous Rocks

Rocks that form when minerals crystallize from a cooling body of magma (molten rock) are referred to as **igneous rocks**—magma that breaches the surface is called *lava*. In general, minerals with the highest melting points crystallize first, which means that only those elements needed for those particular minerals are removed from the cooling magma. As additional minerals crystallize from the cooling magma, a mass of interlocking crystals begins to develop, forcing the magma that remains to occupy the existing space between the mineral crystals. A completely solid rock eventually forms when the final mineral grains crystallize from the remaining liquid. Note that a rock that consists of interlocking mineral crystals is said to have a *crystalline texture*. Recall that texture is important because it holds clues as to the origin and history of a particular rock body. In general, the more time the crystals have to grow before the magma solidifies, the larger the grains become.

Geologists classify igneous rocks based on their texture and chemical composition. Igneous rocks are referred to as *extrusive* if their texture is so fine that individual mineral grains are too small to be seen without the aid of a microscope. Coarse-grained igneous rocks are called *intrusive* and have mineral grains that are clearly visible with the naked eye. Extrusive rocks are fine-grained because they represent magmas that make their way to the surface and cool rather quickly, on the order of perhaps several years (Figure 3.10). Such a short cooling period gives individual mineral grains very little time to grow. Intrusive igneous rocks are coarse-grained since they represent magmas that solidify deep within the crust, taking as much as a million years or more to cool. In some cases when molten rock breaches the surface, the lava can cool so quickly that the atoms making up the melt do not have time to organize into a crystalline structure. The result is the formation of a noncrystalline solid called *volcanic glass*, often given the rock name called *obsidian*. Figure 3.11 illustrates the relationship between the grain size of igneous rocks and the cooling history of magma.

FIGURE 3.10 A lava lake developed inside the crater of Hawaii's Kilauea volcano. When lava is exposed to the surface environment it cools quickly, resulting in fine-grained igneous rocks.

FIGURE 3.11 Volcanic glass (A) is an extrusive rock that cooled from magma so fast that atoms were not able to establish a crystalline structure and form minerals. Other extrusive rocks (B) cool slowly enough that small mineral crystals are able to develop, but are too small to be visible with the naked eye. Intrusive rocks (C) cool much more slowly, allowing mineral grains to grow to the point that they are clearly visible.

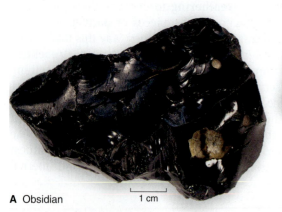

A Obsidian   1 cm

B Basalt   1 cm

C Granite   1 cm

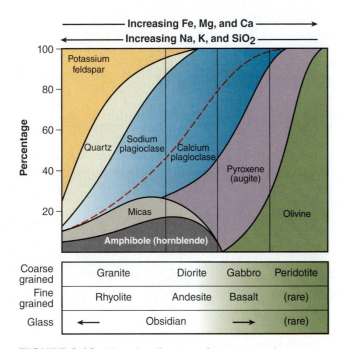

FIGURE 3.12 The classification of igneous rocks is based on texture and mineral composition. Of the various rock types, granites and basalts are the most common igneous rocks in Earth's crust.

In addition to texture, igneous rocks are classified based on their relative abundance of iron and magnesium-rich (ferromagnesian) silicate minerals. As illustrated in Figure 3.12, *peridotite* is the name given to intrusive (coarse-grained) igneous rocks composed entirely of ferromagnesian minerals such as olivine and pyroxene—extrusive rocks with this composition are quite rare. The next category includes the extrusive rock called **basalt** and its intrusive equivalent known as *gabbro*. Both of these rock types contain plagioclase feldspar along with lesser amounts of ferromagnesian minerals as compared to peridotite. As the ferromagnesian content decreases further, amphibole, mica, and plagioclase feldspars increase, creating intrusive rocks called *diorite* and extrusive rocks called *andesite*. The last category in the chart is **granite** (intrusive) and *rhyolite* (extrusive), which contains relatively few ferromagnesian minerals, but is rich in quartz and potassium feldspar.

Of the seven igneous rock types described above, basalt and granite were highlighted because they are the most abundant igneous rocks in Earth's crust. In Chapter 4 you will learn how the oceanic portions of Earth's crust are composed largely of basaltic rocks, whereas the continental crust is dominated by rocks of granitic composition. Moreover, the different ferromagnesian content of basalts and granites creates important density differences that influence the movement of Earth's crustal plates.

Weathering Processes

Before discussing the next major rock type, we need to examine how rocks break down physically and chemically in the process known as *weathering*. As noted earlier in the discussion on the breakdown of ferromagnesian and feldspar minerals, weathering processes generate the materials that form sedimentary rocks and are also critical in the formation of soils. To begin, consider what happens to rocks that are exposed to Earth's surface environment. For many types of rocks Earth's surface is a rather hostile environment. Here rocks are exposed to liquid water, atmospheric gases, biologic agents, and relatively large fluctuations in temperature, all of which tend to cause rocks to disintegrate and decompose over time. Of considerable importance is the fact that weathering rates are highly dependent on the types of minerals a particular rock contains. Another key factor is climate, which, in turn, controls the temperature and availability of water. In the following sections we will explore the two basic types of weathering: *physical* and *chemical*.

Physical Weathering

Scientists use the term **physical weathering** to describe those processes that cause rocks to disintegrate into smaller pieces or particles by some mechanical (i.e., physical) means. The most common way this takes place is when water contained within rocks freezes and thaws in a repetitive manner. As water freezes it increases in volume by about 9%, something which you can readily observe by freezing a bottle of water. When water freezes in a confined space, this volume increase can exert as much as 30,000 pounds per square inch (2,100 kilograms per square centimeter) on its surroundings. Freezing then can literally break rocks apart because the amount of pressure involved far exceeds the strength of even the hardest and most durable rocks. Note that for water freezing within the internal

(pore) space of rocks, this mechanical action generally affects only the outer surface of the rock. However, if the rocks contain planar openings called *fractures,* the effects can extend much deeper into the rock mass. As illustrated in Figure 3.13, a process called *frost wedging* occurs when a fracture filled with water undergoes repetitive freeze and thaw cycles, causing the expanding ice to act as a wedge that eventually breaks the rock in two. Frost wedging is most effective in fractured rocks and in climates where water is abundant and temperatures commonly move above and below the freezing point of water.

Freezing water is not the only mechanical means by which rock fractures can be wedged open. For example, fractures are sometimes filled with fluids rich in dissolved ions. Under the right conditions minerals can precipitate from these solutions, and as the crystals grow they exert sufficient pressure to wedge fractures open. Wedging also takes place when plant roots, particularly tree roots, move down into a fracture and expand in size as the plant grows. In addition to the widening of existing fractures, some mechanical processes actually create new fractures. For example, some deeply buried rocks commonly develop fractures when erosion strips away the overlying rocks, thereby reducing the weight or pressure on rocks in the subsurface. This decrease in pressure allows the rocks to literally expand and begin to fracture. Fractures can also form when surface rocks experience large fluctuations in daily temperature, causing them to expand and contract on a daily basis. Eventually this repetitive expansion and contraction weakens the rocks to the point that they begin to fracture. Regardless of how fractures form, the result is that large rock masses are broken down into smaller particles (Figure 3.13). This process also greatly increases the amount of surface area of the rock where chemical reactions can occur. Therefore, the physical breakdown of rocks into smaller particles can greatly enhance the chemical weathering of rocks.

Chemical Weathering

The term **chemical weathering** refers to the decomposition of the minerals that make up rocks via chemical reactions. Chemical weathering can be thought of as a process where minerals decompose into simpler compounds and individual ions are released into the surrounding environment. Basically, minerals become susceptible to chemical weathering when they encounter an environment different from that in which they were formed. For example, an igneous rock like granite forms deep below the surface in a relatively closed environment where temperature and pressure are stable. When erosion strips away the overlying material, the rocks can then become exposed to the surface environment. Here in this new environment certain minerals within the rock may no longer be chemically stable, particularly in the presence of atmospheric gases and liquid water. These minerals then will begin to chemically decompose. Because the water molecule plays an essential role in many chemical reactions, climate and the availability of water is a key factor in chemical weathering. Also important is the climate's temperature range, since chemical reaction rates typically increase with temperature. In general, chemical weathering is most pronounced in warm climates where rainfall is plentiful, and in rocks with minerals that are chemically unstable under such climatic conditions.

Although there are over 4,000 known minerals on Earth, they are broken down by only three basic types of chemical reactions: *dissolution,*

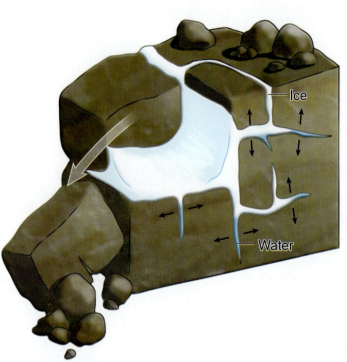

Ice

Water

FIGURE 3.13 One of the ways physical weathering occurs is when water repeatedly freezes and expands within a fracture, slowly wedging the rock into smaller pieces. This causes the surface area of rock body to increase dramatically, thereby increasing the area where chemical weathering can take place.

hydrolysis, and *oxidation/reduction.* Figure 3.14 provides example reactions to help illustrate the chemical changes that are involved in each type of reaction (names of chemical symbols are listed inside the back cover). From this figure one can see that in **dissolution** reactions there is no solid product in the right side of the equation (try not to let the chemistry intimidate you as this is actually quite simple). What this means is that the original mineral will completely dissolve or disassociate in water, leaving only individual ions (charged atoms) in the solution. Perhaps the familiar example is how halite (i.e., common table salt) completely dissolves in water. When you

FIGURE 3.14
Important types of chemical reactions involved in the chemical weathering of minerals. Example reactions are shown for a few of the more common minerals that undergo chemical decomposition.

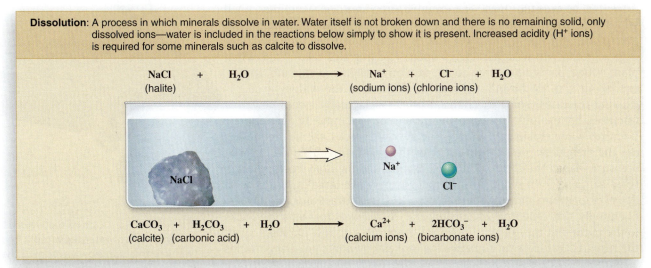

Dissolution: A process in which minerals dissolve in water. Water itself is not broken down and there is no remaining solid, only dissolved ions—water is included in the reactions below simply to show it is present. Increased acidity (H^+ ions) is required for some minerals such as calcite to dissolve.

$$NaCl + H_2O \longrightarrow Na^+ + Cl^- + H_2O$$
(halite)   (sodium ions) (chlorine ions)

$$CaCO_3 + H_2CO_3 + H_2O \longrightarrow Ca^{2+} + 2HCO_3^- + H_2O$$
(calcite) (carbonic acid)   (calcium ions) (bicarbonate ions)

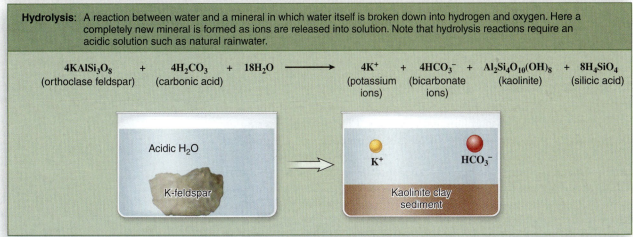

Hydrolysis: A reaction between water and a mineral in which water itself is broken down into hydrogen and oxygen. Here a completely new mineral is formed as ions are released into solution. Note that hydrolysis reactions require an acidic solution such as natural rainwater.

$$4KAlSi_3O_8 + 4H_2CO_3 + 18H_2O \longrightarrow 4K^+ + 4HCO_3^- + Al_2Si_4O_{10}(OH)_8 + 8H_4SiO_4$$
(orthoclase feldspar) (carbonic acid)   (potassium   (bicarbonate   (kaolinite)   (silicic acid)
                                              ions)   ions)

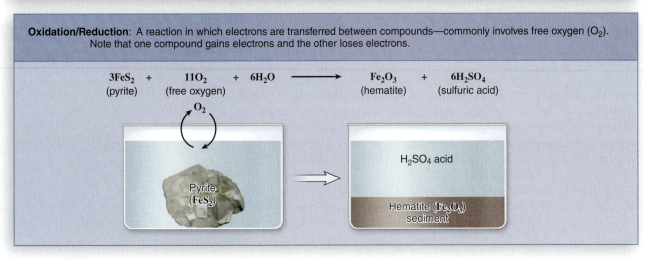

Oxidation/Reduction: A reaction in which electrons are transferred between compounds—commonly involves free oxygen (O_2). Note that one compound gains electrons and the other loses electrons.

$$3FeS_2 + 11O_2 + 6H_2O \longrightarrow Fe_2O_3 + 6H_2SO_4$$
(pyrite) (free oxygen)   (hematite)   (sulfuric acid)

pour halite into a pan of water, the chemical bonds within the mineral begin to break, thereby releasing positively charged sodium (Na^+) ions and negative chlorine (Cl^-) ions into the solution. These dissolved ions (i.e., salts) are what give the water its salty taste.

The other dissolution example in Figure 3.14 is the reaction for calcite ($CaCO_3$). The dissolution of calcite is geologically significant because calcite is the primary mineral in limestone, a fairly abundant rock type as discussed earlier. Note how this reaction requires the presence of an acid in order for calcite to dissolve. This also means that calcite will dissolve faster in more acidic waters. There are many types of acids, but one of the most common is carbonic acid because it forms when water in the atmosphere reacts with carbon dioxide (CO_2). Because CO_2 is well mixed in the atmosphere, rainwater is everywhere naturally acidic. Highly acidic rain forms from sulfur dioxide (SO_2) and is primarily a human-made problem (Chapter 15). The key point here is that since rainfall is naturally acidic, calcite-rich rocks will dissolve more rapidly in climates where rainfall is more plentiful.

The second type of reaction shown in Figure 3.14 is known as **hydrolysis,** which differs from simple dissolution in that water molecules actually take part in the chemical reaction. For example, note in the reaction for the potassium feldspar mineral called orthoclase how the H_2O molecule itself is broken down to create other chemical compounds. Another important difference is that in hydrolysis the original mineral does not simply dissolve away, but rather is transformed into a new mineral—often called a *secondary mineral* or *weathering product*. Note in the feldspar example how potassium ions (K^+) are released into the water and the original mineral is transformed into *kaolinite*—one of the clay minerals mentioned earlier. Similar reactions transform calcium and sodium-rich feldspars into clay minerals, but release calcium (Ca^{2+}) and sodium (Na^+) ions instead. Because feldspar minerals are so abundant in crustal rocks, the breakdown of these minerals by hydrolysis is the primary means by which sodium, potassium, and calcium ions are released into Earth's surface environment. Finally, it is worth noting that the decomposition of feldspars requires acidic solutions. Again, atmospheric CO_2 plays a critical role in generating naturally acidic rainfall, which makes the breakdown of feldspars and calcite possible.

The last type of reaction listed in Figure 3.14 is called **oxidation/reduction,** where individual atoms exchange electrons. When an atom gains or loses electrons, its chemical properties change and it can then form an entirely new mineral or compound. Perhaps the most familiar oxidation/reduction reactions are those in which iron objects begin to rust (i.e., oxidize). In the presence of water and free oxygen (O_2), iron atoms in minerals or compounds can exchange electrons with oxygen atoms, resulting in the formation of an *iron oxide* mineral. Iron oxide minerals are usually reddish, yellowish, or brownish in color. There are a large number of iron-bearing minerals that are susceptible to oxidation/reduction reactions, but the most abundant are the ferromagnesian silicates that are common in rocks like basalt. Because these minerals are susceptible to hydrolysis and oxidation/reduction reactions, the chemical weathering of ferromagnesian-rich rocks will generate sediment consisting of iron-rich clays and iron-oxide minerals.

Perhaps the best example of an oxidation/reduction reaction shown in Figure 3.14 is the one for the iron mineral called pyrite—more commonly known as *fool's gold*. Pyrite is of considerable interest in environmental geology because it is found in a wide variety of rock types and because it oxidizes very rapidly. Note in the chemical reaction for pyrite how it reacts with atmospheric oxygen (O_2) and water to form the bright red iron-oxide mineral known as hematite. Consequently, soils that form from rocks containing even small amounts of pyrite are typically bright red due to the presence of hematite. Also note how the sulfur (S) from pyrite reacts with water to form sulfuric

acid. You will see in Chapter 12 how the mining and processing of rocks containing pyrite is the primary way in which humans release sulfuric acid into the environment. Sulfuric acid is particularly strong and causes significant damage to both the environment and human structures.

Sedimentary Rocks

When rocks are exposed to Earth's surface environment, they naturally undergo physical weathering and break down mechanically into smaller fragments of rock and mineral grains called **sediment**. While this is occurring, minerals that are susceptible to chemical weathering will start decomposing within the rocks. This generates secondary minerals (e.g., clays) and releases electrically charged ions, which attach themselves to water molecules. Both physical and chemical weathering therefore produce sediment. A different process known as **erosion** occurs when sediment and ions are removed from a given area. This takes place whenever rock or sediment is chemically dissolved, physically picked up, or mechanically worn down by the slow abrasive action of moving sediment particles—similar to how sandpaper wears down a solid object. The actual way in which Earth materials are moved from one location to another is called **transportation,** which involves some combination of gravity, running water, glacial ice, and wind. During transportation, material is carried from areas of higher elevation to low-lying areas where it accumulates in a process referred to as **deposition.** In the case of dissolved ions, they are carried away by flowing water and end up in either a body of surface water or become part of the groundwater system (Chapter 11).

Given enough geologic time, the combination of weathering, erosion, and transportation can result in entire mountain ranges being broken down into sediment and ions and then deposited elsewhere. In this section we are interested in how sediment grains and ions are reassembled, layer by layer, to form what geologists call **sedimentary rocks** (Figure 3.15). We will examine the two basic types of sedimentary rocks: those consisting of weathered rock and mineral fragments and those composed of new mineral grains that chemically precipitated from dissolved ions. Note that sedimentary rocks are particularly important in environmental geology because they often contain void spaces between the individual sediment grains. These void spaces then are capable of storing significant quantities of freshwater and petroleum,

FIGURE 3.15 Sediment that forms by the weathering of rocks is normally transported to some other site where it is deposited. Given the right conditions, the deposit may be transported into sedimentary rocks. Example photos showing massive sandstone rock (A) that represents ancient sand dune deposits, active wind transport of sediment (B), and a wind-blown sand deposit (C).

Sandstone

Shale

A Coconino sandstone, Grand Canyon

B Sandstorm approaching a town in Eritrea, Africa

C Death Valley, California

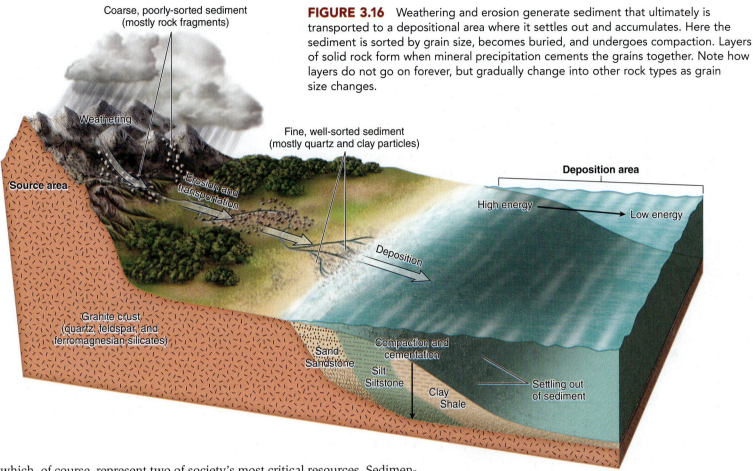

Coarse, poorly-sorted sediment
(mostly rock fragments)

FIGURE 3.16 Weathering and erosion generate sediment that ultimately is transported to a depositional area where it settles out and accumulates. Here the sediment is sorted by grain size, becomes buried, and undergoes compaction. Layers of solid rock form when mineral precipitation cements the grains together. Note how layers do not go on forever, but gradually change into other rock types as grain size changes.

Weathering

Fine, well-sorted sediment
(mostly quartz and clay particles)

Deposition area

Source area

High energy

Low energy

Erosion and transportation

Deposition

Granite crust
(quartz, feldspar, and
ferromagnesian silicates)

Compaction and
cementation

Sand
Sandstone

Silt
Siltstone

Clay
Shale

Settling out
of sediment

which, of course, represent two of society's most critical resources. Sedimentary rocks also serve as the raw materials for making glass, concrete, drywall (sheetrock), and a host of other items of great importance. Consequently, the information on sedimentary rocks that follows will be quite useful in understanding the topics of water resources, mineral and rock resources, and petroleum supplies (Chapters 11, 12, and 13, respectively).

Detrital Rocks

Perhaps the most common type of sedimentary rocks are those known as *detrital sedimentary rocks,* which consist of preexisting rock and mineral fragments, called *detritus,* cemented together to form a solid rock (because geologists refer to the granular nature of this material as having a *clastic texture,* the term *clastic* is sometimes used in place of detrital). As illustrated in Figure 3.16, the process of turning sediment into detrital rocks begins when sediment is deposited in low-lying areas, reaching thicknesses of hundreds or even thousands of feet. Depositional sites include lakes, shallow seas, and near-shore areas of the oceans. Here water naturally fills the void or *pore spaces* that exist between sediment grains. As the sediment is buried progressively deeper, the weight of the overlying material causes the individual grains to rearrange themselves such that the sediment layers become more compact. With increased depth of burial both the sediment and its pore water are exposed to higher temperatures and pressures. In this environment new minerals begin to precipitate from the dissolved ions within the pore water. The new mineral matter acts to bind or cement the sediment grains together, thereby forming solid rock. It is through this process of *burial, compaction,* and *cementation* that layers of loose sediment are gradually transformed into sequences of sedimentary rocks (Figure 3.16).

TABLE 3.2 Common detrital sedimentary rocks. Classification is based on grain size and shape.

Particle Diameter	Sediment	Rock Name
> 2 mm	Gravel	Conglomerate (rounded) breccia (angular)
1/16 to 2 mm	Sand	Sandstone
1/256 to 1/16 mm	Silt	Siltstone
< 1/256 mm	Clay	Shale

A

B

FIGURE 3.17 Detrital sedimentary rock called conglomerate (A) consists of coarse rock and mineral fragments which represent sediment that is young and has not traveled far. As the transport distance increases, feldspar and ferromagnesian minerals in the fragments break down into clays particles, whereas quartz remains unaltered and tends to dominate the grain size called sand. Photo (B) shows a sandstone rock composed almost entirely of quartz grains.

Of interest here is the size and composition of the sediment itself. Earlier you learned that the bulk of Earth's crust is composed of ferromagnesian, feldspar, and quartz minerals. When rocks containing these minerals are exposed to chemical weathering, the ferromagnesian and feldspar minerals are transformed into clay minerals. Quartz, on the other hand, is extremely resistant to chemical weathering, and thus remains largely unaltered. Consequently, when sediment is transported, particularly by running water, the rock fragments not only get progressively smaller due to mechanical action, but the ferromagnesian and feldspar minerals turn into clay minerals. Ultimately the rock fragments break apart during transport, liberating individual quartz grains and clay particles. As time increases and sediment travels farther from its source area, the overall grain size and number of rock fragments will decrease (Figure 3.16). Given enough distance and time, the sediment will eventually be dominated by individual grains of quartz and clay minerals. The sediment will also become better sorted, meaning the fragments are closer to being all the same size.

When sediment reaches its final resting place, such as at the mouth of a river, it is sorted one last time. Here differences in water energy, related to stream velocity or wave action, will sort sediment particles based largely on grain size (Figure 3.16). In the case of sediment consisting of sand, silt, and clay-sized particles, the larger sand grains are deposited in the areas where the water has relatively high amounts of energy. Progressively finer silt and clay particles are deposited farther out as the energy continues to decrease. In this way, fairly uniform layers of sand, silt, and clay are deposited. Note in Figure 3.16 how the grain size of an individual sediment layer is not constant, but becomes finer as one moves laterally in the direction of lower energy.

Because of the way sediment is naturally sorted by size, geologists classify detrital sedimentary rocks based primarily on grain size. As can be seen in Table 3.2, sedimentary rocks that are dominated by fine, clay-sized particles are called **shale;** those consisting of silt particles are called *siltstone*, and those with more coarse sand grains are called **sandstone.** Rocks composed of gravel-sized particles are referred to as *conglomerate* or *breccia*, depending on whether the particles are angular or have been rounded during transport. In general, sediment that has been transported great distances will form sandstones and siltstones that are dominated by quartz grains, whereas shales are composed largely of clay particles. This results from the fact that quartz is highly resistant to chemical weathering, but feldspar and ferromagnesian minerals decompose into clay minerals. On the other hand, conglomerates and breccias typically consist of various rock and mineral fragments, representing sediment that had not traveled very far, hence had less exposure to chemical weathering. The examples in Figure 3.17 illustrate the relationship between transport distance and sediment composition and size.

Chemical Rocks

The other major class of sedimentary rocks is referred to as *chemical sedimentary rocks*, which form when dissolved ions precipitate from water and form new mineral grains. Table 3.3 lists some of the more common chemical rocks—note that coal is composed of altered plant material rather than mineral matter, thus it is not a chemical precipitate. The most common chemical sedimentary rocks are those called **limestone,** which consist chiefly of the mineral calcite ($CaCO_3$). Although calcite can precipitate inorganically from seawater or groundwater, most calcite forms when marine organisms remove calcium (Ca^{2+}) and carbonate (CO_3^{2-}) ions from seawater and then biochemically precipi-

TABLE 3.3   A list of some of the more common chemical sedimentary rocks. The classification for these rocks is based on mineral composition. Also, coal is often listed as a chemical rock despite the fact it consists of altered plant material and does not form by chemical precipitation.

Composition	Rock Name
Calcite ($CaCO_3$)	Fossiliferous limestone
Calcite ($CaCO_3$)	Crystalline limestone
Calcite ($CaCO_3$)	Chalk
Dolomite ($CaMg[CO_3]_2$)	Dolomite
Microcrystalline quartz (SiO_2)	Chert
Gypsum ($CaSO_4 \bullet 2H_2O$)	Rock gypsum
Halite ($NaCl$)	Rock salt
Altered plant material	Coal

FIGURE 3.18   Many marine organisms create body parts made of calcite by extracting certain dissolved ions from seawater. Their skeletal remains can accumulate on the seafloor over time to form fossiliferous limestone as shown here.

tate their own hard parts or protective shells. As these organisms die, their skeletal remains can accumulate on the seafloor and become buried, which slowly raises the temperature and pressure and causes the material to undergo compaction. This, in turn, causes the calcite to recrystallize into solid limestone. Should the original fossil material be preserved, the resulting rock is classified as a *fossiliferous limestone* (Figure 3.18).

In some cases the burial and recrystallization process destroys any visible evidence of the original fossil material, producing a rock consisting of a mass of interlocking calcite grains that geologists call a *crystalline limestone*. The crystalline texture of this type of limestone is similar to that of igneous rocks where individual mineral crystals interlock as they grow. Note that under the right conditions calcite grains will chemically react with magnesium ions (Mg^{2+}) in seawater to form the mineral called dolomite ($CaMg[CO_3]_2$), resulting in a rock called *dolomite*—some geologists use the term *dolostone*.

Although calcite can precipitate inorganically, the most common mechanism is by marine organisms that thrive in what are known as reef systems. As shown in Figure 3.19, reefs typically form in shallow environments where the water is relatively free of suspended sediment. This causes the water to be clearer (i.e., less cloudy), which, in turn, allows

FIGURE 3.19   Fossiliferous limestone rocks typically form where the water column is free of suspended sediment (A), allowing calcite-producing marine organisms to thrive. Limestone forms in shallow seas beyond the point where sediment settles out to form detrital rocks (B) or in near-shore areas where there is minimal sediment influx (C).

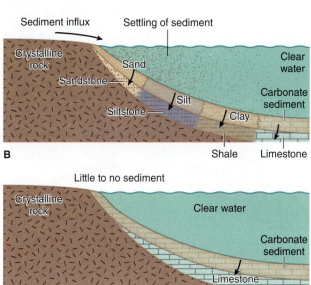

FIGURE 3.20 The exposed limestone rocks of the Guadalupe Mountains in Texas are approximately 250 million years old and were once under the sea as part of an extensive marine reef system.

direct sunlight to penetrate deeper into the water column. Such conditions can occur close to shore in areas where very little sediment is being transported by streams into the marine environment. Relatively clear water can also be found farther offshore, beyond the point where clay particles are settling out from the water. A good example of a major deposit of reef limestone is the Guadalupe Mountains in Texas (Figure 3.20). This limestone sequence, approximately 250 million years old, represents part of an extensive reef system that was over 400 miles (650 km) in length. Finally, there is a special type of fossiliferous limestone geologists call *chalk* (Table 3.3), consisting of tiny calcite shells made by single-celled plants and animals. The Chalk Cliffs of Dover, England are perhaps the most famous example. Because of its fine-grained texture and low hardness, chalk has long been used as a writing tool in classrooms.

The rock known as *chert* is commonly found as nodules within beds of limestone and as discrete layers, and has long been used by humans for making stone tools (Figure 3.21). Chert itself is quite dense and is composed of extremely fine, interlocking crystals of quartz (SiO_2)—often referred to as microcrystalline quartz (Table 3.3). Most chert is thought to originate from tiny marine organisms that biochemically secrete shells made of SiO_2 rather than calcite. After burial, the SiO_2 shell material recrystallizes into a more compact and dense form. Because chert comes in a variety of forms and colors, it has been given specialized names, such as *flint, jasper,* and *agate.* Since chert is quite hard and forms razor-sharp edges when broken, it has played an important role throughout human history, where it has been used for making stone tools, including arrowheads and implements for scraping animal hides.

Another important group of chemical sedimentary rocks form when a body of water evaporates and the concentration of dissolved ions becomes so great that minerals begin to precipitate. The newly formed minerals then settle out of the water column, creating layers of sediment on the seafloor or lakebed (Figure 3.22). Rocks that form in this manner are referred to as *evaporites,* and include valuable mineral deposits of halite (common table salt—NaCl) and gypsum ($CaSO_4 \bullet 2H_2O$). Because various evaporite minerals precipitate under different chemical conditions, various minerals commonly form in different locations with a body of water (Chapter 12). It is by this mechanism that relatively pure deposits of *rock salt* form, consisting almost entirely of the mineral halite. Layers of nearly pure gypsum, known as *rock gypsum,* form in a similar manner. Rock salt is mined and used as a source of sodium and chlorine in the chemical industry as well a deicing agent for roadways. As noted earlier, gypsum is the raw material used in making drywall (sheetrock).

Finally, **coal** is a sedimentary rock composed of altered plant remains, which originally accumulates in swamps where oxygen-poor conditions help preserve the organic matter. As additional plant material accumulates, the underlying material undergoes compaction and is transformed into what is known as *peat.* Should the peat become even more deeply buried, temperature and pressure can increase to the point where the

FIGURE 3.21 Chert is commonly found associated with sedimentary rock deposits. Because of its hardness and ability to create sharp edges when broken, chert has been used for thousands of years to make stone tools, such as these arrowheads.

organic matter undergoes chemical and physical changes. This process results in a more concentrated form of carbon we call coal. Because coal gives off considerable amounts of heat during combustion, and it is quite abundant, modern societies have found coal to be a useful and inexpensive source of energy. The origin and use of coal will be described more thoroughly in Chapter 13 on fossil fuels.

Metamorphic Rocks

In the previous section you learned that when sediment is buried, it begins to compact and experience increases in temperature and pressure. This triggers physical and chemical changes within the sediment that transform it into rock. Similarly, preexisting rocks can be placed in a new environment where they undergo physical and chemical transformations such that new types of rock form. In some cases the temperature and pressure will become high enough that certain minerals within the rocks will become chemically unstable, at which point they begin transforming into new minerals. This results in an altered rock that contains a new set of minerals, all of which are chemically stable under higher levels of temperature and pressure. Geologists use the term *metamorphism* to describe the process where rocks are altered by some combination of heat, pressure, and fluids, producing what are called **metamorphic rocks.** Metamorphism may cause rocks to become so highly altered that the original rock type can no longer be recognized. Other times the temperature may become so high that the metamorphic rocks will begin to melt, forming magma that eventually cools into igneous rocks.

Metamorphism occurs in two basic types of geologic environments. One involves an increase in heat, and the other an increase in both heat and pressure—reactive fluids are common in both environments. As illustrated in Figure 3.23, *contact metamorphism* occurs when magma comes into contact with preexisting rocks. Here just the addition of heat causes minerals to recrystallize into larger grains and/or be transformed into new and more stable minerals. Because any type of rock (igneous, sedimentary, or metamorphic) can undergo metamorphism, the degree to which a particular rock is altered is highly dependent on the original minerals it contains. Also important is the amount of heat the minerals are exposed to and the presence of reactive fluids in the metamorphic environment. For example, in Figure 3.23 one can see that the effect of the magma coming into contact with the igneous rock is minimal since granitic minerals are generally stable at such temperatures. On the other hand, sedimentary rocks often contain minerals that are highly susceptible to being altered during contact metamorphism. In the case of limestone composed of calcite, the mineral grains typically recrystallize into larger grains of calcite, forming the metamorphic rock known as *marble*. Similarly, when a quartz sandstone undergoes metamorphism, the quartz grains will recrystallize and form a more coarse-grained rock known as *quartzite*.

FIGURE 3.22 Death Valley in California once held a freshwater lake that later evaporated to the point where the concentration of dissolved ions became so great that minerals began to precipitate. Shown here is an evaporite deposit of mostly rock salt (halite) covering the valley floor.

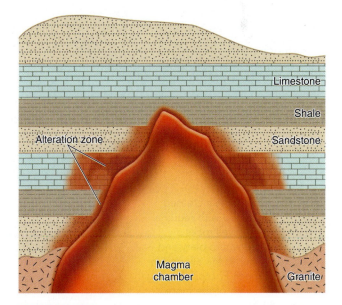

FIGURE 3.23 When magma comes into contact with rocks, the increased heat can cause minerals to recrystallize into larger grains and/or be transformed into more stable minerals. The width of the metamorphic alteration zone depends on how susceptible the original minerals are to higher levels of heat.

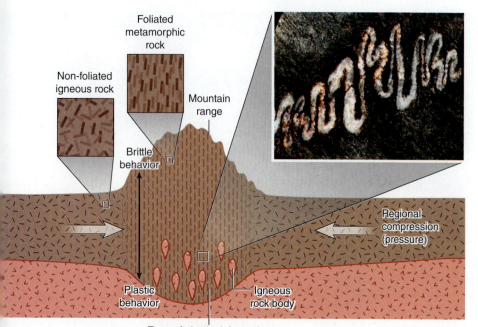

Foliated metamorphic rock

Non-foliated igneous rock

Mountain range

Brittle behavior

Plastic behavior

Igneous rock body

Regional compression (pressure)

Zone of plastic deformation and partial melting of rocks

FIGURE 3.24 Regional metamorphism commonly occurs when deeply buried rocks are subjected to compressive forces. Elevated levels of both heat and pressure cause minerals within the rocks to recrystallize or be transformed into more stable minerals. The directed pressure forces elongated and platy minerals to become aligned, giving the rock a foliated (layered) texture. At higher levels of heat and pressure, rocks may begin to deform by flowing in the solid state (plastic flow) as opposed to fracturing in a brittle manner. At high enough temperatures the rocks can begin to melt and form magma.

The other major type of metamorphism is known as *regional metamorphism,* which is where deeply buried rocks are exposed to elevated levels of both temperature and pressure. As with all metamorphic environments, those minerals that are not stable under the new conditions will either recrystallize or be transformed into more stable minerals. However, as illustrated in Figure 3.24, the increased pressure during regional metamorphism causes elongate and sheetlike (i.e., platy) minerals to reorient themselves in a parallel manner. This parallel realignment of minerals gives the rocks a **foliated texture,** which is similar to how wood fibers align themselves in a parallel manner in sheets of paper. For example, when shales undergo regional metamorphism, their clay minerals are transformed into platy minerals of the mica family, forming a fine-grained and highly foliated rock called **slate** (Figure 3.25A). In the case of granites composed primarily of quartz, feldspar, and ferromagnesian minerals, regional metamorphism produces a strongly foliated rock known as *schist,* which is dominated by micas and recrystallized quartz grains (Figure 3.25B). If the temperature and pressure become great enough, the minerals will start to separate into dark and light bands, forming a rock called *gneiss* (Figure 3.25C). Note that when rocks are exposed to increasing levels of temperature and pressure they tend to develop fewer fractures as they become less brittle. Instead the rocks begin to deform in a plastic manner, where they literally begin to flow, as evident by the deformed gneiss shown in Figure 3.24.

A Slate

2 cm

B Schist

5 cm

C Gneiss

1 cm

FIGURE 3.25 The increased pressure associated with regional metamorphism gives rocks a foliated texture where platy and elongated minerals are aligned in a parallel manner. Photos showing examples of some of the more common types of foliated metamorphic rocks.

The Rock Cycle

Based on what you learned about rocks in Chapter 3 it should be clear that geologic processes can take any rock and transform it into an entirely different type of rock. Igneous rocks, for example, can be transformed into layers of sandstone and shale. These sedimentary rocks can later be altered and become metamorphic rocks, which might even undergo melting and form magma. The magma, of course, would eventually cool and create igneous rock. Geologists refer to this recycling of rocks from one rock type to another as the **rock cycle.** As shown in Figure 3.26, the rock cycle is commonly represented in terms of a flow chart in order to illustrate the various ways in which rocks can be transformed. This simple flow chart is a helpful tool for understanding the fairly complex ways in which rocks can be recycled.

Imagine if we could follow the possible paths that an igneous body of granite might take through the rock cycle. One path would be for the granite to become exposed at the surface. Here weathering and erosion processes would slowly break the rock down into sediment and dissolved ions. If the sediment were to undergo compaction and cementation after being transported to a depositional site, then material would develop into detrital sedimentary rocks. Likewise, some of the ions may precipitate into minerals and form chemical sedimentary rocks. Should these new sedimentary rocks later become exposed to high levels of heat and pressure, they will naturally be transformed into metamorphic rocks. However, should the heat and pressure become too intense, the metamorphic rocks could begin to melt and form magma. The magma of course would eventually cool and form igneous rocks, thereby completing the cycle.

From Figure 3.26 you can see that our original igneous rock body could take another path through the rock cycle; namely, it could simply stay buried and never be exposed to weathering and erosion. In this case the igneous rocks could either remain unaltered for eons of time, or undergo metamorphism and be transformed into metamorphic rocks. Should the heat and pressure continue to increase, the rocks would begin to melt and ultimately turn back into igneous rocks. However, these metamorphic rocks do not necessarily have to melt, but instead may stay buried and experience repeated episodes of metamorphism. The rocks could also be uplifted and be exposed to weathering and erosion, and hence turn into sedimentary rocks. Note that sedimentary rocks themselves can undergo weathering and erosion and be recycled back into sedimentary rocks. Ultimately, any of the three rock types can either be: (a) exposed at the surface and transformed into sedimentary rocks; (b) remain buried and unaltered; (c) go through metamorphism; or (d) experience extreme metamorphism and begin to melt and form magma. The particular path a rock body takes through the rock cycle is entirely dependent upon the type of geologic environments it happens to encounter.

A key question at this point is why do rocks encounter different geologic environments within the rock cycle? In other words, why do some rocks remain buried and potentially subjected to metamorphism, yet others become uplifted and exposed to weathering and erosion? The answer remained a mystery until the 1960s when geologists developed the *theory of plate tectonics*. This new theory explains how Earth's crust is broken up into rigid slabs or plates, which are set in motion by forces associated with heat that is generated in the interior of the planet. These internal forces create the conditions for metamorphism and cause parts of the crust to be uplifted, forming mountains. Interestingly, plate tectonics not only plays a key role in the rock cycle, it also affects nearly every aspect of the Earth system (Chapter 1). Because understanding plate tectonics is central to the study of environmental geology, most of Chapter 4 will be dedicated to the subject.

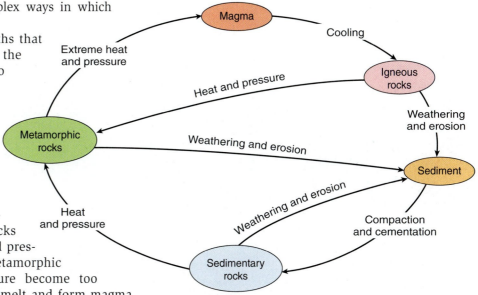

FIGURE 3.26 The rock cycle explains how various geologic processes can cause rocks to be transformed into different types of rocks. The geologic processes that operate within the rock cycle ultimately cause the rocks within Earth's crust to be recycled over time.

FIGURE 3.27 The Gunnison River in Colorado has exposed this ancient body of metamorphic rock, which contains numerous veins of igneous rocks (white colored). These igneous bodies within the metamorphic rocks prove that the temperature during metamorphism was high enough to generate magma.

Rocks as Indicators of the Past

In Chapter 1 you learned how geologists constructed the geologic time scale by classifying rocks based on their relative ages—absolute ages were later obtained by radiometric dating. Because many sedimentary rocks contain fossils, the geologic rock record then provides us with a vast history of life on our planet. The rock record also contains valuable evidence on an array of important topics, such as the movement of continents over time, location of ancient mountain ranges, climatic changes, asteroid impacts, and more. In this section we want to briefly explore how you can take what you learned about rocks in Chapter 3 and make some basic interpretations about Earth's past.

Suppose you find a rock that is strongly foliated and shows evidence that the rock has been deformed and flowed in a plastic manner. Based on what you learned earlier, the foliation and plastic deformation tell us that this rock was once deeply buried and under high levels of temperature and pressure. In other words, it must have experienced regional metamorphism. The exposed rock in the cliff face shown in Figure 3.27, for example, represents a rock body that at one time was deeply buried and subjected to regional metamorphism. This rock is not only highly foliated, but has large veins of light-colored igneous rocks cutting through the entire rock mass. We can conclude therefore that the temperature and pressure must have reached the point where this mass of metamorphic rock underwent partial melting and formed magma. By using radiometric dating techniques we could also determine when these igneous veins cooled into solid rock.

While igneous and metamorphic rocks tell us a great detail about the history of Earth's interior, sedimentary rocks can provide specific information about the surface environment. For example, because sediment and dissolved ions interact with the atmosphere and hydrosphere, scientists are able to interpret past climatic conditions based on the composition of sedimentary rocks. In the case of evaporite deposits such as rock salt (halite) and gypsum, we can infer a warm and arid environment due to the fact these rocks require very high evaporation rates. The presence of coal in a sedimentary sequence tells us the environment must have been warm and humid because coal forms from thick accumulations of organic material in swamp conditions. Sedimentary rocks also contain features that form during deposition which provide valuable information about the environment. For example, sand that is transported by wind has a characteristic type of layering called *cross-bedding* (Figure 3.28A). Because large masses of blowing sand occur in arid climates where there is a lack of vegetation, cross-bedding in ancient sandstones is indicative of an arid environment. Similarly, when *mudcracks*

FIGURE 3.28 Features preserved in sedimentary rocks hold important clues as to the environment where the original sediment was deposited. The so-called cross-bedding of layers in (A) are the result of windblown sand being deposited in shifting sand dunes. The ancient mudcracks in (B) developed in clay-rich sediment of a shallow lake that periodically dried up.

A Zion National Park, Utah

B Glacier National Park, Montana

FIGURE 3.29   A 400-million-year-old fossiliferous limestone from the Great Lakes region in North America proves that life flourished in the marine environment that once existed in the area.

FIGURE 3.30   Image of Mars taken from an orbiting spacecraft showing what appears to be sedimentary rocks and an ancient shoreline.

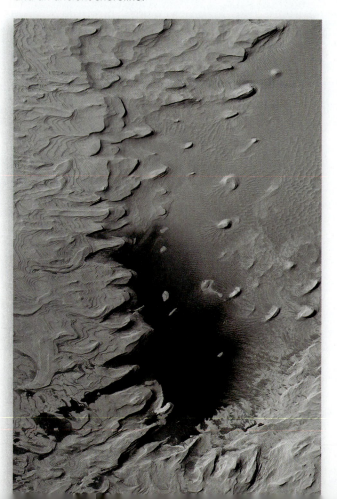

(Figure 3.28B) are found in alternate layers of shale, it tells the story of a shallow lake that periodically evaporated, allowing the muddy lakebed to dry out and develop these characteristic cracks.

Sedimentary rocks also provide key environmental information through the fossils they contain (Figure 3.29). Although land animals may live in a variety of environments, land plants and many marine organisms live in very specific environments. Take, for example, corals, which thrive in clear, warm water that is also rich in microscopic organisms on which they feed. Other marine organisms live only in the turbulent beach zone, and some attach themselves to rocks in the intertidal zone. Still others crawl around on the seafloor in shallow waters, and some have even adapted themselves to the great pressure found in the deep ocean. Most fossils, therefore, reveal at least some information about the environment in which they once lived. In general, sedimentary rocks and their fossils often tell us whether the sediment was deposited in fresh, brackish, or salt water and also some indication of the water depth and temperature.

Finally, because Earth and other bodies in the solar system formed from the same solar nebula, one could expect that some of the minerals and rocks found on Earth might be found on these other bodies. When Apollo astronauts brought samples back from the Moon, indeed most of the rocks turned out to be nearly identical to the basalts and their coarse-grained equivalents found here on Earth. Robotic craft on the surface of Mars have likewise found basaltic rocks, but have also discovered sedimentary rocks that had been deposited in a water-rich environment (Chapter 2). This means that at one time water and gases in Mars' atmosphere had caused these basaltic rocks to break down into sediment and ions through the processes of weathering and erosion. Supporting this interpretation are images from orbiting spacecraft (Figure 3.30), showing what appears to be extensive deposits of sedimentary rocks and ancient shorelines. This means that we can use our knowledge of geologic processes and rocks not only to interpret past environmental conditions on Earth, but on other planets as well.

SUMMARY POINTS

1. Different types of atoms (elements) are created in stars by the nuclear fusion of hydrogen. These atoms can then be ejected into space and incorporated into planets.

2. Minerals are inorganic solids composed of one or more types of atoms that are arranged in crystalline structure. Each mineral is unique and has a unique set of physical properties. Rocks are simply assemblages of one or more types of minerals.

3. Oxygen, silicon, and aluminum make up over 80% of Earth's crust by weight. Of the more than 4,000 known minerals, only a dozen or so make up the bulk of Earth's crustal rocks. These minerals, referred to as the rock-forming minerals, are assembled into three basic rock types.

4. Igneous rocks form when magma begins to cool, which allows different types of ions to be incorporated into the crystalline structure of minerals. As the cooling period increases, minerals have more time to grow, producing an igneous rock with a coarser texture.

5. Physical weathering at the surface breaks rocks down into smaller fragments called sediment. Minerals within rocks that are susceptible to chemical weathering then decompose into ions and/or are transformed into new minerals.

6. Erosion and transportation take place when sediment is removed from an area by wind, water, or ice—ions move primarily by flowing water. The sediment is ultimately transported from its source area to a depositional site where it accumulates.

7. Sedimentary rocks form when accumulated sediment undergoes compaction and cementation, or when dissolved ions precipitate to form new minerals.

8. Metamorphic rocks form when preexisting rocks are altered by heat, pressure and/or chemically active fluids. Contact metamorphism occurs when rocks are altered by the heat from a magma body, whereas regional metamorphism involves both heat and directed pressure.

9. The continual transformation of rocks from one rock type to another is referred to as the rock cycle. The rock cycle is ultimately driven by plate tectonics and Earth's internal heat.

10. Because each rock type forms under certain geologic conditions, rocks provide scientists with a record of past events and environmental conditions on Earth.

KEY WORDS

basalt   74
calcite   72
chemical weathering   75
clay minerals   72
coal   82
deposition   78
dissolution   76
erosion   78

feldspars   72
ferromagnesian minerals   72
foliated texture   84
granite   74
hydrolysis   77
igneous rocks   73
ions   67
limestone   80

magma   70
metamorphic rocks   83
mineral   68
oxidation/reduction   77
physical weathering   74
quartz   72
rock   70
rock cycle   84

rock-forming minerals   71
sandstone   80
sediment   78
sedimentary rocks   78
shale   80
slate   84
transportation   78

APPLICATIONS

Student Activity   Does your car or a friend's car have any rust on it? Look at the rust spot carefully. Is it flakey or powdery? Can you flake it off? Do you think this can happen to rocks?

Critical Thinking Questions
1. Explain the rock cycle. Where does it end and begin?
2. Can an element also be a mineral?
3. Can a rock be made up of only one mineral?
4. What are the different types of rocks? How were each formed?

Your Environment: YOU Decide   Can a rock be both sedimentary and igneous? Igneous and metamorphic? Why or why not?

Chapter 4

Earth's Structure and Plate Tectonics

Earth's continents and ocean basins sit atop crustal plates that move slowly over geologic time in response to forces generated by Earth's internal heat energy. In some places plates move away from one another, increasing the size of ocean basins, such as the Atlantic shown here. In other areas plates come together, creating colossal collisions that push mountains far up above sea level. The moving plates not only reposition the continents and ocean basin over time, but cause most of the planet's earthquakes and volcanic eruptions. Scientists have also shown that plate movements have played a critical role in the evolution of the entire Earth system, including life as we know it.

CHAPTER OUTLINE

Introduction

Deformation of Rocks

Earth's Interior
- Earth's Structure
- Earth's Magnetic Field
- Earth's Internal Heat

Developing the Theory of Plate Tectonics
- Continental Drift
- Mapping the Ocean Floor
- Magnetic Studies
- Location of Earthquakes
- Polar Wandering

Plate Tectonics and the Earth System
- Types of Plate Boundaries
- Movement of Plates
- Surface Features and Plate Boundaries
- Plate Tectonics and People

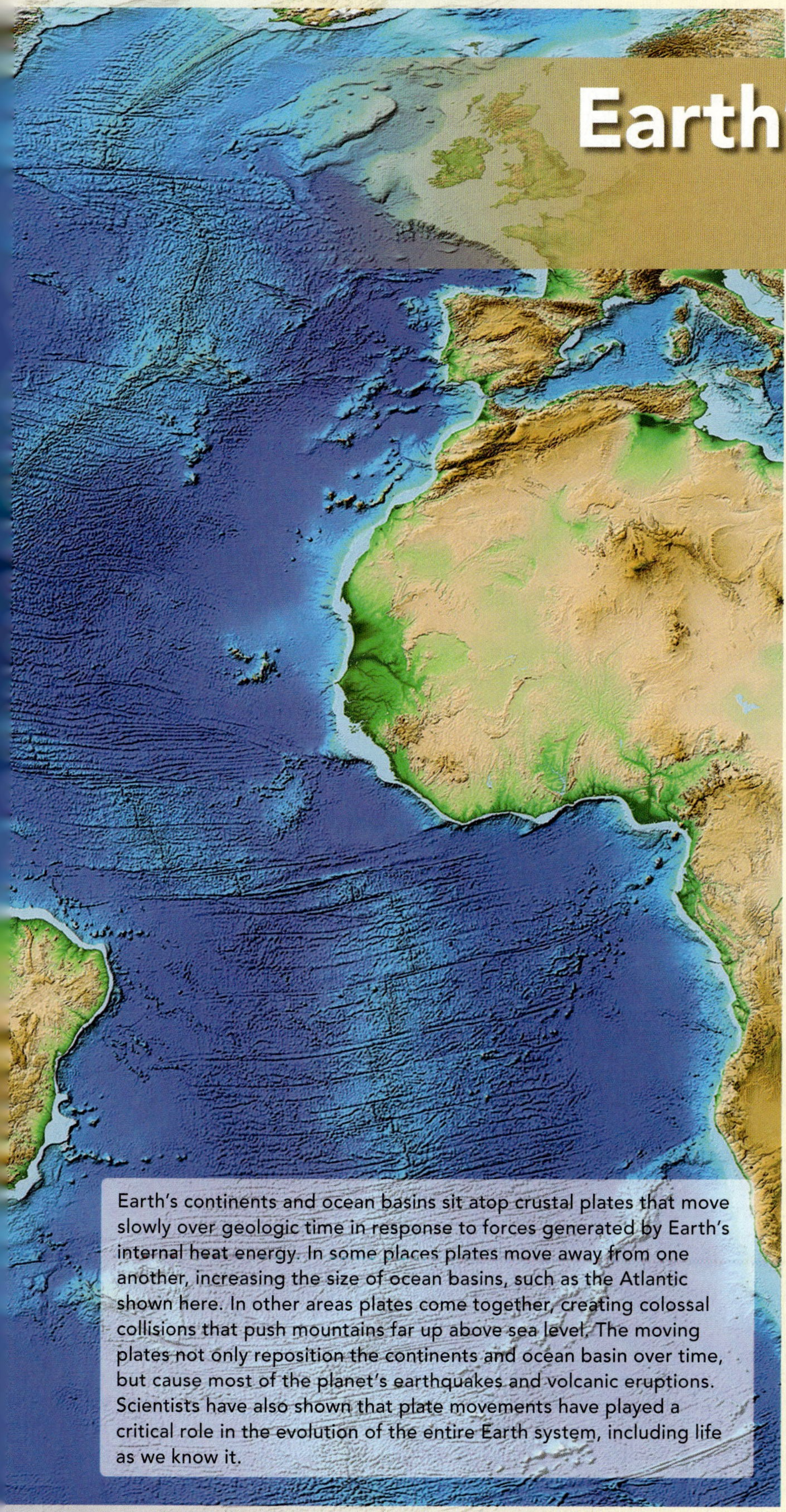

LEARNING OUTCOMES

After reading this chapter, you should be able to:

- Describe the three different forces that deform rocks and explain what happens to rocks when they are deformed beyond their elastic limits.
- Know the different layers making up Earth's internal structure and understand the basic way in which scientists have determined this structure.
- Understand the difference between oceanic and continental crust and be able to describe the difference between the lithosphere and asthenosphere.
- List the two sources of Earth's internal heat and explain how this heat helped create the planet's layered structure and is driving its system of moving tectonic plates.
- Explain how scientists have been able to confirm that seafloor spreading and subduction processes are taking place and how this proved continental drift.
- List the major types of plate boundaries and describe the types of surface features that develop at each one.

Introduction

Unlike Mars, Mercury, and the Moon, the Earth is a very active and restless planet where earthquakes and volcanic eruptions are still relatively common. In addition, vast amounts of freshwater evaporate from Earth's oceans and fall over the landmasses, making it possible for life to flourish in the terrestrial biosphere. Much of the energy that drives this interactive and dynamic system, called the *Earth system* (Chapter 1), comes from the solar radiation that streams outward from the Sun. The other primary source of energy to the Earth system is the heat contained within our planet. It is this internal heat that sets rock masses in motion, causing large-scale metamorphism and uplift of land surface. When the land experiences uplift, it produces rugged and mountainous terrain, thereby exposing greater amounts of rocks to the surface environment. Here at the surface, weathering and erosion processes work to lower the elevation of the landscape, generating sediment that eventually accumulates and forms sedimentary rocks. Earth's internal heat therefore is a major driving force in transforming rocks from one type to another, referred to as the *rock cycle* (Chapter 3). The purpose of Chapter 4 is to explore the critical role Earth's internal forces have on shaping the environment in which we live.

Humans, of course, have long been aware that Earth's landscape is highly varied, from broad flat plains to rugged mountains that are virtually impassible. As with other aspects of our physical world, we learned to use the process known as *science* to try and explain how different landscapes form. This effort resulted in a new area of study called geology, which focused on the different types of rocks, and sediment upon which all landscapes are built. From the study of rocks, early geologists were able to develop hypotheses that explained how volcanic mountains were built by rising magma. However, they found it difficult to provide an adequate explanation for the presence of mountains such as the Appalachians, Himalayas, and Alps, whose strongly deformed rocks show few signs of volcanic activity. Not until the 1960s did geologists come to understand that Earth's outer layer, or *crust,* is broken up into rigid slabs or plates that are in motion due to forces associated with the planet's interior heat. This concept of moving plates eventually became known as the **theory of plate tectonics.**

With the theory of plate tectonics modern geologists could explain how deformed mountain ranges represent giant collision zones between crustal plates, some the size of continents. Plate tectonics not only provides a simple and elegant explanation for the rock cycle and why the land rises, it also explains the occurrence of the vast majority of earthquakes and volcanic eruptions. The theory also explains how internal forces sometimes cause landmasses to break up and begin drifting apart, eventually becoming separated from one another by an ocean. Biologists have discovered that the great diversity of plant and animal life found on the planet is a direct result of both evolutionary processes and this movement of landmasses over time. Similarly, climatologists and oceanographers have found that the reconfiguration of the continents and ocean basins creates important changes in the circulation patterns of both the oceans and atmosphere. This, in turn, has produced changes in the global climate system over eons of geologic time.

Finally, it should be noted here that plate tectonics is referred to as a *theory* rather than a *hypothesis* (Chapter 1) because it acts as a unifying framework that explains a wide variety of natural phenomena. It is no exaggeration to say the theory of plate tectonics has revolutionized the way in which scientists view our planet. The solid earth is no longer seen as a rigid mass of rock, but rather a dynamic system that affects nearly every aspect of the Earth system. Plate tectonics is not only central to the study of geology, but it also represents one of the most profound advances in

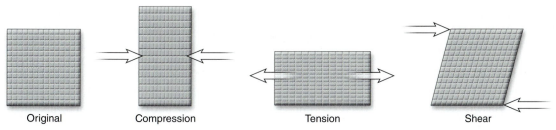

FIGURE 4.1 A two-dimensional representation of the deformation (strain) that would result from different types of stress acting on a square.

Original   Compression   Tension   Shear

modern science. Because plate tectonics is so important, we will devote this entire chapter to the subject. However, it will first be necessary to examine some background material on how rocks deform and the layered structure of our planet.

Deformation of Rocks

For rocks to become deformed, they must be acted upon by some type of force, also called *stress*. This deformation involves some change in shape or volume, technically known as *strain*. As illustrated in Figure 4.1, there are three basic types of stress that can result in deformation. **Compression** pushes on rocks from opposite directions, causing them to be shortened as if they were put in a vise. **Tension** pulls on rocks from opposite directions, resulting in the rocks becoming stretched or lengthened. Finally, **shear** occurs when rocks are being pushed on in an uneven manner, causing the rocks to be skewed such that different sides of a rock body slide or move in opposite directions.

People are often surprised to learn that rocks near the surface are *elastic*, meaning that when a force (stress) that is acting on them is removed, the rocks will return to their original shape. Common examples of elastic materials include rubber bands and tree limbs. However, all elastic materials have what is called an **elastic limit,** which is the point at which they no longer behave elastically and deformation becomes *permanent*. For example, if the wind forces a tree limb to bend (deform) beyond its elastic limit, the limb will break or snap, which of course is permanent. In the case of rocks exceeding elastic limit, deformation can become permanent in one of two ways. As illustrated in Figure 4.2, one way this occurs is by fracturing or breaking, in which case the rocks are referred to as being *brittle*. The other way is to deform by flowing, in which case rocks are called *ductile*. For example, a glass rod is considered brittle because if it bends beyond its elastic limit, it will fracture or snap in two. On the other hand, a steel rod is ductile because if its elastic limit is exceeded, it will literally flow and develop a permanent bend. When it comes to rocks that are ductile, geologists often refer to the deformation as *plastic deformation*.

The reason why glass and steel rods have different elastic limits and deform differently is because of their composition and internal structure. In the case of rocks, it

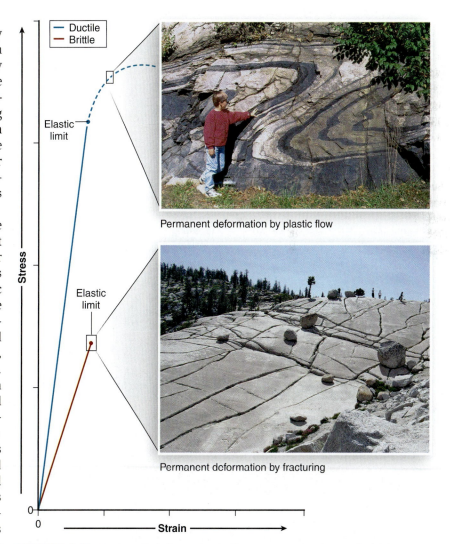

Permanent deformation by plastic flow

Permanent deformation by fracturing

FIGURE 4.2 Rocks will deform elastically up to a point, beyond which deformation becomes permanent. Ductile materials deform permanently by flowing plastically, whereas brittle materials fracture. Rocks near the surface are typically brittle and will fracture, but when buried, the higher temperatures and pressures cause them to become ductile and deform plastically.

TABLE 4.1 Approximate temperature and confining pressure at selected depths below the Earth's surface. Note in this example that there is a 50-fold increase in pressure, whereas temperature increases only 10-fold. Estimates based on average crustal density of 2.8 g/cm³ and temperature gradient of 2.5°C/100 meters; pressure in units of pounds per square inch (psi) and megapascal (MPa).

Depth	Approximate Temperature	Confining/Overburden Pressure
1,000 feet (305 m)	74°F (23°C)	1210 psi (8.4 MPa)
10 miles (16 km)	790°F (420°C)	64,000 psi (440 MPa)

is their texture and mineral composition that helps determine their elastic limits and manner in which they deform (brittle or ductile). Other key factors include temperature, pressure, and time over which the stress acts. For example, if you applied enough heat to our glass rod, it would change from being brittle to being ductile; hence, it would undergo permanent deformation by flowing rather than fracturing—a *fracture* is simply a planar opening or break. On the other hand, if you kept the glass rod at room temperature, but wrapped it tightly with a fabric, the fabric would confine or add pressure to the rod such that its elastic limit would increase. Many materials will experience greater deformation under a relatively small force (stress) acting over a long period of time compared to a stronger force acting rather quickly. Also note that when rocks deform they often slide past one another along a fracture plane, at which point the fracture is called a **fault.** All faults then involve some type of slippage or movement, whereas fractures do not.

These factors that affect deformation are important to our discussion of rocks because as one descends toward the center of the planet, rocks experience progressively higher levels of both temperature and pressure. What happens is that rocks in the subsurface have to bear the weight of the overlying column of rocks, creating what geologists refer to as *overburden* or *confining pressure.* Confining pressure is similar to the pressure you feel when diving to the bottom of a pool—the difference being that one results from the weight of water, and the other the weight of the rocks. When rocks are under greater confining pressure, their elastic limit increases, just as the fabric wrapped around the glass rod made it stronger. Keep in mind, however, that because of Earth's internal heat, temperature also increases along with confining pressure. Consequently, rocks near the surface are under little confining pressure and tend to be quite brittle, hence will fracture when subjected to a stress beyond their elastic limit. As can be seen in Table 4.1, temperature and pressure increase rapidly with depth such that rocks which are deeply buried can easily become ductile and deform plastically. Examples of rocks that have undergone brittle and plastic deformation are shown in Figure 4.2.

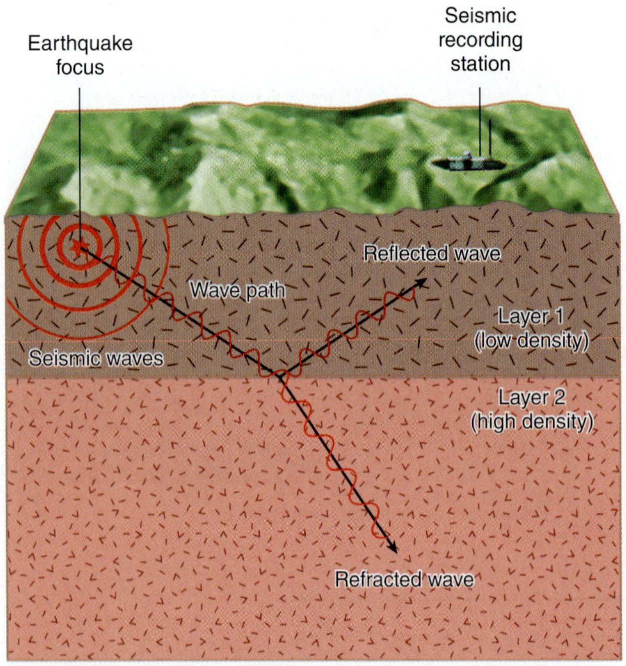

Earth's Interior

As one descends deeper into the Earth, not only does temperature and pressure change, but so too does the composition of earth materials. Recall from Chapter 3 how Earth's crustal rocks are made up of a relatively small number of minerals called the *rock-forming minerals.* Because most rock-forming minerals contain the silicate ion (SiO_4^{4-}), oxygen and silicon atoms end up accounting for 49% and 26% of the crust by weight, respectively. However, if we consider the Earth as a whole, iron becomes the dominant element at 35% by weight, followed by oxygen at 29%. This means that Earth must have a layered structure and that the lighter elements are more abundant in the outermost layers. Likewise, Earth's interior must be denser than the outer shell we call the crust.

Scientists today know with certainty that Earth is layered because of the way earthquake waves, also called *seismic waves* (Chapter 5), change velocity and direction as they travel through the planet's interior. As indicated in Figure 4.3, when seismic waves encounter layers of different density, their velocity changes, causing the wave to both reflect and refract (bend)—similar to how light reflects and bends when passing from

FIGURE 4.3 Seismic waves generated by earthquakes and human-made explosions will reflect and refract when encountering layers of different density. Recording instruments measure the waves that return to the surface, enabling scientists to determine the depth of different layers all the way to Earth's core.

air into a pool of water. By measuring refracted and reflected seismic waves that return to the surface, scientists have been able to locate the boundaries between different materials deep within the Earth. Seismic studies have also shown the Earth to be comprised of four major layers that vary in composition and physical properties. Beginning at the surface, the major layers are the *crust, mantle, outer core,* and *inner core.* Interestingly, the deepest well ever drilled reached 7.6 miles (12.3 km), which is a mere pinprick considering the center of the Earth is nearly 4,000 miles (6,400 km) deep. If one compared the Earth to an apple, this deep well would not even have penetrated the apple's skin! Therefore, were it not for the study of seismic waves, Earth's internal structure would have remained a mystery.

Earth's Structure

From Figure 4.4 one can see that the solid portion of the Earth consists of a metallic sphere surrounded by a rocky shell mostly composed of silicate minerals. The center of the metallic sphere is solid and known as the **inner core,** which is surrounded by a shell of molten metal called the **outer core.** In order to account for the overall known density of the Earth, scientists hypothesize that both the inner core and outer core are composed of an iron-nickel alloy. If the density of the this alloy were combined with the density of the silicate shell, it would produce an average that is consistent with the known density of Earth as a whole. This nickel-iron hypothesis is also supported by the fact that many of the asteroids that strike the Earth are composed of these two metals. Because asteroids represent leftover material from the formation of the solar system (Chapter 2), it follows that Earth's inner sphere is an alloy of iron and nickel. Scientists have also discovered that certain types of seismic waves, called *shear waves,* do not penetrate into the outer core, whereas *compressional waves* do. Because laboratory experiments show that shear waves are unable to pass through liquids, but compressional waves can, scientists infer that the outer core is molten and the inner core is solid.

Earth's metallic center is surrounded by an 1,800-mile (2,900 km) thick rocky shell called the **mantle,** which is composed of iron-rich silicate minerals (Figure 4.4). The mantle rocks are subjected to very high temperatures and pressures because they lie at such great depths below the surface. This, of course, makes the mantle rocks rather susceptible to plastic deformation as described earlier. Note that as you descend progressively deeper into the Earth, temperatures will eventually reach the melting point of different silicate minerals. However, pressure increases with depth at a faster rate compared to temperature (Table 4.1). Because higher pressure raises the melting point of minerals, this means that despite the high temperatures deep within the mantle, silicate minerals are generally unable to reach their melting points due to the tremendous pressure (similar to how water in a pressure cooker is forced to boil at a higher temperature). This explains why the mantle is not molten, but instead is made of solid rock that is very hot and subject to plastic flow.

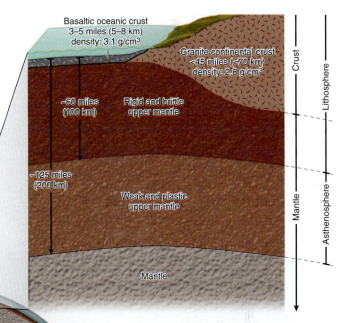

FIGURE 4.4 The Earth has a layered structure consisting of a high-density metallic core surrounded by a lower-density rocky shell of silicate minerals, called the mantle. Near the top of the mantle the silicate minerals are close to their melting points, creating a weak and plastic zone called the asthenosphere. The outermost silicate shell is called the crust, which has the lowest density of all the layers. Geologists refer to the crust and uppermost mantle as the lithosphere since they behave as a single, rigid slab that moves over the asthenosphere. Note that the lithosphere in continental areas contains granitic crust, whereas basaltic lithosphere lies beneath the oceans.

Near the top of the mantle is a zone called the **asthenosphere,** where the combination of temperature and pressure are such that the silicate minerals are near their melting points. This makes the asthenosphere a rather weak layer that is also prone to melting, particularly in areas where the rocks are disturbed by a localized decline in pressure, or by the introduction of water-rich material. One of the reasons the asthenosphere is important in geology is because it is a likely source for the magma that rises and forms the igneous rocks we find near Earth's surface. This magma sometimes of course breaches the surface and results in volcanic activity (Chapter 6). The asthenosphere is also important because of the fact it represents a layer that is very weak, which allows it to flow and deform plastically when it is acted upon by Earth's internal forces. Note that above the asthenosphere, the temperature and pressure become low enough that the mantle rocks become more rigid and brittle.

Earth's outermost layer is called the **crust,** which consists primarily of silicate-rich rocks whose density is even lower than those in the underlying mantle. One of the more striking features of the crust, shown in Figure 4.4, is that it is exceedingly thin compared to the 4,000-mile (6,400 km) depth of the planet. Also striking is how the composition and thickness of the crust varies. Under the ocean basins the crust is made of more dense, basaltic-type rocks that maintain a fairly uniform thickness of around 3 miles (5 km). Under the continents, however, the crust is composed of rocks with an overall composition similar to granite, and can reach thicknesses as much as 45 miles (70 km). Because both the mantle and crustal rocks overlying the asthenosphere are relatively brittle and rigid, they effectively behave as a single layer, which geologists refer to as the **lithosphere.** The fact that the lithosphere is brittle and rigid is important because this allows forces within the Earth (compression, tension, or shear) to break the lithosphere up into individual slabs called **tectonic plates.** Earth's internal forces, commonly called *tectonic forces,* also cause these rigid plates to glide on top of the weak, and easily deformed asthenosphere. We will take a close look at the actual movement of tectonic plates in a later section.

Earth's Magnetic Field

The core is of considerable importance to life on Earth because of its role in creating the planet's strong magnetic field. As indicated in Figure 4.5, scientists think that Earth's magnetic field results from the circulation of metallic ions within the outer core—creating a magnetic force field similar to that around a familiar bar magnet. It is generally believed that the fluid nature of the outer core allows the inner core to rotate slightly faster than the rest of the planet. This, in turn, causes the electrically charged metallic ions in the outer core to circulate, essentially making the Earth a giant electromagnet. For over 700 years humans have made use of Earth's magnetic field for navigational purposes using compasses. Interestingly, biologists now hypothesize that certain types of migrating animals (e.g., birds and turtles) also use the magnetic field to navigate. Most important of all is how the magnetic field acts as a critical shield that helps block out harmful radiation streaming out from the Sun (Figure 4.5). Earth's core then has likely played an important role in

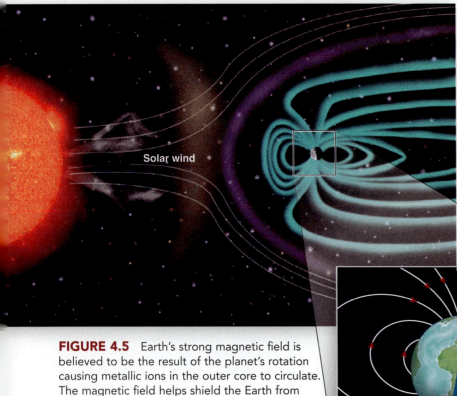

Solar wind

FIGURE 4.5 Earth's strong magnetic field is believed to be the result of the planet's rotation causing metallic ions in the outer core to circulate. The magnetic field helps shield the Earth from harmful radiation streaming from the Sun, hence is important to the biosphere.

making Earth's environment hospitable and also in the evolution of life as we know it. This interaction between the solid earth and biosphere is yet another example of how the entire Earth operates as a system (Chapters 1 and 2).

Earth's Internal Heat

It is well established that the Earth radiates more heat energy into the coldness of space than the planet receives from the Sun. Moreover, temperature is known to increase as you go deeper into the planet. This leads to the obvious question: what is the source of Earth's internal heat? Scientists originally thought that this heat was *residual,* meaning it was left over from the formation of the planet 4.6 billion years ago. When the young Earth was accreting large amounts of material from the solar nebula (Chapter 2), gravity caused the material within the planet to undergo compaction due to the increasing overburden (confining) pressure. As the material compressed, individual particles reoriented themselves which generated frictional heat— similar to heat generated by rubbing your hands together. The planet was also gaining heat energy during this early period from the large number of impacts that were taking place. Scientists now believe that at some point the young Earth was mostly molten except for a relatively thin crust.

Temperatures within the young planet would naturally have been higher near the center and cooler near the crust, where heat was radiating outward into space. Geologists refer to the increase in temperature with depth as the **geothermal gradient,** which today averages about 75°F/mile (25°C/km). Naturally the geothermal gradient today is much lower since the planet has cooled considerably over the last 4.6 billion years. Interestingly, in the mid-1800s physicists attempted to calculate Earth's age based on the difference between the modern geothermal gradient and the gradient believed to have existed when the planet first formed. Assuming Earth was initially molten, they calculated it would take approximately 20 to 40 million years of cooling for the planet to reach its present geothermal gradient. In the early 1900s, however, scientists discovered that the nucleus of certain types of atoms, particularly uranium (U), thorium (Th), and potassium (K), give off heat when they emit energetic particles in a process called *radioactive decay.* This means that minerals containing *radioactive elements* must be giving off considerable amounts of heat deep within the Earth. The fact that radioactive heat is being added to the residual heat tells us that the planet must be cooling at a much slower rate compared to what had been calculated earlier. The Earth therefore had to be much more than 20 to 40 million years old. By using radiometric dating techniques as described in Chapter 1, scientists now calculate the age of the Earth to be about 4.6 billion years old. Recall that this older age is consistent with many other lines of geologic evidence on both the Earth and Moon.

Scientists today believe that early in its history, the Earth became molten as it continued to gain heat energy from a combination of internal compression (friction), impacts, and radioactive decay. During this period the planet was also starting to cool through a process called *conduction,* where in this case the heat is transferred through the atmosphere and into space. Because of its molten state, the planet was also able to cool by *convection* as molten matter transported heat toward the surface where some of it would be lost via conduction through the atmosphere. During convection, the hottest material near the center of the planet becomes less dense, causing it to rise toward the surface—similar to how a hot-air balloon rises. Once near the surface this material loses some of its heat, at which point it becomes denser and starts to sink. As illustrated in Figure 4.6, **convection cells** represent the circular motions of heat and matter, which are driven by temperature-induced changes in the density of the material.

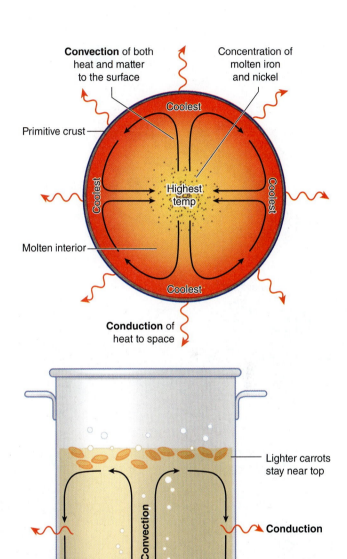

FIGURE 4.6   Soon after Earth formed, enough heat was generated to cause the planet to become molten with only a thin crust. Convection cells developed and began transporting both heat and matter to the surface, some of which is lost to space by conduction—similar to that of a pot of boiling soup. Earth soon developed its layered structure as molten iron and nickel sank to the core and lighter elements tended to rise, eventually forming the crust and mantle.

FIGURE 4.7 The distribution of unique plant and animal fossils on different continents (A) supports the idea of a single supercontinent. The matching fossil assemblages between South America and Africa were first noted by Alfred Wegener. The supercontinent called Pangaea (B) was originally proposed by Wegener, who suggested that the continents slowly drifted to their present position. Modern data show that Pangaea began to break up approximately 225 million years ago.

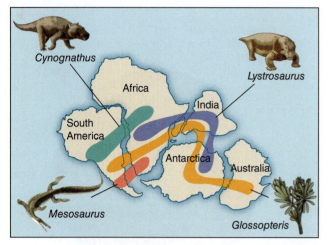

A

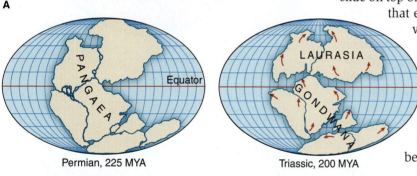

Permian, 225 MYA   Triassic, 200 MYA

Jurassic, 135 MYA   Cretaceous, 65 MYA

B   Present day

The convection process within a hot fluid (Figure 4.6) is important because it can explain how planets can obtain their layered structure. For example, when the Earth was mostly molten, the high density of iron and nickel would cause these elements to sink and accumulate in the core. On the other hand, lighter elements such as oxygen and silicon would tend to rise and form a crust made of silicate minerals. As the planet continued to cool, iron-rich silicates would eventually crystallize and form the mantle. This process of convection and separation of elements and minerals based on density nicely explains Earth's structure as we know it. Although the mantle is now solid rock, its temperature remains hot enough for this material to undergo plastic flow. Scientists generally believe that convection cells continue to operate within the mantle, slowly bringing iron-rich silicate rocks up from great depths, where they then cool and sink back into the mantle. This convective motion in the mantle may help explain why lithospheric plates are slowly moving on top of the asthenosphere. The movement of tectonic plates will be the focus for the remainder of Chapter 4.

Developing the Theory of Plate Tectonics

Recall how plate tectonics is a unifying theory that revolutionized the way in which scientists view our planet. The solid earth is no longer seen as a rigid mass of rock, but rather as a dynamic system with crustal plates that slide on top of a semimolten zone in the upper mantle. Scientists now think that even the mantle is being constantly overturned by deep convection cells. In short, the plate tectonics theory has become central to the study of geology and represents one of the most profound advances in modern science. Although the theory of plate tectonics is now widely accepted, its basic principles had been rejected by most geologists prior to the 1960s. In this section we will briefly examine how this relatively new concept was developed, and the geologic evidence that was obtained which finally caused it to become accepted by the scientific community.

Continental Drift

The idea that Earth's continents actually move is thought to have first been suggested in 1596 by a Dutch mapmaker who proposed that the Americas had been torn away from Africa and Europe by some unknown catastrophic event. Later in the 1850s an American writer, noting how closely the shorelines of South America and Africa fit together (Figure 4.7), also suggested that the continents had been ripped apart in a violent and sudden manner. Then in 1910 an American geologist named Frank Taylor published a paper proposing the continents were once joined, but had separated and reached their present position by slowly plowing through the ocean basins—an idea which later became known as the *theory of continental drift*. Taylor's paper drew little attention until 1922 when a book written by a German meteorologist named Alfred Wegener was translated into several languages, including English. In this new work Wegner had taken Taylor's concept of continental drift and actually supported it with some rather compelling geologic evidence. Supported by evidence, continental drift quickly aroused the interest of the geologic community.

Like others before him, Wegener was intrigued by how South America and Africa fit together like a jigsaw puzzle. However, Wegener was able to show that the two continents, now separated by the Atlantic Ocean, held similar sequences of rocks and unique plant and animal fossils that matched up remarkably well (Figure 4.7A). Moreover, this fossil evidence and the fact that coal had been found in Antarctica strongly suggested that the continents were at one time located in vastly different climatic zones. Because this evidence was difficult to explain other than by continental drift, Wegener proposed an ancient supercontinent called *Pangaea,* which broke apart into separate continents that slowly began to drift apart (Figure 4.7B). Although Wegener's argument was quite compelling, it met considerable opposition by the geologic community. In the end, most scientists rejected Wegener's idea of continental drift largely because of his inability to provide a plausible explanation as to *how* the continents actually moved.

Despite the fact no one could come up with a mechanism to explain how continents could move, additional evidence for continental drift continued to be found. For example, geologists discovered evidence for glaciation on continents where the climate is presently hot and dry (Figure 4.8). This odd occurrence could be accounted for had the continents broken up and drifted into warmer climatic zones. In 1958, geologists studying the continental margins of Africa and South America found that the true edges of the continents were below sea level, thus they actually extended farther offshore. When the two continents were reassembled based on the edge of their undersea margins rather than their coastlines, the fit was now almost perfect (Figure 4.9). Other lines of evidence were soon uncovered that also supported Wegener's idea of continental drift, eventually leading to the more comprehensive theory of plate tectonics.

Mapping the Ocean Floor

Prior to the 1800s very little was known about the ocean floor for the simple reason it was inaccessible to humans. This meant that nearly two-thirds of Earth's surface was completely unknown. The first real data came from depth measurements, called *soundings,* whereby a weighted line was dropped to the seafloor. Based on sounding surveys, the U.S. Navy published a depth or *bathymetric* map in 1855, showing the presence of a submarine mountain in the middle of the Atlantic Ocean. The ability to gather bathymetric data greatly increased after World War I with the development of *sonar.* Here a cruising ship could measure depth along a line, called a *profile,* by bouncing sound waves off the ocean floor and measuring the time it takes for them to return—the sound is the familiar "ping" you hear in movies.

An international effort during the 1950s led to a great number of sonar surveys that gave scientists their first detailed look at the entire ocean floor. The previously mapped mountain in the mid-Atlantic was now shown to be part of a chain of submarine mountains referred to as **mid-oceanic ridges.** As illustrated in Figure 4.10, these mid-oceanic ridges form an extensive network that circles nearly the entire globe. The detailed mapping also revealed the existence of **ocean trenches,** which are narrow, steep-sided depressions running parallel to landmasses. Ocean trenches reach depths as great as 36,000 feet (11,000 m) below sea level, which is deeper

FIGURE 4.8 The mismatch of glacial deposits and present-day climates on several continents helped support the idea of continental drift.

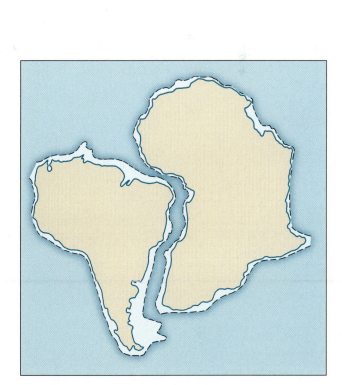

FIGURE 4.9 A near-perfect fit of Africa and South America was obtained when the two landmasses were reassembled using the edge of their continental shelves and opposed to their coastlines.

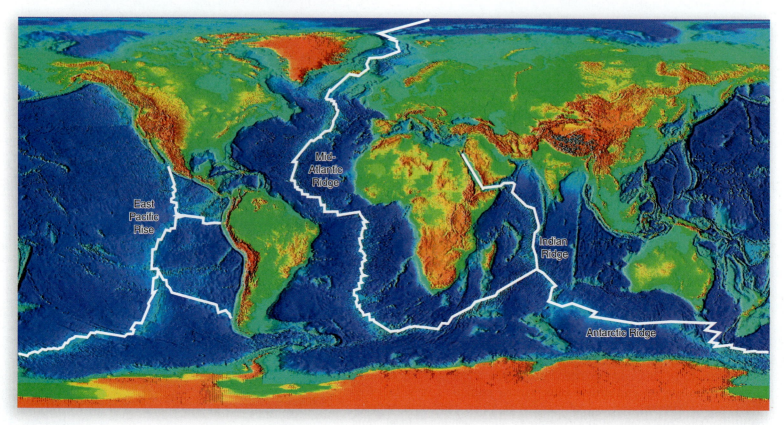

FIGURE 4.10   Modern map showing the topography of both the land and seafloor. One of the striking features of the oceans is the extensive network of mid-oceanic ridges that circle the globe. The oceans also contain narrow trenches that reach depths of nearly 7 miles (11 km).

than Mount Everest is high! Modern maps of Earth's surface (Figure 4.10) show that the ocean floor is not the flat and featureless plain it once was assumed, but rather a complex surface containing long mountain chains and extremely deep canyons.

Magnetic Studies

Certain rocks, particularly igneous basalts, contain appreciable amounts of the iron mineral known as magnetite (Fe_3O_4). Magnetite has the rather rare property of being magnetic, meaning it is naturally attracted to a magnet— similar to a piece of steel. When magnetite crystals cool below 1,075°F (580°C), called its *Curie point,* their iron atoms orient themselves parallel to Earth's magnetic field lines shown earlier in Figure 4.5. In this way grains of magnetite act as tiny compasses frozen in time, provided that metamorphism has not heated the grains above 1,075°F. This property has led to the development of a new field of study called *paleomagnetism,* in which geologists use magnetite-rich rocks (e.g., basalts) as a means of recording historical changes in Earth's magnetic field.

Early paleomagnetic studies found that magnetite grains in many ancient rocks had reversed *polarity,* meaning the magnetic field lines pointed in the opposite direction compared to Earth's current magnetic field. In other words, what was "north" in these rocks actually points toward today's magnetic South Pole. Consequently, geologists classify magnetic rocks as having either normal or reversed polarity. Note that the reason why Earth's magnetic field reverses itself is not well understood, but it is generally believed to be related to periodic changes in the circulation of metallic ions within the outer core.

Up until World War II paleomagnetic studies had largely been restricted to the continents, hence no one had ever systematically studied the magnetic properties of basaltic rocks known to exist on the seafloor. During the 1950s, ships began measuring the magnetic properties of the seafloor by towing an instrument called a *magnetometer* along profile lines. Maps of the seafloor were then constructed and revealed alternating bands of normal and reverse polarity. As illustrated in Figure 4.11, these mysterious bands extended for great distances and were oriented parallel to mid-oceanic ridges. One of the keys to understanding this strange striping pattern was that the bands were of different widths and were symmetrical on either side of a ridge. This meant the patterns were mirror images, hence individual stripes could be paired up with an identical stripe on the opposite side of a ridge. Another key observation was that the rocks along the ridge crest show normal polarity, and that the rocks become progressively older as one moves away from the ridge in either direction. Geologists soon recognized that the magnetic properties of the seafloor were somehow related to formation of the mid-oceanic ridges.

In order to explain this new oceanographic data, geologists developed a hypothesis known as **seafloor spreading,** whereby mid-oceanic ridges represent weak zones along which magma erupts to form new oceanic crust. According to this hypothesis, ocean ridges spread or open up over time and are simply filled with new magma, as shown in Figure 4.11. This process would account for the symmetrical pattern of magnetic reversals and the fact ridge crests all contain rocks with magnetic signatures consistent with Earth's present magnetic field. Seafloor spreading also explains why rocks get progressively older farther away from the ridge crests. The hypothesis gained widespread acceptance in 1968 when a research vessel equipped with a drilling rig systematically collected actual rock samples over much of the Atlantic basin. Results from radiometric dating of the basaltic rocks (Figure 4.12) showed that the entire Atlantic seafloor gets progressively older on either side of the mid-oceanic ridge. Moreover, the oldest parts of the seafloor were found to be 200 million years old, which is quite young compared to 3.8-billion-year-old rocks found on the continents. This was conclusive proof that seafloor spreading was acting like a conveyor belt, carrying newly formed rocks away from mid-oceanic ridges.

Seafloor spreading presented a problem because if new oceanic crust is constantly being formed, then either the Earth must be growing in size or some process must exist that is destroying crustal rocks. Because the age of the Atlantic seafloor (Figure 4.12) proved that the oceanic crust there is expanding, geologists reasoned that perhaps this growth was being compensated for by the destruction of crustal material under the Pacific. In this way the Earth would not be getting larger since part of the Pacific basin would be decreasing in size at the same time the Atlantic basin grew larger. Scientists felt that the most likely place where crustal material is being destroyed is in the deep ocean trenches located around the perimeter of the Pacific Ocean. This idea would eventually be confirmed through the study of earthquake (i.e., seismic) waves, which had also provided information on Earth's layered structure as described earlier.

Location of Earthquakes

We will explore earthquakes in greater detail in Chapter 5, but for now we can define an earthquake as the release of energy that occurs when rocks are deformed beyond their elastic limit, causing them to rupture. This energy then travels outward in all directions in the form of *seismic waves*—the *epicenter* is the point on the surface that directly overlies the point underground where rocks rupture and release their stored energy. In the 1960s a global network of seismic recording stations enabled seismologists to map the location of earthquake epicenters around the world. In the more recent map in

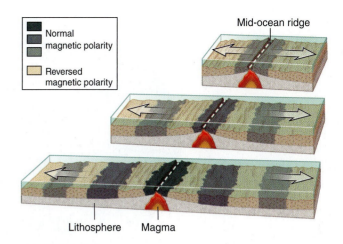

FIGURE 4.11 Illustration showing the development of magnetic striping by seafloor spreading. Magma rises up through mid-oceanic ridges and then cools to form basaltic rocks whose magnetite grains record the orientation and polarity of Earth's magnetic field. As the seafloor continues to spread and new rocks form, a symmetrical pattern of reverse and normal polarity develops on opposite sides of the ridge.

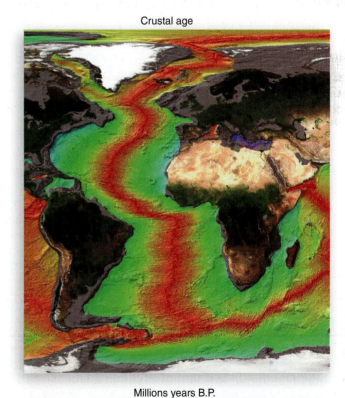

FIGURE 4.12 Map showing both the age and topography of the Atlantic seafloor. Note how the seafloor gets progressively older farther away from the mid-oceanic ridge, with oldest seafloor being about 200 million years old.

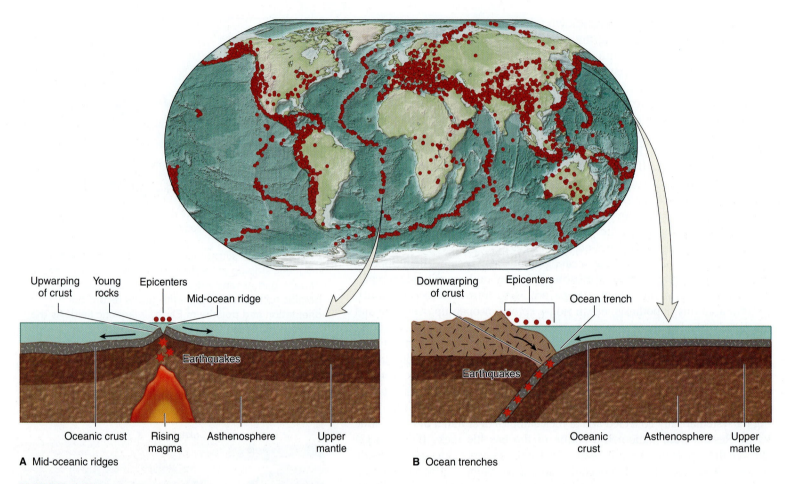

A Mid-oceanic ridges

B Ocean trenches

FIGURE 4.13 Map showing the location of earthquake epicenters from 1963 to 1998. Note how the earthquakes are not randomly distributed. Inset (A) illustrates how rising magma beneath mid-oceanic ridges generates earthquakes whose epicenters lie in a relatively narrow zone at the surface. Insert (B) shows how epicenters near ocean trenches are spread over a wider area due to the way the subducting slab generates earthquakes in an inclined manner.

Figure 4.13, one can clearly see that earthquakes do not occur randomly, but rather are concentrated in distinct zones. Perhaps most striking is just how well the epicenters correlate with the location of mid-oceanic ridges (Figure 4.13A). This can be explained by magma rising vertically upward from the mantle, buckling the crust and thereby forming the ridge. The force of the rising magma also causes rocks to deform beyond their elastic limit, at which point they rupture and release stored energy in the form of seismic waves.

In Figure 4.13B you can also see there are places where epicenters are clustered in relatively wide zones, which happen to coincide with mountain ranges and ocean trenches. When seismologists performed detailed studies near ocean trenches, they found that the earthquakes were originating in an inclined zone that extended several hundred miles down into the mantle. This inclined zone was interpreted to be caused by the collision of two lithospheric plates, generating earthquakes as an oceanic slab of lithosphere sinks down into the mantle (Figure 4.13B). This process of lithospheric slabs descending into the mantle is referred to as **subduction**, whereby the oceanic slabs are believed to eventually undergo melting and become incorporated into the mantle. Subduction then can explain the occurrence of certain types of earthquakes, and it provides a mechanism by which oceanic crust is destroyed. In addition, scientists believe that ocean trenches themselves are the result of the descending slab forcing the lithosphere to buckle or downwarp. Later you will see that melting of the descending slab also explains the volcanic activity that commonly occurs along ocean trenches.

In the end, seismic studies provided convincing evidence that magma is rising up from the asthenosphere at mid-oceanic ridges, forming new oceanic lithosphere. These studies also indicated that oceanic plates are being destroyed along ocean trenches as lithospheric slabs undergo subduction. It was now quite clear that the Earth was not expanding, but rather the Atlantic seafloor was growing at the expense of the seafloor under the Pacific.

Polar Wandering

The last major line of evidence supporting continental drift came as a result of paleomagnetic studies involving continental rocks. Recall that magnetite-rich rocks record the orientation of the magnetic field at the time the rocks form. These data were used to make maps showing the location of the magnetic North Pole at different times in the geologic past. Scientists were at first baffled when the location of the pole took on a strange wandering path, which was referred to as *polar wandering*. Even more intriguing was that the different continents each showed a different polar wandering path. These separate paths made no sense since the Earth has only one magnetic North Pole at any given time. As illustrated in Figure 4.14, the puzzle was solved when scientists took into consideration the possibility that the continents had been moving rather than the poles. Sure enough, when seafloor spreading data were used to reposition each of continents at different times in the geologic past, paleomagnetic data showed a common location for the pole. Therefore, the combination of paleomagnetic and seafloor spreading data provided convincing proof that Wegener's concept of continental drift was indeed correct. Even the most skeptical geologists now agreed that a supercontinent had once existed, and then subsequently broke apart and began moving in different directions as separate landmasses.

Plate Tectonics and the Earth System

Although it took nearly 50 years of gathering data and testing various hypotheses, by the late 1960s scientists had proven that Wegener's basic idea of continental drift was correct. Radiometric age dating and paleomagnetic studies of the oceanic crust left no doubt that seafloor spreading was actually taking place at mid-oceanic ridges. Moreover, earthquake studies showed that magma rises up from the mantle at ocean ridges to form new crust, while at the same time older crust is being destroyed through subduction at ocean trenches. By unraveling the mystery of polar wandering, geologists also proved Wegener's concept that a supercontinent had once existed, which later broke up and began drifting apart. Finally, it was the clustering of earthquake epicenters that allowed geologists to define the boundaries of the rigid slabs of lithosphere, now called *tectonic plates*.

Perhaps the most important outcome of this long process of gathering data and testing different hypotheses was that scientists were able to show that continental drift, seafloor spreading, and subduction are all connected to one another. This resulted in the development of a single, unifying theory (Chapter 1) that explained the interrelationship between these different processes. Geologists referred to this new and more comprehensive theory as the *theory of plate tectonics*, a name that originates from the Greek word *tekton*, meaning "builder." Tectonics is an appropriate term since Earth's major surface features are literally built by the planet's internal forces and moving plates. These tectonic forces help drive the rock cycle (Chapter 3) by causing metamorphism and pushing mountains up above sea level. Weathering and erosion processes then act to wear the uplifted terrain back down to sea level, ultimately transforming the weathered material into sedimentary rocks. In the end, plate tectonics is a critical component of the Earth system and has helped make life as we know it possible. We will begin this section by taking a brief look at the tectonic plates themselves, as well as the types of interactions that occur at the plate boundaries.

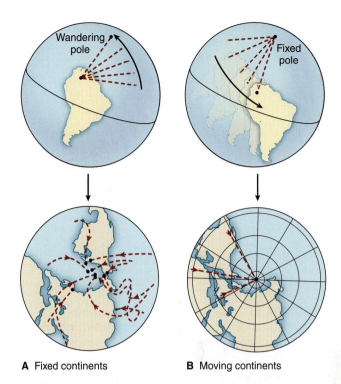

A Fixed continents   **B** Moving continents

FIGURE 4.14   When paleomagnetic studies assumed that the continents remained fixed (A), the position of the magnetic North Pole appeared to wander over time. Moreover, each continent showed a different wandering path for the pole. When seafloor spreading data were used to reposition the continents at different times in the geologic past (B), a single location for the pole emerged.

Types of Plate Boundaries

Based largely on seismic data, scientists have determined that Earth's lithosphere is broken up into seven major plates and several smaller ones as shown in Figure 4.15. This concept of Earth having a rigid outer shell composed of individual plates is analogous to a cracked eggshell. Note that some of the plates in Figure 4.15, such as the Pacific, are covered almost entirely of oceanic (basaltic) crust, whereas others are covered by both continental (granitic) and oceanic crust. For example, notice that the eastern edge of the North American plate does not stop at the Atlantic shoreline, but rather extends all the way out to the oceanic ridge in the middle of the Atlantic. Keep in mind that plate boundaries are usually not sharp, distinct features, but rather are zones that are defined by concentrated earthquake activity.

Another key aspect of plate tectonics is that Earth's internal forces cause these rigid slabs to move over the weak asthenosphere. A useful analogy is how wind or water currents will cause broken slabs of ice to move over a water body, which is naturally weak because it is a liquid. In some areas the currents will cause individual separate slabs of ice to collide, and in other places the slabs may move in opposite directions. The compressive forces generated along collision boundaries can cause the ice

FIGURE 4.15 Map showing the distribution of lithospheric plates. Note how the North American plate is covered by both continental and oceanic crust, whereas the Pacific plate is covered entirely by oceanic crust.

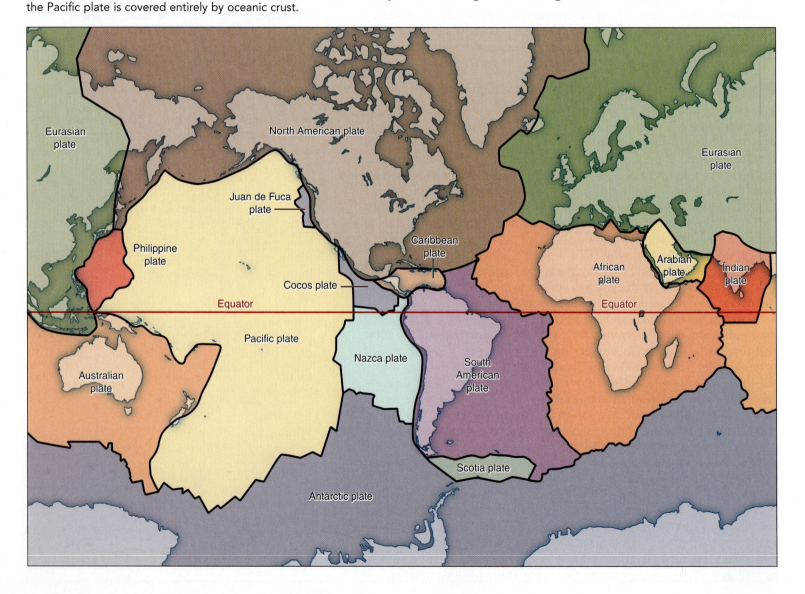

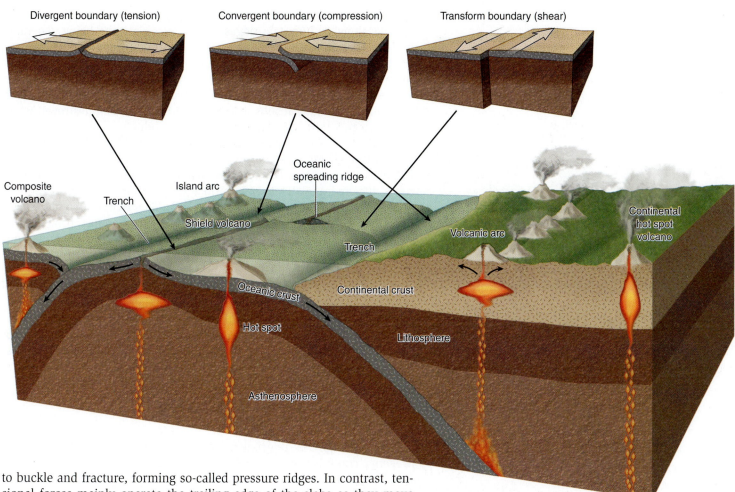

to buckle and fracture, forming so-called pressure ridges. In contrast, tensional forces mainly operate the trailing edge of the slabs as they move apart from one another, resulting in minor deformation. The movement of Earth's lithospheric plates is similar in that the tectonic forces are most prevalent along the boundaries of plates. It is along these plate boundaries that geologists find the vast majority of earthquakes, volcanic eruptions, and mountains ranges. Therefore, we will focus our attention on the way in which plates interact with each other at their boundaries.

The movement of Earth's tectonic plates generates three basic types of forces at the plate boundaries, namely *compression, tension,* and *shear* (see Figure 4.1). Depending on the relative motion of the plates and the dominant type of force that is being generated, plate boundaries can be placed into one of the three categories shown in Figure 4.16. A **divergent boundary** is dominated by tension forces and involves two plates moving away from one another. For example, a mid-oceanic ridge defines a divergent boundary because it marks where two plates are moving in opposite directions. In contrast, a **convergent boundary** is where two plates are moving toward each other and are under compression. An example here is a subduction zone in which the plates collide, forcing one plate to slide under the other. Finally, when two plates slide past each other and are dominated by shear forces, the boundary is called a **transform boundary.** Although we have yet to describe a geologic example of a transform boundary, the concept can be illustrated by imagining two cars moving toward each other. If the cars were to have a glancing impact, as in a side-swipe, it would create shear forces on the cars and be analogous to a transform boundary. A head-on crash would generate mostly compression forces, thereby representing a convergent boundary.

FIGURE 4.16 Illustration showing how divergent, convergent, and transform plate boundaries are under tension, compression, and shear forces, respectively. Note how mid-oceanic ridges and mountain chains are features that form in a parallel manner to plate boundaries. Volcanic hot spots occur away from plate boundaries and are believed to be related to hot plumes of material that rise from deep within the mantle.

One of the key features of Figure 4.16 is how oceanic lithosphere is created along mid-oceanic ridges. Here we see magma being generated in the asthenosphere, which then rises to form a layer of basalt over the rigid rocks of the upper mantle. As the two plates diverge away from the ridge system, notice how the mantle portion of the lithosphere thickens as the upper asthenosphere cools and becomes rigid. Continued seafloor spreading eventually results in the oceanic plates being destroyed when they descend into the asthenosphere at subduction zones. As a slab descends it starts to melt, generating magma that ultimately rises through the overlying plate and forms a volcanic mountain range. This rising magma at convergent boundaries commonly interacts with continental rocks, which are relatively rich in silica (SiO_2). This interaction produces andesitic and granitic magmas that are more SiO_2 rich, but contain fewer ferromagnesian minerals compared to basaltic magmas (Chapter 3). Therefore, the oceanic lithosphere that is destroyed at convergent boundaries ends up producing magma, which in turn forms new continental rocks. At transform boundaries, however, magma is not being generated since the plates simply grind past one another. Also notice in Figure 4.16 how magma rises from deep within the mantle at so-called *hot spots*, creating volcanic landmasses in the middle of tectonic plates. The Hawaiian Islands, for example, lie over a mantle hot spot. We will discuss hot spots in more detail in Chapter 6, but for now simply note that the vast majority of volcanic activity takes place along divergent and convergent plate boundaries.

Movement of Plates

Geologists originally rejected the concept of continental drift because Wegener could not come up with a mechanism to adequately explain how the plates actually move. Although the exact mechanism is not fully understood even today, scientists generally agree that plate movements result from the complex interaction between convection cells within the mantle and the plates themselves. Remember that deep within the mantle the temperature and pressure are extremely high, which allows the silicate rocks to flow as they undergo plastic deformation. As illustrated in Figure 4.17, this hot material tends to rise toward the surface where it begins to lose heat, causing its density to increase. Because of the increased density the rocky material eventually begins to sink, creating large-scale convection cells within the mantle. Notice that near the surface, the tectonic plates are believed to be moving in the same direction as the convection cells. Keep in mind, however, that the asthenosphere is near its melting point and is very weak. Because of this weak layer, it is unlikely that the convective motion in the mantle is very effective in moving tectonic plates over the asthenosphere. Scientists have concluded that additional forces must be operating which are helping to drive the plates over the asthenosphere.

Modeling studies have shown that the density and elevation differences of tectonic plates play an important role in the convective movement within the mantle. For example, at a subduction trench the descending slab of lithosphere is relatively cool and dense compared to the

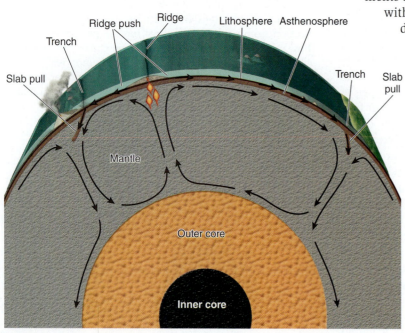

FIGURE 4.17 Relationship between mantle convection cells and plate boundaries. Note how the plates move in the direction of the convection cells.

hot material within the asthenosphere. This density difference causes the slab to sink, which tends to drag or pull the rest of the plate into the subduction zone, a process geologists call *slab pull* (Figure 4.17). In addition, the rising magma at divergent boundaries pushes the lithosphere upward, forming an elevated ridge system. This creates what is known as *ridge push*, which is where gravitational forces cause the elevated rocks to literally push the oceanic plate downward over the sloping surface of the asthenosphere. Recent studies have concluded that slab pull at subduction zones has a greater effect on the overall movement of tectonic plates than does the ridge push mechanism at spreading centers. An example of ridge push can be seen in the snow-covered car in Figure 4.18. In this case gravity caused the elevated section of snow on the windshield to push downward on the sloping surface of the glass. This force caused the entire layer of snow to slide over the surface of the car. Note that this movement was facilitated by unfrozen water on the car's surface, which acted as a weak layer similar to the asthenosphere.

Surface Features and Plate Boundaries

In this section we want to take a closer look at the different types of features and processes that occur at each of the three types of plate boundaries. We want to focus our attention on plate boundaries because this is where processes such as earthquakes, volcanic eruptions, and the uplift of mountains are taking place. These processes, of course, are important in the study of environmental geology because they represent natural hazards. As you read through this section it will be helpful for you to refer back to Figure 4.19 as it shows the boundaries of Earth's major plates, and the direction in which the plates are moving.

FIGURE 4.18 A thick layer of snow on this car provides an example of the ridge push mechanism. Gravity caused the snow to push downward on the sloping surface of windshield, forcing the rest of the snow to slide over the hood. Note that the snow layer behaved as a rigid plate, allowing it to buckle, forming a fold. Also note that unfrozen water was present on the surface of the car, creating a weak layer that facilitated the movement of the overlying snow.

FIGURE 4.19 Map showing the types of movement taking place at the boundaries of Earth's major plates.

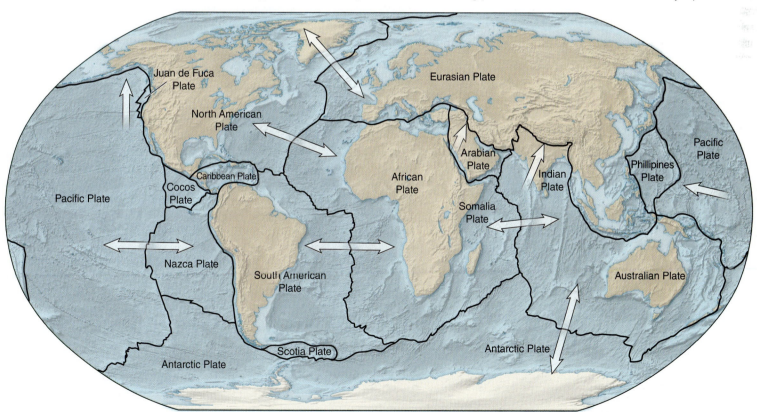

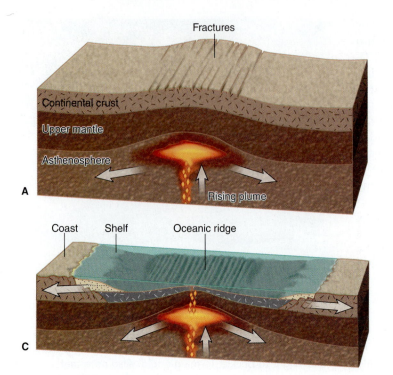

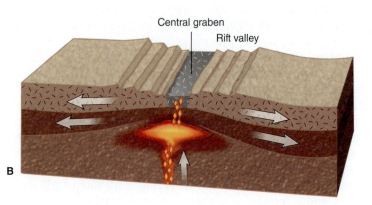

FIGURE 4.20   Sequence of events showing the development of a mid-oceanic ridge system. A rising convection cell (A) allows magma to push upward against the brittle lithosphere, creating tensional forces that cause the plate to fracture. Continued tension (B) leads to faulting and the development of a rift valley where magma forms layers of basaltic rock. As the rift widens it becomes flooded (C), forming a new ocean with a spreading center in the middle.

Divergent Boundaries

Earlier you learned that mid-oceanic ridges are divergent boundaries where the process of seafloor spreading is taking place. Here the tectonic plates on opposite sides of a ridge move away from each other like giant conveyor belts. The sequence of events in Figure 4.20 shows how scientists believe such spreading centers originate. The process begins when a convection cell causes magma to rise up from the mantle and eventually encounters the rigid lithosphere. The upward force of the magma not only causes the lithosphere to buckle upward, it also creates a considerable amount of tensional stress within the relatively brittle rocks. Eventually the rocks are deformed beyond their elastic limit, at which point they fail by fracturing. This fracturing typically leads to the development of a linear fracture zone where individual fractures fill with magma. If the tensional forces continue, then faults can develop along the fracture zone in a stair-step fashion to form a down-dropped featured called a **rift valley,** also called a *graben* (Figure 4.20B). The upwelling of magma then results in basaltic rocks forming in the valley floor. As the basaltic floor continues to widen, the entire valley can become flooded with seawater, at which point a linear-shaped ocean is born with a spreading center located in the middle.

The mid-Atlantic ridge (see Figures 4.10 and 4.19) is perhaps the most well studied site of seafloor spreading. With a length of nearly 10,000 miles (16,000 km), this ridge system is actually the longest mountain range in the world. Based on the age of the basaltic rocks making up the Atlantic seafloor, and their distances from the spreading center, geologists calculate that the seafloor is spreading at an average rate of 2.5 centimeters per year. Although this may seem quite slow to us humans, if this rate continued for another million years of geologic time, then the Atlantic Ocean would become 15 miles (25 km) wider. Likewise, if we could somehow reverse the plate motion and go backward in time, the ocean basin would get smaller. Were we to go back 200 million years, the Atlantic Ocean would be a narrow sea bounded by the African and the American continents, which were just starting to drift away from each other.

One of the fascinating things about geology is that there are examples of geologic processes that we can study in different stages of development. For instance, the mid-Atlantic ridge and ocean basin show us what an active spreading center can do given over 200 million years of geologic time. To observe a spreading center in its initial stages of development, we can look at Africa, where the eastern portion of the continent is currently being torn apart by great tensional forces. From Figure 4.21 one can see that the Arabian peninsula has already broken away from Africa, creating a rift valley that has become flooded to form the Red Sea—this is similar to how the Atlantic once looked. Moving south we find the African Great Rift Valley where the rifting is even younger. Should the spreading continue, then this valley will also flood and form a narrow sea. Note that the Great Rift Valley contains some of the world's oldest hominid fossils and is also well known for its volcanic activity.

Convergent Boundaries

Unlike the tensional forces that form rift valleys and spreading centers, compressional forces associated with convergent boundaries produce complex mountain ranges. Moreover, different combinations of oceanic and continental crust can be found colliding with each other at convergent boundaries. This, in turn, produces different types of mountains. As shown in Figure 4.22, the various combinations of convergent boundaries are as follows: *oceanic-oceanic, oceanic-continental,* and *continental-continental.* In this section we will briefly explore the different types of collisions and the type of landforms that are generated.

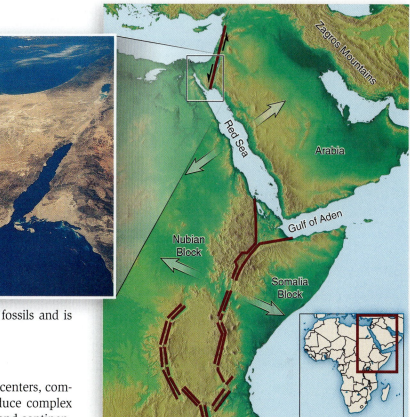

FIGURE 4.21 Map showing the location of a young spreading center in eastern Africa. The northern portion of the rift zone has opened to a point where it has become flooded, forming the Red Sea and Gulf of Aden. To the south is the much younger African Great Rift Valley. Satellite photo showing the Sinai peninsula located at the top of the Red Sea. This peninsula is bounded on the left by the spreading center and on the right by a transform boundary.

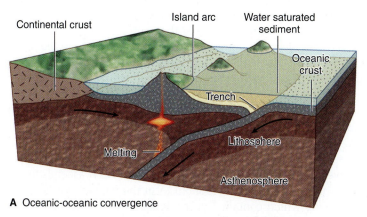

A Oceanic-oceanic convergence

FIGURE 4.22 Different types of convergent plate boundaries produce different landforms. The collision of two oceanic plates (A) results in subduction and the formation of a volcanic island arc. When oceanic and continental plates collide (B), subduction leads to a continental arc system that contains more SiO_2-rich rocks. When two continental plates collide (C) there is no volcanic activity since neither plate undergoes subduction. Here the collision produces a highly deformed mountain belt called a suture zone.

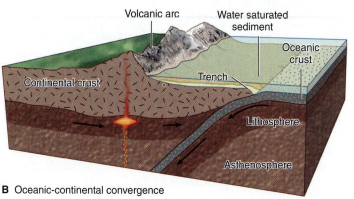

B Oceanic-continental convergence

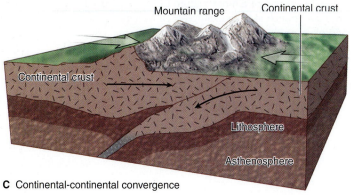

C Continental-continental convergence

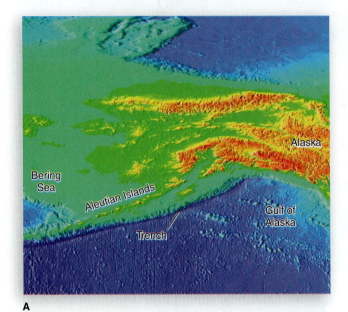

A

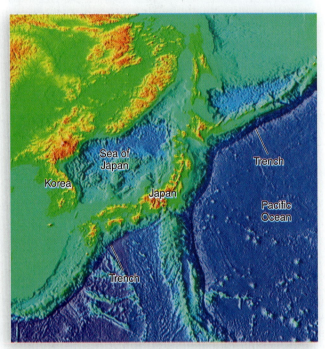

B

FIGURE 4.23 The Aleutian Islands off the Alaskan mainland (A) are a narrow string of volcanic islands that represent a young island arc system. The much larger islands of Japan (B) represent a more mature island arc. Note the deep ocean trench in both examples, caused by subduction of one plate on the ocean side of the island arcs.

Oceanic-Oceanic When two oceanic plates collide at a convergent boundary (Figure 4.22A) one of the basalt-covered slabs will undergo subduction, causing the crust to buckle and form an ocean trench. Water-rich sediment is also carried downward with the descending slab, which acts to lower the melting point of rocks in the upper mantle. This commonly produces magma of basaltic composition, which then rises up through the buckled plate to form a string of volcanic islands called an **island arc.** The Aleutian Islands off the Alaska mainland (Figure 4.23A) are a good example of a relatively young island arc system. Over time, the subduction process not only results in the growth of the islands, but will produce *andesitic* magmas. Recall from Chapter 3 that andesite is an igneous rock that contains more SiO_2 (silica) but less iron and magnesium than basalt. One way andesite forms is when a subducting slab of basaltic crust becomes hot enough that it undergoes *partial melting*—note that water in the subduction zone helps lower the melting point (Figure 4.22). Because the minerals in basalt that have the lowest melting points also happen to be the most SiO_2 rich, partial melting generates andesitic magmas that are relatively rich in SiO_2. Andesite can also form when a rising body of basaltic magma causes silicate minerals in the surrounding rocks to melt. This material is then incorporated into magma, enriching it in SiO_2—a process called *assimilation.*

Over time, the processes of partial melting and assimilation will produce crustal rocks that are more SiO_2 rich, but less dense compared to the basaltic crust being destroyed in the subduction zone. This means that as island arcs get older, they not only become larger, they generally contain more andesite rock. The islands of Japan (Figure 4.23B) are a good example of a relatively large, more mature volcanic arc system that is composed of somewhat less dense crustal rocks.

Oceanic-Continental Convergent boundaries involving the collision of oceanic and continental plates (Figure 4.22B) also lead to subduction and volcanic activity. In this case, however, the resulting string of volcanoes forms on the continental plate, and is therefore referred to as a **continental arc** as opposed to an island arc. Recall that the continental (granitic) lithosphere is considerably less dense than the oceanic plate because the rocks are more enriched in SiO_2 and contain relatively few ferromagnesian minerals. Since these two plates are moving over the asthenosphere, when they collide the more buoyant continental plate overrides the denser oceanic slab. As in the oceanic-oceanic setting, the oceanic plate then undergoes subduction and provides the water that helps initiate melting in the upper mantle. Here the magma must rise through a much thicker continental plate as opposed to the thin oceanic slab. This means the magma is likely to assimilate even more SiO_2-rich material into the melt as it ascends to the surface. The result is a magma that cools to form the igneous rocks called *granite* and *rhyolite,* which contain few ferromagnesium minerals but are rich in quartz (Chapter 3).

One of the best examples of an oceanic-continental plate boundary is along the western coast of South America (Figure 4.24). Here nearly the entire edge of the continent is converging with an oceanic plate, forming an ocean trench and a continental arc system known as the Andes Mountains. This impressive mountain range contains numerous volcanoes and has peaks over 22,000 feet (6,700 m) above sea level. Moreover, the Andes are approximately 5,500 miles (9,000 km) in length, which is greater than the distance from New York City to Rome, Italy. Another example of a continental arc is the Cascade Range, located in the Pacific Northwest region of the United States. Note that Mount St. Helens, which erupted in 1980, is one of the better known volcanoes in the Cascades.

It is important to emphasize that both island and continental arcs are built by magma that is derived from the subduction of oceanic lithosphere. This process generates igneous rocks that become part of the plate that is not subducted, ultimately producing landmasses that are more granitic in composition. The continued recycling of oceanic plates and creation of new continental rocks explains why oceanic crust is generally less than 200 million years old. This also explains why the continents contain ancient rocks as much as 3.8 billion years old. Studies of subduction zone processes have helped scientists unravel Earth's history, and they have led to a better understanding of the geologic hazards associated with earthquakes and volcanic eruptions (Chapters 5 and 6).

Continental-Continental   The last type of convergent boundary is where two continental plates collide (Figure 4.22C). Because both lithospheric plates in this case are composed of low-density granitic material and are both rather buoyant, neither plate will undergo subduction. Instead the two plates will literally plow into each other, forcing rocks to rise vertically above sea level and also push downward into the weak asthenosphere. In this setting the rocks are under so much compressive force that they deform beyond their elastic limits. This causes the rocks near the surface to fail mostly by fracturing and folding, whereas more deeply buried rocks will undergo plastic deformation. As illustrated in Figure 4.25, the result is a linear mountain chain consisting of intensely folded and faulted rocks, most of which have undergone regional metamorphism (Chapter 3). Geologists often refer to such mountain belts as *suture zones* because they act as a bond between two landmasses that have come together.

Numerous examples exist of ancient suture zones around the world, indicating that plate tectonics has been operating since Earth was quite young. The Appalachian Mountains of North America represent a suture zone that formed 250 million years ago, thus what we see today are the erosional remnants of a once-lofty mountain chain. Modern examples naturally consist of mountains that are much higher, such as the Himalayas in Asia and the Alps in Europe. The formation of the Himalayas is of considerable interest to geologists because the collision of India with Asia is a relatively recent event, geologically speaking. By studying the Himalayas, much can be learned about the formation of ancient suture zones.

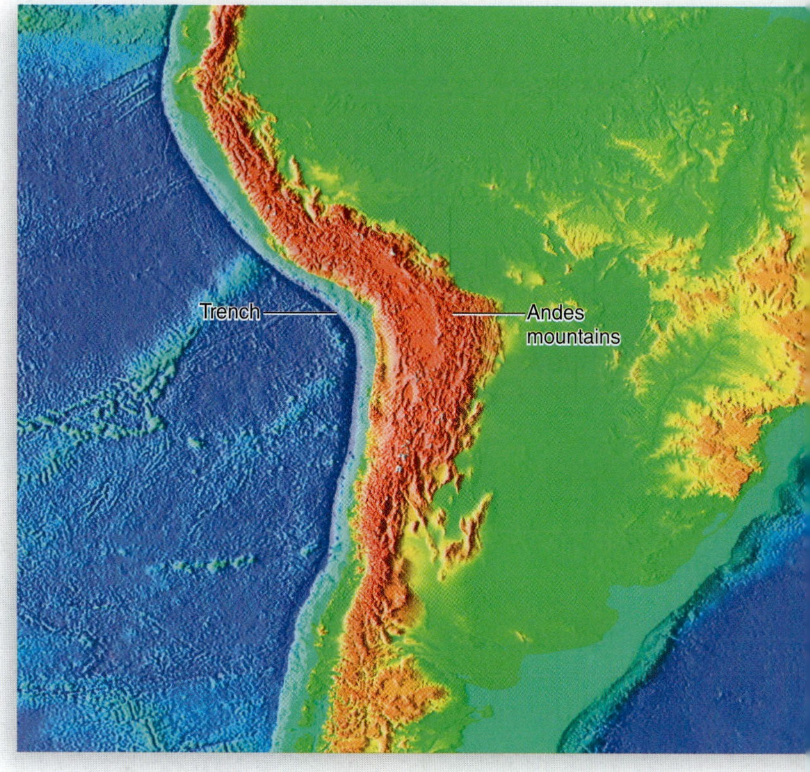

FIGURE 4.24   The Andes Mountains are a continental arc system associated with an oceanic-continental boundary. The boundary of these plates is marked by the position of the ocean trench. The Andes are the result of continued convergence and subduction of the oceanic plate, which has produced volcanic activity and buckling of the plates.

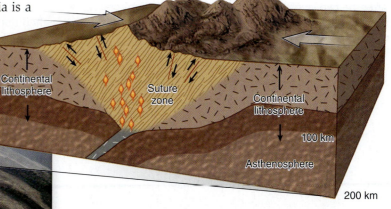

FIGURE 4.25   Subduction does not occur along convergent boundaries involving two continental plates composed of relatively low-density material. Instead, the collision creates a thick zone of highly faulted and deformed mountains where the rocks undergo regional metamorphism. Photo shows the folded Zagros Mountains in Iran.

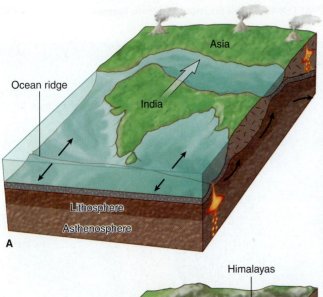

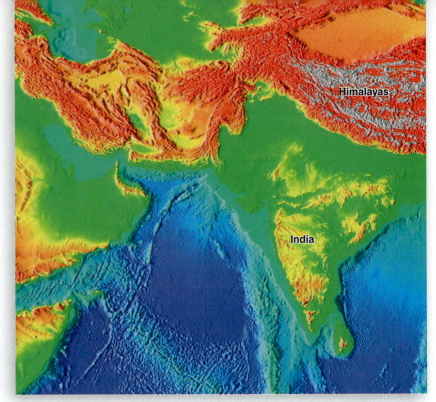

FIGURE 4.26 The Himalaya Mountains are a modern example of the collision between two continental plates. As India moved toward the Asian landmass, the subduction of oceanic crust created a volcanic arc (A). After the two landmasses collided, the subduction zone eventually shut down and volcanic activity ceased. The collision continues to cause uplift and deformation yet today (B).

As illustrated in Figure 4.26, India was once a separate landmass surrounded by oceanic crust that was moving away from a spreading center as if on a conveyor belt. For millions of years the oceanic plate on the leading edge of India was being subducted under the Asian continent, thereby creating a volcanic arc system. Then about 40 to 50 million years ago the two continental landmasses started to collide, and the modern Himalayas began to rise. Eventually the subduction of the oceanic plate stopped, causing most of the volcanic activity to cease. This collision continues today and is actively raising the Himalayas at an incredible rate of more than 1 centimeter per year. Similar to the uplift of mountains associated with subduction processes, the rise of the Himalayas is not necessarily smooth and continuous, but rather occurs in sudden movements related to major earthquakes. Note that the height of all mountains, regardless of their tectonic setting, is controlled by the difference in the rates of uplift and erosion. Mountains naturally rise when uplift is greater than erosion, and they become lower when erosion is greater than uplift. Climate also plays a key role because erosion rates are affected by the amount of precipitation and the formation of glacial ice.

Transform Boundaries

The last of the three major types of plate boundaries we need to consider are *transform boundaries,* which are where *shear* forces dominate between two plates. In this setting the plates simply grind or slide past one another along what are called *transform faults.* These faults were first identified in the 1960s when scientists noticed that mid-oceanic ridges are commonly offset, creating a zig-zag pattern as shown in Figure 4.27. Notice in the areas where the ridges are offset, how seafloor spreading produces a shearing motion along the various transform faults. Because the plates are moving in a horizontal manner, magma does not form since there is no subduction. The lack of subduction and volcanic activity means, of course, that lithospheric material is neither being created nor destroyed. Later we will examine how the shearing forces that develop along transform faults can result in major earthquakes.

Because of their association with spreading centers, most transform faults are within oceanic plates. Consequently, the vast majority of transform boundaries lie beneath the oceans, and relatively few are found on continents. A good example of a transform boundary is the San Andreas fault in California, which

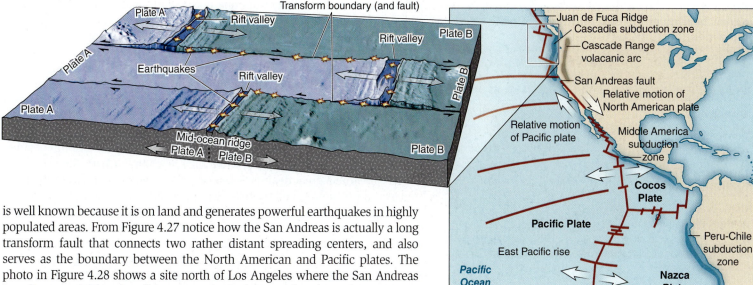

FIGURE 4.27 The San Andreas fault is a transform boundary associated with the East Pacific Rise spreading center. The San Andreas forms the boundary between the Pacific and North Amercian plates. Note that just north of San Francisco, the western edge of North America changes from a transform to a convergent boundary that includes a subduction zone and volcanic arc.

is well known because it is on land and generates powerful earthquakes in highly populated areas. From Figure 4.27 notice how the San Andreas is actually a long transform fault that connects two rather distant spreading centers, and also serves as the boundary between the North American and Pacific plates. The photo in Figure 4.28 shows a site north of Los Angeles where the San Andreas is well exposed. Here the offset stream channels prove that dramatic movement along the fault has taken place relatively recently. Interestingly, a person at this site could stand on one tectonic plate, then walk a short distance and be on a different plate that is literally moving in the opposite direction. This lateral movement along the San Andreas means that Los Angeles, which sits on the Pacific plate moving to the northwest (Figure 4.27), will eventually meet up with San Francisco as it moves to the southeast on the North American plate. Keep in mind that it will take millions of years for the cities to meet. Also, contrary to common folklore, these plate movements are not going to cause Los Angeles to suddenly slide off into the ocean in some giant landslide!

Finally, notice in Figure 4.27 that as one goes north of San Francisco, the tectonic setting changes from a transform boundary to a convergent boundary. Moreover, the presence of a subduction zone and volcanic arc (Cascade Range) tells us that the boundary along the Pacific Northwest involves the convergence of an oceanic plate with the continental plate. As a result, the residents of the Pacific Northwest face both volcanic and earthquake hazards, whereas the primary hazard in Southern California is one of earthquakes. We will explore both earthquake and volcanic hazards in considerable detail in Chapters 5 and 6.

Plate Tectonics and People

In Chapter 4 you have learned how plate tectonics is a comprehensive and unifying theory that is central to the study of geology. This system of moving plates is not only responsible for driving the rock cycle, but it also creates the major surface features of our planet. However, the importance of plate tectonics goes far beyond geology. Because Earth acts as a system, we should expect that anything as significant as plate tectonics will have an important impact on other parts of the system, namely the atmosphere, hydrosphere, and biosphere. In this final section we will briefly touch on some of the ways in which plate tectonics affects the lives of humans and modern societies. This is quite appropriate since the focus of this textbook is on the interaction between environmental geology and people.

Natural Hazards

There are many geologic processes that create natural hazards for humans and society. In this text we will focus on the hazards that are associated with earthquakes, volcanic eruptions, and the downslope movement of earth materials (i.e., landslides), all of which are clearly related to plate tectonics. As you have learned thus far in this chapter, the majority of

FIGURE 4.28 Air photo showing the San Andreas fault in Southern California. Note how the stream channel has been offset due to recent movement along this transform plate boundary.

TABLE 4.2 Relative risk of earthquake and volcanic hazards at the different types of plate boundaries.

Tectonic Plate Setting	Relative Risk of Earthquake Hazards	Relative Risk of Volcanic Hazards
Divergent	Moderate	Moderate
Convergent: oceanic-oceanic	Major	Major
Convergent: oceanic-continental	Major	Major
Convergent: continental-continental	Major	None
Transform	Major	None

earthquakes and volcanic eruptions occur along plate boundaries as does the uplift of mountain ranges, which generates steep slopes and the potential for landslides. Therefore, the level of risk people face from these hazards greatly depends on where they live relative to Earth's plate boundaries.

If we consider just earthquakes and volcanic eruptions, we will find that the relative risk from the hazards varies according to the type of plate boundary, as indicated in Table 4.2. Earthquakes, for example, can happen far from a plate boundary, but most occur along plate boundaries where rocks are subjected to tremendous amounts of stress. However, the amount of energy released in an earthquake is directly related to the amount of stress rocks can store before rupturing when they reach their elastic limit. Because rocks can store a great deal more energy when under compression or shear forces than tension, major quakes are much more common along convergent and transform boundaries (Table 4.2). Volcanic hazards, on other hand, exist only in areas where rocks deep in the subsurface are able to melt and form magma. The generation of magma then is largely restricted to divergent boundaries and those convergent boundaries involving subduction. The reason volcanic hazards are ranked higher along convergent settings in Table 4.2 is because subduction zones typically generate magmas that cause more explosive and violent eruptions.

Natural Resources

Earth naturally provides water and food needed for human survival, but it also provides mineral and energy resources that make modern societies possible. As pointed out in Chapter 1, water is absolutely essential for human bodies to function and also for the production of our crops (i.e., food). Plate tectonics then is tied to both our water and food supplies in part because it influences Earth's climate and precipitation patterns as described above. In many dry regions of the world people depend almost entirely on rivers whose water comes from melting snow. This snow, in turn, accumulates in mountains that had been uplifted far above sea level by tectonic forces. For example, millions of people who depend on such meltwater live in the desert regions below the Himalayas of Asia, the Andes of South America, and the Rockies of North America. Plate tectonics also influences the formation and fertility of the soils in which we grow our crops.

In addition to water and food, modern societies depend on vast array of mineral and energy resources in order to supply the goods and services we enjoy. As you will learn in Chapters 12 and 13, mineral and energy deposits are not randomly located around the world, but rather form under very specific geologic conditions related to the rock cycle and plate tectonics. Consequently, some countries have the good fortune to be located in an area whose tectonic history has left them with valuable mineral and energy deposits. For example, Saudi Arabia is quite wealthy because of its enormous oil deposits, whereas many nearby African countries remain desperately poor in part due to a lack of mineral or energy resources.

Climate

Earth's climate naturally depends on the amount of solar radiation the planet receives from the Sun, but is also governed by a series of complex interactions between the atmosphere, hydrosphere, biosphere, and solid earth. We will look at Earth's climate in more detail in Chapter 16, but for now let us consider how plate tectonics can affect the climate over the course of geologic time. For example, the major ocean currents shown in Figure 4.29 transport vast quantities of both heat and water, thus play a significant role in regulating Earth's climate. As the positions of the continents change over geologic time due to continental drift, so too does the pattern of ocean circulation and Earth's climate. The shifting of tectonic plates also produces large mountain ranges (e.g., Himalayas and Andes) that have a significant impact on precipitation patterns around the globe.

In addition to atmospheric circulation patterns, Earth's climate is influenced by the rates of volcanic gas and ash that are emitted into the atmosphere. For example, volcanic ash has a cooling effect on the climate because airborne particles reduce the amount of solar radiation that can strike Earth's surface. Sulfur dioxide (SO_2) gas has a similar effect, and therefore contributes to global cooling. In contrast, the carbon dioxide (CO_2) emitted during volcanic eruptions causes the planet to warm as this gas traps some of the heat that is radiating out into space. Therefore, varying rates of volcanic activity can have complex and long-term effects on Earth's climate system. Varying rates of mountain-building activity can have a similar effect. As tectonic uplift increases, so too does erosion and the subsequent formation of limestone ($CaCO_3$) in the rock cycle (Chapter 3). The deposition of limestone is important because it removes carbon dioxide (CO_2) from the atmosphere, which traps heat radiating into space.

Development of Life

There is little doubt that one of the most important aspects of plate tectonics for humans is the crucial role it has played in the evolution of life on our planet. As mentioned in Chapter 2, many scientists now believe life may have originated around volcanic vents on the seafloor early in Earth's history. One of the reasons for this is that modern extremophile bacteria are found thriving in the hot gases and fluids discharging from seafloor vents along spreading centers. Whatever the exact origin, more complex life is known to have evolved and colonized the landmasses.

Once terrestrial (land) life was established, the combined effects of continental growth, continental drift, and evolutionary processes led to the great diversity and abundance of life-forms we see today. As you have learned in this chapter, subduction allows continents to grow at the expense of oceanic lithosphere, and mountain building creates various landforms at different elevations. The increase in land area over time then allowed terrestrial populations to increase, and the more varied landscape led to greater species diversification. Diversification was increased further when the supercontinent of Pangaea began to break up around 225 million years ago. As the fragmented landmasses began to drift apart, the terrestrial life-forms they contained were allowed to evolve in isolation, leading to even greater diversification (Case Study 4.1). The increased abundance and diversity of life resulting from plate tectonics proved to be of considerable importance when life had to reestablish itself after periodic mass-extinction events (Chapter 2). In the end, one can safely say that if it were not for plate tectonics, life as we know it would not exist.

FIGURE 4.29 The shifting position of continents affects Earth's climate by altering the circulation of heat and water in the oceans. Climate is also affected by the uplift of mountains and by the rates at which heat-trapping gases are released by volcanic activity and absorbed in the rock cycle.

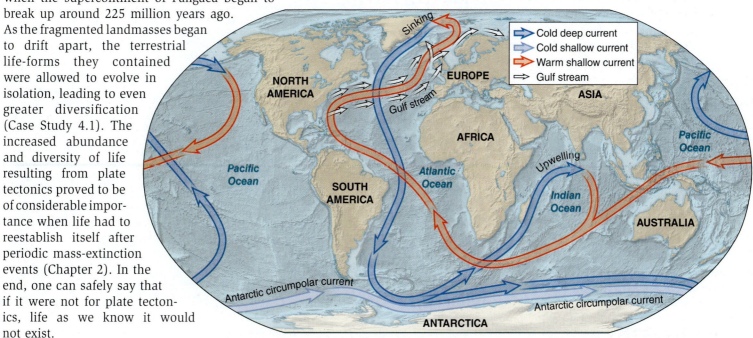

Biogeography and Plate Tectonics

In the mid-1800s a naturalist named Alfred Wallace spent considerable time observing and collecting various plants and animals on islands throughout Southeast Asia. As he sailed along the chain of volcanic islands known today as Indonesia (Figure B4.1), Wallace was struck by the sudden and dramatic change in the types of birds between two islands that were a mere 20 miles (30 km) apart. Moreover, he noted that plants and animals on the islands to the west of this break were clearly related to one another and shared similar characteristics to species found on the continent of Asia. Sailing in the opposite direction toward the east, Wallace found a similar pattern of related species on different islands, but here they all shared common characteristics with those living on the Australian continent. It was as if the two halves of this island chain had life-forms that had come from opposite sides of the world.

In 1859 Wallace published his maps showing *different biogeographic regions,* each of which defined a region where plants and animals were related, but distinctly different from those in other regions. Interestingly, his work was similar to that of Charles Darwin, with both scientists coming to the same, but independent, conclusions regarding evolution and natural selection. Darwin's work, however, received more recognition in part because he was better known and also because he published his work before Wallace. In honor of Wallace's contribution, the sharp boundary he mapped in Indonesia is still referred to as the *Wallace Line.* Modern biogeographers have found that the Wallace Line is more of a zone consisting of several different lines, with each line representing the biogeographic boundary of more specific groups of plants and animals.

Another interesting aspect is that during Wallace's life he could not provide a biological explanation for the differences he observed, but rather suggested that geologic processes were somehow involved. As the study of geology advanced, it became apparent that many of the islands in the region were once joined by land bridges that developed during glacial periods when sea levels were much lower. Once the landmasses were interconnected, various species would then be forced to compete with each other; in other words, they evolved together. Eventually sea level would rise and submerge the land bridges, allowing species on the various islands to once again evolve under conditions of geographic isolation. This, of course, nicely explains why many islands in Southeast Asia today contain different species that share common ancestors. However, it does not account for the sharp break at the Wallace Line where species on either side do not have such a common ancestry.

The key to the Wallace Line remained a mystery until after the development of plate tectonics in the 1960s. Scientists could now show that the Wallace Line coincides with a deep ocean trench associated with a subduction zone. Here the Asian and Australian continents are slowly getting closer together along a convergent plate boundary. Most critical though was the fact that regardless of how low sea level might fall, land bridges between islands could never have formed across the deep trench. The trench then permanently isolated the plant and animal communities on either side, forcing them to evolve separately. The Asian and Australian continents may be getting closer together, but the ocean trench remains an effective migration barrier.

Perhaps the most familiar example of animals forced to evolve in isolation across this ocean trench are the mammals known as *placental and marsupial.* The offspring of placental mammals develop internally (e.g., humans), whereas marsupial young develop externally in a pouch (e.g., kangaroos). It turns out that marsupials are the dominant type of mammal to the east of the Wallace Line, particularly on the Australian continent. On the other hand, placental mammals are dominant west of the Wallace Line, which includes the Asian, African, and American continents.

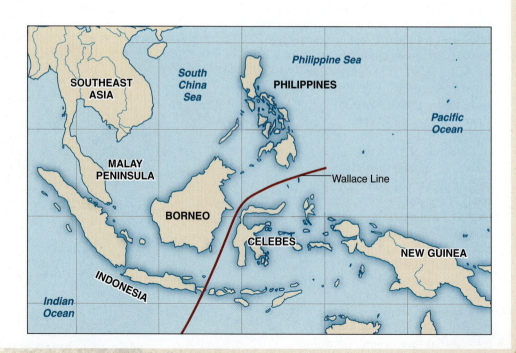

FIGURE B4.1 The Wallace Line in Southeast Asia represents the boundary between distinctly different groups of plant and animal species. This boundary coincides with a deep ocean trench that forced species to evolve in isolation from one another on opposite sides of the trench.

SUMMARY POINTS

1. Rocks deform elastically when they are subjected to tension, compression, or shear forces. If they deform beyond their elastic limit, they either fail by fracturing (i.e., are brittle) or they flow plastically (i.e., are ductile).

2. With increasing depth, rocks are exposed to greater levels of temperature and pressure, making them more ductile and prone to plastic deformation.

3. Earth's internal structure is known through the study of how earthquake (seismic) waves refract and reflect at boundaries between materials with different properties. The planet has a layered structure consisting of the crust, mantle, outer core, and inner core. Density of these layers increases downward, reflecting an increase in iron content.

4. Oceanic crust is relatively thin and composed of basalt whereas continental crust is thicker and composed of granitic rock. The crust and upper mantle act as a single rigid layer and together are called the lithosphere. The lithosphere is broken up into rigid plates that move over the weak, semimolten layer called the asthenosphere.

5. Decay of radioactive elements within the Earth generates heat and helps create a large temperature difference between the core and the crust. This sets in motion large convection cells that transport both heat and plastic mantle material toward the surface.

6. The rising and sinking motion of convection cells explains how Earth obtained its layered structure and movement of its lithospheric plates. Spreading centers develop along rising convection cells, whereas subduction zones correspond to areas where cells descend down into the mantle.

7. Seafloor and seismic studies have confirmed that new oceanic crust forms at mid-oceanic ridges and is eventually destroyed along subduction zones and forms continental crust. This explains why the ocean crust is younger than continental crust.

8. The three basic types of plate boundaries are defined by the dominant forces that exist along the boundary: divergent (tension), convergent (compression), and transform (shear).

9. Major surface features develop along plate boundaries and include: ocean ridges, ocean trenches, rift valleys, island arcs, volcanic arcs, and complex mountain belts. The majority of earthquakes and volcanic eruptions also occur along plate boundaries.

10. In addition to geologic processes, plate tectonics plays a central role in the Earth system by affecting the atmosphere, hydrosphere, and biosphere. For humans, plate tectonics is important because it creates hazards (earthquakes, volcanic eruptions, landslides), regulates our climate, distributes natural resources, and was important in the development of life.

KEY WORDS

asthenosphere 96	elastic limit 93	mid-oceanic ridges 99	tectonic plates 96
compression 93	fault 94	ocean trenches 99	tension 93
continental arc 110	geothermal gradient 97	outer core 95	theory of plate tectonics 92
convection cells 97	inner core 95	rift valley 108	transform boundary 105
convergent boundary 105	island arc 110	seafloor spreading 101	
crust 96	lithosphere 96	shear 93	
divergent boundary 105	mantle 95	subduction 102	

APPLICATIONS

Student Activity   Look on a map. Does the eastern coast of South America seem to fit into the western coast of Africa? How about North America? Does it seem to fit into the northwestern coast of Africa?

Critical Thinking Questions
1. What are the forces that deform rocks?
2. How have scientists determined the interior structure of the Earth?
3. What are the major types of plate boundaries and what features develop?
4. What was wrong with the theory of continental drift?

Your Environment: YOU Decide   Will California ever fall into the Pacific Ocean? Or will something else happen? Explain.

Chapter 5

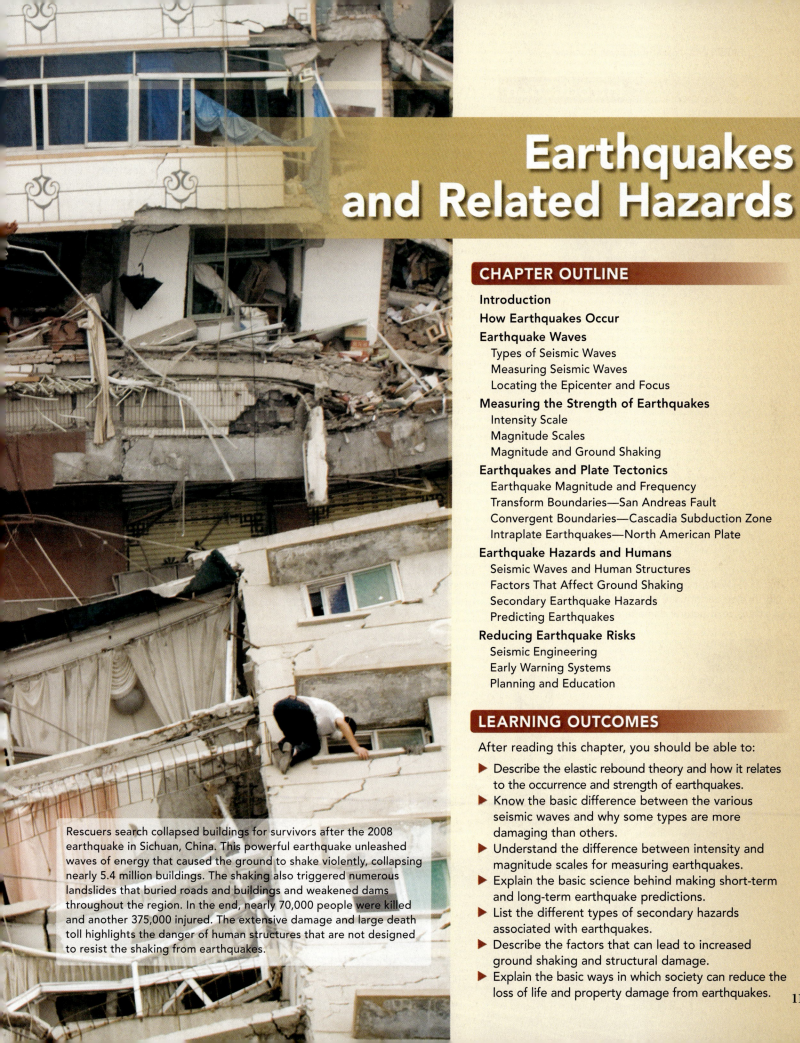

Earthquakes and Related Hazards

Rescuers search collapsed buildings for survivors after the 2008 earthquake in Sichuan, China. This powerful earthquake unleashed waves of energy that caused the ground to shake violently, collapsing nearly 5.4 million buildings. The shaking also triggered numerous landslides that buried roads and buildings and weakened dams throughout the region. In the end, nearly 70,000 people were killed and another 375,000 injured. The extensive damage and large death toll highlights the danger of human structures that are not designed to resist the shaking from earthquakes.

LEARNING OUTCOMES

After reading this chapter, you should be able to:

▶ Describe the elastic rebound theory and how it relates to the occurrence and strength of earthquakes.

▶ Know the basic difference between the various seismic waves and why some types are more damaging than others.

▶ Understand the difference between intensity and magnitude scales for measuring earthquakes.

▶ Explain the basic science behind making short-term and long-term earthquake predictions.

▶ List the different types of secondary hazards associated with earthquakes.

▶ Describe the factors that can lead to increased ground shaking and structural damage.

▶ Explain the basic ways in which society can reduce the loss of life and property damage from earthquakes.

Introduction

At around 4:00 a.m. in the morning of July 28, 1976, Earth's crust suddenly shifted beneath the city of Tangshan, China. The resulting earthquake unleashed waves of vibrational energy, slamming residents into the ceiling of their homes, who were then quickly entombed as the buildings around them collapsed. When it was over, more than 90% of the homes and 75% of commercial buildings in Tangshan had been destroyed. Chinese authorities placed the death toll at 250,000, whereas Western officials estimated 650,000 people were killed in the disaster. In contrast, in 1994 an earthquake of slightly smaller magnitude struck around 4:30 a.m. in densely populated Los Angeles, California, near the town of Northridge, causing extensive damage and 61 deaths (Figure 5.1). Although damage from the Northridge earthquake was considerable, it paled in comparison to the complete devastation in Tangshan. Since both earthquakes released about the same amount of energy and occurred in populated areas during the early morning hours, an obvious question is why did these two events have such vastly different outcomes?

The reason so many more people were killed in the Tangshan earthquake compared to Northridge is largely related to the geologic setting of the two cities. Here the geologic setting affected their level of preparation for such a disaster. Los Angeles (Northridge) happens to be located along a transform plate boundary (Chapter 4) where powerful earthquakes are fairly common. Tangshan, however, is far from a plate boundary and earthquakes are rare. This difference in tectonic setting and corresponding earthquake frequency directly affects the way people prepare for such disasters. Recall from Chapter 1 how humans generally take greater steps to minimize the effects of a natural hazard in areas where the frequency is high and consequences more severe. Because the earthquake risk is high, the State of California requires that buildings and other structures in Los Angeles be constructed in such a way as to minimize the chance of collapse during an earthquake. In contrast, similar earthquake building codes were not required in Tangshan primarily because earthquakes were uncommon. Therefore, it was Tangshan's tectonic setting, and corresponding lower earthquake frequency, that led to the lack of preparedness and enormous death toll.

FIGURE 5.1   Violent ground shaking associated with the 1994 Northridge earthquake near Los Angeles, California, caused extensive structural damage and killed 61 people.

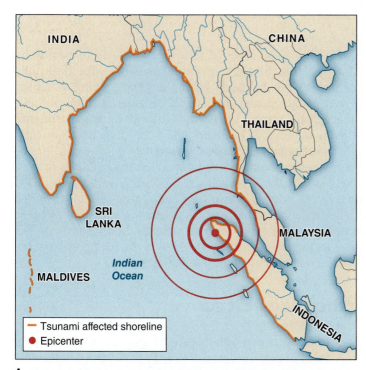

FIGURE 5.2   The map in (A) shows the shorelines around the Indian Ocean that were impacted by the 2004 tsunami that originated from a subduction zone earthquake off the coast of Indonesia. The massive ocean waves swept through low-lying coastal areas, killing an estimated 230,000 people. Before and after photos (B) of the once thriving community of Meulaboh, Indonesia, illustrating how entire communities were literally swept away by the tsunami. Note that concrete foundations mark the remains of buildings.

Although the number of deaths from the 1976 Tangshan earthquake was exceptionally high for a single event, the cumulative worldwide death toll from earthquakes each year is generally quite large. This is partly due to the fact that, in addition to the collapse of buildings, earthquakes create many secondary hazards that also claim lives and damage property. No disaster in modern times illustrates the problem of secondary hazards better than the 2004 subduction zone earthquake along the coast of Indonesia (Figure 5.2). This massive earthquake resulted in the deaths of approximately 230,000 people, the majority of whom were literally swept away by a series of ocean waves known as a *tsunami*. During this subduction zone earthquake the tectonic plates shifted suddenly, displacing a large volume of seawater. The resulting ocean waves then traveled outward, taking the lives of tens of thousands of people on shorelines so far away that the vibrations from the earthquake itself could not be felt. Similar to the Tangshan example, people living along the Indian Ocean were caught unprepared largely because tsunamis are relatively rare in that region. Around the Pacific Ocean basin where tsunamis are more common, millions of residents enjoy the benefits of an early warning system that has been in place for decades.

Regardless of whether the danger is from collapsing buildings or massive ocean waves, earthquakes are a sober reminder that we are often at the mercy of natural forces over which we have little or no control. However, we can use science to better understand the nature of earthquake forces so that we can develop ways of minimizing the hazards. In Chapter 5 we will first explore how earthquakes occur, followed by a discussion on how science can be used to reduce the risk from earthquakes.

How Earthquakes Occur

Prior to modern science, many people believed earthquakes were random events; some even thought they were punishment by gods for evil or immoral behavior. By gathering data and testing various hypotheses, scientists have developed a theory based on the elastic properties of rocks that nicely explains how earthquakes occur. Recall from Chapter 4 that when rocks are placed under a force, also called *stress,* they can become deformed and change their shape or volume, a process known as *strain.* Interestingly, rocks are also considered to be *elastic,* meaning that if the force (stress) is removed they will return to their original shape. Similar to a rubber band, rocks then are capable of releasing the strain they have accumulated and returning to an undeformed state (i.e., zero strain). All elastic materials have what is known as an **elastic limit,** which is the maximum amount of strain they can accumulate before either fracturing or undergoing plastic deformation. Moreover, when *brittle* materials reach their elastic limit they undergo permanent deformation by fracturing, whereas *ductile* materials deform by flowing plastically.

This somewhat complex process of how rocks accumulate strain and deform can more easily be understood if we consider the example in Figure 5.3. Here the wooden rod represents more brittle rocks in the lithosphere, and the iron rod is analogous to ductile rocks that are more deeply buried. Notice that when a force is applied to the rods they both bend and accumulate strain. If the force does not exceed their elastic limits, then this strain can be released once the force is removed, allowing the rods to return to their original shape. However, if the elastic limit of the iron (ductile) rod is exceeded, it will flow plastically and become permanently deformed (i.e., bent). On the other hand, should the force go beyond the elastic limit of the wooden (brittle) rod, the rod will suddenly fracture and break. At the moment the rod fractures, the strain it had accumulated is suddenly released and is transformed into vibrational wave energy. This energy then quickly travels away from the fracture and moves down the two broken pieces. You can experience this yourself by bending a dried tree branch until it snaps, at which

FIGURE 5.3 Both iron and wooden rods will deform and return to their original shape as long as their elastic limit is not exceeded. When an iron rod exceeds its elastic limit it deforms permanently by bending. When a wooden rod exceeds its limit it will suddenly break by fracturing, releasing energy in the form of vibrational waves. Note that when the fractured rod breaks the separate pieces rebound and become straight again.

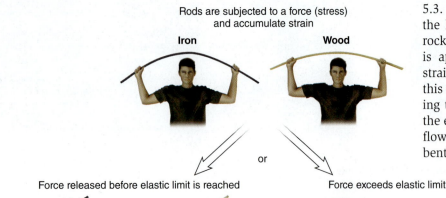

Rods are subjected to a force (stress) and accumulate strain

Iron   Wood

Force released before elastic limit is reached   or   Force exceeds elastic limit

Accumulated strain is released   Iron rod begins to deform plastically and continues to accumulate strain   Wooden rod breaks and accumulated strain is suddenly released as vibrational wave energy

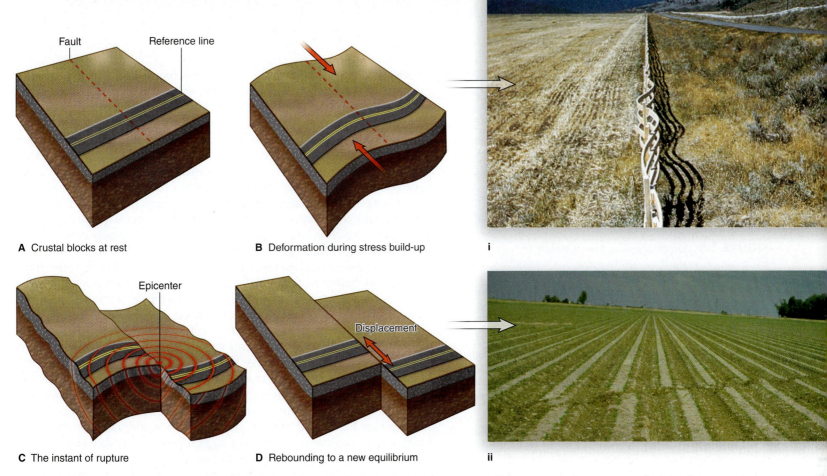

A Crustal blocks at rest   **B** Deformation during stress build-up   **i**

C The instant of rupture   **D** Rebounding to a new equilibrium   **ii**

FIGURE 5.4 When a rock body accumulates strain (A, B) and reaches its elastic limit, it will fail at its weakest point, called the focus. As the strain is suddenly released (C), waves of vibrational energy begin radiating outward in all directions from the focus, causing the ground shaking known as an earthquake. After the earthquake, the rock body becomes displaced (D) on opposite sides of the fault, but is no longer deformed because the strain has been released. (i) Buckled fence is evidence that the underlying rocks are accumulating strain. (ii) Displacement of rows in a farm field along a fault is evidence of a recent earthquake and that strain has been released.

point you should feel a vibration in your hands as the energy waves travel down the wood.

Based on the relationship between stress and strain and the deformation of rocks, earth scientists have developed the **elastic rebound theory** that explains the occurrence of earthquakes. As illustrated in Figure 5.4, this theory holds that earthquakes originate when a force (stress) acts on a rock body, causing it to deform and accumulate strain. Eventually the rock reaches its elastic limit, at which point it *ruptures* or fails suddenly, releasing the strain it had accumulated. This sudden release of strain, lasting anywhere from several seconds to a few minutes, is transformed into vibrational wave energy that radiates outward and causes the ground to shake in what is called an **earthquake.** In addition, the release of energy generally begins at a point called the **focus,** typically located where the rock is the weakest, such as along a preexisting fracture plane. While the strain is being released, rocks on either side of the fracture will begin moving in opposite directions, creating what geologists refer to as *displacement*. Note that when rocks are displaced along a fracture, the fracture technically becomes a *fault* (Chapter 4). Also notice in Figure 5.4 that once the strain is released, the displaced rock body is no longer deformed after the earthquake. It is worth pointing out that when rocks return to their undeformed state, scientists refer to it as *rebound*, hence the name elastic rebound theory.

The elastic rebound theory not only explains how earthquakes form, but it can account for why some areas experience repeated earthquakes. Recall from our discussion on plate tectonics (Chapter 4) that Earth's tectonic forces operate on a large scale, and operate over long periods of geologic time. This means that when a rock body ruptures, the strain may

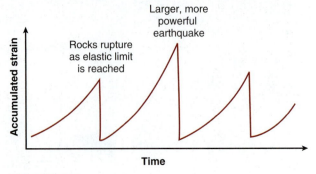

FIGURE 5.5 Graph showing how a steady tectonic force can cause earthquakes to occur in a repetitive manner. Here a rock body accumulates strain until its elastic limit is reached. The instant the rock ruptures, the strain is released as vibrational wave energy, which produces ground shaking. Note that more powerful earthquakes occur when greater amounts of strain are allowed to accumulate before the rupturing event.

be released, but the tectonic forces are still operating. The result is the cyclic pattern shown in Figure 5.5. After each rupturing event, or earthquake, the strain slowly rebuilds until the elastic limit of the rock body is reached, thereby producing repeated earthquakes. Because faults and fractures represent weak zones within a rock body, rupturing generally takes place where the strain becomes greater than the frictional resistance along a particular fault or fracture. Since the rocks in tectonically active areas usually contain numerous faults, the sudden release of strain along one fault can alter the distribution of strain on the other faults. This redistribution of strain commonly produces a series of smaller earthquakes called *aftershocks,* which may continue to occur for days or weeks after the primary earthquake, sometimes called the *main shock.*

Finally, based on the elastic rebound theory, we know that the more strain a rock body can accumulate, the more energy it will release when it finally ruptures. More strain energy will, of course, translate into more vibrational energy, hence larger earthquakes. The key factors that determine the size of an earthquake are the frictional resistance along fault planes and the elastic properties of the rock body. For example, imagine we have two faults that are under the same tectonic force. One fault offers little frictional resistance and is able to slip fairly easily, whereas the other one offers a great deal of resistance. In the first case the rock body will need to accumulate a relatively small amount of strain before the fault slips and releases the strain, thereby generating an earthquake. On the other hand, a great deal more strain must accumulate in order to cause slippage along a fault with more frictional resistance. When this fault eventually slips, the additional strain energy will result in a larger earthquake.

In addition to faults, we also need to consider the elastic property of the rock itself. Recall that when rocks become more ductile (less brittle) they tend to accumulate less strain, and instead undergo plastic deformation (Chapter 4). This is important because for an earthquake to take place a rock body must be rigid enough to accumulate strain. It turns out that earthquakes do not occur deeper than 435 miles (700 km) below the surface because the higher temperatures cause the rocks to become so ductile that they deform only by plastic flow, hence do not rupture.

Earthquake Waves

Today scientists understand that most natural earthquakes are caused by the buildup of strain associated with tectonic forces, in which case they are often referred to as *tectonic earthquakes.* However, there are also many *magmatic earthquakes* that form when magma forces its way up through crustal rocks (Chapter 6). We should also point out that not all earthquakes result from the slow buildup of strain, but rather from the sudden transfer of energy in events such as landslides, large volcanic explosions, and meteor impacts. Nevertheless, the vast majority of earthquakes are tectonic or magmatic in nature, which is why most earthquakes are found along plate boundaries as described in Chapter 4.

The fact that most earthquakes coincide with plate boundaries also means that vast numbers of people in the world will never experience a natural earthquake. However, most everyone has felt the ground shake from small *artificial* earthquakes whereby some human-derived energy source generates vibrational waves that travel through the ground—explosions, passing trains, and heavy construction equipment all generate such waves. To help illustrate this point, imagine you were standing outside when a demolition company brought down a large building using small explosive devices to weaken the structure. As the building collapsed you

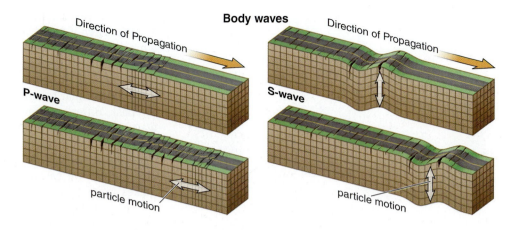

Body waves

P-wave

S-wave

Surface waves

Rayleigh wave

Love wave

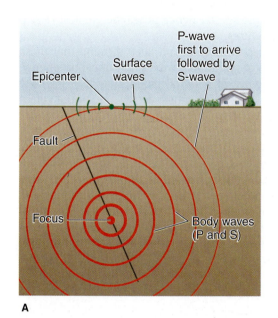

A

FIGURE 5.6   Body waves travel throughout the solid earth (A) and then generate surface waves upon reaching the land surface. Note how the orientation of the particle motion varies relative to the direction the different wave types are traveling. P-waves are the least damaging and travel the fastest, thus are the first to arrive at any given location. Most of the damage to human structures results from the side-to-side and rolling motion of surface waves as indicated by the photograph in (B).

would first hear a great deal of noise, and then very quickly begin to feel a rumbling sensation in the ground. What happens is that as the building falls it generates kinetic energy (objects in motion), which is then transferred to the ground when the debris makes impact. The energy then travels outward in all directions in the form of vibrational waves, eventually reaching the ground where you are standing.

Regardless of their origin, we will refer to vibrational waves that travel through solid earth materials as **seismic waves,** as opposed to the less scientific term of earthquake waves. From Figure 5.6 you can see that in the case of tectonic or magmatic earthquakes, seismic waves originate at the point of rupture called the *focus*—also called *hypocenter.* Note how the waves travel outward from the focus in all directions, similar to how exploding fireworks expand outward as a sphere. A key point here is that as seismic waves travel outward, the energy they contain is spread out over an ever-increasing volume of rock, thus their energy steadily decreases away from the focus. Consider what would happen if there were two earthquakes that released the same amount of strain energy, but were located at different depths. Because wave energy steadily decreases away from the focus, we could expect that the earthquake with the more shallow focus would be more destructive than the one with a deeper focus.

Finally, note that the **epicenter** is the point on the surface that lies directly above the focus, which means it is also the closest point to where the strain energy was released (Figure 5.6). The epicenter then is not only where seismic waves *first* reach the surface, but more importantly it is

B

where the waves contain the *most* energy—similar to being the closest to a bomb when it explodes. This is why the epicenter is commonly the place where the ground shaking is most severe, hence poses the greatest risk to humans and human-made structures.

As we explore seismic waves in more detail, keep in mind that these waves are similar to other types of vibrational waves that transport energy, the basic difference being the types of material (solid, liquid, or gas) the waves must pass through. For example, the music we listen to or the voices we hear both involve wave energy traveling through air, which forces gas molecules to vibrate as the wave passes. Interestingly, when a stone falls into a pond we can actually *see* this process taking place. As the stone makes impact, its kinetic energy is quickly transferred to the water, at which point the energy begins to travel outward in all directions as waves that we can see. Here the passing energy wave is causing water molecules to move (vibrate) instead of gas molecules or rock particles. Similar to seismic waves, water and sound waves also begin at some localized point (i.e., focus) where an energy transfer takes place. The waves then travel outward in all directions, slowly losing energy due to the friction created by the vibrating particles.

Types of Seismic Waves

During the brief period when a rock body is rupturing and being displaced along a fault, the seismic waves leaving the focus are called **body waves.** These waves then travel throughout Earth's interior (i.e., body) as illustrated in Figure 5.6. However, when body waves reach the surface the energy is not transferred to the atmosphere, but instead begins to move along the land surface in what are called **surface waves.** Of considerable interest in environmental geology is the fact that most of the damage in earthquakes is caused by surface waves. Therefore, in this section we want to briefly examine what it is about seismic waves that makes some of them more dangerous to people and their structures.

As with other forms of wave energy, body and surface waves can be classified based on the direction that particles within rocks are forced to vibrate relative to the direction the wave is traveling, or propagating. Basically, particles either vibrate in the same direction the wave is traveling or they vibrate at right angles. For this reason there are two types of body waves shown in Figure 5.6. In **primary (P) waves** the particles vibrate in the same direction the wave is traveling, causing rocks to alternately compress and decompress as successive waves pass through. On the other hand, particles in **secondary (S) waves** vibrate perpendicular to the wave path, which creates a shearing (side-to-side) motion. Although primary and secondary waves travel along the same path, the compressional nature of the P-waves allows them to travel faster, hence they are the first to arrive at any given point. Later in this chapter you will see how the early arrival of P-waves provides the basis for earthquake early-warning systems.

Similar to body waves, there are two basic types of surface waves, but here the motion is more complicated because of the way the waves interact with the ground surface. As indicated in Figure 5.6, the particles in *Rayleigh waves* move back and forth in the direction of the wave, *plus* they move up and down, creating a rolling motion similar to ocean waves. In *Love waves* the ground moves back and forth and side to side at the same time. In essence then, both types of surface waves force the ground to move in two different directions at the same time, something buildings are not normally designed to handle. Consequently, surface waves cause far greater damage to human structures than do body waves. Later we will explore how engineers use this knowledge of how seismic waves vibrate in order to minimize the amount of structural damage in an earthquake.

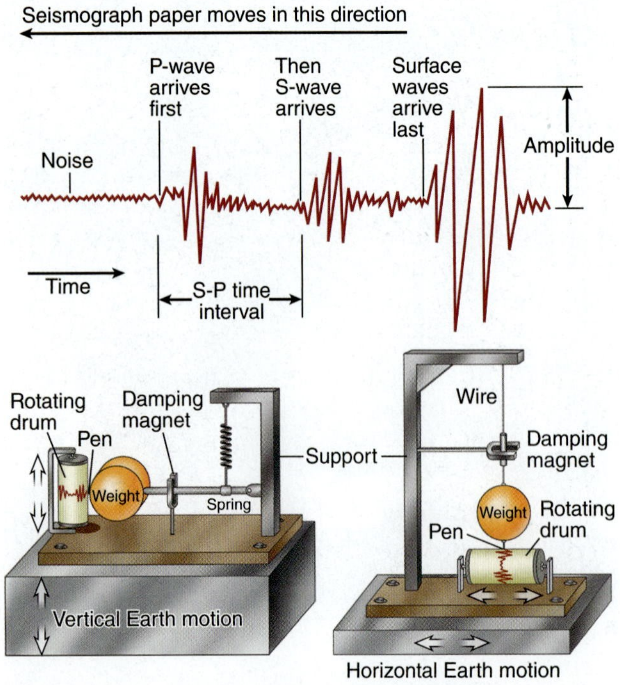

FIGURE 5.7 Seismographs operate on the principle of inertia, where a pen attached to a weight remains stationary whereas a rotating drum moves as the ground vibrates. The ground movements are recorded on the paper attached to the drum.

Measuring Seismic Waves

When body and surface waves pass through an area, everything that is firmly attached to the ground is going to be forced to move in the same direction as the waves are vibrating. However, loose objects like file cabinets and suspended light fixtures tend to remain stationary. This seemingly odd behavior is the result of **inertia,** which is the tendency of objects at rest to stay at rest unless acted upon by some force; objects in motion also tend to stay in motion because of inertia. For example, imagine you are standing on a rug when someone literally pulls the rug out from under you. Because you are just standing (i.e., at rest), you will not move with the rug, but rather will fall straight to the ground. Inertia is important because it helps explain much of the structural damage that occurs in earthquakes.

Inertia is also important because it forms the basis for instruments called **seismographs,** which measure the ground motion during earthquakes. The first known seismograph was built in China in 132 BC, but it was not until 1889 that a seismograph was made that could actually make a record of the ground movement—such a record is called a *seismogram.* As illustrated in Figure 5.7, early seismographs consisted of a pen attached to a suspended weight (mass) that sat over a rotating drum wrapped with paper, all of which was anchored to solid ground. In-between earthquakes the pen simply traces a rough line on the rotating paper. But when seismic waves pass through, the pen and attached weight remain stationary due to inertia, whereas the underlying paper and rotating drum vibrate or shake. The result then is a tracing on the seismogram that records the arrival of different seismic waves over time. You can simulate how a seismograph works by placing a pencil on a piece of paper, and then slowly pull the paper in one direction while keeping the pencil steady. Try it again, but this time move the paper from side to side while pulling the paper past the pencil.

From Figure 5.7 one can see that seismographs can be configured to record different directions of ground motion (e.g., vertical and horizontal). Also, the seismogram is best viewed when the paper is removed from the drum and laid flat. Notice that the seismic waves from an earthquake arrive at a station in the following order: P (primary), S (secondary), and then surface waves. Also, note how the trace of the surface waves has the largest amplitude (height) and body waves the smallest. This means that surface waves create the greatest ground motion and body waves the least, which helps explain why surface waves cause the most structural damage. Finally, note that modern electronics have made mechanical seismographs obsolete. Instead of having drums being anchored to the ground, the movement of the pen is now based on signals from a buried electronic sensor. The signal can also be stored electronically and displayed on a computer screen as opposed to paper.

Locating the Epicenter and Focus

As described earlier, P- and S-waves are generated the instant a rock body begins to rupture. This means that the first set of P- and S-waves will leave the focus at the same time. But since P-waves travel faster, the S-wave will lag progressively farther behind as the waves increase their distance from the focus—similar to how two cars traveling at different speeds continue to get farther apart. Scientists have devised a means of locating the epicenter based on the time interval, or lag time, between the first set of P- and S-waves to arrive at different seismograph stations. As indicated in Figure 5.8, the distance from each station to the epicenter is determined based on their respective lag times. However, because the distance data do not indicate in which direction the epicenter is located, a minimum of three stations is required. By plotting the distances as circles around each station, the epicenter is the point where the circles all intersect.

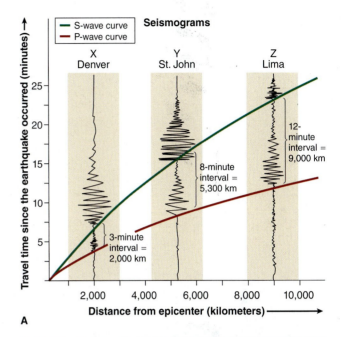

A

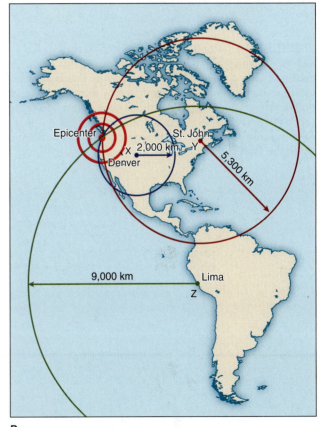

B

FIGURE 5.8   Locating the distance from a seismograph station to the epicenter is found from the difference in arrival time between P- and S-waves from at least three stations. The distances from each station are then plotted as circles, with the epicenter lying where the circles all intersect.

Scientists have also developed methods for locating the actual focus within the subsurface. Although these methods are rather complex, the basic approach involves using multiple seismic stations to compare the time it takes P-waves to travel along different paths in the subsurface. By knowing the location of the focus from multiple earthquakes, scientists can construct three-dimensional views of the fault zones where earthquakes are taking place. For example, Figure 5.9 shows a map and cross-sectional views from a study of the 1989 Loma Prieta earthquake, which occurred just south of San Francisco. From the map view we can see that the epicenters of the main earthquake and its numerous aftershocks are concentrated along the trace of San Andreas fault zone. The cross-sectional views allow us to see how the individual focal points are distributed throughout the subsurface. Of particular interest is the plot (Figure 5.9C) that cuts across the fault zone at right angles. This view clearly shows how the fault plane is inclined at a steep angle within the subsurface.

By using seismic data to locate focal points and epicenters of earthquakes, geologists have a powerful tool for studying Earth's interior. In fact, it was through the analysis of seismic data that scientists were able to define the boundaries of tectonic plates (Chapter 4). Today scientists can even document the position of lithospheric slabs as they descend into subduction zones. From a hazard perspective, seismic studies are important as they help scientists understand how and where rock strain is being relieved along fault zones. This information is very useful in making long-term earthquake predictions.

FIGURE 5.9 Plotting the epicenters of an earthquake and its aftershocks on a map allows geologists to determine the surface trace of a fault zone. The fault plane itself can be defined in the subsurface by plotting the focal points along different cross-sectional views. Note the cross section in (i) is parallel to the fault trace and (ii) is perpendicular. Here each focal point represents the point along the fault zone where rock strain had been relieved.

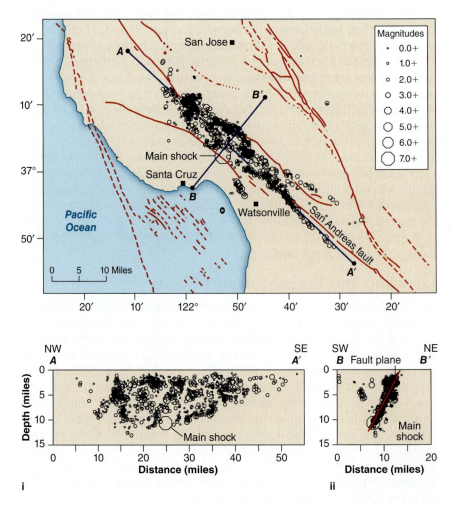

Measuring the Strength of Earthquakes

For thousands of years people have been both fearful and fascinated by earthquakes due to their enormous destructive power and mysterious origin. Prior to the development of the modern seismograph, scientists had no means of quantifying this energy by direct measurements. All that was available were people's written accounts of the damage they observed and sensations they felt during an earthquake. In this section we will explore the two basic methods scientists use to measure the destructive power of earthquakes.

Intensity Scale

In 1902 an Italian seismologist named Giuseppe Mercalli developed a means of comparing both modern and historical earthquakes through the use of firsthand human observations during earthquakes. He created what is known as the **Mercalli intensity scale,** whereby earthquakes are ranked based on a set of observations most humans could report objectively, particularly the type of damage sustained by buildings. A modified version of Mercalli's rankings and standardized observations is listed in Table 5.1. Note how the intensity scale ranks earthquakes from I to XII, with XII representing total destruction.

The way in which the intensity scale is employed for any recent earthquake is basically the same as taking a survey. Immediately after an earthquake, people throughout the region are asked to read the list of observations from the scale (Table 5.1), and then pick the classification which best fits their experience. The individual rankings and locations are then plotted on a map and contoured such that similar rankings are grouped together, as shown in Figure 5.10. Although the intensity scale is a *qualitative* measure

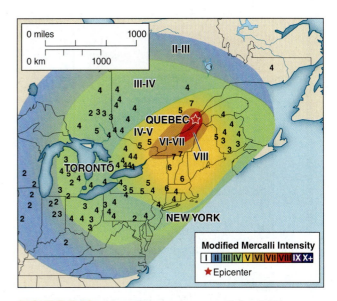

FIGURE 5.10   Mercalli intensity map of the 1925 Charlevoix-Kamouraska earthquake along the St. Lawrence River, in Quebec, Canada. Note the individual intensity rankings from the original survey.

TABLE 5.1   Modified Mercalli intensity scale for classifying earthquake effects based on human observations.

  I. Not felt except by very few people under especially favorable conditions.

 II. Felt only by a few people at rest, especially on upper floors of buildings. Delicately suspended objects may swing.

III. Felt quite noticeably indoors, especially on upper floors of buildings. Vibration like passing truck.

 IV. Felt indoors by many, outdoors by few. Dishes, windows, and doors disturbed; walls make creaking sound. Sensation like heavy truck striking building.

  V. Felt by nearly everyone. Some dishes, windows, and so forth, broken; a few instances of cracked plaster; unstable objects overturned.

 VI. Felt by all; many frightened and run outdoors. Some heavy furniture moved; a few instances of fallen plaster or damaged chimneys. Damage slight.

VII. Everybody runs outdoors. Damage negligible in buildings of good design and construction; slight to moderate in well-built ordinary structures; considerable in poorly built or badly designed structures. Some chimneys broken.

VIII. Damage slight in specially designed structures; considerable in ordinary buildings, great in poorly built structures. Fall of chimneys, factory stacks, columns, monuments, walls. Heavy furniture overturned.

 IX. Damage considerable in specially designed structures. Buildings shifted off foundations. Ground cracked conspicuously. Underground pipes broken.

  X. Some well-built wooden structures destroyed; most masonry and frame structures destroyed. Ground badly cracked and rails bent. Landslides considerable along river banks and steep slopes.

 XI. Few masonry structures remain standing. Bridges destroyed. Broad fissures in ground. Underground pipelines completely out of service. Rails bent greatly.

XII. Damage total. Waves seen on ground surfaces. Objects thrown upward into the air.

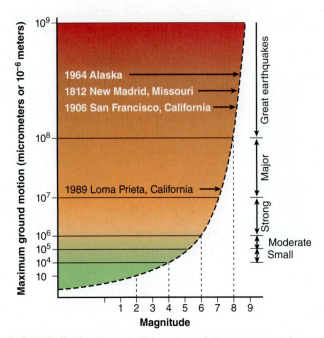

FIGURE 5.11 Graphic illustration of the exponential nature of the Richter magnitude scale where each increase represents a 10-fold increase in ground shaking. Here a magnitude 8.0 quake has 10 times greater ground motion than a 7.0 quake, 100 times greater than a 6.0 quake, and 1,000 times greater than a 5.0 quake.

since it is based on subjective human observations, it still provides a useful means of comparing the strength of different earthquakes. Finally, notice in Figure 5.11 how the intensity scale can be used in fairly populated areas to locate the epicenter with considerable accuracy. Prior to modern seismographs, this ability to accurately locate the epicenter represented a significant achievement.

Magnitude Scales

It was not until 1935 that Charles Richter and his colleague, Beno Gutenberg, devised a means of quantifying earthquake energy and ground motion based on the amplitude (height) of the waves recorded by seismographs. Here they analyzed the records of numerous earthquakes, ranking them according to the amplitude of the largest wave in each record. However, the amplitudes could not be directly compared since the distance between the earthquakes and seismic stations were different. Therefore, they applied a correction factor to the data to account for the different amounts of energy that were lost as the waves traveled to the seismic stations. Because their corrected amplitude data varied over such a wide range (i.e., contained very small and very large values), they used a logarithmic scale and called it the *magnitude scale*. Their original magnitude scale, shown in Figure 5.11, is now known as the **Richter magnitude scale.**

As the number of seismograph stations around the world steadily increased, scientists eventually realized that results obtained using the Richter magnitude scale were not always consistent with one another, particularly for large-magnitude earthquakes. The problem basically is that the Richter scale was developed using data which had a somewhat narrow range of epicenter distances, plus all the earthquake data was obtained from the same geologic setting, namely California. Today most scientists favor the **moment magnitude scale,** which is based on similar types of seismogram measurements as Richter's, but is more accurate over a wide range of magnitudes and geologic conditions. Despite the preference for the moment magnitude scale within the scientific community, the Richter scale is still widely used by the media when reporting earthquakes. In Table 5.2 you can compare the Richter and moment magnitudes of a select number of large, historical earthquakes. Note how magnitude itself is not necessarily a good predictor of death toll—compare the 1989 Loma Prieta and the 1995 Kobe quakes for example.

Magnitude and Ground Shaking

Magnitude scales are useful because they quantify the amount of ground motion during an earthquake, and the energy that was released when the rocks ruptured. Interestingly, when the news media reports the magnitude of an earthquake, the general public typically does not appreciate the logarithmic nature of the scale. Take for example how people often think that the difference between a magnitude 5 earthquake and a magnitude 8 is simply 3 on a scale of 1 to 10. However, as can be seen in the exponential graph in Figure 5.11, the Richter scale magnitudes are the logarithms of the values for ground motion found on the vertical axis. In other words, the magnitude numbers are simply the exponents of the ground motion values. Notice on the graph that a magnitude 5 earthquake causes 10^5 (100,000) micrometers (0.1 meter) of ground motion, whereas a magnitude 8 represent 10^8 (100,000,000) micrometers (100 meters) of movement. While there was only an increase of three units on the magnitude scale, the amount of

TABLE 5.2 Richter and moment magnitudes and corresponding death toll for selected earthquakes.

		Richter Magnitude	Moment Magnitude	Estimated Death Toll
1556	Shansi, China	~8	*	830,000
1811–1812	New Madrid, Missouri	~7.3–7.5	~7.8–8.1	*
1857	Fort Tejon, California	~7.6	~7.9	1
1886	Charleston, South Carolina	~6.7	~7.3	60
1906	San Francisco, California	8.3	7.8	3,000
1923	Tokyo, Japan	*	7.9	143,000
1960	Chile, South America	8.5	9.5	5,700
1964	Anchorage, Alaska	8.6	9.2	131
1976	Tangshan, China	7.6	7.5	650,000
1985	Mexico City, Mexico	*	8.0	9,500
1989	Loma Prieta, California	7.0	6.9	63
1994	Northridge, California	6.4	6.7	61
1995	Kobe, Japan	6.8	6.9	5,502
1999	Izmit, Turkey	*	7.6	17,118
1999	Taipei, Taiwan	*	7.6	2,400
2001	Gujarat, India	*	7.7	20,085
2003	Bam, Iran	6.3	6.6	30,000
2004	Sumatra, Indonesia	*	9.1	230,000
2008	Sichuan, China	*	7.9	70,000

Source: Data from U.S. Geological Survey, www.earthquake.usgs.gov.

Data not available.

ground motion increased by a factor of 10^3, or 1,000. Although a unit increase on the magnitude scale represents a 10-fold increase in ground motion, this corresponds to about a 30-fold increase in energy released at the focus—recall that the release of stored elastic energy is what causes the shaking in the first place.

To give you a better appreciation of what this all means, we will compare the two largest earthquakes in the San Francisco area in modern times, namely the 1989 Loma Prieta earthquake and the great earthquake of 1906. With a moment magnitude of 6.9, the 1989 Loma Prieta event was ranked as a strong to major earthquake, whose shock waves caused extensive damage as much as 60 miles (100 km) away from the epicenter. In comparison, the 1906 San Francisco earthquake had a moment magnitude of 7.8 and was considered to be a strong to great earthquake. We can easily calculate how much stronger the ground shaking was during the 1906 quake than in the 1989 event by subtracting the two magnitudes (7.8 – 6.9) to get 0.9. Since magnitude represents the exponent on the value for ground motion, we take 10 and raise it to the 0.9^{th} power ($10^{0.9}$), which comes out to an eight-fold difference. In other words, the ground shaking in the 1906 earthquake was eight times greater than in the 1989 quake. For anyone who experienced the violent shaking of the 1989 Loma Prieta event, being shaken eight times harder would be difficult to imagine.

TABLE 5.3 Differences in ground motion (i.e., shaking) of selected earthquakes as determined by their differences in moment magnitudes.

Earthquakes		Difference in Moment Magnitude	Difference in Ground Motion
1906 San Francisco 1989 Loma Prieta	$M_m = 7.8$ $M_m = 6.9$	$7.8 - 6.9 = 0.9$	$10^{0.9} = 8\times$
2004 Indonesian 1906 San Francisco	$M_m = 9.1$ $M_m = 7.8$	$9.1 - 7.8 = 1.3$	$10^{1.3} = 20\times$
2004 Indonesian 1989 Loma Prieta	$M_m = 9.1$ $M_m = 6.9$	$9.1 - 6.9 = 2.2$	$10^{2.2} = 158\times$

Finally, we can take our example one step further and compare these two earthquakes to the 2004 Indonesian earthquake that was responsible for the tsunami that took nearly 230,000 lives. With a moment magnitude of 9.1, the Indonesian earthquake was one of the most powerful ever recorded. In Table 5.3 you can see that by doing the same type of calculations as before. The ground motion in the Indonesian earthquake was 20 times greater than the 1906 San Francisco quake, and 158 times greater than the Loma Prieta quake! Clearly, the 2004 Indonesian earthquake was a truly massive event that generated a level of ground shaking that is nearly incomprehensible.

Earthquakes and Plate Tectonics

In Chapter 4 you learned that most earthquakes are not random events, but are intimately related to plate tectonics. For example, the maps in Figure 5.12 show that earthquake epicenters in the United States are clustered along the transform boundary in California and along the convergent boundaries off the coast of Oregon and Washington and Alaska. Another well-defined cluster is associated with tectonic uplift of the Rocky Mountains. However, there are also small clusters and isolated epicenters scattered throughout the continental interior. Geologists refer to earthquakes that occur far from a plate boundary or active mountain belt as **intraplate earthquakes.** Although less well understood, intraplate earthquakes are generally believed to be related to tectonic forces that are being transmitted through the rigid plates. These forces cause crustal rocks to slowly accumulate strain, which is then released along buried fault systems, producing earthquakes in the interior of continents.

In this section we will examine how the theories of plate tectonics and elastic rebound can help explain the location and frequency of most earthquakes. We will pay particular attention to the relatively small number of large-magnitude (>6) earthquakes since these are the ones that pose the greatest risk to people. Understanding the occurrence of strong earthquakes is important in environmental geology because such knowledge can be used to help minimize the damage and loss of life from earthquakes.

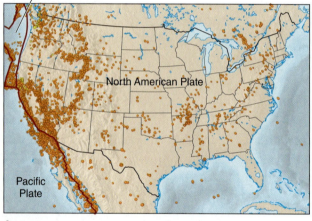

Juan de Fuca Plate

North American Plate

Pacific Plate

A

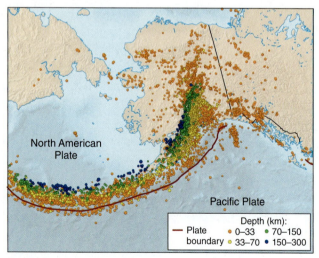

North American Plate

Pacific Plate

Depth (km):
— Plate boundary
● 0–33 ● 70–150
● 33–70 ● 150–300

B

FIGURE 5.12 Earthquake epicenters in the United States from 1990 to 2000. Note that the majority of epicenters are associated with plate boundaries and active mountain belts; the remainder are called intraplate earthquakes as they occur in the interior of plates.

Earthquake Magnitude and Frequency

Seismologists estimate that several million earthquakes occur worldwide each year, with the vast majority going undetected because they are either too small or lie too far from an existing seismograph station (nearly all of the strong earthquakes are detected by an international monitoring network of 128 seismograph stations). Interestingly, of the estimated several million earthquakes each year, only about 100,000 are large enough (magnitude > 3) to be felt by people. From a hazard perspective, our concern is with the 150 or so *strong earthquakes* (magnitude 6 and higher) since these are the ones capable of causing significant damage. Note that there are only about 20 *major earthquakes* each year with magnitudes between 7 and 7.9, and just one *great earthquake* with magnitude 8 or higher. The key point here is that although the number of potentially damaging earthquakes each year is minuscule compared to the several million earthquakes, the destruction and loss of life from these large earthquakes can be enormous.

At this point we need to examine how the theories of plate tectonics and elastic rebound can help explain where strong earthquakes occur, and why they are relatively rare in most parts of the world. From the elastic rebound theory we know that more powerful earthquakes occur when rock bodies are able to accumulate greater amounts of strain energy before rupturing. The key here is the strength of the rock itself and the nature of the faults within a rock body. Recall from Chapter 4 that rocks are much stronger under a compressional force compared to a tensional force. This means that at convergent boundaries where compressive forces dominate, rocks are able to accumulate much more strain before rupturing than at divergent boundaries where tensional forces are dominant. Rocks can also accumulate considerable amounts of strain under the shear forces found along transform boundaries.

The other key factor in the ability of a rock body to store strain is the frictional resistance of the faults. In areas where tensional forces dominate, the friction along faults is naturally low, allowing them to slip in an almost continuous process known as *fault creep.* When a rock body experiences fault creep it obviously cannot build up much strain, which helps explain why large magnitude earthquakes generally do not occur at divergent boundaries. On the other hand, compressional and shearing forces at convergent and transform boundaries tend to create high levels of friction on faults, creating the potential for large magnitude earthquakes. In fact, the resistance may become so great that faults become locked, allowing the strain to build to the point where extremely powerful earthquakes are generated. We will now turn to some examples that illustrate the connection between tectonic setting and the occurrence of large magnitude earthquakes.

Transform Boundaries— San Andreas Fault

The San Andreas fault, shown in Figure 5.13, is one of the few places in the world where a transform boundary is found on land. The San Andreas is actually a large transform fault that separates the Pacific and North American plates, but is often referred to as a *fault zone* due to the network of interlocking faults located on either side. This means that as tectonic forces cause strain to accumulate along the boundary, some of the strain is distributed among the different faults within the fault zone. As indicated by the location of epicenters in Figure 5.13, not only does the San Andreas fault occasionally slip, generating a strong earthquake, but other faults within the fault zone do so as well. Moreover, due to the interlocking nature of the fault zone, strain relieved along one fault can disrupt the delicate balance of relationships within the fault zone, triggering additional earthquakes.

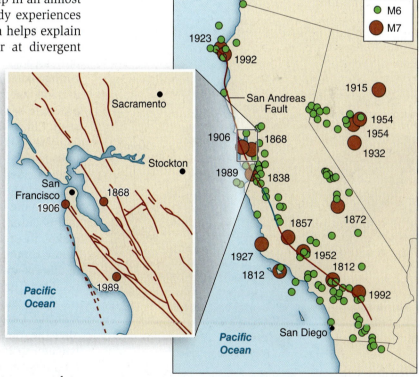

FIGURE 5.13 Location of magnitude 6 and 7 earthquakes along the San Andreas fault zone in California between 1800 and 1994. This transform fault is a boundary between the Pacific and North American plates, but also contains a network of interlocking faults (inset). Fault creep occurs along the blue segment of the main fault, which greatly limits the chance of earthquakes greater than magnitude 7.

From Figure 5.13, notice that the segment of San Andreas shown in blue is an area where fault creep prevents strain from building to the point of generating major earthquakes (magnitude > 7). Along those sections shown in red, the main fault tends to become locked, allowing strain to build to more dangerous levels. In fact, the largest earthquake ever recorded in California, an estimated magnitude 7.9, occurred in 1857 along the southern portion of the San Andreas. Because this section of the fault near Los Angles has remained locked since 1857, the strain energy there has continued to build. Of course, it is only a matter of time before the fault ruptures and releases its strain, generating another major earthquake—Los Angeles residents often refer to such a quake as the *"Big One."* Unlike the 1857 earthquake, however, the Los Angeles area is now a densely populated metropolis. Unfortunately, the damage and death toll from such a quake is expected to be high.

Convergent Boundaries—Cascadia Subduction Zone

In northern California where the San Andreas fault moves offshore (Figure 5.14) the boundary of the North American plate changes from a transform (shear) setting to one of convergence (compression). At this point the North American plate starts to override a series of relatively small oceanic plates along what geologists call the *Cascadia subduction zone.* This subduction zone not only produces the volcanic arc (Chapter 4) known as the Cascade Mountain Range, it also generates **subduction zone earthquakes,** which form when an oceanic plate is overridden by another plate. Subduction zones are important to our discussion because they are capable of generating extremely powerful and devastating earthquakes. For example, of the ten largest earthquakes ever recorded, nine were subduction zone earthquakes, and of these, four were magnitude 9 or higher. The possibility of such an earthquake along the Cascadia subduction zone is rather frightening because a magnitude 9 earthquake will unleash 32 times more energy than one of magnitude 8. Keep in mind that throughout history, earthquakes in the magnitude 8 range have proven to be sufficiently powerful enough to level entire cities, such as Tangshan, China, described at the beginning of this chapter.

The reason subduction zone earthquakes are capable of releasing unusually large amounts of energy is partly due to the way the overriding plate buckles and becomes locked, as shown in Figure 5.14. Another key factor is that the surface area over which the slippage or rupture occurs can be quite large compared to that in other plate settings. Equally important is the fact that the descending oceanic plate is relatively cool, which makes the rocks more brittle and capable of accumulating more strain before rupturing. Finally, in addition to the intense ground shaking, some of this energy can be transferred to the ocean, creating tsunamis that reach heights of 100 feet (30 m) as they crash into coastal areas.

Although geologists have long been aware of the hazards associated with the Cascadia subduction zone, public awareness grew considerably after the massive magnitude 9.1 subduction zone quake and subsequent tsunami in Indonesia in 2004. What is particularly worrisome about the Cascadia subduction zone is that the last major earthquake to occur there was in 1700, which means that over the past 300 years strain may have accumulated to dangerously high levels. Recent studies have also found ample evidence that a large tsunami was associated with this event. To make matters worse, unlike in California where earthquakes are common, there has not been a major earthquake in the Pacific Northwest in more than 300 years. This unfortunately has resulted in relatively few buildings having been designed to withstand the shaking associated with seismic

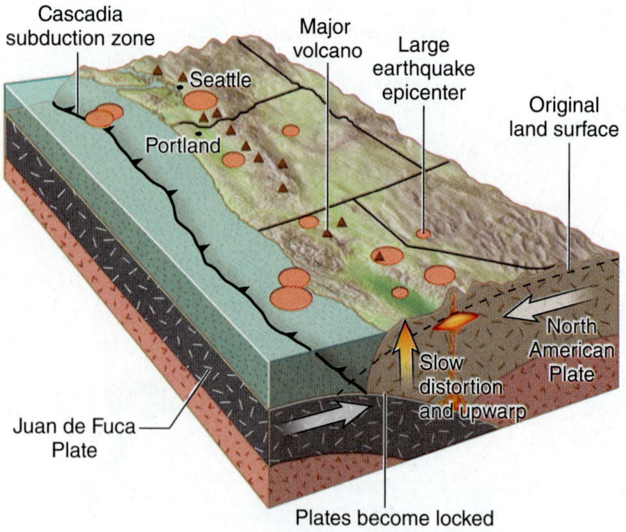

FIGURE 5.14 The Cascadia subduction zone along the Pacific Northwest is not only responsible for the volcanic activity in the Cascade Mountain Range, but also for considerable seismic activity—note the position of epicenters. Subduction zones are notorious for generating powerful earthquakes because of the way the plates lock and the ability of the rock to accumulate large amounts of strain.

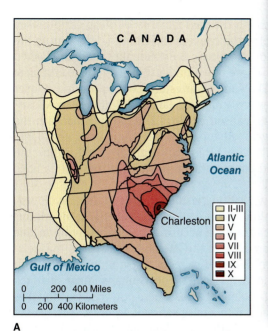

FIGURE 5.15 In 1886 an intraplate earthquake occurred near Charleston, South Carolina, causing severe damage. As indicated by a Mercalli intensity map, this earthquake was felt over a large portion of the eastern United States.

waves. Therefore, this leaves us with the frightening prospect of a magnitude 9 earthquake occurring in a populated area, whose buildings and other infrastructure are relatively unprepared for such an event.

Intraplate Earthquakes—North American Plate

Of considerable interest in the United States are the New Madrid and Charleston seismic zones because they have a history of producing powerful intraplate earthquakes. In 1886 a strong earthquake occurred about 50 miles (80 km) outside of Charleston, South Carolina, causing 60 deaths and extensive property damage throughout the city and surrounding region. As indicated by the Mercalli intensity map in Figure 5.15, the earthquake was felt over nearly the entire eastern part of the United States. Amazingly, structural damage was reported as far away as central Alabama and central Ohio. Although this earthquake occurred prior to the development of modern seismographs, based on damage and other lines of evidence, scientists estimate its moment magnitude was 7.3. Modern studies have shown that this seismic zone is still active, as evident by the clustering of small earthquakes in three distinct areas west and north of Charleston. Based on seismic data, scientists have mapped the position of several faults buried beneath a thick sequence of sedimentary rocks. Despite the fact this region is far from a plate boundary, geologists believe the crust is still accumulating strain, which is then periodically released along buried faults.

In the case of the New Madrid seismic zone shown in Figure 5.16, geologists have been able to link modern earthquake activity there to faults associated with a large, buried rift system called the *Reelfoot rift*. This structure is over 500 million years old and is thought to be similar to the rift currently forming in East Africa, where tensional forces are literally tearing the continent apart (Chapter 4). Geologists now believe that compressional forces within the North American plate have reactivated ancient faults within the Reelfoot rift, generating a clustering of earthquake

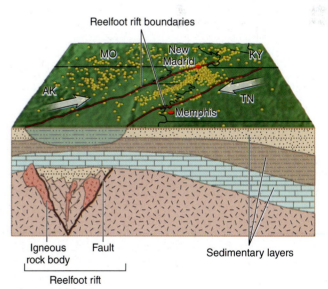

FIGURE 5.16 The clustering of epicenters (most too small to be felt) shows the area of high seismic activity within the New Madrid seismic zone. Other geologic data define an ancient rift system that coincides with the seismic activity. Geologists believe that compressional forces within the continental plate cause strain to build within the rift, which eventually slips and causes earthquakes.

epicenters (Figure 5.16). Although the vast majority of these earthquakes are too small for people to feel, what concerns scientists is that during the winter of 1811–1812 a series of magnitude 8 earthquakes were unleashed within the New Madrid zone. These powerful earthquakes caused damage as far away as Washington, D.C. and Charleston, South Carolina. Closer to the epicenter the seismic waves triggered landslides along the Mississippi River, and in some areas, caused entire islands to sink beneath the river. Water waves even developed on the Mississippi that were large enough to swamp boats and wash others up onto dry land. Since the region was sparsely populated back in the early 1800s, structural damage and loss of life were minimal.

Today, of course, the New Madrid seismic zone is highly developed, including the nearby metropolitan areas of Memphis and St. Louis. Unfortunately, should a powerful earthquake occur again, people living in this region face the same danger as those in the Pacific Northwest in that relatively few buildings have been designed to resist the ground shaking. You will see in the section "Earthquake Hazards and Humans" how this lack of preparedness is a common problem in areas where powerful earthquakes occur infrequently, lulling people into a false sense of security. Perhaps the best example of the phenomenon is the 1976 Tangshan disaster described at the beginning of this chapter. This strong intraplate earthquake occurred in a heavily populated area, which was totally unprepared largely because the people had no memory of a large earthquake ever occurring in the region. The 250,000 to 650,000 people that perished provide a sober lesson for cities located in areas with large but infrequent earthquakes.

Earthquake Hazards and Humans

Based on what you learned so far, it should be clear that earthquakes are a natural consequence of Earth's shifting tectonic plates. Seafloor spreading, subduction, and the uplift of mountain ranges are processes that generally do not occur in a smooth and continuous manner, but rather by the sudden release of strain energy and displacement along faults. Therefore, were it not for earthquakes and plate tectonics, Earth would not have the variety of landscapes or the biodiversity we see today. Simply put, life as we know it would not exist. This brings up an interesting question. If earthquakes are important to the evolution of life, why then do we consider them a "problem"? After all, our ancestors managed to live for hundreds of thousands of years in the presence of earthquakes.

While earthquakes are an important part of the Earth system, the reality is they have always posed a variety of hazards to humans. Prior to the development of agriculture and cities, the primary hazards people had to face from earthquakes were landslides and tsunamis. While these hazards are significant, they are relatively minor compared to problems that developed once people started living and working in buildings, particularly as these structures became progressively taller and heavier over time. Another key element here is that exponential population growth has resulted in many more people living in earthquake hazard zones compared to the past.

In this section we will examine the various types of earthquake hazards, particularly those related to the failure of buildings and other human structures. This is important because structural failure is the leading cause of death and property damage in most earthquakes. Consequently, we will also focus on how society can mitigate the risk of earthquakes by designing structures that are more resistant to ground shaking.

Seismic Waves and Human Structures

There is a common saying among seismologists that "earthquakes don't kill people, buildings do." You can begin to appreciate this statement by examining the collapsed structures in Figure 5.17. This structure was one of an estimated 5.4 million buildings that collapsed during the magnitude 7.9 earthquake in China in 2008, in which nearly 70,000 people died. Most victims in collapsed buildings die from being crushed, but some manage to survive in small void spaces within the pile of rubble. In major earthquakes, however, it is not uncommon for there to be numerous collapsed structures within a city, which completely overwhelms the available rescue personnel. There is also the constant threat that aftershocks will cause the rubble to shift, endangering both rescuers and survivors. To make matters worse, rescue generally requires heavy equipment to remove the overlying debris, a process that may take days or weeks to accomplish. Sadly, most of the uninjured are never rescued from within the rubble, but die within a few days due to hypothermia or dehydration. Therefore, the most effective way of reducing the loss of life in an earthquake is to design buildings so that the chance of collapse is minimized.

In modern cities there are many different types of structures that may fail or become damaged in an earthquake, including homes, office buildings, factories, highways, bridges, and dams. When engineers design a structure they take into account the fact that the structure must be able to withstand a range of different forces, with gravity being the most important. Because all structures have mass, at a bare minimum they must be strong enough to support their own weight against the force of gravity. Because gravity works in the vertical direction, structures are usually the strongest in the vertical direction. Engineers also design for horizontal (lateral) forces such as wind, but this is usually a minor consideration compared to the vertical load or weight. In most places of the world this lack of structural strength in the lateral direction is not a problem, but it becomes one of critical importance in areas where strong earthquakes occur.

FIGURE 5.17 Although most people are killed in the total collapse of buildings, some survive in void spaces within the rubble. The problem is gaining access to the survivors before they die from their injuries, or from hypothermia or dehydration. Rescue is also made more difficult by the unstable nature of the rubble and the constant threat of aftershocks. Photos from the 2008 earthquake (M 7.9) in Sichuan, China.

Construction Design

Recall that when rock ruptures and releases its strain, both body (P and S) and surface (Rayleigh and Love) waves are produced. Of these, surface waves are the most destructive due to the fact they cause the ground to vibrate in a lateral direction, and at the same time, roll up and down like an ocean wave. How well a structure withstands the violent shaking associated with surface waves is highly dependent upon the way in which it was constructed. In the case of homes, most have a wood frame that either sits

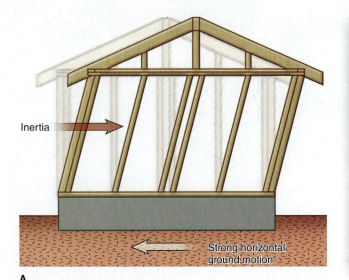

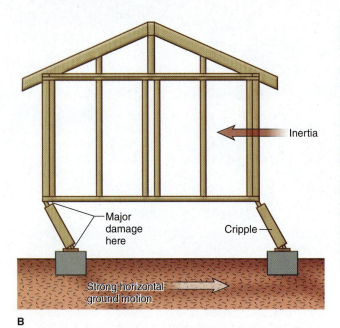

FIGURE 5.18 During an earthquake, buildings are subjected to lateral shear stress due to the horizontal ground motion and their own inertia. This lateral shear force causes structures built on slabs (A) to become skewed after an earthquake. Buildings with crawl spaces (B) or large open areas on the ground floor are inherently weak and prone to cripple-wall failure.

directly on a concrete slab or is built over a crawl space or a garage like shown in Figure 5.18. During an earthquake the base of the structure will move laterally with the ground, but the top tries to remain stationary due to its inertia, which places a lateral shearing force on the structure. Because most houses are not specifically designed for lateral shear, they offer very little resistance to ground vibrations, and consequently they are damaged quite easily. Later, we will discuss how the *probability* of a strong earthquake plays a key role in determining whether buildings are required to be designed for such lateral forces.

Another important aspect of the two types of home construction shown in Figure 5.18 is how they experience different types of damage in an earthquake. In the case of a framed structure built on a concrete slab, the shearing motion during an earthquake typically leaves the building heavily skewed (Figure 5.18A). For those built over a crawl space, the walls surrounding this open space commonly fail since they are the weakest part of the structure. In this case houses tend to stay intact while falling over onto the open space (Figure 5.18B). Note that engineers refer to the collapse of a crawl space as a *cripple-wall failure,* but when the open space is taller

A   **B**

FIGURE 5.19 The mortar in unreinforced masonry walls such as these in Iran, 1990 (A) can easily fail during an earthquake. Oftentimes the entire structure crumbles, leaving a pile of rubble in which few survive. Shown in (B) is one of the many masonry homes to collapse during the 1988 Armenia earthquake (M 6.9), claiming nearly 25,000 lives.

the failure is called a *soft-story collapse.* Soft-story collapse is a common problem in commercial buildings where the first floor has considerable open space for parking or retail shopping.

Generally speaking, the most dangerous types of homes are those constructed of *unreinforced masonry* because they offer very little resistance to lateral shearing motion. In this technique walls are usually constructed of brick or stone bound together with mortar, as opposed to reinforced walls with internal supports of wood or steel—note that most brick homes in the United States have an internal wooden frame with a brick facade. Although unreinforced masonry walls have great load-bearing capacity, the mortar readily fails during an earthquake, leaving the wall in a weakened state (Figure 5.19). When this occurs, continued shaking during an earthquake may cause the entire structure to crumble, crushing its inhabitants. Some of the highest death tolls from earthquakes have occurred in regions where homes were largely built of stone or brick. For example, most of the deaths in the 1976 Tangshan earthquake were attributed to homes being built with unreinforced masonry walls. Compounding the problem in Tangshan was the fact the earthquake struck at night while residents were asleep in their homes.

With respect to multistory buildings, many of which are nonresidential, construction usually involves an interior skeleton made of steel or steel-reinforced concrete. Under normal conditions the entire weight of the building is easily supported by its vertical columns as shown in Figure 5.20. However, during an earthquake the strong lateral forces will cause the structure to sway. In some cases this swaying motion may become so great that some of the floors within the building become detached from the columns, leaving the floors to fall freely. Once a floor becomes free, it naturally falls onto the one below, which can cause additional floors to fail in a cascading manner that engineers call *pancaking.* The result is either a total or partial collapse of the structure in which few people survive.

FIGURE 5.20 Floors in a multistory building can become detached from the supporting columns as the building sways due to lateral ground motion. This can lead to a total collapse, where survival is extremely remote. In the photo from the 1985 earthquake (M 8.0) in Mexico City, note how the vertical column punched through the cascading floors as they fell.

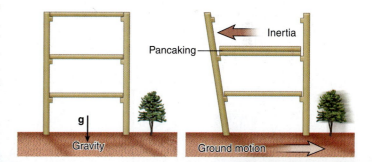

FIGURE 5.21   Brittle failure of steel-reinforced concrete columns occurs when swaying motion causes the columns to reach their elastic limit. Note that the steel rods themselves are not strong enough to support the weight of the structure. Photo from the 1999 earthquake (M 7.6) in Taiwan.

Another important type of structural failure in earthquakes is the sudden rupture of steel-reinforced concrete columns, as shown in Figure 5.21. Such columns are widely used for supporting highways, bridges, and buildings. However, they can fail when the swaying motion of a structure becomes so great that the concrete columns, which are quite brittle, reach their elastic limit and literally explode. Once the concrete shatters, the entire structure can collapse since the steel-reinforcing rods alone are not capable of supporting the weight of the structure.

Natural Vibration Frequency and Resonance

Although all multistory buildings are flexible to some degree, a dangerous phenomenon can develop during an earthquake which causes a building's swaying motion to actually increase. Consider that a building which is swaying back and forth would technically be vibrating—similar to how a guitar string vibrates. Moreover, the building will vibrate at a fixed frequency called its **natural vibration frequency;** frequency is the number of times the motion is repeated in a set amount of time. A key point here is that as building height increases, the natural vibration frequency decreases—similar to how lengthening a guitar string will produce a note with a lower or deeper pitch (i.e., frequency). Therefore in a city with multistory buildings of different heights, some buildings will have a relatively low vibration frequency and others will have a relatively high frequency.

The problem occurs during an earthquake when the natural vibration frequency of a given building matches that of the seismic waves. The matching of frequency then leads to the phenomenon called **resonance,** whereby the amplitudes of the individual waves combine as shown in Figure 5.22A. What happens is that the increased amplitude causes the building to sway even more violently, increasing the chance that its supporting structure will fail as previously described. In some instances the swaying motion due to resonance can be so severe that a building can slam repeatedly into adjacent structures, as was the case for two buildings in Figure 5.22B. However, because seismic waves have a relatively narrow range of frequencies, only those buildings whose vibration frequency falls within this range can experience resonance. It turns out that buildings around 10–20 stories high are

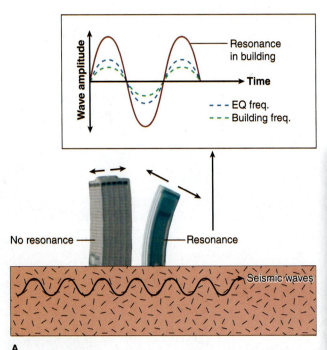

A

FIGURE 5.22   When a building's natural vibration frequency matches the frequency of seismic waves, resonance can occur (A), causing a building to sway more violently. Because vibration frequency varies with height, not all multistory buildings will experience resonance. In the photo (B) the shorter building on the far left experienced resonance and repeatedly slammed into the adjacent hotel, half of which then partially collapsed. Photo from the 1985 (M 8.0) earthquake in Mexico City.

B

most susceptible to resonance since they tend to have natural vibration frequencies that match that of seismic waves. On the other hand, tall skyscrapers are unlikely to experience resonance as their vibration frequency is beyond the frequency range of most seismic waves.

Factors That Affect Ground Shaking

It should be obvious by now that strong ground shaking during earthquakes is the principal cause of structural damage and loss of life. However, ground shaking creates several other hazards in addition to structural failure, topics which we will explore in the following sections. But first we need to discuss the reasons why the level of shaking varies between earthquakes and from one location to another, sometimes even within the same city.

Focal Depth and Wave Attenuation

Recall that magnitude and ground shaking are ultimately controlled by the amount of elastic strain energy that is released when rocks rupture. Moreover, the energy of the resulting seismic waves steadily decreases as they travel away from the focus, a process referred to as **wave attenuation.** The reason the ground shaking is typically the strongest at the epicenter is because it is the closest surface point to the focus, which means wave attenuation is at a minimum. Therefore, the level of ground shaking at any given location depends on the distance to the focus and the amount of energy that was released. This explains why the most dangerous earthquakes tend to be those with a combination of large magnitude *and* shallow focal depth. Keep in mind that it is quite possible for a relatively shallow, low-magnitude quake to generate greater ground motion than a deep, high-magnitude quake.

Of interest here is the fact that seismic waves experience different amounts of wave attenuation, depending on the types of geologic materials the waves must past through. It turns out that loose materials and rocks of lower density will absorb more energy from passing seismic waves compared to rocks that are more rigid and dense. This means that in areas of rigid rocks, seismic waves are able to retain more of their energy as they travel farther. Because the waves undergo less attenuation, they therefore have the potential to cause damage farther from the focus. A good example is the series of magnitude 8 earthquakes that struck the New Madrid area during the winter of 1811–1812. The rigid rocks throughout this region were able to transmit the seismic waves efficiently and with little wave attenuation. In fact, the seismic waves from these earthquakes were able to retain enough energy to ring church bells as far away as Boston, Massachusetts.

Ground Amplification

Although seismic waves lose energy as they travel through the subsurface, their velocity remains relatively constant so long as the density and rigidity of subsurface materials stays the same. However, the subsurface in many regions consists of a variety of rock types and loose sediment, forcing the velocity of seismic waves to change as they encounter these different materials. What is important here is that when seismic waves travel through weaker materials, they slow down and lose energy at a faster rate. This, in turn, causes wave amplitude to increase, creating a phenomenon known as **ground amplification.** In addition, weaker materials can begin to vibrate at the same frequency as that of the seismic waves. This can lead to resonance, similar to buildings as described earlier, but in this case it increases ground amplification even farther. As shown in Figure 5.23, one could expect ground amplification to be most severe in areas with thick layers of loose sediment and where the surface materials are relatively weak.

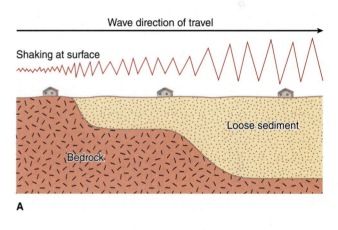

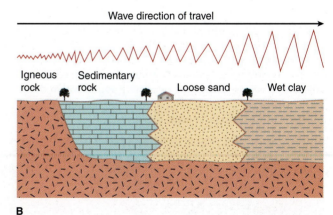

FIGURE 5.23 As seismic waves travel from bedrock into materials of lower density, their velocity decreases, which can cause the waves to amplify. Resonance in loose sediment (A) often leads to ground amplification and is most severe where the sediment is thicker. Ground amplification is also more severe in weaker materials (B), which offer less resistance to seismic waves.

FIGURE 5.24 Map showing levels of ground amplification in the Los Angeles area as predicted by mathematical models. Note that areas of highest risk (red and yellow) are sediment-filled valleys, whereas the lowest risk is in the mountainous bedrock areas (purple).

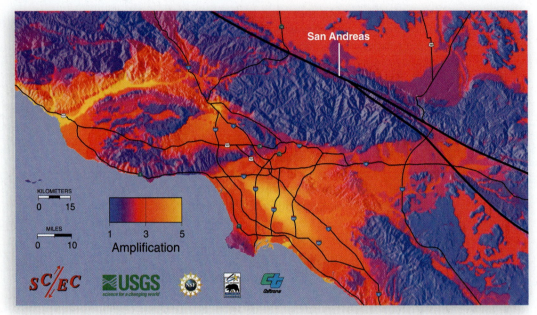

Ground amplification is a serious problem because it greatly increases the level of shaking, which, in turn, increases the risk of structural failure. Consequently, scientists and engineers are keenly interested in identifying those areas at highest risk of ground amplification, as shown in the map of the Los Angeles area in Figure 5.24. From the map notice how the ground shaking is expected to be up to five times greater in the sediment-filled valleys (yellows and reds) than in the mountainous areas (purple) composed of solid rock (bedrock). Also note that the areas of highest population density coincide with the more flat-lying terrain of the valleys, which unfortunately is where geologic conditions create the highest risk of ground shaking.

This brings up another problem, namely, how seismic waves behave in *sedimentary basins,* which are depressions in the crust filled with sediment and sedimentary rocks. As shown in Figure 5.25A, ground amplification can occur when seismic waves enter a basin and begin to amplify since they are forced to slow down in the sedimentary material. Seismic waves can also become trapped within a basin and undergo internal reflection (Figure 5.25B), creating a reverberating effect that extends the duration of the shaking—similar to the way jello continues to vibrate after being set down. This is undesirable since the longer the shaking goes on, the greater the likelihood that structures will fail. Finally, the convex shape of a basin (Figure 5.25C) can cause waves to refract and merge, focusing their energy into localized areas which then experience more intense shaking. All told, the combined effects of amplification, internal reflection, and focusing of seismic waves can

FIGURE 5.25 Ground shaking can increase when seismic waves travel through sedimentary basins: (A) ground amplification occurs as waves slow down when encountering the lower-density material within the basin; (B) seismic waves reflect and become trapped within the basin, prolonging the shaking; and (C) refracted waves can merge and focus their energy.

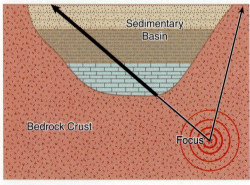

A Wave amplification

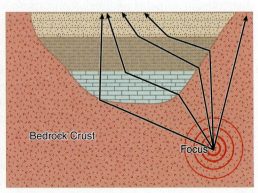

B Internal reflection

C Focusing by refraction

cause ground shaking in a sedimentary basin to be as much as 10 times greater than in the surrounding areas composed of more dense and rigid bedrock.

Secondary Earthquake Hazards

Although the primary hazard associated with earthquakes is the failure of human structures, intense ground shaking often produces secondary hazards such as fires, landslides, and saturated ground that suddenly turns into a liquid. In addition to shaking-related hazards, the displacement of lithospheric plates along subduction zones can generate devastating tsunamis. In this section we will explore some of the secondary hazards associated with earthquakes.

Liquefaction

In areas where saturated conditions are found close to the surface (i.e., a high water table—Chapter 11), ground shaking can cause a phenomenon called **liquefaction,** where sand-rich layers of sediment behave as fluid. As illustrated in Figure 5.26, compacted sand grains are normally in contact with one another, hence are able to support the weight of overlying sediment and human structures. During an earthquake, however, S-waves produce a shearing motion that increases the water pressure within the pore space of the sediment, thereby preventing the vibrating sand grains from making contact with one another. While the grains are in this suspended state the saturated sediment will behave as a fluid, a process often referred to as becoming *liquefied* or *fluidized.* As soon as the shaking stops, the sand-rich material will again behave as a solid as the individual sand grains are able to make contact with each other.

Liquefaction is a serious problem because while subsurface sand layers are in the liquid state, heavy objects sitting on the surface are left unsupported, allowing them to sink or topple over (Figure 5.26B). In hilly terrain, liquefaction can cause slopes to become unstable, triggering different types of landslides geologists refer to as *mass wasting*—a topic that will be covered in detail in Chapter 7. The increased water pressure within the saturated sediment can also cause geysers of liquefied sand to erupt onto the surface, creating what are called *sand blows* (Figure 5.26A). Although sand blows do not present a hazard, they are important since they can be overlain by new sediment and become part of the geologic record. Buried sand blows have provided geologists with a valuable tool for dating ancient earthquakes associated with the New Madrid and Charleston seismic zones discussed earlier.

Disturbances of the Land Surface

In addition to ground shaking and liquefaction, earthquakes can cause disruptions to the land surface that damage buildings and other important infrastructures. For example, recall that when a fault ruptures, the rocks on either side of the fault are *displaced* (i.e., move away from each other). Obviously then, anything built across an active fault plane is at risk of being damaged when the fault slips and the ground becomes displaced, such as the bridge shown in Figure 5.27. Because of the potential for displacement, critical structures like dams, nuclear power plants, underground pipelines, hospitals, and schools should not be built across known faults.

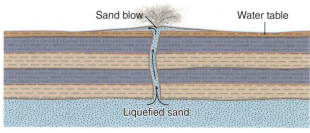
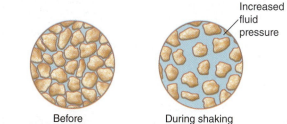

A   Before   During shaking

B

FIGURE 5.26   Liquefaction (A) occurs when ground shaking causes an increase in water pressure within sand-rich sediment. As individual sand grains lose contact with one another, the material behaves as a fluid and loses its ability to support the weight of overlying materials. This lack of support causes structures (B) to sink or topple over. Photo from the 1964 earthquake (M 7.5) in Niigata, Japan.

FIGURE 5.27   This bridge was destroyed when an earthquake caused nearly 20 feet (6 m) of vertical displacement along a fault, which is marked by the newly formed waterfalls. Photo from the 1999 (M 7.6) earthquake in Taiwan.

As shown in Figure 5.28, earthquakes also cause large open cracks called **ground fissures** to form over a wide area of the landscape (faults themselves do not open up like fissures). Ground fissures typically develop close to the surface in loose sediment where there is little resistance to the rolling and stretching motion associated with surface waves. Unlike ground displacement that occurs directly along the fault trace, open fissures have the potential to affect a greater number of structures because they occur over a much broader area. Ground fissures not only can damage surface features such as roads and buildings, but also disrupt underground gas, electric, water, and sewage lines. Damage to transportation links and basic utilities can be particularly disruptive to the overall economy of a region. Moreover, the interruption of utility services, which we often take for granted, creates additional hardships for people who are already struggling to cope in the aftermath of an earthquake. The loss of basic utilities can also lead to disease outbreaks due to a lack of sanitation (sewers) and clean water. Finally, ruptured gas lines pose a serious fire hazard, a topic we will explore separately in the next section.

Earthquakes also provide one of the basic triggering mechanisms for a surface hazard known as *mass wasting,* which is where earth materials move downslope due to gravity (Chapter 7). Familiar forms of mass wasting include landslides, rock falls, and mudflows. Although mass wasting events can occur over a wide area, they are largely restricted to hilly and mountainous terrain where steeper slopes are more common. Here there are many slopes that are inherently unstable, needing only the vibrations from an earthquake in order to fail, sending material downslope. Particularly troublesome are slopes consisting of highly fractured rock, and those covered with thick layers of loose material. One of the worst mass wasting disasters occurred in Peru in 1970, when an earthquake triggered a rock and snow avalanche that killed an estimated 18,000 people. The earthquake itself was responsible for another 48,000 deaths.

Fires

As noted earlier, underground gas lines are easily broken when surface waves roll through a city. All that is needed for a fire is an ignition source, which is readily provided in many cities by sparks from countless electrical shorts in damaged buildings and downed power lines (Figure 5.29). Gas-fed fires can be extremely difficult for fire crews to extinguish, especially when blocked or damaged roads limit access to the fire. To compound the problem, once a crew gets to a site, broken water mains may mean there is no water available to fight the fire. Often the sheer number of fires is so great that local firefighters are simply overwhelmed. Such was the case in the 1995 earthquake near Kobe, Japan, where over 300 fires broke out, most of which were started by gas cooking stoves in residential homes. To make matters worse, blocked streets and a lack of water resulting from an estimated 10,000 water-line breaks, greatly hampered fire-fighting efforts. In the end, over 7,000 buildings were destroyed by fire. Because of the lessons learned from this and other recent earthquakes, some cities in earthquake-prone regions have built systems that will use seawater or underground storage reservoirs (cisterns) for future emergencies.

Perhaps the best known example of an earthquake fire is the one that followed the great 1906 quake near San Francisco. Between 500 and 700 people perished and approximately 20% of the city was destroyed in this 7.8 (moment) magnitude earthquake. Although the earthquake caused heavy structural damage, the fire that swept through parts of the city was responsible for 70–80% of all buildings that were destroyed (Figure 5.29B). Like Kobe nearly 100 years later, broken

FIGURE 5.28 Ground fissures damaged this highway during the 1964 Alaskan earthquake (M 9.2). Open fissures occur in unconsolidated sediment and can cause serious damage to transportation links and various surface structures as well as underground utilities.

A **B**

FIGURE 5.29 Fires are a common secondary hazard associated with earthquakes. Aerial panorama (A) of San Francisco showing the extensive damage caused by the fires that swept through the city after the 1906 (M 7.8) earthquake. Photo (B) showing a gas-fed fire caused by a broken underground gas line. The spark for this fire probably came from the nearby electrical lines.

water mains made it nearly impossible to extinguish the initial fires, which then raged out of control for days. In a last-ditch effort to stop the spreading fire, firefighters resorted to dynamiting buildings to try and create firebreaks.

Tsunamis

A **tsunami** is a series of ocean waves that form when energy is suddenly transferred to the water by an earthquake, volcanic eruption, landslide, or asteroid impact. The majority of tsunamis, however, form during subduction zone earthquakes when crustal plates abruptly move and displace large volumes of seawater. As illustrated in Figure 5.30, compressive forces cause the overriding tectonic plate to slowly buckle until the fault eventually ruptures, at which point the plate lurches upward as the strain is released. The kinetic energy from this vertical displacement is quickly transferred to the water, taking the form of wave energy that travels in both directions away from the subduction zone. Note that a tsunami is similar to seismic waves in that they originate where an energy transfer takes place. The basic difference is that the wave energy travels through water as opposed to rock. Because wave attenuation (rate of energy loss) in water is relatively small compared to rock, a tsunami will transmit energy for considerable distances. For example, tsunamis that originate along the subduction zone of South America can travel across the entire Pacific Ocean and strike Japan, a distance of over 10,000 miles (16,000 km).

While a tsunami is in the deeper parts of the ocean it travels at great speed, around 450 miles per hour (725 km/hr), but its amplitude (height) is usually less than 3 feet. As the waves approach shore and enter shallower waters they are forced to slow down, a process which causes them to become taller. When the waves finally crash ashore they may be as high as 65 feet (20 m). Even taller waves can be generated in bays and inlets since they tend to collect or funnel the waves. Here the energy becomes more focused, producing wave heights that can

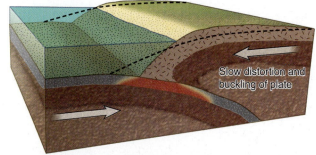

Locked area Overriding plate
Subducting plate

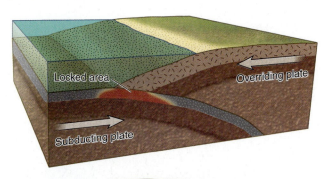

Slow distortion and buckling of plate

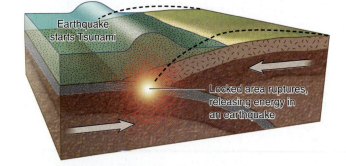

Earthquake starts Tsunami
Locked area ruptures, releasing energy in an earthquake

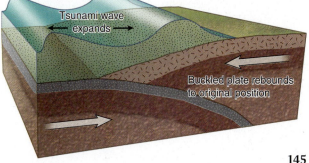

Tsunami wave expands
Buckled plate rebounds to original position

FIGURE 5.30 Tsunamis form at subduction zones when a buckled plate suddenly slips, displacing a large volume of seawater. Kinetic energy from this movement is transformed into wave energy, which then travels outward away from the subduction zone. As the waves approach shore they slow down, which causes an increase in wave height.

145

exceed 100 feet (30 m). Keep in mind what ultimately governs wave height is the volume of rock that displaces seawater. This, in turn, is controlled by the amount of vertical displacement along a fault and the horizontal distance over which the rupture occurs.

To help illustrate this process consider the 2004 tsunami that swept ashore in the Indian Ocean. This event was triggered by an extremely powerful magnitude 9.1 earthquake in the subduction zone off the Indonesian coast. Here the tectonic plates not only shifted 50 feet (15 m) vertically, but shifted for 750 miles (1,200 km) along the length of the subduction zone. The result was the sudden displacement of an exceptionally large volume of rock, which in turn produced an unusually large tsunami. Because wave attenuation in water is relatively low, the tsunami retained much of its energy as it crashed onto distant shores. In fact, of the estimated 230,000 people killed in this disaster, nearly 60,000 were in coastal communities located so far from the epicenter that they never felt the vibrations from the earthquake itself. On the other hand, those living close enough to feel the earthquake had little time to escape, plus they had to face waves that were much higher. In the hardest hit areas close to the epicenter, the waves reached heights of over 100 feet (30 m) and swept ashore in less than 20 minutes.

Finally, Case Study 5.1 provides a good example of how earthquakes can trigger a variety of secondary hazards. Also, please see Chapter 9 for more details on tsunamis and the Indonesian disaster.

Predicting Earthquakes

Many lives certainly would be saved if scientists could predict the occurrence of earthquakes on a short-term basis, say within hours or days of the actual event. Despite the considerable knowledge of how earthquakes occur when a rock body accumulates strain beyond its elastic limit, seismologists are still unable to predict just *when* the rock will rupture. Short-term predictions, therefore, are not yet a reality, and some seismologists believe they never will be. On the other hand, scientists have been successful in predicting earthquakes on a long-term basis using statistical probabilities. Such long-term predictions are similar to weather forecasts in that they give the *probability* that an earthquake will occur within a given time period. In this section we take a brief look at the current status of both short- and long-term earthquake predictions.

Short-Term Predictions

When strain energy is released along a fault, or series of faults, it usually takes place as a cluster of separate events. Small earthquakes called *foreshocks* will sometimes precede the major release of energy, known as the *main shock,* which is usually followed by a series of less powerful *aftershocks.* In addition to the foreshocks, there are phenomena called **earthquake precursors** that sometimes occur just prior to the main shock. Precursors include such things as changes in land elevation, water levels and dissolved gases in wells, and unusual patterns of low-frequency radio waves.

Earthquake precursors can largely be explained by the way strain accumulates in rocks on a microscopic scale. Here we can use the bending of a wooden rod as an analogy. As the wood accumulates strain and gets near its elastic limit, individual wood fibers will start separating and produce a subtle cracking sound. While this is happening the volume of the rod increases slightly as air space develops between the fibers. Ultimately the strain becomes so great that the fibers are no longer able to resist the force, at which point the rod ruptures. Rock behaves similarly in that as strain accumulates, the individual mineral grains begin to separate, which

Secondary Hazards in Anchorage, Alaska

The great Alaskan earthquake of 1964 provides us with a good lesson on how a region's geologic setting plays a key role in creating secondary earthquake hazards. At around 5:30 p.m. on March 27, which happened to be Good Friday, the Pacific and North American plates suddenly slipped in the nearby subduction zone, unleashing an extremely powerful magnitude 9.2 earthquake 75 miles (120 km) from Anchorage. Residents there reported that the shaking lasted between 4 and 5 minutes, which is relatively long for an earthquake, and which probably seemed like an eternity to those who experienced the shaking. In these few minutes some areas experienced as much as 40 feet (12 m) of tectonic uplift, whereas others experienced over 7 feet (2 m) of subsidence (sinking). Although this event brought great destruction, it also provided a lesson in how earthquakes have helped to form the surrounding mountains and shoreline.

The violent ground motion of the powerful seismic waves not only shook people, but shook every single structure in the region, some of which collapsed into open ground fissures created by rolling surface waves. The intense shaking directly triggered secondary hazards in the form of numerous landslides, some of which were themselves triggered indirectly when liquefaction suddenly destabilized slopes. Some of these mass wasting events destroyed entire neighborhoods and several schools, whereas others fell into coastal waters and generated yet another localized hazard, namely tsunamis. Then, approximately 20 minutes after the earthquake, the final blow came in the form of a massive tsunami generated by vertical displacement within the subduction zone. Although it would eventually sweep across the entire Pacific Ocean, the tsunami first crashed ashore in coastal Alaska, with the largest recorded wave reaching an incredible height of 220 feet (65 m). All told, the tsunami was responsible for 122 of the 131 deaths caused by this earthquake—16 died when the waves struck Oregon and California.

Many of the secondary hazards from the Alaskan earthquake can be explained by the aerial photograph of Anchorage in Figure B5.1. Here you can see that like many cities, Anchorage is conveniently located on relatively flat ground with access to the ocean. In the distance lies more hilly terrain, and eventually, very steep and rugged mountains. However, Anchorage itself is built on a large sediment deposit that is quite close to sea level, which is conducive for both ground amplification and liquefaction. Moreover, seismic waves can undergo internal reflection within the sedimentary basin, which, in turn, can prolong the shaking and increase the risk of liquefaction. The loose nature of the sediment at the surface is also conducive to surface waves opening up large ground fissures and generating landslides in the more hilly terrain. Of course, being close to the sea naturally presents a threat from tsunamis.

After the 1964 earthquake, Anchorage officials had the foresight to impose building codes that would help minimize damage from seismic waves and developed programs to help educate the population about earthquake hazards. Despite efforts to minimize the loss of life and property damage in a future earthquake, Anchorage's population has gone from 82,000 in 1964 to over 260,000 in 2000, thereby placing a far greater number of people and buildings in harm's way.

FIGURE B5.1 Aerial view of Anchorage, Alaska. The nearby subduction zone, combined with the fact that the city is built on sedimentary material, puts Anchorage at high risk of hazards associated with wave amplification and liquefaction.

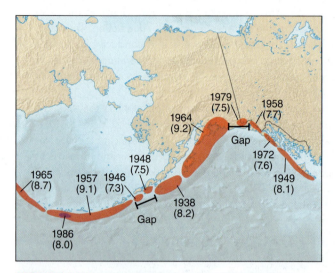

FIGURE 5.31 Map showing the location of seismic gaps along the Alaskan subduction zone. Note how the gaps represent areas where little of the strain along the plate boundary has been released by major earthquakes in modern times. Therefore, the next major earthquake is likely to occur in one of these seismic gaps.

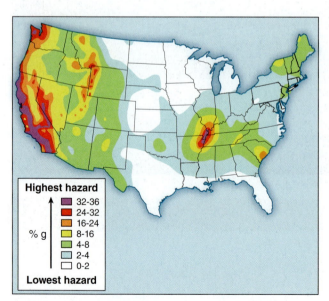

FIGURE 5.32 Seismic hazard map of the continental United States prepared by the U.S. Geological Survey. The map indicates the strength of the horizontal ground motion, relative to gravity, with a 10% chance of occurring over a 50-year period.

develop into microscopic cracks prior to rupturing. This model can explain some of the following earthquake precursors:

1. *Increase in foreshocks*—microcracks forming prior to complete rupture, or main shock.
2. *Slight swelling or tilting of the ground surface*—microcracks increasing the rock volume.
3. *Decreased electrical resistance*—water entering new void spaces that is more conductive than surrounding minerals.
4. *Fluctuating water levels in wells*—water entering new cracks causes water levels to lower; levels rise when voids close again.
5. *Increased concentration of radon gas in groundwater*—new cracks allowing the gas, a radioactive decay product of uranium, to escape from rocks and enter wells.
6. *Generation of radio signals*—changes in rock strain or movement of saline groundwater.

One of the difficulties in using precursors as a predictive tool for earthquakes is a lack of consistent and reliable data. For example, with the exception of foreshocks, all of the precursors just listed can result from things other than the buildup of strain. Therefore, without sufficient data, scientists cannot necessarily attribute a particular precursor to earthquake activity. Another problem is that the triggering of earthquakes is a highly complex process involving the interaction of many different factors, or variables. Moreover, different earthquakes may involve different combinations of variables. This means that earthquakes may be unique in that no two will have the same set of factors leading up to the main shock. Research in this area continues, but at the present time a reliable means of making short-term earthquake prediction simply does not exist.

Long-Term Predictions

Geologists have long noted that in seismically active areas, large earthquakes are more likely to occur as the amount of time increases since the last major event. This observation is consistent with the elastic rebound theory where a continuous force will cause a rock body to accumulate strain and then rupture in a repetitive manner. Geologists use the term **seismic gap** when referring to those sections of an active fault where strain has not been released for an extended period of time. Seismic gaps then can be useful in predicting what areas are most likely to experience a large earthquake. For example, the map in Figure 5.31 shows the areas along the Alaskan subduction zone where strain has been released in modern times by major earthquakes. Based on the location of the ruptured areas, geologists have identified two seismic gaps where a major earthquake is most likely to strike. A similar situation exists along the Cascadia subduction zone in the Pacific Northwest discussed earlier. Here the entire subduction zone can be considered a seismic gap since a major earthquake has not occurred for nearly 300 years.

Although seismic gaps are good indicators of the potential for future earthquakes, statistical estimates are more useful as they give us a better idea of what to expect over a specific time period. One statistical approach involves the creation of *seismic hazard maps,* which incorporate information on past seismic activity, magnitudes, and displacement rates on faults. For example, the seismic hazard map of the United States in Figure 5.32 shows the potential strength of the horizontal ground motion compared to the strength of gravity—sometimes simply called *g-force.* The colors represent different levels of ground motion which have a 10% probability of being exceeded in a 50-year period. To help understand what this means, we will use the New Madrid seismic zone in the Mississippi River area as an example. Here we see that the highest order color (red) is found in the central portion of the New Madrid zone. Therefore, based on the map legend, this area has a 10% chance over the next 50 years of experiencing lateral ground forces that are 32% or 0.32 times as strong as gravity. Despite being statistical in nature, these types of projections can be quite useful to local governments in developing building codes that improve the ability of structures to resist lateral ground motion.

In addition to hazard maps showing potential ground shaking, statistical method can also be used to estimate the probability of earthquakes with a certain *magnitude*. For example, the map in Figure 5.33A shows the percent probability of an earthquake of magnitude 6.7 or larger on individual faults within the San Francisco area between 2000 and 2030. Taken as a whole, seismologists estimate that this region has a 70% chance of a magnitude 6.7 or greater earthquake over the next 30 years. The basis behind these estimates is illustrated in the accompanying graph (Figure 5.33B). Notice how the level of seismic activity along the various faults in the San Francisco area abruptly decreased after the great earthquake in 1906. Probabilities are calculated from the frequency and magnitude of these earthquakes along with the amount of strain the rocks have accumulated. For example, suppose that two magnitude 7 earthquakes occurred in a 100-year period on the same fault. We could then say that statistically, there is a 50% probability of another magnitude 7 event taking place over the next 50 years (i.e., an equal chance of happening or not happening).

One of the problems with earthquake probabilities is that our historical records only go back a few hundred years, plus the number of large earthquakes within this period is typically small. This tends to lower the confidence level statistical projections. In recent years, however, geologists have been able to extend the record farther back in time by finding evidence for large earthquakes within layers of sediment. For example, carbon-14 dating techniques (Chapter 1) can be used to assign specific dates to buried sand blows and layers of sediment that have been offset, both of which form as a result of large earthquakes.

Although scientists' ability to predict earthquakes is improving, it is important to keep in mind that all predictions, short or long term, can only be made for faults where data are available. This lesson was driven home in 1994 when a magnitude 6.7 quake struck the Los Angeles area along a "blind" or concealed fault previously unknown to geologists. Despite these limitations, scientists, engineers, and planners have made great progress over the past several decades in minimizing the risk from earthquakes, which is the topic of our final section.

Reducing Earthquake Risks

A common question asked by students is whether scientists could prevent major earthquakes by somehow creating small earthquakes, thereby preventing strain energy from reaching dangerously high levels. Studies have shown that injecting water into the subsurface can initiate small earthquakes. Although a technique might someday be developed to prevent the buildup of strain, it would be of little use in areas where the strain is already at high levels. This leaves us with the fact that earthquakes are simply inevitable in certain tectonic settings. Our best solution then is to reduce the risks associated with earthquakes by designing more earthquake-resistant structures. Another is to develop early warning systems so that critical preparations can be made before the shaking begins. We will begin by examining some of the engineering designs that have proven effective at minimizing structural damage.

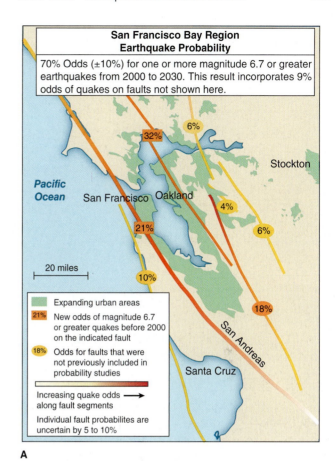

A

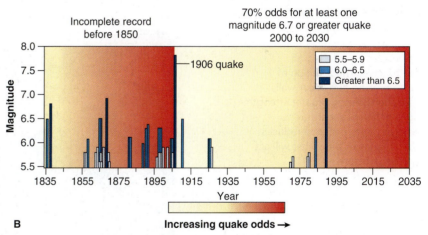

B

FIGURE 5.33 Map (A) showing earthquake probability estimates for individual faults within the San Andreas fault zone near San Francisco. Estimates are based on the frequency and magnitude of earthquakes over time as shown in the graph. Note in (B) how seismic activity abruptly decreased after the great quake in 1906. Overall, the region has a 70% chance of experiencing a 6.7 magnitude or larger quake between 2000 and 2030.

Seismic Engineering

Over the last century engineers have learned how to design structures that are more resistant to the ground shaking associated with seismic waves. This has not only led to a reduction in structural damage and property losses, but it has also reduced the loss of life. The basic approach in this branch of engineering, known as *seismic engineering,* is to: (a) provide greater structural strength with respect to the shear forces generated by lateral ground motion and a structure's own inertia; and (b) reduce the actual amount of shear force that can develop on the structure.

As shown in Figure 5.34, engineers can add a variety of structural elements to a building to make it more earthquake resistant. Of particular importance is the addition of cross-bracing and shear walls, which give a building greater strength against the lateral shearing forces generated during an earthquake. Another important element is *base isolation,* where the amount of shear force that can act on a structure is minimized. Base isolation involves placing rubber bearings (dampers) between the structural skeleton and the foundation supports. Because the skeleton is no longer connected to the ground in a rigid manner, more of the lateral shearing force will be directed on the rubber bearings as opposed to the building. Taken together, shear walls, cross-bracing, and base isolation can greatly reduce the amount of movement within a building's internal skeleton during an earthquake.

Although seismic engineering techniques have proven to be highly effective, there are many structures in earthquake-prone areas that had been built with outdated designs, or worse, built without any seismic con-

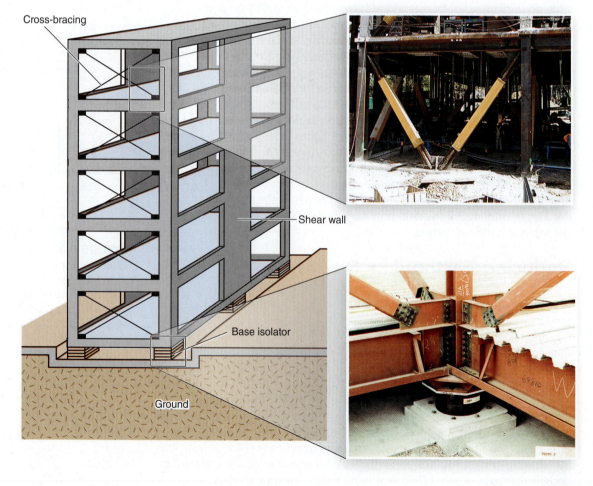

FIGURE 5.34 Seismic engineering involves adding elements to structures that reduce risk of damage during an earthquake. The addition of cross-bracing and shear walls give the skeleton greater strength against lateral shear forces, whereas base isolation reduces the amount of shear that will act on the structure.

Cross-bracing

Shear wall

Base isolator

Ground

trols. A somewhat expensive, but viable option is to retrofit existing buildings with seismic controls as shown in Figure 5.35. Here you can see how an external frame with cross-braces can be attached to the outside of a building, providing strength against lateral shear. Rubber bearings can also be placed between the structural skeleton and its original foundation supports, thereby providing base isolation. Retrofitting projects are most common for those buildings whose continued operation would be critical in the aftermath of a major earthquake. Examples include emergency command and control centers, hospitals, and police and fire stations. Retrofitting is also becoming common in areas where citizens have only recently come to recognize that they lie in areas with a history of major earthquakes. This typically occurs in places that were settled *after* the last major earthquake, thus never saw much need to incorporate seismic engineering. Examples here include the Pacific Northwest and New Madrid seismic zones (see the hazard map in Figure 5.32), where a powerful earthquake has not occurred in living memory.

Recall from earlier how steel-reinforced concrete columns are prone to brittle failure as they are forced to flex due to the lateral ground motion during an earthquake. This presents a major problem since these types of columns are not only used in buildings, but as highway and bridge supports. Failure here of course, can lead to serious disruptions in transportation networks. As shown in Figure 5.36, engineers have developed new designs whereby the vertical steel rods are wrapped in a spiral fashion by a single rod. After the concrete is poured and allowed to harden, the entire column is wrapped with a steel jacket or sleeve. The result is a support where the concrete is far less likely to fail in a brittle manner when it is forced to flex during an earthquake. Note that existing highway columns can be retrofitted with steel jackets, but nothing can be done to the reinforcing rods that are already set in concrete. Finally, notice in Figure 5.36 how the road deck can be tied to the columns using cabling devices.

Early Warning Systems

Although humans are unable to prevent earthquakes, or predict their occurrence on a short-term basis, we are capable of creating early warning systems that can reduce the damage and loss of life. Recall that P- and S-waves leave the focus at the same time, but since P-waves travel faster, they are always the first to arrive at any given location. Moreover, the farther a city is from the epicenter, the greater the time lag between the arrivals of the first P- and S-waves. The basic idea behind an early warning system is to take advantage of this time lag and the fact that P-waves do very little damage. The first P-wave then is simply used as an alert that the highly destructive S-waves and surface waves will soon follow. Depending on the distance back to the epicenter, the warning time may range from a few seconds to around a minute (beyond one minute the epicenter will be far enough away that damage should be minimal). A warning on the order of seconds may not seem significant, but it can be enough for people to seek safety. It can also allow utility and transportation networks to shut down in a controlled manner. For example, only seconds are needed for preprogrammed systems to close valves on gas lines, thereby reducing the risk of uncontrolled fires. Likewise, trains can be programmed to automatically stop. Electric utilities can also shut down critical control systems on electrical grids and at power plants. This not only prevents damage to the systems themselves, but allows electrical service to be restarted much more quickly.

A

B

FIGURE 5.35 Existing buildings can be strengthened against lateral shear by adding an external skeleton (A), whereas the amount of shear force can be reduced using base isolation (B).

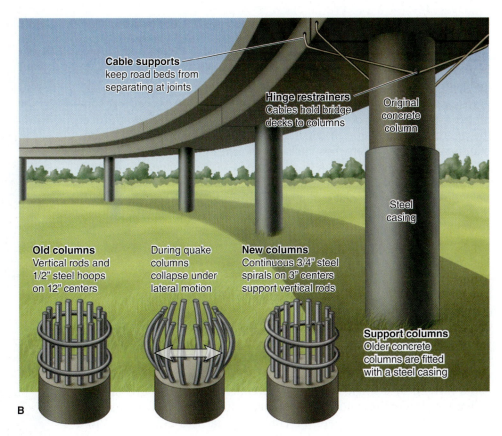

FIGURE 5.36   Lateral motion during an earthquake can cause steel-reinforced concrete columns to flex such that they reach their elastic limit, resulting in failure (A). One solution is to wrap the columns with a steel jacket (B) and use a spiral wrapping technique on the reinforcing rods, thereby reducing the chance of failure.

In the 1980s Japan was the first to begin operation of an early warning system, which was only used for shutting down its high-speed rail system in the event of an earthquake. More advanced systems were later tested and provided early warnings for a select group of government agencies, businesses, and schools. This system makes use of communication stations and a dispersed network of seismographs. When a seismograph detects the ground motion from incoming P-waves that is greater than some preset limit, a signal is automatically transmitted to a processing center, which then sends out an electronic warning. Early warning systems based on this Japanese design are now in operation in select parts of Mexico, Taiwan, and Turkey.

After years of testing and development, Japan began operation of the first comprehensive early warning system in 2006. This system provides coverage for the entire nation as opposed to covering only select areas or key facilities. However, due to problems with false alarms (approximately 10%) and the likelihood of public panic, this system does not send out a general public alert, but rather provides warnings only to critical private and government functions. Officials worry that a general alert would cause panic and create serious hazards, such as human stampedes in crowded places and cars coming to a sudden stop on busy highways. In order to reduce the possibility of panic, the Japanese government is planning to eventually phase in a general alert in conjunction with a public education program. In the United States, in 2006 the federal government began a three-year program in which several different experimental designs are being tested throughout California. Once the testing is complete, officials hope to build a fully operational early warning system.

With respect to tsunamis, the United States first began developing an early warning system in 1946 after a tsunami killed 170 people on the Hawaiian Islands. This system has been continuously improved and updated over the years, particularly after tsunamis struck Hawaii again in 1960, and the Oregon and northern California coasts in 1964. With the cooperation of numerous Pacific nations, the system now utilizes a network of seismograph stations to detect earthquakes with the potential for generating a tsunami. When such an event is detected, an electronic message is sent over the network to various coastal centers around the Pacific. From there, an alert is issued in the form of emergency sirens and public address systems. Note that in 1995 a series of deep-ocean buoys were added to the system that can detect passing tsunamis waves. To increase the effectiveness of the system, a public education program was included so that when citizens hear the warnings, they know to immediately seek higher ground. Efforts are currently underway to develop a similar system to warn of tsunamis in the Indian Ocean (for details see Chapter 9).

Planning and Education

In order to reduce the risk of earthquakes it is critically important that all levels of society, from school-aged children up to emergency management officials, know what to do before, during, and after an earthquake. The first step is to determine the level or severity of risk in a given area. In the United States this task has largely been performed by the U.S. Geological Survey through the construction of various hazard maps, particularly the seismic hazard map shown in Figure 5.32. Based on the hazard assessment, state and local government agencies will typically develop building codes that require appropriate levels of seismic engineering in buildings and other structures. The goal here is to reduce structural damage and the loss of life. Also important are local zoning ordinances that keep homes and key facilities, such as hospitals, schools, dams, and fuel storage areas, from being built across known faults.

In addition to taking steps to minimize structural damage, it is equally important that individual citizens, businesses, and local emergency services know exactly what they need to do both during and after an earthquake. Such planning is critical because unlike most other natural disasters, earthquakes provide no warning. Individual citizens therefore need to know in advance how to seek protection in a variety of situations, such as sitting at work, walking down a sidewalk, or driving a car. Prior to an earthquake people should also take steps that will minimize damage to their home. This may include securing loose objects such as free-standing cabinets, water heaters, and other heavy objects. Family members should also plan on where to best seek shelter in different places within the house, and how to shut off gas and electricity when the shaking stops. Since emergency services will probably not be available, people should be prepared to take care of themselves and family members for a minimum of several days after an earthquake. This means that emergency supplies of food, water, and medicine should be on hand at all times, and that family members should all have first-aid training.

With respect to local government and emergency services, there are a number of things that can be done in advance of an earthquake that will help minimize the loss of life and property. For example, because water main breaks are all but inevitable, some municipalities have built underground storage systems so that water will be available for firefighting efforts. Planning can also help ensure that hospitals are equipped to treat

large numbers of injuries, and that rescue crews have the specialized equipment needed for extracting people from collapsed buildings. Public utilities can make use of devices that detect P-waves that will allow gas lines to shut down automatically, thereby reducing the threat of fire.

Although planning is critical, practice drills are important so that various professionals are more familiar with their assigned tasks in advance of an actual earthquake. Drills are also valuable in that they reveal flaws in the planning, which can then be corrected. Of course, practice drills are just as important for individual family members as they are for emergency personnel. Not surprisingly, in areas where earthquakes are frequent it is common to find high levels of earthquake planning and training as well as building codes that require seismic engineering controls. On the other hand, in places where large earthquakes occur less frequently, the level of planning and preparation is often quite low. Should you discover that you live in an area at risk from earthquakes and want to find out more on how to protect you and your family, just type *"earthquake preparedness"* into any Internet search engine. This should lead you to a number of governmental websites that provide detailed information on the subject.

SUMMARY POINTS

1. A rock body placed under a force will deform and accumulate strain. When deformation exceeds the rock's elastic limit, it will suddenly rupture, releasing the strain in the form of vibrational wave energy called seismic waves.

2. Seismic waves travel outward from the rupture point (focus) and cause the ground to vibrate or shake in what is called an earthquake. Because seismic waves lose energy as they travel, the greatest shaking on the surface is usually at the epicenter, which is directly above the focus.

3. Seismic waves are classified based on how they force rock particles to vibrate. Body waves (P and S) travel through solid earth and then generate surface waves when reaching the land surface. P-waves travel the fastest, but cause very little damage compared to the highly destructive S-waves and surface waves.

4. In tectonically active areas, earthquakes occur repeatedly because of the cyclic manner in which strain accumulates and is then released. Large earthquakes occur when rocks accumulate greater amounts of strain energy before rupturing.

5. The Mercalli intensity scale is a qualitative means of measuring the strength of an earthquake using human observations. Magnitude scales are quantitative measures of the amount of ground motion and are based on measurements taken from a seismograph.

6. Most large earthquakes occur along convergent or transform plate boundaries as rocks are stronger under compressional and shear forces compared to tension. Large earthquakes sometimes take place in the interior of plates when tectonic forces cause strain to accumulate along ancient faults.

7. Seismologists are currently unable to use precursor data to make reliable earthquake predictions on a short-term basis. More successful are long-term predictions using statistical probabilities that make use of historical earthquake frequency and magnitude.

8. Structural failure of buildings due to ground shaking is the single greatest cause of human deaths and property loss in earthquakes. Other related hazards include ground fissures, landslides, fires, and tsunamis.

9. Factors that can lead to increased structural damage due to shaking include magnitude, wave attenuation, resonance effects in buildings, ground amplification, and liquefaction.

10. Buildings and structural supports commonly fail during earthquakes if they do not have adequate strength against lateral shear force. Key elements in seismic engineering include cross-bracing, base isolation, and spiral-wrapped support columns.

11. Loss of life and property damage due to earthquakes can be reduced through seismic engineering controls, early warning systems, and better preparedness within a community.

KEY WORDS

APPLICATIONS

Student Activity

Liquefaction is a very interesting phenomenon, but it is hard to imagine the ground turning to a liquid and then back again. This activity should help illustrate that process. Please note, this can be messy but fun! You will need the following items: 2–3 cups of cornstarch, a bowl large enough to spread your hand across the bottom and deep enough to cover your hand to a depth of approximately 1 inch (a plastic shoe box is great for this), mixing spoon, and approximately 1.5 cups of water. Add 2.5–3 cups of cornstarch to the bowl. The correct consistency is about a 2:1 ratio of cornstarch to water. Pour the water down the sides of your container and begin mixing. As it gets thick, it will be better to use your hand. As the right consistency is reached, it will look like a solid. It will support small structures, army men (toys), and other small items. But if you shake the container or beat the sides, it will turn to a liquid, causing those things to sink. If you try to put your hand in, it will be very hard, but if you beat on the sides or vibrate your hand, it will go right in. If you try to remove your hand, it will seem stuck, but if you vibrate your hand and beat on the sides, it should come out fairly clean. The cornstarch and water make a non-Newtonian fluid that behaves both as a solid and a liquid. To clean up, let the mixture dry and use dry paper towels or sponges. Do not use wet paper towels or sponges, because you will start the mess all over again!

Critical Thinking Questions

1. Do you live in an earthquake-prone area? Do you have an emergency plan in case of a major earthquake? If you don't live in an earthquake-prone area, would you move to such an area, if everything else were to remain the same (salary, climate, etc.)? Why or why not? If no, what would it take to cause you to move to such an area?

2. Why aren't seismographs located on the tops of buildings?

3. Explain how an earthquake high on the Mercalli scale could be low on the Richter scale. Could an earthquake that is low on the Mercalli scale ever be high on the Richter Scale?

Your Environment: YOU Decide

"Earthquakes don't kill people, buildings do. . . The loss of human life and property damage is directly related to the number of structures that fail in an earthquake." These words are quoted in this chapter. There is a lot that can be done to strengthen existing buildings and make new construction safer in these areas, but some of the time this does not occur.

There is a lot of research being done to predict when earthquakes will occur. Some research has been very successful, and some has not. We have a relatively good idea where earthquakes will take place. Of course, this is a complicated issue and deals with many other factors that are not covered in this book, such as the politics of local zoning and how scientists get funding for research. If you were in charge of granting research money, would you give it to earthquake prediction or would you fund improvements in both new and old buildings in those earthquake-prone areas? Explain your reasoning.

Chapter 6

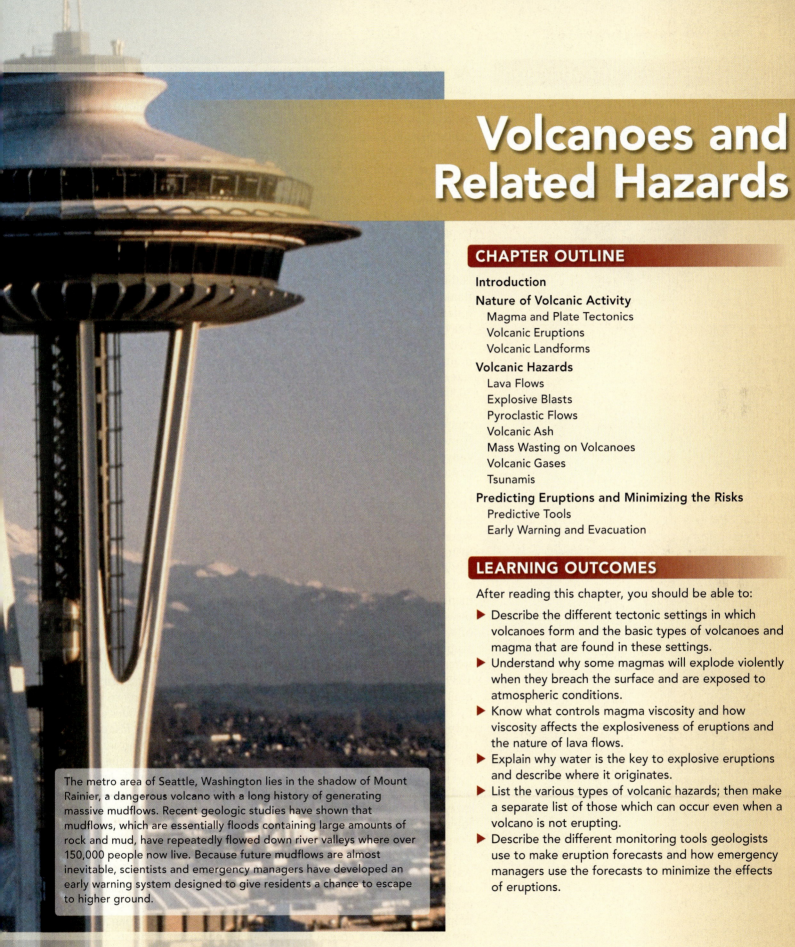

Volcanoes and Related Hazards

LEARNING OUTCOMES

After reading this chapter, you should be able to:

▶ Describe the different tectonic settings in which volcanoes form and the basic types of volcanoes and magma that are found in these settings.

▶ Understand why some magmas will explode violently when they breach the surface and are exposed to atmospheric conditions.

▶ Know what controls magma viscosity and how viscosity affects the explosiveness of eruptions and the nature of lava flows.

▶ Explain why water is the key to explosive eruptions and describe where it originates.

▶ List the various types of volcanic hazards; then make a separate list of those which can occur even when a volcano is not erupting.

▶ Describe the different monitoring tools geologists use to make eruption forecasts and how emergency managers use the forecasts to minimize the effects of eruptions.

The metro area of Seattle, Washington lies in the shadow of Mount Rainier, a dangerous volcano with a long history of generating massive mudflows. Recent geologic studies have shown that mudflows, which are essentially floods containing large amounts of rock and mud, have repeatedly flowed down river valleys where over 150,000 people now live. Because future mudflows are almost inevitable, scientists and emergency managers have developed an early warning system designed to give residents a chance to escape to higher ground.

Introduction

As with earthquakes, people have long been fascinated by the destructive power of volcanoes and the various types of hazards they create. In addition to explosive eruptions and rivers of molten lava, volcanoes generate secondary hazards such as falling ash, mudflows, avalanches, and tsunamis. Like almost all natural hazards, however, people tend to become ignorant or complacent with respect to volcanic hazards, particularly whenever several human generations or so pass between major events. A tragic example where complacency and ignorance resulted in devastating consequences is the 1985 eruption of the 17,681-foot (5,389 m) high Nevado del Ruiz volcano, which buried the Colombian town of Armero, located in the Andes Mountains of South America (Figure 6.1).

After nearly a year of minor activity, on the afternoon of November 13, explosive eruptions began on Nevado del Ruiz, ejecting approximately 700 million cubic feet of hot rock and ash onto its snow- and ice-covered flanks. The eruption also released large quantities of water vapor, which quickly condensed and fell as rain. Together, the ejected material and heavy rains caused rapid melting of the snow and ice, forming mudflows that began traveling downslope and into the network of streams that drain the volcano. As the stream valleys merged, the mudflows grew in size, eventually reaching heights of 130 feet (40 m) and speeds of over 30 miles per hour (50 km/hr). One of these massive mudflows swept down a steep and narrow river valley toward the town of Armero, a seemingly safe 46 miles (74 km) away from Nevado del Ruiz's summit. Here Armero had been built on a large sediment deposit at the mouth of the canyon, where the river empties out onto a broad valley (Figure 6.1). Around midnight, nearly two and a half hours after the eruption began, a giant mudflow reached the base of the canyon and deposited a new layer of sediment on the valley floor, burying most of the city. Of Armero's 28,700 residents, approximately 23,000 were killed and thousands more were injured or left homeless. It was the worst volcanic disaster in the history of Colombia. It also could have been avoided.

Armero is a grim reminder that ignorance and complacency can have serious consequences for humans when it comes to the geologic environment (Chapter 1). Armero was founded by Spanish settlers who selected the site in part because it offered a flat area to build and a steady supply of water from the nearby river (Figure 6.1). However, Nevado del Ruiz erupted in 1595 and again in 1845, killing hundreds of Spanish settlers when mudflows came roaring down the same river valley as it did in the 1985 disaster. Just like the previous disasters then, modern Armero was built on relatively fresh mudflow deposits. The major difference was that by 1985 population growth had resulted in a far greater number of people living in harm's way. Because 140 years had passed since the last mudflow, individual residents of Armero in 1985 must have either been ignorant of the risk they faced or felt the risk was low. Finally, despite the growing level of volcanic activity leading up to the eruption, government officials and Armero citizens tended to believe that the volcano was simply too far away to be much of a threat. No one but the scientists seemed to appreciate the fact that the city was built on old mudflow deposits.

In Chapter 6 we will examine what science can tell us about the various hazards people have to face when living in volcanically active areas. As demonstrated by the Armero tragedy, some of these hazards can reach unsuspecting towns and cities that have grown complacent since the last eruption. Fortunately, there are usually clear warning signs prior to an eruption that if acted upon, can be used as the basis for ordering an evacuation. It is therefore crucial that scientists and government officials work together in educating citizens about volcanic hazards and in developing an effective warning system that allows sufficient time to evacuate. We will

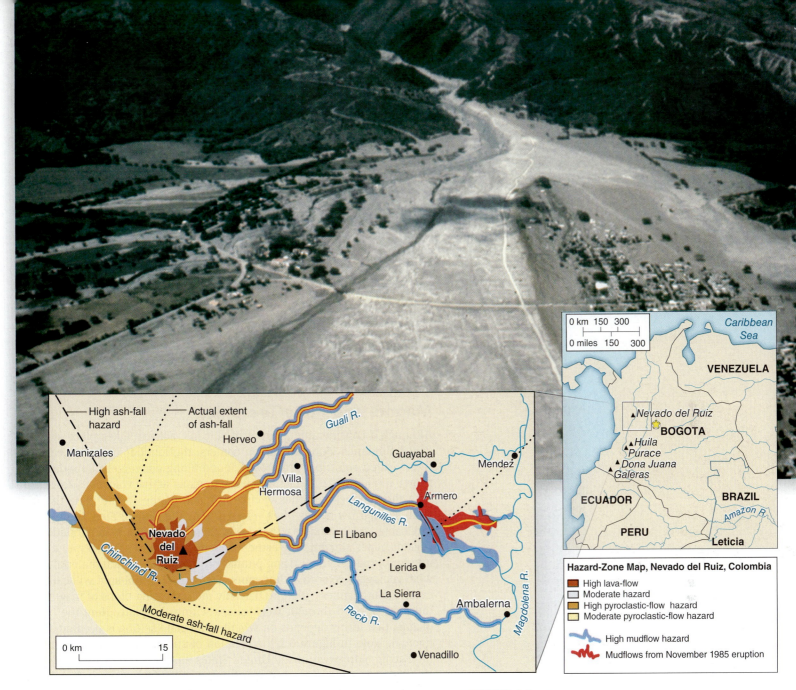

FIGURE 6.1 Three-quarters of the residents of Armero, Colombia, were killed when a massive mudflow came roaring down a canyon and emptied out onto the valley floor where the city was built. The mudflow formed when Nevado del Ruiz, located 46 miles (74 km) away from the town, began to erupt, causing its glacial ice cap to melt.

begin by taking a look at the basic science behind volcanic eruptions in order to better understand the hazards we face.

Nature of Volcanic Activity

In the previous chapters you learned that crustal rocks are continuously being transformed from one rock type to another in what geologists call the *rock cycle.* Moreover, a key part of the rock cycle is Earth's system of moving tectonic plates, which are driven by the planet's internal forces. It is primarily along the boundaries of these plates that molten rock called **magma** rises up through the lithosphere and eventually cools to form igneous rock. From here the newly formed rock can take a variety of paths through the rock cycle. Our interest in this chapter is the magma which makes it through the entire lithosphere and erupts out onto the land surface, at which point it is technically called *lava.* This eruption of lava results in several hazards for us humans, and at the same time, produces a variety of landforms, the most familiar being what geologists call *volcanoes.*

159

Volcanic eruptions have been taking place for Earth's entire history, leading to several important aspects of the planet's environment critical for humans. For example, continued volcanic eruptions along subduction zones have led to the formation of island arcs and volcanic arcs (Chapter 4), and eventually to the continents themselves where most humans live today. The eruption of lava also releases water vapor that enters the *hydrologic cycle* (Chapter 8), which, in turn, is critical to most every life-form on the planet. In fact, scientists believe that most of Earth's vast oceans formed by extensive volcanic activity and comet impacts early in the planet's history. In addition to land and water, volcanic activity often produces fertile soils in which we grow crops, and is also responsible for creating ancient mineral deposits which are the main source of copper for today's society (Chapter 12).

Although volcanic eruptions pose serious hazards for people, life as we know it would certainly not have been possible without volcanoes. However, at this point we will focus our attention on volcanic hazards and leave soil, mineral, and water resources for later chapters. We will begin by reviewing the relationship between the different types of magma and plate tectonics that was described in Chapter 4.

Magma and Plate Tectonics

Although a considerable amount of volcanic activity takes place at divergent boundaries along spreading centers, the familiar cone-shaped landforms called *volcanoes* are mostly found along convergent boundaries as shown in Figure 6.2. In fact, there are so many active volcanoes associated with the subduction zones encircling much of the Pacific plate that this relatively narrow belt is often referred to as the *Ring of Fire*. As described in Chapter 4, the magma that feeds these volcanoes is generated when oceanic plates sink into the asthenosphere. Volcanoes are also found in the interior of plates where rising plumes (columns) of mantle material create what geologists refer to as **hot spots** (Figure 6.2). A hot spot will cause the overlying lithospheric plate to partially melt, creating blobs of magma that then rise upward through weak zones within the plate. Here the type of magma that forms depends on whether the plate setting is oceanic or continental. Note in Figure 6.2 the location of subduction zone volcanoes in the Pacific Northwest of the United States and along the Aleutian Islands in Alaska. So-called *hot spot* or *intraplate* volcanoes are found in Hawaii and in the Yellowstone area of the Rocky Mountains.

The two basic tectonic settings, subduction zone and hot spot (Figure 6.2), produce different types of magma because each involves different kinds of geologic processes. This is important in the study of geologic hazards because some magmas erupt in a highly explosive and violent manner, whereas others are more benign and tend to create mostly lava flows. We therefore need to take a closer look at the relationship between the different types of magmas and the tectonic setting in which they form.

Recall from Chapters 3 and 4 that there are three classes of magma: *basaltic*, *andesitic*, and *rhyolitic*. Basaltic magmas contain the most iron and magnesium and are generated in the upper mantle from ferromagnesian- (iron and magnesium) rich rocks that contain very little water. In some areas basaltic magma rises up through lithospheric plates to form hot spot volcanoes. Other places the basalt comes up along divergent plate boundaries in the process called seafloor spreading, forming new oceanic crust. Eventually lithospheric plates covered with this basaltic crust descend into a subduction zone, where the abundant water acts to lower the melting point of the rocks. As shown in Figure 6.2, this melting

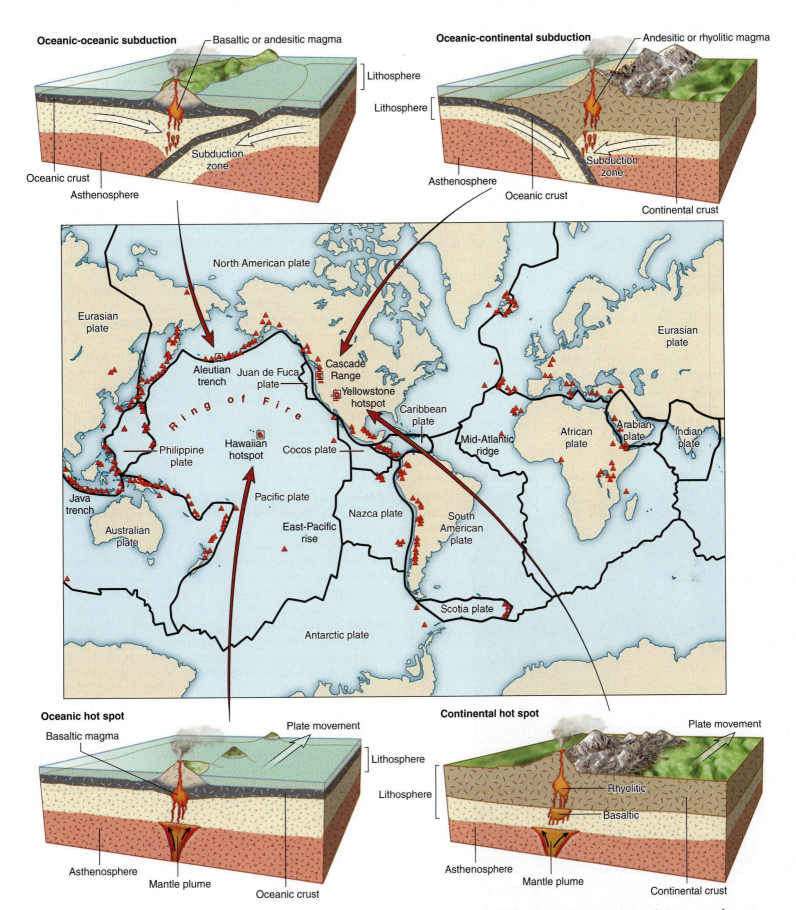

Oceanic-oceanic subduction

Basaltic or andesitic magma

Lithosphere

Oceanic crust

Asthenosphere

Subduction zone

Oceanic-continental subduction

Andesitic or rhyolitic magma

Lithosphere

Asthenosphere

Oceanic crust

Subduction zone

Continental crust

North American plate

Eurasian plate

Aleutian trench

Juan de Fuca plate

Cascade Range

Yellowstone hotspot

Caribbean plate

Mid-Atlantic ridge

Eurasian plate

African plate

Arabian plate

Indian plate

Ring of Fire

Philippine plate

Hawaiian hotspot

Cocos plate

Java trench

Australian plate

Pacific plate

East-Pacific rise

Nazca plate

South American plate

Scotia plate

Antarctic plate

Oceanic hot spot

Basaltic magma

Plate movement

Lithosphere

Asthenosphere

Mantle plume

Oceanic crust

Continental hot spot

Plate movement

Lithosphere

Rhyolitic

Basaltic

Asthenosphere

Mantle plume

Continental crust

FIGURE 6.2 Location of active volcanoes and their relationship to tectonic plate boundaries. Most volcanoes derive their magma from subduction zones or from the upper mantle in what are referred to as hot spots. Hot spot volcanoes in an oceanic setting produce basaltic magma, and those on continents generate more rhyolitic magma. Depending on the tectonic setting, subduction zone volcanoes will erupt basaltic, andesitic, or rhyolitic magmas.

Classification and Flow Characteristics of Volcanic Rocks

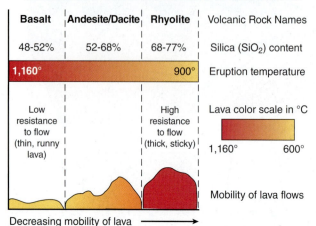

Basalt	Andesite/Dacite	Rhyolite	Volcanic Rock Names
48-52%	52-68%	68-77%	Silica (SiO₂) content

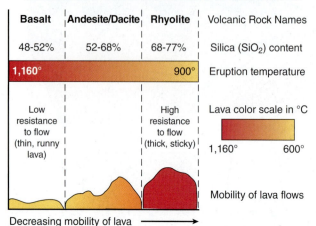

Eruption temperature: 1,160° — 900°

Lava color scale in °C: 1,160° — 600°

Basalt: Low resistance to flow (thin, runny lava)

Rhyolite: High resistance to flow (thick, sticky)

Mobility of lava flows

Decreasing mobility of lava →

FIGURE 6.3   The viscosity of magma increases with increasing SiO₂ and decreasing temperature. Basaltic magmas are the least viscous because they form in the upper mantle where the temperature is high and the SiO₂ content of the rocks is relatively low. Andesitic and rhyolitic magmas form at much shallower and cooler depths and under processes that cause the melt to become enriched in SiO₂.

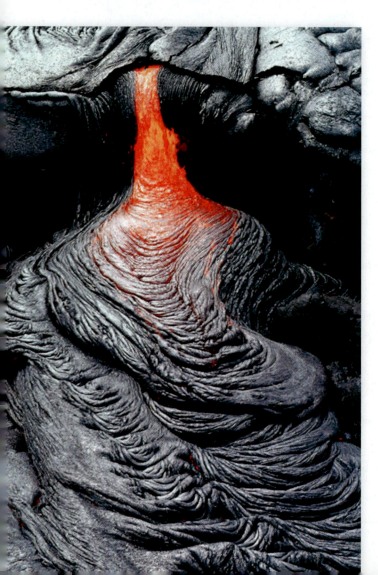

during subduction in oceanic-oceanic settings can lead to basalt magma that eventually erupts and forms volcanic arcs. However, when the rocks undergo *partial melting*, the magma tends to be andesitic as it contains less iron and magnesium but more SiO₂, a compound called *silica*. Remember that andesitic magmas can also form by the process called *assimilation*. As a basaltic magma rises, it will cause silicate minerals in the overlying plate to melt, enriching the magma in SiO₂ such that it becomes andesitic in composition. Andesitic magma is also common in oceanic-continental settings where the magma has to rise through continental plates. Because continental crust is thicker and composed of granitic minerals with even higher proportions of SiO₂, assimilation can lead to rhyolitic magmas that are highly enriched in silica. Rhyolitic magmas also occur at hot spots where mantle plumes rise up through thick, continental plates.

From a hazard perspective, the varying content of silica (SiO₂) in basaltic, andesitic, and rhyolitic magmas is important because it plays a key role in determining the explosiveness of volcanic eruptions. Silica is actually made up of SiO_4^{2-} ions that form strong chemical bonds as the magma begins to cool, which, in turn, increases the internal friction within the magma. Therefore as the SiO₂ (silica) content increases, the ability of a magma to resist flow also increases, a property called **viscosity**. A useful way of understanding viscosity is to think in terms of thick versus thin fluids. For example, water is considered to be a "thin" fluid with low viscosity because it has little internal friction, hence it flows quite easily and can form a thin film or layer. However, if you add flour or cornstarch to water, chemical bonds develop that make it more viscous such that it has greater resistance to flow. In a similar manner, silica is the key ingredient in magma that controls internal friction and viscosity. As shown in Figure 6.3, basaltic magma has the lowest viscosity of the three types of magma primarily because it has the lowest SiO₂ content.

Figure 6.3 also illustrates how there is a progressive decrease in eruption temperature among the three types of magma, with basaltic being considerably hotter than rhyolitic magmas. Part of the reason for this is that basaltic magmas generally form in the upper mantle where temperatures are naturally higher. On the other hand, the water present in subduction zones allows melting to take place at lower temperatures; hence the temperature of andesitic magma is lower than that of basaltic. Rhyolitic magma has the lowest temperatures of all due to the fact that granitic rock within the crust is able to melt at even lower temperatures. The key point here is that as temperature *decreases*, the internal friction within magma *increases*, which means viscosity increases—similar to how maple syrup gets even thicker when it is cooled. Therefore, the reason rhyolitic magmas are the most viscous is because they have the highest SiO₂ content *and* the lowest temperature. In contrast, basaltic magmas are more fluidlike (Figure 6.4) because they contain around 25% less SiO₂ than rhyolitic magmas and are about 250°C (480°F) warmer.

FIGURE 6.4   How easily magma flows depends on its viscosity, which is controlled by its SiO₂ content and temperature. Shown here is high-temperature, silica-poor basaltic magma of relatively low viscosity. Photo taken during the 1989 eruption of Kilauea volcano, Hawaii.

In addition to individual ions (charged atoms), magma also contains various types of gas molecules that form when minerals within a rock body begin to melt. These gases remain dissolved within the magma until the magma gets near the surface, where the decreased pressure allows the gases to escape—similar to how carbon dioxide (CO_2) gas in a soda remains dissolved until the can is opened. The variety and amount of gases involved in this process largely depend on the rock types being melted. This is particularly important at subduction zones where water-rich sediment is incorporated into the melt, generating andesitic magmas rich in water (H_2O) vapor. Rhyolitic magmas typically get their water vapor from the melting of continental crustal material. Table 6.1 compares the gas composition of a basaltic and an andesitic magma, illustrating how H_2O and CO_2 together make up close to 90% of the gas content. We can also see from the table that water is by far the most dominant gas in andesite magma. In the next section you will learn that the *amount* of dissolved gases within magma is the ultimate control on the type of volcanic eruption. At the end of this chapter you will see how the monitoring of dissolved gases escaping from a volcano can be used to help predict future eruptions.

Volcanic Eruptions

Recall from Chapter 3 that when magma stays buried deep within the crust it slowly cools and forms *intrusive* igneous rock, but if it reaches Earth's surface environment, it cools quickly and forms *extrusive* or *volcanic* rock. A volcanic eruption then occurs whenever magma breaks through Earth's surface. Note that once molten rock is on the surface it is referred to as *lava*. In this section we will look at how the gas content and viscosity of magma determines whether eruptions take place in an explosive or nonexplosive manner.

When rock begins to melt, droplets of magma form and eventually coalesce into larger blobs, which then rise because their density is lower than that of rock. Should enough magma accumulate it can form a zone of molten material called a **magma chamber,** as illustrated in Figure 6.5. Because this occurs at considerable depth, the magma chamber must support the weight of the overlying column of rock, referred to as *confining* or *overburden pressure* (Chapter 4). The key point here is that this gives molten material within the magma chamber a tremendous amount of fluid pressure. Therefore, as the pressurized magma rises and encounters less confining pressure, it tries to expand by pushing outward on the surrounding rock. Should the magma get close enough to the surface, its fluid pressure may be sufficient to force open fractures and faults. This not only creates small earthquakes, but can also provide the magma with a pathway to the surface. If the overlying rocks at some point are no longer capable of containing the fluid pressure, then significant amounts of magma can make its way to the surface, resulting in a volcanic eruption.

Once magma reaches the surface it is free to expand since the only confining pressure now is atmospheric pressure (Figure 6.5). This also allows the dissolved or compressed gas within the magma to form bubbles and escape into the atmosphere. However, should the pressurized magma contain significant amounts of dissolved gas, then this rather sudden encounter with the atmosphere will allow large volumes of gas to rapidly decompress in an explosive manner—similar to what happens if you shake a bottle of champagne and then pull the cork. In some

TABLE 6.1   Percent breakdown of gases escaping from magmas at a hot spot and a subduction zone volcano. Water vapor and carbon dioxide are typically the most abundant gases, but notice the dominance of water in andesitic magma that is commonly found at subduction zones. Also note how the andesitic magma is relatively cool compared to the basaltic magma.

Volcano	Kilauea Summit (Hawaii)	Mt. St. Helens (Washington)
Magma Type	Basalt	Andesite
Tectonic Style	Hot spot	Subduction zone
Temperature	1,170°C	802°C
H_2O	37.1	91.6
CO_2	48.9	6.64
SO_2	11.8	0.21
H_2	0.49	0.85
CO	1.51	0.06
H_2S	0.04	0.36
Other	0.16	0.28

Source: R.B. Symonds, Rose, W.I., Bluth, G., and Gerlach, T.M., 1994, "Volcanic gas studies: Methods, results, and applications," in Carroll, M.R., and Holloway, J.R., eds., *Volatiles in Magmas: Mineralogical Society of America Reviews in Mineralogy,* 30: 1–66.

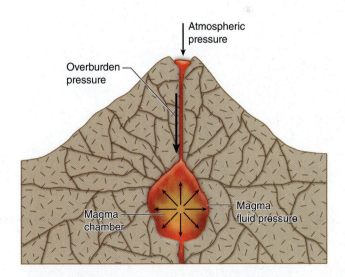

FIGURE 6.5   Molten rock eventually accumulates in what is called a magma chamber. The weight of the overlying column of rock creates overburden pressure, which is offset by the magma's fluid pressure. As magma continues to rise and encounters less overburden, the fluid pressure is able to open fractures, creating possible pathways to the surface. At the surface the pressurized magma is allowed to expand freely.

eruptions the volume of gas is so great that this decompression results in a cataclysmic explosion that literally blows most of the existing volcano up into the atmosphere. For example, consider that when one cubic meter of rhyolitic magma, with just 5% dissolved water, moves from being confined at depth to the surface environment, it will expand to an incredible 670 cubic meters! On the other hand, when magma contains a relatively small amount of gas, like a can of soda gone flat, it will simply flow out onto the surface as a lava flow. These two basic eruption styles, explosive and nonexplosive, are nicely illustrated in Figure 6.6.

It should now be clear that subduction zone volcanoes tend to erupt in an explosive manner because their andesitic magmas usually contain large volumes of dissolved gas—mostly water vapor. Hot spot volcanoes located in the interior of continental plates also erupt violently, as their rhyolitic magmas usually contain abundant water. In addition to the dissolved gas content, the more viscous nature of andesitic and rhyolitic magmas contributes to eruptions being explosive by making it more difficult for the decompressing gases to escape. In contrast, hot spot volcanoes in oceanic settings typically produce low-viscosity basaltic magma with relatively small amounts of gas; hence they generally do not erupt violently. However, any volcano which erupts in an area where water is abundant in the subsurface, like in oceanic settings, can generate a large steam explosion.

From Figure 6.6 one can see that in explosive eruptions most of the lava is ejected into the atmosphere, where it cools into different-sized particles. Because these eruptions also create pulverized rock that is ejected along with the lava, geologists use the term **pyroclastic material.** When referring to these particles of solidified lava and pulverized rock, some geologists prefer the term *tephra.* The coarsest pyroclastic fragments quickly fall from the sky and are deposited on the volcano itself. The finest material, called **volcanic ash,** can travel hundreds, even thousands, of miles before falling back to Earth's surface. In some eruptions ash can reach the upper atmosphere and circle the entire globe.

FIGURE 6.6   Volcanic eruptions occur when pressurized magma breaches the surface. Explosive eruptions (A) are associated with more viscous and gas-charged magmas in which the dissolved gases rapidly decompress, ejecting rock and ash into the atmosphere. Nonexplosive eruptions (B) are associated with hot fluid magma containing less dissolved gas, in which case the eruption generates lava fountains and lava flows.

A Mount St. Helens, Washington

B Kilauea, Hawaii

Low viscosity

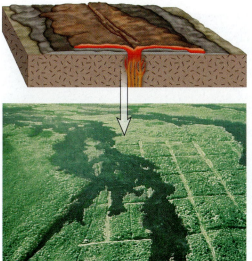

A Basalt, Pu'u O'o, Hawaii

B Andesite, Lassen, California

High viscosity

C Rhyolite, Long Valley, California

FIGURE 6.7 The shape and thickness of a lava flow depends upon the magma's viscosity and slope of the land surface. Low-viscosity basaltic lava flows (A) tend to travel greater distances and spread out into thin sheets in areas where the terrain is more gentle. Andesitic lava is more viscous and creates thicker flows (B) that travel relatively short distances. Highly viscous lava (C) hardly flows at all, but rather builds into a mound-shaped feature called a lava dome.

Volcanic Landforms

In this section we will briefly examine the main types of landforms that are created during volcanic eruptions. This is important because volcanic landforms can provide information as to the relative proportion of lava and pyroclastic material in ancient eruptions, which then tells us something about the gas content and viscosity of the magma itself. In this way geologists can assess the potential hazards humans may face from volcanoes whose last eruption may have taken place prior to human settlement.

Lava Flows

When lava flows out onto the land surface it eventually cools and solidifies into an igneous rock body called a **lava flow.** The shape and thickness of lava flows vary widely due to differences in the viscosity and volume of lava being extruded and the slope of the terrain over which it flows. For example, Figure 6.7 illustrates how low-viscosity basaltic magma can travel long distances and spread into thin sheets in areas where the terrain is relatively flat. On the other hand, more viscous andesitic lava flows are generally thicker and do not travel as far. Rhyolitic lavas are so viscous that they hardly flow at all, which causes the lava to build into steep-sided mounds called **lava domes.** Although highly viscous lavas do not present much of a threat to distant human settlement, a lava dome can act like a plug when it begins to solidify. This, in turn, can allow pressure to build in the magma chamber and result in a more explosive eruption.

In some geologic settings large volumes of basaltic lava will flow onto the surface along large fracture zones, creating vast deposits geologists call *continental flood basalts.* These basaltic magmas are believed to originate within the mantle, then rise up through the lithosphere and erupt onto the land surface. Although this process is similar to the formation of hot spot volcanoes, the volume of magma here is so great that the lava flows occur on a more continuous basis as opposed to periodic eruptions. A good example of flood basalts is the extensive basalt sequence located along the Columbia River in the Pacific Northwest region of the United States (Figure 6.8).

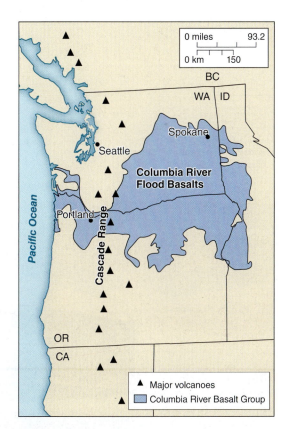

FIGURE 6.8 Extensive lava flows, called flood basalts, form when large volumes of highly fluid basaltic magma are extruded, usually along fracture zones. The map shows the aerial extent of the Columbia River flood basalts in the Pacific Northwest region of the United States. These extensive flows reach a thickness of over 2,000 feet (600 m). The Cascade Range is a volcanic arc that contains subduction zone volcanoes.

165

A Pu' u O'o, Hawaii, 1986

B Mount Etna, Sicily, Italy, 2001

FIGURE 6.9 A volcanic vent can coincide with a fault or fracture, resulting in a linear extrusion known as a fissure eruption (A). In other cases the vent is a single opening whereby the ejected material creates the familiar cone-shaped feature known as a volcano (B).

Volcanoes

In simple terms a **volcano** is the accumulation of extrusive materials around a vent through which lava, gas, or pyroclastics are ejected (Figure 6.9). Sometimes the vent is a fault or fracture, resulting in a *fissure eruption* whereby materials are extruded and accumulate in a linear fashion. Other times the opening is a single pipelike feature called a *central vent* where material is ejected, creating the familiar cone-shaped deposit most people know as a volcano. There are three basic types of volcanic cones: *cinder*, *shield*, and *composite*.

Cinder cones are relatively small features that form when lava is ejected into the air and cools into pyroclastic material called *cinders*, which then fall and accumulate around the vent. In contrast, **shield volcanoes** are exceptionally large landforms composed primarily of basaltic lava flows. Because of the low viscosity of basalt, these lava flows can travel considerable distances from the vent and spread out over large areas. This, in turn, gives shield volcanoes a broad cross-sectional shape similar to a warrior's shield (Figure 6.10), hence the name. From a hazard perspective, note that shield volcanoes do not erupt in an explosive manner because basaltic magma

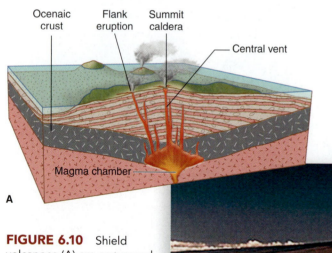

Ocenaic crust
Flank eruption
Summit caldera
Central vent
Magma chamber

A

B

FIGURE 6.10 Shield volcanoes (A) are composed primarily of basaltic lava flows that accumulate over geologic time. Photo shows Mauna Loa (B) in the Hawaiian Islands, which sits over a hot spot, providing a steady supply of magma that has allowed it to grow to 14,400 feet (4,400 m) above sea level.

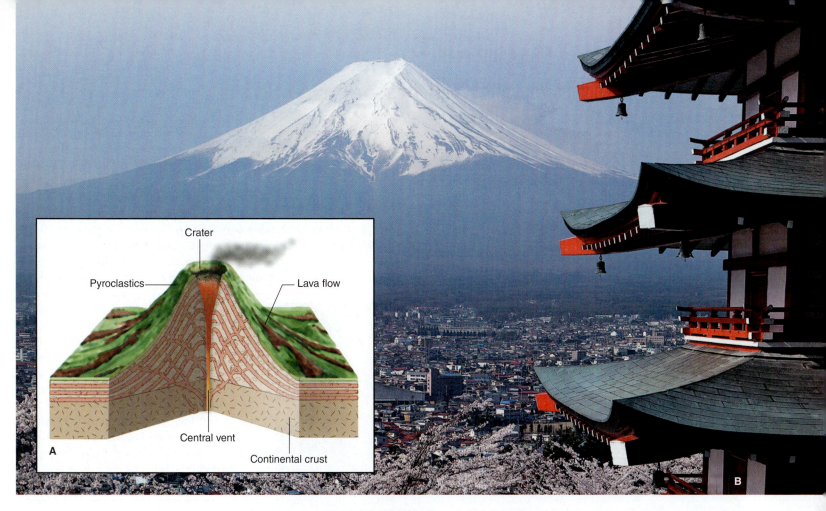

Crater

Pyroclastics

Lava flow

Central vent

Continental crust

A

B

contains only small amounts of dissolved gas; thus, eruptions consist primarily of low-viscosity lava flows. Although these volcanoes have relatively gentle slopes, repeated lava flows over geologic time can allow these volcanoes to grow to great heights. Perhaps the best example is Mauna Loa in Hawaii, which at 14,400 feet (4,400 m) above sea level is the largest shield volcano in the world. Because of its great mass, Mauna Loa has caused the seafloor to become depressed. Altogether, this volcano represents an enormous deposit that is nearly 56,000 feet (17,100 m) high. The reason Mauna Loa is so massive is because it sits over a mantle *hot spot* that provides a steady supply of magma (Case Study 6.1).

The other major type of volcano is known as a **composite cone,** also called a *stratovolcano,* which are cone-shaped features with steep slopes consisting of alternating layers of pyroclastic material and lava flows (Figure 6.11). Composite cones are normally associated with viscous and gas-charged andesitic magmas that generally erupt in an extremely explosive and violent manner. In addition, composite volcanoes typically are found along subduction zones, making up the high peaks in volcanic arc mountain ranges (Chapter 4). In the Andes Mountains of South America, for example, composite cones reach heights of over 20,000 feet (6,000 m) above sea level. The typical eruption sequence begins with an explosive event and the accumulation of pyroclastic material around the vent. This is followed by the extrusion of viscous andesitic lava that forms relatively thick and short lava flows, which then solidify on top of the pyroclastic layers. Because andesitic lavas are rather viscous and resist flowing downslope, the volcano is able to maintain steep slopes and thus attain great heights. Although composite volcanoes along subduction zones are indeed impressive landforms, they are actually quite small compared to shield volcanoes, like the Hawaiian Islands, that develop over hot spots. Figure 6.12 should help you appreciate the vast size difference between composite and shield volcanoes.

FIGURE 6.11   Composite volcanoes (A) are composed of alternating layers of pyroclastic material and lava flows. Viscous andesitic lavas tend to form short and thick flows that enable the volcano to maintain steep slopes and reach great heights. Composite cones are typically quite symmetrical, like Mount Fuji (B) in Japan.

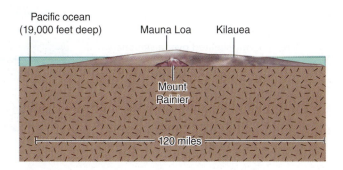

Pacific ocean
(19,000 feet deep)

Mauna Loa

Kilauea

Mount Rainier

120 miles

FIGURE 6.12   Size comparison of one of the larger composite cones in the Cascade Range, Mount Rainier, with the shield volcanoes on the island of Hawaii (See Case Study 6.1).

167

Hawaiian and Yellowstone Hot Spots

Both Hawaii and Yellowstone are similar in that they are known to be active volcanic hot spots related to mantle plumes. However, their eruption styles and associated hazards are entirely different because Hawaii lies in the interior of an oceanic plate and Yellowstone lies within a continental plate. The idea of mantle hot spots originated in the 1960s by geologists who tried to explain why some volcanic centers have been active for long periods of geologic time. The Hawaiian Islands are particularly interesting because they are part of an exceptionally long chain of volcanic islands and seamounts (submerged volcanic remnants) shown in Figure B6.1. Moreover, radiometric dating proved that the oldest islands in the chain are about 70 million years old and get progressively younger toward the Hawaiian Islands, where three volcanoes are currently active on the big island of Hawaii. Geologists generally agree that the explanation for this is that a stationary mantle plume has been generating basaltic magma in the overlying Pacific plate for the past 70 million years. Moreover, as the plate slowly moves in a northwesterly direction, individual shield volcanoes are cut off from their source of magma and become extinct whereas new volcanoes develop over the plume. The shield volcanoes of Mauna Loa and Kilauea on the big island of Hawaii then are simply the latest in a long line of volcanoes to have been situated over the hot spot. Kilauea has been erupting almost continuously since 1983, and the youngest volcano, named Loihi, has not yet risen above sea level (Figure B6.1). Due to the fluid nature and low gas content of the basaltic magma, the primary hazard from these non-explosive shield volcanoes is lava flows.

Yellowstone National Park is a hot spot located in a continental setting and is famous for its geothermal hot springs and geysers (Figure B6.2). To geologists, Yellowstone is also one of the largest calderas on the planet, measuring an impressive 53 by 28 miles (85 by 45 km) in diameter. In addition to geothermal activity, the Yellowstone caldera is seismically quite active, producing a powerful magnitude 7.5 earthquake in 1959 that killed 28 people.

Continental shelf

Aleutian volcanic islands

Emperor Seamount chain

Kuril volcanic islands

Hawaiian Ridge

Hawaii

FIGURE B6.1 The Hawaiian Islands are part of a long chain of islands and submerged volcanic remnants known as the Hawaiian Ridge–Emperor Seamount chain. The active shield volcanoes on the big island of Hawaii are currently situated over a stationary mantle plume, which has been generating basaltic magma for approximately 70 million years as the Pacific plate slowly moves to the northwest.

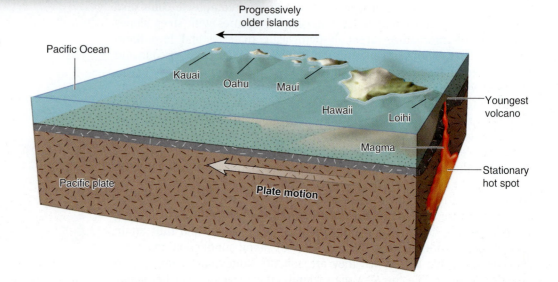

Progressively older islands

Pacific Ocean

Kauai

Oahu

Maui

Hawaii

Loihi

Youngest volcano

Magma

Stationary hot spot

Pacific plate

Plate motion

Although geologists have long known that Yellowstone's geothermal features are related to past volcanic activity, only in the 1990s did it become clear that this giant caldera sits over a mantle hot spot. Today, researchers have been able to identify a series of calderas that get progressively older to the southwest as shown in Figure B6.2. Similar to the tectonic setting in Hawaii, the North American plate is moving over the hot spot which has been active for approximately 17 million years, with Yellowstone representing the site of the most recent series of eruptions.

Although the Hawaiian and Yellowstone hot spots have some common characteristics, at Yellowstone the rising basaltic magma is able to melt continental crust of granitic composition, thereby pro-ducing highly explosive rhyolitic magma as illustrated in Figure B6.2. Eruptions at Yellowstone therefore tend to generate large amounts of pyroclastic material and comparatively few lava flows, which explains why massive shield volcanoes are not found there. Instead, an extremely large caldera developed at Yellowstone when huge volumes of magma were drained from the magma chamber after the last eruption around 640,000 years ago. Based on the size of the caldera and the extensive pyroclastic deposits surrounding it, scientists have concluded that the last major eruption was a truly cataclysmic event. If such an eruption were to happen today, it could have disastrous consequences within the United States and also likely affect the climate of the entire planet.

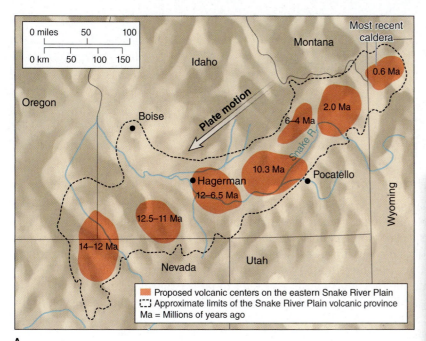

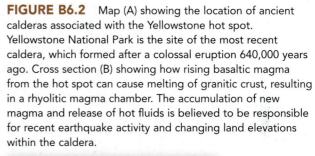

FIGURE B6.2   Map (A) showing the location of ancient calderas associated with the Yellowstone hot spot. Yellowstone National Park is the site of the most recent caldera, which formed after a colossal eruption 640,000 years ago. Cross section (B) showing how rising basaltic magma from the hot spot can cause melting of granitic crust, resulting in a rhyolitic magma chamber. The accumulation of new magma and release of hot fluids is believed to be responsible for recent earthquake activity and changing land elevations within the caldera.

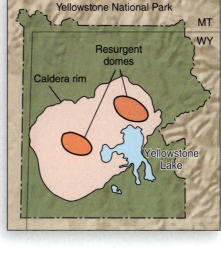

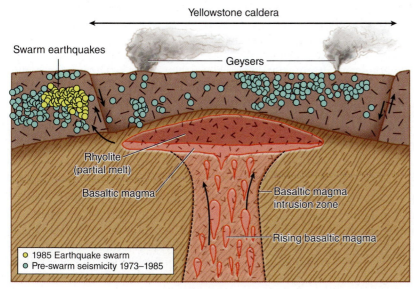

FIGURE 6.13 Calderas form (A) when magma is ejected from a shallow magma chamber, leaving its roof unsupported and eventually causing it to collapse or subside. Photo (B) shows the water-filled caldera of Crater Lake in the Cascade Range in Oregon.

Craters and Calderas

During a volcanic eruption a circular depression called a **crater** will form around the vent where material is being ejected into the air (Figure 6.11A). Especially large craters can form in great explosive events that remove a volcano's summit. A **caldera** is also a circular depression, but one which forms *after* the eruption when rocks in the subsurface begin to collapse or subside (some geologists also refer to large explosive craters as calderas). Basically, a caldera forms when large volumes of magma are ejected from a shallow magma chamber, leaving it relatively empty. The roof of the magma chamber then becomes unsupported and weak, causing it to collapse or subside in on itself as shown in Figure 6.13. The size of a caldera usually depends on the size of the magma chamber that was drained and its depth beneath the surface. For example, the Yellowstone caldera (Case Study 6.1) is exceptionally large, which indicates its last major eruption was enormous. Note that some calderas later become filled with water and form very scenic freshwater lakes, like Crater Lake in Oregon. In some cases, hot, mineralized fluids circulate through the highly fractured rock and deposit valuable ore minerals containing gold, silver, and copper.

Volcanic Hazards

Like most natural hazards, the way in which humans respond to potential volcanic threats depends on several factors, in particular the frequency at which eruptions occur and the availability of habitable living space. For example, people have lived for thousand of years on the flanks of active volcanoes because these areas commonly provide critical resources in the form of fertile soils, lush forests, and supplies of freshwater. The availability of resources plus the fact that several human generations may pass between volcanic eruptions makes the risk of living in such hazardous areas acceptable to most people. Moreover, when enough time passes between eruptions people may become complacent because the probability of an event is low, or even worse, become unaware of the risk altogether.

TABLE 6.2 Some notable volcanic catastrophes in human history and the primary hazard responsible for most of the deaths.

Location	Date	Deaths	Volcanic Hazard
Vesuvius, Italy	79	3,600	Pyroclastic flow
Tambora, Indonesia	1815	80,000	Ash and short-term climate change (starvation from crop failure)
Krakatau, Indonesia	1883	36,000	Tsunami
Mt. Pelee, Martinique (Caribbean)	1902	29,000	Pyroclastic flow
Nevado del Ruiz, Colombia	1985	22,000	Mudflow
Lake Nyos, Cameroon	1986	2,000	Asphyxiation by volcanic gases

Source: Data from the U.S. Geological Survey.

When it comes to volcanic hazards, people most often think of violent explosions and flowing lava, which can lead to the erroneous assumption that the only danger is in the immediate area around a volcano. While the area nearest an active volcano is indeed dangerous, there are a number of secondary hazards that can reach considerable distances. A grim example is the mudflow described earlier that killed over 23,000 residents of Armero, Colombia, which is located a seemingly safe 46 miles (74 km) from the volcano. Other examples of hazards with far-reaching effects include the release of volcanic ash into the atmosphere and large tsunamis resulting from explosive eruptions or the sudden collapse of volcanoes. In addition, some hazards such as mudflows, landslides, and deadly gases can pose serious risks even when a volcano is not erupting. Table 6.2 lists some of the more notable volcanic catastrophes in human history and illustrates the diverse nature of these hazards. In the following sections we will explore the various types of volcanic hazards in some detail.

Lava Flows

Perhaps the most familiar volcanic hazard is a *lava flow,* which occurs whenever magma reaches the surface and begins to move across the landscape. Because lava is a high-density fluid that is very hot, 1,100–2,100°F (600–1,150°C), nearly everything in the path of a flow will either be pushed over, buried, or incinerated (Figure 6.14). As you might guess, how fast and how far a flow will travel is largely determined by the slope of the land surface, volume of lava being emitted at the vent, and the lava's viscosity. Very fluid basaltic lavas, for example, can move as fast as 6 miles per hour (10 km/hr) in steep terrain and extend as much as 30 miles (50 km) from the vent. Note that on flat terrain most people can easily walk around 3 miles per hour, and could easily outrun such a flow. However, in Hawaii where fluid lavas are sometimes confined in steep channels where the heat is retained better, flows have been clocked as high as 35 miles per hour (55 km/hr). On the other hand, more viscous andesitic lavas rarely travel faster than a few miles per hour or extend much more than 10 miles (16 km) from the vent. As discussed earlier, rhyolitic lavas are so viscous that they hardly flow and instead tend to form lava domes.

Lava clearly becomes a problem whenever humans find themselves or their property in the path of an advancing lava

FIGURE 6.14 Lava flows cause considerable damage when they bury valuable real estate and infrastructure, such as the highway and personal property shown in (A). Flows are also destructive when they encounter combustible materials, causing them to catch fire, as in the case of Volcanoes National Park Visitor Center in Hawaii (B).

A Kalapana, Hawaii, 1986

B Volcanoes National Park, Hawaii, 1989

flow. If lava repeatedly erupts from the same vent or fissure, the path is somewhat predictable as it will follow the slope of the surrounding terrain. In areas with steep slopes, flows tend to be thicker since they are more likely to follow the natural stream network, whereas in flat terrain lava tends to spread out and form broad sheets. Note that even the most fluid lava is much more viscous than water, which gives large-volume flows the ability to move up gentle slopes a short distance due to the weight of lava pushing down on the flow from above. Fortunately, most lava moves slowly enough that people can escape an advancing flow simply by walking away briskly. However, as shown in Figure 6.14, permanent structures and various land-use activities (personal use, agriculture, etc.) are completely at the mercy of moving lava.

Various attempts have been made in modern times to stop or divert lava flows to protect valuable structures and property. In some cases explosives have been used as a way to disrupt moving lava so that new directions of flow are created. Ideally this can cause the lava to spread out over a larger area and cool more rapidly, thereby forming a thinner and shorter lava flow. Another method involves placing an earthen barrier in the path of an advancing flow to divert it away from human structures or property. Finally, large volumes of water can be strategically sprayed on the leading edges of a lava flow, thereby creating a chilled zone and a more continuous barrier made of rock. A key factor, of course, in any attempt to divert or stop a flow is the volume of lava involved. In many cases the volume being emitted is so large that the flow simply overwhelms any human effort to contain or divert it.

Perhaps the most famous diversion project was the successful attempt in 1973 on the island of Heimaey in Iceland, where relatively slow-moving basaltic flows were threatening to block the harbor of the country's most important fishing port. Since unlimited seawater was readily available, pumping operations aimed at chilling the leading edge of the flows were able to begin within weeks of the eruption. After laying 19 miles (30 km) of pipe and pumping continuous for nearly five months, the lava flows were finally stopped and the vital fishing harbor was saved.

Explosive Blasts

The most spectacular volcanic processes, and also the most dangerous, are undoubtedly the blast effects associated with explosive eruptions and composite cone volcanoes (stratovolcanoes). Recall that the explosive power typically comes from highly compressed gases (primarily water) dissolved within andesitic and rhyolitic magmas. These gases then violently expand when the magma breaches the surface and encounters the low-pressure conditions of the atmosphere. Even gas-poor magmas can generate large steam explosions should the magma suddenly come into contact with a significant volume of groundwater or seawater.

Photographs like the one in Figure 6.15 help us appreciate the immense power of volcanic eruptions. Here one can see how the rapid decompression of magmatic gases and steam from ground or surface waters can literally blow the summit of a composite cone to bits, leaving only a crater behind. One of the most powerful eruptions in recorded history was the 1883 eruption of the composite cone called Krakatau (or Krakatoa) in Indonesia, where seawater is believed to have entered the magma chamber, creating enormous steam explosions. The eruption of Krakatau was so violent that the 2,600-foot (792 m) mountain, along with most of the island, ceased to exist.

Where this mountain once stood scientists found a hole in the seafloor over 1,000 feet (300 m) below sea level! While this colossal eruption ranks about fifth in terms of known volcanic explosions, it ranks first in the number of humans killed by a volcano. The final series of explosions created several tsunamis, killing over 36,000 people on islands throughout the region (Case Study 6.2).

Although explosive eruptions can generate tsunamis, the most common hazard from these events is pyroclastic material and hot gases that get blasted away from a volcano. What happens is that a pressure or shock wave is created by the rapid expansion of gases that are either released from the magma or form when surface water and groundwater quickly turn to steam. The shock wave then pulverizes rocks making up the volcano into smaller fragments, hurtling them upward and outward at great speed along with superheated gases and blobs of lava. This expanding cloud of pyroclastic debris and hot gases simply obliterates and incinerates everything in its path. For example, during the 1980 eruption of Mount St. Helens, dense forests of 200-foot (60 m) fir trees were completely swept away in an approximately 8-mile (13 km) radius around the volcano. From 8 to 19 miles away these giant trees remained, but were blown down like matchsticks as shown in Figure 6.16 (see Case Study 6.2 for more details on the eruption).

The 1980 eruption of Mount St. Helens provided a sober reminder of the serious risks people face within the blast zones of composite cones along the Cascade Range in the Pacific Northwest. Although the Cascade

People

FIGURE 6.16 A lateral blast during the 1980 eruption of Mount St. Helens obliterated all trees and vegetation within about 8 miles of the volcano. From 8 to 19 miles away the blast leveled 200-foot (60 m) tall fir trees like those shown here—note the people in the lower right for scale.

Explosive Blast of Mount St. Helens

The 1980 eruption of Mount St. Helens in the United States generated considerable public attention despite the fact it was not particularly large in terms of explosive power, ash volume, or outpouring of lava. Even the death toll of 57 was relatively small compared to the tens of thousands killed in other volcanic eruptions (Table 6.2). However, Mount St. Helens was on the U.S. mainland, which made it easy for teams of scientists from the U.S. Geological Survey and other institutions to monitor key indicators prior to the eruption, particularly seismic data (Chapter 5). Because the data allowed scientists to anticipate the eruption, many journalists and photographers were on hand to record the actual eruption. The results (Figure B6.3) helped focus the world's attention on the hazards associated with explosive eruptions, particularly the blast generated by the expansion of gases within the magma.

The explosive blast of Mount St. Helens was especially devastating because it was initially directed in a lateral direction as opposed to being directed upward into the sky (Figure B6.3). This occurred because, as the magma slowly began to rise within the vent, it created a bulge on one side of the volcano (Figure B6.4). As

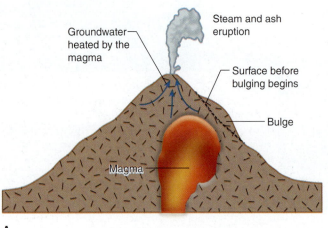

FIGURE B6.3 The eruption of Mount St. Helens was triggered by a small earthquake, which caused a landslide (A) to occur over a bulge that had formed on the side of the volcano. Once the landslide removed the weight of this overlying rock (B), dissolved gases within the highly pressured magma were allowed to rapidly expand. Because of the location of the bulge, the initial blast was directed horizontally (C), devastating the landscape.

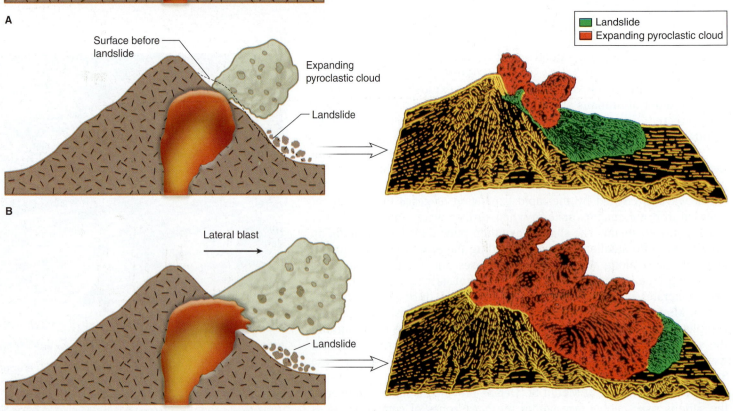

the bulge continued to swell it caused the slope to become steeper, making the bulge less stable. Ultimately, one of the many earthquakes generated by the rising magma caused the bulge and its oversteepened slope to collapse, which created a massive landslide. As the rock mass slid down the mountain it also suddenly removed much of the weight that had been containing the highly pressurized magma, allowing the dissolved gases to rapidly expand and causing the enormous explosion.

Among the people killed in Mount St. Helens' initial blast was a geologist with the U.S. Geological Survey, named Dave Johnston. Dave was at a monitoring station located approximately 5 miles (8 km) from the summit. The site had been carefully chosen and deemed safe from the hot avalanches expected to take place in the pending eruption. Tragically, however, no one anticipated the lateral blast. When the blast occurred it hurled rocks as large as 9 feet (3 m) in diameter toward the monitoring station at an estimated 670 miles per hour (1,080 km/hr). At this rate it took the cloud of hot ash and rocks approximately 30 seconds to reach the station. Of the 57 people killed in this eruption, most were on the side of the volcano that was devastated by the lateral blast. This tragedy caused scientists and emergency management officials to reconsider the size of the evacuation zones for similar volcanoes in the future.

For details about the eruption go to: http://vulcan.wr.usgs.gov/Volcanoes/MSH/Publications/MSHPPF/MSH_past_present_future.html

FIGURE B6.4 A massive bulge (A) developed on Mount St. Helens' northern face as magma within the mountain slowly pushed outward toward one side. It was here that an earthquake triggered a landslide, which then led to the lateral blast that blew out the side of the volcano (B).

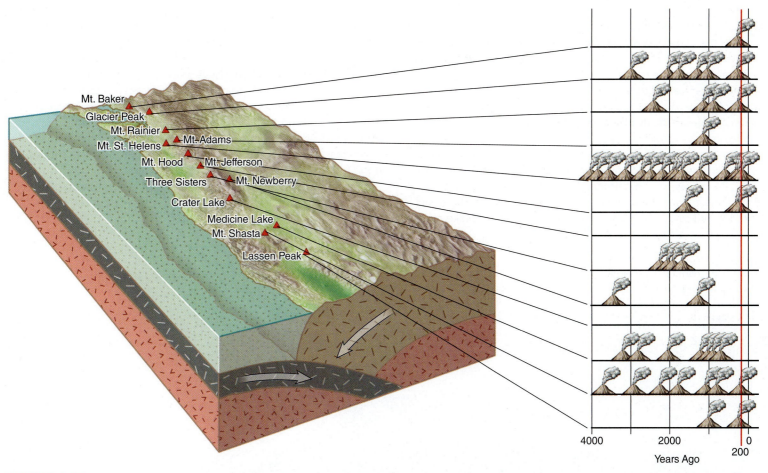

Mt. Baker
Glacier Peak
Mt. Rainier
Mt. St. Helens — Mt. Adams
Mt. Hood   Mt. Jefferson
Three Sisters   Mt. Newberry
Crater Lake
Medicine Lake
Mt. Shasta
Lassen Peak

4000   2000   0
                        200
Years Ago

FIGURE 6.17   The Cascade Range in the Pacific Northwest contains active stratovolcanoes associated with a subduction zone. Most of these volcanoes have erupted in the recent geologic past; some like Mount St. Helens erupt more frequently.

volcanoes are rather active in terms of geologic time (Figure 6.17), the recent event at Mount St. Helens was the first major eruption of such magnitude since European descendants settled the region over 200 years ago. This, in turn, has lulled people into living dangerously close to these ticking bombs. Although major cites like Seattle and Portland are located fairly far from the expected blast zones of their respective volcanoes, many smaller towns in the region are not. Other hazards, like mudflows and ash fall for example, extend far beyond the blast zone of these volcanoes and threaten parts of the urban population centers.

Pyroclastic Flows

A **pyroclastic flow** is a dry avalanche consisting of hot rock fragments, ash, and superheated gas, all rushing down the side of a volcano at great speed. A typical flow consists of two parts: a tumbling mass of large rocks overlain by a turbulent cloud of finer material. Although highly destructive pyroclastic flows can form during explosive or nonexplosive eruptions, they are almost always associated with more viscous, SiO_2-rich magmas. As illustrated in Figure 6.18, one of the ways a pyroclastic flow can form is when a rising eruption column begins to collapse on itself, sending hot gases and fragmental material down the flanks of the vol-

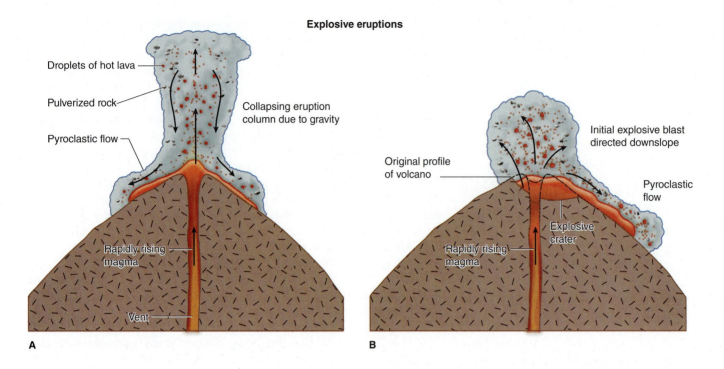

Explosive eruptions

A

Droplets of hot lava

Pulverized rock

Pyroclastic flow

Collapsing eruption column due to gravity

Rapidly rising magma

Vent

B

Original profile of volcano

Initial explosive blast directed downslope

Pyroclastic flow

Explosive crater

Rapidly rising magma

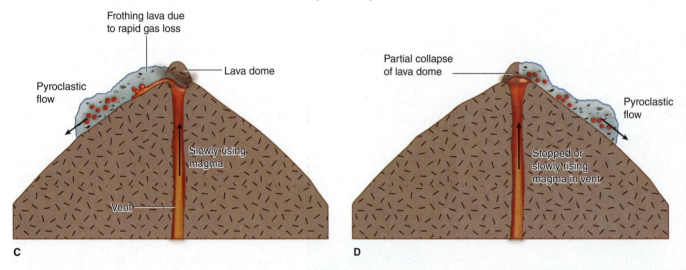

Non-explosive eruptions

C

Frothing lava due to rapid gas loss

Lava dome

Pyroclastic flow

Slowly rising magma

Vent

D

Partial collapse of lava dome

Pyroclastic flow

Stopped or slowly rising magma in vent

FIGURE 6.18   Pyroclastic flows are dry, hot avalanches where large rock fragments tumble along the ground surface and are overlain by a flowing cloud of finer fragments and droplets of lava. Mixed with these materials are superheated gases, creating a flow that will obliterate and incinerate everything in its path. The illustrations show some of the ways pyroclastic flows form during either explosive or nonexplosive events.

cano. Another is during explosive blasts when hot fragmental material is sent rushing along the ground surface away from the volcano. Pyroclastic flows can also develop when gases escape rapidly from erupting lava, creating a frothing mass of semimolten material that rolls downslope. Lastly, flows often form in nonexplosive eruptions when a growing lava dome becomes unstable due to the oversteepening of its slopes. This can lead to partial or total collapse of the dome, sending large volumes of hot rock tumbling down the volcano.

Pyroclastic flows are extremely destructive in part because of the weight of the rocks involved and the fact this material travels on average around 50 miles per hour (80 km/hr), but can reach speeds upward of 450 miles per hour (725 km/hr) depending on the steepness of the slope and relative amounts of gas and rock fragments. Adding to its destructive power is the high temperature of the gases within the flow, which ranges between 400° and 1,500°F (200–700°C). Not surprisingly, a pyroclastic flow will obliterate or bury everything in its path as well as incinerate any combustible materials. Like most avalanches these flows tend to funnel down narrow valleys on the flanks of volcanoes, then form a fanlike deposit at the base where the slope becomes less steep (Figure 6.19). Due to the infrequent nature of volcanic eruption, people will often place themselves in danger by using this relatively flat terrain for building settlements and cultivating agricultural crops—similar to how Armero was built on old mudflow deposits.

Perhaps the most famous example of a pyroclastic flow is the one that struck Pompeii, Italy, a thriving Roman resort city on the flanks of Mount Vesuvius. In 79 AD a series of pyroclastic flows rushed down from this stratovolcano, completely burying the city and its inhabitants. Although Pompeii is recognized today as having been destroyed by pyroclastic flows, this deadly hazard was largely unknown to scientists until 1902, when a series of flows devastated the city of St. Pierre on the island of Martinique (Figure 6.20). Residents of St. Pierre at the time knew that nearby Mount Pelée, a composite cone only 4 miles (7 km) away, had previously been active, but eruptions were rare and none had caused serious damage. Consequently, people were not alarmed when minor eruptions began in April, 1902. Concern grew a few weeks later with the smell of sulfur gas followed by larger explosions and increasing amounts of ash falling on the city. By the morning of May 8, activity on the volcano had subsided and only a thin cloud of smoke was seen rising from the crater. Suddenly a burst of activity began around 8:00 a.m., resulting in an immense pyroclastic flow that rushed down the flanks of the volcano and engulfed the city and seaport in less than two minutes. The flow, consisting mostly of hot gases, incinerated nearly the entire population of 30,000 and instantly ignited everything that was combustible. Only two people from the city itself were known to have survived, along with a few sailors whose ships had been anchored offshore. Twelve days later another flow, one containing more rock fragments, entered the empty city and knocked down many of the walls in buildings that had remained standing. The once bustling seaport of St. Pierre simply ceased to exist.

FIGURE 6.19   Repeated pyroclastic flows funneled down a steep valley, forming a large deposit of debris at the base of Soufrière Hills volcano on the Caribbean island of Montserrat.

Volcanic Ash

In explosive eruptions, parts of the volcano will disintegrate into smaller rock fragments that are blown skyward along with fine droplets of lava, which eventually cool into different-sized particles of rock and non-crystalline material called *glass* (Chapter 3). Moreover, the hot, expanding gases quickly carry this pyroclastic material into the atmosphere as shown in Figure 6.21, creating a towering eruption column up to 80,000 feet (25,000 m) high. The coarser fragments in this column naturally rain down closer to the volcano, whereas upper-level winds may transport finer particles around the globe before they are deposited. Our interest here is in the fine material geologists call *volcanic ash*, consisting of jagged rock and glass fragments less than 2 millimeters in diameter (2 mm is slightly larger than a pinhead). Due to its small size and loose nature, people commonly get the false impression that volcanic ash is soft and lightweight. Ash is actually extremely abrasive because it is made up of tiny rock and glass fragments, which also makes it fairly dense.

FIGURE 6.20   St. Pierre was a thriving seaport on the island of Martinique in the Caribbean. In 1902 a pyroclastic flow raced down nearby Mount Pelée and incinerated the city and its entire population of 30,000.

A Klyuchevskoy 1994, Kamchatka peninsula, Russia

FIGURE 6.21   Explosive eruptions can send a column of volcanic ash (A) as much as 80,000 feet (25,000 m) into the atmosphere, at which point it may be carried around the globe by upper-level winds. Airborne ash poses a serious threat to commercial air traffic, particularly routes over the North Pacific (B) where there is a high concentration of active composite cones. Ash itself consists of pulverized rock and glass fragments, some of which can be extremely fine.

Soufrière Hills volcano, 1997, Montserrat

FIGURE 6.22 Ash clouds can turn day into night. Deposits of ash can add weight to buildings, damage moving parts in mechanical devices, and lead to mudflows during heavy rains. Ash also creates economic losses in agricultural and forestry activities.

There are a variety of environmental hazards associated with volcanic ash, such as inhaling the fine particles, which is particularly dangerous to children or adults with cardiac or respiratory conditions. Other ash hazards occur when it falls from the sky (Figure 6.22) and forms a blanketlike deposit on the landscape. Here the additional weight of the ash can destroy crops and cause buildings to collapse. Ash can also ruin crops when it is impractical to wash off before processing, or when it disrupts pollination and changes the acidity of soils. Pastureland covered with ash can force farmers to either sell their livestock or purchase expensive feed until the grasses can regenerate. Likewise, new sources of water may be needed to replace those contaminated with ash. Another serious problem occurs when ash is washed off the landscape by heavy rains, clogging rivers and streams and making flooding more frequent for years to come. Because ash is so fine-grained it easily finds its way into all types of mechanical equipment where its abrasive properties can cause considerable damage. Since ash also conducts electricity, especially if wet, it can short-circuit electrical equipment and cause entire power systems to shut down.

In 1989 a new volcanic ash hazard was discovered when a commercial airliner flying from Amsterdam to Anchorage, Alaska, flew into what appeared to be a typical layer of clouds, causing all four of its engines to shut down. For nearly five minutes the pilots tried to restart the engines as the plane and its terrified passengers silently fell from the sky. Finally, just a few thousand feet above the ground, the pilots successfully restarted the engines and averted disaster. The plane and its 231 passengers safely landed in Anchorage, but required four new engines at a cost of $80 million. What the pilots thought were normal-looking clouds turned out to be clouds of volcanic ash that had drifted 155 miles (250 km) after being ejected from Alaska's Redoubt volcano the previous day. Later analysis showed that the high temperatures within the jet engines caused ash to melt and resolidify on the engine blades and other moving parts, which ultimately led to engine failure. Although volcanic ash has yet to cause a jet to crash, many planes have encountered similar problems, sometimes from ash that has drifted thousands of miles. Today the airline industry recognizes the seriousness of this hazard and has worked with various scientific institutions to develop an early warning system for pilots. These systems include seismic monitoring of volcanoes and satellite imagery for tracking ash clouds so that pilots can avoid flying into these otherwise normal-looking clouds.

Upper-level winds can carry volcanic ash and gases around the globe, creating a widely dispersed volcanic cloud. The amount of ash in some eruptions is so great that it cools the planet, in part by reducing the amount of sunlight reaching the surface. The 1991 eruption of Mount Pinatubo in the Philippines, which spewed 1.2 cubic miles (5 cubic km) of ash and 22 million tons of sulfur dioxide gas, provided scientists a good opportunity to study the effects of large eruptions on Earth's climate. A group of NASA scientists separated the effects of volcanic eruptions from natural climate fluctuations, and then compared the average response of the climate to the five largest eruptions this century. From this they concluded that the Mount Pinatubo eruption resulted in a small cooling of 0.5°F (0.25°C). Although the Mount Pinatubo eruption was significant, it was small compared to the 1815 eruption of Tambora in Indonesia where 7 cubic miles (30 cubic km) of ash were ejected into the atmosphere. The resulting ash and gas cloud from Tambora circled the globe, dropping global temperature as much as 5°F (3°C) and causing frosts during the summer months in parts of Europe and North America. The year of 1815 was long remembered as "the year without a summer." Most significant was the fact that the frosts had a devastating effect on crops, leading to an estimated 80,000 deaths by starvation.

In Figure 6.23 one can compare the volume of ash ejected by some of the eruptions discussed in this text along with other notable eruptions in the recent geologic past. Note the relatively small amount of ash emitted by Mount St. Helens in 1980 compared to the 1815 eruption of Tamboro. Even more amazing is Yellowstone's eruption 640,000 years ago (Case Study 6.1), which is estimated to have ejected 240 cubic miles (1,000 cubic km) of ash, an amount so large that it was left off the graph due to its scale.

Mass Wasting on Volcanoes

Geologists use the term *mass wasting* to refer to the downslope movement of earth materials due to the force of gravity. Although Chapter 7 is devoted entirely to the topic of mass wasting, in this section we will examine two types of mass wasting processes and associated hazards that are specific to volcanoes. Note that pyroclastic flows were discussed separately due to the fact that hot gases play such a critical role in transporting the volcanic debris, as opposed to mostly gravity.

Volcanic Landslides

Whenever the steep flanks of a volcano become unstable, the underlying material can fail, resulting in rocks, snow, and ice moving rapidly downslope in what is known as a **volcanic landslide**—also called a *debris avalanche*. Since composite cones have much steeper slopes than shield volcanoes, they are more likely to experience volcanic landslides. Moreover, it is important to realize that these landslides do not necessarily take place during eruptions, but are often triggered by heavy rains or earthquakes, as was the case with the massive landslide just before the eruption of Mount St. Helens (Case Study 6.2). Another key factor is the movement of corrosive gases and groundwater within a volcano that can breakdown feldspar-rich rocks into much weaker clay minerals (Chapter 3). This weakens the rocks making up a volcano and tends to destabilize the slopes, thereby making it easier for an earthquake or heavy rain event to trigger a landslide.

Volcanic Mudflows

Unlike a landslide, a **volcanic mudflow** (sometimes called a *lahar* or *debris flow*) is a mixture of ash and rock that also contains considerable amounts of liquid water. Because of its more fluid nature, volcanic mudflows tend to rush down stream valleys that lead away from a volcano, reaching speeds of 20 to 40 miles per hour (30–65 km/hr). The amount of ash and rock in the flow can be as much as 60 to 90% by weight, making some mudflows resemble a river of wet concrete. All this solid material causes mudflows to be fairly dense and viscous, which is why they are so much more destructive than ordinary river floods. In fact, mudflows are so

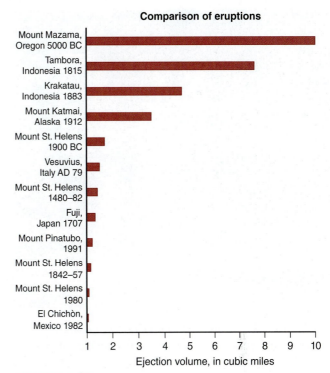

Comparison of eruptions

FIGURE 6.23 Graph comparing the volume of ash ejected in selected volcanic eruptions in the recent geologic past.

FIGURE 6.24   The height of a mudflow generated during the Mount St. Helens eruption is clearly marked on the trees—note the person for scale.

FIGURE 6.25   Mount Rainier is an active composite cone in Washington, whose snow- and ice-capped summit contains more water than all the other Cascade volcanoes combined. This volcano has a history of generating extremely large mudflows, as evident by the map showing ancient mudflow deposits in river valleys leading up to the volcano's summit. Many rapidly growing communities in the Seattle-Tacoma metropolitan area, located within river valleys draining Mount Rainer, are at high risk of volcanic mudflows.

powerful they can easily rip up trees and transport large boulders far downstream (Figure 6.24). This large debris also has a hammering or crushing effect on objects it encounters. As noted earlier, most of the debris within a mudflow will be deposited at the base of the volcano where the stream gradient abruptly flattens out, burying everything in its path like at Armero. In addition to their destructive power, mudflows can travel considerable distances and occur even when a volcano is quiet, hence they can strike without warning.

Volcanic mudflows typically form whenever loose ash lying on the flanks of composite cones is picked up by moving water from either rain events or by rapidly melting snow and glacial ice. Rain-induced mudflows are common in eruptions that release large volumes of water vapor, which then quickly condenses and falls as rain, carrying loose ash off into stream channels. In some areas with a humid climate, heavy rains may occur fairly frequently as part of normal weather patterns or be the result of tropical storms and hurricanes. Note that heavy rainfall events may produce mudflows long after the actual ash eruption. Such was the case around Mount Unzen in Japan, where mudflows destroyed thousands of homes in 1995, more than a year after the ash was deposited.

Mudflows also form by the rapid melting of snow and glacial ice, which are commonly found on the summit of many composite cone volcanoes (Figure 6.25). Ice caps can hold tremendous quantities of water, hence have the potential to produce exceptionally large mudflows. One of the ways this stored water is quickly released is when pyroclastic flows, lava, or hot ash come into contact with the ice. Another mechanism is a sudden slope failure near a volcano's ice-capped summit, resulting in a massive landslide of rock and ice, as shown in Figure 6.26. As the debris travels downslope, some of the ice will begin to melt, transforming the landslide into a mudflow. This type of slope failure can occur when magma creates an unstable bulge or when gases and fluids within the volcano transform solid rock into much weaker clay minerals.

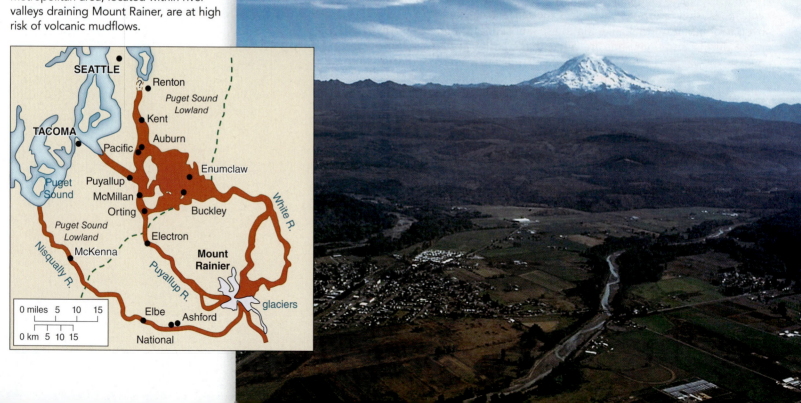

An area that is at serious risk of a mudflow is the Seattle-Tacoma metropolitan area, where nearly 3 million people live in the shadow of Mount Rainier (Figure 6.25). This 14,413-foot (4,393 m) high composite cone contains more water in the form of glacial ice than all the other Cascade volcanoes combined. Moreover, geologic studies in river valleys leading up to Mount Rainier's summit have found evidence of repeated mudflows over the past 6,000 years—some are as recent as 500 years old. The scientific data also show that these flows were enormous, depositing debris in valleys all the way to Puget Sound, a distance of 55 miles (90 km). Of great concern to scientists and emergency management officials is that the geologic conditions which led to past mudflows are still present on Mount Rainier. In addition, due to growth in the Seattle-Tacoma metro area, large numbers of people have moved to communities nestled within the valleys leading up to Mount Rainier (Figure 6.25). Compounding the problem is the fact that geologists have recently found considerable amounts of clay minerals within the ancient mudflow deposits. This indicates that feldspar-rich rocks within the volcano are being transformed into clay minerals that may destabilize the volcano's steep slopes. Consequently, it's quite likely that large portions of the volcano's summit periodically collapse, generating huge landslides that turn into mudflows as the ice rapidly melts. Keep in mind that the volcano does not need to erupt for this to occur, in which case residents would have no warning that giant mudflows were headed their way.

Because of the high risk to residents living in river valleys draining Mount Rainier, a mudflow detection system has recently been installed. This system consists of a network of seismic sensors at five stations located in the upper reaches of the two most vulnerable valleys in the area. The system operates on the principle that mudflows generate seismic waves (Chapter 5) as they flow down the valleys. Therefore, if a sensor records ground vibrations above some preprogrammed limit, a radio signal will be transmitted to an emergency management operations center. Warning sirens will then be sounded in the lower reaches of the valleys, giving some residents as much as 30 minutes to climb or drive to higher ground. This detection and warning system is particularly important since mudflows can occur even when the volcano is not erupting.

FIGURE 6.26 Extremely large volcanic mudflows can form when a landslide develops beneath a glacial ice cap on a composite cone. Such a landslide can be triggered when magma creates an unstable bulge in the mountainside, or when gases and hot fluids weaken slopes by turning solid rock into clay minerals.

Volcanic Gases

When rocks begin to melt deep within the Earth, the resulting volcanic gases are produced and remain dissolved in the newly formed magma. Along subduction zones where water-rich sediment is incorporated into the melt, large volumes of H_2O gas are generated, making andesitic magmas highly explosive. Consequently, the most abundant volcanic gas is commonly water vapor (H_2O), followed by carbon dioxide (CO_2), and sulfur dioxide (SO_2), which together account for over 95% of all volcanic gases. One of the reasons why a volcanic gas cloud is hazardous to humans is simply because it contains no free oxygen (O_2). Therefore, should a volcanic cloud descend into a populated area it poses an asphyxiation (i.e., suffocation) risk to people. Because volcanic gases are typically quite hot, severely burned skin and lung tissue is another life-threatening hazard.

The hazard from volcanic gas clearly should be greatest during or just before the eruption of magma when the release of gas is most common. However, deadly gas clouds are known to have moved down the flanks of a volcano in the complete absence of any volcanic activity. In 1986 clouds of carbon dioxide (CO_2) gas silently moved down the slopes of two water-filled craters in the Oku volcanic field in Cameroon, Africa. The result was the asphyxiation of over 2,000 people and thousands of head of livestock in a 12-mile (20 km) radius around the craters. Officials at first were baffled as to the cause since there was no indication of a volcanic eruption. What they did find, however, was that the once-clear lake water within the craters was now laden with sediment. Scientists later determined that CO_2 gas had been accumulating in the thick sediment lying on the lake bottom. An unusual wind or cooling of the lake surface is believed to have triggered the sudden release of CO_2, which churned up the sediment and overturned the water within the lake as it escaped. Because CO_2 is heavier than air, the resulting gas cloud rolled down the flanks of the crater and into low-lying areas, where people and livestock quickly suffocated.

Tsunamis

In Chapter 5 you learned that earthquakes create tsunamis when the seafloor suddenly moves in a vertical direction, thereby displacing large volumes of seawater. Tsunamis also form when volcanoes explode violently in an oceanic setting. Due to the large number of volcanoes around the world along shorelines, volcanic tsunamis represent a threat that can strike coastal communities far from the volcano itself. As described earlier, the 1883 eruption of Krakatau in Indonesia is notorious for the deadly series of tsunamis created by the colossal eruption which obliterated most of the volcano.

Scientists have recently come to better appreciate the fact that large tsunamis can also form by volcanic landslides, partly because of geologic studies on the Hawaiian Islands. For example, the map in Figure 6.27 shows an extensive debris pile lying on the seafloor north of the islands of Oahu and Molokai. In the photo one can see abrupt *scarps* (clifflike face) on the islands themselves, which are also characteristic of landslides. All this strongly indicates that a massive slide started well above sea level, which then must have displaced a tremendous amount of water, and, in turn, generated a very large tsunami. Such events appear to have occurred on nearly all of the Hawaiian Islands, implying that volcanic landslides may be a common process there. It is quite possible then that similar landslides

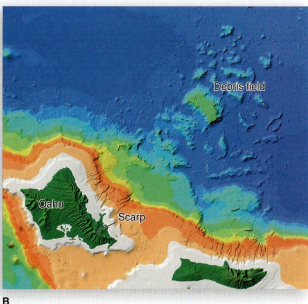

B

FIGURE 6.27   Large scarps (A) and an extensive debris field (B) offshore of the Hawaiian Islands point to an enormous landslide that likely generated a large tsunami. The green areas on the map are above sea level, representing the islands of Oahu and Molokai.

A

may be common on volcanic islands in other places, creating a tsunami threat along unsuspecting coastlines where such waves are not known to have occurred in modern times.

Predicting Eruptions and Minimizing the Risks

Approximately 20% of Earth's population lives in areas at risk of a volcanic hazard. As noted throughout this text, people choose to live in hazardous zones for many reasons, including fertile soils, limited availability of use-able land, economic opportunity, scenery, and recreational value. However, another key factor is complacency. Remember that the time interval between volcanic events is often much longer than the life span of an individual human; many people therefore consider the risk to be low and worth taking. Although many generations may pass while an active vol-cano remains dormant, an eruption is generally all but inevitable. This issue of human complacency is compounded by the fact that as population grows, greater numbers of people are placed in harm's way, increasing the potential loss of life and property. For example, researchers at Columbia University used 1990 census data, satellite images, and geologic informa-tion to study the relationship between population patterns and volcanic hazards. Here they examined 1,410 volcanoes that have been active within the past 5,000 years, which is very recent in terms of geologic time. Of these, 457 volcanoes were found to have populations greater than 1 million

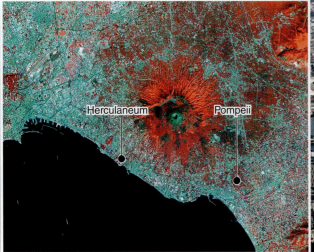

A

B

FIGURE 6.28 False-color satellite image showing Mount Vesuvius and the surrounding urban area of Naples, Italy. In 79 AD, Mount Vesuvius erupted and buried 3,600 residents of the Roman city of Pompeii and its surrounding settlements. Nearly 4 million people now live in the urban area of Naples.

living within a 60-mile (100 km) radius—only 311 active volcanic zones were found to be relatively uninhabited.

An example of a potential megadisaster where people have been crowding into a known volcanic hazard zone is the metropolitan area of Naples, Italy. Here nearly 4 million people now live within 18 miles (30 km) of the summit of Mount Vesuvius, an active and very dangerous composite cone (Figure 6.28). For perspective, consider that the metro area's 18-mile radius is about the same size as Mount St. Helens' blast zone (Case Study 6.2). Moreover, this dense urban area has been built on ash and pyroclastic flow deposits laid down during previous eruptions. Recent geologic studies have shown that over the past 25,000 years Mount Vesuvius has experienced, on average, a major eruption every 2,000 years. The most famous eruption occurred in 79 AD when pyroclastic flows and volcanic ash entombed approximately 3,600 residents of the Roman city of Pompeii and surrounding settlements (Figure 6.28). Archaeological studies have recently found that the 1780 BC eruption also killed thousands of people on the plains surrounding this composite cone. Geologic evidence shows that this eruption was considerably larger than the one that destroyed the Roman cities in 79 AD. Based on the distribution of ash and pyroclastic flow deposits, geologists estimate that a 900°F (480°C) pyroclastic cloud rushed along the ground at 240 miles per hour (385 km/hr), incinerating everything in its path—the cloud would have been hot enough to boil water as much as 10 miles (16 km) from the vent. Of major concern to scientists is the fact that based on its 2,000-year cycle, Mount Vesuvius is due for another major eruption.

Since more people are now living in volcanic hazard zones, an obvious question is what can society do to reduce the risks? One first must recognize that there is nothing humans can do to prevent the hazards themselves. Although we have had limited success in diverting lava flows, the only thing we can do with respect to most volcanic hazards is to simply get out of the way. Unlike earthquakes, where buildings can be constructed to withstand the intense shaking, the forces involved in volcanic blasts, pyroclastic flows, and mudflows are beyond our capacity to protect ourselves. Moreover, there is still the problem of the searing heat associated with many volcanic hazards. What society can do is take advantage of the fact that volcanic eruptions are almost always preceded by fairly reliable warning signs, such as swarms of earthquakes, topographic bulges, and escaping gases. The most effective way then of minimizing many volcanic hazards is for scientists to monitor precursor activity and evaluate the potential for an eruption. Based on this information, emergency managers can decide whether to implement evacuation plans for those who are at risk.

Predictive Tools

In this section we will explore the different tools volcanologists use to assess the potential for an eruption. Note that when it comes to noneruptive hazards (mudflows and landslides), the effectiveness of some predictive tools is rather limited. The best approach for noneruptive hazards is to install sensors to detect the movement of earth materials as they begin to move downslope. Once a slide or flow is detected, a signal is sent to an early warning system that automatically alerts residents farther down the drainage system so they can quickly move to higher ground.

Geologic History

When evaluating potential volcanic hazards, one of the first things volcanologists do is try to learn how a particular volcano has behaved in the past. Because written accounts do not go back very far in terms of geologic time, scientists commonly make use of volcanic deposits since the rocks themselves hold clues to how they formed. The way geologists do this is by performing field studies and making maps of the different types of volcanic deposits found around a volcano. Based on the size, shape, composition, and layering characteristics of its particles, a deposit can usually be identified as to its origin (i.e., lava flow, mudflow, pyroclastic flow, ash fall, or landslide). By using a map to view the way various deposits are distributed around a volcano, geologists can get a pretty good idea of the hazards associated with past events. For example, it was the discovery of ancient mudflow deposits around Mount Rainier that led geologists to voice their concern about the safety of communities located on top of these deposits.

Topographic Changes

Another useful tool for assessing the potential for a volcanic eruption is monitoring changes in the topography, or shape, of a volcano. When rising magma begins to collect in void spaces near the surface, it forms a reservoir called a *magma chamber*. Because the confining (overburden) pressures are low near the surface, the presence of pressurized magma commonly causes the volcano to swell or inflate. After an eruption the pressure naturally decreases and the volcano will deflate. In cases where the magma simply moves, then some parts of volcano will swell while others deflate. Therefore, by accurately surveying changes in the shape of a volcano over time, scientists can get an idea as to the position of magma within the volcano as well as the volume moving into the magma chamber. Note that modern global positioning system (GPS) receivers are now used to survey topographic changes.

On some active volcanoes the conditions can quickly change, making it too dangerous for scientists to revisit and take measurements. In these

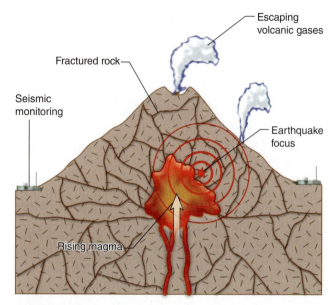

FIGURE 6.29 The monitoring of magmatic earthquakes and gases are key tools used in predicting volcanic eruptions. Portable seismographs record the rhythmic vibrations of magmatic earthquakes and allow scientists to track the magma body as it pushes upward through the fractured rocks. Measuring the chemistry of gas samples collected at the surface helps determine whether the magma is new, hence potentially more explosive.

FIGURE 6.30 A seismograph recording showing numerous earthquakes as measured at a monitoring station located on Mount St. Helens.

situations electronic tiltmeters can be installed that accurately measure very small changes in slope, and then transfer the data in real time to a safe location. Tiltmeters work on the same principle as a carpenter's level, the major difference being that an electronic sensor is used to determine the position of a bubble in a fluid-filled container. By installing numerous tiltmeters at strategic locations around a volcano, scientists can observe, simultaneously and in real time, those parts of the volcano that may be inflating and areas that may be deflating. This information naturally can be quite useful in tracking the movement of magma within the volcano and assessing the potential for an eruption.

Seismic Monitoring

Since earthquake activity almost always increases as magma moves toward the surface, seismic monitoring (Figure 6.29) is an excellent tool for predicting eruptions. Recall from Chapter 4 that earthquakes occur when rocks reach their elastic limit, at which point they fail and the strain they had accumulated is suddenly released in the form of vibrational wave energy. Although most earthquakes result from forces associated with the movement of tectonic plates (i.e., *tectonic earthquakes*), strain also accumulates when rising magma forces its way through crustal rocks, creating what geologists call **magmatic earthquakes** (sometimes called *harmonic tremors*). Like all earthquakes, magmatic earthquakes cause rocks to become displaced along faults, which makes it easier for the magma to move upward (Figure 6.30). A key point here is that as magma pushes its way to the surface, the resulting earthquakes vibrate in a steady and rhythmic (i.e., harmonic) manner. In addition to rhythmic vibrations, magmatic earthquakes have relatively low magnitudes and occur in distinct swarms, which may last an hour or more and consist of tens to hundreds of small earthquakes. This stands in sharp contrast to more powerful tectonic earthquakes that take place very abruptly and last a minute or two at most.

Because tectonic and magmatic earthquakes have such vastly different characteristics, it is a somewhat easy task for seismologists to tell the difference between these two types of earthquakes. Moreover, since magmatic earthquakes occur when magma is on the move, an increase in seismic activity under a volcano is a strong indication that an eruption may occur. In situations where people would be at risk in an eruption, seismologists typically place an array of portable seismographs (Figure 6.29) around a volcano. From this data they can determine the focal depth of each earthquake, and therefore monitor the position of the main magma body as it moves upward. Should the magma continue to move closer to the surface, then the probability of an eruption will naturally increase.

Monitoring of Volcanic Gases

Recall that the dissolved gases in magma primarily consist of water vapor (H_2O), carbon dioxide (CO_2), and sulfur dioxide (SO_2). As a magma body gets closer to the surface, it becomes increasingly likely that some of these highly pressurized gases will escape through fracture systems and be released into the atmosphere (Figure 6.29). Volcanologists typically monitor these gases at the surface on a fairly regular basis, and then look for changes in gas chemistry that may indicate a possible eruption. However, the use of this technique as a predictive tool is complicated by the fact that volcanic gases do not always originate from fresh magma moving up from depth. Other sources include heated groundwater and older magma leftover from a previous eruption. Determining the source is important since leftover magma is likely to have lost much of its gas pressure, which means the chance of an explosive eruption is far less than if the magma is fresh and fully charged with gas. Should the primary source turn out to be heated groundwater, then the worst-case scenario would be a large steam explosion.

Fortunately, the origin of volcanic gases can be determined from careful chemical analysis of samples collected directly at the surface. For example, if the samples consist almost entirely of water vapor, then one can conclude that the primary source is heated groundwater. On the other hand, if the samples contain appreciable amounts of CO_2 and SO_2 in addition to water vapor, then the source is most likely magma. Finally, should the proportion of individual gases within the samples be significantly different from samples taken earlier, then it can be inferred that new magma is moving up from depth. This, in turn, means there is a strong possibility that a major eruption could occur. The interpretation that fresh magma is moving upward can be verified by plotting the depth of focal points obtained from seismic data.

Geophysical and Groundwater Changes

In addition to ground deformation, earthquakes, and release of volcanic gases, rising magma can also change the physical properties of rocks—called geophysical changes—as well as the temperature and chemical composition of groundwater. When measuring various rock and groundwater properties over a period of time, scientists look for changes that may indicate that magma is getting closer to the surface. For example, we would naturally expect the temperature of both rocks and groundwater to increase as magma approaches the surface. Moreover, because magmatic gases and fluids commonly flow outward from the magma chamber along fracture systems, water samples from wells typically show an increase in acidity and sulfur content as magma gets closer to the surface. Finally, the electrical resistance of rocks will often change due to increased temperatures and circulation of conductive fluids and gases within the volcano.

Early Warning and Evacuation

The best way to minimize the risks associated with volcanic hazards is to use predictive tools in order to produce reliable eruption forecasts. The forecasts then provide an early warning so that officials can implement emergency response plans and allow people to safely evacuate. The ideal situation is to monitor a dangerous volcano with a variety of sensitive ground-based instruments, all of which are capable of transmitting data to scientists in real time. Such a system allows baseline (i.e., background) data to be collected that can later be compared to data that arrives when activity on the volcano begins to increase. The ability to compare to baseline conditions is important as it helps scientists to increase the reliability of eruption forecasts, which in turn provides more accurate early warning times for emergency managers.

Although early warning systems have proven to be very valuable, only a small number of the world's volcanoes that threaten populated areas are well monitored and have instruments with real-time data transmission capabilities. Most volcanoes are only lightly instrumented with seismographs whose sensitivity is not adequate for detecting the subtle earthquakes that are typical during the earliest stages of an eruption. Even worse is the fact that some dangerous volcanoes are not being monitored at all. This means that when they become active, scientists and emergency managers lose precious time and data when trying to assess the danger. Eruption forecasts in such cases tend to be less reliable. Because of this problem, the U.S. Geological Survey recently proposed a program for developing adequate monitoring capabilities on each of the United States' 57 volcanoes that the agency identified as being undermonitored.

Ultimately, the issue of having adequate instrumentation on volcanoes that threaten populated areas is important because emergency managers need reliable forecasts when making a decision on whether to order an evacuation. For example, there simply may not be enough time for a successful evacuation should officials wait to issue evacuation orders until scientists are more certain of an eruption. Such a delay could then lead to a large loss of life. On the other hand, should evacuation orders be given and an eruption not occur, then people will be less inclined to evacuate the next time, which may also lead to unnecessary deaths. Finally, one should keep in mind that the decision to evacuate is further complicated by the fact that evacuations are highly disruptive to the local economy and create serious hardships for individual citizens. Evacuations involve more than science; they include sensitive political and economic issues as well.

SUMMARY POINTS

1. Most of the world's active volcanoes are located along convergent and divergent boundaries of tectonic plates where magma is generated. Hot spot volcanoes occur in the interior of plates where a plume of magma rises up from the mantle.

2. There are three basic types of magma: basaltic, andesitic, and rhyolitic (granitic). Basaltic magmas are the dominant type at divergent plate boundaries, whereas andesitic magmas are most common at convergent boundaries along subduction zones. Rhyolitic magmas form when continental crust becomes involved in the melting process at convergent and hot spot locations.

3. The viscosity of magma refers to its ability to resist flowing. Magma viscosity increases with increasing SiO_2 (silica) content and decreasing temperature. Basaltic magmas are the least viscous and rhyolitic the most viscous.

4. When rocks melt into magma, gases are formed and generally remain dissolved within the magma, with water (H_2O) vapor, carbon dioxide (CO_2), and sulfur dioxide (SO_2) being the most abundant gases. Highly gas-charged magmas contain mostly water vapor and generally form in subduction zones where water-rich sediment is involved in the melting process.

5. Magma deep within the Earth is under considerable pressure, forcing magmatic gases to remain dissolved within the magma. When magma breaches the surface and encounters atmospheric conditions, the compressed gases rapidly expand and can create an explosive eruption.

6. The type of landforms that develop when magma reaches the surface largely depends on the relative proportion of lava, gas, and pyroclastics that are extruded. Explosive eruptions generate considerable amounts of pyroclastics, but relatively small volumes of lava. The opposite is true for nonexplosive eruptions.

7. Shield volcanoes have gentle slopes and usually form by the eruption of gas-poor, low-viscosity basaltic magmas, which tend to erupt in a nonexplosive manner. Composite cones have steep slopes and form during explosive eruptions of gas-rich, high-viscosity andesitic magmas.

8. Some volcanic hazards such as explosive blasts, lava flows, and pyroclastic flows typically occur in a relatively small radius around the volcano. Other hazards like mudflows, ash fall, and tsunamis can threaten people considerable distances from the volcano itself. Noneruptive hazards can include volcanic landslides, mudflows, and deadly gases.

9. Explosive eruptions can have global consequence by ejecting large quantities of volcanic ash and gas into the upper atmosphere, causing a cooling effect over the entire planet.

10. As population continues to expand, greater numbers of people are living in areas of volcanic hazards, thereby increasing the potential loss of life and property.

11. Volcanic eruptions are normally preceded by precursor activity, such as topographic changes, swarms of earthquakes, and release of volcanic gases. Scientists can measure these precursors and make fairly reliable eruption forecasts. The forecasts then provide an early warning for implementing emergency response plans and evacuating people, thereby minimizing the effects of an eruption.

KEY WORDS

caldera   170
cinder cones   166
composite cone   167
crater   170
hot spots   160
lava domes   165

lava flow   165
magma   159
magma chamber   163
magmatic earthquakes   188
pyroclastic flow   176
pyroclastic material   164

shield volcanoes   166
viscosity   162
volcanic ash   164
volcanic landslide   181
volcanic mudflow   181
volcano   166

APPLICATIONS

Student Activity   Buy two small bottles of club soda or seltzer water. Do this activity outside. Without shaking, open one bottle and let it stand. Shake the second bottle and open immediately. What happens? There is dissolved gas in both, but the shaken one "erupts" more violently. Shake it up a second time and open. Was the "eruption" as violent as the first time? Shake again and open. There should be a great decrease in the level of eruption. This shows the energy that dissolved gases have in magma. The more violent the eruption, the greater the chances that there are a lot more dissolved gases in it.

Critical Thinking Questions
1. Why are there so many volcanoes in the Ring of Fire?
2. Why are basaltic eruptions more fluid than rhyolitic ones?
3. How does a shield volcano differ from a composite cone?
4. What is a hot spot? Name two.

Your Environment: YOU Decide   If you had to live near a volcano, what kind would you choose?

Mass Wasting and Related Hazards

LEARNING OUTCOMES

After reading this chapter, you should be able to:

▶ List the main factors that affect slope stability as they relate to the balance between gravitational and frictional forces acting on a slope.
▶ Describe the different types of weakness planes found in earth materials and how they can affect the stability of slopes.
▶ Explain the two ways in which water acts to destabilize a slope.
▶ List the four main triggering mechanisms for mass wasting events.
▶ Explain the fundamental differences between the following types of mass wasting: falls, slides, flows, slump, and creep.
▶ Understand why subsidence occurs and the geologic conditions that lead to rapid or gradual subsidence.
▶ Describe the primary ways in which humans can reduce the risk of mass wasting.

In 2005 earth materials moved downslope and severely damaged several homes and a roadway in Laguna Beach, California. This movement, called mass wasting, is a natural process that results from the pull of gravity on sloping surfaces composed of rock or sediment. Certain triggering events, such as heavy rains, earthquakes, and construction activity, can cause less stable slopes to suddenly fail. Buildings and other human structures in sloping terrain are naturally at risk from mass wasting processes.

Introduction

In Chapter 4 you learned that Earth's moving tectonic plates generate long underwater mountain ranges and deep ocean trenches, creating considerable topographic *relief* (elevation differences) in the ocean basins. Plate movements also result in parts of Earth's land-masses being elevated far above sea level. While plate tectonics gives the globe its vertical relief, gravity works in the opposite sense by moving rock and sediment from areas of high elevation to low-lying areas, thereby acting to fill in Earth's rough surface. Gravity alone of course can directly move rock and sediment downslope (i.e., downhill), as in rocks falling from a cliff, or it can indirectly move material through the action of flowing water, ice, and air. Geologists use the term **mass wasting** to describe the general process of earth materials moving downslope due only to gravity, whereas those materials carried by some secondary agent fall under the category of stream, glacial, or wind transport (Figure 7.1). Note that the terms *landslide* and *avalanche* are often used synonymously with mass wasting, but actually imply specific types of movement. In order to avoid confusion, only

Okanagan Valley, British Columbia, Canada

FIGURE 7.1   Photo showing how gravity caused part of a hillside to slide downslope in a process known as mass wasting. Note how the rock and sediment slid into the stream valley where running water will transport the material downstream (toward the top of the photo). Together with sediment transported by water, wind, and ice, mass wasting plays an important role in shaping the landscape.

the term *mass wasting* will be used to describe the general movement of material by gravity alone.

Although mass wasting processes play an important role in many different aspects of the Earth system, our focus in Chapter 7 will be on how it affects humans. The basic problem is that when a body of rock or sediment begins to move downslope, anything in the path of this material is in danger of being destroyed. People and human-made objects typically are no match for the forces generated by moving masses of earth materials, especially when the material is moving quickly. Mass wasting then not only places human lives at risk, but also threatens valuable buildings, transportation networks, and utility lines (water, sewer, and electric).

Similar to other natural processes such as earthquakes and volcanic eruptions, some of the worst mass wasting disasters have occurred when people live and work in a hazard zone. For example, from the photo in Figure 7.2 one can see the effects of an earthquake-triggered mass wasting event in 1970 that buried parts of two cities in Peru, killing an estimated 18,000 people. Note how this event created a fan-shaped deposit at the mouth of a canyon where the cities had been built on more flat-lying terrain. This disaster is very similar to the volcanic mudflow described in Chapter 6 that killed 23,000 residents of Armero, Colombia. Both of these disasters involved mass wasting processes where large numbers of people were living in a hazard zone. Keep in mind that the cities were located in hazard zones because the sites offered relatively flat terrain and an abundant water supply at the mouth of canyons. Moreover, several generations had passed between major events, thus many were either ignorant or complacent regarding the risks they faced. In the end, population growth helped ensure that more people would be living there when earth materials came crashing down the valleys as had happened repeatedly in the geologic past.

While large-scale mass wasting events can have devastating consequences, they are naturally quite rare. Much more common are small-scale movements, whose cumulative effects are rather substantial due to the fact they are so numerous. For example, consider that in the United States alone, mass wasting is estimated to cause $1–2 billion in damage and 25 to 50 deaths, *each* year. One reason why small-scale events are so numerous is that mass wasting can occur even on moderate slopes, which means it is a potential problem over a significant portion of the landscape. In addition to being a common process that is geographically widespread, people often inadvertently trigger mass wasting through routine activity. For example, excavating material from hillsides for construction purposes and harvesting timber and growing crops on sloping terrain are activities that tend to destabilize slopes, and therefore trigger mass wasting. Ironically, society allocates

FIGURE 7.2 Aerial photo showing the source area for a rock and snow avalanche that destroyed much of the Peruvian city of Yungay in 1970—parts of Ranrahirca were also destroyed. Note how the cities were built at the mouth of a canyon that leads up to the source area on Mount Huascarán.

FIGURE 7.3 Photo showing the Beartooth Highway traversing steep terrain northeast of Yellowstone National Park. Construction of the highway required excavating material from the hillside, which made the slopes even more prone to mass wasting—note the scars left by repeated movements of rock and sediment. This photo was taken in 2005 after a series of rockslides and mudslides had cut or blocked the road in 13 separate locations, requiring extensive and costly repairs.

considerable sums of money and resources to address mass wasting problems resulting from routine human actions.

Perhaps the best example of the interaction between human activity and small, but numerous mass wasting events is the construction and maintenance of highways. In order to connect towns and cities, highways naturally have to be constructed across a variety of different terrain. As shown in Figure 7.3, constructing a flat roadbed that winds up and down steep terrain requires that material be excavated from the hillsides. The problem is that removing this material only helps destabilize steep slopes that are already prone to mass wasting—note the scars in the photos where rocks and sediment have routinely moved downslope. This results in costly maintence and emergency repairs that must be performed indefinitely. In some cases cost and safety considerations may be so great as to justify building a tunnel.

In the following sections we will explore in more detail the different types of mass wasting and the ways in which they occur. Of particular importance will be the topic of how human activity can help trigger these events, which, in turn, leads to costly repairs and maintenance. Lastly, we will examine some of the more common engineering techniques used to stabilize slopes and minimize the hazards associated with mass wasting.

Slope Stability and Triggering Mechanisms

Although every rock and mineral grain on our planet is under the influence of Earth's gravitational force, some slopes are inherently less stable than others and are therefore more prone to mass wasting. One of the key factors affecting the stability of a slope is its steepness, which can range from almost horizontal to completely vertical. Figure 7.4 illustrates how the pull of Earth's gravity on a rock creates the force we call weight. Here we see that on a horizontal surface (a zero slope) a rock's entire weight is directed vertically downward, making movement by gravity alone impossible. However, on an inclined surface a portion of the gravitational force, which itself is a constant, will act parallel to the slope. This means that on a sloping surface some of the rock's weight is directed downslope. As the slope increases, so too does the gravitational component acting parallel to the slope, which, in turn, allows more of the rock's weight to be directed in the downslope direction.

Most of us know from experience that rocks and sediment grains are able to remain stationary on a hillside despite the force of gravity. The reason for this is frictional forces acting to keep them in place, as illustrated in Figure 7.4. You can demonstrate this for yourself by placing a block of wood on a board, and then gradually lifting one end until the block begins to slide. What happens is that the gravitational component in the slope direction continues to increase until it finally overcomes the frictional resistance, at which point the block starts to slide. The same thing happens on a hillside where rocks and sediment remain in place until something causes the gravitational force to become greater than the frictional forces.

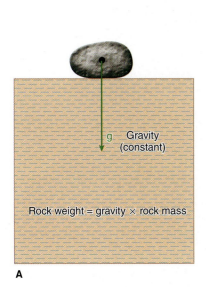
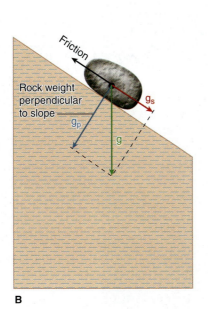
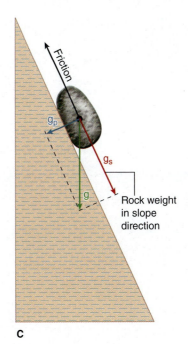

A   **B**   **C**

A   Gravity (constant)

Rock weight = gravity × rock mass

Friction

Rock weight perpendicular to slope

Friction

Rock weight in slope direction

FIGURE 7.4
On horizontal (A) terrain (zero slope), the full weight of the rock is directed downward due to gravity (g). On a hillside (B), part of the gravitational force (g_s) acts parallel to the slope, thereby directing some of the rock's weight downslope; the remaining component (g_p) acts perpendicular to the slope. On steeper slopes (C) more of the rock's weight is directed downslope, which requires greater friction to hold it in place.

When this happens the movement can occur quickly, such as in a rockfall, or so slowly that it is imperceptible to the human eye.

In order for us to better understand slope stability and the potential for mass wasting, we need to examine the various factors that influence the gravitational and frictional forces on a slope. In addition to steepness, these factors include the type of rock or sediment making up the slope, the presence of water or ice, and the amount of vegetative ground cover. We will also need to explore what geologists and engineers call **triggering mechanisms,** which are processes or events that reduce the frictional forces on a slope and/or increase the effect of gravity. Examples of mass wasting triggers include earthquakes, heavy rains, and wildfires that remove vegetation. We will begin this section by considering the types of material that make up the slope itself.

Nature of Slope Material

Some rocks are inherently so strong and homogeneous that they are able to form stable cliffs, such as the granitic rocks of Yosemite Valley, California shown in Figure 7.5. Here slabs of rock separate from the main rock body along fractures, leaving walls that have remained almost vertical for thousands of years. These walls are stable because the minerals in the massive granite are only slightly weathered, forming an interlocking network of crystals with great frictional resistance. In contrast to the towering cliffs in Yosemite, the alternating layers of sedimentary rocks in the Grand Canyon

FIGURE 7.5   This nearly vertical 2,000-foot (600 m) cliff has existed for thousands of years on Half Dome in Yosemite National Park due to the strength of the granite's interlocking mineral grains and resistance to weathering.

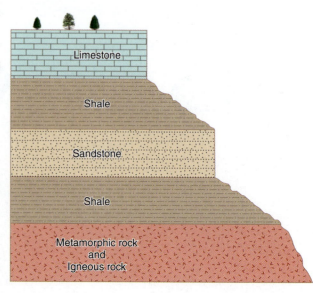

FIGURE 7.6 Differences in strength and resistance to weathering of sedimentary layers control the steepness of slopes in the Grand Canyon. Steep cliffs develop in resistant sandstones and limestones, whereas broad, gentle slopes form in the much weaker shales.

FIGURE 7.7 Planar surfaces such as bedding planes, faults, fractures, and foliation represent weaknesses within rocks that can greatly reduce slope stability. Particularly dangerous situations occur when these surfaces are inclined in the same direction as the slope, creating the potential for blocks of material to slide downslope.

(Figure 7.6) are each composed of different types of minerals that vary greatly in their strength and ability to resist weathering. The weathering of these alternating layers of relatively weak and strong rocks creates a stair-step effect in the slope as opposed to a single massive cliff face. Note in Figure 7.6 that the steep, but relatively thin, cliffs in the Grand Canyon form from individual layers of sandstone and limestone which are resistant to weathering and inherently strong. Between these resistant layers are easily weathered beds of shale whose internal frictional resistance is much lower, allowing only gentle slopes to develop.

From these examples we see that maintaining a nearly vertical cliff requires a material with strong internal friction, such as homogenous rocks with minerals that interlock and are resistant to weathering. Obviously, frictional forces within loose or unconsolidated sediments are usually lower than solid rock, making sediment more prone to mass wasting and less able to form vertical slopes. Depending on water content and the size and shape of the individual sediment grains, loose material typically does not form slopes greater than 35 degrees—geologists call this the *angle of repose*. Generally speaking, large angular fragments generate greater frictional forces, and therefore are capable of maintaining steeper slopes compared to small, well-rounded fragments. In terms of composition, sediments containing considerable water and clay minerals tend to form the least stable slopes because they can behave as a plastic material (Chapters 3 and 4), which allows them to flow and spread out.

We also need to consider the fact that earth materials commonly contain planar or sheetlike features that weaken them considerably. For example, both sediment and sedimentary rocks contain *bedding planes*, which are horizontal surfaces that often represent some change in the sediment during deposition. The important point to note here is that sedimentary materials tend to split or break along these surfaces. Should tectonic activity cause these materials to become inclined, slippage can occur along bedding planes such that overlying sections of rock glide downslope, as shown in Figure 7.7. Common to nearly all rock types are faults and fractures which represent planar breaks or openings in a rock body (Chapters 3)—sometimes these features are even found in semiconsolidated (partially cemented) sediments. Similar to bedding planes, masses of rock can slide along fracture and fault surfaces that are inclined in the same direction as the slope (Figure 7.7). Finally, foliation planes that commonly develop in metamorphic rocks (Chapter 3) can serve as a sliding surface in the same way as do bedding planes, fractures, and faults.

Oversteepened Slopes

On steeper hillsides the component of gravity operating the direction of slope is greater, which, in turn, increases the potential for mass wasting. There are a number of ways, both natural and unnatural, where the steepness of a particular slope can change rather suddenly in terms of geologic time. Therefore, any activity that abruptly increases the slope and leads to mass wasting can be considered a triggering mechanism. Perhaps the most geologically important and common mass wasting trigger is the undercutting of stream banks (called *cutbanks*) due to the natural migration of stream channels as shown in Figure 7.8A. When a stream channel migrates and undercuts its bank (Chapter 8) it creates a highly unstable overhang, which inevitably will fall or slide into the channel. Interestingly, the primary way in which river valleys become wider over time is by mass wasting caused by this process of streams undercutting and destabilizing their banks.

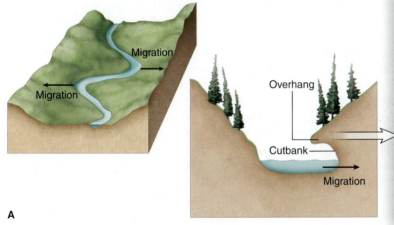

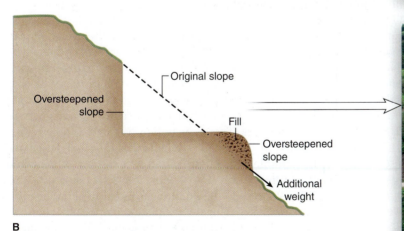

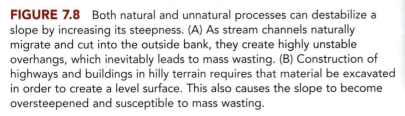

FIGURE 7.8   Both natural and unnatural processes can destabilize a slope by increasing its steepness. (A) As stream channels naturally migrate and cut into the outside bank, they create highly unstable overhangs, which inevitably leads to mass wasting. (B) Construction of highways and buildings in hilly terrain requires that material be excavated in order to create a level surface. This also causes the slope to become oversteepened and susceptible to mass wasting.

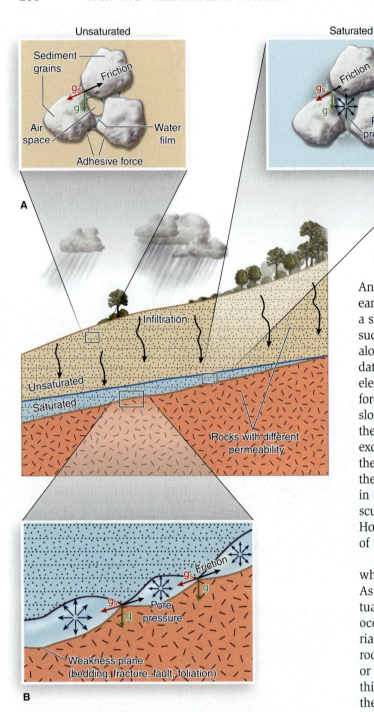

FIGURE 7.9 Rain or melting snow will infiltrate and eventually cause subsurface void spaces to become saturated. The weight of the water in the saturated zone causes the fluid or pore pressure within the voids to increase (A), which reduces the friction between the solids. Downslope movement occurs when the frictional forces become less than the gravitational force in the slope direction. Note the enlarged view (B) shows the irregular nature of most planar surfaces.

In addition to natural processes, human activity often results in oversteepened slopes and costly mass wasting problems. For example, flat surfaces are required for the construction of roads, buildings, and parking lots. In sloping terrain this means that material must be excavated from hillsides as shown in Figure 7.8B in order to create a level surface. This weakens the slope significantly and greatly increases the potential for mass wasting. Note that this process can also trigger mass wasting below the cut due to the additional weight of the excavated material. Later we will examine how engineers try to avoid these problems by building retaining walls above the cut and by hauling the excavated material away rather than placing it on the slope.

Water Content

Another common mass wasting trigger is the addition of water to permeable earth materials because of the way it upsets the balance of forces existing on a slope. As illustrated in Figure 7.9, earth materials commonly contain voids, such as the *pore space* (Chapter 3) between individual grains and the space along faults, fractures, bedding, and foliation planes. In the case of unconsolidated sediment, when a thin film of water forms within the pore space, the electrical attraction between the water molecules and mineral grains (adhesive forces) will make the sediment stronger. This, in turn, acts to strengthen the slope. On the other hand, if the pores become saturated (Figure 7.9A), then the additional weight of the water will tend to destabilize a slope. Here the excess water also reduces the frictional forces between the individual grains, thereby causing the material to become weaker. Water therefore can reduce the stability of a slope because it both increases weight *and* reduces friction in the direction of the slope. A good example is how relatively strong sand sculptures can be made by adding moderate amounts of water to beach sand. However, if too much water is added, then the sculpture will collapse because of the additional weight and loss of friction between the grains.

The process where a hillside is destabilized by water usually begins when rain or melting snow infiltrates the subsurface as shown in Figure 7.9. As the infiltrating water moves downward through void spaces, it will eventually encounter some material with relatively low permeability. When this occurs the water is forced to slow down, and the voids in the overlying material will begin filling with water. The filling of voids is critical because when rock or sediment becomes saturated, the weight of the water generates fluid or **pore pressure** that acts outward in all directions within the voids. Because this fluid pressure is also acting in the opposite direction of gravity, it reduces the weight of the earth material bearing down at the contact points between solid grains and planar surfaces. This reduction in weight naturally reduces the frictional force within the materials, which makes the slope less stable. Consequently, if the saturated thickness within a rock or sediment body keeps increasing, due to heavy rains for example, then the rising pore pressure can reduce frictional forces to the point where solid material begins to move downslope. A particularly dangerous situation can occur when pore pressure reduces friction along a weakness plane (Figure 7.9B), allowing the entire mass of overlying material to begin moving.

Climate and Vegetation

The long-term average weather for a region is defined as *climate,* which is an important factor in slope stability because it ultimately determines how and when precipitation falls. Climate also determines the types of vegetation we see blanketing the various slopes, which influences the fraction of rain or snow that infiltrates into the subsurface. For example, *humid climates* are characterized as having relatively consistent amounts of precipitation (rain and snow) distributed throughout the year (e.g., the Pacific Northwest and southeastern United States). This supports fairly dense vegetation that tends to stabilize slopes as the plant roots help bind together loose particles of rock and sediment. However, during unusually large rainstorms or rapid snow-melts, dense vegetation will increase infiltration since it reduces the ability of surface water to move downslope. Excessive infiltration adds significant weight to a slope and reduces friction through higher pore pressures. There-fore, under normal conditions, dense vegetation helps to stabilize slopes, but during heavy and prolonged precipitation events, the vegetation facilitates infiltration and leads to less stable slopes.

A much different situation exists in warm *arid climates* where relatively small amounts of precipitation produce sparse vegetation (e.g., the south-western United States). Although rainfall is infrequent in these regions, when it does occur it can be fairly intense. Intense rains therefore combined with sparse vegetation makes it easier for loose material to move downslope in arid climates. Note that a similar situation can occur in humid climates where wildfires or logging activity suddenly removes vegetation from a hillside. Because of the rapid way in which the slope is destabilized, the loss of vegetation can be considered as a triggering mechanism that will initiate mass wasting. Once the vegetation is gone, mass wasting may occur during a rainfall event that would not otherwise cause the slope to fail. Vegetation, of course, will eventually reestablish itself, but the slope will remain unstable during the period of regrowth. Note that in California there is a fairly regular pattern of wildfires during the dry summer conditions, followed by mudslides that are triggered by winter rains. This explains some of the state's frequent mass wasting events. Finally, we should men-tion that in cold climates (e.g., the U.S. Upper Midwest and Canada), where freeze and thaw cycles are common, material can move downslope due to the expansion and contraction of water.

Earthquakes and Volcanic Activity

The ground vibrations associated with earthquakes and volcanic eruptions are another common type of triggering mechanism for mass wasting. For example, the massive landslide described earlier that took the lives of 18,000 people in Peru was triggered by a large (magnitude 7.9) earthquake. The 1980 eruption of Mount St. Helens (Chapter 6) also triggered a land-slide of gigantic proportions. Recall that these types of processes release tremendous amounts of energy, which are then transformed into seismic wave energy (Chapter 5) that travels outward in all directions. As the seis-mic waves pass along the surface, the least stable slopes will tend to fail when the ground vibrations suddenly reduce the frictional forces within the slope materials. Clearly, we can expect that stronger earthquakes and volcanic eruptions would cause more slopes to fail due to the higher levels of ground shaking. Also note that passing seismic waves often cause sur-face materials to liquefy, which can immediately destabilize a slope and trigger a mass wasting event.

In addition to seismic waves, volcanic activity can trigger massive mudflows when lava or hot debris causes rapid melting of a volcano's snow and ice cap, such as the Armero disaster described in Chapter 6. Gases and fluids commonly found within a volcano can also weaken the rocks to the point that slope failure occurs. It is now believed that the internal break-down of rocks has played a major role in triggering the large landslides known to have occurred repeatedly on Mount Rainier (Chapter 6).

Types of Mass Wasting Hazards

As noted at the beginning of this chapter, because mass wasting processes play an important role in the Earth system they also help shape the environment on which we humans depend. At the same time, however, these processes pose serious risks to human safety and society's infrastructure, such as buildings, bridges, and highways. Mass wasting is a problem because it involves large masses of earth material, whose downhill motion will obliterate, crush, or bury everything in its path. In some cases the material moves so slowly that the motion is imperceptible to humans, but still powerful enough to destroy or cause considerable damage to human-made objects. Perhaps most familiar, and most common, are situations where the sudden movement of material damages a section of highway, creating an immediate hazard to motorists and requiring difficult and costly repairs.

Because mass wasting presents serious and costly problems for society, geologists and engineers have devised different techniques and strategies for reducing the risk associated with these earth movements. The particular approach that is used largely depends on the type of mass wasting. For example, minimizing the effects of rocks free-falling from a cliff requires an entirely different approach than that used to keep wet, finer-grained sediment from flowing downhill. Scientists and engineers who study mass wasting, and try to mitigate its effects, naturally find it useful to classify the different types of mass wasting. However, due to the number of variables involved, no single classification system has been devised which adequately describes all the different types of mass wasting processes. This, in turn, has resulted in a somewhat confusing and overlapping set of terms.

In order to simplify our discussion of mass wasting processes, we will examine the different types in terms of the type of movement, namely falling, sliding, flowing, and creeping. As the name implies, a *fall* involves material free-falling from a cliff or tumbling down a steep slope. On the other hand, a *slide* is where a body of material moves downslope along some surface, and in a *flow* the material moves with the consistency of a

FIGURE 7.10 Mass wasting can be categorized based on the type of material involved and the way in which it moves downslope. Blanked-out areas indicate those combinations of materials and movement that occur rarely, or not at all.

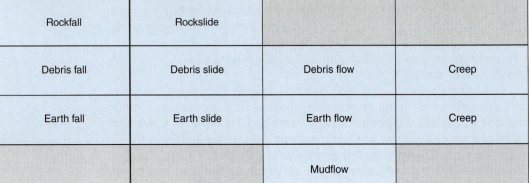

Type of motion

	Fall	Slide	Flow	Creep
Rocks (large blocks of solid rock) 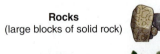	Rockfall	Rockslide		
Debris (mixture of rock, earth, plants, and mud)	Debris fall	Debris slide	Debris flow	Creep
Earth (loose sediment, weathered rock fragments)	Earth fall	Earth slide	Earth flow	Creep
Mud (mixture of water and finer-sized sediment) 			Mudflow	

Type of material

viscous fluid. The last type of movement is *creep,* which refers to the imperceptibly slow movement of loose material (earth or debris). By combining the kind of movement with three basic types of slope materials, namely *rock, earth,* and *mud,* we get the simplified classification system shown in Figure 7.10— the exception to this naming convention is creep. This system is well suited for our discussion in the following sections because the terminology is both informative and easy to understand. Note that some combinations of materials and movement are not represented in Figure 7.10 (e.g., mud fall, rock flow) either because they are rare or do not occur at all. Finally, a very common process called *slump* is not listed in this classification as it is a hybrid that involves both sliding and flowing movements.

Falls

A **fall** is the rapid movement of earth materials falling through air. As illustrated in Figure 7.11, *rockfalls,* sometimes called *topples,* involve relatively small amounts of material that begin when a slab or block of rock becomes detached from a steep wall of solid rock. This process typically includes the repeated freezing and thawing of water within fractures or other planar surfaces in a rock body. When liquid water freezes into ice it naturally expands, which then acts to slowly wedge a block of rock away from the wall until it is allowed to free-fall through the air. The falling block will usually crash into the cliff face before breaking up into smaller pieces as it hits the previously fallen rocks at the base of the cliff. These smaller pieces of rock will then bounce, roll, or slide on top of the existing rock pile, often dislodging individual rocks such that they too start to move downslope. Eventually the entire pile of rock attains a stable slope between 30 and 35 degrees—referred to as the *angle of repose.* Over time this process produces a cone-shaped deposit of rocks called a **talus pile,** which are common features found at the base of exposed rock bodies as shown in Figure 7.11.

Rockfalls are an obvious hazard, but they are also an important geologic process that helps widen valleys and lower mountain ranges, piece by piece over long periods of time. Clearly this material must go somewhere, otherwise talus piles would simply grow until a valley is filled in. What normally happens is that the rocks at the bottom of the pile are slowly incorporated into streams or glaciers, which then carry the material away and deposit it elsewhere. Although most falling debris consists of solid rock, *earth falls* and *debris falls* are common in areas where migrating rivers undercut stream banks made of unconsolidated materials (refer back to Figure 7.8A). In this way earth falls and debris falls help supply sediment to a stream, which then carries it away in a similar manner as rocks from a talus pile.

Slides

As the name implies, a **slide** occurs when material moves in a sliding manner on some zone of weakness, such as along bedding planes, faults, fractures, and foliation planes. Slides that involve masses of rock, earth, or debris (mixtures of rock and earth) are referred to as *rockslides, earth slides,*

FIGURE 7.11 Rockfalls generally result from repeated freezing and thawing of water within fractures in steep exposures of solid rock. Expanding ice creates a wedging effect that eventually pushes the slab outward to the point where it free-falls. Over time this process creates a deposit of broken rocks at the base of the cliff called a talus pile, which can then be carried away by streams and glaciers.

FIGURE 7.12 Rockslides (A) consist of blocks of solid rock sliding on top of a weakness plane such as bedding, foliation, faults, and fractures. Earth and debris slides (B) are less coherent and tend to break up and move as a jumbled mass. Common triggering events include streams undercutting their banks and infiltrating water that increases the pore pressure along less permeable layers.

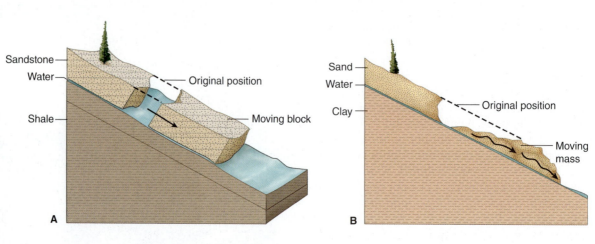

FIGURE 7.13 Photo showing the aftermath of the 2001 debris slide that killed over 1,100 people in the village of Guinsaugon on Leyte Island, Philippines. A period of unusually heavy rains is believed to have played a major role in triggering the deadly slide.

or *debris slides,* respectively. As illustrated in Figure 7.12, slides are more likely to take place when the planes of weakness are inclined in the same direction as the slope of the land surface. In the case of rockslides, it is common for one or more large blocks to move downslope as a single unit. However, in earth and debris slides the material above this weakness zone is more likely to break up and form a jumbled mass that moves downslope on top of a relatively stable and undisturbed layer of material.

The potential for a slide is also high in areas where a permeable layer, such as sand or sandstone, overlies a less permeable layer of shale or clay (Figure 7.12). As described earlier, infiltrating water from prolonged rains or melting snow is forced to slow down once it reaches the less permeable layer. This causes water to accumulate in the overlying layer, which leads to the slope becoming less stable. Here the extra water increases the weight acting in the direction of the slope, plus it raises the pore or fluid pressure such that the frictional forces are reduced along the contact between the two layers. In addition to the effects of water, the slope can be further destabilized by the undercutting action of a migrating stream or highway construction. Slides are a particularly serious problem for society because they commonly involve large volumes of material that move onto highways and into flat-lying areas where humans tend to live and build.

An unfortunate example of the hazards associated with slides occurred in 2006 on Leyte Island in the Philippines. Here over 1,100 people perished when a massive debris slide buried much of the village of Guinsaugon, which was built in a valley at the base of the steep hillside shown in Figure 7.13. Material from the slide reached nearly half the width of the valley, traveling as much as 2.4 miles (3.8 km) from the source area. Scientists later estimated that the slide reached a velocity of 78 miles per hour (126 km/hr), which made it nearly impossible for the residents to escape. It was also concluded that the primary triggering mechanism was a period of unusually heavy rain that ended four days prior to the slide.

Slump

As mentioned earlier, a **slump** is a complex form of mass wasting where both sliding and flowing take place in unconsolidated material (earth and debris materials). From Figure 7.14 you can see that in the upper portion of a slump the material slides along a curved surface whose shape is similar to a spoon. At the very top of the slump is a curved and distinct scar called a *scarp,* which marks where the disturbed material becomes detached from the undisturbed material. Note how at the base of the slump, also called the *toe,* a prominent bulge develops. Because slumps generally occur in unconsolidated materials with a high water content, the bulging mass near the toe tends to flow (see the section on "Flows"), giving the terrain a jumbled or disorderly appearance. Slumps are problematic in that once they form, additional movement is difficult to prevent, particularly if the toe is removed in order to clear debris away from a road or building site. Here the mass within the toe actually helps to support the slide material lying farther up the slope. Therefore, when material is cleared away from the toe, the entire mass becomes destabilized and triggers additional slumping at the top, which then sends more material downslope. The complex slump and flow that occurred at La Conchita, California (Case Study 7.1), provides a good example of the problems associated with recurrent movement.

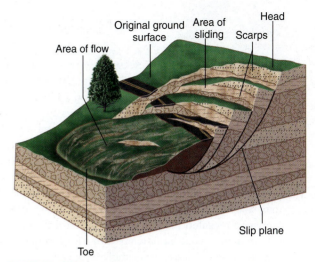

FIGURE 7.14 Slumps are complex events where material moves by sliding along a spoon-shaped surface near the top, but then flows toward the bottom or toe. Note the distinctive scars called scarps at the top and the jumbled terrain near the toe.

Flows

In Chapter 4 you learned that when solid rock, deep within the Earth, is subjected to enough heat and pressure, it will begin to flow and deform in a plastic manner. For our purpose here we will consider a **flow** to be one where surface material moves in a continuous manner due to some external force. Take for example how gravity will cause liquids, such as water or oil, to flow when they are on a sloping surface. A similar process occurs when loose (unconsolidated) material covering a hillside accumulates enough water so that internal friction is reduced, allowing it to behave like a fluid and start flowing downslope. The speed at which the material moves depends on the type of material involved, the amount of water it contains, and the steepness of the slope. Perhaps the most common type of mass wasting involving flow is a *mudflow,* which is composed of mostly fine-grained sediment and enough water to enable it to flow downslope. In mudflows the relative amounts of sediment and water may vary considerably, making some so viscous the entire mass moves rather slowly, whereas others are more fluid and resemble a turbulent river heavily overloaded with sediment.

Recall from Chapter 6 that volcanic mudflows (lahars) can form when heavy rains pick up ash lying on the flanks of composite volcanoes, and then carry it off into stream channels. Volcanic mudflows can also begin near a volcano's summit when glacial ice undergoes rapid melting. Note that mudflows form in nonvolcanic regions as well, particularly in rugged terrain where vegetation is sparse because of an arid climate or recent wildfires. Since sparse vegetation offers less protection against the impact of falling raindrops, heavy rain events can loosen significant amounts of fine sediment, making it easier for running water to carry it away. A key point to note is that regardless of whether a mudflow begins on a volcanic or sparsely vegetated slope, the resulting mudflows are quite similar. From

Recurrent Mass Wasting at La Conchita, California

On January 10, 2005, a debris flow crashed into the small residential community of La Conchita, California, taking the lives of 10 people and completely destroying 23 homes and damaging 13 others. This disaster drew national news coverage in the United States, and then the story quickly took on an added dimension when reporters learned that a similar event occurred at the same site 10 years earlier in 1995. Questions naturally arose as to why people would continue to live at the base of a hillside with a history of mass wasting. Other questions centered on why the local government allowed people to build there in the first place. Television interviews revealed that many of the local residents felt that living in a pleasant seaside community with easy access to the beach and wonderful ocean views was simply worth the risk. To help under-

stand why the local government allowed people to live in a known hazard zone, it will be useful for us to examine the geology of the site and its historical human development.

La Conchita lies on a narrow strip of land located between the Pacific Ocean and a steep hillside (Figure B7.1). These bluffs consist of poorly consolidated layers of marine sediment that have recently been uplifted by tectonic activity. Modern geologic studies have shown that mass wasting has been taking place along these same bluffs for many

FIGURE B7.1 Color infrared photo (vegetation in red) showing the town of La Conchita, California, located below bluffs overlooking the Pacific Ocean. The bluffs consist of uplifted marine sediments that are weak and prone to mass wasting after periods of heavy rain. The inset shows the area where movement occurred in 1995, and then again in 2005.

Scars from previous movement

Ranch Road

La Conchita

thousands of years. In terms of human development, historical records as far back as 1865 note that a wagon trail along this section of the coast was plagued by masses of earth falling down from the bluffs. Later when a railroad was built along the coastal strip, it too experienced problems when earth movements buried the tracks in 1887 and 1889. In 1909 another event destroyed a work train, prompting the railroad company to try to reduce the hazard by removing part of the hillside using bulldozers. The idea was to create a relatively flat area that would collect earth material moving down from the bluffs, thereby minimizing the chance of any material reaching the railroad tracks. However, this operation also created level ground that attracted the attention of real-estate developers, who eventually purchased the property in 1924. A housing development soon emerged with 330 lots with easy access to the ocean, and an additional 47 lots were located along the base of the bluffs themselves (Figure B7.1). La Conchita was thus born, but it was only a matter of time before earth materials would again start moving downslope, posing a threat to this new community.

Small-scale earth movements took place along the bluffs above La Conchita in 1988, 1991, and 1994; fortunately none were large enough to reach the town below. Then on March 4, 1995, a large section of the bluff moved several tens of meters downslope in a matter of minutes, taking the form of a combined slump and earthflow (Figure B7.1). Although this relatively slow-moving mass damaged or destroyed a total of nine homes, no one was injured. Geologic studies later determined that the event was triggered by unusually heavy rain that winter. For example, in the six-month period leading up to the event, this coastal region received nearly 30 inches (761 mm) of rain, which was nearly double the normal average of 15.4 inches (390 mm). Also significant is the fact that 24.5 inches (623 mm) of rain fell in January alone, a month where only 4.3 inches (108 mm) normally falls. In the end, scientists concluded that the primary triggering mechanism of the large slump/earthflow was a rise in pore pressure caused by infiltrating water from the exceptionally heavy January rains.

A mere ten years after the 1995 movement a smaller event occurred on January 10, 2005, claiming the lives of 10 people. Unusually heavy winter rains again served as the primary triggering mechanism, but in this case the movement took on different and more deadly characteristics. In 2005 the movement occurred at a much shallower depth and took the form of a fairly rapid debris flow, leaving residents no time to flee. Scientists later concluded that the differences in speed and volume between the 1995 and 2005 events were related to the depth that water had been able to infiltrate into the slope. From the graphs in Figure B7.2 one can see that the 2005 debris flow occurred right at the end of a 15-day period of heavy rain as opposed to the month-long delay in 1995. This implies that water did not infiltrate as deep in 2005 before rising pore pressures reduced the frictional resistance of the slope materials to the point of failure. This also explains why the failure occurred at a much shallower depth and involved a smaller volume of material. Finally, the fact that water accumulated closer to the surface helps account for the 2005 debris flow being more fluidlike, giving it greater speed and little time for people to get out of the way.

Based on the geologic history and scientific studies of mass wasting near La Conchita, one can expect that future earth movements will be triggered by periods of heavy rainfall. However, it is nearly impossible to predict, with any degree of certainty, exactly where and when any movement will take place and whether it will be rapid or slow. The reason such predictions are so difficult is because individual mass wasting events are all somewhat unique since they often depend upon complex subsurface conditions that are difficult to determine in advance. Further complicating the issue is the knowledge that future movement could be triggered by an earthquake, which unlike periods of heavy rainfall provides absolutely no warning that movement may be imminent. What is known is that future mass wasting activity is almost certain to occur near La Conchita and that people will likely find themselves in harm's way.

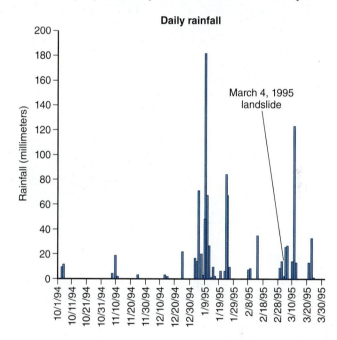

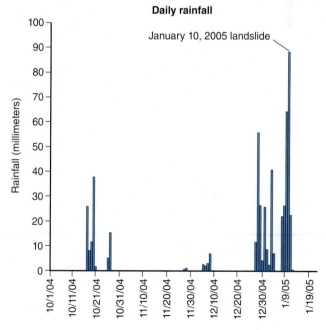

FIGURE B7.2 Graphs showing the different rainfall accumulation patterns that led up to the 1995 and 2005 mass wasting events at La Conchita.

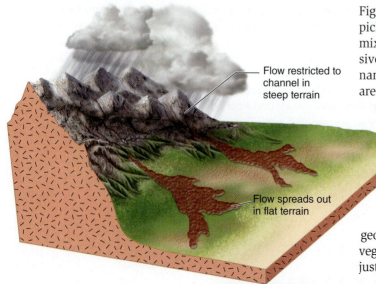

FIGURE 7.15 Mudflow and debris flows typically form in areas where there is an abundance of loose sediment on relatively steep slopes. Water can then carry this material off the slopes, funneling it down steep channels where it forms a fan-shaped deposit near the base of the slope. Debris flows can also occur on heavily vegetated slopes when unusually heavy rains saturate loose material. Photo of a 2007 debris flow that buried several homes and closed a highway near Clatskanie, Oregon.

Figure 7.15 you can see that in both cases large volumes of sediment are picked up by water moving over a fairly rugged landscape. The resulting mixture of mud then moves into small stream channels, which progressively merge into a single channel that funnels the mudflow through a narrow canyon. Eventually the canyon empties out onto a relatively flat area, allowing the mudflow to spread out and form a fan-shaped deposit (Figure 7.15). As was the case with the Armero disaster (Chapter 6), humans have a history of putting themselves at grave risk by building towns and cities on old mudflow deposits because of the flat land and steady water supply.

A *debris flow* is similar to a mudflow; the primary difference is that it contains particles ranging in size from mud and sand to large boulders and trees (see Figure 7.10). Because of the wide variety of materials involved, debris flows naturally form in a greater number of geologic settings. For example, debris flows form on volcanic and sparsely vegetated slopes, consisting of both fine and coarse fragments as opposed to just fine particles. Water flowing down the slopes can then funnel this material (debris) into channels, sending it crashing downhill where it forms a fan-shaped deposit near the bottom of the slope (Figure 7.15). In addition, debris flows occur on heavily vegetated slopes that become unstable after taking on too much water, causing the internal friction within the material to be reduced. This situation is fairly common during unusually large rainfall events in more humid climates with steep terrain (Case Study 7.1). Examples include Pacific storms that move inland over rugged coastal areas and tropical storms and hurricanes that make landfall and move over parts of the Appalachian Mountains. Note that the news media typically uses the term "mudslides" when referring to both mudflows and debris flows.

Unlike mud and debris flows where water picks up loose material, an *earth flow* involves a large section of an unstable hillside, which then flows downslope as a more coherent and viscous mass. In general, earth flows are common on slopes that are underlain by more clay-rich sediment and held in place by thick vegetation. Here the extensive root system helps bind the loose particles together, and therefore minimizes the potential for mudflows. However, during prolonged rains the infiltrating water can saturate the material, which adds weight to the slope and reduces the internal friction within the sediment. The presence of clay minerals is important in reducing friction because their *plasticity* (ability to deform by flowing) greatly increases as they take on water. Note that a special type of earth flow, called *solifluction*, is common in cold climates, such as in northern Canada and Russia, where most of the ground remains permanently frozen—called *permafrost*. The problem occurs when the uppermost zone thaws during the summer causing the material to become saturated such that it begins to flow downslope over the frozen base. Solifluction can be a serious and costly problem for roads and buildings, especially since these features have the tendency to increase the rate of thawing, which, of course, can exacerbate the problem.

Creep

Creep is an exceptionally slow process where repeated expansion and contraction causes unconsolidated materials to move downslope. One of the ways creep takes place is by the freezing and thawing of water within

the pore space of unconsolidated materials. Liquid water naturally expands as it freezes into ice, which forces the overall sediment volume to increase. When the ice melts, the volume decreases. Expansion and contraction also occurs due to the cyclic wetting and drying of clay-rich materials. Here electrical forces allow the microscopic clay-mineral particles to alternately trap and release water molecules. During wet periods the clay particles will take on water, causing the overall volume of material to expand. When the material dries it will shrink—this is similar to a sponge that expands when it becomes wet and then shrinks as it dries.

The effect of repetitive expansion and contraction on unconsolidated material is illustrated in Figure 7.16. From the inset you can see that during expansion the individual particles move upward, perpendicular to the slope. But then, on the contraction cycle, gravity pulls the particles vertically downward. Over time this results in particles taking a zigzagging path downslope. Because the greatest motion occurs closest to the surface, particles will move at different speeds within the sediment profile. It is this difference in speed that causes weathered layers of rock to bend downslope. Keep in mind that even the fastest creep rates are so slow that the resulting movement is imperceptible to humans. As could be expected, the rate of creep in loose materials depends on several factors, such as the amount of water and clay in the sediment, number of freeze/thaw or wet/dry cycles, slope steepness, and presence of plants with deep roots. Although creep is exceedingly slow, it exerts a nearly continuous force that is capable of damaging human structures, such as retaining walls, fences, and buried utility lines, and also cause utility poles to tilt (Figure 7.16). While creep may not be as dramatic as rockslides and falls, it is a widespread and costly problem.

Snow Avalanche

Mass wasting can involve snow rather than rock or sediment, in which case the process is usually referred to as a **snow avalanche** (Figure 7.17). In recent years the number of fatalities and property damage from snow avalanches has increased significantly as mountain resorts have expanded to

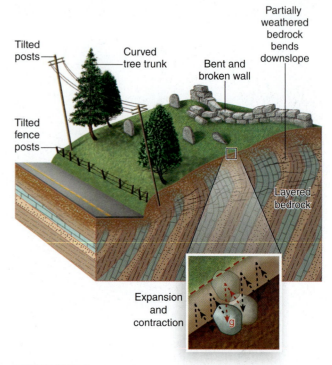

FIGURE 7.16   Creep is the extremely slow movement of unconsolidated materials caused by repeated expansion and contraction of soils that result from freeze/thaw and wet/dry cycles. The inset shows how expansion and contraction results in particles taking a zigzagging path downslope. Because the motion decreases with depth, broken rock layers appear to bend downslope. This movement can cause damage to a variety of human-made structures.

A 2008, San Juan Mountains, Colorado

FIGURE 7.17   Most snow avalanches occur in mountainous areas where weak layers form within the snowpack. Common triggers are heavy snowfall events and human activity that add weight to the slope and overwhelm the frictional forces along a weak snow layer. The avalanche in (A) began when a slab started sliding along a weak layer. Closer view (B) of a detached slab—note the skier for scale.

B 2001, Logan Mountains, Utah

accommodate the increasing number of people who enjoy winter sports. However, long before winter sports became popular, avalanches have posed a serious hazard to railroad and highway traffic in mountainous areas where steep slopes and thick accumulations of snow are common. Under the right climatic and snow conditions, the snow may move downslope in a tumbling mass known as a *loose-snow avalanche.* This type of avalanche is relatively rare and accounts for only a small percentage of all snow avalanche-related deaths and property damage. Much more common is what is called a *slab avalanche,* where a coherent slab (sheet) of dense snow slides along a weakness zone within the snowpack. The key to understanding most snow avalanches then is the formation of slabs and weak layers within a snowpack.

The development of slabs and weak layers is related to the way snow accumulates in mountainous terrain and is then subjected to wide variations in temperature, sunlight, and wind. These variations, combined with the weight of the overlying snow, cause individual snowflakes to recrystallize into a variety of more compact forms. During this process, layers of snow will bond to one another, forming more dense and coherent masses of snow. However, under certain weather conditions the bonding between two layers can be rather poor, resulting in the development of a weak zone or layer within the snowpack. A weak layer often forms, for example, when the surface of the snowpack becomes crusted with ice and is then overlain by a fresh layer of snow. Over time this produces thick slabs of coherent snow that are separated by thin, weak layers as you can see in Figure 7.17. Also important is the fact that these layers are all inclined in the same direction as the slope, which as you recall from Figure 7.7, is the least stable orientation. Clearly, movement is most likely to begin along a weak layer because this is where the frictional resistance is the lowest.

As with other types of mass wasting, a snow avalanche can be triggered by natural processes or human actions that alter the balance between the weight acting in the slope direction and frictional forces along weak layers. Natural triggering mechanisms include heavy accumulations of fresh snow and melting on the surface that enables water to percolate downward and reduce friction along a weak layer. Interestingly, snow avalanches are more common later in winter partly because weak layers tend to form more frequently early in the snow season. This means that there are usually a greater number of weak layers in the lower parts of a snowpack. As snow continues to accumulate throughout the winter, the additional weight makes the slope progressively less stable.

With respect to snow avalanches triggered by humans, most occur when a skier or snowmobiler provides the additional weight necessary for the snow to overcome the frictional resistance along a weak layer. When this happens the overlying slab breaks free and quickly accelerates, reaching speeds of up to 80 miles per hour (130 km/hr), during which time the slab begins to shatter into smaller and smaller pieces. People who get caught in such an avalanche find themselves in a moving fluid whose density is less than a human body, thus they tend to sink into the flowing snow. In some cases people are able to literally swim their way to the surface of the flow and stay there until it stops. Unfortunately, most people remain trapped within the flow. Once the fluidized snow stops, it takes on the consistency of concrete and escape becomes impossible. Although the packed snow can be about 60–70% air, most people die from asphyxiation as carbon dioxide gas builds up around their mouths. In fact, over 90% of buried avalanche victims are found alive if dug out within 15 minutes, but only 20–30% will survive after 45 minutes. In this situation rapid rescue is obviously critical, which is why many skiers and snowmobliers now wear radio transmitters or beacons so they can be quickly located by rescuers.

Submarine Mass Wasting

Although hidden from human view, mass wasting also takes place on the seafloor, especially where sediment accumulates on slopes found along the edge of continental shelves as shown in Figure 7.18. Here thick sediment sequences are naturally saturated and often contain gases, which helps reduce cohesion between the sediment grains. As sediment continues to accumulate on the shelf, it pushes seaward. Eventually the sediment deposit begins to spill over the edge of the shelf, causing the slope areas to become progressively less stable. Mass wasting can then be triggered by an earthquake, hurricane, or passing tsunamis. These processes can cause movement within the sediment, which, in turn, reduces the cohesion or frictional resistance between the grains of sediment. Oceanographers today are obtaining the clearest pictures yet of the seafloor and are finding submarine slides, slumps, and flows similar to those found on land, but at much larger scales, such as the slump shown in Figure 7.18.

As oceanographers continue to study the offshore environment, they are discovering potential hazards associated with submarine mass wasting events. For example, should a submarine slope abruptly fail and involve a large volume of sediment or rock, the movement would suddenly displace an equal volume of seawater, creating the potential for a tsunami (Chapter 5). Just such an event is suspected to be responsible for a 55-foot (17 m) wave that struck the coast of Papua New Guinea in 1998. This tsunami originally was thought to have been triggered by an earthquake, but the source has now been traced to a large slump located 16 miles (25 km) offshore. Other slides and slumps that correlate to known tsunamis have also been found off the coasts of Peru and Puerto Rico.

In addition to tsunami hazards, submarine mass wasting can cause significant damage to offshore oil and gas pipelines lying on the seafloor. For example, in 2004 hurricane Ivan roared up through the Gulf of Mexico, disrupting offshore oil production for months and impacting world oil prices. After the storm, production facilities and over 10,000 miles (16,000 km) of pipeline lay damaged, largely due to submarine slides and flows. Overall, the damage caused a reduction in daily U.S. oil production by almost 500,000 barrels, representing nearly 15% of total U.S. production. Because this disruption came at a time when world petroleum supplies were already tight, the submarine mass wasting events in the Gulf of Mexico were a significant factor in keeping oil prices relatively high throughout the latter part of 2004.

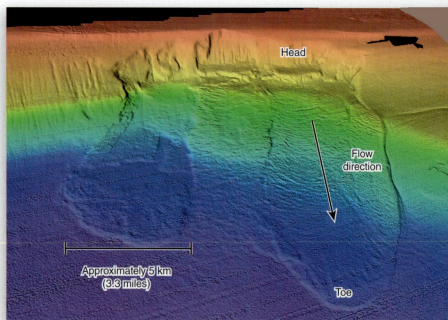

FIGURE 7.18 A high-resolution survey of the ocean floor off Santa Barbara, California, has revealed a nearly 9-mile (15 km) long slump involving approximately 30 cubic miles (130 cubic km) of material. Note the prominent scarp at the top of the slump and the flow pattern at the base, or toe.

Subsidence

In this section we will examine the process known as **land subsidence,** which is the lowering of the land surface due to the closing of void spaces within subsurface materials. Since land subsidence occurs in a strictly vertical sense and does not require a slope, it is technically not a form of mass wasting. The reason subsidence is included in this chapter is because it fits the general theme of earth materials moving downward in response to gravity. In addition, by introducing the topic here, it should provide the reader with a better understanding of the relationship between subsidence and coastal flooding (Chapter 9), groundwater withdrawals (Chapter 11), and extraction of mineral and energy resources (Chapters 12 and 13). Details on these relationships will be provided later in the appropriate chapters.

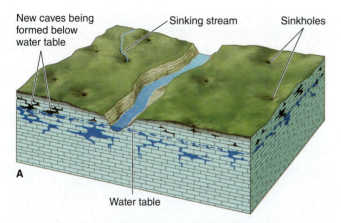

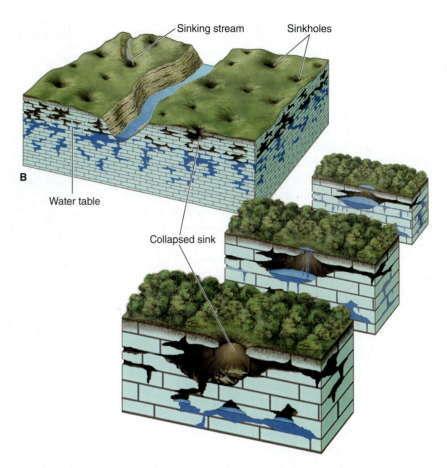

FIGURE 7.19 Groundwater flowing through soluble limestone (A) will create large voids or caverns. As rivers cut downward (B) they cause the water table to lower, leaving void spaces high and dry and in a weakened state. Infiltrating water will cause the cavern roofs to eventually weaken to the point they collapse, forming sinkholes.

The closing of void spaces that leads to subsidence is ultimately related to *overburden pressure,* which we defined in Chapter 4 as the weight of the overlying rock and sediment bearing down on subsurface materials. This means that a subsurface body of rock or sediment must be strong enough to support the weight of everything above it. Should something reduce the ability of subsurface materials to support this weight, then pore spaces and open fractures will tend to close. Subsurface materials then become more compact. Because most of the compaction occurs in the vertical sense, the end result is that the surface naturally sinks or subsides. Note that under some geologic conditions subsidence takes place gradually, but in other situations it happens so suddenly that it is referred to as a *collapse*. In the following sections we will briefly explore both gradual and sudden (collapse) types of subsidence.

Collapse

Sudden or rapid land subsidence typically occurs in areas where unusually large void spaces are found in subsurface materials. Subsurface cavities, commonly called *caves,* can form by groundwater circulating through soluble rock, magma draining from a magma chamber, and underground mining. However, the vast majority of caves form when groundwater slowly dissolves away limestone rock, which is composed of the soluble mineral called calcite (Chapter 3). As illustrated in Figure 7.19, the key to this process is the way cavities are left high and dry as a river slowly cuts downward and lowers the water table. When the cavities were full, the water, of course, helped support the weight of the overlying rock, but once the water is gone, the roofs and walls must bear the full weight. Moreover, infiltrating rainwater will continue to dissolve away the cavern roofs until

they weaken to the point they suddenly fail and collapse, forming what geologists call **sinkholes.** In areas where limestone is relatively close to the surface, the number of sinkholes can be so great that the landscape takes on a pitted or cratered appearance (Figure 7.19). This type of landscape is often referred to as *karst* terrain.

The sudden collapse of sinkholes can be a widespread and serious problem in areas underlain by soluble limestone, as is the case for large portions of Florida and Kentucky. From the photo in Figure 7.20 one can see that any human structure which lies directly above a limestone cavity is at risk of being destroyed should the roof suddenly fail. The formation of sinkholes can also create serious and costly disruptions to transportation and utility networks. Although heavy rainfall events commonly trigger the collapse of limestone sinkholes, human activity can also play an important role. For example, constructing a building directly over a large cavity may increase the overburden pressure (weight) beyond what the cavern roof can safely support. Another common factor is the lowering of the water table by large groundwater withdrawals (Chapter 11), leaving caverns empty and more prone to collapse.

Finally, void spaces created during underground mining for mineral resources or coal (Chapters 12 and 13) can collapse, thereby causing land subsidence. Mining voids that are relatively shallow can collapse and create pits at the surface, which appear similar to limestone sinkholes. In the case of mining voids that are deeper, the collapsed space results in a more gentle sagging of the land surface.

Clearly, the best way for society to minimize the problems associated with collapse features is to avoid constructing permanent and valuable structures over large cavities. However, the location of subsurface voids is easily known in advance, particularly in limestone terrain. In recent years ground-penetrating radar has been successfully used to locate voids that are relatively close to the surface. Should a cavity be found beneath an existing building, engineers will typically try to prevent rainwater from weakening the cavern roof. This can be done by installing drains and covering the surface with impermeable material so as to minimize the amount of surface water that can infiltrate the void.

Gradual Subsidence

Gradual or slow land subsidence is usually associated with the compaction of pore space within a sedimentary sequence. As illustrated in Figure 7.21, compaction occurs in fine-grained sediment when overburden pressure causes individual clay mineral particles to become aligned in a parallel manner. The realignment causes the material to become more compact, which can lead to gradual land subsidence. This process occurs naturally in areas where new sediment is being deposited, such as in the Mississippi Delta. Here additional sediment increases the overburden pressures, causing compaction within the delta. In Chapters 8 and 9 we will take a closer look at how human activity has increased the rate of subsidence in the Mississippi Delta. This, in turn, has increased the risk of flooding along the Louisiana coast, including New Orleans.

Perhaps the most common cause of slow subsidence is a reduction in pore pressure associated with the pumping of large volumes of water or oil from the subsurface. Since fluid pressure acts outward in all directions, any reduction in the amount of water or oil forces the sediment grains to bear more of the overburden pressure or weight (Figure 7.21). This results in compaction and gradual subsidence which cannot only increase risk of flooding, but also causes structural damage to buildings and underground utilities. A dramatic example of pumping-induced subsidence is the area around

Winter Park, Florida

FIGURE 7.20 The sudden collapse of sinkholes is normally associated with large cavities that form in layers of soluble rock that are relatively close to the surface. Sinkholes cause significant damage to buildings, highways, and utility lines.

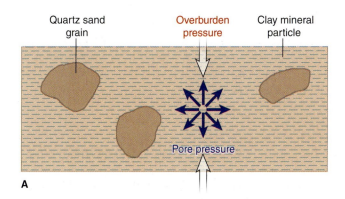

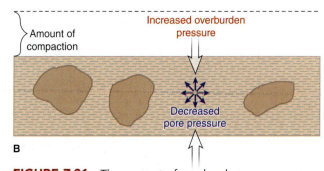

FIGURE 7.21 The amount of overburden pressure, or weight, that sediment grains must bear (A) is offset by the level of pore (fluid) pressure within the sediment. Compaction and subsidence can occur (B) whenever there is a reduction in fluid pressure, or when additional sediment is deposited, allowing more weight to bear down on the grains.

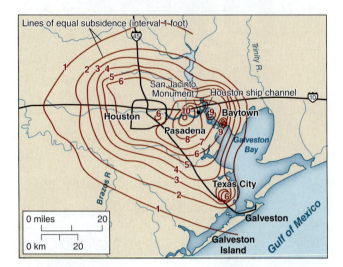

FIGURE 7.22 Gradual subsidence can take place when large volumes of water or petroleum are removed from the subsurface. Compaction occurs in clay-rich layers where individual water molecules being held between the clay particles are removed. Heavy withdrawals of water and petroleum in the Houston area have resulted in as much as 10 feet of subsidence, creating a bowl-shaped depression that has led to serious flooding problems.

Houston, Texas, shown in Figure 7.22. Here the subsurface withdrawal of water and oil has created a bowl-shaped depression that has subsided as much as 10 feet (3 m) since 1906 (some low-lying areas are now submerged). This has made it more difficult for streams to drain water off the landscape, thereby increasing the frequency and intensity of flooding.

Reducing the Risks of Mass Wasting

In the previous sections you learned that tremendous forces are involved when rock or sediment moves downslope. Moreover, people sometimes place themselves in danger by building in areas susceptible to mass wasting. Other times our own actions serve as triggering mechanisms. There are several ways in which society can reduce the risk of mass wasting, the most obvious of which is to simply avoid building or living in hazardous areas. This of course may not be feasible where people are already living in a hazard zone, or where there is a need for a certain activity, such as an important transportation link through mountainous terrain. If it is undesirable or impractical to avoid the hazard, then another option is to try and minimize the risk through engineering strategies. In some situations we may use control structures designed to help stabilize a slope, hence keep material from moving in the first place. Should engineers determine that it is not cost effective to try and stabilize a slope, a structure can be built that will help shield buildings and highways from any material that moves downslope.

In the following section we will examine the risk management approach in more detail, as well as some of the more common types of engineering controls.

Recognizing and Avoiding the Hazard

The first step in minimizing risks associated with mass wasting is to identify those slopes that are unstable. Because slopes tend to fail repeatedly, the easiest way to locate such slopes is to recognize the telltale signs of mass wasting. For example, boulders at the base of a cliff are a clear indication that rockfalls have taken place, and will occur again in the future. Likewise, scarps and jumbled terrain point to past slump activity; curved or deformed walls are clear evidence of creep. Such features can be seen in the field and on aerial photographs, enabling geologists to make hazard maps showing those areas of past movement. In addition, by understanding the science behind mass wasting processes, geologists and engineers can identify unstable slopes where movement has yet to occur, but which have a high potential for failure. Key risk factors include oversteepened slopes, weakness planes oriented parallel to the slope, unconsolidated materials, and materials with a high water content. Note that hazard maps showing high-risk areas are often available through the U.S. Department of Agriculture, federal or state geological surveys, and local planning offices.

Once a high-risk slope is identified, steps can be taken to minimize human activity within the hazard zone, particularly that which may trigger a mass wasting event. Such steps commonly involve zoning laws designed to restrict development, and construction ordinances that prohibit the oversteepening of slopes. In general, it is much cheaper to avoid human activity in areas with unstable slopes than it is to design and construct engineering controls to prevent mass wasting. In situations where a hazard zone is simply unavoidable, detailed surveys can be performed to gather data on slope materials and conditions prior to the design and construction of an engineering solution. Note that this preventative form of risk reduction is highly desirable since the cost involved in avoiding a mass wasting event is often far less than the potential property damage.

A   **B**

FIGURE 7.23
Retaining walls are commonly used both above and below highways to strengthen oversteepened slopes. The retaining wall in (A) helps stabilize an oversteepened slope created when part of the hillside was removed, whereas the wall in (B) is supporting fill material that was placed on the slope to create a flat area for the road.

Engineering Controls

Engineering controls are a key element in most efforts directed at minimizing the hazards associated with mass wasting. In general, engineering solutions involve a number of different controls, and a combination of these controls is usually chosen based on the type of expected movement and the characteristics of the site itself. In some cases the goal is to prevent movement; in others, it is simply to provide protection from movement that is all but impossible to stop. If the objective is to prevent movement, then slope stability can be maximized by simultaneously increasing frictional forces and decreasing the weight acting in the downslope direction. Consequently, most engineering efforts aimed at increasing slope stability will involve one or more of the following engineering controls.

Retaining Walls

A **retaining wall** is an engineering structure whose basic purpose is to strengthen oversteepened slopes. Retaining walls are commonly used whenever a flat or level surface is needed in sloping terrain, such as for roadways, buildings, and parking lots. Recall from our earlier discussion that creating a level surface on a slope requires either removing (i.e., excavating) part of the slope or bringing in fill material and placing it on the slope (refer back to Figure 7.8). Either technique will generate an oversteepened slope, which will likely begin to move unless it is supported by a retaining wall made of steel, concrete, rock, or even wood. For example, the photo in Figure 7.23A shows how a retaining wall was used to stabilize an oversteepened slope that was created by cutting into the slope above the highway. In some situations, however, making such a cut is not desirable because the amount of material above the road would be too great for a retaining wall to support. Engineers therefore can put fill material on the slope and hold it in place with a retaining wall underneath the roadway as shown in Figure 7.23B.

Although many retaining walls are used as a preventative measure, they can also be used to gain control over slides or flows that occur repeatedly. For example, Figure 7.24 shows an active slide in which material from its toe has been removed from the highway. Because the toe provides support for the upper part of the slide, removing material from the

FIGURE 7.24   Photo of a recent slide in which material that had moved onto the highway was removed, reducing support for the remaining material above the road and increasing the chance of continued movement. A retaining wall here can help provide support and prevent additional movement. Note the large amount of material placed below the road in an attempt to help support the highway by forming a buttress.

Mendocino County, California

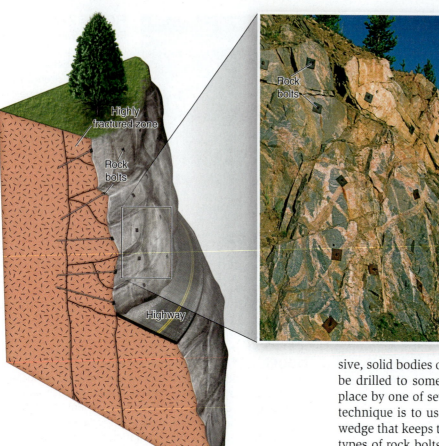

FIGURE 7.25 Rock bolts are used to attach loose rock or slabs to more massive, solid bodies of rock, thereby reducing the chance of rockfalls or small-scale slides along highways and in tunnels.

toe area typically results in additional movement. A retaining wall here would provide additional support and decrease the likelihood of future movement. However, as just mentioned, these structures may not be adequate if there are large amounts of loose material behind the wall. Retaining walls are also not effective if the plane of movement lies beneath it, in which case the wall is likely to become part of the moving mass. Note in Figure 7.24 the large amount of material that was placed below the roadway in an attempt to form a buttress and help support the highway from below, similar to a retaining wall.

Rock Bolts

Rockfalls or slides are common in areas where highly fractured rocks are exposed on steep slopes and where weakness planes are inclined in the same direction as the slope. To minimize the potential for rockfalls or small-scale slides, **rock bolts** are commonly used to anchor loose rocks to more massive, solid bodies of rock (Figure 7.25). Installation requires that a hole first be drilled to some desired depth, and then a bolt is inserted and held in place by one of several different techniques. The most common anchoring technique is to use a tip that expands when the bolt is turned, forming a wedge that keeps the entire assemblage from backing out of the hole. Other types of rock bolts make use of cement grout or simple friction within the borehole to remain fixed in place. Rock bolts are used extensively along highways and rail lines where fractured rocks create a near constant threat of rockfalls. Bolts are also widely used for stabilizing walls and ceilings in tunnels and underground mines.

Controlling Water

In this chapter you have learned that water plays an essential role in a number of mass wasting processes, in particular, excessive water often serves as a triggering mechanism. As water accumulates in the subsurface, the increase in fluid pressure within pore spaces and weakness planes reduces the frictional forces within slope materials. Water not only reduces friction, but it also increases the weight acting in the direction of the slope. Consequently, a highly effective means of keeping a slope stable is to control or limit the amount of water that can accumulate within the slope material. Here a common technique is to install a network of perforated pipes and/or gravel beds in order to drain water from within the slope. For example, Figure 7.26 illustrates how drains can be placed behind retaining walls to prevent the buildup of water. Note in this figure that another useful approach is to combine drains with features that prevent water from entering a slope in the first place. In this technique a berm or channel is used to divert surface water away from the upper portion of an unstable slope. Diverting water is especially important in keeping water from flowing into open fractures at the top of slumps and slides. Under some circumstances it may even be necessary to cover large sections of a slope with impervious plastic sheeting to prevent water from infiltrating into unstable materials. Finally, the buildup of water in some slopes can be minimized by limiting the amount of water being used nearby for landscape or agricultural irrigation.

FIGURE 7.26 Drains can be used to prevent water from accumulating in porous earth materials, thereby minimizing weight in the slope direction and the buildup of pore pressure. Berms and channels can also be used to divert surface water away from unstable slopes.

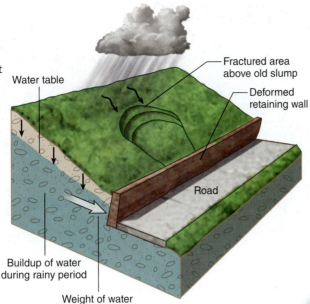

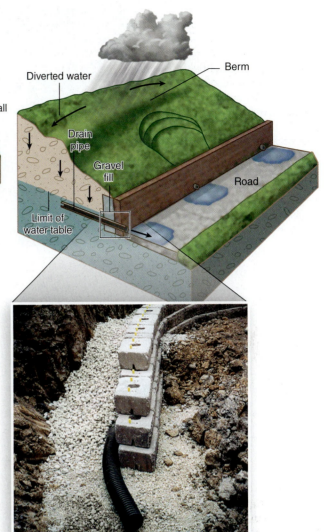

Terracing

As can be seen from the example in Figure 7.27, **terracing** involves creating a series of *benches* (flat surfaces) on a hillside—retaining walls are often used to support the oversteepened portions of the slope. The construction of terraces is an ancient practice in parts of Asia and South America where they provide flat areas for growing food in rugged terrain. Terracing is also common around homes built on slopes in order to create level areas for landscaping and recreation. In addition to creating areas of level ground, terracing is also an effective technique for reducing the risk of mass wasting. For example, multiple terraces can be used to stabilize a slope in situations where it would not be feasible to construct a single, large retaining wall. Also, along highways where rockfalls are difficult to control with rock bolts, an alternative approach is to create a series of terraces by cutting into the rock face in a stepwise fashion (Figure 7.27). This not only decreases the overall steepness of the slope, but also breaks the slope up into shorter segments. Therefore, when a rock does break loose it will travel only a short distance before encountering a bench or step. The rock will most likely come to a stop on the bench as opposed to tumbling down onto the highway.

Covering Steep Slopes

It is not uncommon for slopes to become unstable when construction activity or wildfires remove the vegetation cover. Therefore, covering a bare slope with new vegetation, crushed rocks, or a synthetic mesh or fabric will increase frictional forces within the slope and increase slope stability. Vegetation is particularly effective in stabilizing slopes consisting of sediment because plant roots help bind the loose material together. In many instances the roots will extend through the surficial material and penetrate into fractures within the underlying rock, effectively anchoring the sediment. Generally speaking, vegetation with deeper root systems is more effective at stabilizing slopes. Regardless of root depth, however, should movement occur below the root zone the vegetation will simply move downslope along with the underlying material. Vegetation also helps to stabilize slopes by taking up water, thereby removing water from the soil. This, in turn, helps limit the amount of deep infiltration and buildup of pore pressure, ultimately decreasing the potential for movement within the slope.

FIGURE 7.27 Terraces were built into this road cut in order to keep rocks from falling onto the highway. By breaking the slope into shorter segments, terracing allows rocks to come to rest on the terrace steps as opposed to tumbling down the entire length of the slope.

FIGURE 7.28 Covering slopes with vegetation and synthetic materials increases friction within the slope, thereby decreasing the potential for mass wasting. Photo (A) shows workers applying a hydroseed mixture over a synthetic fabric, which had been draped over the sloped surfaces of a construction site. Photo (B) illustrates how the slopes later became covered with vegetation. Note how large rocks were used to stabilize the slope in certain areas.

A

B

Depending on the slope material and steepness, grass alone may be sufficient or it may serve as a temporary cover until permanent and more deeply-rooted vegetation can take hold. Perhaps the most common way of establishing a blanket of grass is a technique called *hydroseeding*, where a slurry of seed, mulch, and fertilizer is sprayed onto a bare slope (Figure 7.28). On steeper slopes it becomes more of a challenge to reestablish vegetation because of the need for deep-rooted plants, which of course take more time to mature compared to grass. Compounding the problem is the fact that erosion will quickly form gullies on steeper slopes, and reestablishing vegetation once a gully develops is especially difficult. One solution is to fill a gully with new soil and then place deep-rooted plants along small terraces constructed in the hillside. The terraces greatly reduce erosion by decreasing the speed of water flowing downslope, which gives the plants more time to grow.

Reducing Slope Materials

Depending on the geological conditions at a site, it may be more cost effective to actually remove a dangerous slope rather than to try and stabilize it by some physical means. Figure 7.29 illustrates a situation

FIGURE 7.29 Removal of a slope altogether is sometimes the most practical and cost-effective means of reducing a mass wasting hazard.

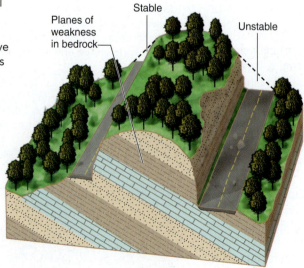

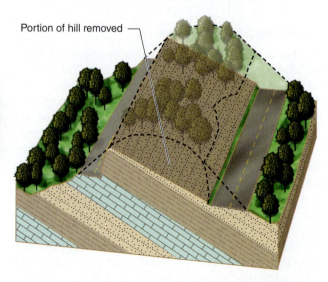

Planes of weakness in bedrock

Stable

Unstable

Portion of hill removed

where weakness planes are inclined in the direction of the slope, creating a serious hazard that can require continuous and costly engineering efforts to prevent a slide. Removing the entire slope has the advantage of eliminating both the hazard and the long-term costs associated with stabilization efforts. Another common situation where slope reduction or removal is a viable option is when it is impractical to use rock bolts to stabilize loose rocks above a highway. In some cases this may simply involve having highway engineers periodically dislodge threatening rocks in a controlled manner while traffic is stopped. Finally, note that slope reduction is the preferred method used by ski resort operators for reducing the threat of snow avalanches. Here explosives are used to initiate snow avalanches such that dangerous accumulations of snow are removed from the surrounding slopes. Avalanche control programs at most ski resorts are typically quite good, which accounts for the fact most snow avalanche fatalities occur outside the boundaries of resorts where such protection is not available.

Protective Structures

There are some situations where it is not cost-effective to physically stabilize or remove materials from a dangerous slope. For example, some slopes may simply contain too much material to remove, while others may require an unreasonable amount of engineering in order to reduce the threat. An alternative approach is to build structures that keep material that is moving downslope from coming into contact with people and/or infrastructure. One technique is to construct a retaining wall (Figure 7.30A) that diverts material away from a building or group of buildings. To protect an entire village or town, large barriers made of reinforced concrete can be placed in a stream valley (Figure 7.30B) that will trap material from flows or slides as it moves downslope. For highways and railroads, special shelters (Figure 7.32C) are used as a cost-effective means of shielding those sections where mass wasting occurs frequently. Shelters not only keep people from being injured, they protect the road or rail line from being damaged and minimize costly repairs.

FIGURE 7.30 In areas where it is not feasible to prevent mass wasting, buildings can be protected by constructing retaining walls (A) that will divert material, or by installing large barriers in valleys designed to trap debris (B). Shelters (C) can also be built that allow material to safely move over transportation routes.

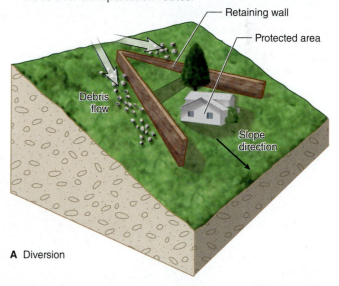

A Diversion

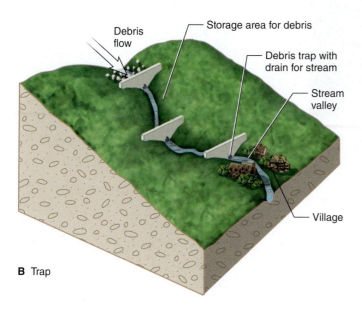

B Trap

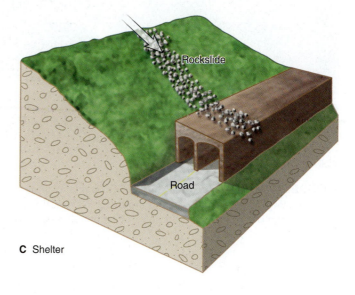

C Shelter

Lahaina, Maui, Hawaii

FIGURE 7.31 A heavy chain-link mesh was draped over a crumbling rock mass in Hawaii in order to keep small-to-medium-size rocks from falling onto the roadway.

Perhaps the most common technique is to drape a heavy chain mesh over an entire rock exposure (Figure 7.31) in order to protect transportation routes from small-to-medium-sized rockfalls. Alternatively, transportation lines can be shielded from rockfalls by constructing a tall, heavily reinforced fence parallel to a road or rail line. Finally, because mass wasting problems can lead to long-term repair and maintence costs, it is sometimes more cost-effective to bore a tunnel (Figure 7.32). In this way troublesome sections of a highway or railroad can be bypassed altogether.

Wolf Creek Pass, Colorado

FIGURE 7.32 This tunnel was constructed as a long-term and cost-effective solution for protecting both the highway and motorists from earth materials moving downslope. Note that the old roadway (to the left of the tunnel) is no longer open to traffic.

SUMMARY POINTS

1. Rock or sediment will begin to move downslope whenever the gravitational force acting on the material in the slope direction becomes greater than the frictional forces keeping the material in place.

2. The main factors affecting slope stability are: steepness of the slope, type of material, presence of water, existence of weakness planes, and amount of vegetative ground cover.

3. A triggering mechanism is anything that destabilizes a slope by reducing the frictional forces on a slope or by increasing the effect of gravity. Earthquakes, heavy rains, removal of vegetation, and oversteepening of slopes are common examples.

4. Most rocks are not homogeneous, but rather contain planar weaknesses such as bedding, fracture, or foliation planes. Slopes are the least stable when these weakness planes are inclined in the same direction as the slope.

5. Water acts to destabilize slopes by adding weight and by reducing the internal friction between sediment grains and weakness planes. When materials become saturated, pore pressure will increase, which then reduces internal friction.

6. Mass wasting can be classified based on the type of movement: *falls*, *slides*, and *flows* and also on the material involved: *rock, debris,*

earth, and *mud*. Slump is a complex type of movement involving both sliding and flowing. Creep is the imperceptibly slow movement due to repetitive expansion and contraction of sediment.

7. Snow avalanches are a form of mass wasting that pose a serious hazard along highways and for ski resorts in mountainous terrain.

8. Land subsidence occurs when void spaces begin to close within subsurface materials. Rapid subsidence (collapse) takes place in limestone terrain and in mining areas where large underground voids are present. Gradual subsidence is common in areas of large water withdrawals that reduce pore pressure, causing clay particles to undergo compaction.

9. Reducing the risk of mass wasting requires that people first identify hazardous areas, then take steps to avoid hazard zones or implement engineering controls.

10. The goal of some engineering controls is to prevent movement by increasing slope stability. Efforts here focus on increasing frictional forces within slope materials and decreasing the amount of weight acting in the slope direction.

11. Other engineering efforts provide protection from mass wasting events that are either impractical or impossible to prevent.

KEY WORDS

creep   208
fall   203
flow   205
land subsidence   211
mass wasting   194

pore pressure   200
retaining wall   215
rock bolts   216
sinkholes   213
slide   203

slump   204
snow avalanche   208
talus pile   203
terracing   217
triggering mechanisms   197

APPLICATIONS

Student Activity   As you are driving around your town or in the countryside, look for mass wasting. Even if you live in the flattest country, you can still see evidence of creep, such as fences tilted and walls leaning. Once you see one example of mass wasting, you can see a lot more.

Critical Thinking Questions

1. What are the main factors that affect slope stability?
2. How does the addition of water destabilize a slope?
3. Are snow avalanches a form of mass wasting?
4. How can the risk of mass wasting be reduced?

Your Environment: YOU Decide

The logging business is a necessary part of modern life, yet the environmental impact of this industry includes human-made mass wasting events. By clear-cutting on very steep slopes, the vegetation cannot grow back fast enough to hold the soil in place. Should the logging companies be held responsible for this problem, or since most of these clear-cuts occur on land that will probably never be developed, just let it go?

Chapter 8

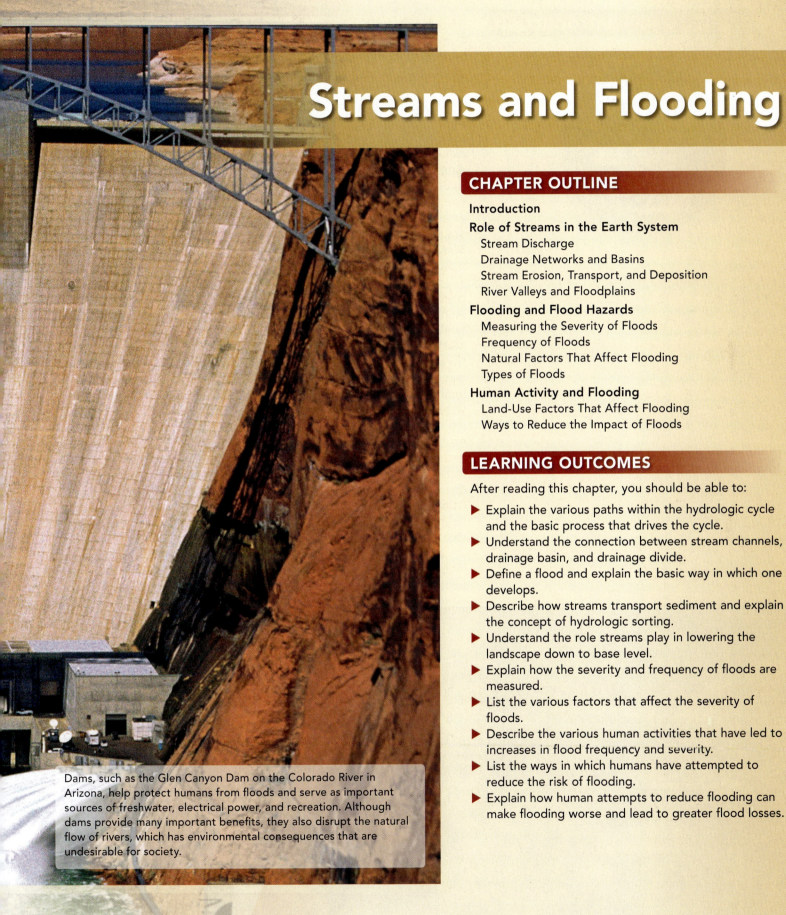

Streams and Flooding

LEARNING OUTCOMES

After reading this chapter, you should be able to:

▶ Explain the various paths within the hydrologic cycle and the basic process that drives the cycle.
▶ Understand the connection between stream channels, drainage basin, and drainage divide.
▶ Define a flood and explain the basic way in which one develops.
▶ Describe how streams transport sediment and explain the concept of hydrologic sorting.
▶ Understand the role streams play in lowering the landscape down to base level.
▶ Explain how the severity and frequency of floods are measured.
▶ List the various factors that affect the severity of floods.
▶ Describe the various human activities that have led to increases in flood frequency and severity.
▶ List the ways in which humans have attempted to reduce the risk of flooding.
▶ Explain how human attempts to reduce flooding can make flooding worse and lead to greater flood losses.

Dams, such as the Glen Canyon Dam on the Colorado River in Arizona, help protect humans from floods and serve as important sources of freshwater, electrical power, and recreation. Although dams provide many important benefits, they also disrupt the natural flow of rivers, which has environmental consequences that are undesirable for society.

Introduction

The basic role of streams and rivers (larger streams) within the Earth system is to drain water off the landscape and to transport sediment. Periodically, however, the ability of a stream to carry water is overwhelmed by the sheer volume of water flowing off the landscape. When this occurs a stream or river may overflow its banks and create what is known as a *flood* (Figure 8.1). Of all the natural hazards facing society, floods perhaps take the greatest toll in terms of human lives and property damage. For example, for the past 50 years, floods in the United States alone have claimed an average of more than 100 lives and $3.5 billion in property damage each year. Although a single volcanic eruption, earthquake, or mass wasting event can result in tremendous losses, these events are relatively infrequent and are usually restricted to relatively small geographic areas. In comparison, rivers and streams are found in all parts of the landscape, and humans tend to concentrate their settlements along waterways. This, combined with the relatively frequent nature of flooding, puts large numbers of people and buildings at risk of flooding.

One of the reasons people have historically built settlements near rivers and streams is because of the basic need for water to survive. Rivers not only serve as important transportation corridors and sources of food (e.g., fish), but the adjacent low-lying areas provide rich topsoil where crops can flourish. As societies advanced, people eventually discovered that waterfalls and rapids were ideal sites for harnessing the energy of falling water, thus mills were built for grinding grain and cutting lumber. While there were many benefits to constructing settlements along waterways, there were also serious risks associated with periodic flooding. To reduce these risks, humans learned to build permanent structures on higher ground as well as levees and dams that provide protection against floods. The use of these engineering controls, however, tends to cause increased development in areas that previously had been avoided due to frequent flooding. A prime example is the 2005 disaster in New Orleans, where engineering controls (levees) allowed development to flourish, but then failed and resulted in a catastrophic flood with extensive damage and loss of lives.

In Chapter 8 our focus will be on the basic way in which humans interact with the stream processes that transport both water and sediment off the landscape. In particular we will examine how human modifications to the landscape have actually led to an overall increase in the frequency and severity of flooding.

— Levee

FIGURE 8.1 Humans have historically built settlements along rivers to take advantage of the water supply, fertile soils, and ability to transport people and goods. Photo taken in 2001 shows how engineering controls (levees) help keep the town of Trempealeau, Wisconsin, dry when the Mississippi River overflows its banks. Note that the town would flood if the engineering controls fail or become overwhelmed by an exceptionally large flood event.

Role of Streams in the Earth System

Recall from Chapter 1 that the *hydrosphere, biosphere, atmosphere,* and *solid Earth* are subsystems that interact with one another, forming the *Earth system.* A key part of the Earth system is the constant transfer of water between each of these subsystems, referred to as the *hydrologic cycle.* The hydrologic cycle is ultimately driven by solar energy (electromagnetic radiation) that reaches Earth's surface. When liquid water is warmed by sunlight it undergoes *evaporation,* sending water vapor into the atmosphere. This vapor eventually cools and condenses into tiny water droplets, which then return to the surface in the form of rain, snow, sleet, or hail, collectively called *precipitation.* In the following sections we will explore how this water moves across the landscape on its journey through the hydrologic cycle.

Stream Discharge

Although nearly 75% of all precipitation returns directly back to the oceans, the remaining 25% that falls on Earth's landmasses can take several different paths through the hydrologic cycle. As illustrated in Figure 8.2, some of the precipitation reaching the land surface moves downslope (under the force of gravity) in thin sheets in a process called **overland flow.** During overland flow, which can be observed moving across parking lots during heavy rains, water eventually accumulates in low areas of the terrain and begins flowing as a discrete body or *stream* of water through a sinuous pathway called a *channel.* **Stream discharge** is the term used to describe the volume of water moving through a channel over a given time interval, commonly measured in units such as cubic feet per second (ft^3/s). Note that hydrologists refer to the process of water flowing through stream channels as *runoff.* Since water flows from areas of higher to lower elevation, most stream discharge (runoff) eventually makes its way to the oceans. Once in the ocean, the water undergoes evaporation and returns to the atmosphere, thereby completing its journey through the hydrologic cycle.

From Figure 8.2 we can see that instead of flowing downslope, some of the precipitation that falls on the land surface will directly infiltrate the soil zone. Once in the soil this water can return back to the atmosphere due to evaporation or *plant transpiration* (i.e., uptake by root systems and then released through plant leaves). However, when soils become wet enough, gravity will begin pulling water from the pore spaces. This allows infiltrating water to reach the water table and enter the saturated zone, at which point it is called *groundwater.* Groundwater then moves slowly through subsurface materials (rocks or sediment) as it follows the slope of the water table, eventually discharging into a stream channel, lake, wetland, or ocean (see Chapter 11 for details). Hydrologists use the term **groundwater baseflow** to refer to this discharge of groundwater into the surface environment. It is important to realize that unlike the sporadic input of water to a stream from overland flow, groundwater baseflow is fairly continuous. Also, groundwater may travel anywhere from a few days to thousands of years before discharging into a stream channel.

A great deal of information about a river or stream can be obtained from a *stream hydrograph,* which is simply a plot of discharge versus time. For example, the hydrograph in Figure 8.3 illustrates how stream discharge is affected by the contributions of water from overland flow and groundwater baseflow. The overland flow component typically occurs during

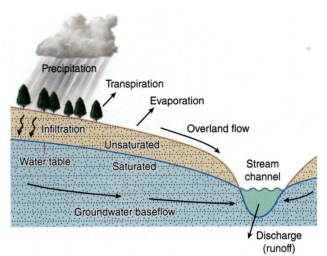

FIGURE 8.2   Precipitation that falls on the land surface can take different paths through the hydrologic cycle. Some of the water moves as overland flow and enters directly into stream channels. Most of the remaining water infiltrates the soil zone where it can return to the atmosphere via evaporation and plant transpiration. If soil moisture becomes great enough, infiltrating water can reach the water table and then flow with the groundwater system and eventually discharge into a stream channel. Stream discharge (runoff) is simply the volume of water moving through a channel.

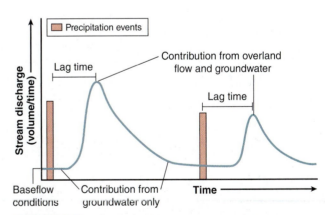

FIGURE 8.3   Stream hydrographs are useful tools in analyzing how stream discharge changes over time. Overland flow from heavy rain events creates spikes in discharge, which can be large enough to cause streams to overflow their banks and flood low-lying areas. Note the lag time between rain events and peak discharge. In contrast, water that infiltrates and slowly makes its way to a stream through the groundwater systems provides a steady supply of water called baseflow. Groundwater baseflow enables streams to keep flowing in between precipitation events.

FIGURE 8.4 Greater infiltration in humid regions causes the water table to be higher than the level of stream channels, forcing groundwater to flow into streams. In more arid regions the water table is commonly below the stream channel, causing water to flow into the groundwater system.

heavy or prolonged rain events, creating a rapid input of water that generates a sudden rise or spike in stream discharge. Because it takes time for water to move across the landscape and into channels, scientists use **lag time** to describe the time difference between a rain event and the resulting discharge peak. Obviously, lag time will vary depending on the distance between where the rain is falling and the particular channel where discharge is being measured. The key point here is that overland flow is sometimes so great that the resulting discharge peak is large enough to force a river or stream to overflow its banks and cause flooding—floods can also be caused by the rapid melting of snow, ice jams, and dam failures. Finally, notice in Figure 8.3 that the continuous input of groundwater baseflow allows streams in many areas to keep flowing at some minimum level, often called *baseflow conditions*. This contribution of groundwater baseflow keeps streams from going dry between rain events, thus is critical in maintaining the health of stream ecosystems.

Although groundwater systems will be discussed in more detail in Chapter 11, we need to mention here how climate affects the relationship between streams and groundwater systems. In humid climates, more water infiltrates to the water table because precipitation is more plentiful. As can be seen in Figure 8.4, this causes the water table in humid regions to be higher than the stream channels, thereby forcing groundwater to flow into streams. Such streams are often referred to as *gaining streams*. In regions with more arid climates there is less deep infiltration, resulting in a water table that is below the level of most stream channels. Under these conditions the flow of water is reversed, meaning that water in the stream will flow into the groundwater system, in which case it is called a *losing stream*. Because of this loss of water, it is not uncommon in arid climates for all but the largest rivers to go completely dry during the summer months. Most of the water in such rivers typically comes from distant mountains where groundwater baseflow, melting snow, and rainfall is able to contribute to the streamflow.

Drainage Networks and Basins

Figure 8.5 illustrates that as a stream flows downhill it eventually merges with other channels, with the smaller of any two merging channels being called a *tributary*. The result is a network of stream channels called a *drainage system*, where merging tributaries form progressively larger streams. Note that the term *river* is often applied to the larger

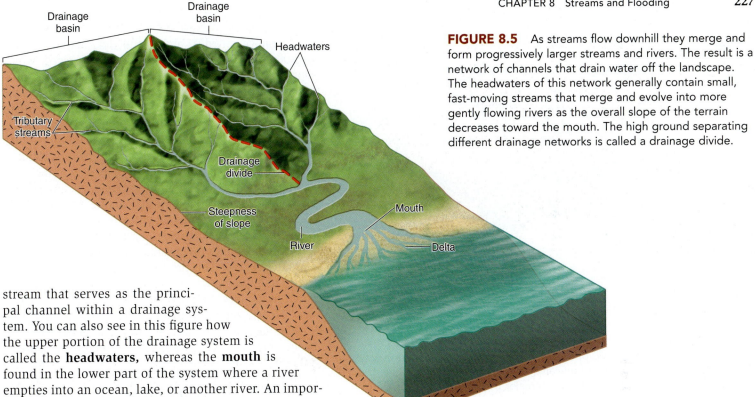

FIGURE 8.5   As streams flow downhill they merge and form progressively larger streams and rivers. The result is a network of channels that drain water off the landscape. The headwaters of this network generally contain small, fast-moving streams that merge and evolve into more gently flowing rivers as the overall slope of the terrain decreases toward the mouth. The high ground separating different drainage networks is called a drainage divide.

stream that serves as the principal channel within a drainage system. You can also see in this figure how the upper portion of the drainage system is called the **headwaters,** whereas the **mouth** is found in the lower part of the system where a river empties into an ocean, lake, or another river. An important characteristic of drainage systems is that there are typically many small tributaries in the headwaters that occupy relatively narrow valleys. Headwater tributaries usually have channels that are gently curved (as viewed from the air) and are relatively steep. As one gets closer to the mouth, the valleys tend to become wider and the steepness and number of channels decreases. In addition, the main channel begins to develop tight, S-shaped curves called *meanders.* Therefore, headwater streams are generally small and relatively fast moving and occupy narrow valleys, but then evolve toward the mouth into gently flowing rivers that occupy wider valleys. Later you will see how the difference in water volume, velocity, and valley width from the headwaters to the mouth affects the nature of flooding as well as the ability of the water to erode and transport sediment.

Another important feature of drainage systems illustrated in Figure 8.5 is that individual systems are separated from one another by a topographic high or crest in the landscape called a **drainage divide.** Rain or snow that falls on different sides of a drainage divide will eventually enter different drainage networks. A good example is the Continental Divide of North America shown in Figure 8.6, which acts as the boundary between surface waters that flow into the Pacific Ocean from those making their way into the Atlantic and Arctic oceans. Note the additional drainage divides that separate the waters flowing into different parts of the Atlantic and Arctic oceans. Another interesting feature of this map is the drainage divide that encloses the Great Basin in the western United States. This area is somewhat unusual in that surface waters here are unable to drain to an ocean, which means the water can exit only through evaporation.

FIGURE 8.6   This map shows the major drainage divides in North America. These divides serve as the boundary between surface waters flowing into separate stream networks. The Continental Divide separates waters flowing into the Pacific from those making their way to the Atlantic and Arctic oceans. The Great Basin is an example of an area where surface streams do not drain into an ocean, but rather exit through evaporation.

FIGURE 8.7   The Mississippi drainage basin represents the land area that contributes water to the Mississippi River, which ultimately discharges into the Gulf of Mexico. The Mississippi basin can also be subdivided into smaller basins, each of which collects water for the river's major tributaries. These smaller basins, in turn, can be subdivided further based on their own tributary network.

Drainage divides are useful for mapping what hydrologists call a **drainage basin** or *watershed,* which represents the land area that collects water for an individual stream or river. For example, Figure 8.7 shows the drainage basin for the Mississippi River—note how part of the basin is bounded by the Continental Divide. Rainfall or snowmelt anywhere within the Mississippi basin will first move into one of the river's numerous tributaries. From there the water will make its way to the main channel of the Mississippi River, and eventually discharge into the Gulf of Mexico at the river's mouth. Figure 8.7 also illustrates how the drainage basin of the Mississippi can be subdivided into smaller basins by following the drainage divides that exist between the river's major tributaries. This process of subdividing a drainage basin into progressively smaller basins can continue down to the level of individual tributaries.

The concept of a drainage basin is also useful in that it helps explain how rainfall can cause flooding in one basin, but yet have no impact on stream discharge in adjoining basins. For example, suppose heavy rains were restricted to the Ohio River basin (Figure 8.7), causing rivers there to overflow their banks. Such an event would raise water levels downstream on the lower Mississippi, but would have no effect on the streams in the Upper Mississippi, Missouri, Arkansas, and Tennessee basins. The drainage basin concept also helps us understand how rivers that drain large tracts of land, such as the Mississippi, tend to carry large amounts of water. This relationship, however, between discharge and drainage basin size is complicated by the role of climate. If two drainage basins were of equal size, then the one located in a more humid region would naturally have greater discharge.

To illustrate the relationship between discharge, drainage basin size, and climate, let us consider the data listed in Table 8.1. Here we see the 10 largest rivers in the world based on the discharge at their mouths. Notice how the Mississippi and Congo drainage basins are comparable in size, but the Congo discharges 2.5 times more water at its mouth than the Mississippi. This is due to the fact that much of the Congo basin lies in a wet

TABLE 8.1   The world's ten largest rivers ranked by discharge.

Rank	River	Continent	Average Discharge at Mouth (ft³/sec)	Size of Drainage Basin (miles²)	Annual Sediment Load (tons)
1	Amazon	S. America	7,750,000	2,670,000	900,000,000
2	Congo	Africa	1,630,000	1,470,000	43,000,000
3	Ganges	Asia	1,570,000	668,000	1,670,000,000
4	Yangtze	Asia	1,110,000	695,000	478,000,000
5	Orinoco	S. America	1,020,000	386,000	210,000,000
6	Paraná	S. America	812,000	1,150,000	92,000,000
7	Yenisey	Asia	681,000	996,000	13,000,000
8	Mississippi	N. America	649,000	1,240,000	350,000,000
9	Lena	Asia	593,000	961,000	12,000,000
10	Mekong	Asia	569,000	313,000	160,000,000

Source: Data from Gleick 1993, Milliman and Mead, 1983, and Zeid and Biswas, 1990.

tropical climate, whereas a considerable portion of the Mississippi basin is semiarid. Also interesting is how the discharge of the Amazon dwarfs that of the other major rivers. The difference here is because the Amazon has such a large drainage basin, and it also happens to be located in a tropical climate. However, if we were to consider sediment load, the Ganges River far outranks even the Amazon. The Ganges transports such a large amount of sediment because it drains the Himalaya Mountains, where weathering and erosion generate tremendous volumes of sediment.

Stream Erosion, Transport, and Deposition

In Chapters 3 and 4 you learned that tectonic forces raise the land surface above sea level, at which point weathering processes begin to break down solid rock into particles called *sediment*. This loose material then makes its way downslope toward stream channels via mass wasting (Chapter 7) and overland flow. Once sediment reaches a channel, it is transported downstream, and is eventually deposited at the mouth of a river or in low-lying areas within the drainage basin (sediment in a stream is often called *alluvium*). In addition to transporting water and sediment, streams play an important role in the process called *erosion*, in which earth materials are removed from a given area. Although erosion and transportation of sediment also occur by the action of wind and ice, our focus in this section will be on the work of running water.

Stream Erosion

One of the key factors in a stream's ability to erode the landscape is the velocity of the water. For example, in Figure 8.8A one can see that when water enters a meander bend it is forced to slow down on the inner part of the bend, but speeds up on the outer part. This velocity increase on the outer bank greatly enhances the ability of the water to cut (erode) into the bank, forming an unstable overhang on what is called the *cutbank*. Eventually the overhang will fall into the stream by mass wasting processes, at which point the material will be incorporated into the sediment already being transported downstream. On the inner bank where velocity decreases, sediment tends to accumulate and form a deposit known as a *point bar*. The combination of erosion on the outer bank and deposition on the inner bank results in the lateral migration of the stream channels over time. As shown in Figure 8.8B, stream migration and mass wasting work together to form progressively wider river valleys.

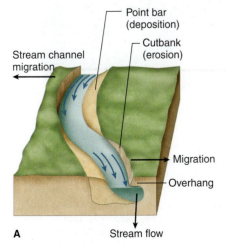

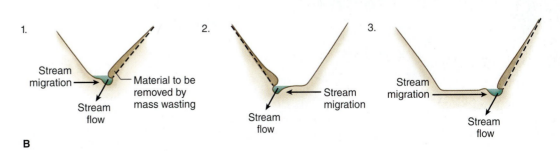

FIGURE 8.8   As flowing water moves through a bend in the channel (A), water velocity increases on the outer bank. This causes erosion and undercutting of the bank, forming an unstable overhang that eventually collapses. The lateral migration of stream channels combined with mass wasting over time (B) produces progressively wider valleys.

A

Sediment filled potholes

B

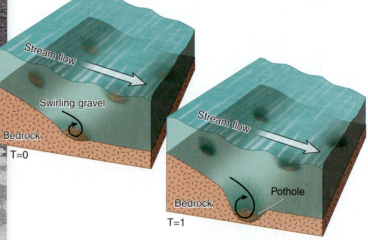

Stream flow

Swirling gravel

Bedrock

T=0

Stream flow

Bedrock

Pothole

T=1

FIGURE 8.9 The abrasive power of sediment within the Yellowstone River (A) has enabled the river to cut downward through solid rock, forming the canyon shown here. Sediment abrasion often takes place within potholes similar to those (B) found in a granite riverbed in Yosemite, California. During high discharge, the swirling action of water causes sediment within the potholes to rotate, slowly grinding holes into the solid rock.

In addition to lateral erosion, running water also cuts downward into the channel or *bed* of a stream. In some cases streams erode vertically through solid rock, forming deep canyons (Figure 8.9A). It is important to realize that the downcutting by streams is not performed by the water itself, but rather by the sediment that physically scrapes or wears away rock in a process called *abrasion*. To illustrate the abrasive power of stream sediment, imagine you were to first rub your skin with a piece of notebook paper, but then switch to coarse sandpaper. Clearly, the grit within the sandpaper would dramatically increase the amount of abrasion on your skin. In a similar manner, sediment moving across the bed gives streams the abrasive power to cut through solid rock. Evidence for stream abrasion in solid rock can be seen in *potholes* that are exposed when water levels are low (Figure 8.9B). Potholes form during periods of high stream discharge when the water column develops a swirling motion called an *eddy current*. These currents cause sediment on the streambed to slowly rotate, such that it grinds progressively deeper holes into the solid rock.

An important factor in a stream's ability to erode is its water velocity. The velocity of a particular stream segment is controlled by the steepness of the channel, called **stream gradient**—also referred to as *grade*. Water velocity therefore decreases whenever the stream gradient decreases, which, in turn, reduces the amount of sediment abrasion. When a river empties into a lake or ocean and the gradient becomes zero, the water stops flowing and erosion ceases. Geologists use the concept of **base level** to describe the lowest level to which a stream can erode. As illustrated in Figure 8.10, sea level is often referred to as *ultimate base level* because the oceans represent the end or low point of most rivers (exceptions include isolated areas below sea

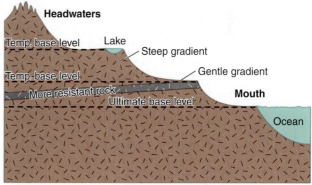

FIGURE 8.10 Base level represents the lowest level to which a stream can cut downward. Sea level serves as the ultimate base level, whereas lakes and resistant rock bodies can create a temporary base level. The photo shows the Grand Tetons, which have been uplifted far above base level, giving tributaries in the mountain headwaters the erosive power to cut down through solid rock. These tributaries eventually merge with the Snake River (foreground), which is approaching a temporary base level, giving the river a much gentler gradient.

level, such as the Dead Sea and Death Valley). Notice that as a river flows toward the ocean it may encounter a series of *temporary base levels* that form when its ability to cut downward is reduced by a resistance rock body, lake, or inland sea. In general, the gradient of a river becomes progressively less steep as the channel gets closer to a particular base level. This results in less downcutting and a tendency for a river to meander more, enhancing its ability to undercut its banks and produce a broader valley.

Stream Transport and Deposition

The ability of running water to transport and deposit sediment is dependent on both the water velocity and the types of particles being transported. Because velocity varies with stream discharge and gradient, rivers and streams tend to transport and deposit sediment in an alternating manner. To better understand how this works, we need to first discuss how solid particles and dissolved ions make up what geologists call a stream's *load*. The term *suspended load* describes the fraction of solid particles that is in a suspended state and moving at the same velocity as the water—suspended material is what makes streams appear muddy. The remaining fraction is called the *bed load*, which consists of sediment particles that roll, bounce, or remain stationary on the streambed. The key point here is that when overland flow causes both stream discharge and velocity to increase, certain types of sediment particles will move from the bed load and become part of the suspended load. Later when discharge and velocity eventually decrease, particles begin to fall out of suspension and return to the bed load. Note that anytime sediment is in motion, individual particles will undergo abrasion, causing them to become smaller and more rounded the farther they travel.

The process where sediment particles move back and forth between the bed and suspended loads depends not just on water velocity, but also on the sediment itself. In Figure 8.11 you can see that when stream discharge and velocity begin to increase, the first particles to be removed from the bed load are the smallest, least dense, and most angular. What remains on the bed are particles that are too large, dense, and rounded to be carried in suspension at that particular velocity. Conversely, when discharge and velocity eventually decrease, the first particles to fall from suspension and return to the bed are the largest, most dense, and most round. This process whereby water separates sediment grains based on their size, shape, and density is called **hydraulic sorting.** Over time, the combination of hydraulic sorting and abrasion will produce sediment that is progressively finer, more rounded, and more uniform in size as one moves closer to the mouth of a drainage basin (Figure 8.11).

An important aspect of sediment transport is that much of the sediment load within a drainage basin remains stationary for extended periods of time, moving only during periodic increases in discharge and velocity. As indicated in Figure 8.11, sediment generally does not move directly from the headwaters to the mouth of a basin, but in a series of steps, each consisting of a period of transport followed by deposition. During this process hydraulic sorting creates sediment deposits where more and more of the grains are similar with respect to their size, shape, and density. Recall from Chapter 3 that the mineral quartz is resistant to chemical weathering, whereas many other common silicate minerals are transformed into various clay minerals. Consequently,

FIGURE 8.11   Abrasion and the selective way sediment rises and falls from suspension results in progressively finer, more rounded, and better-sorted sediment as a river moves toward its mouth. When stream discharge and velocity increase (A), the smallest, least dense, and most angular particles are the first to rise off the bed and be carried in suspension. When velocity begins to decrease (B), the first particles to return to the bed load are the largest, most dense, and most round.

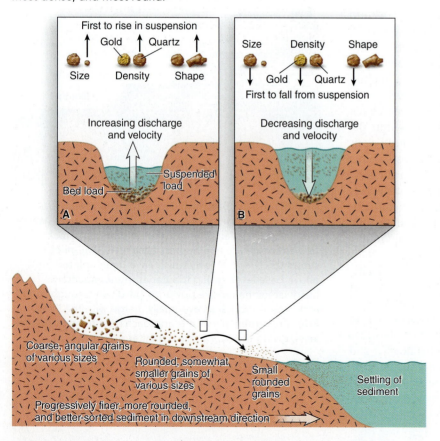

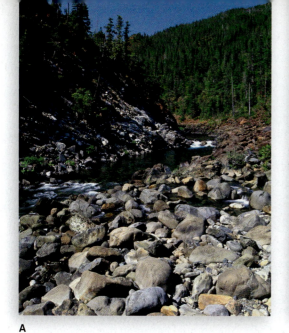

A

B

Point bar

sediment that has been transported great distances is usually dominated by tiny clay particles and rounded grains of quartz. The combination of hydraulic sorting and chemical weathering eventually produces relatively pure deposits of sand and clay, two very important natural resources used in society (Chapter 12). Hydraulic sorting also concentrates high-density particles such as gold, platinum, and titanium minerals, thereby creating valuable ore deposits.

Although describing the various types of stream deposits is beyond the scope of this text, we will briefly mention some of the more common deposits. Perhaps the most familiar are mound-shaped channel deposits called *bars* (Figure 8.12). Individual bars may consist of sorted material ranging in size from boulders to coarse gravel to fine sand. Boulder deposits are generally found near the headwaters of a drainage system, whereas those composed of sand-sized material are more common toward the mouth. Crescent-shaped bars, called *point bars,* develop on the inside of meander bends where water velocity decreases. Another common deposit is a *delta* (Figure 8.13), which forms when a river enters a lake or ocean and splits into smaller channels and begins to deposit sediment due to a decrease in velocity. The weight from this nearly continuous influx of sediment over time can cause the seafloor to sink and cause *land subsidence*

FIGURE 8.12   Hydraulic sorting leads to channel deposits called bars. Bars located near the headwaters (A) are often composed of boulders and coarse gravels, whereas sand-sized material is more common in downstream areas (B) where the gradient is less steep. Point bars are crescent-shaped deposits found on the inside of meander bends. Hydraulic sorting also creates valuable sand and gravel deposits as well as concentrations of gold and other high-density minerals.

FIGURE 8.13   Deltas (A) form at the mouth of rivers where velocity abruptly decreases upon entering a water body, forcing the stream to deposit its sediment. Satellite image (B) shows the Mississippi River as it deposits sediment in the Gulf of Mexico. Brown colors in the image represent sand and silt-sized particles falling out of suspension in the delta; green and blue colors correspond to very fine-grained clay particles that settle farther offshore.

Distributaries

Land

River

Delta surface

A

Accumulated sediment

B

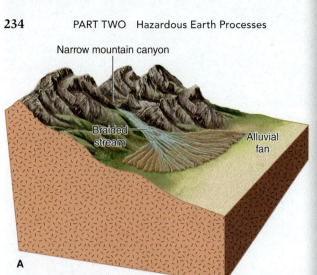

FIGURE 8.14   Alluvial fans (A) develop where steep mountain streams empty out onto valley floors. The abrupt changes in gradient and velocity cause the stream to become choked with sediment. Over time the stream migrates across the entrance to the valley, creating a fan-shaped deposit. Aerial view (B) shows alluvial fans in Death Valley, California.

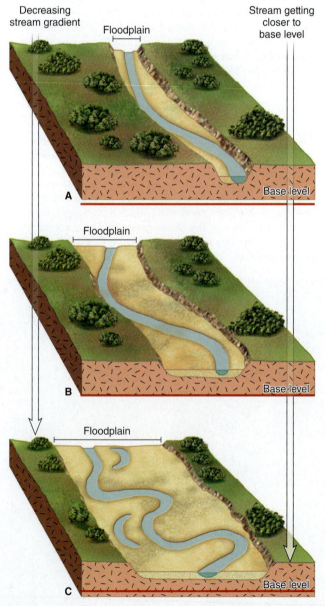

(Chapters 7 and 9). Subsidence allows deltas to become thicker and grow seaward, thereby creating new land area. A good example is the Mississippi Delta (Figure 8.13), where the deposition of vast amounts of sediment has led to subsidence and, until recently, a greatly expanded shoreline.

Also located at the mouth of rivers are *alluvial fans,* which are large fan-shaped deposits that form where steep mountain streams empty out onto valley floors (Figure 8.14). Here the stream is no longer confined by the valley walls, which when combined with the decreased gradient and water velocity, greatly reduce the ability of the water to transport sediment. The result is a channel that is choked with sediment called a *braided stream,* which migrates back and forth across the entrance to the valley, creating a characteristic fan-shaped deposit. As described in Chapters 6 and 7, humans have historically located settlements on alluvial fans because of the availability of water and relatively flat land. However, the potential for large floods, mudflows, and debris slides makes alluvial fans hazardous places to live.

Finally, streams also transport considerable amounts of dissolved ions (charged atoms) in what scientists refer to as the *dissolved load.* The dissolved load originates from the weathering of minerals (Chapter 3), which releases individual ions whose electrical charge causes them to become attached to water molecules. These dissolved ions, often called *salts,* are then carried away and travel with the water as it flows through the drainage system. Ultimately the dissolved load is deposited in an ocean or inland body of water, where the ions stay behind as the water molecules evaporate. Note that dissolved ions are invisible to the human eye because they are on an atomic scale. Therefore, what makes streams, lakes, or oceans appear cloudy is the suspended sediment, not dissolved ions.

River Valleys and Floodplains

Recall that stream gradient, discharge, erosion, and deposition all vary from the headwaters to the mouth of a drainage basin. Here we want to take a closer look at how river valleys change as base level is approached since their shape affects the way in which flooding occurs. As shown in Figure 8.15, river channels tend to meander more as they approach base level and their gradient

FIGURE 8.15   When streams approach base level and their gradient decreases, they tend to meander more, cutting wider valleys with broader floodplains.

becomes less steep. The erosion that occurs along the outside of meander bends produces wider valleys over time, whereas deposition on the inner banks helps to build a flat plain on the valley floor called a **natural floodplain.** When rivers overflow their banks, the first area to be inundated is this flat portion of the valley, hence the name *floodplain*. Floodplains are an integral part of many of Earth's ecosystems, but with respect to flooding, their primary role is to periodically store large volumes of water moving through a drainage basin. Therefore, a floodplain is not separate from a river, but rather is an integral part of a drainage system. This is important because like all natural systems (Chapter 1), changes will occur within a drainage system whenever humans make modifications to the landscape. Later in Chapter 8 we will examine some of the serious consequences that occur when society attempts to disconnect rivers from their floodplain in order to reduce flooding.

Recall that the sediment load in streams nearing base level are typically dominated by sand, silt, and clay-sized particles; overland flow periodically adds decaying organic matter (vegetation) to the sediment load as well. As discharge and velocity increase during the initial stages of a flood, some of the bed load is picked up and carried in suspension. When the water in a river leaves the confines of the channel, it will experience a sudden decrease in velocity as it begins to spread out onto the floodplain. The largest particles, typically sand, will immediately fall out of suspension and be deposited at the edge of the bank where the velocity change occurs. As illustrated in Figure 8.16, this creates a pair of ridges called **natural levees** that run parallel to the bank. The remaining particles of finer silt, clay, and organic matter are carried out onto the floodplain and then slowly fall out of suspension to form a

FIGURE 8.16 Natural levees are typically found along river channels that have well-defined floodplains. When a river overflows its banks, the abrupt decrease in velocity causes the largest grains, commonly sand, to fall from suspension and be deposited along both banks. Finer particles of silt, clay, and organic matter are carried with the floodwaters and deposited on the floodplain. Back swamps develop where floodwaters are unable to return to the main channel.

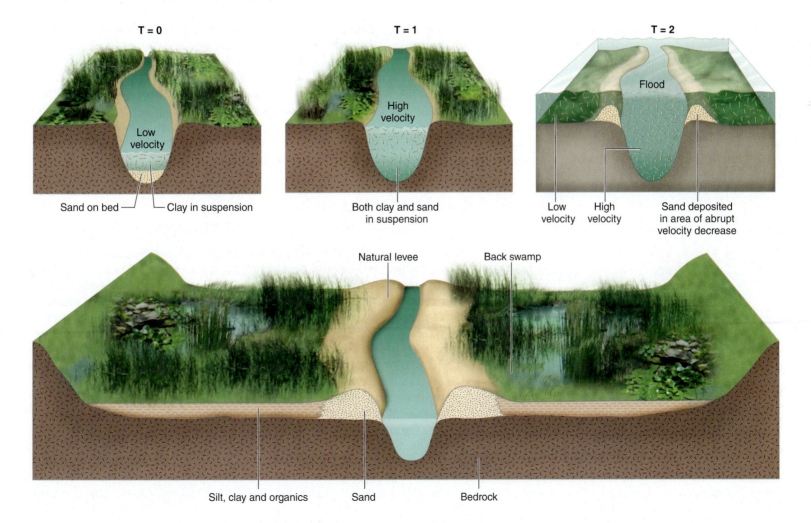

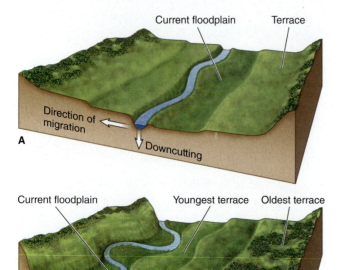

FIGURE 8.17 Terraces are abandoned floodplains left high and dry as the stream channel naturally migrates laterally and cuts downward across the valley floor. Because terraces are flat and less prone to flooding, humans have found them ideal sites for building settlements.

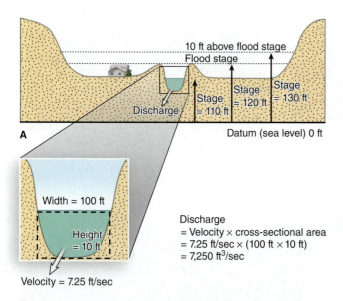

FIGURE 8.18 The severity of a flood is quantified by measuring either stream discharge or stage. Stage is simply the height of the water above some datum (A), whereas flood stage refers to any level where the stream has gone over its banks. Discharge (B) is the product of the water velocity and cross-sectional area of flow. Because the area of flow includes stage or height, there is a direct relationship between stage and discharge.

blanketlike deposit. When discharge eventually decreases, the levees act as a barrier, preventing a portion of the floodwaters from returning to the channel. Draining of the floodplain is further hampered by the fact that it is underlain by fine-grained material that has low permeability. Those areas of the floodplain that are poorly drained are referred to as **back swamps** (Figure 8.16), and can remain wet long after a flood.

When rivers migrate laterally across their valleys they also continue to cut downward, usually at a slow rate, toward base level. As illustrated in Figure 8.17, this lateral movement causes the older deposits in front of the migrating channel to be eroded away while new deposits are laid down behind it. This combination of lateral migration and continued downcutting creates new floodplains at progressively lower elevations in a stairstep fashion. The old floodplains left high and dry as a river migrates are called *stream terraces.* Especially well-defined terraces often develop in response to relatively rapid changes in base level resulting from tectonic uplift or lowering of sea level. Because stream terraces lie above the new flood plain, they are less likely to be inundated during a flood. It is not surprising then that humans have long made use of these geologic features as safe locations for building settlements and developing agriculture.

Flooding and Flood Hazards

Although we defined flooding earlier in this chapter, a more precise definition is now in order. Technically speaking, a *flood* is when normally dry areas of the land become inundated. The most common way a flood occurs is when excessive amounts of overland flow cause discharge to increase to the point that a river channel is no longer able to contain its flow, hence it overflows its banks. With respect to U.S. government regulations, a *floodplain* is defined as any land area susceptible to being inundated. Note that this definition is slightly different from the one given earlier for *natural floodplains* (i.e., flat areas adjacent to river channels). The reason for the government definition is that smaller tributaries lying far above base level often do not have well-developed natural floodplains, but yet still experience flooding. This legal definition therefore is simply used to designate *all* areas at risk of flooding and does not necessarily correspond to natural floodplains formed by geologic processes. In this section we will explore flooding in some detail and the steps humans can take to reduce its impacts.

Measuring the Severity of Floods

Because population density and development are not uniform from one area to another, hydrologists do not use the number of fatalities and property damage as a consistent means of quantifying the severity of floods. In order to compare floods in a quantitative manner, scientists use stream discharge and height. As shown in Figure 8.18A, the term *stage* refers to the height of water in a channel relative to some reference plane, usually sea level. **Flood stage** is the height at which a river begins to overflow its banks. Note that flood stage always refers to some specific site on a river because the channel continually decreases in elevation as you go downstream. Also, when the media reports that a river is 10 feet above flood stage, it simply means the water at that location is 10 feet above the river's bank or natural levee.

The other way hydrologists measure or quantify a flood is by stream discharge, which, as defined earlier, is the volume of water flowing past a given point over some time interval. Rather than using stage to define when a flood occurs, we can also use the amount of discharge required for a river to leave its channel. In Figure 8.18B you can see that discharge is computed by multiplying the cross-sectional area of a river by its water velocity. Because the cross-sectional area includes the water height, this means there is a direct relationship between discharge and stage. As part of its flood-monitoring effort, the U.S. Geological Survey (USGS) measures stage on a nearly continuous basis at approximately 8,500 automated gauging stations located throughout the United States. Discharge is then computed from this stage data and made available to the public in real time over the Internet. To find the water levels in a river near you, type *usgs surface water data* into any Internet search engine, and then follow the appropriate links.

Frequency of Floods

Rivers, of course, do not flood every day. In fact, many do not leave their channels for several years or more at a time. Clearly, if scientists could routinely predict floods, the loss of life and property damage could be significantly reduced. Unfortunately floods are particularly difficult to predict because they are dependent on specific weather events that have a high degree of natural variability. While there are some instances when flooding can be anticipated, as in the landfall of a hurricane, scientists are mostly limited to making projections based on statistical probabilities, similar to that described for earthquakes (Chapter 5). Statistical probabilities are useful because they tell us the frequency at which a river can be expected to reach a particular discharge. Keep in mind that such flood frequencies do not tell us when a flood of a certain magnitude will occur, but rather the probability or chance of it happening. Determining the probability of floods of different magnitudes allows scientists and engineers to quantify the flood risk for land areas adjacent to a river. Of course the lowest areas will have the highest level of risk for flooding.

The flood frequency of a river is determined by first acquiring historical discharge data for a particular river, preferably for as long of a time period as possible. Because the goal is to analyze the probability of flooding, the maximum discharge in a given year is tabulated for the entire record. An example dataset from the Tar River in North Carolina is listed in Table 8.2. Based on the maximum yearly discharge, a **recurrence interval** can be calculated for each value, which represents the frequency a particular discharge value can be expected to repeat itself. For example, a discharge with a 100-year recurrence interval means that on average, 100 years should pass before that same discharge level will occur again. Note that because of the statistical nature of recurrence intervals, a 100-year flood could take more than 100 years to repeat, but could just as easily reoccur in less than 100 years.

The following relationship is used to calculate the recurrence intervals for each yearly discharge value in a historical record:

$$\text{Recurrence Interval } (RI) = (N + 1)/M$$

where N is the number of values in the record, and M is the rank of an individual discharge value. Using the discharge data for the Tar River as an example (Table 8.2), one can see that the record covers a 102-year period, which in this case makes $N = 102$. To determine rank, the discharge values are then sorted from highest to lowest, giving the 1999 event of 70,600 ft^3/sec the highest rank ($M = 1$) and the 1981 event of 3,340 ft^3/sec the lowest ($M = 102$). The recurrence intervals for these two discharge values are calculated as follows:

$$70,600 \text{ ft}^3/\text{sec Recurrence Interval} = (N + 1)/M =$$
$$(102 + 1)/1 = 103 \text{ years}$$
$$3,340 \text{ ft}^3/\text{sec Recurrence Interval} = (N + 1)/M =$$
$$(102 + 1)/102 = 1.0 \text{ year}$$

TABLE 8.2   Maximum yearly discharge over a 102-year period for the Tar River at Tarboro, North Carolina. Flood stage at this particular gauging station near the coast is 19.0 ft above sea level, which corresponds to a discharge of 10,000 ft³/sec. Years in which the river did not flood are highlighted in yellow; years of major floods are highlighted in blue.

Year	Stage	Max. Yearly Discharge	Year	Stage	Max. Yearly Discharge	Year	Stage	Max. Yearly Discharge
1906	23.5	16,600	1940	31.8	37,200	1974	14.8	6,930
1907	21.2	13,400	1941	16.7	8,460	1975	27.1	22,600
1908	29.4	27,300	1942	14.8	7,310	1976	21.7	13,200
1909	19.7	11,500	1943	19.0	10,800	1977	17.5	8,880
1910	27.3	23,100	1944	21.6	13,800	1978	26.4	21,200
1911	17.4	9,210	1945	28.1	24,600	1979	27.0	22,400
1912	17.7	9,480	1946	21.0	13,200	1980	17.6	8,910
1913	21.2	13,400	1947	14.1	6,570	1981	8.9	3,340
1914	19.2	11,000	1948	25.4	19,800	1982	16.0	7,790
1915	19.1	10,900	1949	22.5	15,300	1983	23.7	16,400
1916	16.9	8,770	1950	14.4	6,990	1984	26.4	21,300
1917	23.3	16,200	1951	13.4	6,250	1985	20.8	11,900
1918	21.0	13,100	1952	24.2	17,600	1986	23.4	15,900
1919	34.0	52,800	1953	18.1	9,950	1987	28.4	25,200
1920	19.1	10,900	1954	27.4	23,600	1988	13.2	5,870
1921	17.2	9,030	1955	23.5	16,600	1989	24.2	17,200
1922	26.4	21,400	1956	20.9	13,000	1990	21.7	13,200
1923	22.8	15,500	1957	22.8	15,500	1991	16.4	8,090
1924	20.7	12,700	1958	29.2	26,900	1992	24.5	17,800
1925	33.5	39,800	1959	21.7	14,000	1993	25.7	19,900
1926	19.2	11,000	1960	22.8	15,500	1994	24.5	17,700
1927	17.8	9,570	1961	20.9	13,000	1995	21.6	13,100
1928	30.2	29,200	1962	20.3	12,200	1996	26.6	21,600
1929	25.5	19,800	1963	18.1	9,850	1997	21.2	12,500
1930	27.8	24,000	1964	20.8	12,800	1998	27.6	23,700
1931	17.7	9,480	1965	25.6	20,000	1999	41.5	70,600
1932	20.2	12,100	1966	22.7	15,300	2000	24.5	19,200
1933	16.0	8,050	1967	17.6	8,950	2001	22.5	16,000
1934	22.1	15,900	1968	21.2	12,500	2002	19.0	9,970
1935	27.4	23,500	1969	19.4	10,300	2003	26.3	21,000
1936	25.5	20,200	1970	17.8	9,020	2004	20.3	11,400
1937	26.2	21,500	1971	18.8	9,820	2005	15.4	7,410
1938	21.3	13,500	1972	23.1	15,500	2006	28.0	24,500
1939	27.0	23,000	1973	24.7	18,200	2007	23.0	15,300

Source: Data from the U.S. Geological Survey.

What this means is that a discharge of 70,600 ft³/sec should repeat itself, on average, every 103 years and 3,340 ft³/sec should occur every year. The remaining recurrence intervals in this record range in value between 1.0 and 102. However, the 1999 Tar River flood (70,600 ft³/sec), which was associated with the landfall of hurricane Floyd, was so large that it tends to skew the results of this analysis. Therefore, if we remove the 1999 event from the calculations and plot the remaining discharge values versus recurrence intervals, we get the graph shown in Figure 8.19. From the graph we see that the 1999 flood now has an estimated recurrence interval of over 500 years, perhaps even as high as 1,000 years. Clearly, this was a major flood of historic proportions.

Figure 8.19 also illustrates the fact that low-discharge events are more numerous, and repeat more frequently, compared to high-discharge events. In terms of flooding, the Tar River at this particular site will leave its main channel and begin inundating low-lying areas whenever discharge exceeds

10,000 ft³/sec (flood stage is 19 feet above sea level). From the graph we can see that this minimum discharge for producing a flood occurs on average about every one and a half years. Progressively larger floods, of course, have longer recurrence intervals.

Another useful way of measuring flood frequency is *percent probability,* which is simply the inverse or reciprocal of the recurrence interval (1/*RI*). Note in Figure 8.19 that the recurrence intervals have been converted to probabilities along the top axis. From this, one can see that a 10-year flood has a 10% probability of taking place in any given year, whereas a 100-year flood has a 1% chance of occurring. Recurrence intervals therefore tell us how *often* we can expect floods of a certain size, whereas percent probabilities indicate their *chance* of occurring. It is important to keep in mind that these statistical measures are based on long-term averages. As indicted earlier, 100-year floods do not necessarily happen exactly 100 years apart, but rather they occur *on average* every 100 years. This means that it is entirely possible to have two 100-year floods in back-to-back years, two in a single year, or two separated by 200 or more years.

It should be apparent that the reliability of recurrence intervals is dependent on the availability of historical discharge records. For example, a 100-year discharge record is more likely to contain several major flood events than a 10-year record, thereby producing recurrence estimates that are more reliable. A longer record also helps to smooth out any short-term variations in climate. You can gain an appreciation for the effect of climatic variations by examining the data in Table 8.2. Notice that from 1968 to 1982 there were eight floods in a 15-year period, whereas during the next 15 years (1983 to 1997) there were 13 floods, a 63% increase.

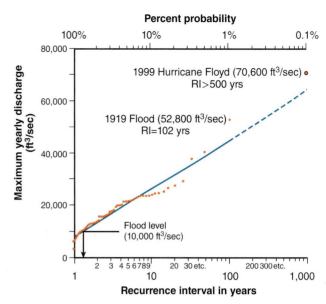

FIGURE 8.19   Plot showing discharge versus recurrence interval and percent probability for the Tar River at Tarboro, North Carolina (graph based on the discharge record in Table 8.2). Note that small discharge events are not only more numerous, but also have shorter recurrence intervals. Major floods such as the ones in 1919 and 1999 are exceptional events with long recurrence intervals.

Natural Factors That Affect Flooding

Earlier we discussed how most floods are related to heavy rainfall events that generate large volumes of overland flow, which then makes its way into a drainage network. However, flooding can also be caused by a number of different mechanisms. For example, coastal areas are commonly inundated by the surge of seawater that pushes ashore as a hurricane makes landfall (Chapter 9). During volcanic eruptions glacial ice caps may undergo rapid melting, producing catastrophic floods called mudflows (Chapter 6). In some regions the rapid melting of snow in the spring will lead to flooding, as will the breakup of river ice and the subsequent formation of an ice dam, causing water to back up such that upstream areas become inundated (Figure 8.20). Because most floods are caused by rain and snow, we will focus our attention in this section on the factors that affect precipitation-induced floods.

Nature of Precipitation Events

There many different types of rainfall events, from light, steady rains that may last for days to heavy, torrential rains lasting anywhere from a few minutes to several hours. The potential for flooding in a given area naturally increases as the *intensity* and *duration* of rainfall increases. Although regional climate determines the annual precipitation and how it is distributed throughout the year, daily weather conditions are most important with respect to flooding because weather governs the rate (intensity) and duration of individual precipitation events. For example, floods are often associated with the intense rainfall from thunderstorms. Such storms are most common in the spring to late summer period when warm air masses are more likely to collide with cool dry air. On the other hand, less intense rains can also produce large volumes of overland flow, provided the rains fall over an extended period of time. In many regions, floods related to these moderately intense rains can occur throughout the year. With respect to snow, its ability to cause flooding largely

FIGURE 8.20   The great 2009 Red River flood in the United States and Canada was caused by a combination of spring rains and the rapid melting of snow that had accumulated over the winter. Note how ice jams can block the flow of water in the main channel, making flooding worse.

depends on how rapidly it melts and how much accumulates before it begins to melt. A good example of a snow-induced flood is the great flood of 2009 along the Red River in Minnesota, North Dakota, and Canada (Figure 8.20). This flood began when spring rains coincided with the sudden melting of large amounts of snow that had accumulated over the winter.

Another important factor associated with rainfall is the *size* of the area over which the rain falls. An isolated thunderstorm for example may generate intense rains, but flooding is generally more localized because such storms are relatively small and tend to keep moving. On the other hand, large regional storms, like those that move inland from the ocean (e.g., tropical storms and so-called nor'easters), can drop tremendous amounts of water over large areas, thereby producing widespread flooding.

Ground Conditions

The ability of the ground to absorb water, referred to as **infiltration capacity,** plays a critical role in flooding because water that is unable to infiltrate is generally forced to move as overland flow. Therefore, rivers are more prone to flooding when large portions of their drainage basins are comprised of areas with low infiltration capacities. The actual rate at which water can infiltrate is determined by the slope of the land surface, type of ground material, and moisture content of the material. Clearly as slopes become progressively steeper, more and more rainwater will move as overland flow, leaving a smaller fraction that can infiltrate. With respect to the material itself, gravel and sand-rich soils have much higher infiltration capacities than soils rich in clay. Also, the infiltration rates of soils are highest when they are dry, then progressively decrease as the pores take on more water. Once soils reach saturation, infiltration will continue at a lower, constant rate provided the land surface is above the water table; otherwise, infiltration will cease completely. Note that other materials have almost no ability to absorb water, which forces rainwater to move almost entirely as overland flow. Examples here include unfractured igneous rock, frozen soils, and surfaces covered with asphalt and concrete (e.g., roads and parking lots).

Vegetation Cover

Vegetation is another critical factor because it helps to reduce the volume of overland flow, thus it tends to reduce flooding. Here vegetation intercepts and stores a certain fraction of the rain, thereby preventing it from reaching the land surface and moving as overland flow. In some heavily forested areas as much as 30% of the annual rainfall is captured and never reaches the ground; instead, it evaporates and returns to the atmosphere. For the fraction of water that does reach the land surface, vegetation restricts its ability to move downslope, thereby giving it more time to infiltrate. Very dense grasses that blanket or carpet the landscape are particularly effective at increasing infiltration and reducing overland flow. Naturally, rates of overland flow are significantly higher in arid climates because of the more sparse vegetation. Later we will examine how human modifications to the landscape (removal of vegetation and creating impervious surfaces) leads to significant increases in overland flow.

Types of Floods

In terms of rainfall and overland flow, the potential for flooding depends on how much and for how long the precipitation rate exceeds the infiltration rate, the size of the area over which the rain falls, and the vegetation density. For example, under even the most ideal circumstances an isolated thunderstorm in a large drainage basin is not going to generate enough overland flow to cause flooding on a major river channel located far downstream. This same storm, however, may be perfectly capable of causing small tributaries to quickly overflow their

banks. In this section we will take a closer look at how flooding hazards vary depending on the size of a channel and its location within a drainage basin.

Flash Floods

Earlier you learned that the headwaters of drainage basins are dominated by numerous tributaries and small rivers. Because these channels are relatively small, even isolated thunderstorms can generate enough overland flow to cause localized flooding. In addition, since small tributaries act as the entry point for overland flow coming into a drainage network, the lag time between the precipitation event and peak discharge is rather short. The result is that small streams and rivers tend to rapidly overflow their banks in what are called **flash floods.** Because small channels are more abundant in the upper parts of a basin, flash floods are also referred to as *upstream floods.* Keep in mind that flash flooding can occur in downstream areas along small tributaries that flow into large rivers.

Although flash floods generally affect only localized areas, they are particularly dangerous due to their sudden nature. What may be a pleasant, gently-flowing stream one moment can quickly turn into a raging torrent the next, leaving little time for escape. Adding to the flash flood hazard is the fact that the water level and velocity can be far greater than that found during normal flow conditions. Recall that channels in headwater areas are typically far above base level, where valleys are narrower and floodplain development is minimal or nonexistent. The general lack of a floodplain essentially means that a stream or small river has a very limited area in which to store excess water during a flood. Thus a narrow valley will force floodwaters to reach greater heights. Moreover, the velocity of the water is very high because the stream gradient in the headwaters is considerably steeper than in channels that are closer to base level. The result can be a raging torrent of water racing down a narrow valley, ripping up trees and moving boulders the size of cars. For people living in narrow mountain valleys, their only means of escape during a major flash flood is to quickly climb to higher ground.

While humans may consider flash floods to be freak events, in terms of geologic time they are a regular occurrence for many mountain streams. The 1976 flash flood along the Big Thompson River near Rocky Mountain National Park in Colorado is a case in point. On July 31 the usual late afternoon thunderstorms began to develop along the rugged Front Range of the Rockies (Figure 8.21). But instead of the normal winds which move the storms eastward out onto the plains, the winds on this day blew in the opposite direction, pushing the thunderstorms westward up against the mountains. This resulted in one particularly large thunderstorm that remained stationary for nearly three hours over the upper reaches of the Big Thompson Valley. Heavy rains began around 6:00 p.m., and over the next four hours as much as a foot of rain fell on the bedrock surfaces in this steep mountain terrain. Naturally very little infiltration took place, forcing nearly all of the rainwater to flow directly into the Big Thompson River. By 8:00 p.m. this normally tranquil mountain stream was transformed into a raging monster, cresting 20 to 30 feet (6–9 m) above normal and reached speeds of up to 50 miles per hour (80 km/hr). In a matter of two hours the Big Thompson flood killed 145 residents and vacationers, destroyed 418 homes and 152 businesses, all of which were washed down the canyon. Before and after photographs in Figure 8.21 attest to the tremendous power of this flash flood.

Deadly flash floods also occur in some desert regions where afternoon thunderstorms are largely restricted to the headwaters of streams in nearby mountain ranges. The combination of heavy rainfall with sparse vegetation and steep terrain create flash floods that sweep down through normally dry channels on the floor of valleys. For example, nearly every year in the western United States people become trapped in dry streambeds by flash floods that form from distant thunderstorms. The rushing water can rise so rapidly that

A

B

FIGURE 8.21 The 1976 flash flood along the Big Thompson River developed when a thunderstorm remained stationary, dumping a foot of rain over the upper reaches of its drainage basin (A) in mountainous terrain. Before and after photos (B) showing boulders that were transported during the flood. These boulders are a testament to the tremendous power and velocity of the floodwaters.

FIGURE 8.22 Photo showing the aftereffects of a failed dam in 1889 that unleashed a catastrophic flash flood, killing 2,209 people in the valley leading into Johnstown, Pennsylvania. As with most river valleys, human development was concentrated here because of the flat terrain, fertile soils, and abundant water resources.

there is not enough time for the victims to scramble to higher ground. In addition to thunderstorm activity, flash floods can also be produced by the sudden failure of dams, either human-made or those comprised of natural debris such as logs and sediment or broken-up ice. In fact the worst flood disaster in U.S. history occurred in 1889 when heavy rains contributed to the failure of a poorly maintained dam, located 14 miles upstream of Johnstown, Pennsylvania (Figure 8.22). The sudden failure of the dam sent a surge of water and debris, nearly 40 feet (12 m) high, roaring down the valley, killing 2,209 residents.

Downstream Floods

As indicated earlier, it takes a much greater volume of water to force a large river to overflow its banks than it does for a small stream located near the headwaters of a drainage basin. A river in the downstream portion of a basin is also more likely to be closer to base level and have a wider valley and more expansive floodplain. Consequently, a **downstream flood** can be defined as one where a river leaves its channel farther down in its drainage basin, flowing out onto its floodplain and inundating large areas of the valley floor. Clearly, the volume of water required to cover the floodplain of a large river will not be generated locally from a single isolated storm. Most downstream floods are caused by regional accumulations of water higher up in the drainage basin. What typically happens is that rain falls for an extended period of time, from days to perhaps weeks, over large areas of the upper basin—in some cases melting snow adds to the volume of water. Discharge from swollen tributaries will then combine, creating ever-larger volumes of flow as the river moves downstream. If the channel is not capable of carrying this combined flow, it will overflow its banks and begin storing the excess water in its floodplain.

In addition to the volume of water involved, downstream floods differ from flash floods in the lag time between the rain event and peak discharge (see Figure 8.3) and the length of time the river remains above flood stage. As illustrated in Figure 8.23, a major rain event in the headwaters will result in water entering the network of small tributaries. Once in the drainage system, the water will collect in progressively larger channels, traveling as a crest or pulse of water that moves downstream through the basin. Also note in the hydrographs how the lag time between the rain event and peak discharge becomes greater as the flood crest moves downstream. Here we see that flash floods form in the upper basin where water levels rise and fall relatively quickly. This stands in contrast to downstream floods where the combined flows of the tributaries cause water levels to rise more slowly as the river's floodplain becomes inundated. The hydrographs also show how the river stays above flood stage in the downstream areas for longer periods of time before eventually returning to its channel. Flash floods are more hazardous than downstream floods because they occur so quickly, which means people have less time to evacuate. On the other hand, downstream floods typically cause greater property damage because they cover much larger areas and can keep buildings under water for days or even weeks.

A good example of a downstream flood is the 1993 flood on the Mississippi River (Figure 8.24), which ranks as one of the largest natural disasters in U.S.

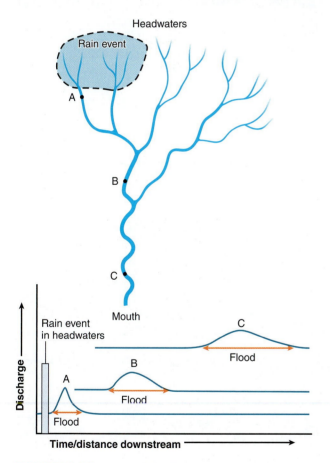

FIGURE 8.23 Idealized hydrographs illustrating how a river will respond to heavy rainfall in the headwaters of a basin. As the flood crest moves downstream, there is a progressive increase in lag time between the rain event peak and discharge; the river downstream also rises more slowly and remains above flood stage for a longer period of time. The hydrograph (A) is characteristic of a flash flood and (C) is characteristic of a downstream flood, with (B) representing a transitional phase.

FIGURE 8.24   Summer weather patterns in the Upper Midwest caused wave after wave of storms to drop rain over the same area. The result was regional flooding as well as flooding far downstream on the Mississippi River. Shown here are satellite images illustrating the effects of the flooding near the confluence of the Mississippi and Missouri Rivers.

history. This flood had its origins in the winter and spring when persistent rains kept the ground nearly saturated, thereby reducing the infiltration capacity of the ground. By June, streams across the Upper Midwest were already filled to capacity, and the heaviest rains were yet to come. During the summer of 1993, rainfall in the Upper Midwest was 200 to 350% greater than normal. One of the hardest hit areas was Iowa, where 48 inches of rain fell from April through August. The problem was related to how weather patterns in the summer cause cool, dry air from the north to collide with warm, moist air coming up from the Gulf of Mexico, producing a belt of thunderstorms that move across the region. In the summer of 1993, however, this frontal system remained fixed throughout June and July, causing wave after wave of thunderstorms to dump rain over the same swath of ground for weeks on end.

The unusual combination of winter and summer rains in 1993 in the Upper Midwest produced major regional flooding as well as downstream flooding on the Mississippi River. The flooding affected nine states where it inundated over 400,000 square miles of land and caused nearly $15 billion in damage. Also impressive was the fact that nearly 150 major rivers and tributaries experienced flooding, with recurrence intervals ranging between 75 to 300 years. With respect to the Mississippi River, it remained above flood stage at Hannibal, Missouri, for an incredible total of 174 days, from April 1 through September 21.

Human Activity and Flooding

The ability of humans to modify the natural environment has had the unfortunate consequence of increasing the risk of flooding. Humans have cleared considerable expanses of land for agriculture, built great cities, and constructed vast transportation networks for moving people and material. While these achievements have allowed us to prosper, they are not without consequences. In recent years scientists have discovered that our extensive modifications to the landscape have altered the hydrosphere such that the frequency and severity of flooding has increased. Ironically, some of our techniques for reducing flooding in one area have led to increased flooding in other areas. Protecting ourselves from floods has also encouraged development in areas where the risk would otherwise be deemed too high. The 2005 disaster in New Orleans (see Case Study 8.1) provides a sober lesson in how flood controls enable development to flourish, but then produce catastrophic losses should the controls suddenly fail.

In this section we will explore how human modifications to the landscape have led to an increase in the frequency and severity of flooding in modern times. We will also examine some of the techniques used to reduce the impact of flooding.

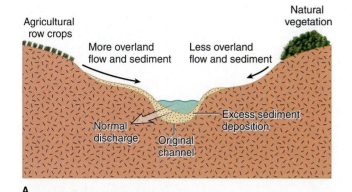

FIGURE 8.25   Sediment pollution (A) occurs when natural vegetation is removed from the landscape, which leads to increased overland flow and erosion that fills stream channels with excess sediment. As channels fill with sediment their capacity to carry water is decreased, making it more likely streams will overflow their banks and cause flooding. Photo (B) showing sediment pollution taking place in a stream. Note that the sediment is being carried off the agricultural field by overland flow.

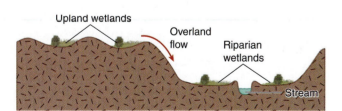

FIGURE 8.26   Wetlands are common along rivers and in topographic depressions in upland areas. Because of their porous nature, wetlands have a great capacity to capture and store water moving across the landscape. The destruction of wetlands has resulted in an increase in flooding as these areas no longer store water.

Land-Use Factors That Affect Flooding

Earlier you learned that when water accumulates on the surface it will either infiltrate or flow across the landscape to the nearest tributary. Many human activities increase the potential for flooding because they tend to decrease the amount of infiltration, which, in turn, increases overland flow. Other activities contribute to flooding by bringing about increased erosion. The result is that stream channels begin to fill with excess sediment, making it more likely that streams will overflow their banks during high discharge events. Flooding can also be exacerbated by the presence of human infrastructure, particularly roads, by restricting the free flow of surface water. Next we will take a closer look at how human activities increase the frequency and severity of flooding.

Removal of Natural Vegetation

Humans have historically cut down forests in order to obtain lumber and create open spaces for growing crops and building settlements. Likewise, vast expanses of natural grassland have been eliminated for the purpose of creating farmland. One of the consequences of the widespread removal of forests and grasslands is that it increases the ability of water to flow downslope, which translates into more overland flow and less infiltration. Note that although agricultural crops replace natural vegetation, overland flow still increases since the density of the crops is typically far less than the vegetation it replaces. The effect on flooding from agriculture is most dramatic when fields are fallow or contain young, immature crops.

Removing natural vegetation from the landscape also leaves soil more exposed to the effects of falling raindrops and overland flow. As raindrops impact the ground, they dislodge soil particles which are then carried into tributary channels by overland flow. This movement of excessive sediment off the landscape and into drainage systems is called **sediment pollution.** Over time sediment pollution can cause channels to become filled with sediment, thereby reducing their capacity to carry water (Figure 8.25). This, in turn, increases the frequency and severity of flooding as streams overflow their banks more easily. Sediment pollution not only exacerbates flooding, it also destroys the ecology of streams. Many types of aquatic species are unable to survive when sediment fills the channel and alters the original stream habitat. Also, fish that depend on relatively clear water to see their prey commonly do not survive when visibility is reduced by the higher levels of suspended sediment.

Destruction of Wetlands

As illustrated in Figure 8.26, wetlands (swamps) are commonly found in topographic depressions and adjacent to river channels, in which case they are called *riparian wetlands*. In many parts of the world wetlands have historically been viewed as wastelands because they were not suitable as building sites and represented obstacles for roads and rail lines. Wetlands were also a source of mosquito-borne disease. It is not surprising then that the practice of draining wetlands became quite common, particularly since they contained organic-rich soils that could be converted into fertile agricultural land. All that was required for wetlands to be drained was to lower the water table by digging a network of ditches. Once the surface dried out people could take uninhabitable "wasteland" and transform it into productive farmland.

Aside from the obvious loss of wildlife habitat and damage to natural ecosystems, the destruction of wetlands has reduced the landscape's ability to store water. Because wetlands are generally quite porous they can cap-

ture and absorb significant quantities of water moving off the landscape, particularly during periods when the organic matter is allowed to dry out. This ability to temporarily store water means that wetlands also help reduce both the rate and volume of water that make it into a stream channel during a heavy rain event. The large-scale destruction of wetlands is a key reason why some areas have experienced an increase in the frequency and severity of flooding in modern times.

Construction Activity

Most construction activity involves removing natural vegetation and regrading the land surface. This activity exacerbates flooding because it increases overland flow and causes stream channels to fill with sediment (Figure 8.27). In many instances the very structures that we build tend to restrict the free flow of water moving through preexisting channels. Most problematic are roadways because of the way these linear features crisscross the landscape. For larger streams bridges are usually constructed for road crossings, but for small streams that flow intermittently, large pipes called *culverts* are typically used. The problem is that the amount of discharge able to flow through a culvert is limited by the diameter of the pipe. During large flood events, culverts are often not large enough to handle the large volume of flow, causing water to back up such that upstream areas become flooded. This problem can be severe in highly developed urban areas with large numbers of culverts.

Urbanization

Of all the changes humans make to the landscape perhaps none has had a bigger impact on flooding than urbanization. Urban areas in developed countries typically have significant portions of the land covered with impermeable surfaces, chiefly roads and parking lots composed of concrete and asphalt. Because these surfaces are almost completely impervious, water is not allowed to infiltrate, but rather is forced to move as overland flow. The roofs of buildings are also impervious, and because urban areas contain large numbers of buildings, they too add to the volume of overland flow. This excess overland flow will make its way to the nearest stream via drainage ditches and storm sewers. As illustrated in Figure 8.28, when urbanization replaces natural vegetation cover with impermeable surfaces, the additional overland flow means that streams will reach flood stage more frequently. Also important is the fact that it takes far less time for overland flow to reach a stream channel than it does water that infiltrates and moves through the groundwater system. Consequently, not only does urbanization cause increased flood heights, it also leads to shorter lag times between precipitation events and peak discharge.

FIGURE 8.27 Photo showing how excessive amounts of sediment have moved off a construction site and into a nearby drainage ditch. Note how the culvert is being clogged with sediment.

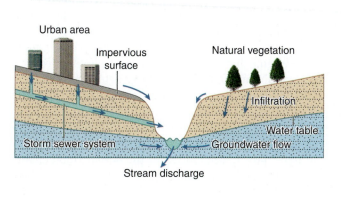

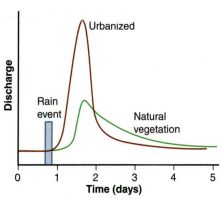

FIGURE 8.28 Urban settings commonly contain large areas of impermeable surfaces where little to no infiltration takes place, generating large volumes of overland flow that rapidly reach the drainage network. The result is more frequent flooding, higher flood crests, and shorter lag time between a rain event and peak discharge.

Ways to Reduce the Impact of Floods

Because flooding has historically presented serious problems, humans have developed a variety of ways to lessen the impact of floods—a process known as *flood mitigation.* Some mitigation techniques require building structures designed to control the flow of water in a channel, and others attempt to decrease the rate of overland flow. Laws have also been passed that are designed to minimize flood losses by managing human activities that tend to exacerbate flooding. In this section we will explore some of the more common flood mitigation techniques.

Dams

When a dam is built across a river it gives engineers the ability to control the discharge of a river. As illustrated in Figure 8.29, by regulating a river's flow engineers can raise or lower the pool or *reservoir* of water behind the dam. Reservoirs benefit society in that they protect against floods, serve as important sources of freshwater and electrical power (Chapters 11 and 14), and offer a variety of recreational opportunities. In general, engineers are able to add water to a reservoir during wet periods when streamflows are high, thereby creating a reserve or stockpile of water. Then during dry periods when the demand for water exceeds the river's discharge, the pool is lowered in order to meet the demand. In the case of dams being used to generate hydroelectric power, engineers raise and lower the pool on a daily basis. Regardless of whether the reservoir height is manipulated on a daily or seasonal basis, managers must ensure that enough water is always being released to meet the needs of ecosystems located downstream.

The storage capacity of a reservoir can also be manipulated for the purpose of preventing flooding downstream of the dam. For this engineers must maintain a sufficient amount of reserve or emergency storage behind the dam (Figure 8.29). This additional storage allows periodic surges of water from upstream areas to be safely contained within the reservoir. Although engineers normally maintain a sufficient amount of emergency storage, unexpectedly heavy or prolonged rains sometimes cause a reservoir to reach its maximum capacity. When this occurs engineers may be forced to release water at such a high rate that it causes downstream flooding, which, ironically, is what the dam is supposed to prevent. The alternative is to risk letting the pool rise to the point that the dam itself is weakened. The dam's structural integrity can also be threatened should water be allowed to flow over the top in an uncontrolled manner. In either case, the structural failure of a dam can generate a truly catastrophic flood, as in the massive flood described earlier that struck Johnstown, Pennsylvania, in 1889.

In addition to the potential for structural failure and devastating floods, dams are also a concern because of their environmental impact on downstream ecosystems. When a river is dammed, the suspended and bed loads are forced to be deposited within the reservoir. Moreover, the water becomes much colder due to the depth of the reservoir. Therefore, what is released from the reservoir is cool, sediment-free water. This can be highly disruptive to aquatic ecosystems that had naturally evolved in the presence of warmer water and a certain amount of sediment moving downstream. Also, because the volume of water that is released from a dam is normally kept within a fairly small range, downstream areas no longer experience a natural range of high and low streamflows. Because many ecosystem functions depend on large variations in streamflow, the highly regulated discharge from dams can have dire consequences for the aquatic food chain. Perhaps the most well-known environmental impact of dams is how they block the migration of certain fish species.

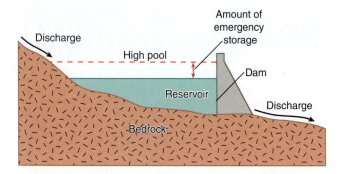

FIGURE 8.29 Dams prevent downstream flooding by intercepting and storing floodwaters in the reservoir. By carefully regulating the release of water, engineers can maintain the reservoir such that it provides a sufficient amount of emergency storage, and enough water to meet the demand for water supply and electricity.

FIGURE 8.30   Artificial levees (A) reduce the frequency of flooding by raising the height a river must reach before overflowing its banks. Most artificial levees consist of a ridge of earthen material constructed parallel to a river bank. Concrete levees called floodwalls (B) are sometimes used along urban corridors. Note how levees also act as bottlenecks that make floods worse in upstream areas.

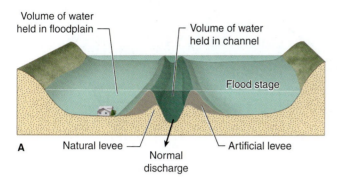

Artificial Levees

Earlier we discussed how *natural levees* form when a river leaves its channel and deposits sediment, forming a pair of sandy ridges running parallel to the banks. **Artificial levees** are those built by humans for the purpose of keeping a river from overflowing its banks and inundating its floodplain. Most artificial levees are constructed of earthen materials, but large concrete panels called *floodwalls* are sometimes used in urban areas. From Figure 8.30 one can see that the increased height of artificial levees reduces the probability that a river will flow out onto its floodplain. While such levees are quite effective in reducing the frequency of flooding, they disrupt the natural drainage system by disconnecting a river from its floodplain. This, in turn, leads to undesirable consequences for society.

The basic problem stems from the fact that a channel lined with artificial levees will hold far less water than what the floodplain is capable of storing. As shown in Figure 8.30, when a river rises the water must now stay within the artificial levees as opposed to flowing out into its floodplain. This means that the water which normally would have been stored in the floodplain is now forced to begin backing up in the channel. In essence then, the artificial levees act as bottlenecks that restrict the flow of a river, resulting in more frequent and severe flooding in areas located upstream. Another undesirable consequence is that the river tends to deposit sediment in its channel rather than in the floodplain. Since the additional sediment takes up space within the channel, it increases the likelihood that floodwaters will be able to reach the top of the levees. The degree of flood protection provided by the levees therefore will slowly diminish over time.

Because levees cause rivers to reach greater heights, upstream property owners are faced with the need for greater flood protection. This creates the incentive for upstream communities to build their own levees. As more of the river becomes confined by levees, floodwaters are pushed to even greater heights. This vicious cycle not only causes more frequent and severe flooding in upstream areas that lack flood protection, the higher water levels also threaten the structural integrity of the levees themselves. As illustrated in Figure 8.31, during periods of high water the levees must bear the additional weight of the water in the

FIGURE 8.31   Earthen levees generally fail or collapse when floodwaters (A) exert greater weight and water pressure on the levee. This also allows water to flow through zones of weakness within the levee and through underlying permeable zones, causing erosion that weakens the structure and its underlying support. Photo (B) showing floodwaters pouring through a broken levee and flowing onto the floodplain.

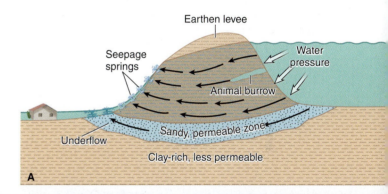

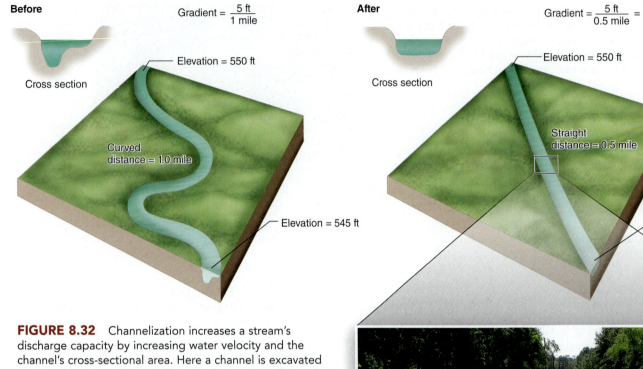

Before

Gradient = $\dfrac{5 \text{ ft}}{1 \text{ mile}}$

Cross section

Elevation = 550 ft

Curved distance = 1.0 mile

Elevation = 545 ft

After

Gradient = $\dfrac{5 \text{ ft}}{0.5 \text{ mile}} = \dfrac{10 \text{ ft}}{1 \text{ mile}}$

Cross section

Elevation = 550 ft

Straight distance = 0.5 mile

Elevation = 545 ft

FIGURE 8.32 Channelization increases a stream's discharge capacity by increasing water velocity and the channel's cross-sectional area. Here a channel is excavated so it becomes more boxed-shaped, plus it is straightened in order to increase the stream gradient. Channelization is effective in reducing flooding, but also destroys the natural habitat for plants and animals, resulting in fewer native species. Note in the photo the steep banks and lack of shade along the channelized stream.

channel. Thus, any weakness within the levee may lead to a collapse and catastrophic breech, allowing floodwaters to inundate areas that were once protected by the levees. Common types of weaknesses include cracks in concrete floodwalls and settling of foundation support due to compaction, whereas earthen levees often have problems with burrowing animals creating tunnels. Earthen levees have the additional problem of weakening when they become saturated during extended periods of high water levels.

Levee failures are commonly triggered when water flows under or through the structure such that some of the material is removed by erosion. This can occur when weakness zones develop within a levee, or when the structure is built across some permeable zone that existed on the original land surface. Notice in Figure 8.31 how flood conditions create additional water pressure that is exerted on the sides and base of the levee. This forces water to preferentially flow along weak zones through the levee itself, and through permeable zones underneath the structure. If the flow or leakage becomes great enough to cause erosion, the structure can weaken to the point where it fails or collapses, thereby allowing flood waters to inundate areas behind the levee.

The sudden failure of artificial levees brings us to perhaps the most serious consequence of levees, namely that once a river is disconnected from its floodplain, people tend to perceive the area as a safe place to live and work. Because the floodplain now contains homes and businesses that otherwise would not be there, failure of the levee can result in catastrophic losses. There is no better example of this so-called *levee effect* than the tragic flooding of New Orleans in 2005, when the levees failed as a result of hurricane Katrina (Case Study 8.1).

Channelization

Another common flood control technique is known as **channelization,** which involves straightening and deepening a stream channel so that its *discharge capacity* is increased. When a section of a stream is allowed to carry more water, the probability that water will overflow the banks is reduced. Channelization makes use of the basic principle that discharge is a function of water velocity and cross-sectional area of the channel. Anything that increases either of these factors therefore will increase a stream's ability to carry water, and reduce the potential for flooding. As illustrated in Figure 8.32, a stream's cross-sectional area flow is enlarged by excavating the channel such that it becomes deeper and more box-shaped. During this process the channel is also straightened by removing the original bends and curves, allowing water to travel a shorter distance while experiencing the same elevation drop. This results in an increase in stream gradient and water velocity. Water velocity can further be increased by creating a smoother streambed so there is less drag or resistance on the flowing water. This can be accomplished by lining the channel with concrete and by removing obstructions such as downed trees and large rocks.

From Figure 8.32 one can see that channelization results in dramatic physical modifications to a stream. Although these modifications have the desired effect of reducing the frequency of floods, the natural stream system also responds in ways humans find undesirable. One of the more serious consequences of channelization is that flooding downstream actually becomes worse. The problem is that in nonchannelized sections, the discharge capacity of the stream remains the same. Consequently, when higher flow volumes move downstream from the channelized segments, the unmodified areas are simply overwhelmed. Another problem is that the increased water velocity in the channelized area causes the stream to begin eroding downward, leaving steeper banks that are more prone to mass wasting.

Perhaps less noticeable, but equally profound, is how channelization affects the ecology of a stream. Natural stream systems provide native plant and animal species with different water depths, velocities, and amounts of shade. Channelization takes this diverse, natural ecosystem and transforms it into an artificial environment with a single habitat consisting of a relatively swift current, uniform depth, smooth bottom, and banks generally free of overhanging vegetation. Particularly harmful is how the reduced shade and more uniform water depth leads to higher water temperatures. Ultimately, many native fish and plants are driven out, whereas the new environment is more favorable to alien, and often undesirable species.

Retention Basins

Earlier we discussed how excessive overland flow from urban areas leads to an increase in both the frequency and severity of floods. The basic problem is that urbanization causes more water to make its way into the drainage network over shorter periods of time. An effective engineering solution is to temporarily store some of this excess water in a series of depressions called **retention basins,** which are constructed within the tributary network as shown in Figure 8.33. During a storm event, water quickly enters a retention basin where the only outlet is a relatively small culvert that restricts the outflow of water. A retention basin is similar to a small-scale dam in that it temporarily stores excess water generated upstream, and then

FIGURE 8.33   Retention basins capture, and then slowly release excess overland flow generated from paved surfaces during storm events. Hydrograph illustrates how a retention basin reduces peak discharge, thereby decreasing the potential for flooding.

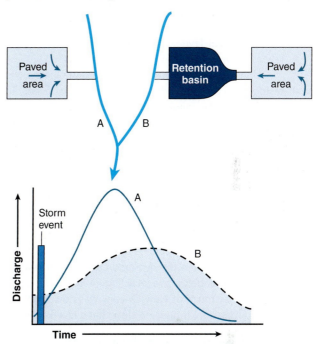

Retention basin

Levees and the Disastrous 2005 Flood in New Orleans

The tragic flood that struck New Orleans in 2005 in the aftermath of hurricane Katrina was more of an engineering failure related to flood control than a true natural disaster. Interestingly, this flood had its origins back in 1718 when the French built a settlement on a natural levee along a bend in the Mississippi River (Figure B8.1). This location gave the French access to a vital transportation corridor into the continental interior, and the levee provided dry ground in an area otherwise surrounded by cypress back swamps. Although the natural levee remained dry most of the year, early settlers found themselves in danger of being swept away by periodic floods. In order to minimize the hazard, French engineers built crude artificial levees. As New Orleans expanded, its protective levees grew in both height and length.

Until the early 1900s the city of New Orleans was pretty much restricted to the high ground along the natural levees. To allow the city to expand further, engineers began draining the back swamps that lay between the river and Lake Pontchartrain to the north. Occupying this land required placing levees along Lake Pontchartrain to hold back periodically high lake levels associated with hurricanes. Because the cypress swamps were near sea level, draining the swamps could only be accomplished by lowering the water table and pumping the water into elevated canals (Figure B8.2). From here the canals could carry the water away from the city. However, once the thick, organic-rich soils in the back swamps dried out, they underwent considerable compaction, which caused the land surface to subside. This created a bowl-shaped depression between the river and Lake Pontchartrain, parts of which eventually sunk below sea level (Figure B8.3).

Because much of New Orleans was now located in this human-made depression, it became clear that a levee failure could lead to a

FIGURE B8.2   Photos showing one the many canals within New Orleans used to help drain the original cypress swamps. Water is pumped into the canals from the adjacent low areas that now consist of residential housing tracts. The canals are lined with floodwalls in order to keep water from flowing out of the canals during hurricanes and flooding the reclaimed low areas.

catastrophic flood in which the depression would fill with water. It was absolutely critical then that the levees be made high enough to hold back potential floodwaters from not just the Mississippi and Lake Pontchartrain, but from the canals as well. In the 1960s the U.S. Army Corps of Engineers embarked on a major program of raising and strengthening the levees. This effort included installing concrete floodwalls along the canals to help prevent the levees from being overtopped by water forced up into the canals during a hurricane. It was this intricate system of levees, pumps, and canals that made it possible for the back swamps to be developed, which ultimately allowed New Orleans to grow into a city of nearly 1.4 million people.

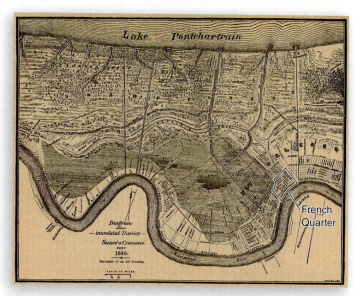

FIGURE B8.1   Map from 1849 showing the growing city of New Orleans and the cypress swamps that once existed north of the city. Note the position of the original French settlement (French Quarter) on the highest part of the levee, and how the swamps once drained to the north into Lake Pontchartrain.

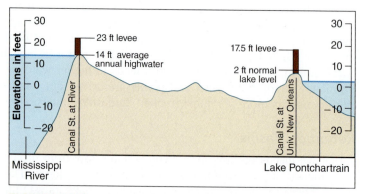

FIGURE B8.3   Draining of the back swamps and subsequent compaction of the organic-rich soils caused the land to subside beneath New Orleans. This has created a bowl-shaped depression, where many parts of the city now sit below sea level. Tall levees are all that keep periodic high water levels in the Mississippi and Lake Pontchartrain from flooding the city.

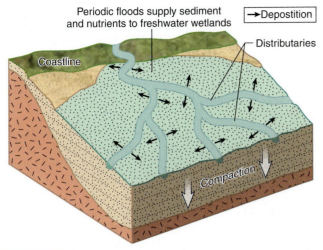

FIGURE B8.4 Fine-grained sediment within river deltas naturally undergoes compaction due to the overburden pressure that develops from the delta's immense weight. The land surface and extensive wetlands remain above sea level since sediment accumulation naturally keeps pace with compaction. In the case of the Mississippi Delta, artificial levees along the main channel have cut the delta off from its supply of sediment and freshwater, causing the entire delta to begin sinking below sea level.

Although it appeared that the engineering efforts had nature under control, the natural system was responding in ways no one had anticipated. In the 1970s geologists were beginning to gather data showing that the entire Mississippi Delta was slowly sinking below sea level. As indicated in Figure B8.4, the immense weight of a delta naturally causes the sediment to compact. Under normal geologic conditions the supply of new sediment being brought into a delta keeps pace with the rate of compaction. Therefore, the land surface does not sink (subside) below sea level. This delicate balance was disrupted in the Mississippi Delta when the artificial levees cut the delta off from its supply of sediment and freshwater. Compounding the problem is the fact that sea level

continues to rise (Chapters 9 and 16). The combined effect of land subsidence and sea-level rise has today caused some parts of New Orleans to be located more than 17 feet (5 m) below sea level.

By lining the Mississippi River with artificial levees and draining its back swamps to allow for greater development, society in the end inadvertently set New Orleans up for a flood of epic proportions. When hurricane Katrina slammed ashore just east of New Orleans in 2005, the city was spared from a direct hit, but the storm pushed water up into the canal system as scientists had predicted. Despite the fact the water levels within the canals remained about 4 feet (1.2 m) below the concrete floodwalls—well within their design limits—the concrete panels failed at nearly 50 locations around the city. Investigators later determined that the panels in these areas had been improperly anchored in their foundations. During the storm, the weakened panels simply fell over due to the weight of water within the canals. This allowed water to rush into the bowl-shaped depression of New Orleans. From Figure B8.5 one can see that while nearly 80% of the city was flooded, certain sections remained dry because some of the canals dividing up the city did not fail. Note that the French Quarter remained dry because the original settlement was located up on the natural levee, which was the highest ground available.

An important lesson learned from this disaster is that while levees do provide protection from floods, they also disrupt natural systems, which may respond in ways society finds undesirable. Levees also encourage development in low-lying areas that otherwise would not be developed. A single event or failure can then result in a major disaster. Because it is only a matter of time before a major hurricane makes a direct strike on New Orleans, questions are being raised as to whether it is a good use of taxpayer dollars to try and upgrade the levee system. This debate is particularly important since the next levee failure could result in losses far greater than those from Katrina. Finally, in light of the fact New Orleans continues to sink while sea level rises, some question whether the money would be better spent relocating the city farther inland on higher ground.

FIGURE B8.5 Photos of New Orleans taken on August 29, 2005, the day Katrina made landfall. Satellite image (A) showing how only certain sections of the city were flooded by failed levees. Aerial view (B) showing flooded neighborhoods and a breached canal levee in background.

releases it at a controlled rate. The slow release of water effectively reduces peak discharge in downstream areas (Figure 8.33), thereby decreasing the potential for flooding. Note that retention basins are commonly located adjacent to parking lots and within residential housing developments in order to capture overland flow from paved surfaces.

Erosion Controls

Because filling stream channels with excessive sediment makes it easier for streams to overflow their banks, techniques that help prevent *sediment pollution* also help reduce flooding. Unwanted sediment primarily comes from agricultural fields and construction sites where soils are bare and readily washed off the landscape during overland flow. There are two basic approaches to reducing sediment pollution. One involves practices that tend to keep soil particles in place so as to minimize the amount of material able to move downslope. The other approach is to use some type of physical barrier to trap sediment before it can enter the drainage network. In agricultural areas where soils are a valuable and irreplaceable resource, farmers quite naturally prefer erosion controls that help keep the soil in place. A good example is the use of vegetation to protect the soil from falling raindrops—see Chapter 10 for a detailed discussion on erosion-control practices.

The use of physical barriers to trap sediment is often required because erosion controls by themselves are not entirely effective at keeping exposed sediment in place. *Stream buffers* are a type of barrier in which vegetated strips line the banks of stream channels, trapping sediment before it can enter the drainage network. Stream buffers are generally required in areas where the land has been cleared by logging or agriculture. Another common barrier system employs temporary **silt fences,** which are made of a synthetic fabric that is fine enough to trap sediment, but yet allows some water to pass. As shown in Figure 8.34, silt fences are placed downslope of construction activity in order to keep exposed sediment from leaving the site. Although these fences are generally effective, the problem is that they can be completely overwhelmed in places where overland flow accumulates, thereby allowing sediment to pour into nearby streams. The last type of barrier system involves the use of *silt basins*, which are ponds constructed for the purpose of trapping any sediment that makes its way into a drainage system. However, heavy equipment must periodically be brought in to dig up the accumulated sediment and haul it away.

Wetlands Restoration

By building artificial levees and disconnecting rivers from their floodplains, humans have drastically reduced the ability of riparian wetlands to store floodwaters. The same holds true for wetlands located in upland areas, most of which have been drained for agricultural and urban development. Scientists now understand that the loss of wetlands has not only had devastating ecological impacts, but has also been a factor in the occurrence of more frequent and severe flooding. Because of these negative impacts, various groups in the United States are actively working to preserve existing wetlands and restore as many as possible back to their native state. Much of this effort involves reconnecting lowlands to their natural water supply by filling in drainage ditches and canals and removing levees. In some cases new wetlands are constructed from scratch so as to counteract the effects of increased overland flow.

Flood Proofing

As described earlier, there are a number of reasons why people build and live on floodplains. In many cases they are aware that their property will periodi-

FIGURE 8.34 Temporary silt fences are used in an attempt to keep exposed sediment from leaving construction sites. Although generally effective, silt fences can be overwhelmed in areas where overland flow accumulates and begins to flow in a channel.

cally be flooded, but choose to live along a river for its beauty, recreational value, or business opportunities. Here people may feel that the benefits simply outweigh the consequences of an occasional flood. Others may consider the risk to be small, and therefore take the chance that a serious flood will not happen in their lifetime, or anytime soon. For example, a property owner may decide that investing in a building with a 50-year life expectancy is worth the risk posed by a flood with a 100-year recurrence interval. Depending on what they feel is an acceptable level of risk, property owners may choose to "flood proof" their buildings or property. As shown in Figure 8.35, a time-honored flood-proofing technique is to raise the building above the expected flood level. Another option is to surround one's property with a permanent levee. For those who do not plan ahead, there is always the possibility of constructing an emergency levee using sandbags. Depending on the particular flood, however, it may not be feasible to build a temporary levee high enough or fast enough. Note also that such hastily built and makeshift levees may not be strong enough to hold back the floodwaters.

Flood Plain Management

In 1968 the U.S. Congress passed the National Flood Insurance Act and created the National Flood Insurance Program (NFIP), providing federally subsidized flood insurance to property owners. This federal program enables people and businesses to obtain flood insurance that would otherwise be difficult and expensive to obtain. To be eligible, a person has to live in one of the nearly 20,000 local communities currently participating in NFIP. Here communities must first perform a hydrologic study whereby a stream's flood characteristics are analyzed, the most important being the height of the projected 100-year flood. By comparing the elevation of the land surface to the projected flood height, a flood map is generated. The map is then used to determine insurance rates for individual land parcels based on the relative flood risk. Communities participating in NFIP are also required to restrict development in those areas lying below the 100-year flood level. As illustrated in Figure 8.36, the area within the 100-year flood zone is called the *regulatory floodplain,* which consists of two parts: the *flood fringe* and *floodway.* The floodway is most critical as this is where floodwaters are the deepest and fastest.

Although NFIP is voluntary, communities that participate are required to manage or restrict development within the regulatory floodplain. This results in what is sometimes called *floodplain zoning* as it works in a similar manner to local zoning ordinances, where commercial, industrial, and residential buildings are restricted to certain areas or zones within a community. In the case of floodplain zoning, regulations prohibit the building of new structures within the floodway (Figure 8.36). The idea here is to keep the floodway free of buildings and other obstructions that would otherwise impede the flow of water, and therefore raise the flood height. Many communities have found that floodways make excellent sites for parks and golf courses because these uses involve few permanent buildings. With respect to the flood fringe, new buildings are allowed, but must be built at or above the 100-year flood stage, which means they must be elevated. The end result is that NFIP provides both affordable flood insurance to communities, and encourages sensible floodplain management strategies that actually reduce flood losses and insurance claims.

Education

Similar to other geologic hazards, educating the public about flooding is a very cost-effective means of reducing the number of

FIGURE 8.35 Elevating homes is one of the oldest flood-proofing techniques and is still commonly used today.

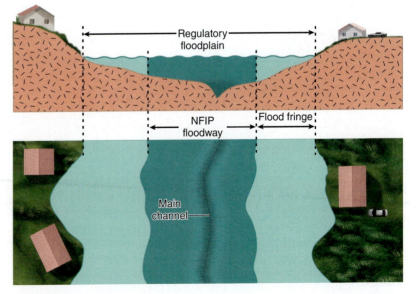

FIGURE 8.36 Floodplain zoning involves identifying areas adjacent to a stream that will be inundated in a 100-year flood. This so-called regulatory floodplain is divided into the flood fringe and floodway, and regulations then restrict the type of development allowed in each of these two zones.

FIGURE 8.37 Being knowledgeable about floods and floodplains can help people avoid placing themselves in high-risk areas. Such knowledge can also be used to take steps to reduce the risk for those who purposely build in flood-prone areas, such as the owners of the homes shown here.

FIGURE 8.38 Most people are not aware of how little water it takes for their vehicle to begin floating, placing themselves at risk of being swept away with the current. Approximately half of all flash-flood fatalities are vehicle related.

fatalities and property damage. Even though the vast majority of people are aware that rivers flood, many are ignorant of the fact that their property may lie in an area that periodically becomes inundated. Of course there are those who are aware of the hazard, but decide to take the chance that a flood will not happen anytime soon. Education then serves to help make people aware of the flood risk, and for those willing to gamble, shows them ways in which they can reduce their risk. For example, in Figure 8.37 you can see that someone *knowingly* built homes directly in the middle of a floodplain, but took steps to reduce the risk by raising the height of the land. While such a strategy might be effective for a typical flood, there's certainly no guarantee it will work for larger, less frequent events. Note that during a flood these people will be living on an island, and in order to access their home they must either have a boat or wait for the water to recede.

Perhaps the most important issue citizens need to understand is the hazard of driving a vehicle through floodwaters. Most flood-related fatalities occur in flash floods, and of these, approximately 50% are vehicle related. This is due in part to the way in which flash floods develop, whereby people quickly find themselves trapped in their vehicles. There are other situations where people will purposely drive through floodwaters. In either case, drivers and their passengers can suddenly find themselves in a life-threatening situation as the water decreases the weight of their vehicle such that it begins to float. When this occurs the vehicle can easily be swept away with the current. As illustrated in Figure 8.38, only 2 feet (0.6 m) of water is needed to float a typical car, and a large SUV will float in as little as 2 to 3 feet. Once a vehicle is swept away, the situation quickly becomes dangerous, because unlike a boat, a vehicle tends to roll over when it floats. Another problem with driving in floodwaters is that the road itself becomes difficult to see. Therefore, it becomes more likely you will drive off the road into even deeper water or into an area where the roadbed has been washed away. In addition to driving hazards, downed electrical power lines pose an electrocution hazard for those who venture out into floodwaters. The safest thing to do then is simply avoid going into floodwaters if at all possible.

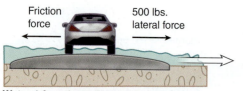

Water 1 foot deep: **Extremely dangerous**

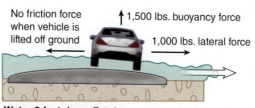

Water 2 feet deep: **Fatal**
Vehicle begins to float when water reaches its chassis, which allows the lateral forces to push it off the road.

Muddy water hides washout: Fatal
Washed-out roadway can be hidden by muddy water allowing a vehicle to drop into unexpected deep water.

SUMMARY POINTS

1. The hydrologic cycle describes the continuous movement of water within the Earth system. Water falling on the land will either infiltrate, evaporate, be absorbed by plants and animals, or move as overland flow downslope into streams. The basic function of streams and rivers is to transport both water and sediment off the landscape.

2. Stream discharge (volume flowing in a channel over time) generally increases as channels merge from the headwaters to the mouth. Stream discharge comes from overland flow (rainfall and melting snow) and baseflow (contributions from the groundwater system).

3. A drainage system is a network of channels that merge to form progressively larger streams, and eventually rivers. A drainage basin is the land area that collects water for a given stream network. The upper portion of a basin is referred to as the headwaters; the lower portion is called the mouth.

4. Streams not only transport water, but play an important role in eroding the landscape down to base level and transporting the resulting sediment to low-lying areas. The abrasive effect of a stream's sediment load enables it to cut down through solid rock.

5. Sediment within a stream channel is transported and deposited in an alternating manner in response to changes in velocity. As velocity increases, streams are able to transport progressively larger and heavier particles; when velocity decreases, the largest and heaviest particles are deposited first. This process results in sediment being sorted based on size, shape, and density.

6. As streams approach base level the following generally occur: (a) sediment size decreases, (b) gradient and velocity decrease, (c) channels meander more, (d) the valley becomes wider, and (e) floodplains, natural levees, and back swamps become more pronounced.

7. Floods originate when heavy rain or melting snow causes water to accumulate on the land surface faster than it can infiltrate. This excess water moves as overland flow and into the drainage network, where discharge may increase such that a river overflows its banks and inundates normally dry areas.

8. The severity of floods is quantified in terms of either discharge or stage. Recurrence interval or percent probability are used to measure the frequency of a given discharge event. Large floods occur less frequently than small events.

9. Natural factors that affect flooding are: (a) intensity and duration of precipitation events, (b) size of the area over which precipitation falls, (c) ground conditions that affect infiltration, (d) speed at which snow melts, and (e) amount and type of vegetation cover.

10. Flash floods generally occur in the upper parts of basins and are characterized by a large increase in discharge over a relatively short time period. Downstream floods occur lower in the basin and develop more slowly, but involve large volumes of water moving out onto a floodplain.

11. Flood frequency and severity have increased in areas where human activities generate large volumes of overland flow and excess sedimentation. Significant activities include: (a) removal of natural vegetation, (b) destruction of wetlands, (c) blockage of small tributaries, and (d) urbanization.

12. The impact of floods can be reduced by dams, artificial levees, and channelization. Such flood controls, however, tend to encourage development in flood-prone areas and may create more problems upstream or downstream. This creates the potential for much greater property losses in the event of a large flood.

13. Flooding can also be lessened by reducing the amount of overland flow and sediment into a drainage system. Common techniques include stream buffers, restoration of wetlands, and the construction of retention basins, silt fences, and silt basins. Other important tools include education, elevating structures, and floodplain management.

KEY WORDS

artificial levees   247
back swamps   236
base level   231
channelization   249
drainage basin   228
drainage divide   227

downstream flood   242
flash floods   241
flood stage   236
groundwater baseflow   225
headwaters   227
hydraulic sorting   232

infiltration capacity   240
lag time   226
mouth   227
natural floodplain   235
natural levees   235
overland flow   225

recurrence interval   237
retention basins   249
sediment pollution   244
silt fences   252
stream discharge   225
stream gradient   231

APPLICATIONS

Student Activity

Have you ever washed a car, windows, or anything fairly large with a bucket and mop or sponge? When you went to clean up the bucket, did you notice the sediment in the bottom? What happens when you just pour the water out of the bucket? How do you get rid of the additional sediment? What if you swirl the bucket as fast as you can then dump it out? Can you get rid of more sediment? Why or why not?

Critical Thinking Questions

1. Describe the hydrologic cycle.
2. What are some factors that affect flooding?
3. How does the amount of sediment a river can carry change as the river's velocity changes?

Your Environment: YOU Decide

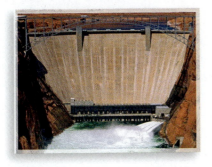

In the 1960s through the 1980s, it was thought that channelizing rivers and filling in wetlands was the best way to control flooding. The new thinking is that we build smaller dams, leave rivers alone, and keep as much of the wetlands as we can. What do you think? A lot of large-scale flooding has happened along channelized rivers and where wetlands have been urbanized. What about people who live in floodplains? Should the government continue to insure them?

Chapter 9

Coastal Hazards

LEARNING OUTCOMES

After reading this chapter, you should be able to:

▶ Explain the fundamental differences between ocean tides, currents, and waves.

▶ Describe how waves form, travel through open water, and change when they approach shore and enter shallow water.

▶ Explain how longshore currents develop and explain the longshore movement of sediment.

▶ Understand the basic way hurricanes form and the three major hazards they pose to humans.

▶ Explain why the number of hurricane-related deaths has decreased in recent years but property losses have increased.

▶ Understand how tsunamis form and why they pose such a hazard to people and property.

▶ Understand the basic cause of shoreline retreat and how humans are making the problem worse.

▶ Describe the different types of coastal engineering techniques, their purposes, and their undesirable consequences.

Aerial view along the North Carolina coast illustrating that when humans build expensive structures too close to the water's edge, their investments are at risk of falling into the sea as the shoreline naturally retreats. Notice how large sandbags were used in a desperate attempt to stop shoreline retreat.

Introduction

Our interest in Chapter 9 is the *coastal environment,* which refers to the unique setting where the terrestrial (land) and marine environments meet. Coastlines, often called *shorelines,* are a critical component of the biosphere because a large number of species are connected in some way or another to the unique habitats found at this interface between the land and sea. In fact, it was along shorelines that marine creatures first ventured onto land, ultimately evolving into the amphibians, reptiles, and mammals that colonized Earth's landmasses. In addition to playing a pivotal role in the evolution of life, coastal environments are immensely important to modern humans, from the food provided by fisheries to the harbors where port cities serve as vital centers of commerce and trade.

Like many other environments in which people live, the natural processes that take place in coastal zones can pose a hazard to human life and property. For example, each year tropical storms make landfall along shorelines around the world, causing major damage and untold human suffering. Moreover, because the coastal environment functions as a system, human modifications to shorelines commonly have unintended consequences and society must contend with them. Perhaps most troublesome is how humans routinely build engineering structures designed to control coastal erosion. These structures interrupt the natural movement of sand along a shoreline, which ironically causes increased erosion elsewhere along the coast. Today rapid population growth in coastal regions (Figure 9.1) is magnifying the problem of coastal erosion as well as the hazards associated with powerful storms. In the United States, for example, 53% of the population now lives in the narrow fringe of coastal counties (including the Great Lakes region), but which represents only about 17% of the nation's land area, excluding Alaska.

In Chapter 9 we will explore some of the natural processes and hazards that occur along coastlines, focusing on how humans interact with this unique

FIGURE 9.1 Plot (A) showing how population density in the United States is much higher in coastal areas, and it continues to grow. Satellite image (B) shows high-density development near Hanauma Bay on the Hawaiian island of Oahu. Note how development is concentrated on low-lying terrain closest to the shore and in valleys leading to the sea. These areas make better construction sites compared to the surrounding rugged terrain. Also note the extinct cinder cones, one of which has been breached, forming a small bay.

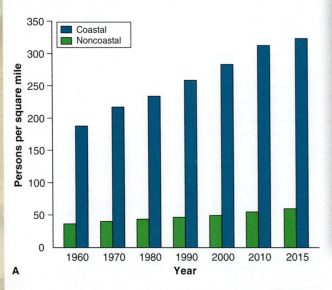

A

B

environment. Although our emphasis will be on marine coastlines, keep in mind that much of our discussion also applies to shorelines along inland seas and freshwater lakes, such as the Great Lakes in North America.

Shoreline Characteristics

As with most physical features on our planet, shorelines have different characteristics due to the dynamic nature of the Earth system. Some shorelines are relatively straight and have broad beaches, while others are more irregular and rugged, where beaches are sometimes restricted to coves. Still others have the appearance of a flooded network of stream channels, with countless inlet tributaries (e.g., Chesapeake Bay). These different characteristics are important in environmental geology because they influence the way humans interact with the coastal environment. In this section we will briefly examine how shoreline characteristics are related to two key geologic processes: plate tectonics and changes in sea level.

Recall from Chapter 4 that mountain ranges form along convergent and transform plate boundaries. In areas where tectonic forces deform and uplift the land, we can refer to the shoreline as an *active shoreline.* These tectonically active shorelines are usually rugged and irregular, with beaches sometimes being restricted to coves and inlets. An example would be parts of the Pacific coast of the United States. In contrast, a *passive shoreline* is one with little to no tectonic activity, commonly resulting in a relatively straight coastline with flat-lying terrain, such as the U.S. Gulf and Atlantic coasts. Figure 9.2 illustrates some of the fundamental differences between active and

FIGURE 9.2   Tectonically active land areas typically have steep terrain that produces irregular shorelines where beaches can be restricted to coves. In passive areas where tectonic uplift is minimal, shorelines generally have broad, straight beaches and low-lying terrain that extends far inland.

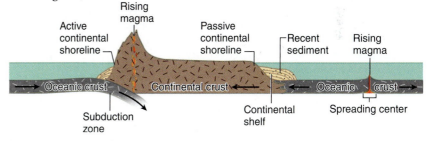

Otter Crest, Oregon

Active shoreline

Pompano Beach, Florida

Passive shoreline

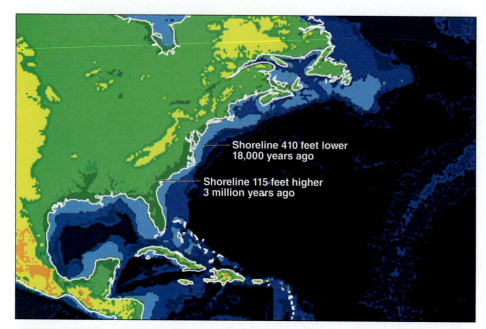

FIGURE 9.3   Climate change and the transfer of water between the oceans and glacial ice over the past 3 million years has led to large fluctuations in sea level and dramatic changes in the position of shorelines. Sea level today is rising at a rate of 0.6 feet (0.2 m) per century, but could increase dramatically should global warming destabilize ice sheets on Greenland and Antarctica.

[Figure labels: Shoreline 410 feet lower 18,000 years ago; Shoreline 115 feet higher 3 million years ago]

passive shorelines. From this figure one can see that mass wasting hazards would be more prevalent along tectonically active shorelines, whereas the low-lying terrain of passive coastlines is conducive to high-density development. While this type of development allows large numbers of people to enjoy living near the sea, the low-lying nature of the terrain can put both people and their associated structures at risk during major storms.

Another important process that affects the nature of shorelines is the relative movement of the shoreline either seaward or landward. Normally such changes take place slowly over geologic time, hence pose few problems for people living along coastlines. Shorelines however can shift much more rapidly, thereby creating serious issues for developed areas, particularly if the shift is landward where the sea begins to drown low-lying areas. One way this occurs is when humans disrupt natural processes such that the land surface begins to sink or subside (Chapter 7). A good example is how the construction of levees have starved the Mississippi Delta of its natural sediment supply (Chapter 8), causing large sections of the Louisiana coast, including New Orleans, to sink below sea level. Another example is the drowning of Venice, Italy, where the withdrawal of groundwater has led to land subsidence (Chapter 10).

Shorelines also shift in response to worldwide changes in sea level that occur when Earth's global climate alternates between cool, glacial periods and warm periods called interglacials (Chapter 16). As the climate cools, a huge volume of water is removed from the oceans and stored on land as glacial ice, lowering sea level and causing shorelines to shift seaward on a global basis. Over long periods of time the climate eventually warms and the water begins returning to the sea as the ice melts, raising sea level and causing shorelines to move inland. As illustrated in Figure 9.3, climatic changes over the past 3 million years have resulted in sea level being as much as 115 feet (35 m) higher and 410 feet (125 m) lower than today. Although the climate system has been quite stable for the past 10,000 years since the last ice age, sea level has continued to slowly rise. The key point here is that sea-level changes are not unusual. Therefore, the familiar position of modern shorelines is not a fixed feature. Throughout Chapter 9 you will see how even moderate rates of sea-level rise is exacerbating the problems of shoreline erosion and the hazards associated with large storms.

Coastal Processes

In this section we will examine several natural processes that play an important role in shaping coastal environments, including tides, currents, waves, and erosion and deposition of sediment.

Tides

If you have ever spent time at an ocean beach, you no doubt have witnessed the landward and seaward movement of the shoreline that occurs as the tide comes in and goes out. This rhythmic movement of the shoreline is caused by the spinning motion of the Earth, combined with the gravitational interaction between the Earth, Moon, and Sun. Because of the way the Earth, Moon, and Sun move within the same plane (i.e., solar plane), this complex interaction creates a net outward force along Earth's equator. As illustrated in Figure 9.4, this net or *tidal* force causes the solid Earth and the oceans to bulge outward along the equator. In this view looking down on the solar plane, notice how the solid portion of the Earth literally moves in and out of the ocean bulges as the planet rotates about its axis. This creates the periodic rise and fall of sea level known as **ocean tides.** Because of the timing of Earth's rotation and the Moon's orbit, most shorelines experience two high tides and two low tides each day (approximately 12 hours and 25 minutes passes between two high tides). Note that tides also occur on large lakes, but the rise and fall is relatively minor compared to that of the oceans.

Scientists use the term *tidal range* to describe the difference in sea level between high and low tides. From Figure 9.4 one can see that tidal range is greatly influenced by the relative position of the Moon and Sun with respect to the Earth. Note how the bulge generated by the Moon is much larger compared to that generated by the Sun. Despite the Moon being so much smaller than the Sun, the fact it lies so close to Earth means that its gravitational influence on the oceans is much greater. Although ocean tides are dominated by the Moon, the Sun can either enhance or reduce the tides, depending on the position of the Moon. For example, when the Moon and Sun periodically line up such that their gravitational effects reinforce one another, the tidal range is maximized in what is called a *spring tide*. Conversely, when the gravitational pull of the Moon and Sun are at right angles, their tidal effects tend to cancel one another, producing a small tidal range known as a *neap tide*. Note that in addition to orbital influences, tidal range varies depending on latitude, water depth, shape of the shoreline, and the presence of large storms. In Chapter 14 we will examine how humans are beginning to harness the tides in order to produce electricity that does not release carbon dioxide, thus does not contribute to global warming.

Currents

Similar to flowing water in a stream, **ocean currents** involve the physical movement of water molecules from one location to another. Ocean currents are driven by various forms of energy, and like all things in motion, currents flow from an area of high energy to one of lower energy. For example, when Earth's rotation brings a high tide into a coastal area, the surface of the sea actually slopes toward land. This difference in elevation of the ocean surface generates mechanical energy that forces water to funnel up into inlets and river channels, creating strong localized currents called *tidal currents*. When the tide goes out, the situation is reversed and the tidal current flows back out toward the sea. In contrast, out in the open water near the surface of the sea, large-scale *surface currents* form that are driven mainly by winds blowing consistently in the same direction. The effect here is similar to a person gently blowing across a pan of water, causing the surface of the water to begin flowing in the same direction.

In addition to wind-driven currents, there are also large-scale *density currents* that form in response to differences in ocean temperature and salinity (i.e., amount of dissolved ions). Because cooler and or more saline water is relatively dense, it tends to sink and flow toward areas where the water is less dense. Oceanographers use the term *ocean conveyor* to refer to the density-driven currents that circulate enormous volumes of water, both vertically and horizontally, in convective manner between tropical and polar regions. Density and wind-driven currents are important components of

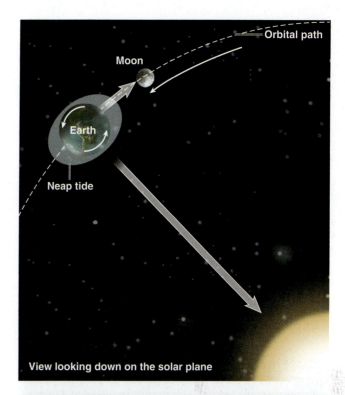

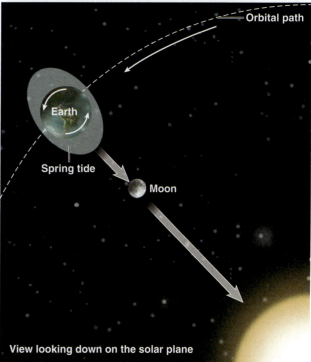

FIGURE 9.4 Earth's oceans bulge outward because of forces created by the planet's spinning motion and gravitational interaction with the Moon and Sun. Ocean tides form as the surface of the Earth rotates in and out of the bulges within the oceans. Note that the Moon has a greater tidal influence because it is much closer to the Earth than the Sun. The maximum tides, called spring tides, occur when the Moon and Sun align such that their gravitational effects reinforce each other.

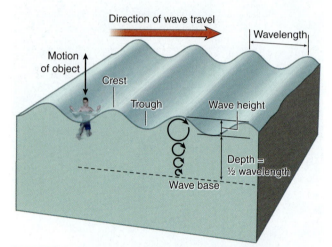

FIGURE 9.5   As wave energy travels horizontally through water, water molecules move in circular paths that get progressively smaller with depth. The level at which all movement stops, called wave base, gets deeper with increasing wave energy. Floating objects do not move horizontally with a passing wave, but rather bob up and down due to the motion of the water molecules.

Earth's global climate system (Chapter 16) since they transfer vast amounts of heat energy from the Tropics toward higher latitudes, providing heat to the cooler parts of the globe. These currents also transport nutrients throughout the oceans, thus are vital to marine ecosystems.

Waves

Recall from Chapter 5 that during an earthquake, that stored energy is released from a point, and then travels outward through solid rock in the form of vibrational (seismic) waves. In contrast, **water waves** transport energy through water such that water molecules move or vibrate in a circular manner as illustrated in Figure 9.5. Water waves not only vibrate differently than do seismic waves, but the frictional resistance of water is far less than rocks. This means that water waves will lose less energy as they travel outward from their energy source, hence they can continue traveling until the shore is reached. Note that water waves transport energy in a horizontal manner, but the circular motion of water molecules causes physical objects to move in a vertical manner. This is why a wave moving through open water will not carry a boat or swimmer along with it, but rather will cause such floating objects to simply bob up and down. Keep in mind that currents are different from waves in that currents transport both water and physical objects, whereas only energy moves in the direction of a traveling wave.

Like all waves, water waves can be characterized by the distance between successive crests (wavelength), the difference in height between the crests and troughs (wave height), and the amount of energy they contain. Notice in Figure 9.5 how the circular paths in the water column get progressively smaller with depth, eventually reaching what is called **wave base** where water molecules are no longer affected by a passing wave. Because the circular movement of water extends to greater depths as waves become more energetic, wave base can be used as a measure of wave energy. It turns out that wave base in a series of waves is equal to about one-half their wavelength (i.e., distance between successive crests). Thus, if wavelength is measured at 100 feet, for example, then wave base would be about 50 feet. Later you will learn how this relationship between energy and wave base is important when waves approach a shoreline.

High-energy waves called *tsunamis* (Chapter 5) form when earthquakes, volcanic eruptions, landslides, or asteroid impacts transfer some of their energy to a body of water. The mechanisms that form tsunamis are relatively rare, particularly compared to how wind routinely forms much lower energy waves on oceans, lakes, and inland seas. These ordinary waves develop when energy from wind (i.e., flowing air) is transferred to a water body. Naturally as wind speed increases, more energy is transferred to the water, producing waves with a deeper wave base. The energy of wind-generated waves is also affected by the amount of contact area between the wind and water, called *fetch,* and the duration of the wind. For example, a wind that blows steadily for 20 hours will generate higher-energy waves compared to the same wind that blows for only 2 hours. Likewise, a 20-square-mile lake will accumulate much more energy than one that is 2 square miles. In general, wave energy is the highest, and wave base the deepest, when high-velocity winds blow for a long period of time over extensive reaches of open water. Such conditions are typically found during tropical storms and hurricanes, a topic we will discuss in a later section.

Wave Refraction and Longshore Currents

When waves travel through deep water they experience little frictional resistance, which allows them to maintain energy and continue traveling until they reach a shoreline. When waves enter shallow coastal waters they eventually begin to drag on the seafloor and lose energy, greatly affecting the shape and velocity of the waves themselves. As shown in Figure 9.6, when wave base comes into contact with the seafloor, the circular water motion begins to encounter greater frictional resistance, which causes the waves to slow down. This decrease in velocity also forces the waves to pile up such that their wavelength decreases, but their height increases. In addition, because the friction is greatest near the seafloor, the bottom part of the waves slows down or decelerates at a faster rate than at the top. This uneven braking action within the water column slows the front part of the waves more than the rear, resulting in waves that become progressively less symmetrical (Figure 9.6). Because the waves continue to rise and become more asymmetric as they approach the shoreline, at some point they literally fall over on themselves, producing what are called *breaking waves*. Once the waves begin to break, water is then pushed up onto the beach, after which it flows back downslope and into the next breaking wave. Note that the *surf zone* refers to the area where the waves break.

Prior to reaching shallow waters, wind-generated waves travel as a series of crests and troughs lined up in a parallel fashion as illustrated in Figure 9.7. Note in this figure that when a particular wave crest moves into shallower water, the end closest to shore will start dragging on the seafloor while the rest of the wave remains in deep water. As the wave continues toward shore, this causes a progressive decrease in velocity along the length of the wave, forcing it to bend in a process called **wave refraction**. This process is similar to how lines of a marching band will begin to turn like spokes in a wheel when band members in a given line walk more slowly the closer they are to the pivot point. Wave refraction is important because as the waves bend, water from the breakers is pushed up onto the beach at an angle as indicated in the figure. But when the water flows back down the slope of the beach face, it follows a path that is generally perpendicular to shore. Overall this process forces water molecules to follow a zigzagging path in the surf zone, ultimately creating a current called a **longshore current** that flows parallel to shore. This longshore current is the reason that swimmers in the surf zone will find themselves slowly drifting away from their blanket and cooler in a direction parallel to shore. In addition to swimmers, the longshore current combined with the zigzagging action of water in the surf zone transports individual grains of sand in a process called *longshore drift*—sometimes referred to as *beach drift*.

Because of longshore drift, the portion of the beach within the surf zone is always moving in the direction of the longshore current, which, in turn, depends on the direction of the wind. A key point here is that as the amount of wave energy changes, so too does the width of the surf zone. Clearly, the wider the surf zone the greater the volume of sand being actively transported at any given time. During storms, when wind and wave energy are at their highest, the surf zone is at its widest and sediment

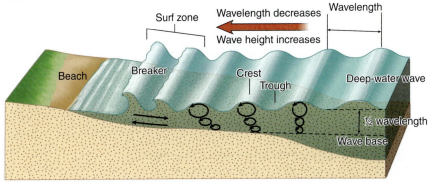

FIGURE 9.6 As waves enter shallow water, the wave base will eventually meet the seafloor, creating friction that causes the waves to slow down. This, in turn, causes the wavelength to decrease as the waves grow in height and become less symmetrical. Eventually the waves become so asymmetric that they fall over on themselves and form breaking waves.

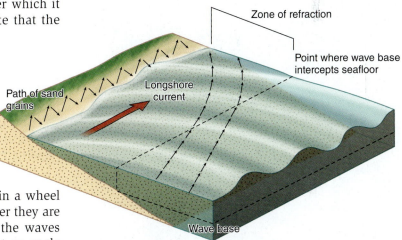

FIGURE 9.7 As a wave approaches land, the end closest to shore encounters the seafloor first, forcing it to slow down while the other end travels at its original speed. This velocity difference results in the wave bending or refracting toward shore. Breaking waves push water up the beach, creating a zigzagging path as the water flows back into the surf zone. This process creates a longshore current that moves both water and sediment parallel to shore.

transport is at a maximum. Moreover, during transportation some sections of the shoreline will undergo a net a loss of sediment, defined as *erosion* (Chapter 3), whereas other sections will experience *deposition*, or a net gain of sediment. Note that along the entire shoreline, erosion and deposition in a natural system will generally be in balance.

Shoreline Evolution

In the previous section you learned how the energy from wind is transferred to bodies of water and then travels as water waves. Upon striking shore, this wave energy is transferred to solid land where it drives both erosional and depositional processes. Similar to the way weathering and stream processes (Chapter 8) slowly wear mountains down, coastlines evolve over time due to the erosion and deposition from breaking waves. This interaction between waves and a landmass can cause the shoreline to slowly move landward, a process referred to as **shoreline retreat.** Landward migration of a shoreline can also occur when there is a rise in sea level, or when the land itself becomes lower due to subsidence (Chapter 7).

To help illustrate how shorelines evolve and retreat landward, consider the example of an irregular coastline shown in Figure 9.8. When waves crash into the exposed rocks it generates tremendous pressure, forcing water into tiny cracks and crevices. This repetitive hydraulic action slowly breaks the rocks apart and forms a notch or undercut within the cliff face. As the notch deepens the overhanging cliff becomes less stable, eventually causing the slope to fail in a mass wasting event (Chapter 7), at which point the cliff face retreats landward. In some instances the hydraulic pressure from crashing waves will slowly bore a hole through a cliff face, forming a *sea arch*.

Notice in Figure 9.8 that an irregular shoreline has what are called coves and points of land called *headlands* that jut seaward. Headlands are important in coastal geology because they are where waves first make

FIGURE 9.8   Headlands are places where waves first make contact with land and with the greatest amount of energy; these are the places where erosion is high. As the waves refract around both sides of the headlands, eroded material is transported into coves via longshore currents where it is deposited, forming isolated beaches.

Headland
(high energy)

Cove
(low energy)

contact with land. This causes waves to refract around both sides of head-lands, generating longshore currents that transport eroded sediment into adjoining coves where it is deposited to form beaches. In the case of tectonically active shorelines, once uplift slows down or stops, the continuous nature of headland erosion and deposition in coves will cause the shoreline to straighten over time. In Figure 9.9 one can see that as headlands are slowly eliminated, isolated beaches eventually begin to merge. At the same time the shoreline is being straightened, inland areas are slowly worn down by weathering and erosion such that the land surface gets closer to sea level (i.e., base level). Over geologic time therefore, a once irregular and tectonically active shoreline will evolve into one with the characteristics of a passive landmass, namely, low-lying terrain with long stretches of continuous beach. Note that during this evolution the shoreline is constantly retreating landward, which is why structures built along a coastline are often at risk of falling into the sea.

Finally, we need to briefly discuss how shoreline evolution affects longshore currents. In Figure 9.9 one can see that as headlands are eliminated and beaches begin to merge, the small cells or sets of longshore currents combine and begin to flow in a single direction with the prevailing wind. Although passive coastlines are relatively straight, sections typically remain that jut seaward and act as headlands where erosion processes dominate. Likewise, there are recessed areas where longshore currents tend to deposit sediment. This means that even along passive shorelines there are places undergoing shoreline retreat (erosion-dominated) and those that are actually growing seaward due to deposition. Later in this chapter you will see that identifying areas dominated by either erosion or deposition is important with respect to human efforts to control shoreline retreat.

Barrier Islands

Many coastlines have elongate deposits of sediment called **barrier islands,** which parallel the shore and are separated from the mainland by open water, lagoons, tidal mudflats, or saltwater marshes (Figure 9.10). In the United States barrier islands are common features along much of the Atlantic and Gulf coasts; worldwide they are found along about 15% of all coastlines. As their name suggests, barrier islands serve as a protective barrier that helps shield the mainland from powerful ocean storms. Despite being the first point of contact for storms, these islands have become highly prized for residential and recreational development due to their wide beaches composed of well-sorted sand. However, as we will explore later in this chapter, the fact that barrier islands are basically narrow ribbons of sand only a few feet above sea level, living here presents a grave risk during major storm events.

Scientists have proposed several different hypotheses as to how barrier islands form, but the exact origin is still somewhat uncertain. What we do know is that barrier islands result from the complex interaction between waves, sea level change, and sediment supply. It is generally believed that these islands originate when wave action causes sand to accumulate offshore, forming shallow sand bars. Enough sand eventually accumulates to where the bar stays above the high tide line, thereby creating a true island. If the sediment supply is sufficient, the island can grow in height as wind piles the sand up into dunes, which are then stabilized by vegetation cover. In general, most barrier islands are no higher than 20 feet (6 m) above sea level, but their shape varies depending on the relative influence of tides and waves. Along coastlines where the effect of tides is greater than that of waves, barrier islands tend to be short and stubby. The constant ebb and flow of the strong tides within the inlets between individual islands

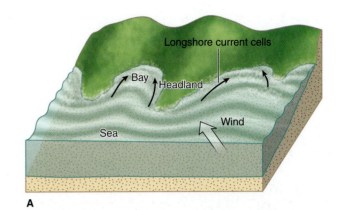

A

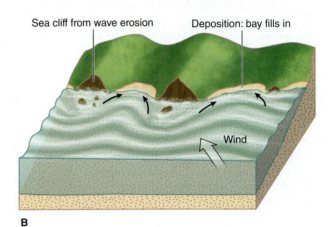

B

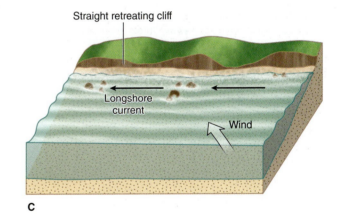

C

FIGURE 9.9 Once tectonic activity ceases, irregular coastlines slowly evolve into passive shorelines with more low-lying terrain and broad, straight beaches. Initially waves break on headlands, forming longshore current cells that transport eroded material into coves. As the headlands become smaller, the beaches and longshore cells will eventually merge, forming relatively straight sections of beach where sediment is transported parallel to shore.

Hutchinson Island, Florida

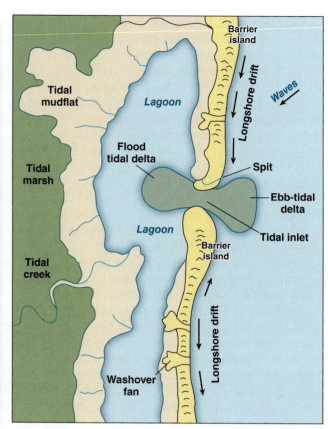

FIGURE 9.10 Barrier islands are elongated sediment deposits separated from the mainland by open water or wetlands. Tides move sand within inlets in an oscillating manner, creating submerged ebb-tide deltas. The islands themselves are highly prized locations for development because of their wide sandy beaches, but their low elevation makes them vulnerable to being overwashed during major storms.

also commonly produces what are called *ebb-tide deltas* (Figure 9.10). In contrast, when wave energy dominates, the longshore currents become stronger, creating longer and more slender islands. Finally, geologists believe that the formation of large barrier islands requires a steady sand supply and a fairly stable shoreline, meaning that the rate of change in sea level or land elevation has been relatively slow.

Although less dramatic than on rugged coastlines, shoreline retreat along passive landmasses can create serious problems for people, particularly on barrier islands. A significant factor here is how continued rise in sea level since the last ice age is forcing barrier islands to retreat landward. Because the rate of sea level rise over the past 7,000 years has been very gradual, these islands have been able to migrate fast enough to keep from being flooded. As illustrated in Figure 9.11, barrier islands move landward during major storm events when the higher sea level combined with high surf and wind erodes sediment from the ocean side of the island. The sediment is then deposited on the back

FIGURE 9.11
Shoreline retreat on barrier islands primarily occurs during storms when sea level increases and sediment is more easily transported over the island by wind and waves, allowing the islands to essentially roll over on themselves.

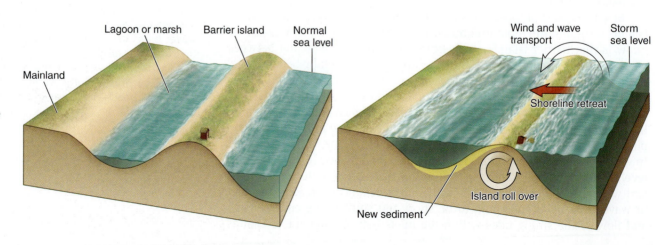

side. This also means that during storms, barrier islands migrate in *two* directions, namely parallel to the coast due to longshore currents and landward as sediment is transported over the islands. A key point to remember is that barrier islands, like most coastal features, remain fairly stationary, but then move in pulses during periodic storm events, when wave energy is at its highest. Later you will see that when humans build directly on the coast, this sudden shoreline retreat often results in expensive structures falling into the sea.

Coastal Hazards and Mitigation

Similar to other geologic processes discussed in this text (e.g., earthquakes, volcanoes, streams, mass wasting), various processes that operate in coastal zones can become hazardous to people and their property. In this section you will learn that most coastal hazards are related to storm events because this is when both wind and wave energy are at their highest. We will also explore some of the steps humans can take to mitigate or reduce the impact of these hazards on society, which is becoming increasingly more important as development continues to expand in coastal zones. We will begin by examining tropical cyclones, commonly known as hurricanes in the Western Hemisphere.

Hurricanes and Ocean Storms

Imagine being on a tropical island on a sunny day with only a gentle wind blowing in off the ocean. You then notice that the waves are getting higher and the surf zone is moving up onto the beach beyond the normal high-tide line. Later the wind begins to pick up and banks of clouds roll in. Intense rain soon follows along with 150-mile-per-hour winds that begin to destroy or damage nearly every human structure on the island. Making matters even worse, the sea soon rises nearly 20 feet, on top of which are large breaking waves that sweep most of the remaining structures off the island. This horrific experience is not over in a minute or two as in the case of an earthquake or tornado, but may last several hours. Should you be lucky enough to survive, it is safe to assume that this is an experience you would not want to repeat. This scenario is not from some science fiction movie, but it is what people actually experience in coastal areas where powerful tropical storms make landfall.

Scientists use the term *tropical cyclones* to refer to large, rotating low-pressure storm systems that originate in tropical oceans. Exceptionally strong tropical storms are called *hurricanes, typhoons,* or simply *cyclones* depending on where they form in the Tropics (Figure 9.12). For the remainder of this chapter we will use **hurricane** when referring to a powerful tropical storm. Of historical interest, the word *hurricane* originates from

FIGURE 9.12 Cyclones, hurricanes, and typhoons (A) are different terms used to describe large, rotating storm systems that originate over warm tropical waters. These storms generally follow curved paths toward higher latitudes and can produce winds in excess of 150 miles per hour and dump torrential amounts of rain, wreaking havoc on coastal areas. Satellite image (B) showing Hurricane Katrina prior to making landfall in Louisiana and Mississippi in 2005.

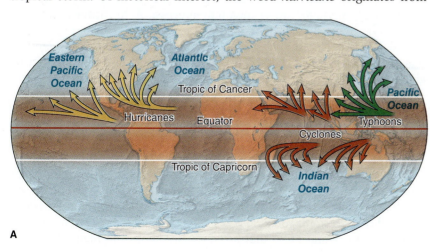

A

B

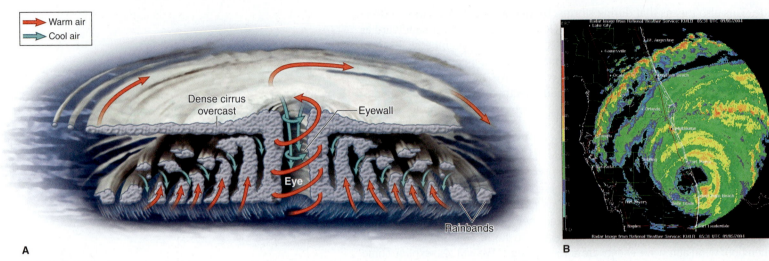

FIGURE 9.13   Hurricanes (A) form around low-pressure disturbances as evaporation removes heat energy and water from tropical waters. The resulting convection combined with Earth's spinning motion produces a rotating system centered about the low-pressure area of the eye. Intense winds, rains, and wave action cause major damage to coastal areas. Radar image (B) of 2004 Hurricane Frances showing spiral bands of heavy precipitation rotating around the eyewall.

the Mayan god of wind and storm called "Hurakan," which the Carib Indians later modified to "Hurican" to describe their god of evil.

Hurricanes develop over the warm tropical parts of oceans where a low-pressure disturbance can become amplified into a gigantic, rotating storm composed of high winds and intense precipitation. As evaporation occurs within the low-pressure disturbance and warm humid air begins to rise, large quantities of heat energy and water are removed from the ocean (i.e., a transfer of *latent heat*). The rising air mass eventually cools to the point where the water vapor condenses, and in the process releases the energy it had withdrawn from the ocean and transfers it to the atmosphere. While the warm air rises and begins to condense, it is replaced by descending air that is relatively cool and dry. This convective movement of heat and water produces heavy rains and thunderstorms. As shown in Figure 9.13, a hurricane forms when the convective movement combines with Earth's spinning motion to produce a rotating system centered about an area of low pressure called the *eye*. In a hurricane this intense convection generates high winds and thunderstorm activity in the wall of clouds surrounding the eye, called the *eyewall*, and also within the spiral bands that rotate around the eye. This rotating type of motion is called cyclonic, hence the scientific name *cyclone*.

The energy that drives a hurricane is extracted from warm seawater (latent heat) through the evaporation process. A hurricane's spinning motion and internal convection is important because it enables the storm to draw even greater amounts of energy from the ocean. Once in motion, this giant heat engine is able to sustain itself and can also further intensify given favorable atmospheric conditions and water temperatures. Hurricanes typically begin to weaken (i.e., lose energy) when they move into cooler waters, or encounter upper-level winds that disrupt their internal circulation. They can also weaken by coming ashore, at which point they become shut off from their basic source of energy, namely the evaporation of seawater. However, a hurricane that makes landfall is clearly an undesirable event for people living in the coastal zone due to the strong winds and high wave energy. Making the situation even more hazardous is the fact that the winds and low pressure within a hurricane produce a rise in sea level called a *storm surge*. This sea-level rise causes flooding and allows heavy surf to pound areas that are normally above the high-tide line. A

TABLE 9.1   The Saffir-Simpson scale is used to rank hurricanes based on their wind speed. Note how wind speed is related to level of air pressure within the eye of a hurricane and its associated storm surge.

Type	Category	Level of Damage	Max. Sustained Wind Speed (miles per hour)	Central Pressure (millibars)	Storm Surge (feet)
Tropical depression	TD		< 39		
Tropical storm	TS		39–73		
Hurricane	1	Minimal	74–95	> 980	4–5
Hurricane	2	Moderate	96–110	965–979	6–8
Hurricane	3	Extensive	111–130	945–964	9–12
Hurricane	4	Extreme	131–155	920–944	13–18
Hurricane	5	Catastrophic	> 155	< 920	> 18

hurricane will also produce heavy rains that commonly lead to river flooding far inland from where the storm makes landfall.

It is important to keep in mind that while hurricanes can be extremely powerful, other types of less powerful ocean storms create similar hazards. Strong storms often develop at higher latitudes over cooler waters and around low-pressure systems. Such storms can form during the winter months, a period in which hurricane activity ceases. Although these cool-weather storms are less powerful than hurricanes, they still produce high winds and heavy precipitation that create serious coastal erosion and inland flooding. For example, winter storms frequently develop in the northern Pacific, bringing heavy surf and rain to the Pacific Northwest coast of North America. On the other side of the continent, so-called *northeasters* (nor'easters) form over cool waters in the Atlantic. Winds from these storms typically blow from the northeast and are notorious for causing coastal erosion and bringing heavy rain and snow to inland areas.

The following discussion on coastal hazards will be broken down in terms of high winds, storm surge, and inland flooding. Although our focus will be on hurricanes because of their extreme nature, keep in mind that the following discussion also applies to the cool-weather storms previously mentioned, the difference being the intensity of the hazards.

High Winds

The unusually strong winds associated with hurricanes result from the circulating air masses within the storm. Because wind speed increases with energy level, scientists developed a scale for measuring a hurricane's intensity or strength that is based on the storm's maximum sustained winds. This scale, listed in Table 9.1, is called the *Saffir-Simpson scale.* Note how different hurricanes are ranked based on their sustained winds, with the lowest category having winds of at least 74 miles per hour—anything less is called a *tropical storm* or *depression*. Although category 1 hurricanes typically cause only minimal damage, their winds are still powerful enough to overturn tractor-trailers (cool-weather ocean storms rarely reach category 1 force winds). At the other end of the scale are category 5 hurricanes, whose sustained winds of over 155 miles per hour are capable of catastrophic damage. Notice in the table that wind speed increases as the air pressure in the eye decreases. The most powerful hurricane on record is the 1979 Pacific typhoon named Tip, whose central pressure was measured at 870 millibars. Based on this pressure, scientists estimate that Tip's sustained winds were an incredible 190 miles per hour!

A

B

FIGURE 9.14   This neighborhood (A) near Miami, Florida, experienced extreme damage from the 145-mile-per-hour winds produced by Hurricane Andrew in 1992. A piece of lumber (B) driven through a tree during Andrew demonstrates the destructive power of airborne debris.

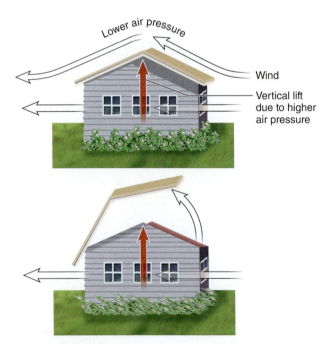

FIGURE 9.15   Hurricanes can destroy buildings with airborne debris, and by high winds blowing through a structure, which increases the amount of vertical lift on the roof such that it is removed.

Hurricanes are clearly capable of producing devastating winds, and the level of destruction naturally depends on building construction and wind speed. For example, mobile homes typically experience the most damage in any hurricane. Winds from a moderate category 2 hurricane can even damage well-built homes by stripping shingles from roofs, causing rain damage to the interior, or by toppling trees that fall onto the structure. At still higher wind speeds it is common for windows to be blown out and poorly anchored roofs lifted off buildings. At the 131–155-mile-per-hour range of a category 4 storm, roof and structural damage becomes pervasive over wide areas as shown in Figure 9.14A. In category 5 storms where maximum sustained winds exceed 155 miles per hour, structural damage is typically so complete that it is classified as catastrophic (Table 9.1).

Finally, we need to examine the actual way in which wind damages buildings and threatens human life. By picking up loose debris, hurricane-force winds typically contain airborne projectiles that not only pose a mortal threat to people left exposed to the wind, but creates a hammering effect that can completely destroy otherwise intact buildings. For example, Figure 9.14B illustrates the penetrating power of lumber traveling in 145-mile-per-hour winds of a category 4 hurricane. Serious structural damage can still occur under less extreme conditions when flying debris penetrates a building's windows, thereby allowing high winds to enter the structure. As illustrated in Figure 9.15, the combination of wind flowing through and over a building causes a difference in air pressure. This pressure differential, in turn, generates an upward force on the roof, which is identical to the vertical lift created by an airplane wing. Therefore, once wind begins flowing through a building and increases the air pressure inside, this lifting force can increase and literally pull the roof off the structure. One of the key reasons for covering windows with plywood or metal sheeting is to prevent wind from entering the structure, thereby keeping both the roof on and the rain out.

Storm Surge

Recall that the rotating eye of a hurricane is an area of abnormally low air pressure. Because air pressure is essentially the weight of the atmosphere pushing downward, the area of low pressure beneath a hurricane exerts less weight on the ocean surface, allowing it to rise. This results in a mound-shaped dome of water centered under the eye of a hurricane as

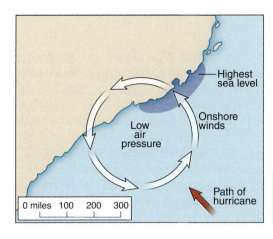

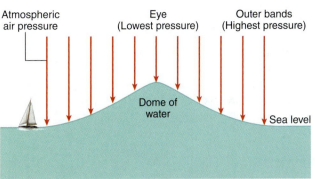

FIGURE 9.16 The decrease in air pressure toward the eye of a hurricane allows the sea surface to rise, creating a dome of water that follows the storm inland. This rapid rise in sea level, called a storm surge, is greatest on the northeastern side of the eye as the counterclockwise rotating system pushes water up against the shoreline.

illustrated in Figure 9.16. This dome of water will literally move or surge up onto land as the hurricane makes landfall, producing a rapid rise in sea level called a **storm surge** that inundates areas above the normal high-tide line. A storm surge moves inland at about the same speed as the hurricane and can produce a 30-foot (9 m) rise in sea level up and down a coastline for as much as 100 miles (160 km). Storm surge is particularly devastating along low-lying coastlines since the water can move inland considerable distances from shore, inundating developed areas.

As indicated in Figure 9.16, the storm surge is enhanced in areas where a hurricane's counterclockwise rotation and winds work together to push water up against the shoreline, creating an even larger rise in sea level. This effect is similar to what happens when a person blows across the surface of a cup of coffee, forcing the drink to pile up on the opposite side. In the Northern Hemisphere, the counterclockwise winds of a hurricane cause the storm surge to be greatest to the east or north of where the eye makes landfall. In contrast, storm surge is reduced to the west or south of the eye as winds blow offshore, thereby reducing the height of the storm surge.

In addition to affecting the height of the storm surge, wind is also directly responsible for the heavy surf that accompanies the storm surge as it moves onto land. It is the addition of large breaking waves that makes a storm surge particularly dangerous and destructive. In fact, more people die from storm surges than any other hurricane hazard. For example, from Figure 9.17A one can see how buildings constructed above the high tide line are not only inundated by storm surge, but are exposed to tremendous forces associated with large breaking waves. Most buildings that take the full impact of such waves are simply demolished, leaving their occupants with little chance for survival. Note in the photos in Figure 9.17B the number of homes that were destroyed by the storm surge from Hurricane Ike, a strong category 2 storm that struck the Texas coast in 2008.

When it comes to a major hurricane, one of the worst places to be is on a barrier island. This is due in part because these islands are where a hurricane first makes landfall, hence they are exposed to the strongest winds. Moreover, since barrier islands are essentially ribbons of sand only a few feet above sea level, the most serious threat comes from the island being completely overwashed by the storm surge. Riding out the surge and its breaking waves, even in an elevated building, is a high-risk and often fatal gamble. For those who decide to take the chance and not evacuate, there is no escape once the leading edge of the storm surge floods the road

FIGURE 9.17 Storm surge (A) not only inundates areas normally above high tide, but also brings breaking waves that demolish structures. Photo (B) of Bolivar peninsula near Galveston Bay, Texas, showing the effects of the storm surge after Hurricane Ike made landfall in 2008. Arrows mark features that appear in both images. Notice how the majority of homes were destroyed.

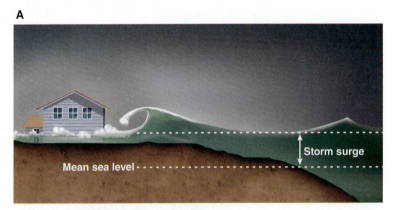

FIGURE 9.18 In 1900 a storm surge from a category 4 hurricane swept over Galveston Island, Texas, killing as many as 6,000 to 10,000 of the city's 35,000 residents. Photo showing the debris pile that formed when breaking waves progressively destroyed city block after city block. Open ocean is to the right in the photo.

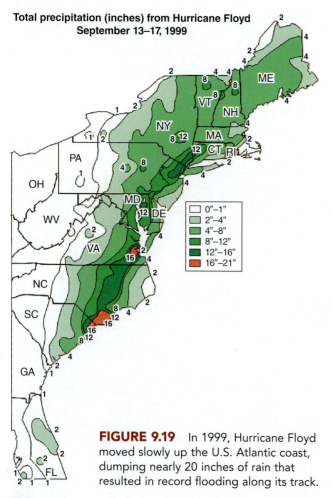

Total precipitation (inches) from Hurricane Floyd September 13–17, 1999

☐	0"–1"
	2"–4"
	4"–8"
	8"–12"
	12"–16"
	16"–21"

FIGURE 9.19 In 1999, Hurricane Floyd moved slowly up the U.S. Atlantic coast, dumping nearly 20 inches of rain that resulted in record flooding along its track.

to the mainland. Prior to modern communications and satellites, people living on barrier islands had no way of knowing that a major hurricane was approaching. They simply had no choice but to go to the top floor of a building or climb a sturdy tree, then hope for the best. Such was the case in 1900 for the 35,000 residents of Galveston, Texas, when the barrier island they were living on was completely overwashed during a category 4 hurricane. As the storm surge progressively destroyed block after block of the city, it created a 30-foot (9 m) tall pile of debris (Figure 9.18) that helped save the remaining buildings from the breaking waves. In the end, an estimated 6,000 to 10,000 people died, making this the worst natural disaster in U.S. history.

A more recent example of a catastrophic storm surge is the one that occurred in 2008 when Cyclone Nargis made landfall on the densely populated Irrawaddy Delta in Myanmar (Burma). At landfall the cyclone had sustained winds of 130 miles per hour (209 km/hr) and produced a storm surge that covered large areas of the low-lying delta. Unfortunately, the government failed to evacuate residents prior to the storm and also inhibited the relief effort of various international organizations in the aftermath of the storm. These factors all contributed to the large death toll, estimated at over 100,000 people. This tragedy is a grim reminder of the risks associated with living in low-lying coastal areas, the need for evacuating people to safety prior to a storm, and for providing adequate relief assistance afterward.

Inland Flooding

Hurricanes and ocean storms naturally remove vast amounts of water vapor from the oceans via evaporation, much of which is eventually returned to the Earth's surface in the form of heavy rain or snow. This intense precipitation commonly leads to inland flooding far from where a storm makes landfall. For example, despite the fact a hurricane becomes cut off from its primary energy source once it makes landfall, large volumes of water still remain in the storm as it tracks over land. It is not uncommon for these storms to produce torrential rainfall rates in excess of 1.5 inches (3.8 cm) per hour, which leads to flash flooding in the steeper parts of a drainage basin as well as downstream flooding (Chapter 8). One of the largest recorded rainfall events occurred in 1966 on Réunion, an island near Madagascar in the Indian Ocean, where a cyclone deposited an incredible 3.8 feet (114 cm) of rain in just 12 hours.

In addition to a storm's precipitation rate, another key factor affecting the level of flooding is a storm's forward speed as it moves over land. Slow-moving storms—less than 10 miles per hour (16 km/hr)—are particularly dangerous since more rain will fall on a given area compared to a storm that passes more rapidly. Consider Hurricane Floyd for example, which in 1999 moved slowly up the U.S. Atlantic coast, depositing as much as 19 inches (48 cm) of rain along its track as shown in Figure 9.19. The result was record flooding, the worst of which was along the Tar River in North Carolina as described in Chapter 8. Notice how heavy rainfall from Floyd was not limited to the immediate area where the hurricane made landfall, but extended far inland along the storm's track. In some cases remnants of a hurricane will merge over land with other weather systems, causing even more intense rainfall.

Rainfall intensity will also increase when an ocean storm moves inland and encounters rugged or mountainous terrain, forcing the humid air masses within the storm to gain elevation. The rapid elevation gain results in faster cooling rates, which, in turn, increases the condensation and precipitation rates. This topographic effect is a major factor in the severe flooding that commonly takes place when hurricanes move inland over the rugged coastlines along parts of Central America and the Gulf of Mexico. Similarly, the Appalachian Mountains in the United States help intensify precipitation rates from ocean storms moving inland from either the Atlantic or Gulf of Mexico. In addition to flooding hazards, the heavy precipitation associated with ocean storms moving inland commonly results in major agricultural losses, in terms of both crops and livestock. In mountainous and hilly terrain, the heavy rainfall often triggers numerous mass wasting events (Chapter 7).

Mitigating Storm Hazards

Similar to other hazards discussed in this textbook, ocean storms have been operating throughout most of Earth's history, thus are part of the natural environment on which humans depend. Although these powerful storms present many hazards to people, they have also played a key role in the evolution of coastal ecosystems, which ironically are what help draw people to live along the sea and in harm's way. Because humans have found it both desirable and beneficial to live close to the ocean, societies have learned how to *mitigate* or minimize the hazards associated with ocean storms. In this section we will explore some of the more common ways of reducing the risk of coastal hazards, particularly with respect to the more powerful and dangerous hurricanes.

Perhaps the oldest mitigation strategy goes back to ancient cultures, where they avoided locating large settlements directly on the coasts where hurricanes were known to frequently make landfall. If one considers how coastal development has been booming in the United States in recent years, it is apparent this strategy is no longer being employed. This modern trend can be explained in part because relatively few major hurricanes struck the U.S. Atlantic and Gulf Coasts during the post World War II construction boom. For example, from Figure 9.20A one can see that in a given year, the percent probability of a major hurricane (category 3–5) making landfall is actually quite low for most sections of the Gulf and Atlantic shoreline. This means that in many coastal areas individuals could likely live their entire lives without experiencing a major hurricane. The low probability of a major hurricane therefore tends to make people complacent with respect to living and building structures along a coastline. In terms of geologic time, however, it is all but inevitable that each section of the coast will sooner or later experience a major strike. Note also that because the strike probabilities are statistical in nature, it is possible for a given area to experience more than one major strike in a single year. Such was the case in 2004 when the paths of three major hurricanes (Charley, Frances, and Jeanne) crossed the same area of South Florida within a six-week period (Figure 9.20B). A fourth hurricane (Ivan) was present in the Gulf of Mexico during this same period, but made landfall in Alabama rather than Florida.

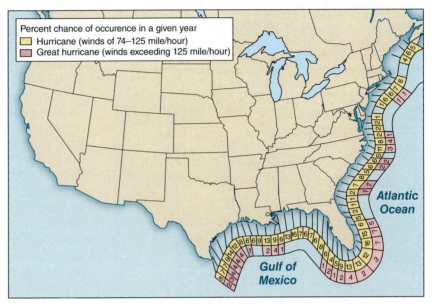

A

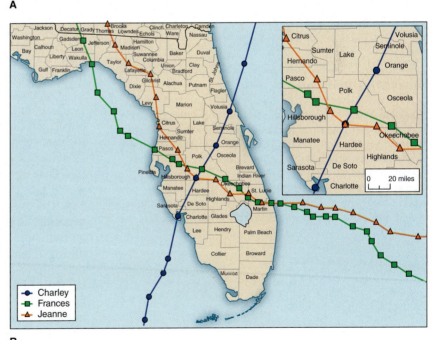

B

FIGURE 9.20   (A) Yellow areas show the percent chance of a moderate hurricane (category 1–2) striking sections of the U.S. coast in a given year. Red shows the percent chance of a major strike (category 3–5). (B) Because strike probabilities are statistical, multiple strikes are possible in a single year, as was the case in South Florida in 2004 where three hurricanes struck the same region.

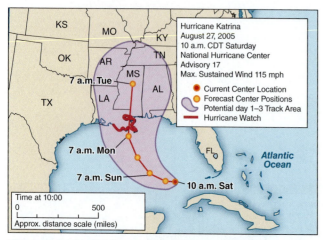

48 hours before landfall (10 a.m., Saturday, August 27, 2005)

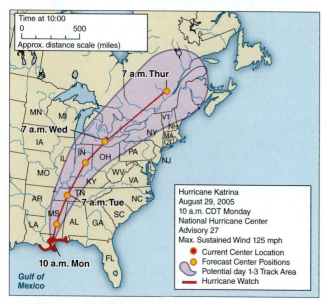

Landfall (10 a.m., Monday, August 29, 2005)

FIGURE 9.21 Computer models can accurately predict hurricanes paths, as illustrated by these three-day forecasts for Hurricane Katrina in 2005. The projected path takes the shape of a cone because the storm's position becomes less certain as distance from the eye increases. The center line within the cone represents the most likely position at any given time. Note how the 48-hour forecast for where Katrina would make landfall was very close to the actual location.

In addition to human complacency, another contributing factor in the modern boom in coastal development was the creation of early warning systems. These systems originated in the early 1900s when ocean-going ships made use of radio technology to begin reporting weather conditions back to land-based stations. After World War II, the U.S. Air Force started flying aircraft into hurricanes to record atmospheric data and to track the storms' positions. Today, weather satellites continuously track the storms' location, which when combined with aircraft data allow scientists to use computer models to predict the path of hurricanes with impressive accuracy. For example, Figure 9.21 shows the computer-predicted paths for Hurricane Katrina on two different dates in 2005 as the storm approached land. Note how the forecasted position of the storm takes the shape of a cone that grows outward from the eye. The cone shape reflects the fact that the uncertainty in the storm's future position keeps getting larger with increasing distance from the eye. As in the case of Katrina, forecasting models can now commonly predict out to 48 hours in advance where a hurricane will make landfall and do so with a fairly high degree of accuracy.

At some point before a hurricane makes landfall, emergency managers must make the decision to order an evacuation and get people to move to a safe location. The goal, of course, is to minimize the loss of life, but managers are also under pressure not to evacuate too early, in part because of disruptions to the local economy. However, as coastal population continues to grow, progressively more time is needed to evacuate, forcing emergency officials to make earlier decisions as to when to evacuate. Because of the need for more evacuation time, officials must begin the evacuation process when the hurricane is relatively far offshore, which is also when forecasters are far less certain of the storm's path (Figure 9.21). This, in turn, forces emergency managers to evacuate even longer stretches of coastline, most of which will experience little impact from the storm. The end result is that many people will have evacuated unnecessarily, making them less likely to evacuate in a future storm and leading to a higher death toll. Should officials decide to wait to order an evacuation until the storm's path is more certain, then they run the risk of not having enough time to carry out the evacuation. This, too, could lead to an unnecessary loss of life.

A good example of the evacuation dilemma emergency officials must face was in 2005 when Hurricane Katrina was far offshore in the Gulf of Mexico. As Katrina grew into an extremely dangerous category 5 storm, forecast models showed it was on track to make a direct strike on New Orleans. This was the nightmare scenario scientists had feared for years (see Case Study 9.1). Emergency planners had previously estimated that it would take 72 hours (3 days) to evacuate the 1.3 million residents in the New Orleans metropolitan area. The long evacuation period was deemed necessary in part because the city has only three escape routes; in addition, an estimated 100,000 residents relied on public transportation and had no means of leaving on their own. Despite the existing plans, city officials did not order a mandatory evacuation until 24 hours before Katrina made landfall. Although most residents were able to flee to safety, many simply had no way to leave. Fortunately, Katrina tracked slightly to the east, sparing the city the worst of the storm surge and high winds. Despite the more favorable storm track, numerous levees within New Orleans broke, sending water pouring into the city. With much of the city flooded, tens of thousands of people became trapped in the Superdome and on the roofs of their homes, many waiting to be rescued for days in the oppressive heat. Had evacuation orders been issued earlier, plans could have been implemented to evacuate those without transportation, and the human tragedy that followed Katrina would likely have been far less.

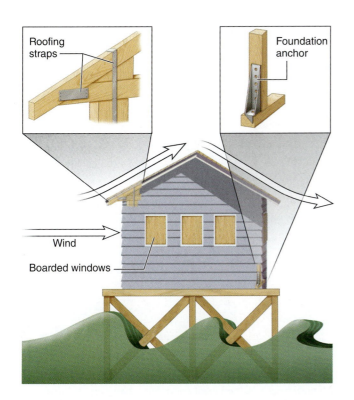

FIGURE 9.22   Structural damage from hurricanes can be greatly reduced by elevating a building above the storm surge so that wave energy can pass underneath. Boarding up windows and strapping the roof and frame help keep the roof from being lifted off the structure. The building can be strengthened further by anchoring the frame to the underlying structure.

Although early warning systems give people the chance to flee to safety, buildings and other types of property, of course, must stay behind and face the storm. In the United States, for example, early warning systems have greatly reduced the number of hurricane-related fatalities. However, the increased safety has also encouraged greater population growth in coastal zones, ultimately resulting in increased property losses. To help reduce property losses, engineers have developed improved construction techniques that minimize the amount of structural damage from hurricanes. As illustrated in Figure 9.22, elevating a building above the expected storm-surge level allows wave energy to pass beneath the structure rather than smashing it completely off its foundation. Note that wave energy can still cause erosion around the foundation supports such that the building collapses into the water. Another key design element for minimizing damage is to strengthen the structure against hurricane-force winds. Here metal straps are used to secure the roof to the main structure, which in turn is bolted to the foundation. Also, windows should be covered with plywood or metal sheeting prior to the storm. This prevents high winds from entering a building and lifting off the roof.

Finally, there is growing concern among scientists that as ocean temperatures continue to rise in response to global warming (Chapter 16), hurricanes will occur more frequently and with greater intensity. Insurance companies and emergency managers are particularly concerned about this issue due to the boom in U.S. coastal development that has occurred since World War II. Recent data show that hurricane frequency has indeed increased in the Atlantic (Figure 9.23). However, this trend does not appear to be worldwide, leading some experts to believe the frequency changes are simply part of natural oscillations within the climate system. Others believe that global warming may be affecting upper-level winds such that it inhibits the formation of these complex storms. While the question of increased frequency remains open, new studies have shown a measurable increase in maximum wind speed and duration of hurricanes on a worldwide basis. Researchers have also found a worldwide increase in the number of category 4 and 5 storms, as opposed to tropical storms of all sizes. Both of these types of studies therefore have shown an overall increase in the power of hurricanes, which certainly suggests a link to global warming.

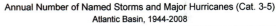

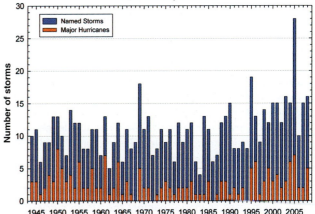

FIGURE 9.23   Histogram shows how the frequency of hurricanes (blue) and major hurricanes (red) in the Atlantic have increased in recent years. Hurricane power has increased measurably on a worldwide basis, which many researchers believe is linked to warmer ocean temperatures caused by global warming. Insurance companies and emergency managers are concerned that hurricane activity may be entering a more active and dangerous phase.

New Orleans and the Next Hurricane Katrina

When Hurricane Katrina approached the Louisiana coastline in 2005, it strengthened into a category 5 storm and was projected to make a direct strike on New Orleans (Figure B9.1). Katrina's 170-mile-per-hour (274 km/hr) winds were clearly capable of causing catastrophic damage and creating a storm surge that would inundate the entire city. This was the nightmare scenario that scientists and emergency managers had been warning of for years, yet it seemed to take government officials by surprise, both before and after the storm. In part because local officials delayed the implementation of a mandatory evacuation plan, tens of thousands of citizens without their own transportation had no means of escape. Fortunately, Katrina tracked slightly to the east and weakened into a category 3–4 hurricane just before making landfall, sparing the city the worst of the storm surge and high winds. Although Katrina was still a disaster for New Orleans, it was by no means the nightmare scenario many had predicted. Since the Federal Emergency Management Agency (FEMA) had long anticipated the worst case, many people found it difficult to understand why the agency's response in the aftermath of the storm was so inadequate.

The disaster in New Orleans that resulted from Hurricane Katrina leads us to two important questions. Why did Katrina have such a devastating impact on the city despite the fact it tracked to the east and weakened before making landfall? Also, is it possible or practical to protect the city from the worst-case scenario, namely a direct strike by a category 5 hurricane? The answer to the first question is that the Katrina disaster in New Orleans was basically a flood caused by an engineering failure of the U.S. Army Corps of Engineers levee system (Chapter 8). Here poorly installed concrete levee panels simply fell over as the levees filled with water from the storm surge. Moreover, the Mississippi River Gulf Outlet, built by the Corps in the 1950s and 1960s to provide ships a shortcut into New Orleans, acted as a conduit that brought the storm surge directly into the city (Figure B9.2). Experts estimated that this shipping canal raised the storm surge within the levee system an additional 3 feet (0.9 m). Despite the increase, the water level remained below what the levee system was designed to handle. This means that the flood protection system for the city had failed for a category 3–4 storm. Therefore, it is highly unlikely that the current system could protect New Orleans in the worst case scenario, namely the storm surge from a category 5 hurricane.

The magnitude of the disaster in New Orleans is directly related to the use of artificial levees to provide flood control and to allow for the expansion of developed areas (Chapter 8). Although the levees

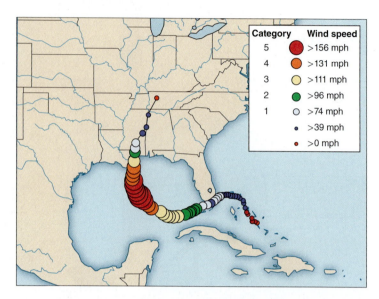

FIGURE B9.1 Map showing the intensity of 2005 Hurricane Katrina in terms of wind speed. Note how the storm developed into a category 5 hurricane, but then weakened into a category 3–4 just before making landfall.

A

constructed along the lower Mississippi River have been effective in controlling floods, they have also cut the delta off from its natural sediment supply. This in turn has caused much of the delta to experience land subsidence as sediment within the delta continues to compact. New Orleans, therefore, continues to sink farther and farther below sea level at the same time sea level is rising at an accelerated rate. To make matters even worse, the subsidence and decreased sediment supply is resulting in severe erosion of the barrier islands that ring the delta

(Figure B9.2). Historically these islands have helped shield both New Orleans and the surrounding wetlands from hurricanes and wave action.

This leads us to our last question of whether it is practical, or even possible, to protect New Orleans from a direct strike by a category 5 hurricane. Taking into account how the city continues to sink while sea-level rise accelerates and its protective marshes and barrier islands disappear, it is obvious that safeguarding the city from a major hurricane will become increasingly difficult. Although it may be possible to build a flood-control system capable of holding back a category 5 storm surge, the cost may be prohibitive, particularly in times of dwindling financial resources. Moreover, there is always the chance that a strengthened flood-control system will fail, leaving the entire city submerged, with the nearest dry ground located tens of miles away. Building higher levees will never erase the fact that most of the city lies below sea level and will always be at serious risk from a major hurricane. In the long run, the most prudent and cost-effective solution may be to relocate much of New Orleans farther inland, and choose to defend the original parts of the city located on high ground along the Mississippi's natural levees.

FIGURE B9.2 The Mississippi Delta (A) is subsiding because human levees have cut off its sediment supply, plus the Gulf Outlet shipping channel is bringing saltwater into its marshes. The lack of sediment and subsidence is causing the protective ring of barrier islands to undergo significant erosion as shown by the photos (B) taken before and after Hurricane Katrina.

Tsunamis

Earlier we defined *tsunamis* as unusually high-energy waves that form not from the wind, but rather by the transfer of energy from earthquakes, volcanic eruptions, landslides, or asteroids to a body of water (*tidal wave* is sometimes used to describe tsunamis, but it is a poor choice because tidal forces are not involved). Due to the large amount of energy contained in tsunamis, the circular motion of water molecules extends to a much deeper level, a level we previously defined as *wave base*. Also, recall that the depth of wave base is about half the distance between wave crests (i.e., wavelength). Because tsunamis traveling through deep ocean waters have exceptionally long wavelengths, typically from 6 to 300 miles (10–500 km), their wave base can be rather deep. Moreover, the height of these waves in deep water is quite small (less than a meter), but they travel at speeds of over 500 miles per hour (800 km/hr). Interestingly, despite the great speed of a tsunami, their small amplitude allows them to pass unnoticed by ships operating in deep waters.

Although harmless out in the deep ocean, tsunamis turn deadly as they approach shore and their wave base starts to encounter the seafloor. This interaction with the seafloor causes the fast-moving tsunamis to quickly decelerate, at which point their enormous energy is translated into progressively taller waves, a process scientists call *run-up*. Depending on the amount of wave energy and shape of the coastline, run-up can produce wave heights of 100 feet (30 m) or more. The towering waves eventually break and push water far beyond the normal surf zone, which, of course, is where people typically build permanent structures along a shoreline. As illustrated in Figure 9.24, buildings and other structures have little chance of withstanding the tremendous forces involved, particularly in the areas where the waves break into surf. To make matters worse, development is usually concentrated in protected bays, which are particularly dangerous locations since the shape of the shoreline acts to funnel the waves into a smaller area, thereby increasing run-up.

Tsunamis are capable of great death and destruction in part because these high-energy events occur relatively infrequently, causing humans to

FIGURE 9.24 Aerial view (A) of Indonesia's coastline where towns and villages once stood, but were obliterated by the 2004 tsunami. Development along this rugged coastline was concentrated on small strips of level ground adjacent to the sea. Notice in the photo how the shape of the shoreline would have helped funnel the waves, thereby increasing the wave height. Photo taken on the ground (B) illustrating how the powerful waves ripped buildings off their foundations, leaving only the foundations themselves and steel-reinforcing rods that were once embedded in concrete walls.

A

B

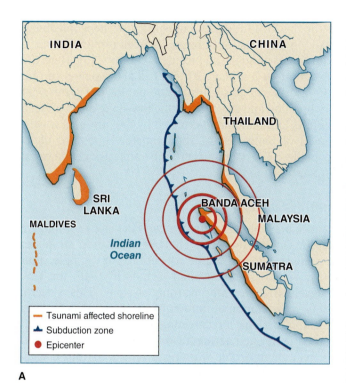

A

Before

After

B

FIGURE 9.25 In 2004 a magnitude 9.1 earthquake off the Indonesian coast (A) generated a tsunami that swept across the Indian Ocean, killing an estimated 225,000 people. Before and after photos (B) of the city of Banda Aceh, which was the closest to the epicenter, provide a dramatic testament to the devastating power of the waves.

become ignorant or complacent regarding the hazard. Because several generations or more may pass between major tsunamis, people tend to build towns and cities along highly vulnerable strips of low-lying terrain next to the sea (Figure 9.24A). For example, consider the tsunami associated with the 1883 eruption of the Indonesian volcano called Krakatau (Chapter 6). This eruption was so violent that the 2,600-foot (792 m) mountain was blown skyward, leaving in its place a hole in the seafloor over 1,000 feet (300 m) deep. Some of the energy from this colossal explosion was transferred to the surrounding water, generating a series of tsunami waves that swept throughout the region, killing over 36,000 people in low-lying coastal towns and villages. People had lived next to the sea for generations and likely had never experienced a tsunami of this magnitude.

Although the 1883 Krakatau tsunami was a catastrophe, an even greater tragedy took place 120 years later when another major tsunami formed, but this time the cause was an earthquake. On December 26, 2004, a magnitude 9.1 earthquake occurred in the subduction zone located off the Indonesian coast of Sumatra (Figure 9.25). As described in Chapter 5, this massive quake caused the seafloor to suddenly shift upward nearly 50 feet (15 meters) along 550 miles (950 km) of the subduction zone. This vertical shift of the seafloor displaced an immense volume of water, creating a series of waves that began to silently travel outward in all directions. The tsunami eventually swept over low-lying coastal communities around the Indian Ocean, killing an estimated 225,000 people and leaving another 2.2 million homeless. This disaster highlighted the grim fact that tsunamis can travel great distances and still pose a serious threat. Consider that of the 225,000 victims, approximately 60,000 lived in coastal communities so far from the epicenter that the people never even knew an earthquake had occurred. On the other hand, those living close enough to feel the ground shake had little time to escape, and they had to face much higher waves.

Those hit hardest by the 2004 Indian Ocean tsunami were people living in cities and villages along the coast of Indonesia's Aceh province, a mere 60 miles (97 km) from the earthquake's epicenter (Figure 9.25). Here the waves came crashing ashore in less than 20 minutes and reached run-up heights of 100 feet (30 m). As can be seen from the photos in Figures 9.24 and 9.25, nearly every building and structure in the low-lying areas of Aceh province was ripped off its foundation and crushed by the powerful waves. In the provincial capital of Banda Aceh, approximately one-third of its 320,000 residents were killed. This tsunami certainly ranks as one the worst natural disasters in human history.

Mitigating Tsunami Hazards

Although humans cannot prevent tsunamis, we can take steps to mitigate or minimize the hazard. One of the most effective mitigation strategies relies on an early warning system and public education. For example, after a tsunami killed 170 people on the Hawaiian Islands in 1946, the U.S. government began developing an early warning system. This system, which today includes the cooperation of numerous Pacific nations, utilizes a network of seismograph stations for detecting subduction zone earthquakes that have a potential for generating a tsunami (Chapter 5). Should such a quake be detected, an electronic warning is sent to various coastal centers, which then alerts residents via emergency sirens and a public address system. This cooperative program also includes a public education component that teaches citizens to immediately seek higher ground whenever a tsunami warning is issued. Over the years this system has been improved and updated, particularly after tsunamis struck Hawaii again in 1960, and the Oregon and northern California coasts in 1964. Most recently, deep-ocean buoys were added to the system in 1995 in which passing tsunami waves are detected based on their unusually long wave length and high velocity. When the buoys detect a tsunami, a message is transmitted via satellite to a ground station, which then sends an alert to the various coastal centers of the system.

After the 2004 Indian Ocean tsunami, it quickly became apparent that a key reason for the horrific death toll was that the region lacked an early warning system similar to the one developed for the Pacific Ocean. This was partly due to the fact that subduction zones are less common in the Indian Ocean compared to the Pacific, which is almost completely lined with subduction zones (Chapters 4 and 5). Since most large tsunamis are associated with subduction zone earthquakes, countries surrounding the Indian Ocean have experienced comparatively few tsunamis in modern times. Although geologists were aware of the potential for a major tsunami in the Indian Ocean, governments there did not develop an early warning system. This was partly due to their limited resources and the fact tsunamis are less frequent in the Indian Ocean. After the 2004 disaster, an international effort led by the United Nations began the process of developing a comprehensive tsunami-warning system for the Indian Ocean. This system, which uses technology similar to the Pacific system, first became operational in 2006. Since that time additional capabilities have been added, including warnings for tropical storms and cyclones.

Finally, it should be noted that Japan, which has a long history of being struck by tsunamis, has developed various engineering controls designed to mitigate the effects of both tsunamis and storm surge associated with tropical storms. These controls include large gates that can be closed to block the flow of water into rivers and harbors, and walls designed to keep large waves from breaking farther up onto the shore.

Rip Currents

A serious risk for people on beaches is getting caught in a strong current that flows away from shore called a **rip current,** sometimes inappropriately referred to as a *rip tide.* Recall from earlier how waves break onto shore, resulting in backwash that flows back down toward the sea. Once at the water's edge this backwash will generally flow parallel to shore until it can escape through a break in an underwater sand bar. As illustrated in Figure 9.26, the water can then funnel through this break and create a narrow, but powerful current that flows toward deeper waters where it eventually spreads out and dissipates. Rip currents are particularly strong, hence dangerous, when the surf becomes higher as this creates greater volumes of backwash that must exit the beach via rip currents. It is esti-mated that in the United States alone, an average of 100 people drown each year by getting caught in rip currents. In fact, officials believe the actual number is far higher, because many deaths are listed simply as drowning and are not reported as being caused by rip currents. Whatever the true number, the estimate of 100 deaths is still much greater than the average of six fatalities each year from shark attacks. Although the media focuses a great deal of attention on shark attacks, swimmers clearly face a much higher risk from rip currents.

Rip currents are dangerous because of the way they drag swimmers into deeper water. Even strong swimmers often drown after quickly becom-ing exhausted in a futile attempt to swim back to shore against these powerful currents. Rip currents sometimes also take the lives of nonswim-mers who are simply wading in shallow water, but get knocked down by a wave and are then carried into deep water by the current. Should you ever find yourself being swept out to sea in a rip current, the best approach is to stay calm and swim parallel to shore. Once you get beyond the narrow zone where the current is operating, it then becomes quite easy to swim back to shore. For people who cannot swim, it is best to stay out of the water during periods of heavy surf. As shown in the photo in Figure 9.26, the location of a rip current can often be rec-ognized by the way the surf is disrupted and the absence of foam floating on top of the water.

Shoreline Retreat

Earlier we defined *shoreline retreat* as the landward migra-tion of the shoreline. Although shoreline retreat occurs whenever sea level rises or the land becomes lower due to subsidence, our focus here will be on retreat that is caused primarily by erosion. Recall that when waves strike a shoreline, wave energy is transferred to the land where it drives both erosional and depositional processes along a coastline. In areas where there is a net loss of material, the shoreline will migrate landward. During storm events wave energy, of course, is at its highest, which means erosion and deposition rates are also at a maximum. Storms there-fore can produce dramatic changes along a shoreline, sometimes literally overnight. In some instances large vol-umes of sediment are deposited in places that society finds undesirable. For example, sediment deposition in shipping channels reduces the depth required (draft) for heavily

FIGURE 9.26 Rip currents (A) form when backwash from the surf zone funnels through a break in underwater sand bars. Photo (B) showing a rip current flowing back out to sea through the surf zone in the Monterey Bay area of California. Note that the rip current can be recognized by how it disrupts breaking waves within the surf zone.

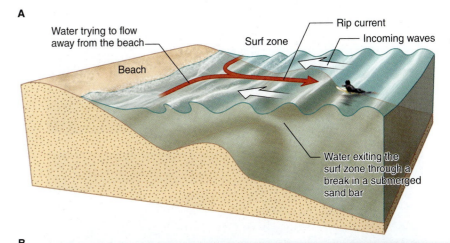

FIGURE 9.27   Photo sequence of Dauphin Island near Mobile Bay, Alabama, showing how shoreline retreat occurs in pulses during major storm events. As the island retreats, homes become closer to the surf zone. Note in the bottom photo the missing homes and the oil rig that came ashore during the storm.

loaded vessels. The sediment must ultimately be removed via expensive dredging operations. Sediment is also frequently deposited on land during major storms, covering roads and other valuable property that adds to the cleanup costs.

Although sediment deposition can interfere with human activity, shoreline erosion and retreat generally cause far more serious and costly problems. Perhaps the biggest problem occurs when people place buildings too close to the water's edge, putting their investments at risk of falling into the sea as the shoreline retreats. Clearly, the level of risk is directly related to the rate of shoreline retreat, which itself is ultimately controlled by the frequency and intensity of storms. Other important factors include the rate of sea-level rise and disruptions in the supply of sediment that moves with the longshore current. In the following sections we will take a brief look at the factors that control the rate of shoreline retreat as well as the engineering controls used to mitigate the problem. We will also examine other engineering controls and how our modifications to the coastal environment cause coastal systems to sometimes respond in ways society finds undesirable.

Increased Frequency of Storms

Because large ocean storms are relatively rare, coastal erosion rates are generally low on a day-to-day basis. This means that a given stretch of a shoreline may remain relatively stationary for several years, perhaps even decades, before retreating suddenly during the next major storm. It is during the periods of relative stability that humans tend to become complacent, leading them to build structures in places that may be located in the surf zone after the next storm. The sequence of photos in Figure 9.27 illustrates how shoreline retreat occurs in pulses, causing homes to move closer to the surf zone after each major storm.

Because shoreline retreat is directly related to storm activity, any increase in the frequency or intensity of storms will cause the overall rate of retreat to accelerate. The frequency and intensity of storms is naturally a function of ocean temperature and climatic patterns, both of which can change over time scales of decades or more in a complex and cyclic manner. There is growing concern among scientists and insurance companies that global warming (Chapter 16) is causing ocean storms to become more frequent and more intense. Some experts, however, believe the changes taking place are simply part of natural oscillations within the climate system. Regardless of the cause, if present trends continue, more frequent and intense storms will certainly cause shoreline retreat to accelerate, putting buildings at even higher risk of falling into the sea.

Effects of Sea-Level Rise

Sea level began to rise rapidly after the end of last glacial period about 18,000 years ago; then around 6,000 years ago the rate began to slow and has remained fairly gradual ever since. Today, there is growing evidence that sea-level rise is starting to accelerate due to global warming. Accelerated sea-level rise is a concern because it would exacerbate the problems associated with shoreline erosion and retreat, and also increase risk from large storms.

Consider that from 1900 to 2000 sea level has increased a total of about 0.6 feet (0.2 m) worldwide, but over the next 100 years climate experts project that sea level will rise an additional 0.6 to 1.9 feet (0.64 m). While this may not seem like much, even a foot of rise would cause inundation problems for low-lying areas that are already close to sea level. There is also the possibility of even higher rates of sea-level rise. Today many climatologists are concerned over growing evidence that large portions of the ice sheets covering Greenland and Antarctica are becoming unstable. This could lead to a collapse of the ice sheets, raising sea level an estimated 33 feet (10 m) over the course of a few centuries. Should this occur, many coastal cities would simply have to be abandoned—see Chapter 16 for more details on climate change and sea-level rise.

Although no one is certain as to the exact amount sea level will rise in the future, what is certain is that the problems associated with shoreline erosion and retreat will worsen. The basic problem is that as sea level continues to increase, it becomes easier for storm waves to reach the top of the active beach. Along more rugged shorelines this means that waves are able to pound away at the sea cliff on a more frequent basis. The slope is then undercut more rapidly such that mass wasting puts human structures at greater risk of falling into the sea. A similar process takes place along low-lying coastlines as higher sea level makes it easier for waves to reach the upper part of the active beach and nearby buildings.

In some areas around the world the problems associated with sea-level rise and shoreline retreat are compounded by the fact that the land is also sinking (Chapters 7 and 11). Perhaps nowhere in the United States is this more of an issue than in coastal Louisiana, where an average of 34 square miles (88 km²) of land has been lost to the sea each year for the past 50 years. The basic problem is that artificial levees lining the lower Mississippi River have cut the delta off from its natural supply of freshwater, nutrients, and sediment, plus a vast network of canals has allowed salt water into the delta (Chapter 8). Because sediment deposition is no longer able to keep pace with natural compaction within the delta, the land surface is now sinking below sea level. The subsidence problem is exacerbated as the freshwater marshes continue to die due to the increased presence of salt water. The combination of low-lying terrain, land subsidence, and accelerated sea-level rise gives coastal Louisiana the unfortunate distinction of having one of the fastest rates of shoreline retreat in world. Figure 9.28 illustrates the dramatic loss of land expected for this area with a relative sea-level change of only 3 feet (0.9 m) over the next century. Clearly, shoreline retreat in coastal Louisiana not only means that an entire culture and way of life will be lost, it also means that New Orleans will be at far greater risk from hurricanes and storm surge (Case Study 9.1).

Disruptions in Sediment Supply

Another key factor in the rate of shoreline retreat is the amount of sediment moving with the longshore current. Should something disrupt the supply of sediment, beaches will naturally become narrower. When this occurs, storm waves will reach the top of the active beach more frequently, accelerating shoreline retreat and putting human structures at greater risk of falling into

FIGURE 9.28 Map showing the amount of shoreline retreat in coastal Louisiana expected from a 3-foot (0.9 m) relative rise in sea level. This change is due to land subsidence within the delta and accelerated sea level rise, both of which are directly related to human activity.

FIGURE 9.29 Map (A) showing the shipping channel leading into the port of Savannah, Georgia. Dredging of the channel has prevented the river's sediment from entering the coastal environment, and has stopped the southward movement of sediment in the longshore system. The nearby barrier island, Tybee Island (B), is consequently starved of sand and experiencing serious erosion problems.

the sea. One of the ways beaches become starved of sand is by dams, as these prevent rivers from transporting sediment to the sea, which leaves less material to be distributed onto beaches by longshore currents. The degree to which dams choke off the sediment supply to a coastline depends upon the number and the location of dams within a drainage system. More dams clearly mean less sediment reaching a coastline, whereas a single dam located near the mouth of a river can prevent nearly the entire sediment load from reaching the sea. Because rivers in the more developed parts of the world are heavily dammed, coastlines there are generally starved of sand and are retreating at an accelerated rate.

In some cases the sediment that does make it to a river's mouth is prevented from entering the longshore current system. Consider how the levees along the lower Mississippi River (Chapter 8) have kept the river from migrating across the delta, causing much of its sediment to be deposited in deep water far beyond any longshore currents. This has resulted in a reduced supply of sediment along the Louisiana coast, and as noted in Case Study 9.1, is contributing to the severe erosion problem on barrier islands that form a protective ring around the delta.

Perhaps the most common way sediment longshore currents are cut off from their sediment supply is the process of *dredging,* where sediment is removed from the bottom of a river or harbor. Dredging is normally done to create a deep channel so that large ships can gain access to port facilities. Due largely to cost considerations, dredged material has historically been used to create human-made islands or to fill wetlands adjacent to the channel rather than moving it to the coastline where it can enter the longshore system. A good example is the deep shipping channel in Figure 9.29A that leads into the Savannah River and the port of Savannah, Georgia. This port is one of the oldest in the United States and its shipping channel has a long history of dredging, which has allowed progressively larger ships to enter the port. While the dredging has contributed positively to the expansion of both the port and Savannah's economy, it has unfortunately stopped sediment from entering the coastal environment. Moreover, because of the way the channel extends offshore, the dredging has also interrupted the southward movement of sediment in the longshore system. Not surprisingly, the barrier islands immediately south of the shipping channel (Figure 9.29B) are experiencing serious erosion problems because they are being starved of sand.

Mitigating the Effects of Shoreline Processes

When humans place buildings and other infrastructures directly on the edge of a coastline, erosion and shoreline retreat eventually force the decision to either let our investments fall into the sea or take steps to stop shoreline retreat. In other situations sediment may be accumulating in places we find disruptive, like shipping channels, which causes us to begin dredging. Other times we simply want a quiet area protected from waves in order to dock boats. Because society faces different types of issues in coastal environments, humans have developed various engineering solutions to help mitigate the problems. In this section we will explore some of the basic engineering techniques and the problems they are designed to address. We will also examine how the techniques themselves further disrupt natural processes operating along a shoreline. Here you will learn that in our attempt to solve one problem, we often create yet another set of problems, all of which involve significant sums of money.

Seawalls   When shoreline retreat threatens valuable real estate or buildings, one solution is to install a **seawall,** which is a physical barrier made of concrete, steel, or large rocks built against the shore (Figure 9.30)— seawalls are also called *bulkheads* and *revetments.* Because of the way seawalls physically prevent waves from directly impacting the shoreline, this technique is often referred to as *hard stabilization* or "armoring a shoreline." Although seawalls are effective in reducing shoreline retreat, there is a rather undesirable side effect in that the beach is eventually lost. What happens is that under natural conditions, storm waves are able to remove sand from the upper part of the beach, which then gets deposited just offshore. Later when wave energy is low, the sand is brought back and the beach is rebuilt. However once a seawall is installed, there is no landward supply of sand to draw upon. As a result, during high tide waves reflect off the seawall and end up transporting some of the sand back

FIGURE 9.30   Seawalls (A) are physical barriers that keep waves from impacting the shoreline. Over time, waves reflecting off the seawall will cause the beach to be redeposited offshore. As shown in the photos of Jekyll Island, Georgia (B), this movement of sediment ultimately results in a usable beach being present only during low tide.

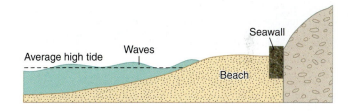
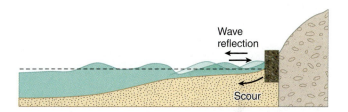
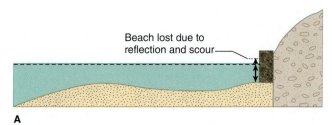

A

B

Lake Michigan, Lake Bluff, Illinois

FIGURE 9.31 Groins are built perpendicular to shore in order to trap sand moving with the longshore current. As the beach widens on the up-drift side of a groin, shoreline retreat is reduced. Note however that down-drift beaches become starved, causing shoreline retreat to accelerate.

Manasquan River, New Jersey

FIGURE 9.32 Jetties normally come in pairs and are placed at the mouth of inlets to help keep the longshore movement of sand from clogging navigational channels. This unfortunately also causes down-drift areas to experience rapid shoreline retreat as beaches there are starved of sand.

offshore. As shown in Figure 9.30B, the beach then gets smaller over time until it can only be found at low tide; at high tide the water is directly up against the sea wall. Once the beach is gone, wave action will begin to undermine the seawall, and ultimately cause the structure to collapse. The irony of building seawalls in order to save buildings is that we end up losing the recreational use of the beach, which is the very thing most people come to the coast to enjoy.

Groins   In areas where shoreline retreat is a problem because of beach starvation, an alternative to building seawalls is to widen the beach by trapping sand that is moving with the longshore current. This method involves installing a barrier called a **groin,** which is made of large rocks or steel sheets and is built seaward as shown in Figure 9.31—a group of groins is called a *groin field.* Because a groin interrupts the longshore current and corresponding movement of sand, the beach will become wider on the up-drift (up-current) side of the structure. This is desirable, because as the beach becomes wider the rate of shoreline retreat will decrease. Eventually the beach will grow to the point where sand is able to go around the groin and continue moving down-drift with the longshore current. This means that once a groin fills, the overall movement of sand is about what it was prior to the structure being installed. However, the problem is that while the groin is filling, the down-drift areas will experience beach starvation and increased erosion and retreat (Figure 9.31). Long-term erosion problems can occur if a groin is built too long, in which case the sand at the tip of the structure moves out into deeper water and is lost from the longshore system. While groins are effective in doing what they are designed for, namely reducing shoreline retreat, they have a rather undesirable side effect in that they generally cause greater retreat in down-drift areas.

Jetties   The sand that moves with longshore currents must eventually cross various inlets located along a coastline. The degree to which these inlets are kept open is due in part to the flushing action of tidal currents and rivers emptying into the sea. Throughout history humans have found inlets valuable in that they provide access to areas where ships can anchor and remain protected from the waves of the open ocean. As described earlier, engineers soon found it necessary to dredge sediment from inlets in order to create deeper shipping channels for bringing larger ships into port. Keeping navigation channels open requires periodic and costly dredging operations. Barriers made of large rocks called **jetties** are often installed at the mouth of an inlet (Figure 9.32) to keep sediment from clogging channels, and thereby reducing dredging costs. Note that jetties normally come in pairs because longshore currents will periodically reverse from their dominant direction.

Although jetties are used for navigation purposes rather than erosion control, they function in a similar manner as do groins. For example, because jetties interrupt the longshore current, beaches get wider on the up-drift side of the structures and are starved in the down-drift direction (Figure 9.32). In addition, because of the need to prevent sand from flowing around the structures and clogging the channel, jetties are generally long so that sand is forced out into deep water. This means that most of the sand is lost from the longshore system, resulting in beach starvation down-drift. Although jetties are effective in keeping shipping channels open, an undesirable side effect is that they create severe erosion and shoreline retreat in down-drift areas.

Breakwaters   Because some coastlines have relatively few natural harbors, large linear structures called **breakwaters** are placed offshore to keep waves from breaking onto land, thereby creating a protected area as shown in Figure 9.33. Breakwaters have traditionally been made of large rocks placed on the seafloor and for the purpose of creating a quiet area to moor boats. These structures have been employed in other situations to help reduce coastal erosion. Regardless of the application, the low-energy environment created by a breakwater also causes the longshore current to abruptly decrease. This results in sand accumulating behind the breakwater, which for a marina is undesirable because it eliminates the space needed for docking boats. Moreover, because breakwaters disrupt the longshore current, they also increase shoreline retreat in down-drift areas because the beaches there become starved of sand.

To help remedy the erosional and depositional problems associated with traditional rock breakwaters, floating systems are now being deployed in some areas. Because floating breakwaters can be set at different water depths, engineers can adjust the system and reduce the impact on the longshore currents, thereby minimizing the problems associated with erosion and unwanted deposition. Floating breakwaters have another advantage in that they give engineers greater flexibility in designing wave-protected areas.

Beach Nourishment   In many cases the only real solution to beach starvation is to manually add sand to the beach in a process called **beach nourishment.** The most cost-effective way of nourishing a beach is usually by pumping sand up onto the beach from offshore sand deposits (Figure 9.34); in some situations trucks are used to transport sand from land deposits. Widening the beach not only reduces erosion, but also enhances the recreational use of the beach, which typically means more tourists and tourist dollars. Another desirable attribute of beach nourishment is that it does not disrupt the longshore current, and therefore does not contribute to downdrift erosion as do groins. The problem, however, is that adding sand to the beach does not address the underlying cause of why the beach is being starved in the first place. This means that the new material being added will eventually be lost, making beach nourishment not just an expensive option, but one that must be repeated over an indefinite period of time. Another drawback is that the sand supply for nourishment projects typically comes from offshore sand deposits, whose texture is different from that of natural beaches. For example, offshore sand commonly contains abundant shell fragments that are relatively coarse, which makes walking barefoot or lying on the beach somewhat uncomfortable.

Because it is expensive, beach nourishment is not always an economically feasible option. The cost effectiveness depends on how frequently a beach needs to be renourished, the value of the property involved, and the amount of tourist revenue gained from the recreational use of the beach. In cases where the local economy is highly dependent on tourists, the cost-benefit analysis usually comes out in favor of beach nourishment. For example, in the 1950s and 1960s Miami Beach in Florida developed into a major tourist destination, but then declined in popularity due in part to the slow loss of its beach. To help reverse the decline, $64 million was spent from 1976 to 1981 on a major renourishment project that is largely credited with bringing tourists back and revitalizing the area. Beach attendance reportedly increased from 8 million in 1978 to 21 million in 1983. In 2001 it was estimated that on an annual basis visitors were spending $4.4 billion, making the original $64 million investment well worth the cost.

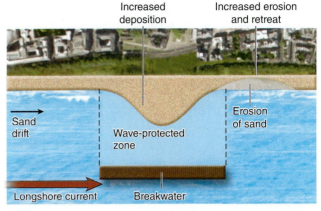

FIGURE 9.33   A breakwater is a barrier placed just offshore and is used to keep waves from directly impacting the shoreline. Although effective in reducing erosion and providing a quiet area for mooring boats, breakwaters also disrupt the longshore current and create unwanted deposition behind the structure and increased erosion in down-drift areas.

FIGURE 9.34   Beach nourishment involves moving sand from offshore deposits and spreading it on an eroding beach. Although expensive, this is often the only solution for bringing back a recreational beach in areas of chronic erosion.

South Amelia Island, Florida

Natural Retreat   Because erosion controls and beach nourishment projects are expensive, they are typically not cost effective in areas with a small economic base and high erosion rates. This leaves one last option, and that is to let the shoreline retreat naturally. Here existing structures would either be allowed to fall into the sea or be relocated farther inland, and new buildings would have to be set back a certain distance from the shoreline. This type of management strategy has been used sparingly in the United States, and is found mainly on barrier islands designated as state or federal parks. Consider for example the Cape Hatteras Lighthouse on the Outer Banks of North Carolina. The present lighthouse there was built in 1870 and located 1,500 feet (460 m) from the shoreline. But by the 1930s, shoreline retreat had placed the structure at the edge of the active beach. A series of erosion control efforts held the retreat in check, but then in the 1990s the decision was made to move the lighthouse to safer ground. In 1999 this historic landmark was moved 2,870 feet (875 m) to its present position. Although this move was successful, it is not feasible for most buildings to be moved along heavily developed coastlines simply because of the lack of vacant land.

For undeveloped coastlines, particularly those on barrier islands, some people propose restricting development of overnight accommodations and other facilities, forcing most visitors to return to the mainland in the evening. In addition to saving money on expensive erosion control and beach nourishment projects, the combination of natural retreat and limited development allows people to enjoy the beach and adjacent wooded areas in a more natural setting. Clearly this is not a desirable option for those who like the convenience of beachfront hotels or condominiums, nor would it be popular among investors who wish to continue developing our coastlines. However, as government budgets become tighter, property owners can no longer rely on receiving federal tax dollars for erosion control and beach nourishment projects. In many areas such projects are now being funded by local property and user taxes. As sea level continues to rise, society will ultimately have to decide where to spend its limited financial resources on trying to stop shoreline retreat, leaving the rest to retreat naturally.

SUMMARY POINTS

1. Shorelines are unique in that they are where Earth's two most fundamental environments meet: the terrestrial (land) and marine (ocean), forming a desirable habitat for humans. Because population growth is much higher in coastal zones compared to inland areas, more people are exposed to coastal erosion and hazards such as hurricanes and tsunamis.

2. Ocean tides are caused by the gravitational pull of the Moon and Sun on Earth's oceans, whereas ocean currents move in response to winds, differences in water density, and wave action along coastlines.

3. Wave energy travels horizontally and causes water molecules to move vertically in circular paths. As a wave enters shallow water, the moving water molecules begin to drag on the bottom, causing the wave to decelerate. This causes the wave to increase in height and to become more asymmetric, eventually curling to form a breaking wave.

4. When waves crash onto shore at an angle, water is pushed parallel to shore in what is called a *longshore current*. Sediment moves with the current and is referred to as *longshore drift*.

5. Irregular shorelines with more isolated beaches are commonly found in tectonically active areas and in places where sea level is rising relative to the land surface. Erosion and deposition from wave action slowly causes shorelines to evolve into ones with longer and wider beaches. At the same time, weathering and erosion of the landscape tend to produce more low-lying terrain.

6. Hurricanes are a serious coastal hazard as they generate powerful winds, storm surge, and heavy rains. Satellite early warning systems have greatly reduced the number of fatalities, but increased coastal development has caused property losses to escalate.

7. Tsunamis most commonly form during subduction zone earthquakes as water is displaced by movement of the seafloor. When reaching shallow water, the tremendous wave energy translates into tall waves that break far beyond the normal surf zone, causing death and destruction along developed coastlines.

8. Rip currents pose a serious risk to swimmers as water from the surf funnels back out to sea through breaks in shallow sand bars. People are unable to swim back to shore against the strong currents, but can get out of the current by swimming parallel to shore.

9. The interaction between waves and a landmass can cause a shoreline to naturally retreat landward. The slow migration of a shoreline can also occur when there is a rise in sea level, or when the land itself becomes lower due to subsidence. Accelerated retreat is occurring in many areas due to human activity, increasing the hazards associated with ocean storms.

10. Humans attempt to protect their property and reduce shoreline retreat through engineering techniques such as seawalls, groins, and beach nourishment. Jetties are used to keep navigational channels free of sediment, and breakwaters provide quiet areas by keeping waves from impacting on the shoreline. Some techniques result in beach starvation and accelerated retreat in down-drift areas.

KEY WORDS

barrier islands   265
beach nourishment   287
breakwaters   287
groin   286
hurricane   267
jetties   286

longshore current   263
ocean currents   261
ocean tides   261
rip current   281
seawall   285
shoreline retreat   264

storm surge   271
water waves   262
wave base   262
wave refraction   263

APPLICATIONS

Student Activity

How many sides of the Moon do we see? Do we see the same side of the Moon? How can this happen? To show how, you will need two coins. Place one coin in the center, and place another face up, with the face pointing at the coin in the center. If you do not rotate the face-up coin as you move it in a circle around the center coin, you will see all sides of the face-up coin. If, however, you rotate the face-up coin one quarter of a turn for each quarter of the circle around the center coin, you will only see the face of the rotating coin. This is what happens to the Moon; it rotates once for every time it goes around the Earth, so we see only one side.

Critical Thinking Questions

1. How are tides created?
2. What causes waves?
3. How can you escape a rip tide or rip current?
4. Why has property damage from hurricanes increased but deaths have decreased? (only in the United States)

Your Environment: YOU Decide

You live on the coast and have a nice waterfront property. Your neighbor (up current) has just installed a groin to prevent his beach from eroding. This will cause your beach to erode. What will you do?

Chapter 10

Soil Resources

LEARNING OUTCOMES

After reading this chapter, you should be able to:

▶ Describe how soils are derived from rocks and why soils are composed primarily of quartz and clay mineral particles.

▶ Explain the process by which soil horizons develop and why the number of horizons typically increases as the soil evolves.

▶ Understand how soil color can be used to indicate the presence of organics and the drainage characteristics of a soil.

▶ Describe the five soil-forming factors and how they control the type of soils that develop.

▶ Describe how quartz and clay mineral particles are different and how this affects soil properties.

▶ Understand how the weathering process of silicate rocks leads to aluminum-rich minerals.

▶ Explain the relationship between soil erosion, soil loss, and sediment pollution.

▶ Describe why soil loss is a problem for humans and list some of the ways it can be reduced.

▶ Understand why salinization and hardpans are problems for agriculture.

The soils that cover the landscape are a critical component of the Earth system and have made it possible for the land plants and animals that we see. In terms of society, soils are most critical as they provide the basis for most of the world's food supply. However, human use of the landscape has upset the natural balance between soil formation and erosion, leading to the steady loss of soils worldwide. The loss of such a critical resource will present challenges as humans attempt to increase food production to meet the growing population.

Introduction

FIGURE 10.1   Photo showing water carrying valuable topsoil off a farm field in Tennessee after a heavy rain. Agricultural activity commonly leads to increased erosion and a net loss of soil because row crops offer far less protection against falling raindrops and flowing water compared to natural vegetation. If left unchecked, soil loss will ultimately lead to a reduction in worldwide food production.

Soils are a unique part of the Earth system in that they are where the atmosphere, hydrosphere, terrestrial biosphere, and solid earth all interact with one another. For example, consider how atmospheric gases and precipitation cause weathering of rocks and minerals, generating loose sediment that blankets the landscape (Chapter 3). Plants and a whole host of organisms then thrive on the water and nutrients that exist within the sediment, ultimately producing a life-sustaining body of natural material called *soil*. Moreover, some of the water that reaches the land surface and infiltrates through the soil zone eventually replenishes the groundwater system, whereas some of the remaining water flows over the soil as overland flow (Chapter 8). The types of soil and plants covering the landscape, therefore, play an important role in determining the portion of water within the hydrologic cycle that infiltrates versus that which flows over the landscape. Because of their influence on the biosphere and hydrosphere, soils are fundamental to the Earth system and life as we know it.

Although soils may not capture people's attention the way other environmental issues do (e.g., earthquake hazards and energy resources), soils are actually far more important because they are critical to our human existence. Were it not for soils there would be no land plants, and, in turn, no food to eat except for what comes from the sea. Soils therefore form the basis for nearly our entire food supply. Because of this fundamental connection, people throughout history have been keenly aware of the difference between fertile and poor soils since most people had to grow their own food in order to survive. In fact, after each fall harvest it was common for people to worry whether they had stored enough food to make it through the winter. Much has changed in today's modern societies where food is grown on a scale that would have been unimaginable just a hundred years ago. This has also resulted in a dramatic decrease in the number of people living on farms in developed nations, causing many of us to lose sight of the connection between soils and our food supply.

Soil, like water, is a natural resource that is absolutely essential for human survival. However, unlike most of our water supply which is replenished fairly quickly due to the relatively rapid rate of movement within the hydrologic cycle, soils form much more slowly over hundreds to tens of thousands of years. This presents a problem in that human activity commonly disturbs the natural vegetation covering the landscape in such a way that soil erosion increases. Humans therefore tend to upset the natural balance between soil formation and erosion, which, in turn, leads to a net loss of soil (Figure 10.1). This means that soils can be considered as a *nonrenewable resource* since they can be lost at a much greater rate than which they form. When Earth's human population was small there was little concern when rich topsoils were lost by erosion or depleted of nutrients by repeatedly growing the same crop. When soils became lost or depleted people simply moved on and cleared new land. As population grew, it became increasingly more difficult to find new land, forcing people to learn how to grow crops on the same soil year after year. This continuous use of the land combined with the practice of exposing (i.e., baring) soils so crops can be planted in rows has led to severe soil loss in many areas. Although farmers are now adopting practices designed to reduce soil erosion, it remains a serious problem in agricultural regions around the world. For society, the continued loss of valuable topsoil means that food production cannot keep increasing forever. Ultimately then, soil loss and food production are on a collision course with our exponentially expanding population (Chapter 1).

In addition to serving as the basis for our food supply, people have found important uses for soils based on the physical and chemical properties of the minerals they contain. For example, some soils are composed of minerals that are ideal for making bricks to construct homes. Others contain minerals that serve as important raw materials, such as those used for making aluminum, a metal which has numerous applications in modern society. Some soils contain so much organic matter that they have historically been used as an energy source for heating and cooking. Soils are also important to people because they help determine the rate at which fluids are allowed to infiltrate into the subsurface. This ability to transmit fluids affects flooding (Chapter 8) and has implications in such things as civil engineering projects and the disposal of human and animal wastes. Soils are even used by geologists to determine the frequency of earthquakes, volcanic eruptions, floods, and other sudden events. For this, geologists use radiometric-carbon dating techniques (Chapter 1) to determine the age of organic matter within buried soils.

We will begin Chapter 10 by examining how soils form, and then take a brief look at some of their properties that are important to society. Because soils are intimately tied to food production, we will pay particular attention to the issue of soil erosion and the ways in which the problem can be minimized.

Formation of Soils

The land surface in most places is covered with broken-down rock fragments, some of which may be quite coarse and others so fine that the individual grains can only be seen with a powerful microscope. In areas where solid rock or *bedrock* is exposed (Figure 10.2), one can often see how this fragmental debris covers the Earth's surface, similar to how snow blankets the landscape. Although the everyday term for this loose material is simply *dirt*, scientists and engineers use different terms depending on the field of study. For example, geologists refer to fragmented material as

FIGURE 10.2   Photo showing a soil that developed from the breakdown of the underlying rock into individual particle grains. Notice how the soil covers the landscape as a thin blanket of loose material.

regolith, but call it *sediment* if it has been transported by wind, water, or ice. Engineers use the term *soil* to describe any type of broken-up material that lies above bedrock. On the other hand, soil scientists define **soil** as a natural mixture of mineral and organic material that is capable of supporting plant life. Because our focus is on soils as they relate to plant growth and food production, we will use the soil science definition throughout this chapter. We will begin by examining how weathering processes break rocks down into smaller particles. For a thorough discussion on weathering, refer to Chapter 3.

Weathering

The origin of soils ultimately begins when rocks physically disintegrate and chemically decompose in a process known as *weathering.* Recall from Chapter 3 that the rate of weathering is generally the greatest at Earth's surface because this environment is relatively hostile compared to the environment in which many types of rocks form. Rocks therefore tend to break down at the surface because they are now exposed to liquid water, atmospheric gases, biologic agents, and relatively large fluctuations in temperature. Geologists use the term *physical weathering* to refer to those processes that cause rocks to disintegrate into smaller particles by some mechanical means. For example, solid rock is commonly broken down mechanically due to the force water exerts as it freezes and expands within pore spaces and in fractures. Plants are also effective at breaking rocks into smaller fragments as their roots force their way into existing fractures. In addition to tectonic activity, fractures can form near the surface as erosion reduces the amount of confining pressure on underlying rocks (Chapter 4). Even wildfires can produce fractures when extreme temperature changes cause the rock to expand. Finally, note that climate is an important factor in physical weathering because of the way many mechanical processes are affected by the presence of water.

An important consequence of the physical weathering of rock into smaller particles is that it causes a significant increase in surface area, exposing more of the rock to chemical reactions—similar to how grinding salt into finer grains makes it dissolve more rapidly. Geologists use the term *chemical weathering* to refer to the process where individual mineral grains within a rock decompose due to chemical reactions (Chapter 3). Climate also plays an important role in chemical weathering because many of the chemical reactions involving rock-forming minerals are affected by temperature and the presence of water. Another important factor determining weathering rates is the types of minerals a particular rock contains. For example, the mineral calcite ($CaCO_3$), which makes up limestone and marble, is highly susceptible to chemical weathering as it will slowly dissolve when exposed to naturally acidic water near Earth's surface. As calcite-rich rocks undergo dissolution, calcium (Ca^{2+}) and carbonate (CO_3^{2-}) ions get carried away with the water, but the insoluble impurities within the rock are left behind to form a layer of soil.

Because of their sheer abundance, the most important rock-forming minerals are those rich in aluminum (Al) and silicon (Si), commonly called *aluminosilicates* or simply *silicates* (Chapter 3). Recall how the silicate mineral named *quartz* (SiO_2) is a major constituent in granitic rocks. Moreover, because quartz is highly resistant to chemical weathering, it is a major constituent in many types of sediment and sedimen-

A Stone Mountain, Georgia

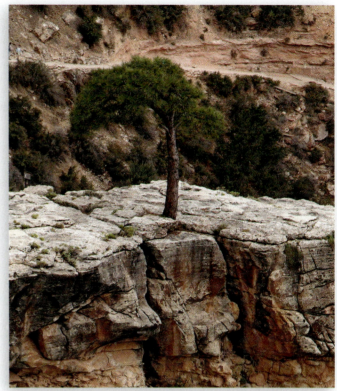

B Grand Canyon

FIGURE 10.3   Photo (A) showing a bowl-shaped depression in solid granite that has been filled with soil formed from the weathering of the rock itself. The roots of the tree in (B) have grown into fractures within the rock, extracting moisture and nutrients from soil within the cracks.

tary rocks. Other common silicate minerals, however, are rather susceptible to chemical weathering. Of particular interest are feldspar and ferromagnesian silicate minerals that make up the bulk of the rocks in Earth's crust. When acidic water near the surface comes into contact with these minerals, they react chemically and decompose into a variety of extremely fine-grained silicate minerals collectively known as *clay minerals.* As feldspar and ferromagnesian minerals are transformed into clay minerals, ions such as sodium (Na^+), potassium (K^+), magnesium (Mg^{2+}), and calcium (Ca^{2+}) are released and get carried away with the water. Because these ions are slowly lost, the clay minerals that are left behind become enriched in aluminum and silicon. Note that the chemical weathering of both ferromagnesian and iron pyrite (FeS_2) minerals release iron ions (Fe^{3+}) into the water. However, these ions will quickly combine with free oxygen (O_2) to form iron-oxide minerals such as hematite (Fe_2O_3). Soils therefore not only contain residual deposits of clay minerals, but if free oxygen is available, they also typically appear bright red or yellow due to the presence of iron-oxide minerals.

Development of Soil Horizons

As rocks undergo physical and chemical weathering and generate soil particles, there are other processes taking place within the soil which result in the formation of horizontal layers called **soil horizons.** For example, from the photos in Figure 10.3 one can see how vegetation will begin growing on small patches of soil that recently forms from the weathering of solid rock. Once vegetation becomes established, organic matter is naturally incorporated into the uppermost portion of the soil, creating a layer that is compositionally different from the soil below. This uppermost soil horizon that is enriched in organic matter is generally known as *topsoil.* In this section we will examine how topsoil and other soil horizons develop over time.

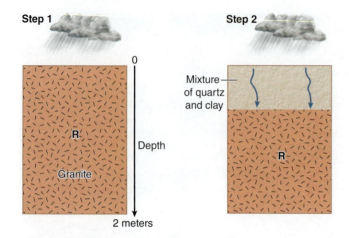

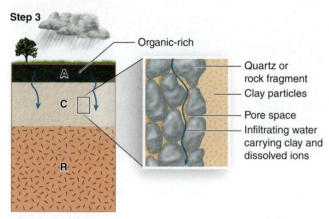

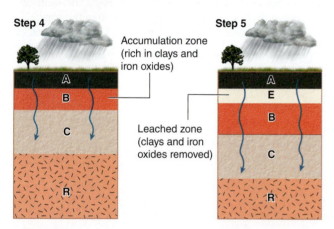

FIGURE 10.4   A time sequence illustrating the order in which soil horizons will develop when granite bedrock becomes exposed to weathering processes on Earth's surface. Note how clay minerals, dissolved iron, and other elements are carried downward with infiltrating water and then accumulate in the B horizon.

We can use a sectional or *profile* view of a block of granite, shown in Figure 10.4, to examine how soil horizons can develop over time from the weathering of bedrock exposed at the surface. Soil scientists refer to unweathered rock that is within a few meters from the surface as the *R horizon,* which, in our example, is granite composed mainly of quartz and feldspar minerals. Notice that once the granite is exposed to the atmosphere, physical and chemical weathering processes transform feldspar minerals into clay minerals near the surface of the rock. Because clay minerals are rather soft and weak, the uppermost part of the granite will literally crumble, resulting in a layer of quartz and clay mineral grains and broken rock fragments (Step 2 in Figure 10.4). At the same time this rudimentary soil starts to form, plants will establish themselves and begin extracting water and nutrients. As time progresses, organic matter is incorporated into the uppermost portion of the soil zone through the decay of leaves, stems, and roots from the plants, resulting in the two distinct soil horizons (step 3). Soil scientists call the uppermost organic-rich zone the *A horizon* (i.e., topsoil), and remaining mixture of weathered material the *C horizon.* In very young soils such as this, the primary difference between the A and C horizons is that the A horizon contains significant amounts of organic matter.

In the fourth step of our time sequence (Figure 10.4), notice how continued weathering has lowered the original bedrock surface, producing an older and thicker soil zone. During the time it took for this to occur, a new layer forms called the *B horizon,* which is enriched in minerals such as different types of clay and iron and aluminum oxide minerals. Because clay particles are extremely small, infiltrating water is able to carry the particles downward through the pore spaces that exist between the much larger grains of quartz and rock fragments. Eventually a B horizon forms as the clay particles accumulate between the A and C horizons. While the clay minerals are accumulating, minerals near the surface will undergo chemical reactions that release iron and other ions (electrically charged atoms) into the infiltrating water. As the dissolved ions move down through the soil, iron quickly oxidizes and forms iron-oxide minerals. These iron minerals, as well as those containing aluminum, tend to accumulate in the B horizon along with the clay particles. Because the B horizon is enriched in clay and other minerals, it is often referred to as the *zone of accumulation.*

The final sequence in our series (step 5) shows the development of a new layer called the *E horizon,* also known as the *zone of leaching,* where clay and other minerals have been flushed from the upper soil zone by the infiltrating water. Finally, note that in low-lying and poorly drained areas with lush vegetation, the uppermost soil layer is referred to as the *O horizon* (not shown) as it is exceptionally rich in organic matter. Because the ground in these conditions is typically saturated with water, oxygen levels within the soil zone are at a minimum. This lack of oxygen slows the decay of organic matter down to the point where it can accumulate faster than it decays, allowing for the development of an O horizon.

Soil Color, Texture, and Structure

In addition to horizons, other important soil characteristics include color, texture, and structure. These characteristics can best be seen by digging a trench to obtain a vertical view called a *soil profile.* For example, from the two soil profiles in Figure 10.5, you can see that the horizons can be distinguished in part by their color. Soil colors are the result of different types of pigments or coloring substances. For example, organic matter gives soil a blackish to brownish appearance and iron oxide minerals generate yellowish to reddish colors. Note that it takes a relatively small amount of pigment to give soil a color—similar to how very little pigment is needed to turn white paint into colored paint.

Because A and O horizons form at the surface, they contain decaying plant matter which gives these horizons a brownish to blackish color. O horizons are much deeper in color because they are chiefly composed of organic matter, whereas A horizons are lighter because they are dominated by mineral and rock fragments with relatively small amounts of organic material mixed in. In contrast, E horizons appear whitish or blonde because they lack pigmenting materials. This lack of color in E horizons results from organic matter and various oxide minerals being leached and transported downward into the soil profile by infiltrating water. B horizons, however, exhibit a range of colors because they represent the zone where various oxide minerals accumulate within the soil profile. The specific color of B horizons varies depending on the presence of free oxygen (O_2) within the pore space of soils. For example, in well-drained soils where oxygen is readily available, the dissolved iron in infiltrating water will combine with the oxygen to form iron oxide minerals, giving the B horizon a yellowish or reddish color. In areas where oxygen cannot enter the pore spaces because water saturates the soil, iron oxides generally do not form, producing a grayish-colored B horizon. Note that because of the relationship between iron oxide minerals and availability of free oxygen, the color of B horizons is a useful indicator as to the level of drainage and aeration within a soil.

Soil scientists also classify soils based on *texture*, which refers to the amount of sand, silt, and clay-sized material within a particular horizon (note that geologists define these sizes somewhat differently). As Illustrated in Figure 10.6, there are 12 textural classes based on the percentage of each grain size within a particular soil. For example, a soil that is composed of 40% sand grains, 40% silt, and 20% clay would be classed as a *loam soil*. A more sand-rich soil (60% sand, 30% silt, and 10% clay) would be called a *sandy loam*. Also notice in the figure how there is a vast difference in particle size, with clay being exceedingly small

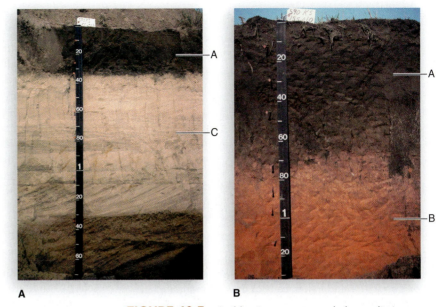

A   **B**

FIGURE 10.5   Soil horizons commonly have distinct colors due to the presence or absence of pigmenting materials. The organic content of the A horizon in (A) gives it a black color, which is in marked contrast to the C horizon that is light colored because of its lack of pigments. The older, more developed profile in (B) shows a much thicker A horizon that overlays a B horizon that is reddish in color due to the presence of iron-oxide minerals.

FIGURE 10.6   Scientists break soils down into 12 textural classes based on the percentage of sand, silt, and clay-sized particles. Texture is important because it helps determine the drainage and fertility characteristics of a soil. Note that soil scientists define the size range for sand, silt, and clay differently than do geologists.

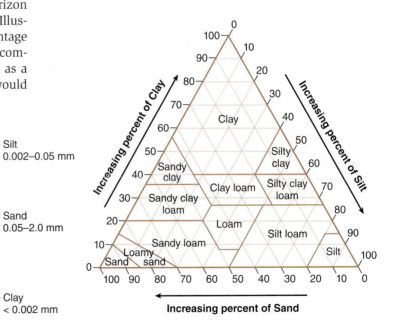

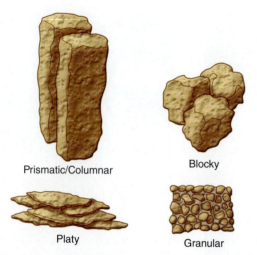

Prismatic/Columnar

Blocky

Platy

Granular

FIGURE 10.7   Illustration showing various shapes of soil peds (aggregates). The size and shape of peds determine a soil's structure and influence root development and infiltration of water.

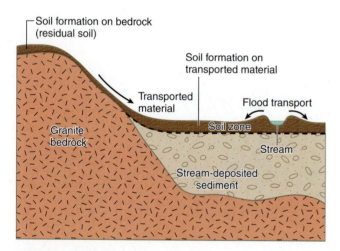

Soil formation on bedrock (residual soil)

Soil formation on transported material

Transported material

Flood transport

Granite bedrock

Soil zone

Stream

Stream-deposited sediment

FIGURE 10.8   Illustration showing how some soils form on parent material that is derived from weathering of the underlying bedrock, whereas other soils form on transported sediment that bears no relationship to the bedrock. Soils that form on river-transported material are commonly quite fertile due to the abundance of organic matter that is deposited during periodic floods and that which grows under the moist conditions.

compared to sand grains. Although sand and silt-sized grains commonly consist of the mineral quartz, and clay is typically composed of various types of clay minerals, note that this classification system is based strictly on size, not composition. The distribution of grain sizes within a soil is important since it plays a key role in determining a soil's permeability, ease of tillage, drought resistance, and fertility. These properties, in turn, help determine the type of crops that a particular soil can support, thus are of great importance in agriculture. For example, sandy soils may have good drainage, but are generally not very fertile and do poorly in droughts because they do not retain water very well. Clay-rich soils on the other hand may be more fertile and drought resistant, but are difficult to work due to the plastic and cohesive nature of clay particles.

Finally, soils can also be characterized based on *soil structure,* which refers to the way in which soil particles are arranged. When previously undisturbed soil is dug up, it will naturally break up into separate clumps, something soil scientists refer to as *peds* or *aggregates.* As illustrated in Figure 10.7, individual peds of soil can take on different shapes, including clumps that are granular, flat and platy, blocky, and more elongated, similar to columns and prisms. Soil structure is important because the size and shape of the peds greatly influences the ability of water to infiltrate and the ease at which roots can penetrate into a soil. Thus, soil structure is important to farmers as it helps determine the productivity of their soil. Farmers therefore should try and avoid running heavy equipment in their fields when they are wet so as to avoid compacting the soil, altering its structure such that infiltration and root development is inhibited.

Soil-Forming Factors

Earth's various landscapes naturally contain many different types of soils. For example, in some areas the soil will have only A and C horizons, whereas there may be well-developed A-E-B-C horizons in other regions. Soils may also be thick in one area and thin in another. Clearly, there must be controlling factors that determine why soils have different characteristics. There are actually five soil-forming factors recognized by soil scientists: parent material, organisms, climate, topography, and time. Moreover, these factors are not independent of one another, but rather commonly work in conjunction with one another. For example, climate strongly influences the weathering rates of minerals and the types of plants and animals found in a given area. Soils therefore are commonly thought of as a system comprised of various components all working together, somewhat analogous to the Earth system described in Chapter 1. In this section we will briefly explore the five soil-forming factors and how they affect soil development.

Parent Material

Of all the different types of soils, approximately 99% are derived from the by-products of weathered rock; the remaining 1% develops from thick accumulations of organic material. Soil scientists define **parent material** as the C horizon, which consists of the original weathering product or organic material from which soil horizons develop. Although bedrock is ultimately the source of weathered material from which soils form, the underlying bedrock in a given area is not necessarily related to the parent material of the soil. As illustrated in Figure 10.8, soils also develop on sediment that has been transported and deposited by streams. Clearly, the parent material in this case bears no relationship to the underlying bedrock. Note that soils which develop on river sediment, called *alluvium,* are commonly quite fertile due in part to the abundance of organic matter that is deposited along with the sediment during periodic floods. Fertile soils also commonly develop on sediment that has been deposited by wind, called *loess,* and by glaciers.

In contrast to parent material that has been transported, so-called *residual soils* develop from parent material (C horizon) that forms from the weathering of the underlying bedrock. Here the type of soil that develops is strongly influenced by the mineral composition of the bedrock. For example, consider the situation in Figure 10.9, where the land surface is underlain by three different rock types: limestone, sandstone, and granite. As discussed earlier, the granite would weather and form a blanket of loose rock, quartz, and clay-mineral fragments. However, limestone is composed primarily of calcite and varying amounts of clay and other impurities. During weathering, the calcite simply dissolves away, leaving a relatively thin layer of insoluble material rich in clay minerals. In the case of the sandstone composed mostly of quartz, weathering produces a parent material that is dominated by quartz sand grains (sand refers to particle size).

Recall that it is only after the parent material or C horizon becomes available that plants will establish themselves. Once plants start to grow, an A horizon will form as organic matter is incorporated into the soil. Here the type of parent material can strongly influence how the soil develops. For example, let us consider what would happen when rain falls on the three different parent materials shown in Figure 10.9. Clearly the highest infiltration rate would be in the sandy material that developed over the sandstone, whereas the lowest rate would be over the clay-rich material that formed from the limestone. Pine trees and other plants whose roots require well-drained soils would preferentially grow on the sandy material. Likewise, oak and other hardwood trees and plants that that prefer poorly drained conditions would tend to grow on the clay-rich material. An even different assemblage of trees and plants would develop on the mixture of quartz and clay minerals blanketing the granite terrain. The important point to note is that the differences in vegetation would, in turn, affect the amount of organic matter that is incorporated into the A horizons. This means that the A horizons in our example would all be of different thicknesses. Also, the high infiltration rate of the sandy material means it would experience the greatest flushing action of water moving downward through the soil zone. This then would increase the rate at which any fine particles and dissolved ions are able to move down into the B horizon.

Organisms

It should be clear that parent material helps determine the plant communities that develop on a given soil. However, soils also contain other living organisms, such as burrowing animals, insects, and microbes that can impact soil development. Soil then not only supports life on the surface, but within the soil itself. This is why some scientists consider soil to be a living system (Figure 10.10). Similar to plants, other organisms contribute organic matter to soil, help break down minerals, and create passageways which allow oxygen and water to circulate more freely within the soil. Consider, for example, how the familiar earthworm creates burrows and processes organic matter within a garden. Likewise, certain types of insects and animals bring material up to the surface and create mounds, a process which overturns soils and aids in the development of the A horizon.

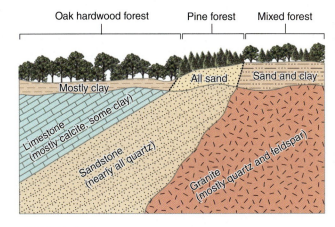

FIGURE 10.9 Because rocks contain assemblages of minerals, the weathering of different rock types can produce parent material with varying proportions of quartz and clay minerals. This, in turn, causes residual soils to vary in their drainage and water storage properties, ultimately leading to the preferential growth of different plant communities.

FIGURE 10.10 Soil can be thought of as a living system, supporting life on both on the surface and in the subsurface. Organisms aid in soil development by adding organic matter, overturning the soil, and providing passageways for air and water.

Climate

Another important factor in soil development is climate, because rainfall and temperature help determine the abundance and diversity of organisms, and also the weathering rates of rocks and minerals. Rich, productive topsoils generally contain abundant organic matter, which, in turn, requires moderate temperatures and adequate amounts of water. Therefore, in areas of extreme temperatures and or limited rainfall, rich soils are uncommon due to a decrease in the number and types of organisms. Also important are how physical and chemical weathering rates increase with temperature and rainfall. This results in faster rates of soil formation in warm and humid climates since mineral and organic matter tend to break down more quickly. However, note that exceptionally warm and humid climates generally have rather poor soils due to more extreme rates of leaching and chemical decomposition. In areas where the parent material remains permanently frozen or dry, soil development simply stops until there is a change in climate to more favorable conditions.

Climatic zones also vary in terms of rainfall patterns. Recall from Chapter 8 that flooding in some regions is related to weather patterns which tend to produce sporadic but intense rainfall events. Because the rainfall is infrequent, vegetation in these regions is more sparse, which, in turn, leads to higher erosion rates during the occasional, but intense rains. The higher erosion rates naturally make it more difficult for thick A horizons to develop because exposed topsoil is easily washed away.

Topography

Topography refers to the configuration of the land surface, including the amount of slope and vertical relief (elevation difference between high and low points). Earth's landscape, of course, is highly variable, ranging from flat plains to rugged mountains. In terms of soil formation, topography is important because it helps control infiltration, erosion, and chemical decomposition rates as well as the types of organisms that inhabit the landscape. To help understand these relationships, scientists often examine a particular location in terms of its *position* on a slope, *steepness*, and orientation of the slope toward the sun—called *aspect*. With respect to position, we can see from Figure 10.11 that the water table is relatively deep on the upper portions of a slope, and shallow in the flat-lying areas at the bottom of the slope. Water therefore drains more readily through the soil zone in topographically high areas, whereas the drainage is rather poor in low areas. This results in greater leaching of the soils in the uplands areas where the infiltration rate is high. The flushing action of the water here also transports clay minerals and dissolved ions deeper into the soil profile.

The varying depth of the water table throughout the terrain also plays an important role in determining the amount of organic matter that accumulates in the topsoil. Notice in Figure 10.11 that vegetation growth is more lush in low-lying areas due to the greater availability of water. In addition, soils here tend to be saturated, which helps block free oxygen (O_2) from entering the pore space within the soil. Because free oxygen is more limited, the decay of organic matter by oxygen-dependent or *aerobic* bacteria is greatly reduced. This slower decay rate combined with more lush vegetation helps produce thick, organic rich soils in topographically low areas. Such areas make for highly productive agricultural lands.

From Figure 10.11 we can also see that soils tend to be thinner on steeper slopes. As a slope becomes steeper, a greater fraction of rain and meltwater will move directly down the slope, which means less water can infiltrate. This

FIGURE 10.11   Soils on topographically high areas generally contain less organic matter because of better drainage and higher rates of chemical decomposition. Soils in low areas commonly have more organic material due to more lush vegetation and poorer drainage, which tends to preserve organic matter. On steeper portions of a slope, soils are thinner due to the higher rates of erosion.

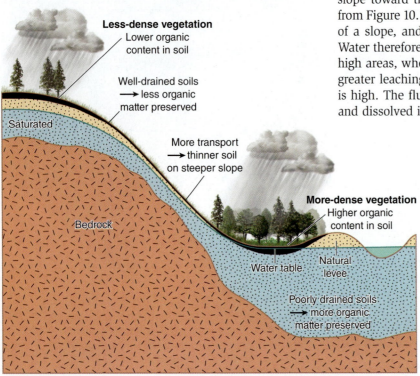

Less-dense vegetation
Lower organic content in soil

Well-drained soils
→ less organic matter preserved

Saturated

More transport → thinner soil on steeper slope

Bedrock

More-dense vegetation
Higher organic content in soil

Water table   Natural levee

Poorly drained soils → more organic matter preserved

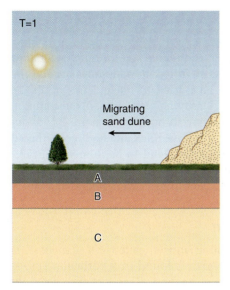 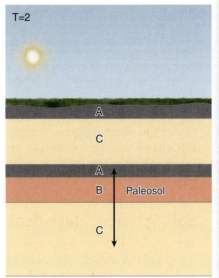

A

B

FIGURE 10.12 Illustration (A) showing how paleosols form when new sediment is quickly deposited over an existing soil sequence, creating an important time marker that can be dated by radiometric carbon-14 techniques. Photo (B) shows a paleosol in Finland that formed when wind-blown sand was deposited over the existing landscape.

produces greater erosion and lower rates of chemical weathering within the soil profile, both of which lead to thinner soils. Finally, we need to consider the orientation of a slope toward the sun. In the Northern Hemisphere, south-facing slopes receive more direct sunlight, hence are naturally warmer and retain less moisture compared to those facing north. The result can be significant differences in vegetation cover and chemical decomposition rates, both of which influence the types of soils that develop.

Time

Because the various physical and chemical processes involved in soil formation operate slowly, time is clearly a factor in the development of horizons. As indicated earlier, these processes can speed up or slow down depending on the manner in which the other soil-forming factors (parent material, plants and animals, climate, and topography) interact with one another. For example, soil horizons develop much more rapidly in warm, humid climates and where slopes are gentle and covered with lush vegetation. To provide some perspective of the time involved, under suitable conditions a simple soil sequence of just A and C horizons may take a hundred years or less to form. On the other hand, several hundred years are normally needed for the development of an A-B-C sequence. More deeply weathered soils with an A-E-B-C sequence require about 5,000 to 10,000 years to form, whereas intensely weathered tropical soils that are highly enriched in aluminum need approximately 100,000 years.

At any point in the development of a soil sequence certain geologic events, such as floods, volcanic eruptions, and migrating sand dunes, can quickly bury the soil with new sediment, generating what is referred to as a **paleosol** (Figure 10.12). Once the old soil is buried, a new sequence of horizons will begin to form, essentially resetting the clock on soil formation back to zero. Because paleosols represent distinct time events, they are quite useful in many types of geologic investigations. Moreover, because the burial helps preserve organic matter contained within the A horizons, scientists are able to use carbon-14 and radiometric dating techniques (Chapter 1) to accurately date the burial event. Although the upper limit of carbon-14 dating is only about 50,000 years, paleosols are used by environmental geologists to help determine the age of relatively recent hazardous events, such as floods, earthquakes, and volcanic eruptions. By dating events that occurred prior to written history, scientists can more accurately determine the recurrence interval (Chapter 8) of certain types of hazardous processes.

Classification of Soils

Due to the many possible combinations of the five soil-forming factors, we find a wide variety of soil types found on different landscapes around the world. Both scientists and engineers have tried to make sense of all the different soil types by grouping or classifying them based on common characteristics. For example, some soils have A horizons that are exceptionally rich in organic matter, whereas others have thick and highly weathered B horizons. Because scientists and engineers tend to be interested in different soil characteristics, they have developed separate classification systems. In this section we will briefly explore these two systems.

Soil Science Classification

The classification system used by soil scientists, called *soil taxonomy,* is based on the characteristics of the horizons found in a particular soil as well as its temperature and moisture regime. This system, listed in Table 10.1, breaks soils down into 12 different categories called *orders.* Although this system may appear complicated due to the large number of orders, it is actually rather simple since the characteristics of each order are related to the five soil-forming factors. For example, histosols are char-

TABLE 10.1 Simplified version of the soil classification system used by soil scientists. Soils are broken down into 12 major categories called orders, which are based on the characteristics of different horizons found within a soil. Although the dominant soil-forming factor(s) is listed for each soil order, all five factors are involved in the development of any soil.

Order	Simplified Description	Dominant Soil-Forming Factor(s)
Alfisols	Soils that are not strongly leached and have a subsurface horizon of clay accumulation. Common in forested areas where the climate is humid to subhumid.	Climate and living organisms
Andisols	Soils that form in volcanic ash and contain aluminum-rich silicates that actively bind with organic compounds.	Parent material
Aridisols	Soils that form in dry climates with low organic matter and that often have subsurface horizons with salt accumulations.	Climate
Entisols	Young soils lacking subsurface horizons because the parent material recently accumulated or because of constant erosion. Common on floodplains and steep mountain terrain.	Time and topography
Gelisols	Weakly weathered soils that contain permafrost in the profile. Common in higher latitudes.	Climate
Histosols	Soils with a thick organic-rich O horizon that contains very little mineral matter (e.g., quartz and clay). Common in poorly drained areas.	Topography
Inceptisols	Soils with weakly developed subsurface horizons because they are either young or the climate does not promote rapid weathering.	Time and climate
Mollisols	Soils that are not strongly leached and have an organic-rich A horizon. Common in grasslands where the climate is semiarid to subhumid.	Climate and organisms
Oxisols	Very old, extremely leached and weathered soils with a subsurface accumulation of iron and aluminum oxides. Common in humid tropical climates.	Climate and time
Spodosols	Soils that have a well-developed B horizon rich in iron and aluminum oxides. Form in cold, moist climates under pine vegetation and sandy parent material.	Parent material, organisms, and climate
Ultisols	Strongly leached soils (but not as strong as oxisols) with subsurface accumulation of clay. Common in humid tropical and subtropical climates.	Climate, time, and organisms
Vertisols	Soils that develop deep, wide cracks when dry due to the presence of swelling clays.	Parent material

Source: Modified after Brevik, *Journal of Geoscience Education,* v. 50, n.5, 2002.

acterized as having an uppermost horizon that is unusually rich in organic matter (i.e., an O horizon). This requires saturated conditions that promote lush vegetation and helps preserve the organic matter, conditions that are commonly found in low-lying areas of the terrain. Topography therefore is the dominant soil-forming factor in the development of histosols. Another good example is the soil order called entisols, which are characterized by a general lack of subsurface horizons. This means that entisols typically have only a simple A and C sequence, hence must be fairly young soils. Entisols are very common along floodplains since repeated flooding continues to deposit new sediment, thereby preventing the development of older soils. Consequently, both time and topography are listed as the dominant factors for entisols.

The relationship between soil orders (Table 10.1) and the five soil-forming factors is nicely illustrated by the maps shown in Figure 10.13. Perhaps the most obvious relationship here is the one between climate and the extensive belt of gelisols found in the high latitude regions of North America and Asia. Because gelisols are characterized by having permafrost within the soil profile, temperature is clearly the most dominant factor in the development of these soils. Another good example is the relationship between deserts and aridisols, which are notable in that they have a low organic content. The major control here is how the lack of rainfall greatly limits vegetation growth, hence the low organic content of aridisols. Finally, note how ultisols cover almost the entire southeastern portion of the United States, which is where the landscape is geologically old and the climate is quite warm and humid. This fits nicely with the fact that ultisols are highly leached and have thick B horizons, characteristics which require considerable amounts of both time and rainfall.

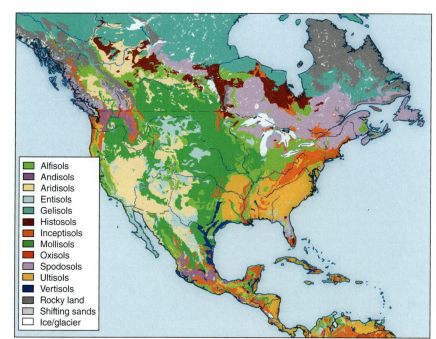

FIGURE 10.13   Map showing the distribution of soil orders in North America. Many of the patterns shown here are related to variations in climate and geologic history, both of which strongly influence soil formation.

Engineering Classification

Although scientists are interested in the relationship between specific types of horizons present in a soil and how they form, engineers are more interested in the physical properties of a soil. For example, engineers find it useful to know how well soil particles stick together and how easily water will flow through a soil. Consequently, the system developed by engineers, called the *unified soil classification system,* is based on physical properties. A simplified version of this engineering system is listed in Table 10.2. Note how soils are classified based primarily on the proportion of gravel, sand, silt, and clay-sized particles. Also notice how the soil names are quite descriptive. Take clayey sand for example, which indicates that the soil consists mostly of sand-sized grains, but yet contains a significant amount of clay particles. On the other hand, the well-graded modifier means that the soil contains a diverse range of particle sizes, whereas poorly graded implies more uniform-sized particles (note that geologic term *sorting,* used to describe the distribution of grain sizes, has the opposite meaning of the engineering term *grading*).

TABLE 10.2   Simplified version of the Unified Soil Classification System used in engineering. Here soils are grouped primarily on the proportion of different grain sizes (gravel, sand, silt, and clay).

Major Divisions	Subdivisions	Soil Name	Symbol
Coarse-Grained Soils (>50% of grains visible with naked eye)	Gravels	Well graded gravel	GW
		Poorly graded gravel	GP
		Silty gravel	GM
		Clayey gravel	GC
	Sands	Well graded sand	SW
		Poorly graded sand	SP
		Silty sand	SM
		Clayey sand	SC
Fine-Grained Soils (<50% of grains visible with naked eye)	Silts	Plastic silt	ML
		Nonplastic silt	MH
		Organic silt	OL
	Clays	Low-plastic clay	CL
		High-plastic clay	CH
		Organic clay	OH
Organic Soils	Organic Matter	Peat	PT

In addition to being descriptive, engineers find this classification system practical since many other soil properties are directly related to its grain-size distribution. Therefore, once a soil has been properly classified, an engineer can get a pretty good idea of its general properties simply from its name. For example, one can predict that a poorly graded (i.e., well sorted) sand would be more permeable than a clayey sand. In the next section we will take a closer look at how soil properties are important in terms of human activity.

Human Activity and Soils

Earlier we stressed how soils are critical to humans because they form the basis of our food supply and provide the raw material for producing aluminum metal. However, the importance of soils goes beyond just food and aluminum. For example, nearly every construction project involves excavation, or digging into the soil. The properties of a particular soil will determine how hard it will be to dig a hole, as well as the hole's tendency to collapse and the speed at which it fills with water. Soil properties are also an important consideration for building foundations, which must be designed so that the weight of a building can safely be supported by the underlying soil. One of the first tasks for an engineer in any large construction project is to determine key soil properties at the site. Measuring soil properties is also important at contaminated sites where it is necessary to know the rate and direction in which pollutants will migrate (Chapter 15). In this section we will examine some of the key soil properties and how they affect the ways in which people interact with soils.

Soil Properties

Recall that Earth's crust is largely composed of silicate minerals, many of which are transformed into various clay minerals during chemical weathering. An important exception is the mineral quartz. Quartz is not only very abundant in granitic rocks, but it is also highly resistant to chemical weathering, which is why most sand and silt-sized particles are composed of quartz. Therefore, because of the weathering characteristics of silicate rocks, soils tend to be chiefly composed of quartz and clay mineral particles along with lesser amounts of organic material.

Also recall that clay-mineral particles are exceedingly small compared to sand and silt-sized grains of quartz. In addition to size, quartz and clay minerals are also quite different in terms of their crystalline structure or internal arrangement of atoms. Because the atoms in clay minerals are arranged in complex sheetlike structures, individual clay particles are plate-shaped and have a negative electrical charge on their outer surface. The surface of quartz grains on the other hand is electrically neutral. Later you will see how the size and electrical properties of quartz and clay minerals play a major role in many soil properties.

Porosity

As illustrated in Figure 10.14, sediment naturally contains void or *pore space* that exists between the solid particles. Geologists use the term **porosity** to describe the fraction of sediment or rock that consists of void (air) space. In terms of total volume, soils average about 45% mineral matter and 5% organics; hence, porosity averages around 50%. Because water can move into a soil and fill the pore spaces, this means that about 50% of a soil's total volume can be taken up by water. Note that when the pores become completely filled with water the soil is said to be *saturated*, but when pores

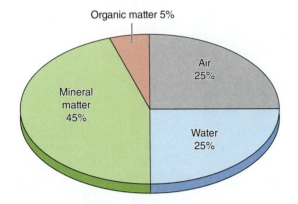

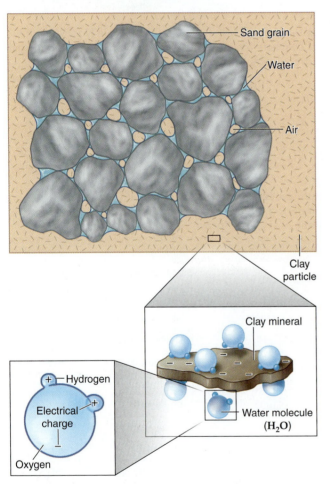

FIGURE 10.14 Soils consist of about 45% mineral matter and 5% organics, with the remaining 50% being void space that is filled with varying amounts of air and water. Within the pores, dipolar water molecules are strongly attracted to the extremely small clay-mineral particles, whose crystal structure creates a negative charge on the outer surface of the particles.

contain both air and water a soil is considered *unsaturated*—rarely are soils completely dry. One of the reasons porosity is an important property is because it indicates the amount of water that is potentially available for plants, particularly agricultural crops.

Soil Moisture and Drought Resistance

Some soils naturally lose water from their pore space more easily than others, making certain crops more susceptible to being lost or damaged due to drought. The so-called *drought resistance* of soils is controlled by the mineral composition of the soils and the dipolar nature of the water molecule (H_2O). As illustrated in the inset for Figure 10.14, the orientation of hydrogen and oxygen atoms in the water molecule causes one side of the molecule to be slightly positive and the other side slightly negative. Although electrically neutral overall, this uneven distribution of electrical charge results in individual water molecules being attracted to one another—scientists use the term *cohesive* force to describe this attraction between similar molecules. Water molecules are also attracted to other types of molecules, such as a solid surface, in which case the attraction is referred to as an *adhesive* force. For example, suppose you take a hot shower and let the resulting water vapor build up in the room until water droplets begin falling from the ceiling. Initially a thin film of water will form on the ceiling as water molecules attach themselves through adhesive forces. Additional molecules would then become attached to this film of water by cohesive forces. As the number of layers of water molecules continues to grow, gravity pulls the layers downward, creating a drop of water. Eventually gravity overcomes the cohesive forces, at which point the drop falls, leaving some water molecules firmly attached to ceiling by adhesive forces.

The simple example of water droplets falling from a ceiling is useful because it is analogous to how water moves through pore spaces within a soil. In the case of soils, however, the different size and electrical properties of quartz and clay mineral grains play a critical role. For example, imagine we have two soils that are completely saturated, but one is composed entirely of clay minerals and the other all quartz. Because the surfaces of the clay particles are electrically charged, whereas those of quartz are not, the adhesive forces between water and the clay particles are much stronger. Moreover, clay particles are extremely small compared to quartz grains, which provides significantly more surface area within a soil for water to be held by adhesive forces. Within the large pores of our quartz sand soil, much of the water is held in place by cohesive forces between individual water molecules, and the adhesive forces between water and the grains are not particularly strong. In contrast, water within the tiny pores of the clay soil is held in place mostly by very strong adhesive forces.

Now imagine that we allow our two saturated soils to suddenly drain by the force of gravity that pulls the water downward. Because water in a quartz sand soil is held rather weakly, much of it will simply drain away, making the soil susceptible to drought. On the other hand, very little if any of water would drain from a clay soil due to the fact it is held so tightly. However, a clay soil is also susceptible to drought because water is held so tightly. Here plants commonly begin to wilt when they are unable to pull water from the clay particles, even though considerable amounts of water remain. One would think that an ideal soil in term of drought resistance would be one with an even mixture of sand and clay. It turns out though that the most drought resistant and highly prized agricultural lands are soils with a high percentage of silt and moderate amounts of sand and clay. Such soils are ideal because they retain more water for plants than a pure sand soil, but yet give up their water more freely than one composed of all clay.

Permeability

The concept of water being held by cohesive and adhesive forces within a soil is also useful for understanding the property called **permeability,** which is the ability of a fluid to flow through a porous material. Soils rich in clay minerals have low permeability since the strong adhesive forces within the soil greatly inhibit the flow of water. In contrast, the permeability of sand-rich soils is high because much of the water is under relatively weak cohesive forces, thus is able to flow and drain much more freely. Therefore, anything that increases the fraction of water being held by the weaker cohesive forces will also cause permeability to increase. Consequently, the most permeable soils are those with large pores (e.g., coarse sands) and those with grains of uniform size such that smaller grains do not plug the pore space and restrict the flow. Although they involve similar forces, drought resistance refers to the *amount* of water in the soil that is available to plants, whereas permeability refers to the ability of water to move through the soil.

In terms of human interactions with the landscape, permeability is important because it determines how fast soils are able to drain. For example, suppose a heavy rain saturates the upper soil horizons in a farmer's field. If the water does not drain quickly enough so that air can get back into the soil, certain types of crops may drown. Clay-rich soils are more prone to this type of crop damage because they drain more slowly. Excessive water can also lead to pest and disease problems. The permeability of soils also affects the proportion of rainwater that is forced to flow across the landscape (overland flow), which is a major factor in the flooding characteristic of a drainage basin (Chapter 8). Likewise, soil permeability is a factor in many mass wasting events (Chapter 7) where infiltrating water decreases the stability of slopes by reducing the frictional forces within the subsurface.

Plasticity

As discussed in Chapter 4, **plasticity** is the ability of a material to deform without breaking when a force is applied—similar to how modeling clay can be shaped into various forms, yet not break. In terms of soil, plasticity generally increases with clay content depending on the specific type of clay minerals present. On the other hand, plasticity always decreases as sand content increases, with pure sand having essentially no plasticity. Another important factor that affects plasticity is water. When some fine-grained soils take on too much water, they can start to flow like a liquid. On the other hand if allowed to dry completely, many clay-rich soils will behave more like solid rock, in which case plasticity approaches zero (similar to how modeling clay becomes hard when it dries). Water therefore has the ability to drastically change the plasticity of soils, which, in turn, can have profound implications in engineering. For example, the weight of a large structure, such as a tall building, dam, or bridge, exerts a tremendous vertical force on the subsurface. If the plasticity of subsurface materials increase because of additional water, they could begin to flow, thereby threatening a structure's integrity. Therefore, when designing foundations for structures, engineers take great interest in soil properties and moisture conditions around the site.

Strength and Sensitivity

The **strength** of soil refers to its ability to resist being deformed or, in other words, how well the particles stick together. For example, if you were to dig a trench, the walls would be less likely to collapse in a strong soil

compared to a soil that is weak. Note that the strength of a soil depends on the cohesive and frictional forces that exist between the grains. Of interest here is the fact that some soils experience a loss of strength when they are suddenly disturbed. Engineers therefore use the term **sensitivity** when referring to how easily a soil will lose its strength when disturbed. For example, if a saturated sandy soil is shaken during an earthquake, the individual sand grains can easily lose contact with one other. When this occurs the grains become suspended in the water, causing the material to behave as a liquid in a process called *liquefaction* (Chapter 5). During the time the soil is liquefied it has virtually no strength, allowing heavy objects to sink into the subsurface. Note that although water is tightly bound in most clay minerals, certain types of clay minerals known as *quick clays* are highly sensitive to vibrations and will readily undergo liquefaction.

Compressibility

We can define **compressibility** as the ability of a material to compact and reduce its volume when placed under a force or load. In general, the compressibility and volume reduction of soil increases as the clay-mineral content increases. As shown in Figure 10.15A, when a load is placed on a soil containing only quartz grains, the grains will reorient themselves so that the pore space is at a minimum. Once this occurs the sand will be nearly incompressible because there is no further room for movement, plus the quartz grains themselves are strong and incompressible. In contrast, when a heavy load is placed on a clay-rich soil where the plate-shaped particles are more randomly oriented (Figure 10.15B), the particles will reorient themselves into a parallel arrangement. This process is facilitated by water in the soil as it exerts pressure within the pore space, which tends to keep the grains apart, making it easier for them to rearrange and become more tightly packed. Compaction in clay-rich soils therefore can be quite significant, resulting in significant volume reductions that can lead to settling problems at the surface. Perhaps the most famous example is the Leaning Tower of Pisa in Italy (Figure 10.15C). Variations in the clay content of the soils underlying the tower, combined with the weight of the structure, led to different amounts of compaction, which, in turn, caused the tower to lean.

Clearly, compressibility and compaction of soils is important to engineers when designing the foundation supports for a heavy structure. The compressibility of soils is also of great concern to farmers, particularly since compaction is essentially irreversible. Agricultural fields can undergo compaction due to the use of heavy farm equipment, which permanently reduces the ability of soils to drain and makes it more difficult for roots to penetrate the soil. Compaction, of course, is more of a problem in wet, clay-rich soils. In order to minimize soil compaction, farm vehicles are equipped with extra wide tires, and farmers generally try to avoid running heavy equipment when their fields are overly wet.

FIGURE 10.15 When a heavy load is applied to a soil the individual grains will attempt to rearrange into a more tightly packed configuration. Sandy soils (A) have low compressibility because the reduction in volume that can occur when rounded grains are rearranged is relatively small. Clay-rich soils (B) are highly compressible because the random orientation of small clay particles allows for a significant reduction in volume. The uneven settling of the Leaning Tower of Pisa (C) was caused by differences in compaction that were related to variations in the clay content of the soils.

A

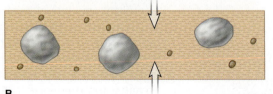

B

C

Dry clay mineral

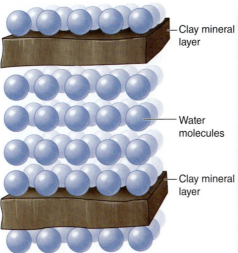

Clay mineral layer

Water molecules

Clay mineral layer

Expansion due to absorption of water

Camera

FIGURE 10.16   The number of water molecules that can exist between the sheetlike structure of clay minerals varies among different types of clays. Expanding clays have a great capacity to take on water, causing soils called vertisols to increase in volume. When vertisols are allowed to dry, they shrink considerably, creating familiar cracks often found in dried-out lake beds as in the photo shown here—note the yellow camera for scale.

Shrink-Swell

As with many types of materials, soils swell or expand when they take on water, then shrink as they dry out—similar to how a sponge expands and contracts. Clay-rich soils generally have the greatest capacity to shrink and swell due to the small particle size of clays and the fact that water molecules have such a strong attraction to clay minerals. As noted earlier, clay minerals have a sheetlike atomic structure dominated by aluminum, silicon, and oxygen atoms. Because the atomic structure of these sheets varies among the different types of clay minerals, it turns out that each clay mineral has a different ability to store water between its sheets. The term **swelling** or **expanding clays** refers to those clay minerals capable of incorporating large numbers of water molecules within their structure, thereby producing significant volume changes as illustrated in Figure 10.16. Note that the presence of expanding clays is a characteristic feature of soils known as vertisols (refer to Table 10.1).

Most of the swelling clays found in vertisols belong to a family of clay minerals referred to as *smectites*, notable for their great capacity to take on water and expand. For example, a pure deposit of the smectite mineral called *montmorillonite* can expand up to 15 times its original volume. Because of their ability to absorb water, humans have found a variety of applications for expanding clays. Perhaps the most familiar application is the dried material used in the litter boxes for cats and for cleaning up spills involving oil, antifreeze, and other liquids. Smectite clays are also commonly used as an additive that increases the density of drilling fluids used during the boring of holes for water and oil wells. After the drilling is complete, expanding clays are then used to create an impermeable seal around the metal or plastic casing of the well.

Ironically, the same properties of expanding clays that people find useful can sometimes cause serious problems in society. For example, the map in Figure 10.17 shows areas in the United States where vertisols contain significant amounts of expanding clays, hence the ground has the capacity to shrink and swell as soil moisture conditions change. Note that the different colors in this map indicate the *shrink-swell potential* of the ground, which is related to both the amount and type of expanding clays present in the soil. It is not uncommon for vertisols in areas with the highest potential to experience a 25–50% volume increase. This level of volume

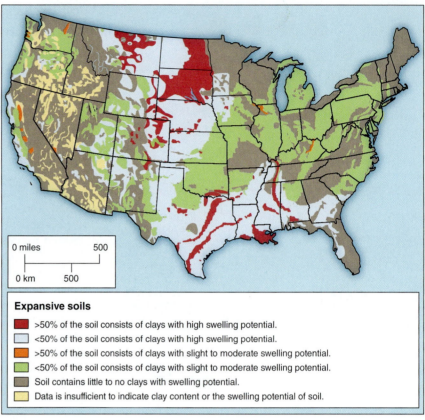

Expansive soils

- >50% of the soil consists of clays with high swelling potential.
- <50% of the soil consists of clays with high swelling potential.
- >50% of the soil consists of clays with slight to moderate swelling potential.
- <50% of the soil consists of clays with slight to moderate swelling potential.
- Soil contains little to no clays with swelling potential.
- Data is insufficient to indicate clay content or the swelling potential of soil.

B

A

FIGURE 10.17 Soils known as vertisols contain significant amounts of expanding clay minerals, which can cause serious structural damage should the soil go through repeated drying and wetting cycles. Photos (A) illustrate the types of damage that can occur when the underlying soil expands and contracts. Map (B) showing areas in the United States where soils have a high swelling potential.

change is highly significant, especially considering that engineers typically require specialized foundations for changes greater than 3%. Problems occur because the expanding soils can generate as much as 20,000 pounds per square foot of force on various types of infrastructure, including sewer and utility lines, highways, and buildings (Figure 10.17B). In fact, damage from expanding and contracting soils in the United States is estimated at nearly $2 billion per year.

Although vertisols can be found across broad portions of the landscape, fewer problems occur in areas with a humid climate since soil moisture remains relatively constant. The most severe problems are found in drier climates where soils typically contain little water. The soils are then able to undergo significant expansion during occasional wet periods, or when the landscape is put under irrigation. Structural damage associated with expanding soils can largely be avoided if a proper soil investigation is undertaken prior to construction. Should the investigation uncover a potential problem, specialized construction techniques can be used to avoid damage when the soil eventually does expand. Another effective technique is to ensure that any landscaping around buildings is done in such a way that soil moisture changes are kept to a minimum, thereby reducing the potential stress on the structure.

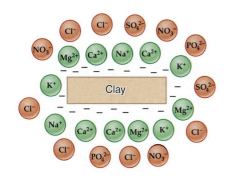

A

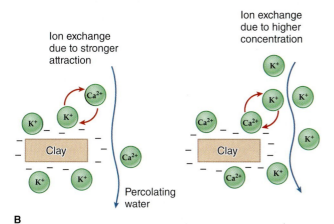

Ion exchange
due to stronger
attraction

Ion exchange
due to higher
concentration

Percolating
water

B

FIGURE 10.18 Positively charged ions naturally attach themselves in layers (A) around the negatively charged surfaces of clay minerals and organic matter within a soil. Notice that negative ions, in turn, surround the positive ions. As percolating water (B) carries additional ions through the soil, they are exchanged with those on the soil particles in a selective manner depending on the attraction and concentration of ions in the water.

Ion Exchange Capacity

Recall that when minerals undergo chemical weathering, charged atoms and groups of atoms called *ions* are released into the water. Some of the ions, such as potassium (K^+) and calcium (Ca^{2+}), are important in soils because they serve as essential nutrients for plants. Other important nutrients include nitrate (NO_3^-) and ammonium (NH_4^+) ions that form not by chemical weathering of minerals, but by the interaction of atmospheric gases and biological processes within the soil. **Ion exchange** is the process whereby dissolved ions attach themselves to soil particles, and are then removed in a selective manner by growing plants and by water moving through the soil zone. This process where various ions are stored and then removed from soil particles is critical to plant growth and crop production.

As shown in Figure 10.18, dissolved ions will attach themselves in a layered manner around the clay mineral and organic particles in a soil. Notice that when water flows through a soil there are two basic ways ion exchange can take place between the water and soil particles. The first is when water contains ions that have a stronger attraction than those ions already attached to the soil. For example, because of their smaller size and greater charge, calcium ions (Ca^{2+}) have a stronger attraction to soil particles than do potassium ions (K^+). Therefore, when calcium-rich water flows through a soil, potassium ions are removed from soil particles and are replaced by calcium ions. Ion exchange can also work in reverse, whereby a strongly held ion (e.g., calcium) is removed from the soil and replaced by an ion with a weaker attraction (Figure 10.18). For this to occur, the water must contain a very high concentration of the weaker ion such that their overwhelming numbers force the more tightly held ions from the soil particles.

Finally, note that scientists use the term *ion exchange capacity* to describe the ability of a particular soil to exchange ions with water. Soils with a high ion exchange capacity naturally contain significant amounts of electrically charged clay and organic particles. In the next section we will examine how ion exchange plays a critical role in soil fertility and agricultural production. In Chapter 15 we will examine how this exchange process is important in the immobilization of contaminants within the soil zone, thereby minimizing the spread of pollution.

Soil as a Resource

Earlier in this chapter we mentioned that land plants and the biosphere as we know it would not exist were it not for soils. Of particular importance to humans are agricultural plants such as wheat, rice, and corn since they make up the bulk of the world's food supply. In this section we will explore the connection between soils and our food supply as well as some other important resources we obtain from soils.

Agricultural Food Production

The single most important use of soils in society is for growing food. Soils are commonly referred to as being *fertile* in areas where crops grow particularly well and infertile where farming is not very productive. In addition to a certain range of water and temperature conditions, different types of plants require a particular set of elements in order to function and grow properly. **Soil fertility** is the term used to define the ability of a soil to supply the elements necessary for plant growth. The most critical elements are often referred to as **essential nutrients,** which include nitrogen (N), phosphorous (P), potassium (K), calcium (Ca), magnesium (Mg), and

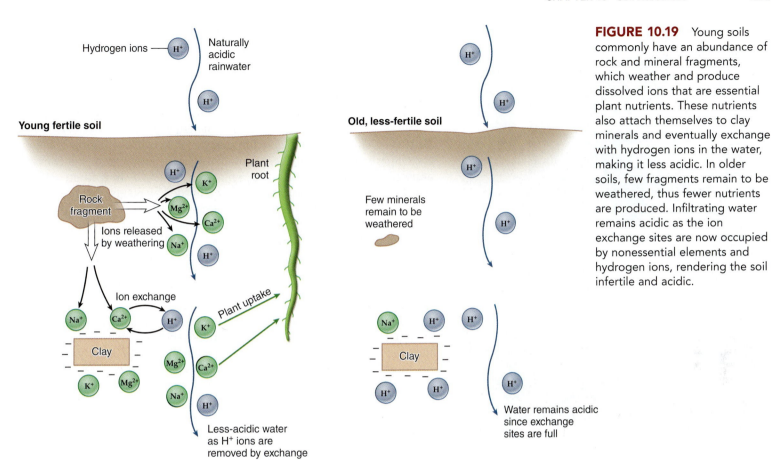

Hydrogen ions — H⁺

Naturally acidic rainwater

Young fertile soil

Rock fragment

Ions released by weathering

Plant root

Ion exchange

Na⁺ Ca²⁺ H⁺

Clay

K⁺ Mg²⁺ Na⁺

Plant uptake

Less-acidic water as H⁺ ions are removed by exchange

Old, less-fertile soil

Few minerals remain to be weathered

Clay

Water remains acidic since exchange sites are full

FIGURE 10.19 Young soils commonly have an abundance of rock and mineral fragments, which weather and produce dissolved ions that are essential plant nutrients. These nutrients also attach themselves to clay minerals and eventually exchange with hydrogen ions in the water, making it less acidic. In older soils, few fragments remain to be weathered, thus fewer nutrients are produced. Infiltrating water remains acidic as the ion exchange sites are now occupied by nonessential elements and hydrogen ions, rendering the soil infertile and acidic.

sulfur (S). These elements come from the weathering of minerals and the interaction of atmospheric gases and biological processes within the soil. The elements are then stored within the soil in the form of electrically charged ions, which plant roots selectively remove depending on the needs of a given plant. For example, because some plants have a relatively high demand for calcium, they will selectively remove calcium ions and leave the remaining types of ions attached to the soil particles.

The ion exchange capacity of a soil is clearly tied to soil fertility as it helps determine whether essential nutrients in the form of charged ions are available in the soil. As indicated in Figure 10.19, the key is the amount of nutrients being produced by weathering and biological activity and the amount of electrically charged clay and organic particles within the soil. Soils that are rich in organic matter and have a moderate amount of clay are naturally fertile as they contain a large number of ion exchange sites for storing nutrients. However, if there are few nutrients moving through the soil, then fertility will be low regardless of the number of ion exchange sites. Older soils are generally not very fertile because chemical weathering has progressed to the point where few minerals remain that can release important ions such as potassium, calcium, and magnesium. When certain nutrients are no longer being released, their continued removal by plants combined with the flushing action of rainwater, which is naturally acidic, eventually leaves the exchange sites occupied by nonessential elements and hydrogen (H⁺) ions (Figure 10.19). Old soils therefore, such as oxisols and ultisols (refer to Table 10.1), are characterized as being both infertile and acidic (i.e., contain numerous H⁺ ions).

Long before scientists learned about ion exchange and soil nutrients, people recognized there was a relationship between the crop production and the amount of organic matter in the upper soil horizon, called topsoil.

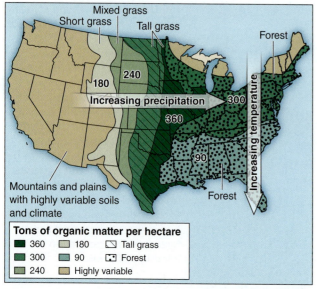

FIGURE 10.20 Generalized map (A) showing the amount of organic matter (metric tons per hectare) in soils in the United States. The trends are strongly related to climate, vegetation types, and amount of weathering. Note the abrupt change in organic content marked by the boundary of tall grass prairie and forest cover. Photo (B) showing the high organic content of an exceptionally fertile topsoil in Iowa.

Because clay minerals alone can provide a sufficient number of ion exchange sites, organic material must play some additional role in soil fertility. Modern scientists have learned that organic matter acts as a food source for bacteria and fungi within the soil, which, in turn, are needed to break down complex organic materials into simpler substances plants can use directly. Although the amount of organic material in topsoils may be relatively small, it is critical because it helps produce certain essential nutrients that are not available through the chemical weathering of minerals. Of particular importance is how biological activity within soil converts atmospheric nitrogen (N_2) into ammonium (NH_4^+) and nitrate (NO_3^-), which are the essential forms of nitrogen needed by plants. The breakdown of organic matter within a soil also produces nitrogen as well as sulfur and other important nutrients. Consequently, soils simply will not support much plant life without organic matter. In Figure 10.20 one can see how the organic content of soils varies across the United States and how the trends correspond to agricultural productivity. These trends are ultimately related to the soil-forming factors of temperature, precipitation, vegetation, and amount of weathering (time).

When farmers repeatedly grow crops on the same field each year, the essential nutrients within the soil are removed at a much faster rate than they can be replaced through weathering and biological activity. Soil fertility therefore quickly decreases. Early on in the agricultural revolution humans discovered that plants grew much better when manure or ashes were spread on the soil. Because this practice increased the fertility of soils, manure and ashes were eventually called **fertilizers.** The use of fertilizers became very important as it allowed people to continue working the same land, plus it could make relatively poor farmland agriculturally more productive. Although modern farmers rely mostly on inorganic or synthetic fertilizers, the basic idea is still the same, namely to ensure that soils contain the essential nutrients needed for maximum crop production.

Another useful technique used for maintaining soil fertility is to rotate crops so that a different crop is grown each year on a given field. Because different crops remove different combinations of essential elements from the soil, crop rotation allows certain nutrients to build back up while others are used by the current crop. Modern farmers do not always find it necessary to rotate crops since they can maintain proper nutrient levels in the soils through the use of fertilizers. However, this more intensive use of fertilizers and other chemicals commonly leads to the contamination of water supplies (Chapter 15).

Mineral and Energy Resources

In addition to providing a fertile medium in which plants thrive, soil-forming processes also result in several important mineral and energy resources that have been important to society. Although mineral and energy resources will be discussed in Chapters 12 and 13, we will take a brief look at some of those that are directly related to soil-forming processes.

Aluminum   Modern societies have found many uses for aluminum metal because it is strong, lightweight, and readily conducts both heat and electricity. Aluminum is particularly valuable in applications where weight and strength are an issue, such as in airplanes and fuel-efficient cars. Deposits of aluminum ore are geologically unique in that they represent the end product of the intense chemical weathering of igneous rocks that are relatively rich in aluminum. Recall that most of Earth's crust is comprised of *aluminosilicate* minerals, the most common being the feldspars (Chapter 3). When feldspar minerals undergo chemical weathering and break down into clay minerals, their aluminum atoms are retained within the atomic structure of clay minerals. Consequently, the clay minerals within the soil zone are enriched in aluminum compared to the original feldspar minerals. Given sufficient amount of time for chemical weathering, as much as 100,000 years, the clay minerals will be transformed into new minerals that are even more enriched in aluminum. To get a sense of the amount of enrichment, consider that the potassium-rich feldspar mineral contains only 10% aluminum by weight, but weathers into the clay mineral called *kaolinite* with an aluminum content of 21%. Kaolinite, in turn, will chemically weather into *gibbsite,* a mineral that is 35% aluminum by weight.

Geologists use the term *bauxite* when referring to the collection of gibbsite and similar minerals that are highly enriched in aluminum. Deposits of bauxite that are economical to mine are generally found in tropical climates where feldspar-rich rocks have underdone long periods of intense chemical weathering. As illustrated in Figure 10.21, the solid rock eventually weathers into a thick soil zone that is rich in clay minerals. Abundant rainfall and warm temperatures cause intense chemical weathering where the clay minerals are transformed into highly enriched aluminum minerals. Running water and other geologic processes then concentrate these minerals into economical deposits of bauxite.

Kaolinite Clay   Under the right climatic and geologic conditions, relatively pure deposits of the clay mineral known as *kaolinite* can develop from the weathering of feldspar-rich rocks. Kaolinite is a highly prized mineral used in a variety of products due to its unique set of physical properties, including being extremely soft and fine-grained and having a white color. For example, these properties make kaolinite useful in the production of high-quality ceramics (porcelain or china) and glossy paper found in magazines. Kaolinite is also used to make the fine, smooth paste found in women's makeup and as the base or thickening agent in a variety of products, including water-based paint. Moreover, its consistently white color makes it possible to add pigments so that products such as makeup and paint can come in any conceivable color. Finally, it's worth noting that kaolinite is used as the base for many medicines, the most familiar perhaps being Kaopectate® and Pepto-Bismol®.

Peat   In some parts of the world where wood is scarce, people have for centuries been using organic-rich soils, commonly known as *peat,* as their primary source of energy for heating and cooking. In these areas the upper soil horizon is composed almost entirely of organic matter, which is defined as the O horizon of soils called histosols (refer to Table 10.1). These peat deposits originate in lakes or wetlands where dissolved oxygen levels are low enough so that organic matter will accumulate faster than it decays. As organic material accumulates, the open water is eventually replaced by saturated ground, at which point the site is referred to as a *peat bog* (Figure 10.22). To extract the peat, selected portions of bogs are drained so that the soil can be cut into squares and stacked on dry land and allowed to dry. Although the use of peat as an energy source has declined in modern times due to the availability of fossil fuels, it is still widely used as gardening mulch and in potting soil. Note that coal forms from peat that becomes buried and transformed by the higher levels of heat and pressure (Chapter 13).

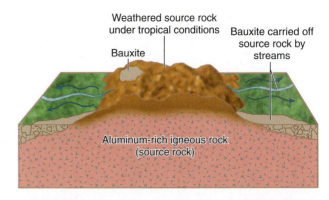

FIGURE 10.21   Economical aluminum deposits form when feldspar-rich igneous rocks undergo long periods of intense chemical weathering. As the aluminum-rich feldspar minerals are transformed into clay minerals, the weight percentage of aluminum increases. Over time this process creates a bauxite deposit containing minerals that are highly enriched in aluminum.

FIGURE 10.22   Organic-rich soils, known as peat, form in bogs where it can be extracted and dried, then used as a fuel for heating and cooking. Peat is also used as gardening mulch and in potting soils. The photo is from a bog in Scotland.

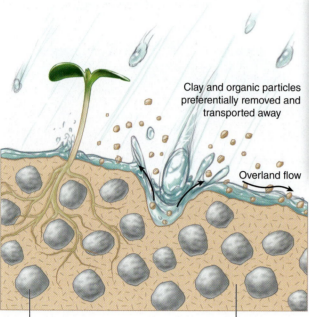

Quartz sand grain   **Clay and organic particles**

Clay and organic particles preferentially removed and transported away

Overland flow

FIGURE 10.23 The impact of raindrops on an unprotected soil creates an explosive effect that preferentially ejects clay and organic particles onto the surface. Soil erosion occurs when the loose particles are transported by wind or water moving downslope as overland flow. Over time this process reduces soil fertility.

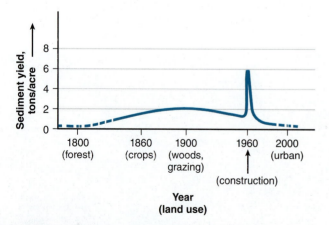

FIGURE 10.24 Graph showing how sediment loss changed over time in response to different land uses at a site in the eastern United States. Note the significant increases in soil loss that accompanied changes to agriculture and a construction boom.

Soil Loss and Mitigation

Although soil-formation processes provide humans with a variety of mineral and energy resources, by far the most important is the production of fertile soils that serve as the basis of our food supply. As mentioned earlier, the problem is that agricultural practices tend to increase soil erosion that leads to the irreplaceable loss of topsoil. Soil loss is a serious problem as it could lead to decreased food production while world population continues to expand. In this section we will explore the various factors involved in soil loss and the ways in which humans can minimize the problem.

Soil Loss

When rain falls on a natural landscape covered by forest or dense grasses, few if any of the raindrops will ever make direct contact with the underlying soil particles. Likewise, wind that moves over the surface rarely makes contact with the soil on heavily vegetated landscapes. We can define *cover material* as anything lying on top of soil that provides protection, similar to a shield. Should anything cause this blanket of cover material to be removed, the bare or unprotected soil will be exposed to the direct impact of raindrops and wind. Cover material is particularly important when it comes to rain because of the explosive effect raindrops have when they make impact with the soil. As illustrated in Figure 10.23, when a raindrop hits bare soil it results in a small impact crater. The ejected soil particles are then easily picked up by wind or water flowing across the landscape. This movement or transport of soil particles away from their place of origin is called **soil erosion.** Moreover, water and wind preferentially transport the smaller and lighter organic and clay mineral particles within the soil, leaving behind the much larger sand grains. This preferential loss of organics and clay minerals is a serious problem, as these particles are the key to soil fertility.

Soil erosion is actually a natural process that has been taking place for hundreds of millions of years. Long before people began inhabiting the planet, there would always be certain parts of the landscape experiencing severe erosion as fire or disease destroyed the natural vegetation cover. But in many parts of landscape, soil formation and erosion were in balance such that there was no net loss of soil for long periods of time. Today, however, the natural vegetation cover over large portions of the landscape has either been removed or disturbed due to human activities that include farming, logging, and construction. Our use of the landscape, in turn, has caused soil erosion to increase to a point that it far exceeds the rate of soil formation in many areas, resulting in a net loss of soil called **soil loss.** From the graph in Figure 10.24 you can see how erosion and soil loss vary over time in response to land-use change in a particular area. Although soil loss can result from a variety of human and natural causes, our interest in the next section will be on those related to human activity. In particular our focus will be on agricultural land use as it leads to the most serious consequence, namely a loss of food production.

A Southern Iowa   **B** Cass County, Iowa

FIGURE 10.25 Photos showing the irreplaceable loss of topsoil from agricultural fields. Soil is actively being removed during a rainstorm in (A), whereas (B) shows a small drainage channel where eroded topsoil has acccumulated.

Consequences of Soil Loss

Soil loss presents farmers with a difficult dilemma, namely that the very act of growing crops requires that natural vegetation be removed from the landscape, allowing topsoil to be washed off their fields as shown in Figure 10.25. From the map in Figure 10.26 one can see that soil loss on agricultural land in the United States is not evenly distributed due to variations in the steepness of the terrain and the intensity of agricultural activity. For example, much of the Great Plains experiences relatively low rates of soil loss simply because agriculture is limited by the dry climate. In these areas much of the land is used for grazing livestock, so soil erosion is relatively low since the cover material generally remains in place. Growing wheat is also common, which also tends to generate lower rates of soil erosion because this particular crop provides fairly good soil protection. On the other hand, the eastern half of the country experiences more severe soil loss because of the types of crops and the more intense level of agricultural activity. Here corn and soy beans are common row crops, both of which provide far less protection against erosion than does natural vegetation. Note how soil loss is not uniform in states such as Iowa and Illinois (Figure 10.26), indicating that factors other than the level of agricultural activity and types of crops are involved. The primary reason for the variation in these areas is the steepness of the terrain. Agriculture fields located on hilly terrain naturally have higher rates of erosion and soil loss compared to fields on flat-lying terrain.

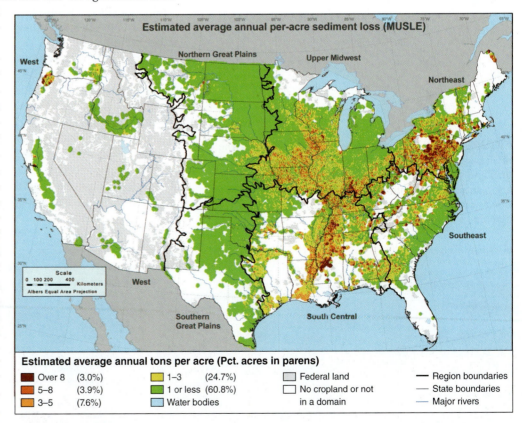

FIGURE 10.26 Map showing the estimated average rate of soil loss per acre on agricultural land in the United States. Variation in soil loss is largely due to the level of agricultural activity and the steepness of the terrain.

Obion River in Central Tennessee

FIGURE 10.27   Photo showing how soil loss leads to sediment pollution where stream channels become choked with excess sediment. Sediment pollution increases the likelihood of flooding and destroys the natural ecology of streams.

FIGURE 10.28   Contour plowing (A) involves planting rows parallel to the slope of the land, reducing overland flow and soil erosion. In crop stripping (B) different crops are planted in alternate strips to trap sediment moving downslope—note in the photo how crop stripping is combined with contour plowing.

Another important consequence of soil loss is *sediment pollution*, which is the choking of drainage systems by excessive sediment moving off the landscape (Figure 10.27). As described in Chapter 8, both water and sediment are carried off the landscape and into stream channels in a process called overland flow. Once in a channel, sediment is then transported downstream through drainage networks. The problem is that in areas of extensive soil erosion, the stream network simply cannot transport all the additional sediment, in which case the channels become choked or filled with sediment. This, in turn, increases the frequency of flooding because there is less room in the channels for water. Sediment pollution also causes problems in rivers that are used as transportation corridors, as the channels must be dredged more often in order to keep the river open for shipping. Finally, the natural ecology of streams can be destroyed when channels become choked with sediment. For example, fish such as bass and trout are no longer able to see their prey in the sediment-laden water, plus their spawning areas of coarse sediment become buried by finer material. Moreover, the aquatic life that once formed the basis of the stream's food web is also greatly impacted. Sediment pollution therefore creates a rather sterile environment in which only a small fraction of the original species survive.

Mitigating Soil Loss

Because both soil loss and sediment pollution are detrimental to society, various techniques have been developed over the years that minimize the problem. These techniques are generally intended to either keep the soil in place so it does not move downslope, or trap the sediment before it can enter the drainage network. In many situations these basic approaches are used in conjunction with one another in order to minimize both soil loss and sediment pollution. In this section we will briefly explore some of the more common techniques. Although most of the following techniques are associated with agriculture, some also address problems resulting from construction and logging activities.

Contour plowing is an old practice whereby farmers plant rows of crops parallel to the contours of the land surface. Although contour plowing originally came about because people naturally found it was easier to plow parallel to the slope, it also happens to be very effective in reducing soil erosion. From Figure 10.28A one can see how the rows reduce the ability of water to flow

A Southwest Iowa

B Iowa-Minnesota border

directly down the slope, thereby helping to keep soil particles from being carried off the fields. Contour plowing is a particularly important technique in areas of rolling or hilly terrain where cultivated fields have more of a slope.

Crop stripping is where crops are planted in strips within the same field (Figure 10.28B) and is often performed in conjunction with contour plowing. Crop stripping originated from the practice of rotating crops on a particular field each year in order to minimize nutrient depletion. Because the foliage and root system of some crops are more effective in reducing soil erosion than others, rotating crops on sloping fields can result in erosion being relatively high one year and low the next. Rather than rotating the entire crop on fields in rolling terrain, farmers today rotate in strips so the crop that is more effective in reducing soil erosion will help trap sediment moving down from the less effective crop. The result is an overall reduction in soil erosion compared to rotating entire fields.

No-till farming is a somewhat new technique where farmers leave the remains of the previous crop standing in their fields rather than tilling (plowing) the residue into the soil. In the past, fields were routinely tilled prior to planting to help control weeds and aerate the soil, but exposing the soil generates high erosion rates. From Figure 10.29 one can see that with no-till farming the new crop is simply planted between the rows of the previous year. The old crop residue then acts as a cover material, preventing raindrops from making direct impact with the ground and dislodging soil particles. The crop residue further reduces erosion by helping to hold the water in place, thereby increasing infiltration and minimizing the amount of water moving downslope as overland flow. Although no-till farming is highly effective in reducing both soil loss and sediment pollution, some farmers have been slow to adapt the practice because of the need to purchase new equipment and additional chemicals for controlling weeds. Note that because no-till farming requires additional chemicals, this technique represents a trade-off between two environmental problems, namely soil loss and pollution of streams from the runoff of agricultural chemicals (Chapter 15).

Grassed waterways are naturally low areas within a field where a farmer plants grass rather than crops. These low areas, also called *swales*, are where water from overland flow starts to collect, and therefore represent the very beginning of small channels within a drainage network. When farmers plant rows of crops across these swales, water flowing through the swales during overland-flow events will form gullies as shown in Figure 10.30A. Therefore, not only is a portion of the field no longer

FIGURE 10.29 Rather than plowing the remains of the previous crop into the soil, in no-till farming the old residue is left in place to help shield the soil from the direct impact of raindrops.

FIGURE 10.30 Overland flow naturally collects in the low areas or swales of a field, easily forming gullies (A) in the exposed soil, creating obstacles for machinery and reducing crop production. By planting grass in the swales rather than crops (B), grassed waterways prevent the formation of gullies.

A Ionia County, Michigan   **B** Missouri

FIGURE 10.31 Terracing allows agriculture to take place on steep hillsides that would otherwise be impossible due to severe soil erosion. Note how the rice fields shown here follow the contours of the land surface.

producing crops, but the gullies are now obstacles that farm machinery must go around. This, of course, makes it more difficult and time-consuming for farmers to work their fields. Farmers have learned that the small amount of extra production they get from growing crops in drainage swales is simply not worth all the problems it creates. Besides, the production is eventually lost in those areas anyway because of soil loss. Today, many farmers wisely choose simply to keep these areas covered with grass (Figure 10.30B) so that soil erosion and gully development is minimized.

Terracing is an ancient technique used in areas of steep terrain, whereby some type of retaining wall (Chapter 7) is constructed in order to create flat surfaces for growing crops (Figure 10.31). This technique has long been used in parts of Asia to grow rice on steep hillsides. Here a relatively high population density combined with limited natural farmland forced people to develop agriculture on rather steep slopes. Clearly, without terraces, sustaining agriculture on such slopes would be virtually impossible since the soil would be quickly lost. Note that in the case of rice, terraces also enable farmers to flood their fields during the critical planting season.

Stream buffers involve leaving strips of grass or forested areas between cultivated fields and stream channels (Figure 10.32), thereby minimizing the amount of sediment that can be carried into a drainage

Story County, Iowa

FIGURE 10.32 Stream buffer zones reduce sediment pollution by trapping sediment moving off fields before it can enter the drainage system. Most stream buffers consist of leaving a combination of grass strips and uncut forest lying adjacent to stream channels.

A Statesboro, Georgia

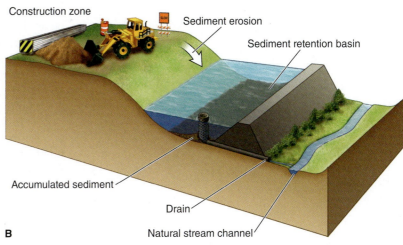

B

FIGURE 10.33 Silt fences (A) help prevent sediment pollution by keeping sediment from construction sites from entering a drainage network. Retention basins (B) are used to collect or trap sediment before it can enter a stream channel.

network. In the case of logging operations, stream buffers are created by leaving uncut strips adjacent to the stream network. In agricultural areas, farmers often plant grass along the edge of their field to form a buffer. Both types of stream buffer are effective in reducing the amount of sediment entering a drainage network, which, in turn, minimizes sediment pollution and its negative impacts on stream ecology and flooding (Chapter 8). Note, however, that because stream buffers do not prevent the downslope movement of soil on a field, this technique does not address the problem of soil loss.

Silt fences are another means of preventing sediment from entering a drainage system and causing sediment pollution problems. However, because silt fences are not designed to be permanent, they are primarily used around construction sites where land disturbances are temporary. Silt fences are perhaps most commonly seen along highways (Figure 10.33A) during construction projects where they are used to keep sediment from filling nearby drainage channels. Because of the 1972 Clean Water Act (Chapter 15), construction projects in the United States are now required to install silt fences to help prevent sediment pollution. Although developers and contractors are generally not fond of having to incur the costs associated with installing silt fences, the costs are minor compared to what taxpayers must bear in terms of increased flooding and destruction of stream ecosystems. Silt fences are an example of how environmental regulations are often perceived as being a burden on society, but in reality help protect people and their property from damage caused by someone else's use of the land.

Retention basins are human-made depressions excavated within a drainage system and designed to store excess water and/or sediment (Figure 10.33B). Although most retention basins serve as a flood-control measure (Chapter 8), they are sometimes built with the dual purpose of trapping sediment in order to prevent sediment pollution. Retention basins are typically located within residential developments and adjacent to large parking lots, both of which have sizable areas where impermeable asphalt and concrete cover the surface. Because the first phase of residential and commercial development projects typically involves the drainage system, retention basins are already in place prior to the major construction phase when erosion rates are highest. As retention basins accumulate sediment, heavy equipment can then periodically be used to remove the material and haul it away.

Slope vegetation cover is a simple and effective way to reduce sediment pollution. Many construction projects involve excavating and regrading land surface, leaving the soil exposed and highly vulnerable to erosion. Covering the sloped areas with vegetation immediately following the regrading is critical to prevent the loss of valuable topsoil. If the topsoil is lost, then it becomes very difficult for vegetation to take hold, which, in turn, allows for continued erosion and the eventual development of deep gullies. As described in Chapter 7, there are several techniques used to keep soil in place while permanent vegetation takes hold, including hydroseeding and a variety of synthetic meshes and fabrics.

Salinization of Soils

Recall that all water within the subsurface contains dissolved ions due to the chemical weathering of minerals. During the summer months when high evaporation rates cause soils to lose moisture, the dissolved ions within the soil zone can combine and form new minerals, commonly called salts. In more humid climates where freshwater is flushed through the soil zone fairly regularly, the mineral salts are eventually redissolved and carried downward through the soil profile and into the groundwater system (Chapter 11). In arid climates where the flushing of freshwater is much more infrequent, the mineral salts are able to accumulate in the soil zone. However, even in arid climates there is usually enough rainfall to flush the salts out of the uppermost soil zone. The result is a soil profile consisting of topsoil with relatively few mineral salts overlying horizons that are relatively salt rich—soil scientists call these aridisols (see Table 10.1).

FIGURE 10.34 Irrigating poorly drained desert soils that naturally contain mineral salts often leads to salinization. Poor drainage allows the irrigated water to accumulate and dissolve the salt minerals. The resulting saline water then moves up into the root zone, reducing crop production.

Cache Valley, Utah

Healthy vegetation

Non-irrigated conditions

Topsoil

Dry salt-rich horizon

Stressed vegetation

Irrigation water

Saline water

Capillary action

Poor subsurface drainage

Deserts have historically been areas with little agricultural activity due to a lack of water. Today, modern irrigation practices have allowed agriculture to flourish on certain desert soils that would otherwise be unproductive. Agriculture in some arid zones can be highly profitable in part because the climate allows crops to be grown year-round. For example, the combination of fertile soils, climate, and irrigation have made the Great Central Valley in California one of the most productive agricultural regions in the world. However, heavy irrigation on poorly drained desert soils can also lead to **salinization,** a process in which the salinity of soil water increases to the point that plant growth is reduced. As illustrated in Figure 10.34, the presence of a horizon with low permeability can prevent the downward movement of water through the soil profile. As the water accumulates it dissolves some salts, making the soil water highly saline. Eventually, this saline water moves up toward the surface via capillary (i.e., wicking) action into the root zone of the plants, reducing their ability to grow. To reduce the effects of salinization, farmers can install drainage systems within their fields (consisting of perforated pipes) so that water can flush through the soil zone, thereby preventing the buildup of saline water. Another strategy is to more closely monitor soil moisture conditions in order to avoid overirrigating the fields so that the water level does not rise into the root zone.

Soils with Hardpans

Some soil profiles contain what soil scientists refer to as a **hardpan,** which is a layer whose physical characteristics limit the ability of either roots or water to penetrate into the soil (Figure 10.35). Hardpans commonly consist of dense accumulations of clay minerals, or layers where the soil particles have been cemented together by minerals, such as calcite or iron oxides that have precipitated from water within the soil. In some cases a hardpan will develop after repeated use of heavy farm equipment causes compaction of clay-rich soils. As noted earlier, farmers try to avoid creating a hardpan, sometimes called a *plowpan,* by staying out of their fields when they are wet and most susceptible to compaction.

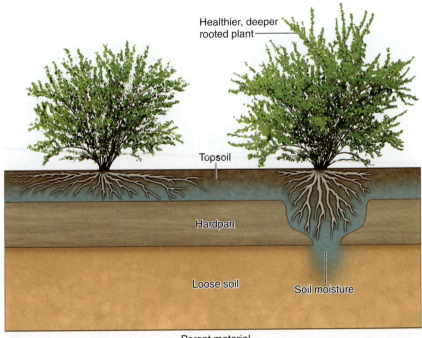

Healthier, deeper rooted plant

Topsoil

Hardpan

Loose soil

Soil moisture

Parent material

FIGURE 10.35 A hardpan is a soil layer rich in clay or cemented together by minerals, making it difficult for roots or water to penetrate the soil. Plants are forced to have shallow root systems, making them more susceptible to wilting during droughts, and drowning during wet periods. When planting trees or shrubs it is best to dig through the hardpan to provide better drainage and room for root growth.

Hardpans are a problem in agricultural areas because they inhibit the movement of water and air in the soil and limit the growth of plant roots. Because a hardpan forces plants to develop shallow root systems, plants are more susceptible to wilting during dry periods when soil moisture becomes depleted. Moreover, the lack of drainage can cause the upper soil zone to become saturated during wet periods, causing plants to drown due to the lack of oxygen in the soil. To remedy the problems, farmers are sometimes able to break up a hardpan through deep plowing. However, for a homeowner wishing to plant a tree or shrub, it is best to dig out a bowl-shaped depression within the hardpan, making it large enough to accommodate the root system when the plant is fully mature (Figure 10.35). It is also important to penetrate the entire hardpan to prevent the depression from filling with water and drowning the tree or shrub. Finally, note that the impermeable nature of hardpans is actually desirable for growing rice since it allows farmers to maintain standing water in their fields, a condition necessary for planting young rice seedlings.

Permafrost

In higher latitudes it is quite common for a soil profile to contain a **permafrost** layer, which is a horizon that remains frozen throughout most or all of the year—the frozen zone can extend downward as much as several thousand feet before encountering a sufficient amount of Earth's internal heat. As shown earlier in Figure 10.13 and listed in Table 10.1, permafrost layer is characteristic of soils known as gelisols, which cover a considerable portion of Earth's land area. Depending largely on latitude, the entire soil zone remains frozen year-round in some regions, but in other areas the uppermost soil layers will thaw during the summer period. Human activity, such as the construction of roads and buildings, can alter the heat balance of the soil zone, thereby causing the permafrost layers to thaw. When this occurs, subsurface materials can begin to compact and flow, damaging buildings and infrastructure (Figure 10.36). Of much greater concern is the impact of global warming. As the planet continues to warm, scientists expect that permafrost will begin to thaw over vast stretches of the Arctic, releasing great quantities of greenhouse gases. We will discuss this topic in more detail in the chapter on global change (Chapter 16).

FIGURE 10.36 Construction of this highway in Alaska caused a portion of the underlying permafrost to thaw, creating a subsurface void that then collapsed into a sinkhole. Damage caused by human-induced melting of permafrost is common in polar regions.

SUMMARY POINTS

1. Soils are natural mixtures of organic matter and rock and mineral fragments that are capable of supporting life.
2. Soils originate from the physical and chemical weathering of solid rock, which generates finer particles of rock and minerals. Most soil particles are composed of quartz and various clay minerals.
3. Soil horizons (layers) develop when organic matter is incorporated in the uppermost horizon and infiltrating water carries clay-mineral particles and dissolved ions downward where they accumulate in subsurface layers.
4. The color of soil horizons is an indicator of organic content and drainage characteristics. Organic matter gives topsoil a blackish and brownish color, whereas iron oxides impart a yellowish and reddish color, indicating a well-drained soil.
5. Five soil-forming factors determine the type of soil horizons that will develop for a particular soil: parent material, organisms, climate, topography, and time.
6. Soil properties are largely a function of water content and the relative proportions of clay, sand, and silt-sized particles. The small size and electrical charge of clay minerals play a key role in determining properties such as drought resistance, permeability, plasticity, strength and sensitivity, compressibility, and shrink-swell.
7. Soil fertility depends upon the ion exchange capacity of the soil and the availability of essential nutrients. The decomposition minerals and organic matter in the soil produce important nutrients, and clay minerals and organic matter help store these nutrients for later use by plants.
8. Soil loss occurs when soil erosion becomes greater than soil formation and is triggered when vegetation cover is removed from the landscape. The loss of topsoil is a concern because it ultimately reduces worldwide food production.
9. Excess soil erosion also causes sediment pollution of streams, which increases flooding and destroys the natural ecosystems of streams.
10. Soil loss and sediment pollution can be minimized by a variety of techniques that keep soil from moving downslope or by preventing what does move from entering drainage channels.
11. Irrigation on poorly drained desert soils containing horizons rich in mineral salts can lead to salinization, causing decreased crop production.
12. Hardpan soil layers restrict the ability of plant roots and water to penetrate the soil, making crops more susceptible to both drought and excessive soil moisture.

KEY WORDS

compressibility 307
essential nutrients 310
expanding clays 308
fertilizers 312
hardpan 321
ion exchange 310

paleosol 301
parent material 298
permafrost 322
permeability 306
plasticity 306
porosity 304

salinization 321
sensitivity 307
soil 294
soil erosion 314
soil fertility 310
soil horizons 295

soil loss 314
strength 306
swelling clays 308

APPLICATIONS

Student Activity   Grab a handful of soil. What color is it? Was there anything growing in it? Does it feel sandy? When you roll it together in your hands, does it stick together in a rope? If you can, then there is a lot of clay in it. If it feels sandy, then there is a lot of sand in it.

Critical Thinking Questions
1. Does the color of the soil have anything to do with the organic content of it?
2. Why are plants and animals important to soil formation?
3. Why is soil considered a resource?
4. How do soil horizons develop?

Your Environment: YOU Decide   From reading this chapter, you have learned how soil is formed. Do you think that soil itself can be considered living?

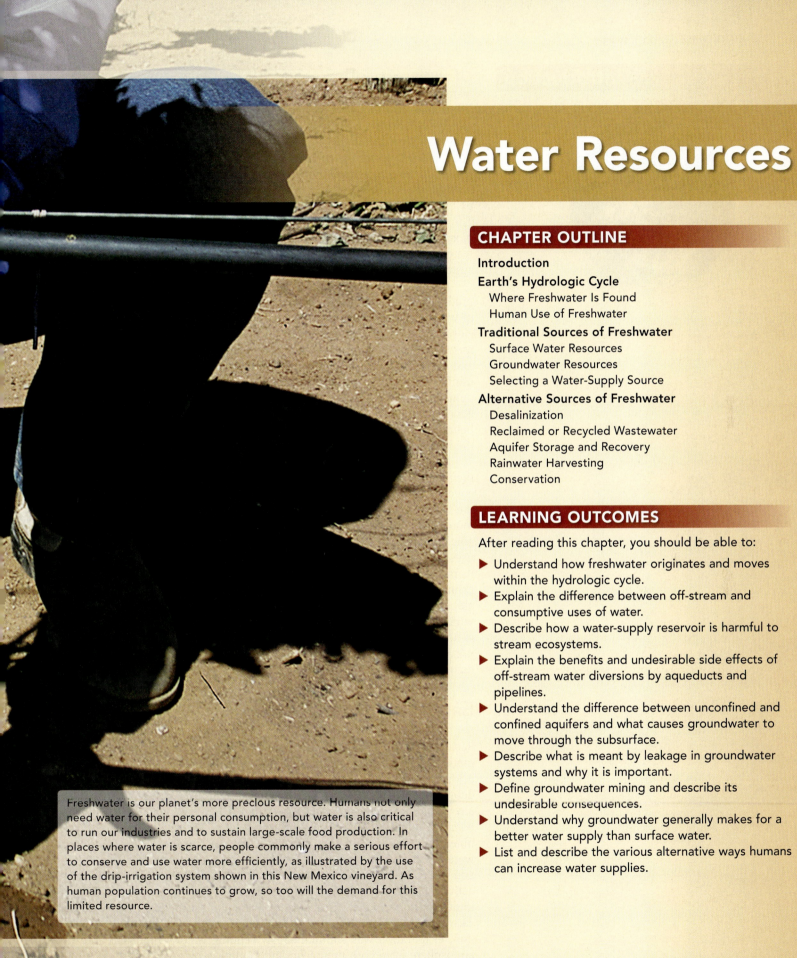

Water Resources

LEARNING OUTCOMES

After reading this chapter, you should be able to:

▶ Understand how freshwater originates and moves within the hydrologic cycle.
▶ Explain the difference between off-stream and consumptive uses of water.
▶ Describe how a water-supply reservoir is harmful to stream ecosystems.
▶ Explain the benefits and undesirable side effects of off-stream water diversions by aqueducts and pipelines.
▶ Understand the difference between unconfined and confined aquifers and what causes groundwater to move through the subsurface.
▶ Describe what is meant by leakage in groundwater systems and why it is important.
▶ Define groundwater mining and describe its undesirable consequences.
▶ Understand why groundwater generally makes for a better water supply than surface water.
▶ List and describe the various alternative ways humans can increase water supplies.

Freshwater is our planet's more preclous resource. Humans not only need water for their personal consumption, but water is also critical to run our industries and to sustain large-scale food production. In places where water is scarce, people commonly make a serious effort to conserve and use water more efficiently, as illustrated by the use of the drip-irrigation system shown in this New Mexico vineyard. As human population continues to grow, so too will the demand for this limited resource.

Introduction

Of all the natural resources, the most important to people are air, water, and food. In terms of human survival an individual can live only a few minutes without oxygen, one to two weeks without water, and approximately a month without food. Although slightly less critical than oxygen, water is essential for the body to function properly (the average human body is 70% water by weight). Interestingly, Earth's water is directly related to our oxygen and food supply. The basis for our food supply is land plants, which depend on rainfall to grow as well as the soil that forms by chemical weathering reactions between water and rocks (Chapter 10). The oxygen we breathe is a by-product of plant photosynthesis. The entire biosphere then, including humans, operates as a system in which liquid water is the key that links together the various components of the Earth system (Chapter 1).

Another key aspect of the Earth system is how global climate patterns produce uneven amounts of precipitation. Because water is critical to life as we know it, the abundance and diversity of species within the biosphere closely corresponds to the distribution of rainfall. Likewise, human settlement patterns have historically followed the availability of water. Hunters and gatherers eventually formed permanent settlements in areas where rainfall was sufficient to support agriculture. This, in turn, led to the development of more complex societies where specialized workers produced goods and services for other members of society. Rivers were then soon being used for water supply and transporting goods and people deep into the interior of continents.

In modern societies water is used for a wide array of activities, from the manufacturing and processing of consumer goods to the production of concrete and vast amounts of electrical power. Individuals use water in their homes to take showers, wash clothes, flush toilets, landscape, and more. The modern use of water has certainly made our lives more comfortable and pleasant. However, human population growth coupled with more uses for water is pushing the available water supply in many regions past its natural limit. At the same time our supplies are being stretched, climate change (Chapter 16) is threatening to exacerbate the problem by altering global precipitation patterns. Some societies will likely find themselves without adequate supplies of water, severely restricting future growth, and in some cases, threatening their very existence. Because water is unquestionably our most precious natural resource, we will begin by examining its origin and distribution.

Earth's Hydrologic Cycle

Recall from Chapter 2 that the origin of Earth's water is related to the formation of the solar system. There is strong evidence that Earth experienced a Mars-sized impact early in the solar system's history, sending ejected debris into Earth's orbit that eventually formed the Moon. The evidence also suggests that this impact caused the Earth to melt, and the subsequent volcanic activity released large volumes of gas. These volcanic gases combined with countless comet impacts to form a dense atmosphere, rich in water vapor. As the planet cooled, water condensed from the atmosphere and fell as precipitation, resulting in an ocean or layer of water. Tectonic activity later caused landmasses to rise above sea level, which allowed rainwater to begin flowing across the young landscape and into rivers that carried the water back to the ocean. While water was falling as precipitation, evaporation was simultaneously sending water vapor back into the atmosphere from both the ocean and landmasses, thereby creating a continual and cyclic transfer of water between the ocean, atmosphere, and land.

Today scientists refer to the cyclic movement of water within the Earth system as the **hydrologic cycle.** As illustrated in Figure 11.1, the hydrologic cycle is driven by solar radiation that causes water to evaporate from the oceans and land surface. The water vapor rises into the atmosphere, where it cools and eventually falls to the surface as precipitation (rain, snow, sleet, or hail). Some of the rain and melting snow and ice will move downslope and into stream channels and return directly back to the ocean. Some of the precipitation also infiltrates permeable sediment and rocks and then enters the groundwater system, eventually discharging into a stream, lake, or ocean. However, a fraction of the infiltrating water is removed from the soil zone by evaporation and plant transpiration and then returned to the atmosphere. Note that in Chapter 8 our focus was on the periodic flooding that takes place as streams transport water back to oceans. The emphasis in this chapter is on how humans make use of precipitation that moves through the hydrologic cycle.

An important aspect of the hydrologic cycle is how water varies in terms of its *salinity,* which is the amount of electrically charged atoms called dissolved ions, or salts, within water (Chapters 3 and 10). Low-salinity water with very few dissolved ions is referred to as **freshwater,** whereas highly saline water is called *salt water.* Freshwater is produced whenever liquid water undergoes evaporation, whereby individual water molecules (H_2O) move into the vapor state but the dissolved ions remain with the original liquid. As indicated in Figure 11.1, the majority of Earth's freshwater is produced by the evaporation of salt water in the oceans. Notice how some of the freshwater precipitation falls on the landmasses where it eventually flows across the landscape or infiltrates the subsurface. Here the freshwater chemically reacts with minerals, which release various ions that act to increase salinity.

The production of freshwater within the hydrologic cycle, of course, is critical to Earth's terrestrial or land-based life-forms, including humans. These organisms have evolved in the presence of freshwater over long periods of geologic time and cannot survive on salt water. In fact, every terrestrial organism has a certain limit or tolerance with respect to the salinity of the water it can safely consume. In the United States the maximum salinity level for human drinking water is set at 500 milligrams of salt per liter of water (mg/l). Chickens, on the other hand, can tolerate water up to 3,000 mg/l and cattle up to 7,000 mg/l. For comparison, seawater averages around 35,000 mg/l of dissolved salts, which is fatal if consumed by any terrestrial organism in sufficient quantity.

In addition to the terrestrial environment, much of the marine ecosystem is dependent on the discharge of freshwater by rivers into coastal estuaries. Here the freshwater reduces the salinity of seawater, producing a zone of so-called *brackish* water. Some marine organisms depend on brackish water during certain periods of their life cycle, particularly reproductive and growth phases. In fact, coastal estuaries with their brackish water serve as critical nursery grounds for nearly 70% of all commercial fish and shellfish in the United States. Therefore, not only does the terrestrial biosphere depend on freshwater, but a significant portion of the marine ecosystem does as well.

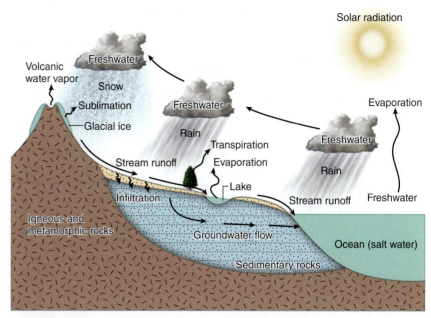

FIGURE 11.1 The hydrologic cycle describes the cyclic movement of water through the Earth system. The cycle is driven by solar energy that causes water to evaporate from the oceans and land surface and allows for plant transpiration. Evaporation of seawater produces large quantities of freshwater, which is vital for humans and the terrestrial biosphere. Humans' primary sources of freshwater are streams and lakes and groundwater systems. Note that groundwater occurs in fractured igneous and metamorphic rocks, but the vast majority is contained in porous sedimentary material.

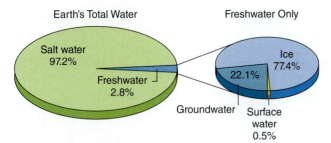

FIGURE 11.2 Breakdown of water within Earth's hydrosphere.

Where Freshwater Is Found

Earth is often called the "blue planet" because of its abundance of water, but as illustrated in Figure 11.2, only 2.8% of the planet's total supply consists of freshwater. If we examine the breakdown of this relatively small amount of freshwater, we see that 77.4% is stored as glacial ice and 22.1% as *groundwater* or subsurface water. This leaves about 0.5% as *surface water*, which is the term hydrologists use for the water in rivers, lakes, and soils. To give you a better sense of the scale involved, examine the detailed breakdown of Earth's water resources listed in Table 11.1. Most striking perhaps is how the rivers in the world account for a mere 3/1,000th of a percent of all freshwater. Although rivers and other forms of surface water represent a tiny fraction of the planet's freshwater, surface water is what makes life possible for humans and other terrestrial organisms. Later in Chapter 11 we will explore how society has expanded its supply of freshwater by learning how to extract groundwater, a much larger store of freshwater (Figures 11.1 and 11.2)—note that despite being considered fresh, groundwater is often too saline for direct human consumption.

In addition to being rather small in terms of overall volume, Earth's freshwater supply is not evenly distributed across the landmasses because of variations in precipitation and subsurface geology. From the map in Figure 11.3 one can see how average annual precipitation varies widely across the United States. As with all large landmasses, the variation in precipitation is related to the circulation of humid air masses within the global climate system and their interaction with topographic features, such as mountain ranges. Freshwater supplies are naturally higher in areas with greater amounts of precipitation because there is more water available in soils, rivers, and lakes. Keep in mind, however, that since precipitation varies seasonally, the amount of available surface water varies throughout the year.

Groundwater systems typically provide large volumes of freshwater in areas of abundant precipitation, provided that there are layers of permeable materials within the subsurface. Here infiltrating water fills the pore spaces and fractures within subsurface materials, which can later be extracted by drilling or digging a well. Because groundwater systems are replenished by infiltration, the water being withdrawn today in many desert regions represents ancient precipitation that entered the subsurface under climatic conditions that were much more humid. This brings up the fact that Earth's climate system, and corresponding precipitation patterns, change over time scales ranging from decades to thousands of years. Therefore, the surface

TABLE 11.1 Distribution and occurrence of Earth's water.

	Volume (Cubic Miles)	Volume (% of All Water)	Volume (% of Freshwater)
Oceans and Inland Seas	317,000,000	97.22	
Ice Caps and Glaciers	7,000,000	2.15	77.35
Groundwater	2,000,000	0.61	22.10
Fresh Lakes	30,000	0.009	0.33
Soil Moisture	16,000	0.005	0.18
Atmosphere	3,100	0.001	0.03
Rivers	300	0.0001	0.003
TOTAL	326,049,400		

Source: Data from the U.S. Geological Survey.

Mean Annual Precipitation (1971–2000)

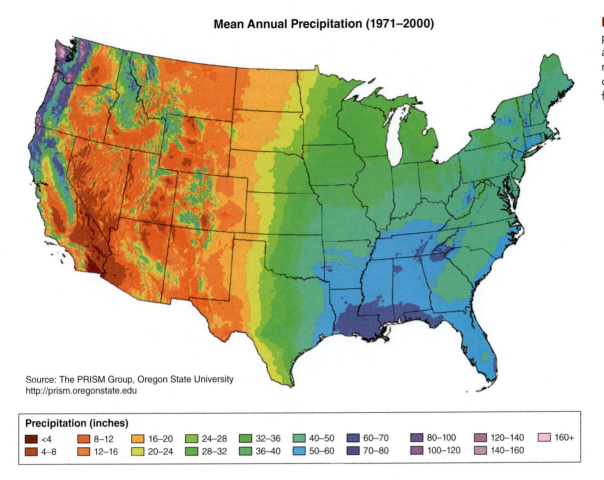

Source: The PRISM Group, Oregon State University
http://prism.oregonstate.edu

Precipitation (inches)

■ <4	■ 8–12	■ 16–20
■ 4–8	■ 12–16	■ 20–24

■ 24–28	■ 32–36	■ 40–50	■ 60–70	■ 80–100	■ 120–140	■ 160+
■ 28–32	■ 36–40	■ 50–60	■ 70–80	■ 100–120	■ 140–160	

FIGURE 11.3 Average annual precipitation varies widely across the United States, resulting in significant differences in the availability of freshwater supplies.

and subsurface supplies of freshwater that we depend on today are not constant, but continuously change over geologic time. Naturally this can have serious implications as much of Earth's human population is located in areas where freshwater is currently abundant. Later in Chapter 16 you will learn that one of most serious consequences of global warming is the likelihood that precipitation patterns will change, causing a redistribution of freshwater supplies around the world.

Human Use of Freshwater

The use of freshwater by humans greatly increased after the Industrial Revolution when societies began requiring water for activities such as manufacturing, energy production, and flushing toilets. Demand for water also progressively increased as agricultural irrigation became more widespread. These changes are reflected in Figure 11.4 which shows the freshwater usage in the United States from 1950 to 2000. Perhaps the most obvious feature is the dramatic increase in total freshwater withdrawals in the period leading up to 1980, with most of the increase being associated with irrigation and electrical production—here water is primarily used to make steam and to cool mechanical systems. Total withdrawals after 1980 then declined and remained steady due to the more efficient use of water across nearly all sectors of society. If we examined the data from 2000 in terms of percentages, we would find that public water supply, often called *municipal* or *city water,* accounted for only 11% of all freshwater withdrawals. This amount is rather small compared to the 48% and 34% that is withdrawn for electrical generation and irrigation, respectively. Note that

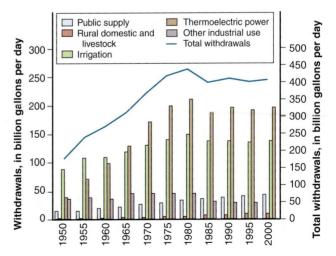

FIGURE 11.4 Freshwater withdrawals in the United States between 1950 and 2000.

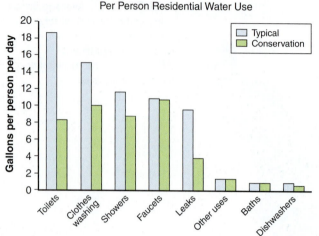

FIGURE 11.5   Breakdown of the personal rate of water usage inside the average U.S. household.

in the United States, surface-water sources make up nearly 75% of the nation's freshwater supply, whereas only 25% comes from groundwater.

Statistics on total water withdrawals are useful, but it is also interesting to examine the rate at which a society uses water. Here it is customary to take the withdrawal volume and divide it by the population to get the *per capita* (per person) consumption rate, thereby eliminating the effect of population growth on the data. In terms of per capita consumption in the United States, the average citizen uses about 1,200 gallons (4,542 liters) of water per day. However, because this represents the water being used to support all aspects of society, it does not provide a good measure of the *personal* consumption rate in our homes. For example, the personal consumption rate for a typical American averages about 69 gallons per day (262 liters/day), which is far less than the per capita rate of 1,200 gallons per day that includes industrial and irrigation use. From Figure 11.5 we can see that personal indoor water use in the United States is dominated by flushing toilets (27%), followed by clothes washing (22%). Notice that when leaks are fixed and more efficient appliances are installed, especially toilets, the daily use of water in the home can be reduced by approximately 35%.

In addition to the total volume and rate of water usage, hydrologists also examine how much is lost and whether it returns to its origin or some other source. The term **consumptive water use** refers to those activities where water is lost or consumed. Examples include irrigating crops because water is lost due to evaporation and plant transpiration. Washing dishes, on the other hand, is a *nonconsumptive* use since nearly all of the water can be returned to a water supply source. The term **off-stream use** describes those situations where water is removed from one supply source, but then returned to a different source after being used. An example is where groundwater is being withdrawn for a public water system, but later discharged into another supply source, such as a stream. In contrast, an *in-stream use* would be if a city removed water from a river, but then discharged its wastewater back into the same river.

Consumptive and off-stream uses are important because they can lead to serious problems in society's sources of freshwater. In the case of a river, substantial water losses may reduce discharge to the point where ecosystems begin to collapse and downstream users are no longer able to withdraw adequate supplies. Because of the nature of subsurface withdrawals, most groundwater is used in an off-stream manner, which almost invariably has a negative impact on groundwater systems.

Traditional Sources of Freshwater

In this section we will explore society's traditional sources of freshwater and how increased human population, combined with higher per capita consumption rates, is pushing some natural systems beyond their ability to supply water. This unsustainable use of water is not only causing harm to the very systems that supply us water, but it is beginning to limit the growth and development of population centers around the world. Compounding the problem is the fact that precipitation patterns are changing in response to global climate warming (Chapter 16). Because of climate change, areas that have been pushing the limits of their water supply may soon find themselves without adequate freshwater supplies. When faced with a water shortage, government officials are generally left with the following options: (1) increase storage capacity by building reservoirs; (2) transport water from another source via pipelines and aqueducts;

(3) institute conservation programs; (4) develop nontraditional sources (e.g., desalinized seawater); or (5) voluntarily limit growth. We will examine these options as we take a closer look at the different sources of freshwater, beginning with rivers and lakes. Note that stream processes were described in considerable detail in Chapter 8; hence, only a brief overview of the subject will be provided here.

Surface Water Resources

Rivers and Streams

Recall that rivers and streams collect freshwater that falls on Earth's landmasses as precipitation, and then transport most of this water to the oceans. The remainder returns to the atmosphere by evaporation and plant transpiration or infiltrates and enters the groundwater system. Here the term *water table* refers to the depth where subsurface materials are completely saturated. In more humid climates where rainfall is plentiful, the water table is normally higher than the nearby stream channels, which causes infiltrating water to eventually flow into streams as *groundwater baseflow* (Chapter 8). Consequently, streams in humid areas not only receive water from periodic rainfall events, but also receive a fairly continuous supply of groundwater. The contribution of groundwater is very important because it allows streams to keep flowing even during prolonged dry periods. In more arid climates where water tables are relatively deep, streambeds may go dry during the summer. Note that some rivers are able to flow year-round across desert landscapes because much of the water comes from melting snow or summer rains in distant parts of the drainage basin. A good example is how snowmelt in the Rocky Mountains helps sustain the Colorado River as it flows across the desert Southwest region of the United States.

In terms of water supply for human activity, humid environments naturally have a greater number of streams that can reliably provide water on a year-round basis than do arid landscapes. This is why human settlements in arid climates have historically been restricted to narrow corridors along major rivers—such as the Nile flowing through the desert of Egypt and Sudan and the Tigris and Euphrates of Mesopotamia (modern Iraq). In the humid regions of Europe and North America, settlements were less restricted geographically because greater precipitation and higher water tables produced a large number of rivers and streams that flow year-round.

Although streams serve as reliable sources of freshwater, human population growth can result in the demand being greater than the streamflow, particularly during summer months when flow is naturally low. Rather than limiting growth and development, societies have often been able to increase supply by diverting water from distant sources via aqueducts and canals. Such diversions would be considered an off-stream use. Perhaps the most famous example is the system of aqueducts the Romans built throughout their empire (Figure 11.6A). This aqueduct system not only increased the supply in places it was needed, but the elevated nature of the system allowed Roman cities to have running water and sewers. A modern example is the California aqueduct system (Figure 11.6B), where massive amounts of water are transferred to the Los Angles basin. This large-scale diversion of water has allowed Los Angeles to grow into a great metropolis of over 17 million, which is far beyond what the local water supply could have supported in this arid environment.

Whenever water is removed from a stream or river, the impact on natural ecosystems will depend on the volume that is removed and the quantity and quality of what is returned. In the case of large off-stream withdrawals where the water is not returned, such as aqueduct systems or where water

A

B

FIGURE 11.6   Roman aqueducts (A) were off-stream diversion projects that allowed areas with inadequate water supplies to flourish. The modern aqueduct system serving the greater Los Angeles, San Diego and San Francisco areas (B) provides water to millions of residents living in an arid environment.

is used for irrigation, stream discharge can be reduced to the point where entire ecosystems collapse. Reduced flows also mean that other users in the drainage basin where the water is being removed are unlikely to be able to expand their supply, limiting future economic growth and development.

With respect to large in-stream withdrawals, the impact on the natural environment largely depends on the quality of the water that is returned to a river or stream. For example, municipal (public) supply systems typically collect the bulk of their wastewater from homes and businesses, and then process the water at a sewage treatment plant before discharging it back into a river (Chapter 15). Although this process is effective in removing harmful bacteria, nutrients such as nitrogen are not removed and can negatively impact a stream's water quality. Another important in-stream use of water involves electrical power plants, which are typically located along rivers because of the need for large volumes of water. Here the issue of water quality involves the discharge of warm water into the river which causes thermal contamination. To minimize this problem, plant operators are now required to recycle water within the plant and to lower its temperature before discharging it back into the river.

Lakes and Reservoirs

Although a natural lake can be an excellent source of freshwater, they are relatively rare features in many areas because they are generally restricted to geologically young landscapes. For example, the Great Lakes and surrounding countless smaller lakes all formed after the last glacial ice sheet began its final retreat about 12,000 years ago. As the ice retreated, lakes began to form because there was no established network of stream channels for carrying surface water off the newly exposed landscape. Similar to all lakes, those in the Great Lakes region will slowly fill in with sediment and organic matter as the terrain and drainage system matures. In addition to glaciated regions, lakes are common in areas of recent tectonic uplift and where streams have been blocked by debris from mass wasting events (Chapter 7).

Perhaps the most common technique for creating a large and reliable source of freshwater is the damming of rivers to create artificial lakes called *reservoirs*. As shown in Figure 11.7, a dam can stockpile an enormous amount of water simply by capturing part of a river's flow during periods of elevated discharge—the volume of water in some reservoirs is greater than the river's total annual flow. By creating a surplus store of water, a reservoir can be used to support additional population growth and irrigation of agricultural land. In fact, some of the Colorado River water in the reservoir shown in Figure 11.7 is sent to Los Angles via the aqueduct system described earlier. Note that the primary purpose of many modern dams is not just to increase water supplies, but also as a means of flood control (Chapter 8) and generating clean hydroelectric power (Chapter 14). Another important benefit is the recreational use of the reservoir for boating and fishing.

Although there are many benefits associated with dams, they also have negative consequences for both society and the natural environment. For example, when a reservoir fills with water and drowns a river valley, it almost invariably results in the displacement of people, businesses, and agricultural activity. In terms of the natural environment, much greater evaporative losses of water take place since the surface area of a reservoir is much greater than the section of the river it replaces. Other problems occur when a once free-flowing river encounters the still waters of a reservoir. Here the sediment being transported by the river is deposited behind the dam. All reservoirs therefore have a certain life expectancy as they will

FIGURE 11.7 In addition to creating a large reservoir of freshwater, modern dams are also built to control flooding and for generating electricity. Shown here is Lake Powell behind the Glen Canyon Dam on the Colorado River in Arizona.

eventually fill in with sediment. Also, the water temperature decreases significantly due to the depth of the reservoir. The water, therefore, that is being released from the dam is both sediment-free and considerably cooler compared to the natural system. This can be devastating to downstream ecosystems that have evolved in the presence of warmer, more sediment-laden water. Another problem is that the volume of water that is released from a dam is highly regulated such that downstream areas no longer experience periods of exceptionally high and low streamflows. Because many aquatic ecosystems depend on large natural variations in streamflow, the controlled discharge of water from dams can cause serious disruptions in the aquatic food chain, particularly those in coastal estuaries where freshwater mixes with seawater.

Groundwater Resources

In addition to surface water, the other major source of freshwater available to society is **groundwater,** which is defined as water that resides within the void or *pore space* of subsurface materials. There are three major factors that determine whether groundwater is a viable source for a given area: (1) the quantity that exists; (2) the ease at which it can be withdrawn; and (3) the quality of the water. When these three conditions are favorable, groundwater will often make a better source of freshwater than will a stream or lake. Keep in mind that the conditions that make groundwater a suitable source vary from user to user. For example, a small supply of groundwater may be fine for a single household, but be totally inadequate to meet the irrigation needs of an agri-

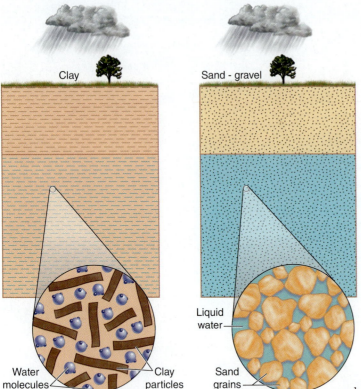

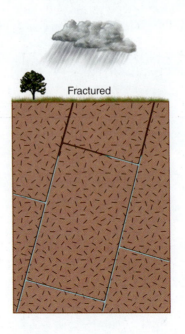

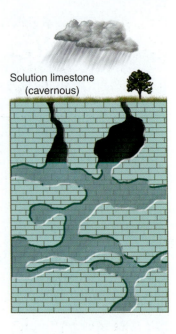

Clay

Water molecules

Clay particles

Sand - gravel

Liquid water

Sand grains

Fractured

Solution limestone (cavernous)

FIGURE 11.8 Porosity determines the volume of groundwater that can be held in subsurface materials. Sedimentary materials composed of clay, sand, or gravel-sized particles normally have high porosity, whereas crystalline rocks have little porosity and usually contain water only in fractures. Water moving through certain types of limestone can create high porosity by dissolving the rock to form passageways and caverns.

cultural user. Likewise, groundwater whose salinity makes it unfit for human consumption might be suitable for poultry and cattle.

From Figure 11.8 you can see that the volume of groundwater beneath a given area is directly related to the amount of void space within subsurface materials, a property scientists refer to as *porosity* (Chapter 10). Igneous and other types of rocks with a crystalline texture (interlocking mineral grains) have virtually no porosity since there is very little void space between the mineral grains. Consequently, areas underlain by crystalline rocks normally hold only small amounts of groundwater, and what does exist occurs in fractures. Large volumes of groundwater are generally found in areas with thick sequences of sediment or sedimentary rock. Here porosity can be quite high in granular material composed of sand, gravel, or clay-sized particles. Note that the porosity is highest in granular material that is unconsolidated, and then decreases when grains undergo compaction and become cemented together, forming sedimentary rock such as sandstone, shale, and limestone. Because limestone contains the soluble mineral calcite, groundwater circulating through fractures can dissolve away the rock and form large passageways and caverns. Areas underlain by such rocks, which geologist refer to as *solution* or *karst limestone,* often contain large volumes of water held in open channels or conduits (Figure 11.8).

Although groundwater volume is an important factor in terms of selecting a water supply, how easily it can be withdrawn is equally important. Recall from our discussion on soils (Chapter 10) that most sediment is composed of rock and quartz and clay-mineral fragments. Clay particles are also extremely small and have a strong electrical attraction to the dipolar water molecule, making it very difficult for water to move through clay-rich materials. Much coarser sand and gravel-sized sediment typically consist of quartz and rock fragments that are electrically neutral. Here the large pore size and lack of electrical attraction allows water to move rather freely through gravels and sands. The property known as **hydraulic conductivity** describes the ease with which a material will transmit a fluid—much like how electrical conductivity relates to the flow of electricity. Note that hydraulic conductivity takes into consideration the transmission properties of the material (permeability) as well as the properties of the fluid itself (density and viscosity). In Chapter 13 you will see how fluid and material properties affect our ability to extract oil and gas from the subsurface.

Geologists use the term **aquifer** to describe a layer of rock or sediment that readily *transmits* water; in other words, a material whose hydraulic conductivity is relatively high. Suitable aquifers therefore normally consist of layers of relatively coarse sediment (sand and gravel) or granular sedimentary rocks (sandstone and limestone). Fractured crystalline rocks can also serve as aquifers, but provide only limited amounts of water due to their low porosity. Conversely, materials with low hydraulic conductivity are considered **aquitards,** such as layers of clay-rich sediment, shale, and unfractured crystalline rocks, including certain types of dense limestone. Figure 11.9 illustrates how hydraulic conductivity varies greatly among the more common types of rocks and sediment. Note that each mark on the scale represents a tenfold increase in conductivity. From this you can see that gravel is about a hundred times more conductive than sand and a billion times more conductive than clay.

Types of Aquifers

Recall from Chapter 3 that sedimentary sequences generally consist of varying amounts of quartz, clay, and carbonate (e.g., calcite) minerals and rock fragments. Because there can be vast differences in the hydraulic conductivity among these layers, sedimentary sequences usually contain multiple aquifers and aquitards. From the example illustrated in Figure 11.10, one can see that an **unconfined aquifer** is a highly conductive layer that is open to the atmosphere and surface waters. In contrast, a **confined aquifer** is overlain by an aquitard that helps to seal it off from the surface environment. Note that aquitards are commonly called *confining layers* because of the way they restrict the vertical movement of water between different aquifers.

There are several important differences between confined and unconfined aquifers that we will examine using the wells shown in Figure 11.10. When a well is drilled into an unconfined aquifer, the hole will eventually reach the **water table,** which is the depth where all the pore spaces are completely filled with water. The area above the water table is referred to as the *unsaturated zone* because pores contain both air and water, and the area below is called the *saturated zone*. Notice how the water table is not necessarily flat, but mimics the topography of the land surface. On topographically high areas, infiltrating water has greater potential energy due to its elevation, thereby forcing water to flow through the unconfined aquifer. Here the groundwater will eventually discharge into the surface environment in topographically low areas such as lakes, wetlands, and free-flowing streams. Note that the water table will rise and fall depending on the amount of infiltrating water that makes its way into the groundwater system, a process called *groundwater recharge*.

The confined aquifers in Figure 11.10 behave quite differently because of the way the aquitards (confining layers) limit the ability of water to move between aquifers. This, in turn, allows water within confined aquifers to become pressurized—similar to how rubber allows a tire to be pressurized. Whenever a well is installed into a confined aquifer, the pressurized water will rise up into the well. Because it is a measure of the amount of potential energy within a confined aquifer, hydrologists use the term **potentiometric surface** to refer to the height water will rise above the aquifer. Note that groundwater within a confined aquifer flows from areas where the potentiometric surface is high to areas where it is lower.

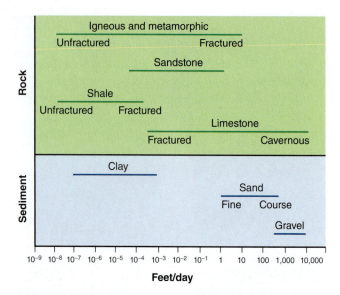

FIGURE 11.9 Graph showing the range of hydraulic conductivity for various geologic materials. Each mark on the scale represents a tenfold change in conductivity. Some materials have a wide range of hydraulic conductivity, and thus can be regarded as an aquifer in some instances and an aquitard in others.

FIGURE 11.10 An unconfined aquifer is open to the surface environment and has a water table that marks the top of the saturated zone. A confined aquifer has an overlying aquitard that limits the vertical movement of water, causing the aquifer to become pressurized. When a well penetrates an aquitard, water rises to the potentiometric surface, which represents the amount of pressure within a confined aquifer.

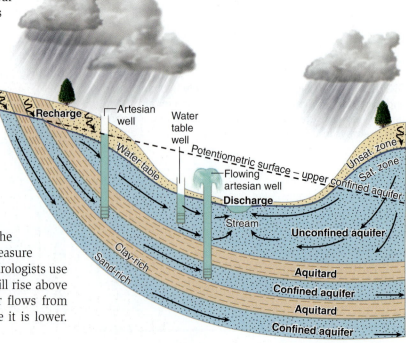

Also, the potentiometric surface has no relationship to the water table or the surface topography. Finally, notice in Figure 11.10 that wells in unconfined aquifers are known as *water-table wells,* and those in confined aquifers are called *artesian wells.* A free-flowing artesian well is one where water freely flows to the surface because the potentiometric surface is higher than the land surface (Figure 11.11).

Movement of Groundwater

Recall that groundwater always flows from areas of higher potential energy to areas of lower energy. Although the energy that causes groundwater flow can involve changes in density and temperature, flow in most situations is due to differences in elevation and water pressure. Because both elevation and pressure are affected by gravity, hydrologists measure the potential energy within an aquifer system in terms of **hydraulic head,** which in practical terms is the height of the water table or potentiometric surface. Figure 11.12 shows how a measuring tape is used to record the depth to the water level in a well. To get the hydraulic head (distance) above some datum such as sea level, the depth to water is subtracted from the elevation at the top of the well.

Measuring hydraulic head is important because it allows hydrologists to determine the direction groundwater is flowing in the subsurface. As shown in Figure 11.12, the **hydraulic gradient** is defined as the slope or steepness of a water table or potentiometric surface, and is calculated based on the difference in head and distance between any two points. Here the direction of groundwater flow is always from higher to lower head. Moreover, the steepness of the gradient indicates how much potential energy is available for driving groundwater flow. Groundwater velocity therefore would be expected to be higher in places where the gradient is steeper because more energy is available. However, porosity and hydraulic conductivity also play a role. Remember that groundwater flows more readily through high-conductivity layers, such as gravel, as opposed to a layer rich in clay minerals. On the other hand, higher values of porosity cause groundwater to move more slowly through granular material because of the greater proportion of pore space.

FIGURE 11.11   A free-flowing artesian well occurs when the potentiometric surface of a confined aquifer is higher than the land surface, allowing pressurized water to escape to the surface.

FIGURE 11.12   Illustration showing how hydraulic head and hydraulic gradient in an unconfined aquifer are determined using two water wells. Note that groundwater flow is always in the direction of the hydraulic gradient.

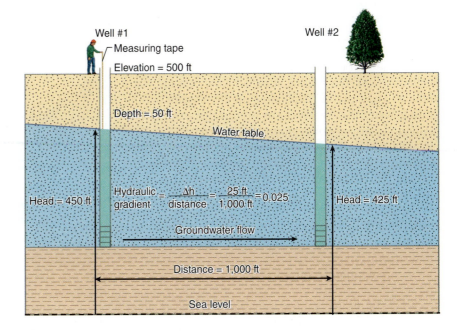

TABLE 11.2 Average groundwater velocities computed from Darcy's law for representative layers of sand and clay and under a typical hydraulic gradient. Travel time is based on horizontal flow through material of uniform composition.

	Hydraulic Conductivity	Porosity	Hydraulic Gradient	Average Groundwater Velocity	50-mile Travel Time
Sand	10 ft/day	0.25	0.025	1.0 ft/day	723 years
Clay	0.00005 ft/day	0.50	0.025	0.0000005 ft/day	1.4 billion years

In 1856, a French engineer named Henri Darcy experimentally derived a mathematical law that describes the relationship between hydraulic gradient and conductivity and the flow of groundwater through granular material. By including porosity, the following version of *Darcy's law* can be used to compute the average velocity of water flowing through a granular body.

$$\text{Groundwater velocity} = \frac{\text{Hydraulic conductivity}}{\text{Porosity}} \times \text{Hydraulic gradient}$$

Calculating the overall average groundwater velocity allows hydrologists to predict the amount of time for water to travel a given distance through porous material. For example, Table 11.2 shows the calculated velocities for water flowing through average sand and clay layers in response to a typical hydraulic gradient. Although the velocity of one foot (0.3 m) per day in the sand may seem slow, it is actually quite rapid in terms of geologic time. Most striking perhaps is just how much faster groundwater moves through the sand compared to the clay. Note that the vast difference in velocity is primarily due to the difference in hydraulic conductivity between these two materials—the effect of porosity is comparatively minor. We can also see that because of the exceedingly slow water velocity in clay, the travel time is very large.

It should be clear that clay and other types of aquitard materials act as barriers to water that naturally flows between different aquifers. Although they inhibit the flow of water, nearly all aquitards allow some water to pass in a process called **leakage,** as illustrated in Figure 11.13. Note that the actual amount of leakage depends on the aquitard's hydraulic conductivity and the difference in hydraulic head between the two aquifers.

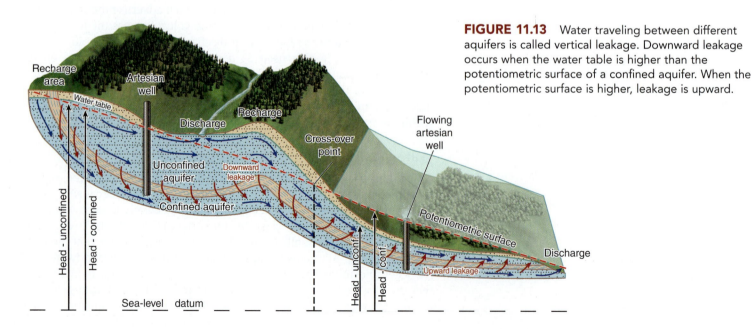

FIGURE 11.13 Water traveling between different aquifers is called vertical leakage. Downward leakage occurs when the water table is higher than the potentiometric surface of a confined aquifer. When the potentiometric surface is higher, leakage is upward.

Moreover, the direction of leakage can either be upward or downward depending on the hydraulic head within the aquifers. Notice in the figure how the relative position of the water table and the potentiometric surface switches at the crossover point. Above this point the hydraulic head in the unconfined aquifer is greater, which forces water to leak downward and recharge the confined aquifer. Below the crossover point, it is the confined aquifer that has the higher head, which means it is discharging toward the surface through upward leakage. Later you will learn how aquitards play a critical role in determining the movement and spread of contaminants in groundwater systems (Chapter 15) and in keeping oil and gas trapped within the subsurface (Chapter 13).

Springs

A **spring** is simply a place where groundwater discharges in a concentrated manner at the surface. As shown in Figure 11.14, springs can discharge directly on the land surface or through the base of a surface-water body (streams, lakes, and oceans). From a historical perspective springs gave humans unique access to groundwater long before the advent of modern water wells. Moreover, in desert climates a spring can create a natural *oasis* that is often the only available source of water. Also interesting is how society has used springs for different purposes, depending on the geology of the springs themselves. For example, springs that discharge from relatively shallow aquifers usually have low salinity because the water is fairly young and has not acquired many dissolved ions. The low salinity combined with the fact that groundwater contains very little sediment and bacteria means that shallow springs make exceptionally good sources of drinking water. However, such springs are also more prone to going dry during extended droughts than those that originate from a deeper source.

Springs that are fed by deep aquifers commonly have such high salinity levels that the water is unfit for human consumption. However, high-salinity springs, sometimes referred to as *mineral springs*, have long been a source of valuable salts for cooking and preserving food. In addition to salinity, the temperature of these springs is usually much warmer than those that originate from shallow aquifers due to Earth's internal heat (Chapter 4). Depending on the actual temperature, the term *hot spring* or *warm spring* is often used to describe saline spring water. Hot springs have long been used to fill pools at resort spas, which advertise the mineral-rich water as having therapeutic value. In some volcanic areas the temperature of the groundwater is so high that it discharges as steam or boiling water. Hot groundwater is also used as a source of clean geothermal energy (Chapter 14). Lastly, because the chemistry of spring water is often quite different from the surrounding surface water, springs generally create a unique environment that serves as a critical habitat for certain species within an ecosystem.

Water Wells and Drawdown Cones

Other than springs, the only access people had to groundwater prior to modern drilling equipment was through hand-dug wells. Because a person can safely dig a hole only so deep, wells were generally located where the water table was relatively close to the surface. Once the hole was completed, the well was usually lined with stones or bricks to prevent the walls from collapsing. All that was required then to extract the groundwater was a bucket and rope. The problem with shallow wells, even today, is that they are easily contaminated by chemicals and animal and human wastes. Hand-dug wells pose a special hazard because their diameter is large

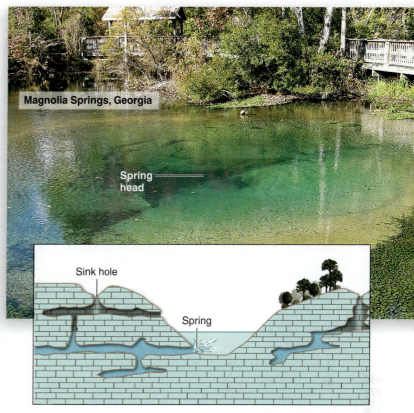

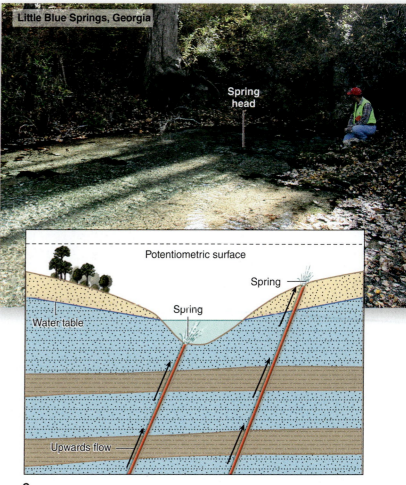

FIGURE 11.14 Springs occur when groundwater discharges at the surface in a localized area. The geology of the site determines the depth of the water source, which in turn influences the spring's salinity, temperature, and consistency of flow. The spring in (A) forms when water becomes trapped in the unsaturated zone, then flows laterally until it discharges along a hillside. In (B) the spring discharge is from a solution passageway in limestone. Example (C) illustrates how water from deeper aquifers can flow to the surface along faults or fractures.

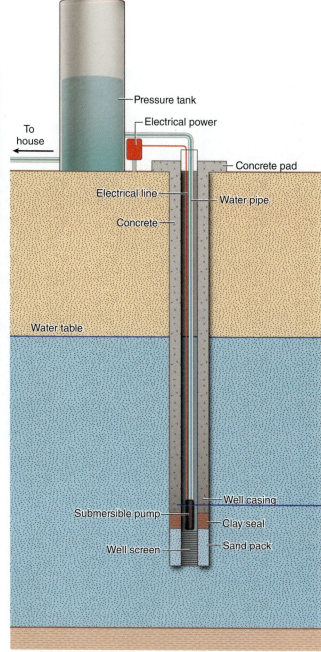

FIGURE 11.15 Diagram showing the construction of a modern water well. Note the clay seal and concrete placed around the well casing to prevent surface contaminants from washing into the well. Photo showing a completed well and pressure tank at the surface.

enough for a person to fall to the bottom. Abandoned wells of this type are particularly dangerous because they are commonly covered with boards, which eventually rot and become obscured by a blanket of leaves.

Modern water wells are usually installed with a truck-mounted drilling rig, where a rotating drill bit bores a small-diameter (about 10 inches) hole into the subsurface. After the drilling is finished a metal or plastic pipe called *casing* is lowered into the hole—the bottom section has a screen with fine slits to keep sediment from entering the well. Figure 11.15 shows a cross section of a modern well used for supplying a home or small development. Note how sand is packed around the screened portion of the casing, followed by a clay seal and concrete to prevent surface contaminants from reaching the screen. To bring water to the surface, a submersible pump is placed in the well, which is then connected to a pressurized storage tank

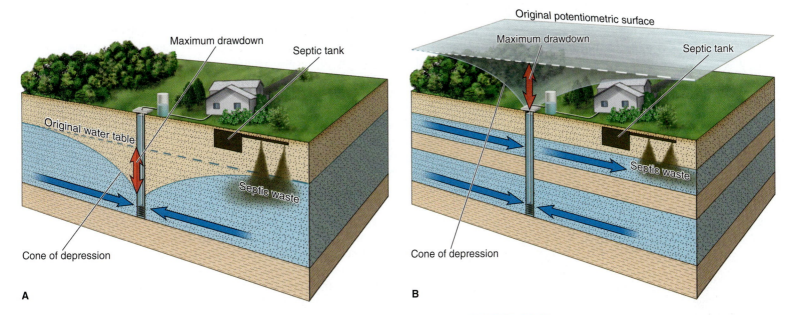

FIGURE 11.16   A pumping well in an unconfined aquifer (A) draws in water and creates a cone of depression in the water table. A well in a confined aquifer (B) creates a cone in the potentiometric surface, which in this case lies above the land surface. These drawdown cones can pull contaminants into a well and reduce the flow of water to nearby rivers or springs.

and electrical power on the surface. Most modern wells are installed at depths ranging from 50 to 1,000 feet (15–300 m), so they are less likely to become contaminated compared to shallow hand-dug wells. Keep in mind that confined aquifers normally make for the safest groundwater supply because the aquitard helps block the downward movement of contaminants.

When a pump is turned on and begins to push water to the surface, the water level (hydraulic head) within the well is lowered. The decreased hydraulic head in turn causes groundwater to flow into the well in a radial or 360-degree manner. As shown in Figure 11.16, this pumping or "sucking" action of a well creates a cone-shaped depression called a **cone of depression** in the water table or potentiometric surface of the surrounding aquifer. Notice in the figure how the cone of depression actually reverses the original direction of the hydraulic gradient in the area down-gradient from the well. This means that a well can easily draw in contaminated water that would have otherwise flowed away from the well.

Impacts of Groundwater Withdrawals

Society has benefited greatly from the use of modern water wells to extract large volumes of freshwater from the subsurface. Perhaps the most significant benefit is that groundwater has allowed population and agricultural production to blossom in areas where surface water is naturally scarce. Consider for example how naturally pure groundwater has made it possible for individual families to live in rural settings, far from any river and water filtration plant. Farmers are also now able to use groundwater to help ensure consistent harvests, particularly in areas facing a high risk of crop failure due to scarce and unreliable rainfall. Although the availability of groundwater has been a tremendous asset to society, this resource is being used in an unsustainable manner in many regions around the world, which is leading to a host of serious issues. The underlying problem is that groundwater levels are falling because withdrawals are being made at a faster rate compared to that at which aquifers are being replenished by natural recharge. Because subsurface water levels are not something people can readily see and appreciate, overuse typically continues until wells start to go dry or other serious problems begin to develop.

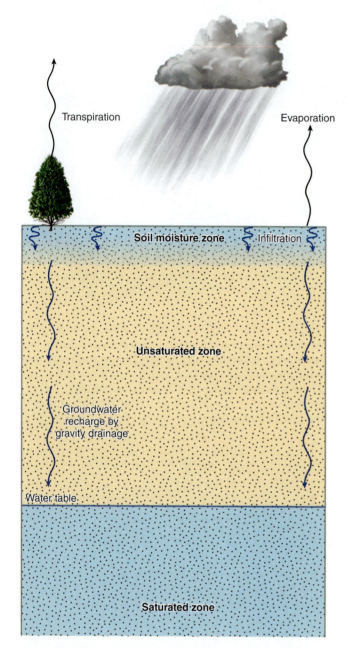

FIGURE 11.17 Groundwater recharge occurs when soil moisture builds to the point where water begins to drain due to gravity. Recharge is less common during hot summer months as most of the water entering the soil zone is quickly returned to the atmosphere by evaporation and transpiration.

FIGURE 11.18 Graph showing how groundwater levels in a semiconfined aquifer in Georgia rise and fall with the seasons. Between 2000 and 2001 the system was in balance as recharge replaced most of the water lost due to discharge. Drought conditions the following year disrupted the system's natural equilibrium, causing water levels to fall. Groundwater levels can return to their normal maximum provided there is a wet year or series of years with enough recharge to make up for the deficit of water in the system.

Before examining the problems associated with the overuse of groundwater, we need to take a closer look at how aquifer systems are replenished. Hydrologists define *groundwater recharge* as the water that leaves the soil zone and makes it way to the water table. Figure 11.17 shows that the recharge process begins when rain or melting snow infiltrates the soil zone. However, unless the water table is at the surface, infiltrating water can only reach the groundwater system if soil moisture increases to the point where water can be pulled from the pore spaces by gravity. During the hot summer months, evaporation and plant transpiration normally prevent soils from accumulating enough water for gravity drainage to take place. Groundwater recharge typically occurs on a seasonal basis during periods when temperatures are cooler and precipitation is more plentiful. The cooler temperatures naturally reduce evaporation and transpiration rates such that soil moisture can accumulate.

Because of climatic controls, groundwater recharge occurs periodically and on a more seasonal basis, whereas discharge back to streams, lakes, and wetlands is a fairly continuous process. Most groundwater systems tend to discharge continuously; therefore, water levels within an aquifer will rise only when the recharge rate becomes greater than discharge. In this sense groundwater recharge is similar to a tire that leaks air. The tire pressure goes up only if you can pump air in faster than it is leaking out. If we consider a groundwater system over the long term, we would see that the system reaches a state of *dynamic equilibrium* in which the overall rate of recharge equals the discharge rate. This means that while water levels may go up and down on a daily or seasonal basis, they remain fairly steady from year to year so long as recharge equals discharge. However, this natural equilibrium can be disrupted by changes in climate or by the excessive withdrawal of groundwater by humans.

The dynamic relationship between recharge and discharge can been seen in Figure 11.18, which shows the water level in a semiconfined aquifer 75 feet (23 m) below the surface. Notice how groundwater levels consistently rise during the winter months when recharge is greater than discharge, and then reach a peak prior to the hot summer months. Despite some significant rains, water levels decline over the summer since recharge is minimal due to the high rates of evaporation and transpiration. Of particular interest is how recharge during the spring of 2001 was able to replace nearly all of the water that had been lost from the system the previous year, returning water levels to their previous maximum. This means that between 2000 and 2001 the system was in an overall state of equilibrium. However, this equilibrium was disrupted the following year due to drought conditions that prevented recharge from returning the system to its previous state. This concept of dynamic equilibrium

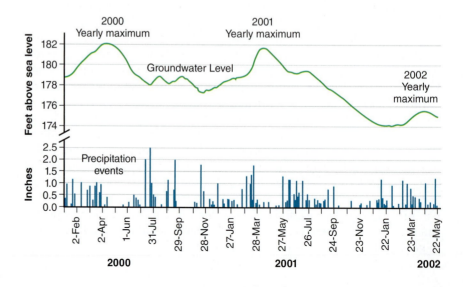

in groundwater systems is important in understanding why the human withdrawal of groundwater is unsustainable in some situations, but sustainable in others. In the following sections we will examine some of the problems that develop from excessive groundwater withdrawals.

Groundwater Mining   Hydrologists use the term **groundwater mining** to describe situations in which water levels in an aquifer get progressively lower because the system is unable to replace all the water being removed by pumping. When this occurs water is no longer considered to be a renewable resource, which makes groundwater withdrawals similar to mining in that what is being removed is not replaced. Groundwater mining is a serious problem because the water resource itself is slowly being depleted, potentially leaving future users without a water supply. Arid regions are particularly susceptible because recharge rates are naturally low, and in some cases nonexistent. In fact, the groundwater beneath many arid regions entered the subsurface as recharge under different climatic conditions during the last ice age, which ended about 12,000 years ago. Such aquifers essentially act as storage reservoirs that are not being refilled. Here water levels rapidly decline and eventually reach the point where the quality declines or the cost of pumping becomes so great that the aquifer is no longer usuable.

A good example of groundwater mining is Tucson, Arizona. This city of over a half-million people (Figure 11.19) lies in a harsh desert environment

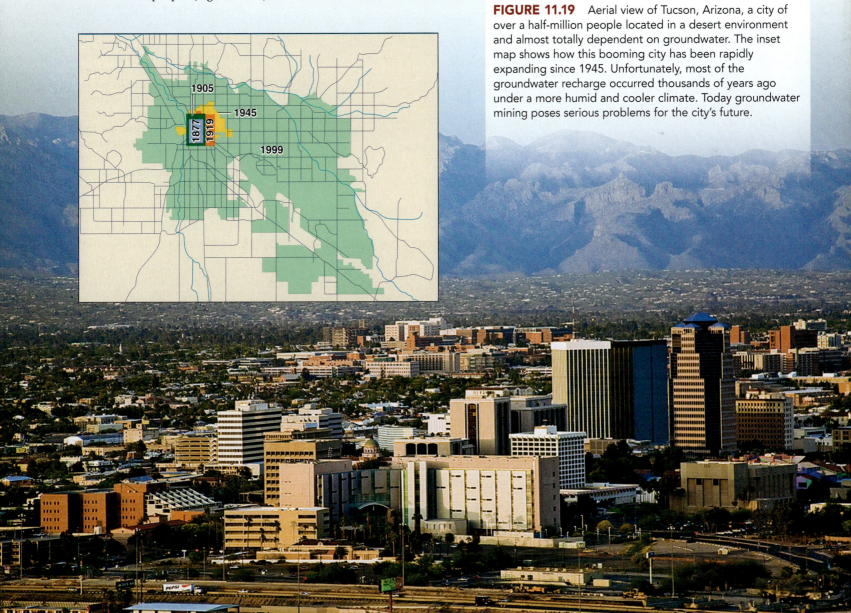

FIGURE 11.19   Aerial view of Tucson, Arizona, a city of over a half-million people located in a desert environment and almost totally dependent on groundwater. The inset map shows how this booming city has been rapidly expanding since 1945. Unfortunately, most of the groundwater recharge occurred thousands of years ago under a more humid and cooler climate. Today groundwater mining poses serious problems for the city's future.

Groundwater Mining in the Breadbasket of the United States

The High Plains region in the central United States (Figure B11.1) is well known for its extensive grasslands and roaming herds of buffalo and Native Americans that used to inhabit the area. Beneath the Plains lies a thick sequence of sedimentary rocks containing vast deposits of sand and gravel known as the Ogallala Aquifer. This enormous aquifer system has allowed farmers to transform the High Plains into one of the world's major grain producing regions, often referred to as the breadbasket of North America. Unfortunately, groundwater mining is threatening the continued high rates of agricultural production.

As with all groundwater mining, the basic problem in the High Plains is that groundwater recharge is less than the withdrawal rate by people. Rainfall across this semiarid region ranges from less than 16 inches (41 cm) per year near the Rocky Mountains and West Texas, to nearly 28 inches (71 cm) in central Kansas. Because evaporation and transpiration rates nearly equal the amount of annual precipitation, soil moisture rarely reaches the point where groundwater recharge can occur. Similar to other aquifers in the American West, much of the groundwater being removed today represents recharge that occurred during cooler and more humid conditions that prevailed after the end of the last ice age 12,000 years ago.

Prior to the 1880s, settlers found the sparse rainfall of this semiarid region suitable mostly for grazing cattle. However, the climate pattern was one of alternating periods of unusually wet years followed by several dry years. Farmers soon discovered that they could grow bumper crops of wheat during consecutive wet years. Eventually they plowed large sections of the Plains to grow wheat, only to experience plummeting production during dry years. With the advent of modern drilling techniques and electric pumps, farmers began ensuring bountiful harvests in the 1930s by tapping the groundwater in the Ogallaha Aquifer. Grain became so plentiful that it was used to feed cattle; soon the High Plains accounted for 40% of all grain-fed cattle in the United States. By 1980 approximately 170,000 irrigation wells (Figure B11.2) were supplying water to 13 million acres of cropland, whereas only 2 million acres were irrigated in 1949.

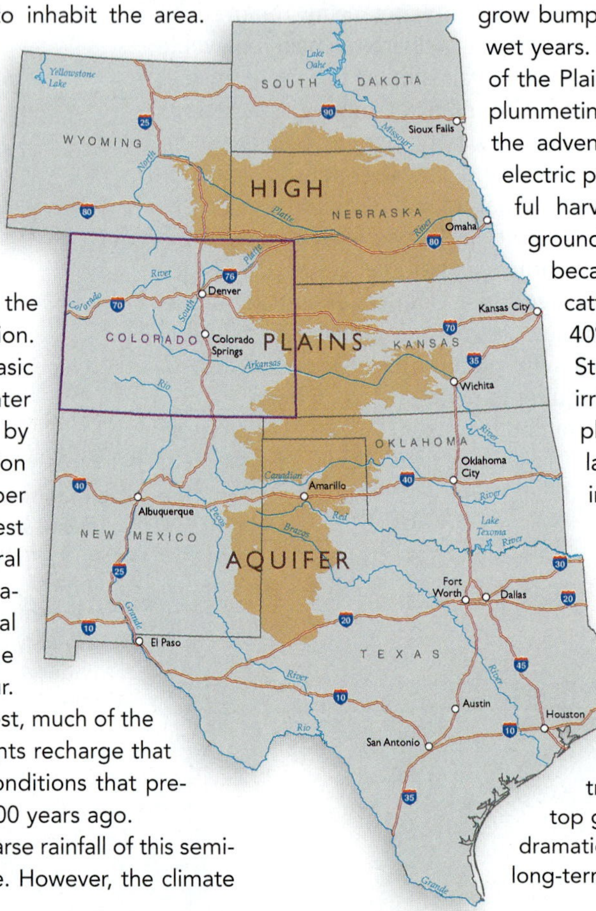

FIGURE B11.1 Map showing where the Ogallala Aquifer lies beneath the surface in the semiarid region of the United States known as the High Plains. Groundwater withdrawals from this vast and complex aquifer system have transformed the region into the nation's top grain producer, but have also resulted in dramatic water-level declines, threatening long-term agricultural production.

where streams are small and flow only intermittently, making groundwater the only reliable supply. The problem in Tucson is that the groundwater recharge rate is very low, while the withdrawal rate is quite high. Officials estimate that approximately 10% of the city's groundwater supply has been mined since 1945, which was when the city began experiencing explosive growth. While a considerable volume of groundwater remains in storage, the stark reality is that once this supply is gone, it is gone forever. Because the city's very existence is at stake, Tucson officials have wisely adopted various strategies designed to extend the water supply farther out into the future.

Groundwater mining is also a serious problem in terms of society's ability to maintain its current high rate of food production. In many regions around the world farmers have been able to make substantial increases in crop productivity by using groundwater irrigation to supplement natural rainfall—in some arid environments, irrigation is the only water the crops

A

FIGURE B11.2 Groundwater irrigation uses a center-pivot system (A) where a large-capacity well, located in the center, supplies water to a wheeled piping system that travels in a circle. In arid and semiarid climates, little if any of the applied water recharges deep aquifers due to the high evaporation and transpiration rates. False-color satellite image (B) over part of Kansas illustrates the large number of irrigation systems. Such extensive irrigation withdrawals in arid and semiarid climates are simply not sustainable.

Although irrigation has transformed the High Plains into a highly productive agricultural region, groundwater levels have dropped as much as 200 feet. Similar to other areas of extensive groundwater irrigation in arid climates, high rates of evaporation and transpiration allow very little if any of the applied water to return to the deep aquifers. In the case of the Ogallala Aquifer, the unsustainable use of groundwater means it is highly unlikely that farmers will be able to maintain today's high rate of agricultural production. Ultimately, grain production can be expected to fall, followed by higher food prices.

B Garden City, Kansas

receive. As the unsustainable use of groundwater causes water levels to continue falling around the world, society is faced with the prospect of decreased food production at the same time population keeps expanding (Case Study 11.1).

Increased Well Costs In areas where water levels continue to fall due to groundwater mining, well owners at some point are forced to incur the cost of lowering the pumps located inside their wells. Should water levels drop below the well's intake screen, then the well itself must be deepened, which is considerably more expensive than lowering the pump. Another problem is that lower water levels result in increased operating costs as the pumps require more electricity to lift the water greater distances to reach the surface. Although water will always be withdrawn for drinking purposes pretty much regardless of the electrical costs, the same

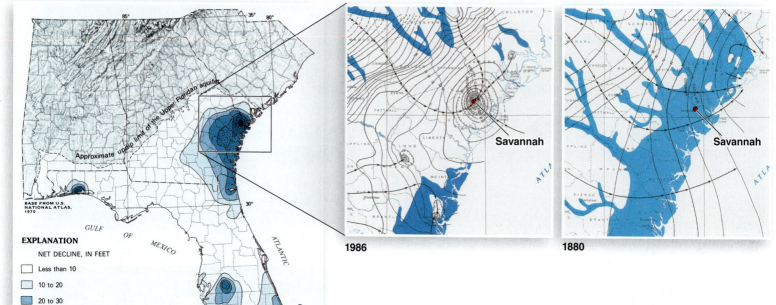

FIGURE 11.20 Large pumping withdrawals from a confined aquifer along the Georgia coast have resulted in a massive cone of depression. Contours show the potentiometric surface before and after major pumping began. Note how the areas of upward artesian flow (in blue) have been dramatically reduced.

does not apply for irrigation. At some point farmers will find that irrigation is no longer profitable, forcing them to rely on natural rainfall. This in turn will result in lower crop production, or land being removed from production altogether.

Reduced Stream and Spring Flow   In regions with a humid climate, it is more likely that large groundwater withdrawals will be sustainable because of greater recharge. Here a cone of depression will typically expand until it captures enough recharge to equalize what is being withdrawn by a well or series of wells. The cone will then stabilize and water levels will stop falling because the amount of recharge equals the discharge. Although the system reaches equilibrium, the hydraulic head of the aquifer is now lower than before the pumping began. This means that the amount of water discharging into streams and springs will decrease. Such reductions in discharge can prove detrimental to many aquatic species, especially during dry periods when streams are sustained solely by groundwater discharge. Reduced flows also have the potential to impact important commercial and recreational fisheries. In some cases groundwater withdrawals can cause springs and rivers to go completely dry, particularly in more arid climates.

Coastal Georgia provides a good example of how heavy pumping has reduced groundwater discharge to the surface environment. Here over 60 years of industrial and municipal withdrawals from a confined aquifer have resulted in a massive cone of depression located near the city of Savannah, Georgia (Figure 11.20). Although the cone today has stabilized, the potentiometric surface is now as much as 135 feet (41 m) lower compared to prepumping conditions. When the potentiometric surface was above the land surface (blue areas in Figure 11.20 insets), the hydraulic gradient was upward, causing water from the confined aquifer to leak through fractures within the aquitard and discharge into the surface environment. This upward flow of artesian groundwater has now largely been shut off because of the drawdown cone. Researchers are just now beginning to study how this reduction in groundwater flow may have impacted the ecology of the coastal region.

Salt Water Intrusion   Large groundwater withdrawals along a coastline can reverse the hydraulic gradient and cause **saltwater intrusion,** where saline water migrates into the freshwater portions of an aquifer. From the

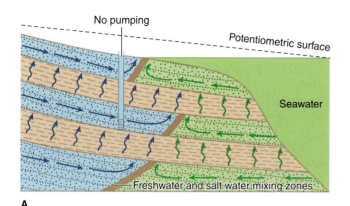

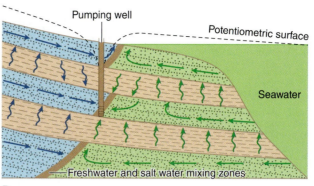

A

B

example in Figure 11.21 you can see how pumping reverses the natural flow of groundwater around a well, drawing seawater into the coastal aquifer. This situation is very undesirable as the intruding seawater can raise the salinity of the aquifer to the point it becomes unusable. Because a pumping well draws in water from all directions, a coastal aquifer can also become contaminated when salt water leaks upward deep along faults and fractures that cut across aquitards. Note that a deep saline aquifer can cause salt water contamination of freshwater aquifers located far from any ocean. Regardless of how it occurs, saltwater intrusion is a serious problem because it permanently degrades the water quality of an aquifer.

Land Subsidence   As described in more detail in Chapter 7, the water in saturated material exerts pressure within the pore spaces that helps support the weight pressing down from the overlying column of rock or sediment. When water is pumped from the subsurface there is a corresponding decline in pore pressure, forcing solid material to support more of the weight. This causes aquifers and aquitards to undergo compaction and experience a reduction in volume, which in turn causes the land surface to slowly sink or subside as shown in Figure 11.22. Because clay minerals are much more compressible than quartz or rock fragments, subsurface layers rich in clay will experience considerable volume reductions when

FIGURE 11.21   Under natural conditions (A), coastal aquifers contain both fresh and salt water that flow toward a mixing zone, the position of which depends on the hydraulic head and water density in the various aquifers. Large pumping withdrawals (B) will alter the hydraulic head within the system, causing the position of the mixing zone to move toward the well, resulting in salt water contamination of the water supply.

FIGURE 11.22   Illustration (A) showing how land subsidence occurs when pumping in a confined aquifer creates a cone of depression, causing leakage and a reduction in pore pressure within the system. Most of the subsidence is due to compaction of the highly compressible, clay-rich aquitards. Photo (B) shows the casing from a well in Mexico City that became exposed when heavy pumping withdrawals across the city caused the land surface to subside.

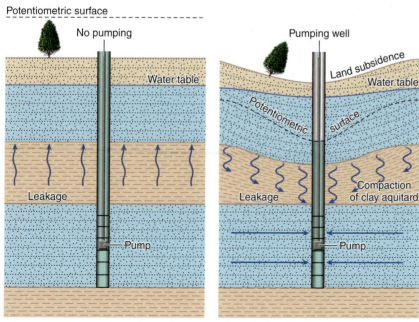

A

B

pore pressure is reduced. Therefore, compaction of the aquifers themselves is relatively minor compared to that of aquitards. Problems associated with water withdrawals and land subsidence are typically more severe in groundwater systems with thick, clay-rich aquitards. Note that once compaction occurs, it is largely irreversible; hence, the resulting subsidence becomes permanent.

One of the best examples of land subsidence is Mexico City, a metropolis of over 20 million people where nearly 70% of the water supply comes from a thick sequence of sand and clay layers. Because the withdrawal rate in this arid environment is greater than the rate of recharge, groundwater levels have been falling by about 3 feet per year. The corresponding loss of pore pressure has caused the clay aquitards to compact, resulting in land subsidence of nearly 30 feet (9 m) over the last century (Figure 11.22). As the unsustainable use of groundwater continues, land subsidence is forcing Mexico City to cope with considerable damage to the foundations of buildings, subway tunnels, and underground utilities, such as water, sewer, and electric lines.

Selecting a Water-Supply Source

People living in the urban parts of more developed countries normally get their freshwater from municipal supply systems, whereas rural residents typically have their own groundwater well—often called a domestic well. As illustrated in Figure 11.23, the supply for municipal systems will either consist of surface or groundwater, or some combination of both. The particular choice is dictated by the quantity and quality of the water that is available. Quality is always a key factor because of the costs involved in treating the water to ensure it is both safe and does not have an unpleasant taste or appearance. Although the quality of a particular source may be high, it might not be capable of supplying a sufficient quantity of water. For example, the Atlanta metro area is underlain by fractured igneous and metamorphic rocks that contain high-quality groundwater, but the volume is nowhere near enough to meet the needs of the 4.2 million residents. Many cities face similar situations in which a suitable aquifer does not exist, forcing them to rely on surface reservoirs where water can be stored in sufficient quantities.

Given a choice between surface and groundwater, cost is always an important factor. To minimize both short- and long-term expenses, municipal supply systems generally use the nearest and purest source possible that can provide water in the quantities that are needed. The ideal source economically is usually groundwater because it involves far less treatment and infrastructure costs compared to surface water. Water wells are relatively inexpensive to install and maintain, plus they can often be placed in or near a city. Because groundwater is typically free of sediment and disease-causing organisms (pathogens), it requires very little treatment prior to human consumption. In fact many homeowners with private wells deeper than 50 feet (15 m) deep do not have to disinfect the water as harmful bacteria and viruses are usually not found beyond this depth.

In contrast, using a river as a supply source requires the construction of a filtration and treatment plant (Figure 11.23). Here the water is first run through settling tanks and filters that remove the particulate matter. Then the water is treated with a chlorine-based disinfectant that kills the bacteria and viruses that are commonly present in surface waters. These processing

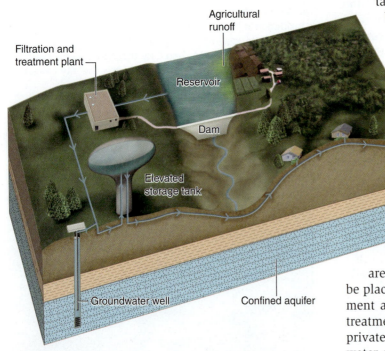

FIGURE 11.23 Municipal water supply systems can use a combination of surface and groundwater sources, with surface water requiring far more filtration and disinfection. After being treated, drinking water is pumped into a storage tank whose elevation creates the water pressure (hydraulic head) necessary to move water through the distribution system.

plants not only require an initial capital investment, but include substantial operating costs on a permanent basis. Moreover, should the city's demand for water be greater than what the river can provide, a dam and reservoir must be constructed, which, in turn, means even greater capital investment and long-term operating costs. The costs are further compounded in cases where a suitable site for a reservoir is located far from a city, thereby requiring miles of pipeline or an aqueduct. Finally, note that regardless of the source, water in a municipal supply system is usually pumped into an elevated storage tank. This provides the hydraulic head (water pressure) that is needed to drive water through the distribution system and into homes and businesses.

Another advantage of groundwater over surface water as a supply source is that it is generally less susceptible to contamination. Confined (artesian) aquifers are the most desirable because the overlying aquitard(s) provide protection from surface waters that can become contaminated by leaking storage tanks, industrial spills, agricultural fertilizers, pesticides, and other pollutants. On the other hand, contaminants can be carried directly off the landscape and into surface streams during heavy precipitation events (Figure 11.23). Managers at water filtration and treatment plants therefore must continuously monitor the quality of surface water coming into the plant, making adjustments to their treatment processes as water quality changes. During unusual events, such as an industrial or sewage spill that takes place upstream, plants will often have to shut down until after the contaminated water moves downstream. Note that water pollution will be described in detail in Chapter 15.

Alternative Sources of Freshwater

In many parts of the world people are using water in an unsustainable manner and are pushing natural systems beyond their ability to provide freshwater. Unfortunately, the demand for Earth's limited freshwater supplies will keep growing as long as the population continues to expand. As supply problems worsen, one traditional approach is to build aqueducts to tap distant sources. There comes a point, however, where this strategy of transporting water great distances is no longer feasible or desirable. While redistributing water resources may increase supply in one area, it can severely affect ecosystems and limit the economic growth within the basin where the water is being removed.

Another traditional approach is to simply extract more water from existing sources. This too has limits and undesirable consequences. For example, there is only so much water that can be removed from a river or stored behind a dam before natural ecosystems begin to collapse. In the case of groundwater, excessive withdrawals result in groundwater mining and saltwater intrusion, both of which threaten the very existence of the water supply and the human activity it supports. Because traditional water supplies are being stretched to their limit in many areas, people are beginning to look at alternative methods for obtaining additional water. We will explore some of these alternatives in the following sections.

Desalinization

Earth has an unlimited supply of saltwater, but before it can be used for most human applications the bulk of the dissolved ions or salts must be removed, a process referred to as **desalinization.** To get a sense of what this entails, recall that seawater contains 35,000 milligrams of salt per liter of water (mg/l), whereas the safe limit for human consumption is set at 500 mg/l. There are basically two ways to desalinize water. One is through the process of *distillation* in which salt water is boiled, vaporizing water molecules but leaving the dissolved ions behind in the remaining liquid. This freshwater vapor is then captured and allowed to condense into what is known as *distilled water*. The other desalinization method is called *reverse osmosis*. Here a semipermeable membrane is used to separate two tanks of water; one side is filled with salt water and the other with freshwater. When the salt water side is pressurized, the membrane allows water molecules to pass through, but the dissolved ions, which are larger, are forced to stay behind. As the amount of freshwater in this process increases, the volume of water on the salt water side decreases and becomes more saline.

Both distillation and reverse osmosis processes create a wastewater disposal problem, because the salt content is about two times greater than that of seawater. When discharged directly into the marine environment, the extreme salinity of the wastewater can be harmful to sensitive ecosystems. One successful solution is to blend the wastewater with the freshwater discharge from a nearby sewage treatment plant. In addition to the wastewater issue, desalinization requires significant amounts of energy, making it considerably more expensive than extracting and processing conventional water resources. It currently costs about 80 cents to desalinate 250 gallons (950 l) of seawater, whereas a traditional supply system can produce the same volume of freshwater for around 15 cents.

Because of the high energy costs, desalinization is currently being used only in those areas with acute water-supply problems and by nations with the necessary financial resources. For example, desalinization is being utilized in some of the Persian Gulf nations where freshwater is extremely limited, but their oil wealth enables them to pay the high costs. In fact, the world's first large-scale desalinization plant was built in Kuwait in 1965, whereas Saudi Arabia currently accounts for nearly 25% of the world's production of desalinized water. Chronic water-supply problems also prompted Israel to build a large reverse osmosis plant (Figure 11.24). Completed in 2005, this plant has the capacity to produce 72 million gallons (270 million liters) of freshwater per day. In the United States the first large-scale desalination plant was built in Tampa Bay, Florida, where population growth had outstripped the natural water supply. This plant operated intermittently between 2003 and 2005, but is now fully operational and producing 25 million gallons (95 million liters) of freshwater per day—enough to meet the average household consumption for 360,000 residents.

Desalination is likely to become a more attractive option in the future as demand for freshwater continues to grow. Despite being tied to rising energy costs, technological improvements are making desalination more economical. Reverse osmosis is the most promising since it can be made more cost-effective by utilizing water of moderate salinity. In some areas brackish water from an estuary or deep aquifer is being used instead of straight seawater. However, even with increased efficiency, desalination is unlikely ever to be a cost-effective means of producing water for agricultural irrigation.

FIGURE 11.24 Aerial view showing Israel's reverse osmosis plant at Ashqelon, which is the largest in the world. Plants such as this one commonly discharge highly saline wastewater into the ocean, where it mixes with normal seawater, potentially disrupting marine ecosystems.

Reclaimed or Recycled Wastewater

Municipal Wastewater Recycling

In addition to filtration and treatment plants for drinking water, municipal governments in the United States also operate sewage treatment facilities that break down organic waste in the wastewater before discharging it into the environment (Chapter 15). In areas with limited water supplies, municipalities are discovering that supplies can be extended by using reclaimed wastewater for certain applications rather than using highly treated drinking water. Moreover, it is far cheaper to expand water supplies in this manner than by using the traditional approach of building new reservoirs or installing additional wells. Wastewater today is increasingly being viewed as a valuable resource as opposed to a financial liability.

Some sewage treatment plants in the United States are also employing ultraviolet (UV) disinfection techniques capable of treating wastewater to the point where it meets drinking-water standards. Despite the fact that reclaimed water can be made safe to drink, the idea of drinking water that had once been flushed down toilets is unappealing to most people. Because of public perception and the potential for operator error that could result in the contamination of drinking-water supplies, reclaimed wastewater is generally not placed directly back into the supply system. Reclaimed water is instead being used for a variety of nondrinking applications, such as irrigating golf courses and parks and for various industrial operations. The use of reclaimed water is currently being limited by the capital costs municipalities must pay to install distribution lines from the treatment plant out to potential users. Consequently, reclaimed waterlines are usually first extended out to the largest users, such as golf courses and large industrial plants.

Due to high population growth and limited water supply, Florida has been a leader in developing a comprehensive program to promote nondrinking uses of reclaimed wastewater. Some counties in California with similar issues have taken the next step and propose putting treated wastewater back into the drinking-water system. Here the treated water would be used to recharge aquifers or piped into a water-supply reservoir. In this way treated wastewater would be mixed with the normal water supply and undergo natural purification processes before entering the filtration and treatment plant. In 1990 Los Angles County officials announced a plan to use reclaimed water to recharge water-supply aquifers. The proposal, dubbed "toilet-to-tap" by critics, met stiff public opposition that eventually forced officials to shelve the project in 2001. When nearby Orange County officials announced a similar plan, they did so in conjunction with an extensive public education program. This effort was successful and resulted in a $481 million plant that is now processing 70 million gallons (265 million liters) of water per day, which is equivalent to the freshwater production at the world's largest reverse osmosis plant.

Industrial and Domestic Recycling

Similar to municipal governments, industrial plants and homeowners have found it cost-effective to recycle their wastewater, particularly in areas where municipal water rates are high. For example, when water was relatively cheap and plentiful, factories would often use water in certain applications and then promptly discharge it as waste. Today, there are financial and political incentives for businesses to reuse water whenever possible. A good example is the reuse of so-called *noncontact cooling water,* which is water used only for cooling and does not come into direct contact with the heat source. Other than being at a higher temperature, the quality of noncontact cooling water is unchanged. Industrial plants therefore have found it economical to store the water so it can cool and be used again.

Reclaimed wastewater can also be used in the home in order to reduce demand on the drinking-water supply. Here the plumbing system is typically modified so that all the wastewater in the home is collected, with the exception of wastewater from toilets. Because this water has never been in contact with human waste it is called *graywater*, which distinguishes it from sewage. Graywater is mostly used to irrigate outside plants and gardens, but can also be used in any number of applications that do not require drinking water.

Aquifer Storage and Recovery

A management technique known as **aquifer storage and recovery** (ASR), sometimes referred to as *water banking*, involves storing surplus surface water in aquifers, and then removing (i.e., recovering) it for use at a later time. Aquifers that are well suited for this are generally composed of granular materials that have both high porosity and hydraulic conductivity. Here surface water is either injected into the aquifer using wells or spread out on the surface and allowed to infiltrate in specially constructed recharge beds or basins. The water is then removed by pumping wells during periods of high demand. In ASR an aquifer functions like a surface reservoir in that it simply stores water for later use.

Although ASR and surface reservoirs function in a similar manner, there are important differences that make storing water in aquifers inherently risky. Some ASR systems use treated water obtained from rivers during seasonal periods of elevated discharge. Careful management is required since withdrawing too much water can damage critical stream ecosystem functions that depend on periods of high discharge. Other ASR systems use treated effluent from sewage treatment plants (Chapter 15). Although in both cases the source water is carefully treated, there is a potential that contaminated water could inadvertently be introduced into the aquifer. Because of the porous nature of most aquifers, once they are contaminated, it is extremely difficult, if not impossible, to remove the contaminants. ASR therefore could permanently render a water-supply aquifer unfit for human consumption. Another potential problem stems from the fact that surface water has a much higher dissolved oxygen content than does most groundwater. Injecting such oxygen-rich water into an aquifer can result in undesirable chemical reactions (e.g., precipitation of iron) and/or bacterial growth.

Despite the potential risks, ASR is being used in many parts of the world, including California and Florida. By injecting water into an aquifer, ASR not only produces a useful stockpile of water but helps to rebuild water pressure (hydraulic head) in the subsurface. This, in turn, can help remediate the effects of saltwater intrusion and land subsidence problems associated with excessive groundwater withdrawals.

Rainwater Harvesting

Collecting and storing rainwater is an ancient practice that is commonly referred to as *rainwater harvesting*. A typical rainwater harvesting system for a home or business (Figure 11.25) consists of a set of gutters to collect water off a roof and a piping system for carrying it to a storage tank called a *cistern*. The building's plumbing system is then modified to make use of the cistern water. Some homeowners choose to have the cistern supply

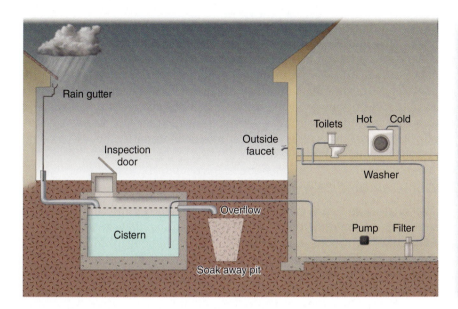

Austin, Texas

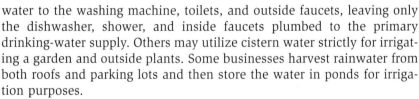

water to the washing machine, toilets, and outside faucets, leaving only the dishwasher, shower, and inside faucets plumbed to the primary drinking-water supply. Others may utilize cistern water strictly for irrigating a garden and outside plants. Some businesses harvest rainwater from both roofs and parking lots and then store the water in ponds for irrigation purposes.

People are beginning to return to this age-old practice because of the need to reduce the demand on their primary water supply. For example, many rural homes must rely on an aquifer that is either thin or composed of fractured rocks, and is therefore not capable of supplying large volumes of water. This means they may not have enough water for such things as washing multiple loads of clothes, irrigating a garden, or even frequent toilet flushing. In other cases homeowners may simply want to reduce their use of municipal water because of its high cost or restrictions on outdoor water use.

FIGURE 11.25 Rainwater harvesting systems involve collecting rainwater from a roof, then storing it either above or below ground in a tank called a cistern. By modifying the building's plumbing system, cistern water can be used in place of drinking-water for certain applications.

Conservation

Alternative water supplies essentially involve adding new or additional freshwater to a region's overall supply. Water conservation is different in that it is about using an existing supply more efficiently so that some of the water becomes available for other uses. Although conservation does not increase the total supply, it has the same net effect. This is analogous to a person who saves money by reducing the amount they spend on nonessential items, such as going to the movies and going out to eat. The savings are then used to purchase more essential items, like gasoline for getting to work. Similar to being fiscally conservative, water conservation means you have freed up some of your water supply for more essential uses.

All water conservation efforts fall into one of two categories. The first relates to *engineering practices* in which plumbing systems, appliances, and operational procedures are modified so less water is used during some process or application. An example would be a more efficient dishwasher. The other category involves *behavioral practices* where people change their water-use habits. Here an example would be spending less time in the

A Denver, Colorado

B Las Vegas, Nevada

FIGURE 11.26   Large reductions in home water use can be realized through landscaping changes. Xeriscaping involves using native plants (A) rather than nonnative vegetation (B) that requires extensive irrigation.

shower. Although conservation is an effective means of increasing water supply, it often faces resistance because it requires people to change personal habits or spend money to purchase more efficient hardware (toilets, showerheads, etc.). This resistance is usually overcome when: (a) the price of water becomes sufficiently high; or (b) there is a shortage and a user has no other means of increasing supply. Next, we will explore some of the more common conservation techniques, with the most desirable generally being those that save the greatest volume of water for the least amount of hardship and capital investment.

Domestic and Commercial Users

The single largest use of water by residents and commercial businesses in the United States is for irrigating lawns, shrubs, flowers, and gardens. Substantial reductions in water use therefore can be realized by changing the way in which we landscape around homes and businesses. People have grown accustomed to using nonnative shrubs, flowers, and grasses for landscaping despite the fact these plants commonly require extensive watering. In general, these plants require additional water because they are not in their natural habitat and are unable to survive in drier settings. Landscape irrigation has not been much of a problem since water has historically been cheap and plentiful in many areas. As water supplies are being pushed to their limits, many local and state agencies are now raising water rates and imposing restrictions on outdoor watering. This has led to the increased popularity of what is called **xeriscaping,** which is a type of landscaping where native plants are used because they require little to no irrigation (Figure 11.26). People are finding that native vegetation not only saves them money by dramatically reducing their water use, but can also be aesthetically pleasing.

With respect to indoor water usage, the most important activity is toilet flushing. It is estimated that the average American uses nearly 7,000 gallons (26,000 l) of water a year just to flush toilets. Traditional toilets use between 3.5 to 5 gallons (13–19 l) per flush and account for approximately 27% of indoor domestic use. In order to conserve water, so-called *low-flow toilets* that use only 1.6 gallons (6 l) per flush have been required in new U.S. homes since 1994. The potential savings with low-flow toilets is significant considering the volume of water being used and the fact that the new toilets use 2 to 3 times less water than traditional models. Also important is showering, which accounts for 12% of indoor water use. Because older shower heads usually release far more water than necessary, replacing them with highly efficient low-flow shower heads is a simple and inexpensive way of conserving water.

In addition to physical changes, there are many behavioral practices people can change which can reduce water usage. For example, running your dishwasher only when it is full can save 10 to 20 gallons (38–75 l) per day per household. Another simple method is to turn off the water while shaving or brushing your teeth—some even choose to take shorter showers and turn the water off while soaping. Similarly, when washing dishes by hand, water can be saved by not allowing the faucet to run continuously—this has the additional benefit of saving energy to make hot water. While laundering, one can also adjust the water level in the washing machine so it matches the size of the load being washed. Behavioral changes are important outside of the home as well. Examples here include turning the hose off while washing a car and using a broom instead of a hose to sweep off driveways and sidewalks.

Although the individual savings from any single appliance or personal habit may seem small, added together they can reduce household consumption rates by 30% or more. Moreover, if every household in a city conserved at this same rate, then the municipal supply system would have 30% more water for additional users. You will see in later chapters how the collective power of conservation can play an important role in our mineral and energy supplies.

Municipal Supply Systems

In order to reduce the overall demand for water, municipal supply systems across the United States are developing educational programs designed to encourage water conservation. Because low-flow toilets and shower heads are relatively inexpensive and make a significant impact on consumption rates, many municipalities have instituted retrofit programs that offer homeowners rebates for installing more-efficient units. In some cases owners are reimbursed for the entire cost of the retrofit. Local governments have discovered that increasing water supplies through conservation programs is often less costly than physically expanding their supply system, which requires large capital investments.

Education is helpful, but water utilities commonly find it difficult to get people to conserve when the price of water is low. Considering that municipal water averages only 30 cents for 250 gallons (950 liters), there is clearly little financial incentive for people to conserve. To help discourage excessive water use, many municipalities are changing from a flat water rate to a tiered rate in which users pay progressively higher rates as the volume they use increases. Another effective conservation approach is repairing leaks in municipal supply systems. Leaks develop over time as water mains deteriorate with age or crack when the ground undergoes compaction and settles. Reducing the amount of water being lost through leaks means that additional water can be made available for new users. Another benefit is that municipalities do not have to pay the necessary processing and treating costs for water that simply leaks into the ground.

Agriculture

Recall that agricultural irrigation is one of largest uses of freshwater. Moreover, modern irrigation techniques are one of the primary reasons agricultural productivity has increased in recent years and has enabled the world to feed an ever-growing population (see Case Study 11.1). As water supplies continue to become stretched, improving the efficiency of irrigation will become more critical, particularly in areas of groundwater mining. For example, depending on temperature and wind conditions, as much as 30% of the water leaving traditional irrigation systems (Figure 11.27A) is caught by the wind and evaporates before ever reaching the ground. Because of the poor efficiency of many systems and the sheer volume involved, minimizing irrigation losses can save enormous amounts of water.

One of the simplest and least expensive ways of improving irrigation efficiency is for farmers to employ techniques that cause soils to retain more moisture—the more moisture that is retained the less the fields need to be irrigated. Recall from Chapter 10 that no-till farming is where the previous season's crop is left standing so as to reduce soil erosion. By keeping the soil covered, this crop residue also reduces the evaporative loss of water from soils, and has the same effect as placing mulch around

A

B

FIGURE 11.27 In older irrigation systems (A) as much as 30% of the water evaporates and never reaches the ground. Systems can be retrofitted (B) so that water is directed toward the ground, which greatly reduces evaporative losses.

FIGURE 11.28   Drip-irrigation techniques are the most efficient as water is applied only to the root zone of each plant. However, irrigating large fields in this manner is labor intensive and involves considerable material costs.

home flower beds. Furrowing and contour plowing are erosion control practices that decrease overland flow, but have the additional benefit of increasing infiltration and soil moisture. Another inexpensive technique is to gather data on soil moisture, soil properties, and climatic conditions so that irrigation schedules can be fine-tuned for specific crop types. The idea here is to avoid irrigating fields when moisture conditions are high and the crops are unable to benefit from the additional water—this is similar to how watering your lawn right after a heavy rain does nothing but waste water.

Finally, irrigation efficiency can be increased by making physical modifications to an existing irrigation system or by installing an entirely new system. Either approach results in a farmer using less water to achieve the desired soil moisture conditions for a particular crop. A good example is shown in Figure 11.27B, where a spray-irrigation system has been modified so that water is directed downward to reduce evaporative losses. Under certain conditions a farmer may find it cost-effective to install a *drip-irrigation system,* where a piping system applies water to the base of the plants one drop at a time (Figure 11.28). Drip-irrigation systems are the most water efficient, but are also quite labor intensive. Although physical improvements in irrigation systems typically involve considerable capital investments, farmers can eventually recover the cost through increased crop yields made possible by the additional water that was saved.

SUMMARY POINTS

1. Earth's abundance of liquid water is what makes human life possible and our planet so unique within the solar system.

2. The hydrologic cycle creates freshwater through evaporation. Freshwater eventually falls on landmasses where it flows into rivers and lakes and infiltrates the subsurface. Humans tap into this freshwater as it moves through the surface and subsurface portions of the hydrologic cycle.

3. Consumptive water use results from human activities where water is lost and cannot be used again. Offstream usage is where water is removed from a source, but discharged elsewhere after being used.

4. Off-stream diversion of surface water via aqueducts provides increased water supplies for those receiving the water, but can adversely affect water supplies, ecosystems, fisheries, and other activities in the basin where the water is being removed.

5. Dams are valuable for stockpiling water, but also discharge cooler water and trap sediment which can damage downstream ecosystems and fisheries.

6. Aquifers are earth materials that easily transmit water. Granular materials like gravels and sands are most desirable because they can transmit larger volumes of water. Fractured rocks and solution limestones also serve as aquifers. Clay-rich materials and unfractured crystalline rocks do not transmit much water, hence are called aquitards.

7. Unconfined aquifers are open to the atmosphere whereas confined aquifers are overlain by an aquitard and are pressurized. The water level in an unconfined aquifer is defined by the water table, whereas the potentiometric surface represents the pressure in a confined aquifer.

8. Hydraulic head is a measure of water's potential energy within an aquifer and is related to the elevation of the water table or potentiometric surface. Groundwater always flows toward areas of lower hydraulic head, thus can move both horizontally and vertically in the subsurface.

9. Groundwater mining occurs when the withdrawal rate from an aquifer is greater than the natural recharge rate, causing water levels to progressively lower. When this occurs water is no longer a renewable resource and can lead to saltwater intrusion and land subsidence.

10. With respect to a water supply, groundwater is generally more desirable than surface water because of its higher purity, resulting in lower treatment and infrastructure costs.

11. Alternative ways of obtaining freshwater are needed in areas where traditional supplies are no longer able to meet demand or where withdrawals are creating environmental problems.

12. Desalinization and reclaimed wastewater are alternative means of increasing existing supplies, whereas various conservation techniques make more water available by using existing supplies more efficiently.

KEY WORDS

APPLICATIONS

Student Activity

Do you know where your drinking-water comes from? If you do not know, call your local water department. If you do not have a water department, then you probably have a well. Most municipal water comes from reservoirs or wells.

Critical Thinking Questions

1. What is the difference between confined and unconfined aquifers?
2. Which kind of aquifer would be better for drinking-water?
3. What is groundwater mining?
4. What is off-stream use and consumptive water use?

Your Environment: YOU Decide

Do you conserve water? Even if you live in an area where there is abundant water, do you think it is necessary to conserve? What do people mean by saying "it is not oil we should worry about, it is potable water"?

Mineral and Rock Resources

LEARNING OUTCOMES

After reading this chapter, you should be able to:

▶ Understand the degree to which people in industrialized societies rely on mineral resources.
▶ Explain why certain minerals are used in specific applications.
▶ Understand the difference between a mineral reserve and mineral resource and what is meant by an economic deposit.
▶ Describe the key geologic processes that concentrate minerals into economic deposits for the basic rock types: igneous, sedimentary, and metamorphic.
▶ Describe two mining-related factors that determine whether a deposit will be economical to mine.
▶ Understand the basic steps in ore processing.
▶ Explain why mineral resources are unevenly distributed on Earth's continents and why some minerals are called strategic minerals.
▶ Describe the two primary ways in which humans can meet future demand for finite mineral resources.
▶ Explain how sulfide minerals are related to acid mine drainage and acid rain.

Workers removing a large block of granite from a stone quarry in Georgia. The granite will be cut into smaller blocks and slabs, and then used for such things as cemetery stones and decorative tiles and countertops. Modern society requires tremendous quantities of mineral and rock resources that must be extracted from the Earth, creating a host of environmental issues. As population continues to grow, the demand for these resources will increase, presenting additional problems as some mineral resources are already in short supply.

Introduction

Nearly all of us who live in modern societies enjoy a lifestyle that was unimaginable a mere 100 years ago. Today there are highly engineered automobiles and jet airliners that can take us anywhere we want to go in climate-controlled comfort, plus we have the convenience of cell phones and MP3 players. We are also able to live and work in comfortable buildings and enjoy a fantastic array of foods and consumer goods. This modern lifestyle is entirely dependent on two key resources: energy and minerals. Minerals are needed for everything we build, from pencils to farm equipment to the very factories that make all these things. Energy resources such as oil, gas, and coal provide the power that keeps everything running, particularly the lights and heating and cooling systems in buildings, as well as our vast transportation network. Moreover, energy is what makes it possible for us to mine the minerals from which most things are built. Although water resources are essential for life itself (Chapter 11), our modern way of life simply would not exist without Earth's abundant, but limited mineral and energy resources. Our focus in this chapter will be on the human use and extraction of minerals, whereas Chapters 13 and 14 will be devoted entirely to energy resources.

One of the more interesting facts about mineral resources is the amount of minerals required to support the modern lifestyle we enjoy. For example, Table 12.1 lists the amount of different mineral resources used by the average American citizen on a yearly basis. Keep in mind that the per capita consumption rate of over 20,000 pounds of stone, sand, and gravel, for example, does not mean you are using this amount each year at your house, but rather it represents your portion of all this material being used in constructing the nation's highways, buildings, factories, and more. We can therefore think of each person as having their own *mineral-resource footprint* on our planet, representing the amount of minerals needed to support their lifestyle—this is similar to the concept of an ecological footprint (Chapter 1) and a carbon footprint (Chapter 16). Clearly, a person living in a modern industrialized nation would have a larger mineral footprint compared to a person in a developing country.

Another interesting aspect of Earth's mineral resources is that unlike water, people rarely see minerals in their original state, thus few of us make the connection between the goods and services we use and the minerals they require. Moreover, people generally do not appreciate the effort that goes into finding, extracting, and processing minerals. For example, we all use glass everyday, lots of glass. But, do you know what glass is made of or what it takes to make it? Glass is actually made by melting quartz, which is one of the dominant minerals found in rocks and sediment near Earth's surface. Every time you drink from a glass, look through a window, or enjoy the comfort of a well-insulated building, you are making use of the same quartz grains you see on a beach or in a river. To help remind people that their everyday lives are dependent on minerals, mining geologists often use the phrase "if it can't be grown, it must be mined."

Although people commonly lose sight of the fact that minerals have made their modern lifestyles possible, they are generally aware that the mining and processing of minerals historically has led to environmental devastation. Despite the fact that recent environmental regulations have greatly reduced the negative impacts of mining,

TABLE 12.1 Average yearly per capita consumption rates of various mineral resources in the United States.

	Mineral Resource	U.S. Yearly Per Capita Consumption	Percent of All Mineral Resources
	Stone, sand, and gravel	20,668 lb (9,383 kg)	87
Nonmetals	Cement	841 lb (382 kg)	3.5
	Salt	365 lb (166 kg)	1.5
	Phosphate rock	254 lb (115 kg)	1.1
	Clays	247 lb (112 kg)	1.0
	Iron	380 lb (173 kg)	1.6
	Aluminum	73 lb (33 kg)	0.31
	Copper	17 lb (7.7 kg)	0.072
Metals	Lead	12 lb (5.4 kg)	0.051
	Zinc	9 lb (4.1 kg)	0.0380
	Gold	.022 oz (.63 g)	0.00001
	All other minerals	869 lb (395 kg)	3.7
	Total	**23,735 lb (10,776 kg)**	**100%**

Source: Mineral Information Institute, 2008.

people often oppose the opening of any new mines near their homes. This gives rise to the concept of NIMBY, or "not in my back yard." While it is understandable why a person would not want a mine to open up next to their house, it is important to realize that we all enjoy the benefits of minerals and therefore create the demand to which mining companies are responding. Therefore, each of us is partly responsible for any negative consequences associated with the mining and processing of minerals. This also means that as citizens we all need to do our part to help ensure that these precious resources are used in a wise and efficient manner, and that they are extracted as safely as possible.

In Chapter 12 we will explore the way in which geologic processes concentrate minerals into deposits that make them economical for people to mine. We will also examine the basic mining and processing techniques used to turn these resources into goods and services, along with the environmental problems that result from these activities. Let us first take a closer look at how society uses minerals and why certain minerals are used in specific applications.

Minerals and People

Recall from Chapter 3 that a *mineral* is a naturally occurring inorganic solid where individual atoms are arranged in an orderly manner (i.e., have a crystalline structure). Some minerals, such as diamond (C), are composed of only one type of atom, but most contain different atoms, as in the case of the copper-rich mineral called bornite (Cu_5FeS_4). Also recall that a *rock* is simply an assemblage of one or more minerals. We can think of a **mineral resource** as any rock, mineral, or element that has some physical or chemical property humans find useful. Mineral resources can range from rocks that contain scattered grains of a specific mineral to a rock body composed entirely of a valuable mineral, such as layers of common table salt or halite (NaCl) shown in Figure 12.1. Our definition of a mineral resource also includes sorted river sands and gravels and crushed rocks used as fill material in construction projects. Certain types of rocks are also cut into slabs and used for decorative countertop and flooring materials or as cemetery stones. Notice in Table 12.1 that geologists typically subdivide mineral resources into metallic and nonmetallic resources, and that rocks are categorized as nonmetallic.

Certain minerals have physical or chemical properties (Chapter 2) that have practical applications. A good example is

FIGURE 12.1   Mineral resources include a variety of different rocks, minerals, and elements. This underground mine is in a layered sedimentary deposit composed almost entirely of the mineral halite. Finely ground halite is used as common table salt, whereas coarsely ground halite (inset), called rock salt, is used for de-icing roadways in the wintertime.

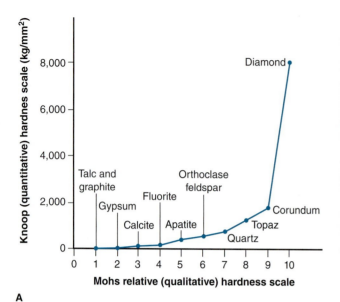

A

B

FIGURE 12.2 Graph (A) comparing the quantitative and qualitative hardness of common minerals illustrates the vast difference in hardness between diamond and the next hardest mineral, corundum. Diamonds are well known as gemstones, but their greatest use is in cutting tools (B) where their extraordinary hardness allows humans to cut through any type of material.

how we use the mineral *diamond* because of its hardness, which is the physical ability to resist scratching. From Figure 12.2 you can see that diamond is over four times harder than the next hardest mineral (corundum), making diamond by far the hardest known substance. People have learned to make use of diamond's extraordinary hardness by placing tiny industrial-grade diamonds on the edge of rotating saw blades or drill bits, thereby creating tools that will literally cut through anything. Metallic copper is another good example of how humans utilize specific minerals because of their properties. Copper metal is widely used for wiring because it easily conducts electricity. Steel, which is made from iron, is also a good electrical conductor and is considerably cheaper than copper. Why then is copper used for electrical wiring and not steel? The answer is that copper has an additional physical property of being *malleable*, meaning it is able to deform without breaking. Malleability is critical in that it allows copper wiring to be tightly bent or wrapped, something that would cause steel wiring to break or crack, stopping the flow of electricity.

Table 12.2 lists some of the more commonly used metallic and nonmetallic mineral resources and is provided here to give you a better appreciation for the connection between mineral properties and their use in society. A good example is how the mineral quartz is the raw material for making glass. Recall that a natural form of glass, obsidian, forms when silica-rich (SiO_2) magma cools so quickly that the atoms within the melt are unable to arrange themselves into the crystalline structure of minerals. People long ago learned how to produce their own glass by melting pure quartz sand, which is composed entirely of silica (SiO_2). When quartz is melted it results in a viscous liquid that can be molded into practically any shape, which can then be preserved by allowing the liquid to cool and harden. By mixing small amounts of other minerals in with quartz sand, glass with slightly different properties can be made.

Humans obviously have found a large number of applications for glass because of its ability to take on any shape and its unique set of physical properties, which are similar to those of the mineral quartz. In addition to making various storage containers, glass is molded into the small fibers that make up *fiberglass*. Because a blanket of fiberglass contains considerable air space, it inhibits the transfer of heat, making fiberglass an ideal insulating material for buildings and plumbing systems. Another key prop-

TABLE 12.2 Applications and properties of selected metallic and nonmetallic mineral resources.

Metallic and Nonmetallic Mineral Resources	Important Applications	Key Physical and Chemical Properties
Gold (Au)	Electronics, jewelry, currency, bullion	Electrical conductor, noncorrosive, malleable
Silver (Ag)	Electronics, jewelry, photographic films	Electrical conductor, malleable
Copper (Cu)	Electrical wiring, plumbing, coins, alloys	Electrical conductor, malleable
Lead (Pb)	Batteries, solder, bullets, weights	High density, soft, low melting point
Zinc (Zn)	Rust-proofing steel, paint, alloys, coins	Corrosion resistant
Iron (Fe)	Iron, steel, yellow to brown pigments	High strength
Aluminum (Al)	Aluminum metal, chemicals	Lightweight, high strength, corrosion resistant
Titanium (Ti)	White pigment, metal for aircraft, ships, human joint replacements	Lightweight, high strength
Graphite (C)	Dry lubricant, graphite compounds, pencil leads	Extremely soft
Diamond (C)	Cutting tools, gemstones	Extremely hard
Quartz (SiO_2)	Glass, sand for mortar and cement, watch crystals	Transparent, hard, chemically resistant
Calcite ($CaCO_3$) (limestone)	Main ingredient of Portland cement, concrete, agricultural lime	Chemically reactive
Gypsum ($CaSO_4\ 2H_2O$)	Sheetrock (dry wall), plaster of Paris	Low density
Kaolinite clay	Paper filler/coating, filler and extender in paint, rubber, plastics, cosmetics, and medicine, ceramics	Extremely soft, white color, absorbant

erty of quartz-based glass is that it is highly *transparent* (i.e., readily transmits light). Consequently, modern societies use large amounts of quartz for making thin sheets of crystal-clear window glass. The fact that quartz-based glass is also quite hard gives it a distinct advantage in certain applications over other transparent materials that are relatively soft. For example, plexiglass (made from plastics) is rather soft and is never used for automotive windshields since it would become so heavily scratched that it would be difficult to see through. Note that quartz-based glass is also chemically inert, which means that glass products do not break down when exposed to most chemicals. Note that many of the minerals and properties listed in Table 12.2 will be referred to throughout this chapter.

Economic Mineral Deposits

In Chapter 3 you learned that there are over 4,000 known minerals on our planet, but only a dozen or so minerals make up the bulk of Earth's crust and mantle. Even more surprising is that 98.3% of the crust by weight is composed of just eight elements (oxygen, silicon, aluminum, iron, calcium, sodium, potassium, and magnesium). All the remaining elements and their compounds therefore are actually quite rare in terms of the rocks we find near Earth's surface. Fortunately, most of these less abundant elements are not evenly dispersed in trace amounts throughout the crust, but rather are often found concentrated in mineral deposits that are formed by geologic processes. Because it takes energy to extract earth materials, the concentration of a particular deposit is a major factor in determining whether or not it is economical to mine. A more concentrated deposit means there is less waste material to remove, which in turn makes it more profitable since less

TABLE 12.3   The average crustal concentration of some commonly used metals and the enrichment factor needed to create an economical ore deposit.

Metallic Element	Average % Concentration in Earth's Crust	% Concentration Needed for Economical Mining	Approximate Enrichment Factor (economic concentration/ crustal concentration)
Aluminum (Al)	8	35	4
Iron (Fe)	5	20–69	4–14
Titanium (Ti)	0.57	32–60	56–105
Copper (Cu)	0.0063	0.4–0.8	80–160
Lead (Pb)	0.0015	4	2,500
Gold (Au)	0.0000004	0.001	2,500

Source: Data from U.S. Geological Survey Professional Paper 820, 1973.

energy is required to extract the resource. Geologists use the term **enrichment factor** to describe the degree to which a mineral resource is concentrated above its average concentration in the crust. A deposit may be deemed economical to mine if it reaches a certain concentration factor. Table 12.3 lists the average crustal composition of several important metals and the amount of enrichment typically required to create an economical deposit. Notice how an element such as gold can be mined profitably at a much lower concentration than aluminum. The different enrichment factors reflect both the value of these resources to society and the energy required to extract them.

Since society has developed applications for a relatively small number of Earth's 4,000 or so minerals, we need some means of describing those minerals which are economically valuable. Although the term *ore* is often associated with metals, for our purposes we can define *ore minerals* as those which contain an element or compound that has some value to society. Likewise, an **ore deposit** will be defined as a body of rock or sediment whose concentration of ore minerals is sufficiently high so that it is economically feasible to extract. Note that the terms *low-grade* and *high-grade* refer to the enrichment level of ore deposits. Later in this chapter we will examine how geologic processes cause crustal elements to become concentrated and form ore deposits.

Resources and Reserves

The first task for any mining operation is to locate a mineral deposit. High-grade deposits in the past were oftentimes located on the surface because they looked noticeably different from the surrounding rocks. Deposits remaining today are generally more difficult to locate because they either lie below the surface or contain fine-grained minerals that are dispersed throughout a rock body. Many mineral deposits are not economical to mine. The volume of a deposit may be too small, or its mineral concentration too low to justify the cost of mining and processing. Keep in mind that changes in technology or supply and demand can alter the economics of mineral extraction and processing. Therefore, deposits that are considered uneconomical to mine may become profitable in the future.

In order to quantify the value of their mineral resources, mining companies and governments will typically classify their holdings based on economic status. For example, the U.S. Geological Survey uses the term *mineral resource* to describe those deposits that are *feasible* to mine using existing technology. This definition simply means the technical ability to extract the minerals exists, but says nothing about its profitability. A **mineral reserve,** on the other hand, is a deposit that is economical to extract under current conditions. As shown in Figure 12.3, this classification makes it possible for geologists to quantify what is referred to as *total mineral reserves*, which represents all the known deposits currently economical to mine. The remaining known deposits are then combined with some estimate of deposits yet to be found, providing what is called the *total mineral resource*. This concept is important because it allows economists to estimate how much of a particular resource is available to the global marketplace at any given time. The ratio of reserves to resources, of course, will vary as supply and demand affects the market price for a particular resource. Should demand decrease or supply increase, prices will fall and many marginally economical mining

	Identified deposits	Undiscovered deposits	
		Known mining areas	Unknown mining areas
Economic	Reserves		
Marginally economic	Marginal reserves	Hypothetical resources	Speculative resources
Subeconomic	Subeconomic resources		

FIGURE 12.3   Mineral resources are either known or undiscovered. Those that have been identified and are profitable to mine are called reserves. Those believed to exist in areas of known deposits but have yet to be located are considered hypothetical resources; all other mineral resources are speculative.

operations will shut down, causing the total reserves to decrease. The opposite will occur when demand goes up or supply goes down. Oftentimes the driving force behind these changes are updated or new technologies that allow lower-grade deposits to be mined.

Geology of Mineral Resources

In this section we will explore some of the more common geologic processes that result in crustal elements becoming concentrated, thereby forming enriched mineral deposits. These enrichment processes operating within the Earth system will be broken down into igneous, metamorphic, sedimentary, and weathering processes.

Igneous Processes

Diamond Pipes

Diamond deposits are unique in that they are found associated with an unusual type of igneous rock that forms from magmas originating at depths of 75 to 125 miles (120–200 km) in the upper mantle. As indicated in Figure 12.4, the pressure and temperature at this depth can cause carbon atoms to arrange themselves into a stable atomic structure (Chapter 3), forming individual crystals of the mineral we call diamond. Closer to the surface, the stable mineral structure for carbon atoms is that of graphite—one of the softest substances, whereas diamond is the hardest. Once formed, the diamond crystals can be carried by magma that forces its way toward the surface through fissures. Because the magma is highly pressurized, it decompresses suddenly and violently (Chapter 6) when encountering a weak zone in rocks near the surface. The result is a small, but very explosive volcanic eruption that produces a carrot-shaped crater known as a diamond or *kimberlite pipe* (named after the town of Kimberly in South Africa). The pipes themselves are filled with magnesium-rich volcanic rocks that generally are quite different from the surrounding rocks.

Diamonds are relatively rare because the formation of kimberlite pipes is geologically somewhat uncommon, plus not all pipes contain diamonds. Interestingly, the volcanic eruptions that form kimberlite pipes create circular deposits of volcanic debris that may contain diamonds (Figure 12.4). It is believed that the first diamonds ever found were in this volcanic debris lying on the surface. Diamonds were later discovered by gold prospectors in stream gravels that had been carried off the landscape by running water. Eventually people recognized that diamonds were associated with the circular volcanic deposits. This association led to the discovery of new deposits and the mining of diamond-bearing rocks from within the kimberlite pipes themselves. Note that kimberlite pipes tend to occur in swarms or clusters, which is why diamond mines are usually found within small geographic areas, often called *mining districts.*

FIGURE 12.4   Diamonds form in the upper mantle under conditions of high temperature and pressure, and then carried to the surface by magma. Near the surface the highly pressurized magma explodes violently, creating a pipe-shaped crater. Diamonds are found in the volcanic rocks filling the crater, ejected material, and nearby stream gravels.

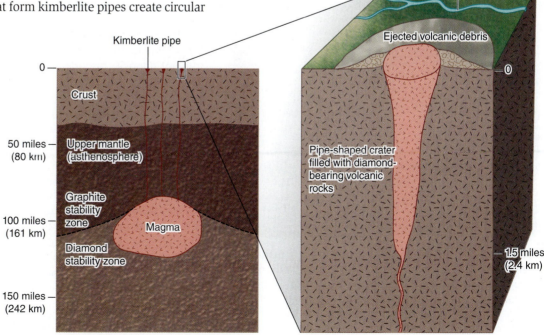

Early stage

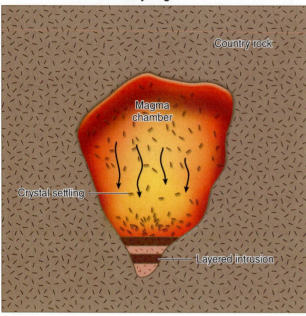

A

FIGURE 12.5 Dense minerals that are the first to crystallize can settle to the bottom of a magma chamber and form a layered ore deposit (A). The photo (B) shows layers of chromium-rich minerals that are part of a layered intrusion in Bushveld, South Africa.

Late stage

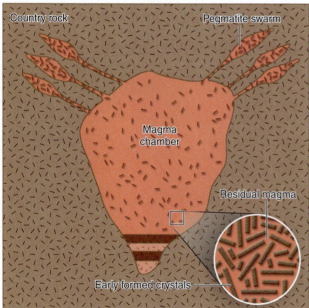

FIGURE 12.6 In the later stages of cooling, a magma chamber will consist mostly of mineral crystals, with the remaining magma residing between the crystals. This residual magma is enriched with certain elements that can form valuable mineral deposits if injected into surrounding rocks, forming small intrusive bodies called pegmatites.

B

Intrusive Deposits

Another important class of mineral deposits is associated with large intrusive bodies of igneous rocks (Chapter 3). Recall that magma forms when the combination of temperature and pressure within the Earth are such that the minerals within a particular rock body begin to reach their melting points—presence of water is also a factor in the melting process. As the minerals melt, the chemical bonds between atoms break, releasing charged atoms called *ions* (e.g., K^+, Ca^{2+}, Cl^-, SiO_4^{4-}) that make up the magma. When the magma eventually begins to cool, the ions will be incorporated into new minerals in a sequential manner according to the specific crystallization temperatures of different minerals. Because minerals crystallize at different temperatures, this means that during the early phases of the cooling process the magma chamber will contain both molten material and solid mineral grains.

Because certain types of minerals are crystallizing from the magma before others, this creates the potential for the development of an economic mineral deposit, provided that the minerals in question have some value to society. What needs to occur is for the early-formed minerals to somehow become separated from the remaining magma before the cooling is complete. One way this can happen is through a process called *crystal settling*, whereby dense, early-formed minerals fall or settle to the bottom of the magma chamber. As shown in Figure 12.5, this process can create layered ore deposits that geologists refer to as **layered intrusions.** Layered ore deposits typically contain metallic minerals that are valuable sources of chromium, titanium, and vanadium.

Igneous ore deposits can also form when magma migrates away from newly formed minerals, as opposed to the minerals settling out of the melt. This process is believed to occur later in the cooling process when the remaining magma resides between the mineral grains—similar to how water exists in the pore space of sediment. Pressure changes can then force this residual magma to leave the magma chamber where it is then injected into fractures within the surrounding rock. As illustrated in Figure 12.6, the magma then cools and forms elongated bodies of igneous rocks. These small intrusive bodies sometimes contain valuable ore minerals because the crystallization process leaves the residual magma enriched with certain elements. One type of mineral deposit that forms in this manner is unusually coarse-grained deposits of silica-rich rock called *pegmatites*. These pegmatites are similar to granite in composition, but will sometimes contain concentrations of minerals composed of rare elements (e.g., lithium, beryllium, niobium, and tantalum), which are normally found dispersed throughout an igneous rock body. Some pegmatites are mined solely for their exceptionally large grains of quartz and feldspar.

Hydrothermal Deposits

In Chapter 6 you learned that all magmas contain varying amounts of water. When magma slowly cools beneath the surface, some of the water will become incorporated into the crystalline structure of certain minerals, and some will remain with the magma during the final stages of crystallization. Water is also commonly present in the rocks surrounding the magma, which geologists refer to as *country rocks*. Naturally, the intense heat from the magma body raises the temperature of any groundwater that is present in the country rocks. The combination of these waters results in hot, mineral-rich fluids that transport ions and chemically react with rocks in a zone around an igneous intrusion. Minerals that crystallize from these highly enriched fluids form what are referred to as **hydrothermal deposits.** Because sulfur (S) is a highly mobile element common in groundwater and magmatic water, hydrothermal deposits typically contain valuable minerals where the sulfide ion (S^{2-}) is bonded to metals such as copper (Cu), lead (Pb), and Zinc (Zn). Gold (Au) and silver (Ag) ores are also common.

Hydrothermal fluids will normally deposit minerals in concentrated masses or in a dispersed manner throughout a rock body as shown in Figure 12.7. For example, *vein deposits* occur when ore minerals crystallize from the hot fluids and fill fractures and small fissures within rocks. Vein deposits are commonly considered to be high-grade deposits because the ore minerals are highly concentrated. In addition to filling fractures, hydrothermal fluids will also move through the country rocks in a diffuse manner, resulting in low-grade deposits where the ore minerals are widely dispersed in what are known as **disseminated deposits.** Note in Figure 12.7 that

FIGURE 12.7 Vein and disseminated ore deposits result from hot, mineral-rich fluids that chemically react with minerals in an igneous intrusion and surrounding rocks, and then transfer elements within a zone around the igneous intrusion. The photo shows a vein deposit containing valuable tungsten and tin minerals in a Portuguese mine.

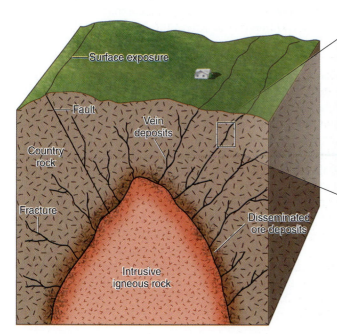

Surface exposure
Fault
Vein deposits
Country rock
Fracture
Disseminated ore deposits
Intrusive igneous rock

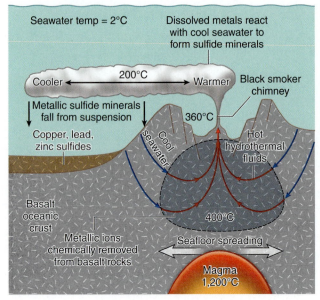

A

B

FIGURE 12.8   Massive sulfide deposits (A) form when hydrothermal fluids discharge from mid-oceanic ridges and then mix with cool seawater. Here metallic ions bond with sulfur, forming sulfide minerals that eventually accumulate on the seafloor. Note how heat convection pulls cold seawater into the ridge, where it reacts chemically with basalt to form hydrothermal fluids. Photo (B) shows sulfide minerals precipitating as hydrothermal fluids escape from vents on the seafloor.

disseminated ores can be found in both the intrusion and the surrounding country rocks. Some of the world's largest surface mines involve removing massive bodies of rock where the disseminated copper minerals are so small that they cannot be seen with the naked eye.

Another type of hydrothermal deposit is associated with divergent plate boundaries where active volcanism occurs along mid-oceanic ridges. Recall from Chapter 4 that in areas of active seafloor spreading how basaltic magmas create new oceanic crust along extensive ridge systems on the ocean floor. As illustrated in Figure 12.8, heat from shallow magma bodies causes hydrothermal fluids to flow upward through fractured rocks within the ridge; the fluids are replaced by cold seawater flowing downward along the flanks of the ridge. This convective motion is believed to cause strong chemical reactions between seawater and the basaltic rocks, generating hydrothermal fluids rich in copper, lead, and zinc. As the hot fluids discharge along the crest of the ridge, the dissolved metals react with cold ocean water to form sulfide minerals, producing a thick mineral deposit called a **massive sulfide deposit.** Notice in Figure 12.8 that sulfide deposits called *black smokers* or *chimneys* form where the hydrothermal fluids discharge from the seafloor.

Massive sulfide deposits have long been important sources of copper, lead, and zinc. These deposits are quite ancient and there is much geologic evidence to suggest that some found on land today have originated along oceanic spreading centers. It is believed that some sulfide deposits are now exposed on land because of simple tectonic uplift, whereas others may have been welded onto an existing landmass when an oceanic plate descended into a subduction zone. A good example is the island of Cyprus in the Mediterranean, where a massive sulfide deposit has been an important source of copper for nearly 4,000 years. In fact, the word *copper* itself is derived from the island's original Greek name of *Kupros.* Cyprus was one of the ancient sites where people made copper implements from pure copper metal found at the surface. After exhausting the supply of pure metal, people continued to obtain copper by placing copper-sulfide minerals into a large fire. The intense heat would cause the chemical bonds within the minerals to break, resulting in molten copper draining into a pit at the base of the fire.

Metamorphic Processes

When rocks undergo metamorphism (Chapters 3 and 4) they often experience important physical and chemical changes that can lead to the development of valuable mineral deposits. However, because igneous activity is commonly associated with metamorphism, it is sometimes difficult to classify an ore deposit as being strictly metamorphic or igneous in origin, particularly when hydrothermal fluids are involved. In this section we will briefly explore the role metamorphic processes play in generating economic mineral deposits.

Regional Metamorphism

Recall from Chapter 3 that *regional metamorphism* takes place when rocks become deeply buried or involved in a mountain-building episode, exposing them to intense levels of heat and pressure. During this process many types of minerals are transformed into more stable minerals. Also, plate-shaped minerals are reoriented such that the mineral grains become parallel to one another (i.e., foliation). A good example is how this metamorphic process transforms clay-rich sedimentary rock (shale) into slate (Figure 12.9A). The transformation of shale into slate creates a much harder and stronger rock that easily breaks along foliation planes, creating thin sheets of slate that are useful for roofing shingles (Figure 12.9B) and flooring tiles. Today slate shingles have largely been replaced by less durable, but less expensive asphalt shingles. Another common use of slate is for making high-quality pool tables—those made of plywood tend to take on water and become warped.

Regional metamorphism also produces *marble,* a rock whose color and attractive crystalline texture makes it highly desirable for use as ornamental building stones, flooring, statues, and so forth. Marble forms from beds of limestone rock during regional or contact metamorphism. Here elevated temperatures cause calcite ($CaCO_3$) grains to recrystallize, creating a new rock with a much coarser texture. Although still composed of calcite, marble is usually white or pinkish in color as the impurities in the original limestone tend to migrate into other rock layers. Unfortunately, marble monuments and building stones, particularly those in Europe, have undergone serious decay due to acid rain (Chapter 15).

Yet another important mineral resource related to regional metamorphism is the collection of fibrous silicate minerals known as *asbestos.* These minerals have widespread applications in industry and in construction materials due to their fibrous nature and ability to resist extreme levels of heat and chemical breakdown. Tragically, the very same properties that made asbestos so versatile also resulted in fatal lung and stomach diseases among workers who had breathed in airborne asbestos fibers. Asbestos not only resulted in the loss of lives, but led to a costly and controversial effort to reduce the health risk by removing asbestos from buildings (Case Study 12.1).

Contact Metamorphism

Contact metamorphism occurs when rising magma comes into contact with rocks, thereby subjecting the rocks to higher temperatures, but not necessarily higher pressures (Chapter 3). Hydrothermal fluids are also commonly present, which chemically react with many of the minerals within the surrounding country rocks. In essence then, rocks that were once in a relatively cool and stable environment suddenly find themselves being baked and infiltrated by hot, corrosive fluids. The result is what geologists refer

A Slate outcrop in Antarctica

B

FIGURE 12.9 Slate (A) forms when shale undergoes regional metamorphism. Because slate breaks into thin sheets that are hard and durable, it has long been used for roofing shingles (B) and flooring tiles.

Asbestos: A Miracle Fiber Turned Deadly

Asbestos is the general name applied to a family of fibrous minerals that have an extraordinary ability to resist heat and chemical breakdown (Figure B12.1). The earliest documented use of this unique set of minerals goes back nearly 5,000 years; in fact, the term asbestos originated from a Greek word meaning inextinguishable. Both the Greeks and Romans were known to weave asbestos fibers into fireproof cloth. Roman restaurants were said to have used asbestos tablecloths, which were cleaned by throwing them into a hot fire! It was not until the Industrial Revolution, however, that the use of asbestos became widespread. The ability of the fibrous minerals to withstand extreme amounts of heat made asbestos ideal as an insulating material for steam boilers, kilns, and ovens and for making fireproof clothing (Figure B12.2). Later applications included thermal insulation for hot water pipes, fireproofing of structural steel supports in buildings, automotive brake pads and gaskets, and as a binder for holding together ceiling and flooring tiles. By the 1970s this so-called miracle fiber was being used in thousands of products. Even in the home it could be found in products ranging from glue to toaster ovens to hair dryers.

FIGURE C12.1 Photograph taken with an electron microscope showing needlelike fibers of the asbestos mineral called anthophyllite. Asbestos fibers can easily become lodged in lung and stomach tissue and cannot be broken down chemically, leading to scarring of the lungs and fatal lung and stomach cancers.

Unfortunately, modern society ignored earlier warnings that these miracle fibers can create serious human health problems, something which was known as far back as Roman times. It was not until the 1960s that medical studies proved that workers who had been exposed to airborne asbestos in the 1930s and 1940s developed high rates of fatal lung and stomach diseases linked exclusively to certain types of asbestos. For example, in one study of shipyard workers who had installed asbestos insulation, over 50% were found to have contracted *asbestosis*—a progressive scarring of lung tissue. Also shocking was the fact that over 10% of the workers' spouses had contracted the same disease, presumably by washing contaminated work clothes.

Medical research showed that the properties which made asbestos so versatile, particularly its fibrous nature and chemical resistance, also made it quite deadly to humans. The needlelike shape of certain types of asbestos (Figure B12.1) and chemical resistance made it nearly impossible for the human lung to expel individual fibers or to break them down chemically. Researchers found that both *asbestosis* (scarring of lung tissue) and *mesothelioma,* a cancer in the lining of

the lungs, were caused exclusively by asbestos. Lung and stomach cancers were also linked to asbestos, but here the link was statistical in nature because these cancers had other causes, such as cigarette smoking. Another important aspect of asbestos-related diseases is that the time period between exposure and the onset of disease, called *latency period,* is rather long, ranging from 10 to 40 years. This, of course, delayed recognition by the public that a medical crisis was slowly building. Between 1940 and 1979, approximately 27 million workers were exposed to airborne asbestos. Based on insurance claims, 23–25% of the workers exposed to high levels of asbestos are estimated to have died of asbestosis, 7–10% of mesothelioma, and 20–25% of lung cancer.

Although definitive medical studies did not take place until the 1960s, U.S. companies and government agencies had documented the risks of asbestos to U.S. workers as early as the 1930s. It was not until 1971, however, that the U.S. Occupational Safety and Health Administration (OSHA) began to regulate exposure levels of asbestos in the workplace. The use of asbestos greatly declined after 1989 when the U.S. Environmental Protection Agency (EPA) banned its use in most commercial products. The combination of environmental regulations and threat of liability lawsuits quickly led to the widespread and controversial removal of asbestos-containing materials from schools and commercial buildings (Figure B12.3). This costly effort was seen by many as not only a poor use of scarce funds, but one that could create an unnecessary health risk. At issue was the fact that asbestos was firmly contained within many types of building materials, and therefore presented very little risk of becoming an airborne hazard. Removing these materials could actually create a greater risk compared to keeping them in place. Moreover, it was discovered that the most serious health threats were primarily related to just two types of asbestos minerals (amosite and crocidolite), which lowered the risk even further for many types of materials that were being removed.

The modern use of asbestos presents an interesting example of how a highly useful mineral was abruptly dropped by society due to its threat to human health. While liability concerns and panic helped drive the wholesale removal of asbestos from buildings—an action that was not necessarily justified in terms of health hazards—

the fact remains that certain forms of asbestos can be deadly. Due to the environmental regulation passed in the 1970s and 1980s, about the only people at risk from asbestos in the workplace today are maintence and construction workers who periodically disturb materials containing asbestos. The key to minimizing even these lower levels of exposure has been for businesses to survey their buildings and identify those materials that contain asbestos, and then train their staff to avoid disturbing these materials. Should maintence or construction work require that the asbestos be removed or disturbed, workers must wear protective respirators and take steps to ensure that asbestos fibers are contained and therefore not released into the building (Figure B12.3).

FIGURE B12.2 Photo showing World War II British firefighters training in asbestos suits.

FIGURE B12.3 The removal of asbestos from buildings is costly and requires that workers and building occupants be protected from airborne fibers. Photo shows asbestos fireproofing being removed from steel support beams within a building.

White marble

Black basalt

B

FIGURE 12.10 Magmatic heat and hydrothermal fluids commonly create a zone of altered rocks (A) surrounding an intrusion in which ore minerals are deposited. Some rock types are more reactive than others, and thus have wider alteration zones and accumulate different types of minerals. Photo (B) showing a marble alteration zone surrounding a basaltic intrusion into limestone beds in Glacier Park, Montana.

Labels in figure A: Ore minerals, Alteration halo, Vein deposit, Disseminated deposit, Magma chamber, Granite

A

to as an *alteration zone* or *halo* that surrounds the intrusion, as shown in Figure 12.10. Note that the width of the alteration zone largely depends on the susceptibility of the minerals in the country rocks to the heat and hydrothermal fluids. Also notice how vein and disseminated ore deposits can occur in selective rock types within the alteration zone. Here, complex chemical reactions between the mineral-rich hydrothermal fluids and surrounding rocks play an important role in determining where the ore minerals precipitate. Limestone rocks typically have deep alteration zones composed of marble, as the mineral calcite ($CaCO_3$) easily recrystallizes into much coarser grains.

Sedimentary Processes

As is the case with the formation of igneous and metamorphic rocks, the various processes that form sedimentary rocks tend to concentrate certain types of minerals, many of which society has found useful. In Chapter 3 we discussed the two basic types of sedimentary rocks: *clastic rocks* composed of fragmental material and *chemical rocks* formed by the chemical precipitation of minerals from water. In this section we will examine some of the minerals from these two basic groups of sedimentary rocks and their applications in society. Although coal forms by sedimentary processes, it is used as an energy resource and will therefore be discussed in Chapter 13 on conventional energy resources.

Sand, Gravel, and Clay

Recall from Chapter 8 that *hydraulic sorting* involves the separation of sediment grains by flowing water based on the size, shape, and density of the individual grains. This sorting actually occurs whenever the velocity of a fluid (water or wind) increases such that sediment grains are selectively picked up and then deposited in a sequential manner as the velocity decreases. This natural enrichment process is important because it provides humans with valuable deposits of relatively pure gravel, sand, and clay. Depending on the source material, hydraulic sorting from wind or water (streams and beaches) can produce sand deposits that are composed of nearly 100% quartz (SiO_2) grains. Humans have long sought such deposits of quartz sand as the raw material for making glass.

The hydraulic sorting that occurs when fine particles settle out in lakes and ponds often results in sediment that is composed almost entirely of clay minerals (Chapters 3 and 10). In addition to sand, clay is one of the most basic types of mineral resources. For example, because of its low permeability clay is used to line ponds and earthen dams so that water can be stored for irrigation or even power generation. Historically one of the greatest uses of clay has been in making clay pots and bricks. Here moist clay is molded into various shapes and then dried in the sun to drive off the water, causing the material to harden. By placing molded materials in a kiln, the temperature can be raised to the point where clay minerals are transformed into stronger and harder minerals, thus producing more durable clay implements; this process is analogous to how shale is transformed into slate during metamorphism.

In terms of weight, stream-deposited sands and gravels are by far the most widely used mineral resources in modern societies. As listed in Table 12.1, these deposits represent 38% of all the mineral resources used in the United States, exceeded only by crushed stone (49%). Huge quantities of sand and gravel are mined for use in construction projects and for making building materials. Some of the more common applications of sand and gravel include aggregate for strengthening mortar and concrete and as fill material that forms the base for roadways and building foundations.

Placers

When rocks containing valuable minerals are exposed to weathering, they naturally begin to break down. In some cases the ore minerals are resistant to chemical weathering and are eventually incorporated into the sediment load of nearby streams, as illustrated in Figure 12.11. Here flowing water will cause any dense and chemically resistant minerals to be sorted hydraulically from the rest of the sediment load, forming a concentrated mineral deposit called a **placer deposit.** Gold-bearing stream placers are perhaps the most well known, but placers can also contain valuable platinum, tin, and titanium minerals as well as diamonds. Historically, some prospectors were able to locate rich hydrothermal vein deposits, which they called the *mother lode,* simply by walking upstream and testing the sediment for traces of gold (Figure 12.11). Note that rich placer deposits are also found in stream terraces (Chapter 8), which are older stream deposits located on ground that is higher than the sediment currently being deposited.

Because most streams ultimately transport their sediment load to the ocean or large lakes, wave and wind action along shorelines will also hydraulically sort minerals that have survived the journey, forming what are referred to as *beach placers.* In fact, the major source of titanium comes from the mining of beach placers, where several types of titanium minerals are found because of their resistance to chemical weathering—on active beaches

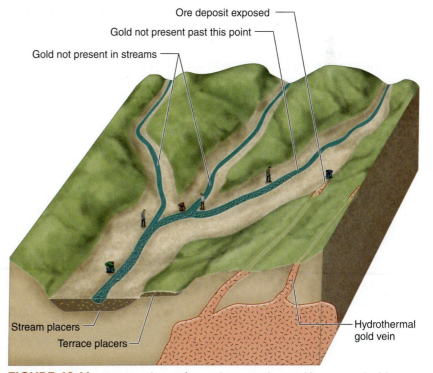

FIGURE 12.11 Stream placers form when weathering liberates valuable minerals from the primary ore deposit, then erosion carries the minerals to a stream where they become concentrated through hydraulic sorting. By following traces of gold in the sediment upstream through the drainage system, prospectors could sometimes locate the primary deposit, referred to as the mother lode.

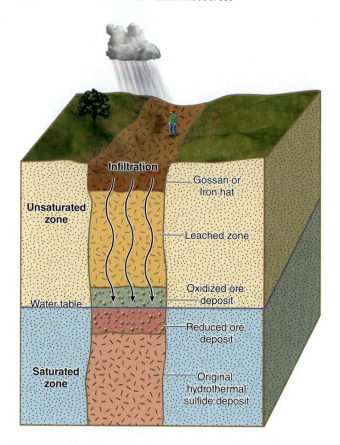

FIGURE 12.12 The weathering of hydrothermal veins causes sulfide minerals to break down, allowing water to carry metallic ions downward where they recombine to form more stable minerals. This commonly results in two enriched ore deposits, one in the oxidized zone above the water table and the other in the reduced zone below, where oxygen levels are low. A low-grade deposit of residual iron minerals lies at the surface, which helps prospectors locate the underlying enriched zones.

these titanium minerals often appear as thin layers of black sand. In southern Africa, beach placers are mined for gold as well as for high-quality diamonds. Keep in mind that most of the beach placers currently being mined are located inland and formed during periods of higher sea levels.

Residual Weathering Products

In Chapter 10 you learned how soils develop from the physical and chemical weathering of the minerals that make up rocks. Minerals that are resistant to chemical weathering, such as quartz, will remain in newly formed soils and the less resistant mineral will break down into secondary minerals, often called *weathering products*. This chemical transformation of minerals into secondary weathering products results in the release of ions or charged atoms that get carried downward with infiltrating water. The secondary minerals within the soil column therefore become enriched in those elements that are not lost, which ultimately may represent minerals that are valuable to society. Likewise, those ions that are carried away with the water may recombine and form yet another economical mineral deposit.

In the case of rocks such as granite and basalt, recall that many of the aluminum- and silicate-rich minerals (i.e., *aluminosilicates*) will chemically break down into various types of clay minerals that are enriched in aluminum (Chapters 3 and 10). As this process continues, the resulting clay minerals become progressively enriched in aluminum. Given a long period of time and a climate with considerable rainfall, the weathering of silicate rocks can lead to a deposit of highly enriched aluminum minerals known as **bauxite.** Through the mining and processing of bauxite ores humans are able to obtain pure aluminum metal, which of course has found numerous applications in modern societies. Aluminum is particularly useful in applications where both weight and strength are an issue, such as in airplanes and fuel-efficient cars. Note that soil deposits rich in iron oxide minerals, referred to as *laterite*, can also form from the intense weathering of silicate rocks (Chapter 10). Laterite deposits have historically been used for making bricks that are relatively strong and easy to cut.

The weathering of low-grade sulfide deposits also produces highly enriched metallic ore deposits. For example, Figure 12.12 shows a hydrothermal vein, which contains disseminated metallic sulfide minerals that have been exposed at the surface and have undergone extensive weathering. Above the water table the vein lies in the unsaturated zone (Chapter 11) where it is exposed to both free oxygen (O_2) and infiltrating water. In this zone the original sulfide minerals, such as pyrite (FeS_2), are chemically unstable and will oxidize relatively quickly. When this occurs many of the metallic ions (e.g., iron, copper, lead, and zinc) will be stripped or *leached* from the sulfide minerals and transported downward with the infiltrating water. During this movement some of the metals will then form new minerals by combining with more stable negative ions such as oxygen (O^{2-}), carbonate (CO_3^{2-}), or sulfate (SO_4^{2-}). The result is an enriched deposit of oxide, carbonate, and sulfate minerals lying just above the water table.

Some of the metallic ions that are leached from the primary sulfide minerals will remain soluble, particularly copper (Cu^{2+}), and reach the water table where oxygen levels are commonly lower and reducing conditions prevail. Once in this reducing environment the metals will commonly precipitate back into sulfide minerals, such as bornite (Cu_5FeS_4). This means that the weathering of the original hydrothermal vein can result in two separate high-grade metallic ore deposits; one lying above the water table consisting of oxide, carbonate, and sulfate minerals, and the other one composed of sulfide minerals located below the water table. Note that the *gossan* or

Red chert layers

Gray hematite layers

A

B

iron hat in Figure 12.12 is a surface deposit consisting of iron oxide minerals and quartz, which represents the final weathering products of the sulfide minerals in the original hydrothermal vein. Although the iron hat is a rather low-grade deposit, the iron oxide minerals give it a strong yellowish-brown color, a feature which prospectors used for locating the much more profitable ore deposits lying below.

Banded Iron Formation

The ability of humans to mine and process large quantities of iron ore was, of course, one of the key factors that made the Industrial Revolution possible. Steel made from iron is used for making the basic machinery that runs our factories, which, in turn, make nearly everything we use. The trucks, trains, and ships of our transportation system are all made from iron as is the farm equipment that produces our food. Iron-based steel also provides the critical strength necessary to support bridges and large buildings.

Interestingly, the vast majority of Earth's iron deposits formed over a billion years ago when the planet's atmosphere contained very little free oxygen (O_2). A sizable portion of the world's iron reserves (Figure 12.13) consists of alternating layers of quartz and iron oxide minerals, particularly hematite (Fe_2O_3) and magnetite (Fe_3O_4). Geologists generally agree that these deposits represent chemical sedimentary rocks that formed between 2.6 and 1.8 billion years ago as iron began to chemically precipitate out of shallow seas. During this period plant life was quite limited, which means very little free oxygen was being produced by photosynthesis. Therefore, the iron that was being released from the weathering of iron-rich silicate rocks would have remained in solution because of the lack of atmospheric oxygen. Later, as photosynthetic algae began to proliferate, the concentration of O_2 in the atmosphere and oceans

FIGURE 12.13 Banded iron deposits (A) are believed to have formed between 2.6 and 1.8 billion years ago when free oxygen became abundant in the atmosphere. Photo showing gray bands of the iron mineral hematite separated by alternating layers of red, fine-grained chert (SiO_2)—note that the layers have been deformed since they were deposited. Photo (B) shows a large open-pit iron mine in the Upper Peninsula of Michigan.

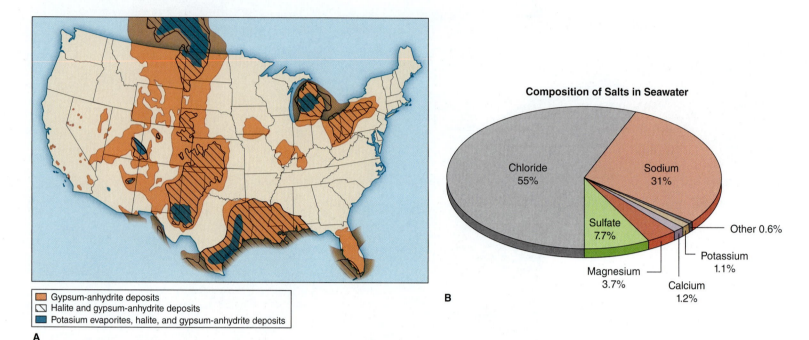

Gypsum-anhydrite deposits
Halite and gypsum-anhydrite deposits
Potasium evaporites, halite, and gypsum-anhydrite deposits

A

Composition of Salts in Seawater

Chloride 55%
Sodium 31%
Sulfate 7.7%
Other 0.6%
Potassium 1.1%
Magnesium 3.7%
Calcium 1.2%

B

FIGURE 12.14   Map (A) showing both marine and freshwater evaporite deposits in the United States. Marine deposits consist chiefly of salts based on chloride and sulfate ions, which are the dominant negatively charged ions in seawater (B).

increased, causing dissolved iron to precipitate and form iron oxide minerals. It is thought that these minerals then fell out of suspension in shallow seas and accumulated in layers. Although the ancient iron deposits could have formed in other ways, this mechanism nicely explains the sheer volume of the deposits within the geologic rock record and the fact that they coincide with the proliferation of photosynthetic algae and free oxygen.

Evaporites

Another important class of economic sedimentary minerals forms when water bodies undergo evaporation. As water molecules vaporize and enter the gas state, dissolved ions or salts will stay behind in the water body, raising its salinity. Depending on the rate of evaporation versus the inflow of any new water, the remaining water, called *brine,* may become so concentrated that mineral grains begin to precipitate from the dissolved salts. These mineral or salt grains then fall to the bottom of the lake or seabed and form layers of chemical sedimentary rock known as **evaporites.** Should a water body evaporate entirely, then all of its dissolved salts will precipitate and form evaporite minerals—you can try this yourself by taking a glass of salt water and letting it evaporate in the sun.

There are two basic types of evaporite deposits found in the world: marine and nonmarine (Figure 12.14). In the case of marine evaporites, the minerals themselves reflect the chemical composition of seawater, which is dominated by chloride and sodium ions. Most economical marine deposits consist of minerals such as halite (NaCl), gypsum ($CaSO_4 \cdot 2H_2O$), potassium chloride (KCl), and calcium chloride ($CaCl_2$). These minerals are important because they serve as the raw materials for the chemical industry and are required in the processing and preservation of foods. Perhaps the most familiar mineral is halite since it is used for common table salt and for keeping highways free of ice during the winter. Because evaporite beds have extremely low permeability, they serve as confining layers (Chapter 11) within the subsurface that are nearly impermeable. These types of confining layers have played a key role in the accumulation of oil and gas in various sedimentary basins around the world, including those in Texas and Louisiana (Figure 12.14A).

Marine evaporites typically form when tectonic activity causes a shallow body of seawater to become restricted, thereby limiting its connection with the

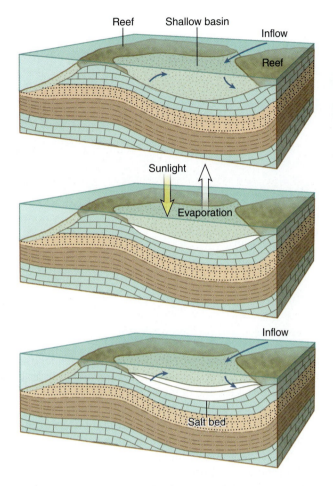

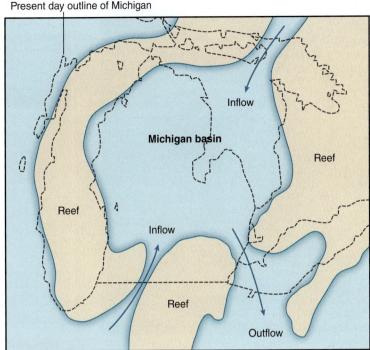

FIGURE 12.15 Evaporite deposits form when surface-water bodies undergo evaporation, increasing salinity to the point where dissolved salts precipitate and fall to the bottom to form layers of salt. Around 400 million years ago a restricted inland sea near the present state of Michigan underwent subsidence and intense evaporation, resulting in thick evaporite and reef limestone deposits that are currently being mined.

open ocean. The restricted circulation allows evaporation to raise the salinity to the point where minerals precipitate (Chapter 3). Keep in mind that the different evaporite minerals do not all precipitate at the same place and time, but in a sequential manner as salinity varies throughout a basin. For example, gypsum will precipitate first in areas of lower salinity followed by halite where the salinity is higher. This results in geographically separate and relatively pure deposits of different salts. To get thick accumulations requires steady salinity levels and a basin that is both actively subsiding and receiving a steady supply of ions to replace those being removed by the deposition of salt. Note that thick evaporite deposits are relatively rare due to the need for maintaining steady salinity levels over long periods of geologic time.

The area known as the Michigan basin is an excellent example of where steady geologic conditions allowed for the deposition of thick evaporite deposits. As illustrated in Figure 12.15, a shallow inland sea covered much of present-day North America during the Silurian period around 400 million years ago. Here a large barrier reef system surrounded what is now the state of Michigan, which created a shallow and highly restricted marine environment. Intense evaporation along with a subsiding seafloor allowed deposits of halite to reach thicknesses of nearly 1,600 feet (490 m) in the middle of the basin. As shown in Figure 12.15, gypsum and reef limestone (calcite) deposits surround the halite in a concentric manner, forming an extensive and important evaporite sequence that is actively being mined.

Because the chemical composition of freshwater is considerably different from seawater, lakes produce certain types of evaporite minerals not commonly found in marine deposits, such as those based on the borate (BO_3^{3-}) and nitrate (NO_3^-) ions. Of particular interest are borate minerals because they serve as the source of boron used in making glass and ceramics as well

FIGURE 12.16   Satellite photo of the Great Salt Lake in Utah. This lake (shown in black) represents the remains of a large body of freshwater that once occupied a series of tectonic valleys. White indicates areas where the lake has completely evaporated, leaving vast salt deposits on the now exposed lake bed.

as lightweight metallic alloys. Examples of freshwater evaporite deposits can be found in tectonic valleys in the western United States. As little as 13,000 years ago at the end of the last ice age many of these valleys, which formed due to extension of the crust (Chapter 4), contained deep lakes whose water was relatively fresh. Since then most of the lakes have completely evaporated as the climate has become more arid, leaving evaporite deposits on the now exposed lake beds. One of the few remaining lakes is the now highly saline Great Salt Lake in Utah, shown in Figure 12.16. This lake lies in a topographic depression that receives water from streams, but the water can only escape through evaporation. Because the overall evaporation rates are relatively constant, water levels today in the Great Salt Lake rise and fall primarily in response to yearly changes in stream runoff.

Phosphorites

Similar to the element nitrogen, phosphorus (P) is one of the essential plant nutrients found in soils (Chapter 10). Consequently, economical mineral deposits containing phosphorus are important in the production of modern fertilizers. Phosphorous-bearing minerals, however, are normally widely dispersed within igneous rocks, and are therefore not economical to mine. When rocks undergo chemical weathering, however, the phosphorus is released in the form of the phosphate ion (PO_4^{3-}), which is then transported to water bodies where it accumulates. Once in aqueous form, animal life is able to extract both phosphate and calcium ions to form teeth and bones composed mainly of the mineral apatite ($CaPO_4$). As a result, most limestone rocks that contain the skeletal remains of marine organisms are relatively rich in phosphorus compared to other rock types. In situations where unusually large amounts of skeletal matter accumulate, the material can represent an economical deposit of phosphate that geologists call *phosphorite*.

Phosphorus has also been mined from deposits of bird and bat excretions called *guano*. Because these excretions are rich in both phosphate and nitrate, guano was once extensively mined for making gunpowder and fertilizer. The mining, unfortunately, was highly disruptive to bird and bat colonies, which experienced sharp population declines. The demand for guano dropped considerably with the advent of modern gunpowder and mining of phosphorite rock deposits.

Mining and Processing of Minerals

The mining and processing of mineral resources has a long history of environmental devastation. The basic problem is that mining by its very nature is destructive to the surface environment. Both ore and waste rock must be removed from the earth, leaving underground passageways or vast open pits at the surface. In some cases entire mountaintops are removed. In addition to the scars, mining creates large volumes of noneconomical rock and processed ore called **mine tailings,** which must be placed on the land surface. All this can lead to poisoned water supplies and serious air pollution problems. In addition, miners can be exposed to hazardous dust and fumes and fatal accidents involving tons of rocks.

As was the case with many other industries in the United States, environmental regulations implemented since the 1970s have greatly increased worker safety and minimized the impact of mining on the environment. For example, the fatality rate in U.S. mines is now about 28 workers per 100,000, which represents a 45% decline between 1990 and 2006, making mining a much safer occupation than fishing (147 deaths per 100,000) and even farming and ranching (37 deaths per 100,000). Despite the fact that mining operations in the United States are a dramatic improvement over those prior to major government regulations, many people still object to new mines being opened near their homes or communities. Part of the problem is that many of the older mines are now abandoned and continue to cause problems, creating a poor impression of modern mining. People also object to new mines because of concerns over lower property values, noise, dust, heavy truck traffic, and, of course, pollution.

Although many of these concerns over mining are valid, we must keep in mind that mineral resources make it possible for us to enjoy the comforts and conveniences of modern society. Like it or not, we are all very much dependent on minerals and help create the demand for mining. It would seem reasonable that as citizens, we help ensure that the impact of mining on the environment is kept at a minimum by practicing conservation and supporting effective environmental regulations. We will first examine the basic mining and processing techniques and then discuss some of their impacts on the environment. A thorough discussion of pollution and environmental regulations can be found in Chapter 15.

Mining Techniques

Prior to the development of modern machinery, the mining of minerals was done using simple hand tools to dig open pits or tunnels into the earth. People would typically begin digging where an ore body was exposed at the surface, and then continued down into the subsurface; an example of this can be seen in Figure 12.17. Similar

FIGURE 12.17   Photo shows the trace of a gold-bearing hydrothermal vein that was mined in the late 1800s in Central City, Colorado. Mining began as a surface operation, but later developed into an underground mine as the ore body was followed into the subsurface. Note the piles of mine tailings (waste material) along the trace of the vein.

techniques are used today, but with large powerful machines and computerized equipment on a massive scale that makes mining much more efficient. Because increased efficiency means greater profitability, mine operators are always looking to use the most efficient means possible for extracting a mineral deposit. Improved mining techniques have also made it possible to extract lower-grade deposits that were once considered uneconomical.

Similar to other businesses, the profitability of mineral extraction is dependent on the market value of a particular mineral resource and the costs associated with getting it out of the ground and to market. One of the major costs involves removing undesirable rock or sediment called **overburden** in order to gain access to a mineral deposit. The problem with overburden is that it requires energy (money) to physically remove it from the earth, plus it must be stored somewhere nearby on the land surface. If the amount of overburden becomes too great, then clearly at some point a mining operation will not be profitable. Another important factor is the concentration or grade of the mineral deposit itself. A higher-grade deposit will enable a mine operator to handle more mine tailings (waste material) and still make a profit. The most desirable situation then would obviously be a high-grade deposit with very little overburden. Other important costs that factor into the economic feasibility of mining include costs associated with transportation, labor, and environmental regulations.

For those who work in the mining industry, an unfortunate aspect has been its long history of so-called boom-and-bust cycles. Fluctuations in the market value of a particular mineral, due to supply and demand changes for example, can make a marginally economic deposit suddenly profitable to mine. Likewise, once-profitable mining operations may be forced to shut down rather abruptly. Because mineral extraction typically requires large amounts of energy, especially in the form of diesel fuel, rapidly rising energy prices can have a significant impact on mining operations around the world. As with most industries, higher energy prices means that consumers pay more for mineral-based products, which basically includes anything that cannot be grown.

Surface Mining

Whenever the amount of overburden is such that a mineral deposit is economical to mine, surface-mining techniques are almost invariably chosen over underground mining, because surface mining has lower operational costs and fewer occupational hazards. There are two basic types of surface-mining operations: open-pit and strip mines. As shown in Figure 12.18, an **open-pit mine** nor-

FIGURE 12.18 Open-pit mines (A) typically involve irregularly shaped, low-grade deposits where large volumes of ore are removed. Terracing allows greater depths to be achieved and minimizes mass-wasting hazards. The Bingham Canyon copper mine (B) in Utah is the world's largest excavation. The inset showing railroad cars on a single terrace step provides a sense of scale.

A

B

mally involves excavating large volumes of a low-grade deposit, such as disseminated ores. In order to minimize mass-wasting hazards, the excavation typically proceeds in a stair-step fashion that creates a series of terraces (Chapter 7). The terraces are also used as transportation routes for trucks and trains that carry material out of the pit. Despite the use of terraces, the overall slope of the excavation can be made only so steep. This means that as the pit becomes deeper, its diameter at the surface must increase. For deep mines this results in a truly enormous hole in the ground. The Bingham Canyon open-pit copper mine (Figure 12.18B) near Salt Lake City, Utah, is the largest pit ever excavated by humans, measuring nearly 2.5 miles (4 km) in diameter at the top and 3,000 feet (900 m) deep.

A **strip mine** is one where the excavation follows a mineral deposit that lies parallel to the surface. These deposits commonly involve sedimentary rocks where the overburden is scraped away in layers using cranes equipped with large dragline buckets (Figure 12.19). Because the deposit is more sheetlike and parallel to the surface, strip mining tends to disrupt a much larger land area compared to underground or open-pit mines. A special form of strip mining, called *mountaintop removal,* involves removing a large hill in order to extract a mineral deposit—coal (Chapter 13) is commonly mined this way. Mountaintop removal is a rather controversial technique in part because it scars the landscape, and because the vast amount of waste material stored on the surface disrupts the natural drainage system and is a potential source of pollution.

Underground Mining

Prior to modern mining equipment that efficiently excavates large volumes of overburden, the only means of accessing deeply buried ore deposits was through underground mining. Today underground mining is used primarily in situations where the ore deposit is so deep that removing the overburden would be too expensive, or where land-use issues makes extensive surface operations impractical. As illustrated in Figure 12.20, underground mining typically involves creating a vertical or inclined shaft by drilling and blasting through solid rock. The ore body is then accessed through a series of horizontal tunnels bored through the solid rock. Depending on the terrain, the ore can sometimes be accessed directly by a horizontal tunnel as opposed to a shaft.

Underground mining clearly creates the potential for deadly rockfalls and cave-ins. Moreover, the blasting of quartz-rich rocks generates fine particles of silica (SiO_2). Long-term exposure to silica dust can lead to the sometimes fatal disease called *silicosis* in which the lungs become inflamed and scarred. Modern safety regulations in the United States have greatly reduced the hazards associated with underground mining. Although underground mining is still more hazardous and has higher operating costs compared to surface operations, it does have an advantage in that it generates a much smaller volume of mine tailings (waste material).

Perhaps the best examples of underground mines are the prolific gold mines of South Africa, which account for nearly 40% of the world's gold reserves. In this region the gold is found in ancient sedimentary rocks that

A

B

FIGURE 12.19   Photo (A) showing a manganese strip mine in South Africa where flat-lying sedimentary beds are overlain by a thin overburden layer. Material is typically scraped up by a large bucket that is pulled along the ground (B), and then dumped into large trucks.

are inclined in the subsurface at a 30–45° angle (similar to the vein in Figure 12.20). After nearly 120 years of following the inclined beds into the subsurface, mining is now taking place at nearly 13,000 feet (3,900 m) vertically beneath the surface—and the inclined distance is much greater. Due to Earth's internal heat (Chapter 3), the temperature of the rocks at this depth is about 140°F (60°C). This requires a massive ventilation and cooling system in order to bring the temperature down to around 90°F (32°C) to create a bearable working environment. The enormous operating costs of these mines are offset by the high market value of gold, and by the fact that the ore body is layered and very high grade which minimizes the amount of waste material that must be removed.

Placer Mining

Placer deposits associated with streams and beach sediments are mined using a technique called *dredging* in which a floating platform is used to scoop up ore-bearing sediment. Once the material is brought onto the dredge, high-density ores such as gold are hydraulically separated from the sediment in much the same way a change in stream velocity naturally sorts sediment. The waste material is then piled onto dry land using a conveyor. Ancient placers are often found above the stream in terrace deposits. During the California gold rush of 1849, miners made extensive use of *hydraulic mining* techniques to mine rich placers located in dry terraces. Mining companies built elevated wooden flumes from the placers to an upstream location where the flume could collect water from the stream. This setup allowed the miners to force tremendous amounts of water through relatively small nozzles, generating high-pressure streams of water. By directing the water onto the terraces, placer material would be washed into wooden sluice boxes (Figure 12.21) where the gold would then be hydraulically separated from the rest of the sediment. Although hydraulic mining

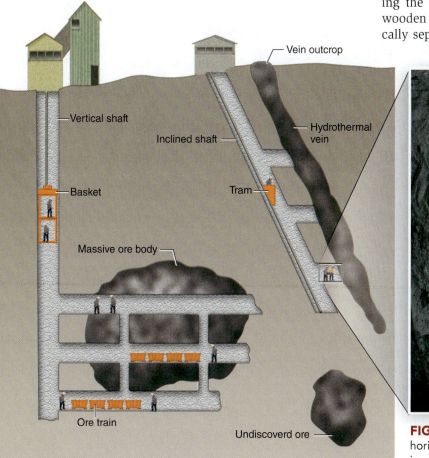

Vein outcrop

Vertical shaft

Inclined shaft

Hydrothermal vein

Basket

Tram

Massive ore body

Ore train

Undiscoverd ore

FIGURE 12.20 Underground mining involves blasting shafts and horizontal tunnels to access mineral deposits. Although safety has greatly improved, surface mining is generally preferred over underground mining as it presents fewer hazards and lower operating costs.

was very effective, it was banned in 1884 because it created serious sediment pollution (Chapters 8 and 10) that degraded drinking-water supplies and increased the risk of flooding. Interestingly, California's ban on hydraulic mining represents one of the earliest pieces of environmental regulation in the United States.

Mineral Processing

When mineral resources are removed from the earth a certain amount of physical and chemical processing is normally required to obtain the desired material. A resource that requires a relatively small amount of processing is crushed stone and sand and gravel used as aggregate (see Table 12.1). To obtain crushed stone, dynamite is used to blast solid rock into pieces that are small enough to be placed on a truck. The material is then hauled to a large mechanical crushing machine located on-site, which physically crushes the rock into appropriate sizes, from the fine gravel found in parking lots to the extremely coarse material used for drainage along highways. In contrast, stream deposits of sand and gravel have the advantage of already being composed of particles of different sizes, so that these only require sorting. Here the sand and gravel is first excavated using earth-moving equipment, and then hauled to a nearby facility where it is sorted by size using different sized screens. After being sorted, the material is washed. In some cases crushing is required to obtain certain sized material that was not present in the natural deposit.

Many mineral resources will contain specific types of ore minerals that are mixed together with noneconomical minerals and rock. In order to extract the desired minerals, the mixture is first crushed and then sieved so it has a uniform grain size. After this process, a variety of techniques can be used to physically separate or extract the valuable minerals. One approach is to make use of the fact that each mineral has its own specific density, which means any mixture is going to contain materials of different densities. Because of the differences in density, flowing water is often used to hydraulically separate the different minerals, which works particularly well for high-density ores such as gold. In contrast, relatively light minerals can be extracted using high-density fluids whereby the minerals rise to the surface of the fluid and literally float. Another means of physical separation makes use of powerful magnets for removing iron ores from the waste material.

After physically separating a desirable mineral, it is sometimes necessary for the material to undergo chemical processing. In many cases, what society finds useful is not necessarily a mineral itself, but rather some specific element it contains. For example, copper (Cu) is the desired element in the mineral bornite (Cu_5FeS_4), but is chemically bonded to iron (Fe) and sulfur (S). The oldest known means of chemically breaking down minerals is the process known as **smelting,** in which metals are obtained by heating certain minerals to the point where their chemical bonds are broken—smelting was used in ancient times to extract copper metal from sulfide minerals. In a similar manner, limestone rock is heated in order to chemically transform calcite ($CaCO_3$) into *lime* (CaO). Note that there are many applications for lime because of its acid-neutralizing ability, but perhaps its most significant use is as the raw material for making cement products.

FIGURE 12.21 Hydraulic mining was developed to facilitate the removal of gold-bearing terrace placers. An elevated flume was used to collect water from upstream, which was then forced through nozzles to create a high-pressure stream of water. The sediment would then be washed into sluices where the gold was separated hydraulically. Photo from 1890 showing hydraulic mining in Nevada County, California, in apparent violation of the 1884 ban of the practice.

Another common chemical processing technique is called **leaching,** in which a solution is allowed to permeate through crushed ore and the resulting chemical reactions liberate the desired element(s). At this stage the mineral resource consists of the ions that are dissolved in the solution, normally referred to as *leachate.* The leachate is then collected and piped to a processing plant where these ions are chemically removed. A good example is the use of cyanide to dissolve metals such as gold and silver from crushed ore. Although cyanide solutions have been used for over 100 years, not until the 1980s did mine operators begin applying the method to what is now called *cyanide heap-leaching.* As illustrated in Figure 12.22, heap-leaching involves spraying a cyanide solution on top of a large pile or heap of crushed ore. The solution then percolates down through the pile and chemically dissolves metallic ore minerals. Buried at the bottom of the

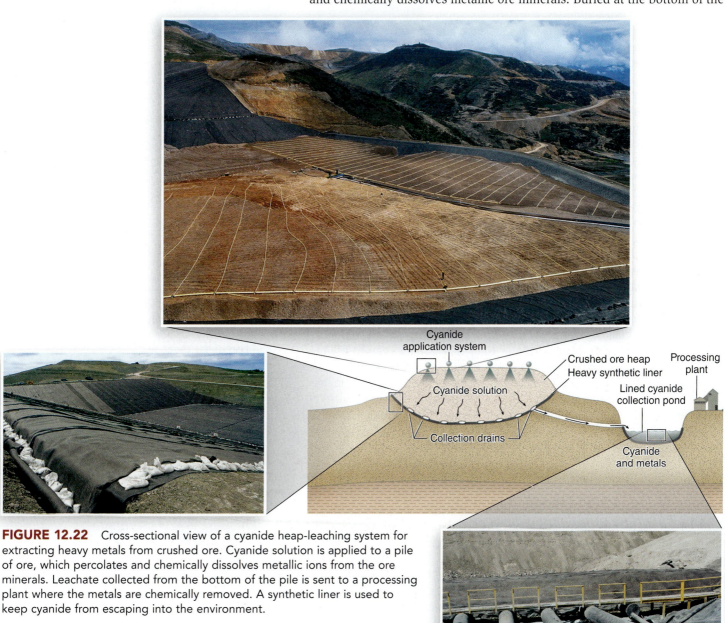

FIGURE 12.22   Cross-sectional view of a cyanide heap-leaching system for extracting heavy metals from crushed ore. Cyanide solution is applied to a pile of ore, which percolates and chemically dissolves metallic ions from the ore minerals. Leachate collected from the bottom of the pile is sent to a processing plant where the metals are chemically removed. A synthetic liner is used to keep cyanide from escaping into the environment.

heap is a system of pipes that collects the resulting leachate and drains it into a pond; both the pond and heap are lined with a heavy synthetic material designed to prevent leachate from escaping. From there the leachate is sent to a nearby processing plant where the valuable metals are removed, after which the cyanide solution is returned back to the heap and used again.

Cyanide heap-leaching has proven to be very effective in dissolving fine-grained ore minerals that are difficult to process by other means. Moreover, this technological advance has now made it possible to mine low-grade deposits that were once considered uneconomical. This process has also made it feasible to recover valuable metals from abandoned mine tailings. Although cyanide heap-leaching represents a significant improvement in our ability to obtain certain metals, it is also rather controversial because of its potential threat to the environment.

Distribution and Supply of Mineral Resources

Humans clearly expend a great deal of energy extracting and processing mineral resources. Also, the different types of mineral deposits are not just found anywhere, but rather in very specific geologic environments where natural processes concentrate certain minerals. For example, the map in Figure 12.23 shows the location of disseminated copper deposits that are associated with igneous intrusions. Note that many of the deposits along the Pacific coasts of North and South America correspond to convergent plate boundaries and igneous activity of the Ring of Fire described in Chapters 4 and 6. Other deposits on this map represent very old intrusions that formed in the distant geologic past when the configuration of plate boundaries was considerably different from today. The key point here is that mineral deposits

FIGURE 12.23 Disseminated copper deposits associated with igneous intrusions are unevenly distributed on the planet, as they are closely tied to plate tectonics. Note how these deposits closely correspond to the convergent plate boundaries along the Pacific coasts of North and South America.

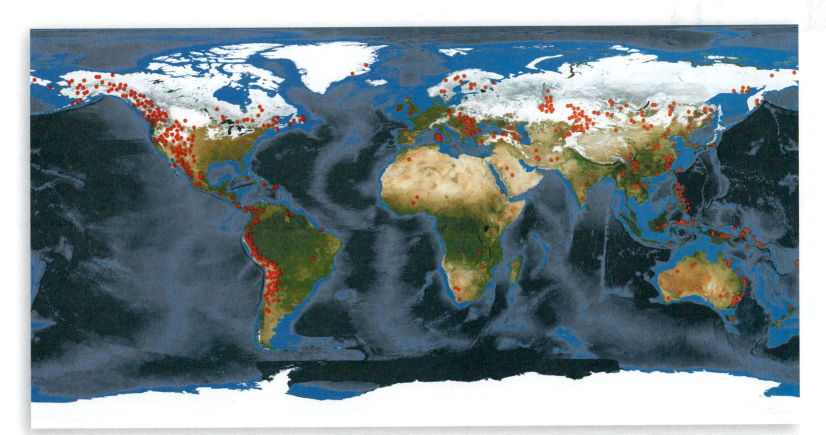

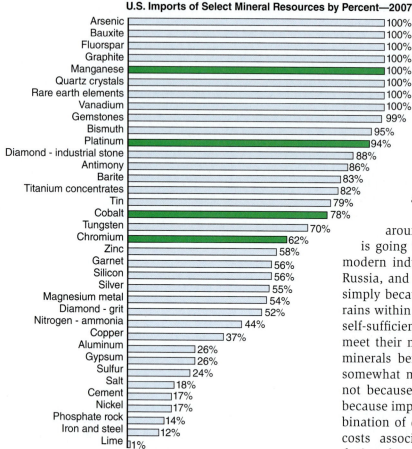

U.S. Imports of Select Mineral Resources by Percent—2007

Mineral	Percent
Arsenic	100%
Bauxite	100%
Fluorspar	100%
Graphite	100%
Manganese	100%
Quartz crystals	100%
Rare earth elements	100%
Vanadium	100%
Gemstones	99%
Bismuth	95%
Platinum	94%
Diamond - industrial stone	88%
Antimony	86%
Barite	83%
Titanium concentrates	82%
Tin	79%
Cobalt	78%
Tungsten	70%
Chromium	62%
Zinc	58%
Garnet	56%
Silicon	56%
Silver	55%
Magnesium metal	54%
Diamond - grit	52%
Nitrogen - ammonia	44%
Copper	37%
Aluminum	26%
Gypsum	26%
Sulfur	24%
Salt	18%
Cement	17%
Nickel	17%
Phosphate rock	14%
Iron and steel	12%
Lime	1%

FIGURE 12.24   List of selected mineral resources and the percentage that the United States imports (green denotes strategic minerals). Some minerals are imported simply because it is less expensive than mining existing U.S. deposits.

are closely tied to the rock cycle and Earth's tectonic history. Therefore, the search for new mineral resources by modern mining companies requires trained geologists who understand both the natural processes that concentrate minerals, and the tectonic history of a particular region.

Because mineral deposits are not evenly distributed around the globe, it is highly unlikely that an individual country is going to have all the different minerals that are needed to run a modern industrial society. Large countries such as the United States, Russia, and China have a greater proportion of the minerals they need simply because they are large and have a wider variety of geologic terrains within their borders. Even so, these large countries are not entirely self-sufficient and have to import certain mineral resources in order to meet their needs. For example, Figure 12.24 lists percentages of select minerals being imported into the United States. The percentages are somewhat misleading, as many of the mineral resources are imported not because the United States has no deposits of its own, but rather because importing is less expensive. This generally occurs due to a combination of deposits that are lower grade, and higher domestic mining costs associated with better wages and environmental regulations designed to protect citizens from pollution.

Since modern societies must rely on imports to meet part of their mineral needs, there is a certain level of risk that a supply disruption will develop for a particularly vital mineral. The risk obviously increases as the percentage being imported increases and when the supply largely comes from politically unfriendly or unstable countries. This brings up the concept of **strategic minerals,** which are those considered critical by a given country, but must be imported in significant quantities. In the United States, strategic minerals include chromium, cobalt, manganese, and platinum because they are vital to civilian and defense industries. For example, the U.S. military depends on high-performance jet engines with operating temperatures that are considerably higher than conventional engines. Such engines are made possible by special alloys containing chromium and cobalt, most of which is imported (Figure 12.24). According to the Center for Environmental Education, Natural History and Conservation, just one of these high-performance engines requires 1,600 pounds of chromium and 1,000 pounds of cobalt. In addition to the military, cobalt alloys have important applications in medical implants because of their low rejection rates by the human body.

Meeting Future Mineral Demand

As with all natural resources, the demand for minerals continues to increase along with population growth and the fact that more countries are becoming industrialized and are developing consumer societies. For example, the graph in Figure 12.25 illustrates the dramatic rise in the consumption

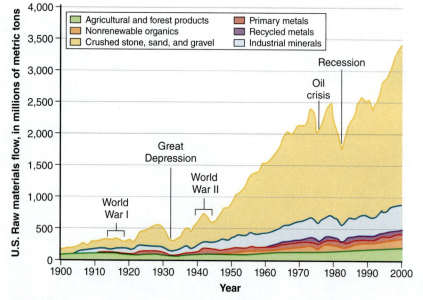

FIGURE 12.25   Growth in the yearly consumption of minerals and other materials in the United States from 1900 to 2000. Note the correlation between economic activity and resource consumption.

of metallic and nonmetallic minerals in the United States as the growing nation built a large industrial base and consumer society. The sharp declines associated with the Great Depression in the 1930s and the oil-related recessions of the 1970s and 1980s demonstrate the close relationship between economic activity and mineral demand. China and other Asian nations are currently undergoing rapid industrialization and are following consumption patterns similar to those of the United States. Since minerals are a finite resource, this exponential rise in consumption is simply not sustainable. An obvious question then is how long will the world's mineral reserves last? To help answer this question, take a look at Table 12.4, which shows current worldwide production rates and total reserves for some of the more common mineral resources. Notice how the life expectancy of different mineral reserves varies, ranging from as little as 13 years for silver to over 300 years for platinum. Keep in mind that these estimates are based on current consumption rates. Should worldwide demand continue to rise, then the life expectancy of the reserves will decrease.

For years scientists have warned about the prospect of important mineral reserves being depleted due to exponential growth in demand. However, depletion has not occurred because total reserves have kept increasing. One reason for this is that geologists have improved their ability to locate new reserves through more sophisticated exploration techniques. Another factor is that technological advances in mining and processing have allowed many low-grade deposits to become economical to extract, thereby causing more mineral resources to be classified as reserves (see Figure 12.3). This does not mean that all mineral extraction can go on indefinitely. History shows that it is becoming harder to locate new deposits, and those that are found are usually of lower grade. Perhaps an even bigger issue is that mineral extraction and processing typically requires large amounts of energy. With rapidly rising energy prices, particularly for diesel fuel, less profitable mining operations will likely begin to shut down, thereby reducing total reserves. The remaining reserves will not necessarily be depleted any faster since higher energy costs will eventually lead to an economic slowdown, followed by a reduction in demand.

It should be mentioned here that due to the critical nature of strategic minerals, manufacturers and governments will sometimes develop stockpiles. Although stockpiling is not a long-term solution, it does provide a means of minimizing the impacts caused by a short-term disruption in supply. In fact, ever since World War I the United States government has routinely stockpiled strategic minerals for possible emergency use.

Recycling and Reuse

At some point humans are not going to be able to afford to keep on expending greater amounts of energy to mine ever deeper and lower-grade mineral deposits. This means we can expect that some mineral resources

TABLE 12.4   World mineral production and projected lifetime of estimated reserves. Reserves represent mineral deposits that are economical to extract under current conditions.

Mineral	2007 Production (thousands of metric tons)	2007 Reserves (thousands of metric tons)	Estimated Life of Reserves (years)
Iron ore	1,900,000	150,000,000	79
Aluminum ore (bauxite)	190,000	25,000,000	132
Phosphate rock	147,000	18,000,000	122
Gypsum	127,000	n/a	
Chromium	20,000	n/a	
Copper	15,600	490,000	31
Manganese	11,600	460,000	40
Zinc	10,500	180,000	17
Titanium concentrates	6,100	730,000	120
Lead	3,550	79,000	22
Nickel	1,660	67,000	40
Tin	300	6,100	20
Cobalt	62	7,000	112
Silver	21	270	13
Gold	2.5	42	17
Platinum group	0.2	71	309

Source: Data from U.S. Geological Survey Mineral Commodity Summaries, 2008.

will become increasingly scarce, which in turn will drive up their market price. As prices increase there naturally will be a greater incentive for manufacturers to try and develop substitute materials and encourage recycling. From Table 12.5 one can see that the recycling rates for metals in the United States is generally around 40%, so there is still considerable room for improvement. Recycling is actually quite desirable due to the high cost of mining and processing of metallic ores. For example, obtaining metallic aluminum from recycled scrap requires only 5–8% of the energy that is required to produce new metal from bauxite ore. Although aluminum and other scrap metal is easy to recycle, many people unfortunately still throw recyclable metals in the trash. A quick look in most garbage cans will likely reveal significant numbers of aluminum cans.

In addition to energy savings and a reduction in environmental impacts, recycling is desirable in that it allows society to create a nearly inexhaustible supply of certain resources. Of course not all materials are so easily recycled; in fact, some are not recyclable at all. Similar to our discussion on water (Chapter 11), certain mineral applications are *consumptive* in nature, meaning the mineral resource is dispersed into the environment or incorporated into some material from which it is difficult to recover. For example, once salt is used on a highway to melt ice, it becomes dispersed, making it impractical to try and recover the dissolved salts. In the case of manganese used to harden steel, once it is alloyed with iron it becomes too difficult to try and extract it from steel scrap. This brings up the concept of *reuse* as opposed to *recycling*. Here manganese in steel may not be recyclable, but a steel beam salvaged from a building can indeed be reused for some other purpose. Yet another option is for society to *reduce* consumption of a particular mineral resource through more efficient use. Take for instance the consumptive use of aluminum foil in the home. By minimizing the amount of foil used in covering or wrapping food, one could make a small but noticeable reduction in their personal use. Note that public education programs aimed at improving conservation commonly use the *3 Rs* concept, which stands for *reduce*, *reuse*, and *recycle*.

Mineral usage can also be reduced as manufacturers develop substitute materials for certain applications. A good example is how the

TABLE 12.5   Metal recycling rates in the United States. Although manganese is a strategic metal used in steel, it is not economical to recover and is not listed.

Mineral Resource	2006 Supply (thousands of metric tons)	2006 Recycling of Old & New Scrap (thousands of metric tons)	Percent Recycled
Iron and Steel	136,000	65,600	48
Aluminum	8,160	3,510	43
Copper	3,000	968	32
Lead	1,570	1,150	73
Zinc	1,390	341	25
Chromium	645	235	36
Nickel	252	108	43
Tin	55	13	25
Titanium	n/a	25	47

Source: Data from U.S. Geological Survey Minerals Yearbook, 2006.

strength and light weight of graphite-based compounds is making this material an ideal replacement for metals in applications ranging from tennis rackets to spacecraft.

Environmental Impacts and Mitigation

As emphasized throughout this chapter, mineral resources are absolutely critical to maintaining our modern societies, of which mining is a necessary part. Although mining activity disrupts a rather small fraction of Earth's total land area, a legacy of poisoned water supplies and devastated ecosystems demonstrates that the impact of mining can be severe. Much of the controversy over mining in the United States is related to the **General Mining Act** passed by Congress in 1872—better known as the 1872 Mining Law. This act governs the mining of precious metals (excluding oil, gas, coal, and nonmetallic ores) on nearly 270 million acres of public lands, mostly in the western United States and Alaska. Under this law a prospector who found a mineral deposit would be able to purchase the land for no more than $5 per acre, and then extract the resource without having to pay any royalties to the government. The intent in the late 1800s was to provide an incentive to those willing to develop the newly acquired lands out West. Some people take issue with the fact that even today, the 1872 Mining Law allows mining to take precedence over all other land-use activities, outside of national parks. In addition, mining companies still only have to pay just $5 an acre to purchase public lands that often yield hundreds of millions of dollars' worth of minerals.

The 1872 Mining Law certainly served its purpose in that it encouraged the mining of precious minerals that were needed as the United States underwent rapid industrialization. Unfortunately, environmental controls on mining or other industrial activity were basically nonexistent until the passage of the Clean Air Act in 1970 and the Clean Water Act in 1972 (Chapter 15). The result was the uncontrolled release of highly toxic metals and other compounds into the environment, leaving streams and rivers throughout the western United States severely impacted even to this day. The pollution from modern mines and processing plants has been minimized due to the regulations enacted since the 1970s and the use of improved mining techniques. Problems and accidents still occur, but the overall amount of pollution from today's mining operations is far less than in the past.

Despite the reduction in pollution, the controversy over mining has not ended in the United States. There are some who view environmental regulations as a costly and unnecessary burden, causing jobs to shift overseas to countries where lax regulations result in lower operating costs. In a global market system, U.S. mines oftentimes simply cannot compete. Others see regulations as a way of forcing companies to incorporate the cost of controlling pollution into their operating expenses as opposed to passing on the much higher cleanup and medical costs onto the taxpayers. Moreover, many point to a lack of enforcement and persistent pollution problems as evidence that the regulations designed to protect people and the environment are falling short of their goals.

The issue of pollution versus jobs is complex and there are no easy answers. We will discuss pollution and environmental regulations in more detail in Chapter 15, but for now our focus will be on some of the specific problems, both past and present, which are associated with the mining and processing of minerals. We will also examine some of the mitigation and prevention techniques that are used to minimize impacts.

Heavy Metals and Acid Drainage

The mining and processing of nearly all mineral resources produces waste material (tailings) where the economic mineral concentration is too low to justify further processing. Depending on the type of deposit, the tailings volume can be significantly larger than the mine excavation because of the way broken-up rock will occupy more space. In the case of open-pit mining of low-grade ores, the vast majority of the material is removed, creating tailing piles whose dimensions are truly staggering (Figure 12.26). Although the permanent storage of mine tailings creates what some view as an eyesore, a more serious problem occurs when melting snow or rainwater percolates through the material and chemically reacts with the remaining ores and other minerals, thereby producing leachate. Tailings from metallic ore deposits are of particular concern as they normally contain sulfide minerals that weather easily, creating leachate that is rich in metallic ions such as lead, copper, zinc, arsenic, cobalt, nickel, selenium, and cadmium. This toxic brew of metals eventually drains from the bottom of the tailing pile, where if left unchecked, will enter both surface streams and groundwater systems.

The sulfide minerals that are commonly present in mine tailings not only release heavy metals as they break down chemically, but they also release sulfide ions (S^{2-}) that react with oxygen and water to produce sulfuric acid (H_2SO_4). The result is leachate that is both extremely acidic and contains highly toxic metals. This same type of leachate also forms in the workings of underground mines where sulfide minerals are exposed to oxygen and water, which can then drain into streams lying below the level of the mine. The term **acid mine drainage** is used to describe acidic leachate, normally rich in heavy metals, that drains from either tailings or underground mines. As shown in Figure 12.27, when acid mine drainage encounters the more oxidizing conditions in a stream, some of the heavy metals will form a precipitate that covers rocks and sediment in the streambed. Because dissolved iron is present in most leachate and easily oxidizes, streams affected by acid mine drainage typically have a characteristic reddish or yellowish color from iron-oxide precipitates. A bluish or greenish color due to copper precipitates is much less common.

In addition to the toxic effect of dissolved metals, acid mine drainage is a serious problem because it can greatly increase the acidity (i.e., lower the

FIGURE 12.26   The volume of mine tailings can be staggering, particularly from the open-pit mining of low-grade copper deposits. For scale, note the size of buildings in these photos.

Twin Buttes, Arizona

Bingham Canyon, Utah

A Silverton, Colorado

B Jerome, Arizona

FIGURE 12.27   Leachate draining from underground mines and from beneath tailings is commonly highly acidic and contains heavy metals due to the chemical interaction of water and sulfide minerals. Dissolved iron quickly precipitates to form iron oxides, giving impacted streambeds a characteristic reddish and yellowish color (A), whereas a bluish or greenish color from copper (B) is less common.

pH) of surface waters. In some areas the pH of a stream can fall from around 6 to 3, which corresponds to a thousand-fold increase in acidity. Together the highly acidic conditions and heavy metals can have devastating effects on the ecology of streams, producing sections that are completely devoid of life for miles downstream. Even less severe cases of acid mine drainage can disrupt an ecosystem by creating conditions where aquatic species that form the base of the food web (e.g., insects) are unable to survive. This sets off a chain reaction that ends in the death of numerous species throughout the food web.

Most of the problems associated with acid mine drainage in the United States are associated with abandoned mines that operated prior to the environmental regulations of the 1970s. Current laws governing mine tailings are complex due to the different ways the waste is classified, which in turn has led to numerous exemptions. In general, most mines attempt to minimize the effects of acid drainage by trying to limit the amount of leachate that is generated, and then collect and treat what does form. For example, the volume of leachate can sometimes be reduced by placing material with low permeability, such as clay, on top of the tailings so that infiltration is reduced. Whatever leachate that does form is then collected and run through a treatment process that reduces acidity and removes heavy metals.

There are techniques used to treat acid-mine waters, but perhaps the most common involve what are known as *constructed wetlands*. As illustrated in Figure 12.28, the first step often involves forcing the acidic water to flow through a bed of crushed limestone, which lowers the acidity as the water dissolves calcite ($CaCO_3$). After this initial treatment the water flows into a constructed wetland where high levels of dissolved oxygen are maintained, causing iron and manganese to be removed as these metals precipitate and form hydroxide minerals (e.g., $FeO(OH)$). From there the water moves into an oxygen-poor wetland where bacteria convert sulfate

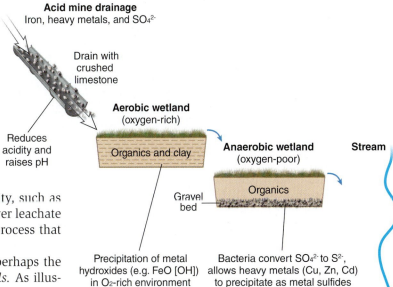

FIGURE 12.28   Remediation of acid mine drainage includes neutralizing the acid by letting it react with crushed limestone, followed by having the leachate flow through a series of constructed wetlands where dissolved oxygen is removed. Here heavy metals are removed by plant uptake and by precipitation under the reducing conditions.

FIGURE 12.29 A cyanide heap-leaching operation at a historic mining area in southern Colorado began developing leaks, contaminating nearby streams that fed into the Alamosa River. The U.S. EPA took over the operation, preventing a potentially catastrophic release of cyanide and heavy metals into the surrounding environment.

(SO_4^{2-}) into sulfide (S^{2-}) ions, which causes heavy metals (e.g., copper, zinc, cadmium) to precipitate as sulfide minerals. Note that special plants in the wetland also aid in removal of the heavy metals.

Processing of Ores

In addition to leachate from waste material stored in tailings piles, the processing of ore minerals can release a variety of harmful toxins and compounds into the environment. One of the concerns with the cyanide heap-leaching method is that the impermeable liner beneath the pile or holding lagoon could develop a leak, releasing cyanide and heavy metals into surface streams and the groundwater system. Such fears were realized in 2000 when a dam holding mine tailings from a Romanian cyanide operation failed, releasing 260 million gallons (100,000 m³) of contaminated liquid into a tributary of the Danube River. This created a massive fish kill and poisoned the drinking water of more than 2 million people. A similar dam failure occurred in 2006 near the town of Miliang, China, resulting in a landslide that killed 17 people and injured 130, and again poisoned water supplies.

Cyanide operations in the United States have good safety controls, but leaks and accidents do happen. Critics often point to how a disaster was narrowly averted in 1992 near Summitville, Colorado, when the U.S. Environmental Protection Agency (EPA) took over a cyanide heap-leaching operation from a mining company that had declared bankruptcy (Figure 12.29). In addition to earlier cyanide leaks from beneath the heap-leach pad and piping systems, the facility's treatment system at the time the EPA took over was in immediate danger of being overwhelmed by snowmelt. The result could have been a catastrophic release of cyanide and heavy metals into the headwaters of the Alamosa River. The potential disaster was fortunately averted, but with the company bankrupt, the estimated remediation and cleanup cost of $120 million had to be paid from the U.S. Government trust fund known as *Superfund* (Chapter 15). Due to the previous leaks, concerns were also raised over the uptake of heavy metals by agricultural crops irrigated by water drawn from the Alamosa River. Later studies revealed that a relatively small portion of the heavy metals found in the irrigated soils was related to the mining operation. Most of the metals are believed to be from the natural weathering of ore-bearing rocks in the region, which were then transported downstream and deposited in the ancient sediment. As in similar areas, the stream here was "naturally polluted" long before humans began mining.

Mercury at one time was extensively used in processing gold and silver ores and is now a persistent problem in streams that drain historic mining districts throughout the western United States. Mercury is a metal that has several unique properties, the most notable being that it exists as a liquid in the temperature and pressure conditions found at Earth's surface. Because mercury is so dense, even denser than iron, most earth materials will literally float on liquid mercury. Moreover, mercury will chemically react with gold and silver to form a paste-like alloy called *amalgam*, which is commonly used in dental fillings. These properties can be used to extract gold and silver by mixing liquid mercury with either placer material or crushed ore. Because gold and silver are denser than mercury, only gold and silver-bearing particles will sink into the mercury and chemically combine to form amalgam. The mercury can then be driven off by heating the amalgam or dissolved in nitric acid, resulting in pure gold and silver—the mercury can be recovered and used again.

Although mercury-amalgam techniques are effective, not all of the mercury is recovered. For example, during the historic gold rushes in the western United States, an estimated 10–30% of the mercury was not recovered and ended up in stream sediment. In California alone, 3–8 million pounds (1.4–3.6 million kg) of mercury are believed to have been lost between the 1860s and early 1900s. The problem with mercury is that it is highly toxic and persists in the environment, plus it can be easily absorbed through the skin, inhaled, or ingested. Mercury also combines with carbon to form methyl mercury, which is ingested by aquatic organisms and then moves up through the food chain and into humans, primarily through the consumption of fish (Chapter 15). Unfortunately, the problems associated with the past use of mercury in the United States are now being repeated throughout South America where recent gold rushes have resulted in the unregulated use of mercury. This has resulted in serious health problems among workers as well as mercury contamination of streams.

In the case of sulfide minerals, ores are typically processed by *smelting*, which is an ancient method where ore is heated to the point that chemical bonds within the minerals are broken, liberating both sulfur and metal ions. Once the sulfur atoms are free, they quickly react with oxygen to form sulfur dioxide (SO_2) gas. The sulfur dioxide gas rises into the atmosphere where it reacts with water to form sulfuric acid (H_2SO_4), eventually falling as *acid rain* that is devastating to land plants and aquatic life in areas downwind of the smelting operations. Note that acid mine drainage forms from the same type of sulfide minerals. In addition to the sulfur dioxide, some of the metallic ions rise up the smokestack, falling back to earth as particulate matter that contaminates soils. Acid rain and heavy metal fallout associated with smelting operations have largely been eliminated in the United States as regulations now require that sulfur dioxide and particulate matter be removed before emission gases are released into the atmosphere (Chapter 15).

FIGURE 12.30 Aerial photo showing numerous pits on the land surface that were caused by the collapse of shallow underground coal mines in Wyoming.

Collapse and Subsidence

Underground mining obviously creates void spaces in the subsurface. As described more thoroughly in Chapter 7, these voids tend to close in on themselves in response to the tremendous weight of the overlying column of rock. Sometimes the voids close slowly and other times they collapse quite suddenly. The result at the surface can be large cracks, a collapsed sinkhole, or a lowering of the land surface itself in a process known as *subsidence.* These surface disruptions can cause serious localized damage of buried utilities and other infrastructure, including homes. Mine-related movement of surface materials is more likely to occur in situations where large workings are relatively close to the surface and in rocks that are rather weak. As shown by the example in Figure 12.30, coal mines are notorious for collapsing, which can lead to subsidence problems at the surface.

Abandoned Mine Hazards

Every year people and animals are killed and injured in abandoned mines, most by falling into quarries and open pits, or down the vertical shafts of underground mines (Figure 12.31). Because mine excavations are permanent features, fences and barricades need to be maintained on a long-term basis in order to keep people from entering, either inadvertently or by intentional trespassing. Unfortunately, there are many historic mines whose owners have long since gone out of business, leaving it up to state agencies (i.e., taxpayers) to erect the barriers. Even with modern mines in the United States, where mining companies are required to restrict access, there is the question of who will maintain such barriers far into the future.

FIGURE 12.31 Abandoned mines pose deadly hazards to humans and animals, such as the vertical shaft shown here at this long-abandoned site in New Jersey. Note the fence is being erected by state officials at taxpayers' expense.

SUMMARY POINTS

1. Nearly every aspect of modern societies depends on Earth's mineral resources. Because mineral extraction is costly and can cause severe pollution, society needs to practice conservation and ensure that minerals are extracted safely.
2. Society uses certain minerals for specific applications because of their physical and chemical properties. Various geologic processes concentrate minerals into deposits that can be mined.
3. Enrichment refers to the degree to which mineral deposits are concentrated. Those deposits that are economical to mine become part of the overall mineral reserves.
4. Economic mineral deposits formed by igneous activity include diamond pipes, layered intrusions, pegmatites, and hydrothermal (vein and disseminated) deposits.
5. A variety of valuable metamorphic deposits result from both contact and regional metamorphism.
6. Important sedimentary deposits include sand, gravel, clay, placers, banded iron formations, bauxite, evaporites, and phosphorites.
7. Mining techniques vary depending on the mineral resource and can be classified as underground or surface (open-pit, strip, and placer). Mining economics depends on the amount of waste material to be removed and the energy this requires. Other factors include ore grade, mining technique, and labor costs.
8. Ore processing involves crushing and physical or chemical separation of ores from waste materials. Common methods include hydraulic sorting, floatation, magnetic separation, smelting, and leaching.
9. Mineral deposits are associated with specific types of geologic environments and are therefore not evenly distributed on Earth's continents. Locating deposits is further complicated as these environments shift over geologic time due to tectonic activity.

10. Countries with large land areas generally contain a greater diversity of geologic terrain and mineral resources. Strategic minerals are those vital to a nation's economy and whose supply is naturally limited.

11. Minerals are finite resources, so future demand will best be met by recycling and conservation, and technological improvements that allow more low-grade deposits to be mined economically. Rising energy costs, however, will make mining less economical.

12. Numerous environmental problems arise from mining and processing, particularly acid mine drainage and release of cyanide and mercury. Pollution from modern mine operations has been minimized due to regulations and improved mining techniques.

KEY WORDS

acid mine drainage   390
bauxite   374
disseminated deposits   367
enrichment factor   364
evaporites   376
General Mining Act   389
hydrothermal deposits   367

layered intrusions   366
leaching   384
massive sulfide deposit   368
mineral reserve   364
mineral resource   361
mine tailings   379
open-pit mine   380

ore deposit   364
overburden   380
placer deposit   373
smelting   383
strategic minerals   386
strip mine   381

APPLICATIONS

Student Activity

Go to a large supermarket. Go down the spice aisle and look for salt. You will find standard iodized salt, but what else do you see? Kosher salt? Sea salt? Where are these from? Do the labels say how these salts were "made"? What other kinds are there? Mediterranean sea salt? Hawaiian red? French grey? Celtic sea salt? What are the price ranges? Do you think that salts can taste different? Buy one, and taste-test it against normal iodized salt. Do you taste a difference? Do you think it would make a difference in cooking?

Critical Thinking Questions

1. What is the difference between a mineral resource and a mineral reserve?
2. List some of the strategic minerals. Why are these minerals strategic?
3. What are some important mining methods? How are they bad for the environment?
4. How has salt gone from a resource that had to be kept under lock and key to being so cheap that it is free at most restaurants?

Your Environment: YOU Decide

Diamonds come from many places in the world. Most come through official channels and have a set, fixed price. There are also diamonds that come from areas of the world that are experiencing uprisings and civil war. The diamonds that come from these areas are termed "conflict" or "blood" diamonds. They end up in regular shops, but not through the official channels. These diamonds can be priced lower than the diamonds from official sources. If you were in the market to buy a diamond or a piece of diamond jewelry, would you make sure it was a conflict-free diamond (as some are now certified) or would you go with the cheapest price, not wanting to know its history?

Chapter 13

Conventional Fossil Fuel Resources

LEARNING OUTCOMES

After reading this chapter, you should be able to:

- List the basic forms of energy and describe some of the common transformations between different energy forms.
- Describe why petroleum was favored over coal to eventually become the dominant resource it is today.
- Explain how organic matter accumulates geologically and is transformed into coal and petroleum.
- Explain the basic process by which petroleum migrates from a source rock and accumulates in a reservoir rock.
- Know where major reserves are located and how that impacts energy security.
- Describe the peak oil theory and how it may affect the world economy.
- Explain why it will take years to scale up production of nonconventional oil resources to make up for declining supplies of conventional oil.
- Describe why conservation and efficiency must play a key role in limiting the impact of future shortages.

Our modern way of life is highly dependent on a finite supply of fossil fuels, which represent stored sunlight that accumulated as organic matter over millions of years. Shown here is an offshore drilling platform in the North Sea oil fields of Great Britain and Norway.

Introduction

In previous chapters you learned that the very existence of human civilization depends upon fertile soils and adequate water resources. From Earth's mineral resources we have made steel, concrete, and various materials that form the basis of our cities and factories, which, in turn, produce machines and countless types of products. This has allowed us to grow large quantities of food on highly mechanized farms and build vast transportation networks that move goods and people around the globe. What makes all these activities possible are energy resources. Although there are different forms of energy, about 88% of the energy consumed by humans comes from the burning of coal, oil, and natural gas, fuels that are collectively referred to as **fossil fuels** as they are derived from remains of ancient plants and animals. In recent years the world has discovered that its demand for energy is outstripping the supply of inexpensive oil and gas, creating price instabilities that threaten to cripple the global economy. Equally troubling is the fact that our massive use of fossil fuels is contributing to the growing problem of global warming.

The world's energy problem is not necessarily that we are running out of oil, but rather that we are running out of *cheap oil*. For over the past 125 years humans have extracted oil that was the easiest to locate and pull from the ground. The viscosity (resistance to flow) of this oil is relatively low, which makes it easier to extract, plus it produces more of the highly desirable transportation fuels such as gasoline, diesel, and kerosene (jet fuel). Most of what remains in the ground today is more expensive because it is more viscous (heavy oil), or is located in areas that are difficult to access (e.g., offshore and in the Arctic). Because nearly every aspect of our society depends on this highly versatile resource, the rising price of oil can have a ripple effect through the entire global economy. For example, food prices will rise along with the cost of diesel fuel because it is used to operate farm equipment and irrigation pumps as well as the trucks that transport the food to market. Higher oil prices then make it more expensive to drive to work or school in our vehicles that run on gasoline. We also feel the impact in the cost of air travel because jet aircraft use kerosene as fuel. In the end, higher oil prices mean that each of us spends progressively more of our income on the basic necessities of life, leaving less to spend on a house, vacation, or new consumer items.

In order to minimize the rising price of oil, society needs to both increase supply by finding new oil deposits, and decrease demand through serious conservation efforts. However, because of the threat of global warming (Chapter 16), burning even more fossil fuels is not a long-term option. Ultimately we must quickly increase the production of energy from sources that do not contribute to global warming, while at the same time find ways to reduce the impact of burning fossil fuels. For us to succeed, today's younger generation will need to become knowledgeable on the science behind energy and global warming issues. After all, it will be young people who will help guide society through the necessary transformations in the coming decades. Keep in mind that although the intertwined problems of energy and global warming pose serious challenges for society, they also present an enormous number of business and job opportunities.

Chapter 13 will focus on the *conventional fossil fuels* (coal, oil, and natural gas) that have provided the vast majority of the world's energy needs since the Industrial Revolution. In Chapter 14 we will examine the nonconventional, or alternative, energy sources that can help meet the world's future energy needs. For the purpose of this book, *alternative energy sources* will be defined as any source that does not involve the types of fossil fuels currently making up the bulk (88%) of our energy supplies. Using this definition, alternative sources will include the remaining (12%)

currently in use, such as hydro, nuclear, wind, and solar power, and biofuels. Alternative energy sources will also include heavy oils and synthetic oils made from tar sands, oil shale, and coal, which presently make up a small fraction of world supply.

Human Use of Energy

Energy is a term we often use when referring to such things as the electric and gas bills for our homes. We may also describe ourselves as being "low on energy," and then go out and buy an energy drink to help stay awake in class. To a scientist or engineer, **energy** is the capacity to perform work or transfer heat, whereas **work** involves moving an object (mass) a certain distance against some force (i.e., work = mass × distance). For example, you do work at the gym when lifting weights against the force of gravity—hence the term "workout." The energy that enables you to do this work comes from the food you eat, which itself is solar energy that has been stored by plants and animals. The key point here is that we are literally surrounded by different forms of energy, and as individuals, are constantly making use of this energy. Because the remaining topics in this textbook require a fundamental understanding of energy, we will next take a brief look at the basic forms of energy.

Electrical energy results from the movement of electrons through a material called a conductor, such as a copper wire or a circuit board. In modern times humans have learned to construct electrical generators that produce a steady flow of electrons, which move through wires in what we call *electricity*. Electrical power is one of the most useful forms of energy in society today, running everything from MP3 players to air conditioners to high-speed trains.

Chemical energy comes from the energy stored in the chemical bonds between the different atoms making up a compound. All chemical reactions either release heat energy (exothermic) or require heat (endothermic). When we eat, acids and enzymes in the stomach break down the chemical bonds within the food, resulting in the release of heat energy and the formation of new compounds. The heat keeps us warm, whereas the new compounds are absorbed into the blood and used by various body functions, such as muscle movement. Another example of chemical energy is when you switch on your cell phone and an electrical connection is made between two compounds stored in the battery. Once the connection is made, a chemical reaction occurs that allows electrons to begin flowing through the phone's electrical circuits.

Thermal energy, often called *heat energy,* comes from the vibration of individual atoms or molecules. What we call temperature is a measure of the frictional heat produced by these vibrations. For example, a metal bar pulled from a fire is hot because its atoms are vibrating fast and generating a considerable amount of frictional heat. When electrons flow through a wire, the wire becomes warm because there is a certain amount of resistance that forces the electrons to slow down.

Kinetic energy is the form of energy that results from the velocity of an object (mass) in motion, and is computed by taking half the object's mass and multiplying it by the square of its velocity (kinetic energy = $\frac{1}{2}mv^2$). Heavier and faster objects therefore contain greater amounts of kinetic energy. For example, an asteroid a mile in diameter has an enormous amount of kinetic energy, whereas a fist-sized asteroid traveling at the same velocity has comparatively little energy. Conversely, if two asteroids were the same size, the faster one would naturally have more kinetic energy.

Potential energy is simply the energy stored in an object that is not in motion, but capable of moving. Familiar forms are *gravitational, elastic,* and *electrical* potential energy. Take for example how a dam creates an elevated pool of water called a reservoir. Because of the water's increased height, its gravitational potential energy has increased. When the water is released and allowed to fall, the stored energy immediately begins converting into kinetic energy, which can be used to spin a generator and make electricity. A practical example of elastic potential energy is a bow and arrow. When an archer pulls on a bow, it bends and stores elastic energy. As the bow is released, this energy is quickly transferred to the arrow and transformed into kinetic energy. Another example is the sudden release of stored elastic energy in rocks that results in the vibrational wave energy of an earthquake (Chapter 5).

Nuclear energy is the energy that binds or holds the nucleus (protons and neutrons) of an atom together—in contrast to chemical energy which involves the bonds *between* atoms. Nuclear energy is a form of potential energy in that it is stored energy, but is released only when the nucleus undergoes one of two types of reactions. *Fission* reactions involve the splitting of the nucleus and *fusion* reactions are where two separate nuclei combine, both of which release tremendous amounts of thermal (heat) energy. Note that fusion is the dominant reaction in the Sun, whereas modern nuclear power plants rely on fission. Also, Earth's interior has remained hot due to the continued release of thermal energy from the fission decay of radioactive elements.

Radiant energy is the general term used to refer to electromagnetic energy traveling in the form of waves (Chapter 2). Radiant energy is associated with the different wavelengths in the electromagnetic spectrum (radio, microwave, infrared, visible light, ultraviolet, X-ray, and gamma ray). Radiant energy is commonly called *solar energy* because the various electromagnetic wavelengths are all produced in the nuclear furnaces we call stars, which naturally includes our Sun.

Energy Conversions

You may have noticed how one form of energy often converts to another. Chemical energy stored in batteries is converted into electricity used to run your cell phone; a stove converts electrical into thermal energy for cooking. In fact, almost everything we do is a series of energy conversions. For example, consider all the energy conversions shown in Figure 13.1 that are involved in just growing and cooking food. Our food originates from the conversion of nuclear energy to solar radiation, which is then converted into chemical energy by plant photosynthesis. Our digestive system converts this chemical energy into thermal (heat) energy and other chemical compounds, some of which are used to move our muscles. Whenever we use our muscles to walk or lift things we are ultimately converting some of the original food into kinetic energy—often called *mechanical* energy. This kinetic or mechanical energy can then be used to do *work*, such as lifting sacks of groceries or pushing a lawn mower. Prior to the Industrial Revolution and the widespread use of fossil fuels, physical work was mostly performed by people (manual labor) and animals, powered solely by chemical energy derived from plants.

Today's modern societies are powered mainly by the conversion of fossil fuels into different forms of energy we

FIGURE 13.1   Humans constantly utilize energy conversions to suit their needs. Consider the number of energy conversions involved in cooking and growing food, a form of chemical energy.

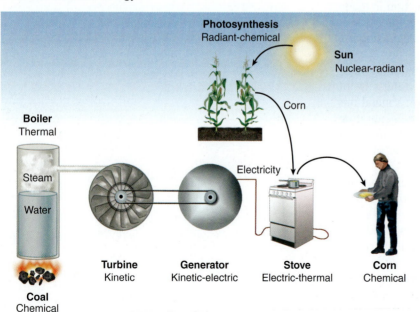

find useful. During this process a certain portion of the energy is converted into forms that are not useful, which means part of what we pay for is wasted. The reason for this is related to the *first law of thermodynamics,* which states that energy cannot be created or destroyed, only transformed from one form to another. Moreover, during every energy transformation there are at least two conversions, one of which is thermal energy, or simply *heat.* A good example is when you flip a switch and electricity begins to run through a standard light bulb, causing the filament inside to begin to glow. Here the electrical energy is being converted into visible light, which is just radiant energy in the visible part of the electromagnetic spectrum. However the bulb also gets hot, telling us that an electrical to thermal conversion is also taking place.

Energy efficiency is a useful measure of how much of the original energy actually goes toward its intended use; the unwanted heat is referred to as *heat loss.* Table 13.1 lists the energy efficiency of some common applications. In the case of a standard incandescent light bulb, only 5% of your electricity is being converted into what you want, namely visible light. Compact fluorescent bulbs are far more efficient as nearly 20% of the electricity goes to producing light. Notice that gasoline engines are only 25% efficient, which means that for every dollar you spend on gas, seventy-five cents goes to making heat that is lost through the exhaust pipe and radiator. Because energy costs each of us money, it is clearly in our own interest to try and be as energy efficient as possible.

Renewable versus Nonrenewable Energy

Although there are many different forms of energy on our planet, solar radiation is essential to the biosphere and life as we know it. Through the process of *photosynthesis,* land plants and the ocean's phytoplankton take radiant energy flowing from the Sun and convert it into chemical energy. The result is the growth of cell tissue and the formation of what is called *biomass.* Plants are a critical form of stored chemical energy since they serve as the primary food source for both land and marine animals. In addition, some of this organic matter from the biosphere becomes buried and is then transformed by geological processes into the more concentrated forms of chemical energy we call coal, oil, and natural gas. These fossil fuels are often referred to as "stored sunlight" since they are derived from the conversion of solar energy into biomass.

Ever since the Industrial Revolution began over 200 years ago, society has been relying on the Earth's supply of highly concentrated fossil fuels, which have been accumulating over the past several hundred million years. Similar to mineral resources, fossil fuels are still forming, but are considered finite or *nonrenewable* because they are being consumed by people at a faster rate than which they form. Society's current reliance on fossil fuels is clearly not sustainable, eventually forcing us to turn to alternative supplies of renewable energy such as hydro, wind, and solar power. Our basic problem is that no other energy source is as concentrated as fossil fuels, or provides the types of liquid fuels being used to run our current transportation systems. Note that electricity is considered to be a *secondary energy resource* because it has to be produced from some primary source of energy, such as coal or nuclear, wind, and solar power.

Historical Energy Usage

For much of history people depended on sunlight for warmth, but eventually learned how to make wood (biomass) fires for cooking and keeping warm, even in cold climates. About 5,000 years ago humans expanded their

TABLE 13.1   Efficiency of some of the most common energy conversions used in modern societies.

Equipment	Desired Conversion	Efficiency (percent going to desired conversion)
Incandescent light bulb	Electrical to radiant	5%
Compact fluorescent light bulb	Electrical to radiant	20%
Gasoline engine	Chemical to kinetic	25%
Diesel engine	Chemical to kinetic	35%
Electric engine (motor)	Electrical to kinetic	70%
Coal power plant	Chemical to electrical	30%
Home gas furnace	Chemical to thermal	90%
Electrical heating	Electrical to thermal	100%

Source: Data from Alternate Energy Guide and International Panel on Fissile Materials.

FIGURE 13.2 Prior to steam engines, wind and water mills were the dominant means of generating mechanized power. The kinetic energy from wind or falling water was used to turn shafts that provided power to perform tasks such as grinding grain, cutting lumber, and making textiles.

energy supply beyond wood by domesticating animals, using them to perform work. By the 1800s, large numbers of mills had been constructed for harnessing the kinetic energy of wind and water (Figure 13.2). These flowing forms of energy provided mechanical power that could be used to perform a variety of useful work, including grinding grain, cutting lumber, and running textile looms. Mills operating on renewable sources of wind and water were later replaced by steam engines as the primary source of mechanical power. Here steam engines converted *stored* chemical energy (mainly wood) into the kinetic energy needed to operate machinery. This was a major technological advance because factories could now be located almost anywhere; no longer were they restricted to waterways or places where the wind was reliable. Steam engines were also put to use powering ships and trains, a step that revolutionized transportation and commerce.

As indicated in Figure 13.3, the rapid development of steam power during the 1800s greatly increased the United States' consumption of wood as an energy source. By 1875, coal was mined in increasing quantities and soon replaced wood as the dominant fuel source. In addition to being more abundant, coal had the advantage of giving off more heat per ton compared to wood because it is a much more concentrated form of chemical energy. This meant that trains and ships could now carry more cargo and less fuel to travel a given distance. Another important development during this period was the switch from whale oil to kerosene as a fuel for lanterns used for indoor lighting. Kerosene was refined from unprocessed or *crude* oil that was collected at seeps where the oil would naturally rise and leak out onto the land surface. This switch not only saved whales from near extinction, but the increased demand for kerosene led to the first U.S. well being drilled in 1859 in Pennsylvania. Oil was soon being extracted in great quantities directly from the subsurface and refined into a variety of products, including grease, lubricating oils, kerosene, and gasoline.

The consumption of oil started to increase significantly after 1900 as the internal combustion engine was put to use in a variety of applications, particularly cars, trucks, tractors, ships, and airplanes. This, in turn, led to the development of large-scale oil exploration (Figure 13.4) and the oil refining industry. As can be seen in Figure 13.3, the dramatic rise in oil consumption was slowed only by the Great Depression of the 1930s and the oil crises of 1970s. Oil eventually eclipsed coal as the dominant energy resource because it contains more energy and its liquid nature makes it more versatile. For example, a relatively small internal combustion engine could produce tremendous amounts of power by burning refined gasoline or diesel fuel. The fact these fuels are liquids makes them easy to transfer and store, plus their highly concentrated nature means that a vehicle does not have to carry much weight in fuel. These characteristics presented a distinct advantage over the heavy and dangerous steam engines, whose fuel supply of coal was bulky and more difficult to transport. Coal also lost its place as the primary fuel for heating buildings, overtaken by natural gas, which soon followed a similar consumption trajectory as did oil (Figure 13.3). The advantage of natural gas over coal is that it burns much cleaner and is easily transported over land via pipelines. The consumption of coal finally began to increase again after 1960 due to the growing demand for electricity, which in the United States is produced mainly by burning coal. This increased demand for electricity was brought on in part by the widespread use of air conditioning.

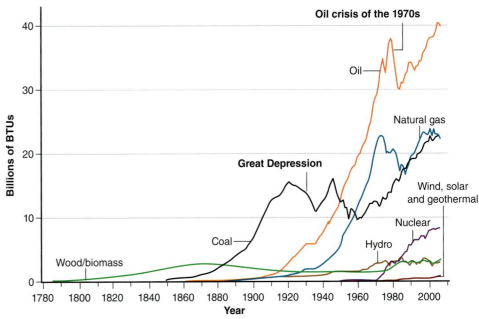

FIGURE 13.3   Plot showing annual U.S. consumption of different energy sources since the Industrial Revolution. Coal usage declined after 1945 as some applications switched to oil and gas, but then increased again after 1960 due to a greater demand for electricity. Note the relatively small contribution of renewable sources compared to the nation's overall energy needs.

Perhaps the most striking aspect of Figure 13.3 is the explosive growth in overall energy consumption that accompanied the Industrial Revolution. The amount of energy that had been harnessed and put to use by burning wood was clearly minuscule to what is now being obtained from fossil fuels. By unlocking this enormous reservoir of chemical energy, human civilization has been able to undergo a phenomenal transformation. The world went from largely agrarian societies to those with high-tech service economies, whose citizens enjoy unprecedented comforts and consumer products. Unfortunately, the conventional fossil fuels that made all this possible are finite. Consider that these fuels had accumulated over several hundreds of millions of years of geologic time, but it will take humans a mere 200 years or so to have consumed the bulk of this irreplaceable resource. This massive use of fossil fuels has released tremendous quantities of carbon dioxide (CO_2) gas into the atmosphere, which, in turn, is affecting Earth's global climate system, a subject we will explore in Chapter 16.

The follow sections cover the basic geologic processes that lead to the formation of conventional deposits of coal, oil, and natural gas. We begin with coal, in part because it was the first fossil fuel to be used on a large scale, and because it forms in a fundamentally different manner than does oil and gas.

Coal

Geologic studies have shown that the world's coal deposits range in age from 2 million to 350 million years old, and formed from ancient plants that thrived under swamplike conditions in humid climates. The plant material eventually became buried, at which point geologic processes transformed the material into sedimentary deposits of coal. Although individual beds of coal can be less than an inch thick, economical deposits generally start at

FIGURE 13.4   Oil's liquid nature and high energy content, combined with the advent of the internal combustion engine and well-drilling technology, helped make oil the most desirable energy source. Shown here is the booming Spindletop oil field in Texas in 1903.

around 6 inches (25 cm), with some reaching over 100 feet thick (30 m). It is estimated that approximately 3 to 10 feet of compacted organic matter is required to produce a foot of coal. Special geologic conditions, however, are required for organic matter in a swamp to be transformed into an economical coal deposit. First, a swamp needs to remain saturated for a considerable period of time. Because saturated conditions help block out free oxygen (O_2), the microbial breakdown of dead plant matter is greatly reduced. The abundance of water also promotes lush vegetation growth, which when combined with the slow decay rates leads to thick accumulation of partially decayed plant material. Next, this organic matter must become buried in order for it to be preserved within a sequence of sediment.

From the types of sedimentary rocks found associated with coal, geologists have determined that coal originates in environments similar to today's river deltas (Chapter 8). As illustrated in Figure 13.5, deltas are ideal because they contain expansive back swamps with lush vegetation. Here the low flow rates help produce stagnant water conditions with low levels of dissolved oxygen (O_2), which greatly slows the decay of dead plant material and allows more organic matter to accumulate. Moreover, large deltas typically experience subsidence as the tremendous weight of sediment presses down on the crust. This progressive sinking allows for even greater amounts of plant matter to accumulate, which, in turn, compresses the underlying organic material into a more compact form called *peat*. At some point the river channels may migrate or sea level will rise, burying the peat with younger sediment. This burial seals the peat off from atmospheric oxygen, increasing the chance that it will be preserved.

FIGURE 13.5   Coal is a sedimentary rock that originates in the back swamps of large river deltas where dead plant material accumulates, eventually compacting into layers of peat. Because large deltas typically undergo subsidence, thick accumulations of peat can develop, and eventually become buried by shifting stream channels or rising sea level. If peat becomes deeply buried, the increased heat and pressure can turn it into coal.

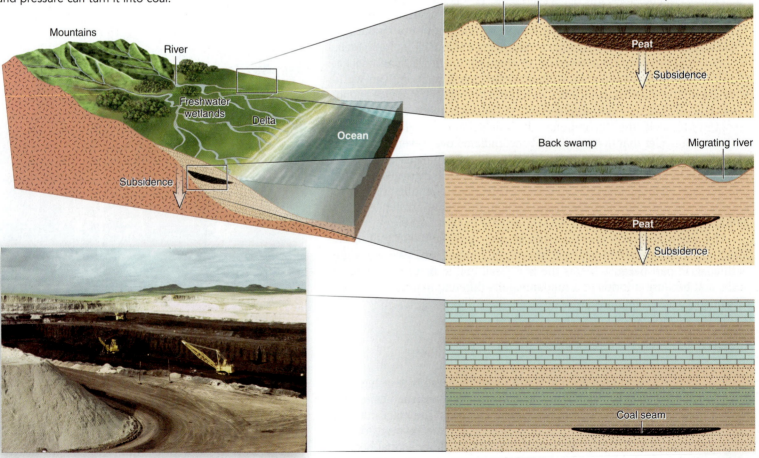

Rawhide mine, Wyoming

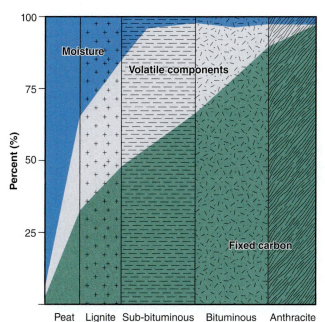

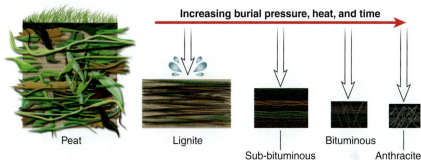

FIGURE 13.6 As peat becomes more deeply buried, the higher temperatures and pressures drive off progressively greater amounts of water and other volatiles, leaving behind a deposit more concentrated in carbon. Higher ranks of coal are generally the most desirable as they are more dense and release greater amounts of energy.

As the delta continues to receive new sediment and subside, layers of peat are buried even deeper and are exposed to progressively higher levels of heat and pressure. Eventually the peat may reach the depth where very low-grade metamorphism transforms it into the combustible sedimentary rock known as **coal.** During this transformation the original plant material undergoes physical and chemical changes that drive off water and other volatile compounds (easily turns to a gas), leaving behind solid coal that is more concentrated in carbon. Because this process is a function of temperature and pressure, the carbon concentration can increase over time should the coal become more deeply buried. As illustrated in Figure 13.6, coals are ranked based on the amount of carbon and volatiles they contain. Lignite is the lowest grade, with progressively higher ranks called sub-bituminous, bituminous, and anthracite. Notice how higher ranked coals are generally more compact, which is a reflection of being buried at greater depths and the volatile compounds being driven off.

In general, the most desirable coals are bituminous and anthracite because they have the highest carbon concentration, and thus release the most energy. Anthracite is relatively rare and is considered to be a metamorphic coal because of the high temperature and pressure at which it forms—beyond anthracite, the coal will metamorphose into the mineral graphite. Figure 13.7 shows the distribution of different coal deposits in the United States. Notice how coals in the eastern United States are generally higher rank than those in the West, and that anthracite is restricted to a narrow belt within the Appalachian Mountains. The differences in rank are due to differences in geologic conditions (temperature and pressure) under which the coals formed.

Environmental Impacts of Mining Coal

Most of the bituminous coal in the United States was historically mined in the Appalachian region using underground techniques (Chapter 12). In this region, layers of coal were found exposed along numerous stream valleys that cut through relatively flat-lying sequences of sedimentary rocks. Mining typically began where coal seams were exposed at the surface, after which the miners followed the seams horizontally into the subsurface. Vertical

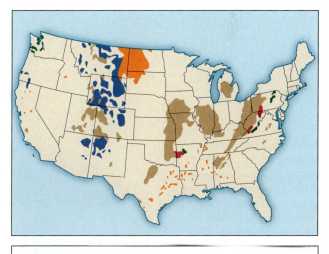

FIGURE 13.7 Major coal deposits in the United States. Although western coals generally have lower energy content than eastern coals, they also contain less sulfur, which reduces the acid rain problem associated with smokestack emissions. This has resulted in a mining boom for western coals and a more depressed market for eastern coals.

FIGURE 13.8   Underground coal mining creates large voids that can slowly close or suddenly collapse, causing land at the surface to sag or develop pits. Subsidence is more common in areas where the layers of coal lie close to the surface and where the mining leaves the roof of the mine with insufficient strength to withstand the weight of the overburden material. Photo from 1976 showing collapse features above a Wyoming coal mine.

shafts were then used to access layers of coal buried deep within the hillsides, creating a network of tunnels and shafts. Here the underground miners faced a high risk of rockfalls and cave-ins in the relatively soft sedimentary rocks. Other common hazards included explosions and fires associated with the buildup of methane gas and coal dust. Long-term exposure to coal dust also caused high incidences of lung cancer among miners, more commonly known as *black lung disease.* Today's modern safety regulations and mining techniques have dramatically reduced the number of fatalities and injuries in underground coal mines in the United States.

In addition to the risks faced by miners, underground mining creates void spaces that tend to close in on themselves due to the tremendous weight of the overlying column of rock (overburden pressure). Sometimes the voids close slowly, and other times they collapse suddenly. As shown in Figure 13.8, the closing of subsurface voids can cause the land surface to develop pits or sag in a process known as *subsidence* (Chapter 7). Land subsidence is undesirable as it can damage buried utility lines, roads, and buildings and adversely affect streams and ponds. Subsidence is a particular problem above shallow coal mines because of the way layers of coal are extracted, leaving the roof of the mine unsupported. For example, in the technique called *long-wall mining* (Figure 13.9A) a cutting machine extracts coal in long strips while hydraulic jacks support the roof of the mine. Then, as the machine moves farther into the coal bed, the jacks are eventually removed and the roof is intentionally allowed to collapse. A much older technique, known as *room-and-pillar mining* (Figure 13.9B) involves removing the coal bed in a rectangular pattern, creating large voids (rooms) surrounded by columns (pillars) of coal that support the roof. In order to extract as much coal as possible, some mine operators systematically remove the coal pillars in a process called *retreat mining.* This practice not only leads to greater subsidence problems at the surface, but puts the miners at much higher risk of cave-ins.

Most of the coal mined in the United States today is done at the surface using *strip mining* (Chapter 12). Cranes equipped with large dragline buckets are used to expose the coal by scraping away the overlying layers of

FIGURE 13.9
Underground coal mining leaves large void spaces that tend to collapse due to the overburden pressure created by the overlying rocks. Long-wall mining (A) is an efficient method where the roof is supported by jacks as a cutting machine removes coal in long strips. The jacks are then removed as the machine moves down the coal bed, allowing the roof to collapse. In room-and-pillar mining (B) columns of coal are left behind in order to support the roof. Collapse can occur when too few pillars are left behind for support.

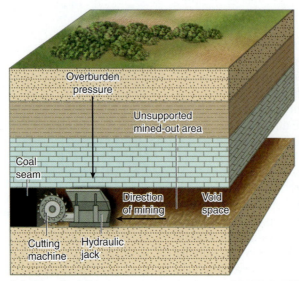

A Long-wall mining

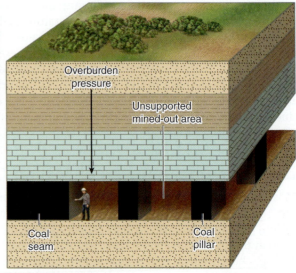

B Room-and-pillar mining

unwanted rock, called overburden. Strip mining is being used extensively in the western United States where it is highly economical because the coal seams are relatively close to the surface and as much as 125 feet thick. Until recently, strip mining has been less economical in the Appalachian coal fields due to greater overburden and thinner coal beds. The development of modern cranes and oversized dragline buckets has changed the economics of strip mining, making it the dominant form of coal mining throughout the United States. This has led to a very controversial form of strip mining in the Appalachians called *mountaintop removal,* where extremely large amounts of overburden are removed to extract multiple layers of coal. As illustrated in Figure 13.10, this technique levels the landscape and places the resulting waste material in adjacent valleys. Once the coal is removed, new vegetation and other improvements are made such that the reclaimed land can be used in a variety of ways, including housing developments and parks. However, many residents and communities vigorously oppose mountaintop mining. They cite the profound changes to the natural landscape and a host of environmental problems, not the least of which are permanently altered streams and ecosystems.

FIGURE 13.10   Mountaintop removal is a controversial form of strip mining that replaces underground coal mining. Here successive coal seams are accessed by removing massive amounts of overburden, causing large-scale disruptions to the land surface. Reclamation efforts can make the area usable for such things as housing developments and parks, but the original mountain streams and ecosystems are permanently lost.

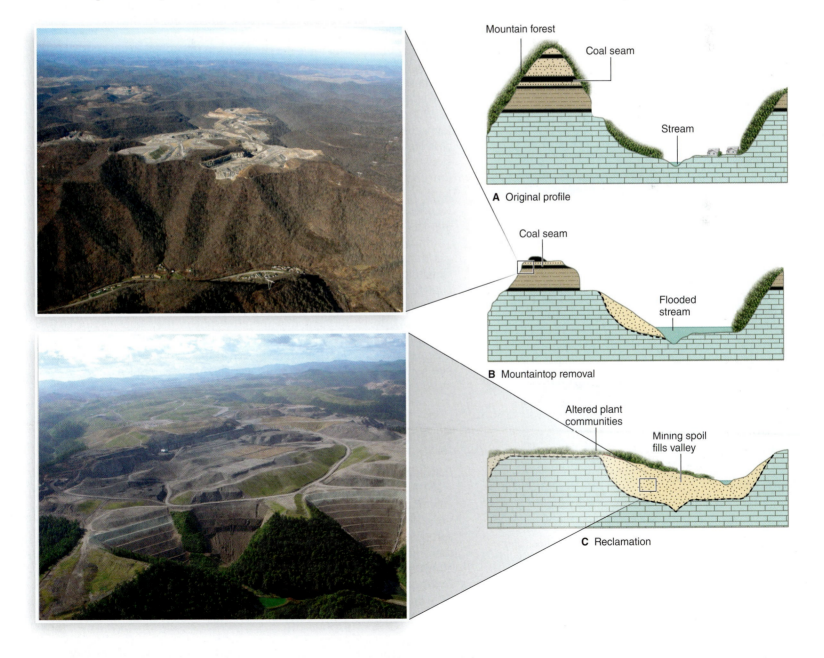

Environmental Impacts of Using Coal

In addition to mining issues, the use of coal presents a number of serious environmental problems, many of which are related to the fact that coal contains metallic sulfide minerals (Chapters 3 and 12). Due to the low oxygen environment in which coal forms, heavy metals bond with the sulfide ion (S^{2-}) in a swamp to form sulfide minerals. When coal undergoes combustion, such as in a power plant, the mineral bonds are broken, allowing the sulfide ions to combine with oxygen to produce sulfur dioxide gas (SO_2). After the sulfur dioxide leaves the smokestack, it reacts with atmospheric water to form sulfuric acid (H_2SO_4), which later falls as *acid rain* along with solid particles containing heavy metals. Acid rain is a problem because it damages ecosystems and has a corrosive effect on bridges, highways, and other structures composed of concrete. With respect to the heavy metals, mercury is a particular concern because it can move up through the food chain into humans and cause neurological damage. Similar problems occur with *acid mine drainage,* except that the sulfide minerals are chemically broken down by rainwater rather than by combustion. Here acid and heavy metals are released into the environment when sulfide minerals are broken down by water percolating through mines and piles of waste material (Chapter 12).

Many of the lower ranked coals found in the western United States (see Figure 13.7) contain relatively few sulfide minerals compared to the Appalachian coals. When environmental regulations designed to reduce the impact of acid rain from coal emissions were passed in the United States in 1970, power companies suddenly found it more cost-effective to burn low-sulfur western coals. Despite producing less heat, using western coals cost less than installing expensive equipment to reduce the sulfur emissions from eastern coals. This ultimately created a mining boom for western coals and decreased the demand in the East, imposing economic hardships on a region long dependent on coal. Refer to Chapters 12 and 15 for more details on the problems and remediation efforts associated with mercury and acid rain and drainage.

Finally, burning coal also releases carbon dioxide (CO_2), a greenhouse gas that affects the amount of heat trapped by Earth's atmosphere. Most climatologists agree that the additional input of CO_2 is affecting the planet's climate system and is a major factor in the current global warming trend. Because the worldwide consumption of coal is causing tremendous quantities of CO_2 to be released into the atmosphere, the future use of coal is uncertain. To make coal more viable as a long-term energy source, technologies are being developed that are designed to capture CO_2 and inject it into permeable rock layers in the subsurface. This approach, called *carbon sequestration,* will be described more thoroughly in Chapters 14 and 16.

Petroleum

Petroleum is a general term geologists use to describe both oil and natural gas. Oil and gas are composed of similar types of organic molecules and are commonly found together in the subsurface under similar geologic conditions. These fuels are similar to coal in that they both contain combustible carbon that originates from organic matter. The primary difference is that the organic molecules in oil and gas contain considerably more hydrogen atoms than those in coal. In fact, oil and natural gas are often referred to as **hydrocarbons** because their molecular building blocks are composed chiefly of hydrogen and carbon atoms. As shown in Figure 13.11, **natural gas** is a mixture of different hydrocarbon gases, the lightest being methane (CH_4)

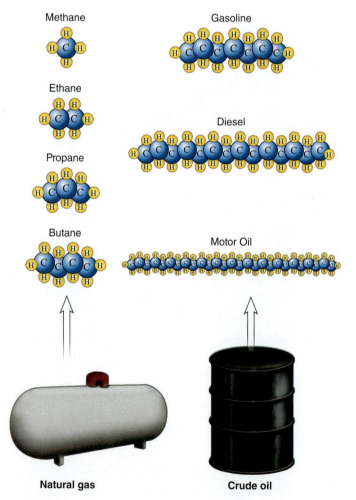

FIGURE 13.11 Petroleum consists of natural gas and crude oil, both of which are composed of hydrocarbon molecules. The refining process produces various petroleum products by separating the gaseous and liquid molecules based on their differences in density. Note that refining produces many more products than those shown here.

because it contains the fewest number of atoms. When natural gas is extracted from the subsurface it is sent to a refining plant where the gases are separated based on their different densities. From this we get familiar products such as propane used in gas grills and butane for disposable lighters. The natural gas that is piped into our homes is composed mostly of methane. Because methane is odorless, refiners add small amounts of a distinctive smelling compound to make it easier to detect leaks.

The term **crude oil** refers to unrefined oil that is composed of a mixture of hydrocarbon molecules existing in the liquid state. Note in Figure 13.11 how the molecules making up crude oil consist of long chains of hydrogen and carbon atoms. Similar to natural gas, crude oil is processed in a refinery where the various molecules are separated into different petroleum products based on their differences in weight or density. Gasoline is one of the lightest and most volatile of the liquid hydrocarbons, whereas diesel fuel and motor oil are progressively less volatile and contain heavier hydrocarbon molecules. In the following section we will take a closer look at how petroleum forms, followed by a discussion on the way geologists find subsurface deposits of this most vital resource.

Origin of Petroleum

The transformation of organic matter into petroleum is a complex process where oil and gas commonly originate in one place, then migrate and accumulate somewhere else. Although there have been several hypotheses developed over the years, the general consensus today is that most petroleum originates in organic-rich marine shales, often referred to as petroleum *source rocks*. Modern examples of places where this type of sediment is accumulating are relatively quiet waters where marine plankton (open-water plants and animals) thrive in the sunlit portions of the water column. As these microscopic organisms die, they fall from suspension and produce a steady rain of organic particles that accumulate on the seafloor. In areas where oxygen levels are low, the decay process slows to the point where the dead plankton tend to be preserved. This results in layers of sediment rich in organic matter.

Should petroleum source rocks become buried under progressively thicker piles of sediment, the corresponding increase in temperature and pressure can cause the original plankton to be chemically transformed, or fossilized, into simpler organic compounds. When the rocks reach a depth of around 7,500 feet (2,300 m) they enter what geologists call the **oil window.** Here temperatures are high enough to cause the organic compounds to begin transforming into oil and natural gas. As illustrated in Figure 13.12, oil will continue to form down to a depth of approximately 15,000 feet (4,600 m). Beyond this depth is the **gas window,** where higher temperatures cause the remaining oil to break down into the simpler molecules that make up natural gas. If the source rocks are buried deeper than 40,000 feet (12,200 m), the gas molecules will be transformed into the mineral graphite. Note in Figure 13.12 that *biogenic gas* forms in the near-surface environment and is mostly methane (CH_4). This type of gas, sometimes referred to as *swamp gas,* forms when nonfossilized organic matter breaks down in low-oxygen (anaerobic) environments. Biogenic gas is common in wetlands and human landfills (Chapter 15).

In order for petroleum source rocks to reach the oil window (Figure 13.12) there must be some geologic mechanism that allows them to be buried by significant amounts of younger sediment. Similar to the formation of coal, geologic evidence suggests that as sediment accumulates the additional weight causes the seafloor to subside. This, in turn, allows even more sediment to accumulate, sending source rocks to progressively greater depths over long periods of geologic time. Note that because coal and

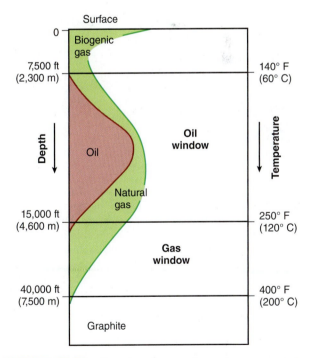

FIGURE 13.12 For oil and gas to form, the source rock must lie within a relatively narrow depth and temperature window in the subsurface. Beyond a depth of approximately 15,000 feet (4,600 m) the temperature reaches the point where oil begins to break down into gas. At 40,000 feet, the temperature is high enough to cause the gas molecules to metamorphose into the mineral graphite.

Oil wells

petroleum form in similar depositional environments, it is not uncommon for both types of fossil fuels to be found in the same sedimentary basin. Despite their similarities, oil and gas contain more volatile hydrogen-rich compounds because they are derived from marine plankton, whereas coal is less energy intensive because it is derived from land-based plants.

Migration and Accumulation

Because oil and gas originate in sedimentary environments, geologists searching for oil and gas deposits focus on areas with thick sedimentary sequences. As illustrated in Figure 13.13, hydrocarbon molecules will form in a sedimentary basin provided there is a suitable source rock that had at one time been buried within the oil and gas windows. Hydrocarbon molecules are less dense than water, so they will migrate toward the surface through the rock's pore spaces, similar to how gravity allows air bubbles to rise in a swimming pool. The migrating oil and gas molecules will eventually reach a more permeable layer within the sedimentary basin, such as a sandstone or limestone. Recall from Chapter 11 that these permeable rocks also serve as aquifers because they readily transmit water. This means that as the hydrocarbons enter an aquifer they can be transported laterally with the flowing groundwater. Here the less dense hydrocarbon molecules will rise and tend to flow on top of the water within an aquifer.

In order to form an economical deposit in a sedimentary basin, oil and gas molecules must somehow become trapped and accumulate; otherwise, they will disperse and continue to rise slowly toward the surface. For example, in Figure 13.13 one can see that the original sedimentary layers have become inclined and folded—due to tectonic forces (Chapter 4). As groundwater and petroleum move along the permeable layer and encounter the fold, the low-density hydrocarbons become trapped, whereas the water continues to flow. Geologists use the term **petroleum trap** to describe any configuration of rocks that allows hydrocarbons to accumulate, and **petroleum reservoir** for those permeable rocks where oil and gas are being stored. Depending on the history of the source rock and its position within the oil and gas windows, a petroleum reservoir may contain oil, gas, or both oil and gas. Whenever both are present, natural gas will lie on top of the oil because it is less dense. In some cases, the pressure is such that the gas will be dissolved within the oil itself—similar to how carbon dioxide is dissolved in a soft drink.

One of the more common types of petroleum traps and reservoirs are those where sedimentary rocks are folded into a dome or arch geologists call an *anticline*, such as the example in Figure 13.13. Notice in the aerial photo how the pattern of layered rocks at the surface clearly reflects the domelike structure that extends into the subsurface. Figure 13.14 provides examples of other common types of traps and reservoirs. Salt domes are particularly common in the Gulf Coast of the United States where evaporite deposits of halite (Chapter 12) flow upward due to the plasticlike nature of halite. Some traps (e.g., buried reefs and unconformities) have little or no expression at the surface, and are therefore more difficult to locate. Interestingly, many of the early oil fields were associated with domes because geologists could easily identify the telltale pattern of rocks at the surface.

An important feature of all petroleum traps is the presence of a **cap rock,** which is a low-permeability rock layer that overlies the trap. A suitable cap rock, typically composed of clay or evaporite minerals, is critical because it limits the ability of oil and gas to escape from the reservoir (Figure 13.14). A good trap and cap rock, however, does not necessarily

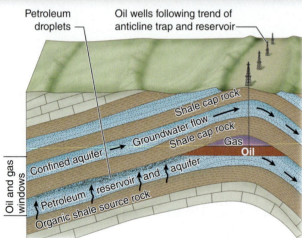

FIGURE 13.13 Hydrocarbon molecules form when organic-rich source rocks are buried to the point where temperatures reach the oil and gas windows. The molecules rise until encountering a permeable reservoir rock, then flow laterally with the groundwater until reaching a trap where the lighter hydrocarbons accumulate on top of the water. Aerial photo shows the characteristic pattern that a dome trap makes at the surface. Note the individual wells that are extracting petroleum trapped in the rocks below.

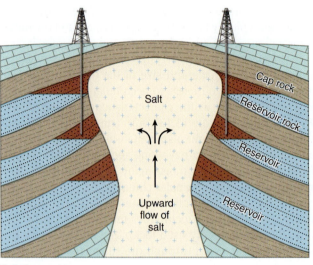

Salt dome

Unconformity

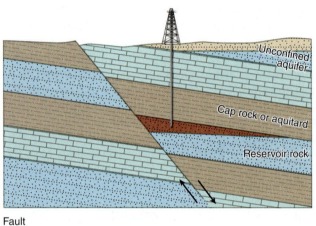

Fault

Limestone reef

FIGURE 13.14
In addition to folded domes, there other common types of petroleum traps and reservoirs. All traps must be overlain by a low-permeability cap rock in order to limit the ability of oil and gas to escape over time. The reservoir rock itself must be quite porous in order to store significant quantities of petroleum, and also permeable enough to allow petroleum to be extracted.

mean there will be economical amounts of oil or gas. For example, any one of the following circumstances will lead to a reservoir being filled with water as opposed to hydrocarbons:

1. a suitable source rock did not exist;
2. a source rock was present, but never reached the oil window;
3. temperature went beyond the gas window, destroying all hydrocarbons;
4. a leaky cap rock allowed petroleum to escape over time;
5. petroleum developed and migrated before the trap formed.

Consequently, traps that contain economical amounts of petroleum are relatively rare from a geologic perspective. Note that because of new technology, natural gas is now being extracted from thick sequences of organic-rich shales, found in various sedimentary basins around the world. The volume of this so-called *shale gas* is quite large, and is expected to greatly increase the global supply of natural gas.

Exploration and Production Wells

Like water wells described in Chapter 11, oil and gas wells are drilled through solid rock using a diamond-tipped cutting tool attached to a rotating steel pipe. This method also allows geologists to determine the types of rocks being encountered by the drill bit simply by examining the chips of rocks that move up to the surface during the drilling. This basic technology was used over a

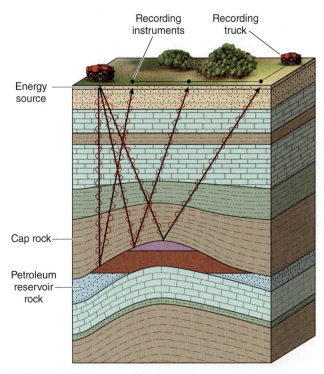

Energy source

Cap rock

Petroleum reservoir rock

Recording instruments

Recording truck

FIGURE 13.15 Seismic exploration techniques use an energy source to generate vibrational waves that reflect and refract off rock layers of different densities. Based on data obtained by instruments that record those waves that return to the surface, detailed views can be obtained of subsurface rock structure that may contain oil and gas.

hundred years ago for installing the first successful oil wells. After a successful strike, additional wells would typically be drilled in a pattern around the initial well. By taking the information on the subsurface rocks and combining it with maps showing the location of successful versus nonproducing wells, it became possible to determine the shape and size of the oil reservoir. This type of mapping soon revealed that oil tends to accumulate in dome traps, which geologists could easily identify by the characteristic rock pattern these structures make at the surface (see Figure 13.13). Airplanes were soon put to use in identifying other domes that might also contain oil. Nearly all of the traps with an obvious surface expression have long ago been identified and tested. What remains are mostly hidden traps, many of which are located offshore on the continental shelves and in polar climates.

To meet the difficult challenge of locating hidden traps, oil companies have developed a variety of geophysical techniques for finding deeply buried geologic structures. The most widely used technique makes use of the same types of *seismic waves* that travel through the subsurface during an earthquake (Chapter 5). As shown in Figure 13.15, *seismic exploration* uses an explosive device or a special vibrating truck as an energy source for generating seismic waves. The waves undergo reflection and refraction as they encounter rock layers of different densities, then a series of instruments, called *geophones,* record the waves that return to the surface (a similar approach is used by ships for exploring offshore areas). From these data, geophysicists are able to construct cross-sectional views of the different rock layers in the subsurface. By repeating this method in long parallel strips, oil companies are able to obtain detailed, three-dimensional views of sedimentary basins on land and on continental shelves.

Once an oil company performs a seismic survey in a sedimentary basin, teams of geologists and geophysicists will determine which traps have the greatest probabilities of containing oil and gas. Despite all the advanced seismic technology, the only sure way to know if a prospective trap is commercially viable is to drill one or more *exploration wells.* Naturally, the prospects with the highest probability of being successful are drilled first—over 50% of all exploration wells do not find economical amounts of petroleum. The exploration process also provides the opportunity to perform porosity and permeability tests on the actual reservoir rock, and chemical analysis on any hydrocarbons that are found. Data from these tests allow engineers to estimate how much petroleum is stored in the reservoir and how much can ultimately be extracted or recovered. In the case of oil, the recovery rate ranges from 5 to 80%, with the highest rates being associated with more permeable reservoir rocks and relatively light, low-viscosity crude oils.

A petroleum reservoir will typically be developed if the exploration data show that there is enough recoverable oil or gas to justify the cost of getting it to market. This can involve enormous sums of money for the installation and operation of long-term *production wells,* processing facilities, and pipelines for sending oil or gas to a refinery. According to the U.S. Department of Energy, the average cost of just a single production well in 2006 was $2.2 million. As shown in Figure 13.16, offshore sites require large and very expensive production platforms that are towed out to sea and then allowed to settle on the seafloor. Once in place, multiple production wells are installed using directional drilling techniques in which the drilling bit is steered through the subsurface by technicians on the platform. Directional drilling is also used on land, particularly in areas that are environmentally sensitive or difficult to access.

Because all petroleum reservoirs contain a finite amount of oil or gas, it is only a matter of time before the production rate reaches a maximum and begins to decline. In the case of large oil reservoirs, the fluid pressure within the reservoir rock is often high enough that oil will flow freely to

A

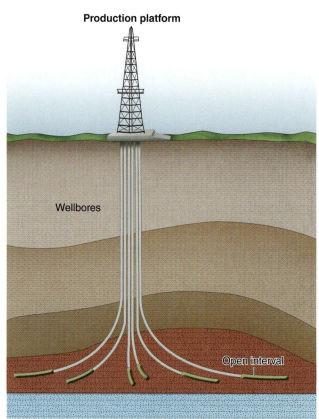

B

FIGURE 13.16   Large production platforms (A) are towed out to sea and then placed on the seabed where oil reserves have been confirmed by exploration wells. Modern land and offshore operations utilize directional drilling technology to install multiple production wells (B) from a single platform.

the surface based on the same principle as artesian water wells (Chapter 11). As the reservoir pressure eventually declines, production costs go up as pumps must be used to extract the oil (reservoirs containing only natural gas all have free-flowing wells). To keep the reservoir pressure up, engineers will often inject water or unused gas into the reservoir rock. This technique also drives the oil in a more uniform manner toward the pumping wells and increases the overall recovery rate. Should the reservoir contain heavy, high-viscosity crude oil, recovery rates can be increased by reducing the viscosity of the oil through steam injection systems.

Although injection and other techniques can improve recovery rates, production will ultimately decrease to the point where the cost of removing oil from the ground is greater than its market value. In its early stages, a production oil well is normally quite profitable since it produces mostly crude and lesser amounts of water. As time goes on it will produce progressively greater amounts of water and less oil, making its operation less profitable. Eventually the costs of operating the well will no longer be justified, forcing the owners to stop production despite the fact a certain amount of oil still remains in the reservoir. Of course, if the market price of crude increases, it may be profitable to continue extracting the remaining oil for a longer period of time, thereby increasing the overall recovery rate.

Petroleum Refining

Petroleum refining utilizes a distillation process similar to how freshwater is made by heating salt water, then collecting the water vapor and letting it cool and condense into a liquid (Chapter 11). Unlike water however, crude oil contains various types of hydrocarbon molecules, each having a different density and corresponding temperature at which they condense.

As illustrated in Figure 13.17, the refining process begins when crude oil is heated to the point where most of the hydrocarbon molecules vaporize into gases, and then rise up into a *distillation tower*. Temperatures within the tower are naturally hotter near the bottom, and become progressively cooler toward the top. Because the heaviest hydrocarbon molecules (tars, greases, oils, etc.) will condense at higher temperatures, they are removed from the lower parts of the tower and then piped to separate holding tanks. Moving upward in the tower where temperatures are lower, progressively lighter and more volatile molecules are collected and condensed into their respective liquids (e.g., kerosene, aviation fuel, and gasoline).

Remember that some crude oils are considered to be relatively *heavy* and more viscous compared to the so-called *light* crudes. The basic difference is that heavy crudes have a higher proportion of long chains of hydrocarbon molecules, making them denser and more viscous. As you might suspect, more gasoline can be refined from a barrel of light crude since it contains more of the shorter molecules that make up gasoline. Because gasoline is in high demand as a transportation fuel, light crude oils command a higher market price compared to heavy crudes. Refiners are able to increase the amount of highly profitable gasoline by using a process

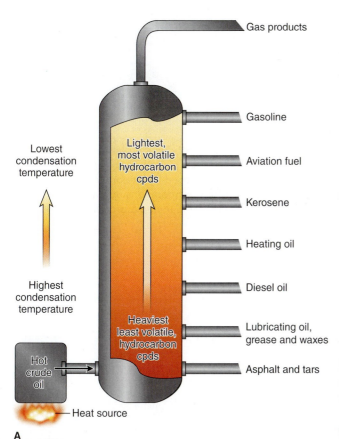

A

FIGURE 13.17 Refining of crude oil into various products (A) begins by heating the crude, then sending the vapors into a distillation tower. At the bottom of the tower, where temperatures are highest, the heaviest hydrocarbon molecules are able to condense, at which point they are collected. As the temperature in the tower gets cooler toward the top, progressively lighter and more volatile molecules will condense. Shown here are just a few of the more common types of products obtained from crude oil. Aerial view (B) of the British Petroleum refinery in Texas City, Texas. Note the numerous distillation towers and storage tanks.

B

called *cracking*, which breaks long, heavy hydrocarbon chains into shorter and lighter molecules. Lighter crudes are the most desirable not only because they produce more transportation fuels, but also because they are easier to extract from petroleum reservoirs and have higher recovery rates. Therefore, most of the crude that oil companies have been extracting for the past 100 years has been relatively light. This has resulted in the bulk of the world's remaining untapped deposits being composed of more viscous and heavy oils, which of course are far more difficult to extract.

Finally, it is interesting that crude oil not only provides the gasoline and diesel fuels that run our cars and trucks, it also provides the tar and asphalt that pave the highways we drive on. There are also countless consumer products that originate from refined petroleum, forming the basis for what is known as the *petrochemical industry*. Perhaps the most familiar products are those made of nylon and various types of plastic, from your cell phone, water bottles, clothes, and tennis shoes, to the interior and exterior of your home and car. Even more important are the agricultural pesticides and nitrogen-based fertilizers that are made from oil and natural gas. In addition to diesel-based farm equipment, these fertilizers and pesticides have largely been responsible for the huge increases in food production over the past century. Without petroleum, it simply would not have been possible for Earth's population to reach its current level of over 6 billion people.

Environmental Impacts of Petroleum

Whenever oil and gas are burned (i.e., undergo combustion), a variety of pollutants are released into the atmosphere, many of which are hazardous to human health. Although some of these combustion by-products such as soot can be seen with the naked eye, most are invisible gases. During this combustion process hydrogen and carbon atoms will bond with oxygen to form water (H_2O) and carbon dioxide (CO_2). Because the combustion is never 100% complete, this process also produces unburned hydrocarbons and carbon monoxide (CO) gas. Unlike carbon dioxide, carbon monoxide is similar in size to the oxygen molecule (O_2) and can be taken in through the lungs and absorbed in the blood. Carbon monoxide is extremely poisonous and often kills people who sleep in cars with the motor running and live in homes with faulty chimneys. Another category of air pollutants includes various nitrogen oxide (NO_x) gases that form when atmospheric nitrogen (N_2) is pulled into the combustion chamber of gasoline and diesel engines (a more thorough discussion of emissions and air pollution can be found in Chapter 15).

In addition to air pollution, society's use of oil has led to an assortment of other environmental problems, particularly oil spills that can damage ecosystems and contaminate water supplies. The drilling of oil wells has historically led to spills, particularly when a high-pressure zone is encountered and operators lose control of the well. When this occurs, oil or gas will escape under intense pressure in what is known as a *blowout*. As shown in Figure 13.18, a blowout not only creates an oil leak, but can produce a very

FIGURE 13.18   Blowouts can occur on wells when operators encounter zones of unusually high pressure, allowing oil or gas to escape in an uncontrolled manner. Shown here is a well blowout in the Bay of Campeche, Mexico, 1979, creating both an oil slick and fire.

intense and dangerous fire that is difficult to extinguish. In 1969 a blow-out occurred on an offshore platform near Santa Barbara, California, releasing nearly 3 million gallons (71,000 barrels) of crude oil into the marine environment. Some of the oil washed ashore along the scenic coastline, covering 35 miles (22 km) of beaches with as much as 6 inches of oil, severely disrupting the sensitive marine ecosystem. Transporting oil can also result in spills, such as the one that occurred in 1989 when the supertanker *Exxon Valdez* ran aground in Alaska's Prince William Sound, discharging 11 million gallons (260,00 barrels) of crude oil into the pristine waters (Figure 13.19). Damage to the delicate ecosystem was extensive and subsequent cleanup costs reached over $2 billion. Although the beaches today look normal, unweathered crude still lies buried along the shoreline. Recent studies indicate that some species have still not recovered, likely due to persistent low-level pollution.

The oil spill near Santa Barbara, California, came at a time when increased restrictions were being placed on offshore oil exploration in the United States. Then in 1990, in the aftermath of the *Exxon Valdez* disaster in Alaska, Congress and President George H. W. Bush established moratoriums on offshore drilling on the U.S. continental shelf, with the exception of parts of Alaska and the existing oil fields in the central Gulf of Mexico. The recent rise in the price of crude, however, has put considerable political pressure to open the continental shelf to exploration. Another hotly contested area is the Arctic National Wildlife Refugee (ANWR) on Alaska's North Slope, originally established by President Eisenhower in 1960 and

FIGURE 13.19   In 1989 an oil tanker ran aground in Prince William Sound in Alaska, polluting beaches and killing wildlife along a vast stretch of Alaska's irregular shoreline. Despite years of cleanup efforts, beaches still contain buried oil that continues to be a source of low-level pollution.

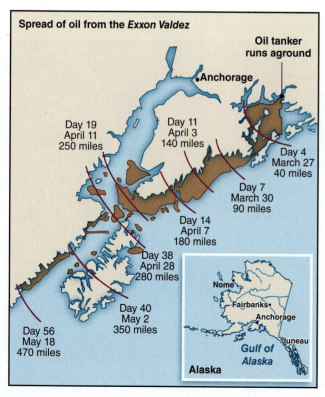

set aside for oil exploration by Congress in 1980 (Case Study 13.1). Some view ANWR and the continental shelf as vital for expanding the nation's oil supplies. Others see drilling in these areas as an unnecessary distraction that keeps the country from focusing on developing alternate energy supplies, and as a potential ecosystem disaster.

Current Energy Supply and Demand

Earlier in this chapter you learned that the Industrial Revolution has combined with rapid population growth to produce today's tremendous demand for energy. This demand is presently being met primarily through the burning of fossil fuels that are both inexpensive and highly concentrated forms of energy. The problem is that fossil fuels are finite, which means continued economic growth based on these resources is simply not sustainable. Another serious issue with fossil fuels is that large quantities of carbon, which were stored in the subsurface for millions of years, have suddenly been released into the atmosphere in the form of carbon dioxide. This is affecting Earth's climate system and is contributing to the problem of global warming. We will examine climate change in Chapter 16, whereas our focus in the remainder of this chapter will be on the dilemma society faces with respect to its limited supply of fossil fuels.

Economic Development and Energy Demand

As discussed throughout this text, the per capita consumption of resources (water, minerals, and energy) depends largely on a nation's level of economic and technological development as well as its efficiency and conservation measures. Developed countries with large populations naturally consume the most energy. This relationship largely explains the uneven distribution of energy consumption by nations around the world. For example, from Figure 13.20 one can see that the United States, with its larger population, consumes a much greater percentage of the world's energy compared to other developed nations (e.g., Russia, Japan, Germany). On the other hand, China and India have very large populations, but relatively low per capita energy consumption rates because they are still developing, hence their total energy usage is still less than that of the United States.

This brings up a very important point: relative to their populations, more-developed countries consume a disproportionate share of the world's energy resources. Consider how the United States consumes 23% of the world's energy resources with just 4.5% of the world's population. In contrast, China has 20% of the planet's population, but consumes only 18% of the energy supply. The problem is that China's per capita consumption rate has been increasing as the country undergoes rapid industrialization. Given its large population and current rate of industrialization, China is expected to surpass the United States in terms of total energy consumption. Consider that China put 25,000 additional cars on the road each day, and doubled its consumption of coal between 2000 and 2008. India is another country with a large population that is also undergoing rapid industrialization. As China, India, and other nations strive to develop consumer societies similar to the United States, the additional demand on the world's energy resources will be enormous. The question is, will the energy supply be able to meet demand, and if so, for how long?

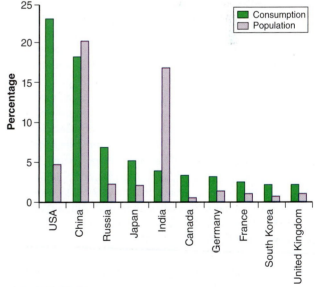

FIGURE 13.20   Plot showing the top 10 energy users in terms of their percent consumption of world energy supply along with their percentage of world population. Considering population size, the United States and other developed nations use a disproportionate amount of energy compared to developing countries such as China and India.

Controversy Over Drilling in the North Slope of Alaska

In 1923, the U.S. government set aside 23 million acres for future oil development in northwestern Alaska. Originally the oil was intended as a fuel supply for the U.S. Navy, but later the area become part of the National Petroleum Reserve. During a survey of the region's natural resources in the 1950s, government scientists identified Alaska's northeast corner (Figure B13.1) as North America's best example of an arctic/subarctic ecosystem. The coastal plain here contains exceptionally diverse and abundant wildlife that are part of a highly integrated ecosystem, including polar bears and great flocks of birds and migrating herds of caribou (Figure B13.2). Due to its unique nature, nearly 9 million acres of the coastal plain and adjoining mountains in northeastern Alaska were designated by President Eisenhower in 1960 as a protected area for wilderness conservation. Then in 1968 the largest oil field in North America was discovered in Prudhoe Bay (Figure B13.1). An 800-mile (1,300 km) long pipeline was later built from the oil fields to the port of Valdez in southern Alaska.

After considerable debate, in 1980 Congress expanded the original wilderness area from 9 to 19 million acres, renaming it the Arctic National Wildlife Refuge (ANWR). However, the same reservoir rocks that contain petroleum at Prudhoe Bay also extend eastward into ANWR. Congress therefore set aside 1.5 million acres of the most promising part of the refuge, known as the 1002 Area, for future oil and gas development. Any development, however, would require congressional approval. During the mid-1980s various seismic and geologic studies were conducted in the 1002 Area by government agencies and private companies. Based on the available data, in 1998 the U.S. Geological Survey (USGS) estimated that for oil to be economically recoverable, crude oil prices would have to be over $16 per barrel. Using the price for crude at the time of $24 per barrel, the USGS estimated that the 1002 Area contains between 2 and 9 billion barrels of oil (for comparison, total recoverable oil from Prudhoe Bay is expected to be 15 billion barrels). Assuming the upper range of 9 billion barrels of recoverable crude oil, the 1002 Area would supply America with 1.2 years' worth of crude at the country's 2008 consumption rate of 21 million barrels per day. The amount of recoverable oil is expected to be higher because crude oil prices are now over $24 per barrel. Keep in mind that until data becomes available from actual exploration wells, no one knows just how much oil and gas will be found. The media often reports that the 1002 Area contains 16 billion barrels, but this includes estimates from nonrestricted sites located offshore and on adjacent private lands.

FIGURE B13.1 Map showing Alaska's North Slope and the extensive coastal plain and offshore areas that are available for oil and gas development. Development in the Arctic National Wildlife Refuge (ANWR) has been prohibited by Congress. The debate over drilling involves the 1002 Area, which happens to contain the best oil and gas prospects and the most sensitive parts of the coastal plain ecosystem.

The debate to open ANWR's 1002 Area to oil and gas development has become more heated in recent years as crude oil prices have increased sharply. Drilling advocates generally believe that the additional oil and gas supplies are vital, particularly since the United States is becoming progressively more dependent on imported oil. But the 1002 Area contains ANWR's most important and diverse ecological habitats. The primary environmental concern over development is that the installation of several hundred production pads and associated pipelines will disrupt the migration of large herds of caribou. This, in turn, would have a major effect on the entire ecosystem. Supporters of oil and gas development tend to believe that environmentalists are exaggerating the effects of crisscrossing roads and pipelines on the migrating caribou herds. Moreover, because of the boglike conditions of the tundra in the summer, exploration activity and construction of permanent production pads will take place during the winter on temporary ice roads. Pipelines will be elevated in order to minimize the impact on the caribou migrations.

To help their cause, drilling advocates frequently state that only 2,000 acres, a mere 0.01% of ANWR's 19 million acres, will be developed. Environmentalists maintain that this statistic is misleading since the development will only occur over the 1.5 million acres of the 1002 Area, not the entire 19 million acres of ANWR. In addition, the supposed 2,000 acres that will be affected represents the actual footprint of the production pads and pipelines, not the size of the ecosystem that will be fragmented by the interconnected production wells and pipelines. This means that oil and gas development will not be confined to a mere 2,000 acres of a vast refuge, but spread out over 1.5 million acres that represent the most sensitive parts of ANWR's complex ecosystem.

Although it remains to be seen what impact oil and gas production would actually have on the ecosystem in the 1002 Area, drilling opponents maintain it is not worth the risk. They argue that even with the most generous estimates for recoverable oil, production rates from ANWR will be quite small in terms of world supply. Because crude oil prices in the United States are not controlled by domestic production, but rather by the global oil market, oil from ANWR would have only a minor effect on world oil prices. Drilling opponents believe that it does not make sense to damage this important ecosystem for a minor impact on crude prices. They also argue that it would be better to invest in clean and renewable sources of energy, as world production of petroleum is expected to decline relatively soon, thereby forcing the United States to move toward these alternative sources anyway. One thing is certain: as long as the price of oil continues to climb, the controversy over drilling in ANWR will continue.

FIGURE B13.2 Large caribou herds are part of a diverse and highly integrated arctic/subarctic ecosystem in the 1002 Area of ANWR.

FIGURE 13.21
Breakdown of U.S. energy consumption in 2007 by sector (A) and by source (B).

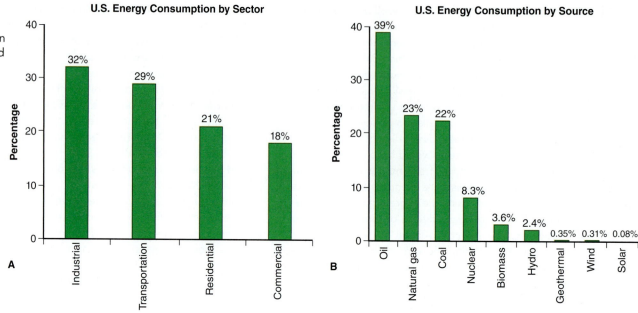

A   U.S. Energy Consumption by Sector

B   U.S. Energy Consumption by Source

Types of Energy We Consume

Because a relatively small number of nations consume the vast majority of the world's energy resources, it is important to understand how these nations use energy. Using the data in Figure 13.21A from the United States as an example, we see that energy usage can be broken down into four major sectors: industrial, transportation, residential, and commercial. Notice that 29% of all energy consumed in the United States goes toward moving people and goods throughout the transportation system. From the breakdown of energy use by source (Figure 13.21B), we see that oil makes up the largest share of the U.S. energy consumption. Interestingly, 70% of all oil is consumed by the transportation sector.

The use of oil for transportation is a good example of how society converts energy from one form to another to accomplish certain tasks. Here an internal combustion engine first converts gasoline or diesel fuel (chemical energy) into heat energy. The heat is then transformed into the kinetic energy that propels our cars and trucks down the highway. Electricity is another very useful form of energy, and is widely used throughout the industrial, residential, and commercial sectors of society. As shown in Figure 13.22, most electricity is made by first transforming fossil or nuclear fuels into steam (heat). The steam is then converted into kinetic energy via a turbine that spins a generator, which actually creates the flow of electrons we call electricity. Electricity can also be made by using falling water or wind to spin a turbine and generator—the conversion of sunlight into electricity will be described in Chapter 14. Notice in Figure 13.22 that burning coal is by far the most common way of making electricity in the United States.

In terms of U.S. energy consumption, the combined use of fossil fuels (oil, gas, and coal) accounts for about 90% of the energy being used across all sectors of the economy (Figure 13.23). China, the world's second largest energy consumer, gets 93% of all its energy from fossil fuels, showing an even greater dependency than the United States. Some nations, however, are less reliant on fossil fuels and use considerably more nuclear and hydro power to meet their energy needs. France, for example, has invested heavily

FIGURE 13.22
Electricity in the United States is made primarily through the burning of coal, followed by natural gas and nuclear fuel. These energy sources are transformed into heat and used to produce steam, which then spins a turbine and electrical generator. Electricity is also made by using water or wind to spin a generator.

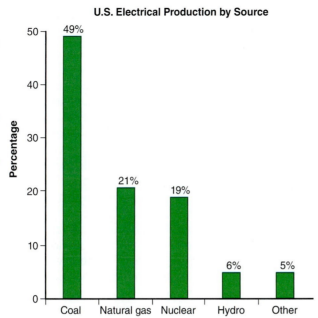

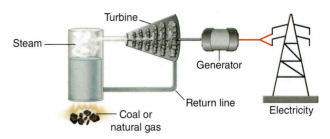

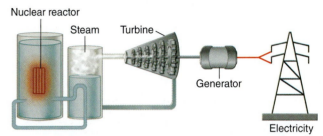

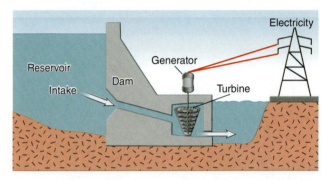

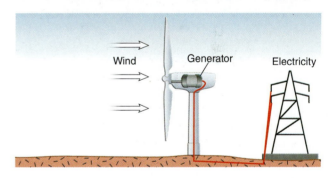

in nuclear power, whereas Sweden has large contributions from both nuclear and hydro. Norway, Canada, and Sweden have all been able to take advantage of their geologic terrain and climate to make significant amounts of hydro power. Another factor here is that after the global oil crises of the 1970s, many countries embarked on large-scale programs for reducing their reliance on fossil fuels.

Unfortunately, the United States dropped its alternative energy program in the early 1980s when oil prices crashed. When China began undergoing rapid development during this same period, it turned to the least expensive form of energy, namely fossil fuels. The result today is that the world's top two energy consumers are burning tremendous quantities of fossil fuels. Moreover, China's growing demand for oil is pushing the already tight global production of oil to its limit. An even more significant problem is that the United States and China are responsible for the bulk of the carbon dioxide being released into the atmosphere, which is affecting our climate system (Chapter 16).

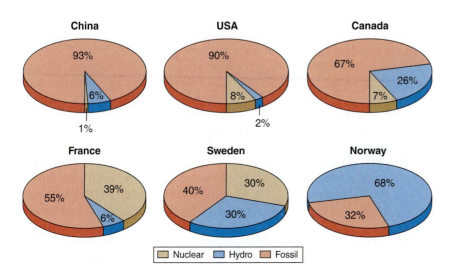

FIGURE 13.23   Countries rely on different combinations of fossil, nuclear, and hydro resources to meet their needs. China and the United States rely heavily on fossils fuels, whereas Canada, France, and Sweden make greater use of nuclear and hydro sources. Norway's geology allows it to utilize hydro power to a much greater degree than most countries.

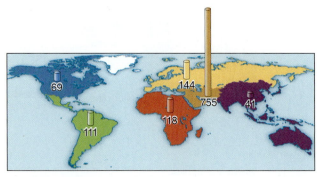

Oil (billions of barrels)

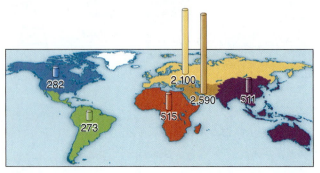

Gas (trillion cubic feet)

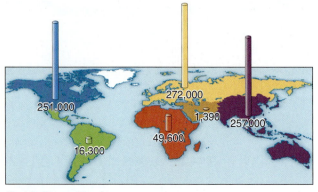

Coal (millions of tons)

FIGURE 13.24   Proven reserves of fossil fuels are not evenly distributed on Earth because economic deposits form only under favorable geologic conditions. Note that the locations of large oil and gas reserves do not necessarily correspond to those of major reserves of coal.

Where Fossil Fuels Are Located

Similar to mineral resources, fossil fuels are not evenly distributed around the globe because economic deposits require certain favorable geologic conditions in order to form. Geographically large countries, such as the United States and Russia, have a higher probability of containing sedimentary basins where conditions were just right for petroleum or coal deposits to form. Large geographic size, however, is no guarantee that sizable deposits will be found. Such is the case for Australia, where only minor amounts of oil are found. It just so happens that the Middle Eastern countries in the Persian Gulf are located in an area where ideal geologic conditions resulted in the world's most prolific oil and gas reservoirs. To get an appreciation for how much petroleum this region contains, compare the oil and gas reserves of the Middle East in Figure 13.24 with those of the rest of the world. Just six countries in the region, Saudi Arabia, Iran, Iraq, Kuwait, Qatar, and the United Arab Emirates, hold 60% of the world's proven oil reserves and 40% of natural gas reserves. Although its petroleum reserves are enormous, the Middle East has essentially no coal, whereas the coal reserves of North America, Eastern Europe, and Asia are quite large. The reason for this is that petroleum and coal form under different geologic conditions.

Due to the economic dependence of modern societies on fossil fuels, the uneven geographic distribution of fossil fuels plays an important role in global politics and the national security of individual countries. Energy security is a serious concern for developed nations whose own energy reserves are insufficient to meet their needs. For example, the United States was once self-sufficient in terms of oil and was the world's leading exporter. Today however the United States has depleted many of its original oil fields and consumes such a large amount of oil that it cannot possibly hope to become self-sufficient again. Japan is in an even more difficult situation as it has virtually no fossil fuel resources of its own, including coal. This sets up the situation where Persian Gulf nations, with their large petroleum reserves, are able to generate enormous sums of money and wield great political power.

The Energy Crisis

As population continues to expand and crude oil production struggles to keep up with demand, the global market price of oil can be expected to increase. Because crude oil is the basis for making gasoline, diesel, and jet fuel, as well as agricultural chemicals and plastics, rising crude prices will have a ripple effect through entire economies. Consumers will not only pay more for a gallon of gasoline, but for nearly everything they buy, including food. As people spend more of their paycheck for the basic necessities, they will have less to spend on various consumer goods and nonessential activities. Ultimately this can lead to an economic recession and the collapse of various sectors of the economy. Because of the severe consequences, countries that are not self-sufficient in terms of oil production can be expected to compete more aggressively for existing supplies. This can lead to a greater potential for armed conflicts over oil resources.

History provides several important examples of economic and political crises that were triggered in part by limited oil supplies. In 1941, the United States, then the world's leading oil exporter, imposed an oil embargo on Japan over its war with China. Japan responded five months later by attacking U.S. and British forces throughout the Pacific. This allowed Japan, a nation with no petroleum reserves of its own, to seize the oil fields in the Dutch East Indies, thereby eliminating their oil crisis. Germany was in a similar position prior to the war, and had even resorted to producing gasoline and diesel fuel from coal. One of the reasons Hitler made the fateful decision to invade the Soviet Union (Russia) was to capture its oil fields in the Caucasus.

After World War II, the United States remained the world's leading oil exporter until 1970, when domestic production reached its peak. To meet its growing demand, the United States had to begin importing oil. As shown in Figure 13.25, crude prices remained quite stable until 1973 when the oil-rich countries of the Middle East imposed an oil embargo (refused to sell) on Western nations because of their support for Israel in the 1973 Yom Kippur War. The oil embargo resulted in an immediate shortage and a surge in crude oil prices, sending shock waves through the Western economies that triggered a global recession. Eventually the embargo collapsed and prices stabilized. Then in 1979 yet another oil crisis emerged when the Iranian revolution disrupted that nation's considerable oil exports, sending oil markets into a panic and prices skyrocketing. By 1985 prices tumbled as supplies increased due to the increased production that resulted from the previously higher prices. Major new production from Alaska and from fields in the North Sea then helped create an oil glut. The world then enjoyed relatively low prices, fueling an extended period of economic prosperity until 2000, at which point prices began to climb. Note that in 2008 crude oil prices spiked at around $150 per barrel, but then fell to around $40, producing the yearly average of $99 shown in Figure 13.25.

There are several important lessons we can learn from the oil crises of the 1970s and early 1980s. Although the embargo in 1973 created a real oil shortage in Western nations, the world was still capable of producing far more oil than the market demanded. The consequences were more than just long gas lines, many people lost jobs in the resulting recession. In the second crisis of 1979 there was never an actual shortage since oil production continued to meet demand. Prices were driven upward by the *fear* that unrest in Iran could spread and lead to a significant loss of production. The market then was being driven more by speculation rather than actual supply and demand. The question we want to explore next is, what caused the sharp rise in oil prices between 2000 and the 2008 peak? Are we really running out of oil or are the high prices simply the result of fear driving the market as it did in 1979?

Peak Oil Theory

During the early part of last century, there was a dramatic increase in the rate at which new oil and gas fields were being discovered in the United States. A boom in production followed, but lagged years behind the discoveries due to the time required for installing production wells and pipelines. Oil and gas production from individual fields would steadily increase and reach a peak, after which production would steadily decline as the reservoir became depleted. This natural depletion did not affect overall supply since new fields were continually being discovered. However, finding new reservoirs became progressively more difficult, forcing oil companies to develop new technologies for locating hidden traps. Companies also began searching overseas, which eventually led to the discovery of giant oil and gas fields in the Middle East, most of which were found in the 1930s and 1940s.

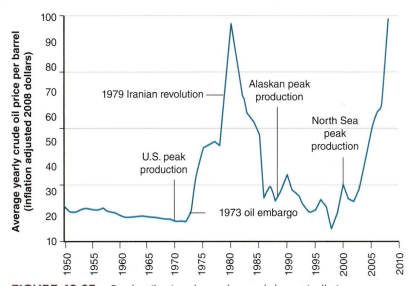

FIGURE 13.25   Crude oil prices have changed dramatically in response to changes in supply and world events. The major price increases of 1973 and 1979 sent shock waves through the economies of developed nations. Additional production caused prices to fall and eventually stabilize at a lower level until 2000. In recent years crude oil prices have risen because of increased demand and limited production capacity. The sharp increase and subsequent decline that occurred in 2008 are not seen here because the data represent yearly averages.

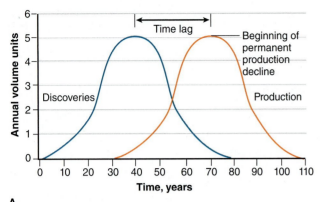

A

FIGURE 13.26   Hubbert's 1956 model (A) used a statistical approach based on the idea that discoveries of new fields would eventually peak, causing production to peak about 30 years later. Using this statistical model, Hubbert predicted the United States would reach peak production in 1970. A graph from a 2004 study (B) showing that U.S. production peaked around 1970. Note the production peaks that followed in other oil-producing countries.

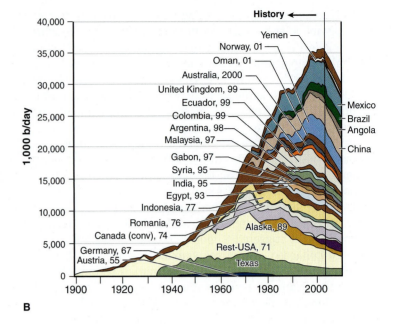

B

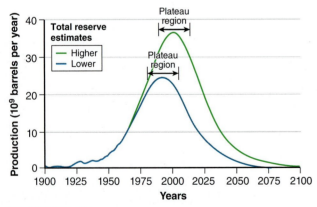

FIGURE 13.27   Based on world production trends and estimated total reserves, Hubbert used his model to project peak world production, with the larger reserve estimate placing the peak at the year 2000. Note how production will reach a plateau before beginning a permanent decline.

In the 1950s a petroleum geologist named M. King Hubbert noticed that the rate at which new oil fields were being discovered in the United States had peaked in the 1930s, and then began to decline. Based on the fact that oil is a finite resource, Hubbert developed a model predicting that exploration and production trends would follow the bell-shaped curves shown in Figure 13.26A. Hubbert's predictive model, often referred to as the **peak oil theory,** was based on a statistical analysis of historical exploration and production data. In 1956 he published his work showing that new discoveries were declining and predicted that U.S. oil production would peak in the early 1970s. To many people at the time, the idea that the world's leading oil producer would reach peak production in the 1970s was simply inconceivable— that is until 1970, when U.S. production actually peaked, and began following a slow downward decline just as Hubbert's model predicted.

After U.S. production peaked, experts watched as country after country followed similar production declines (Figure 13.26B). The message was now clear: conventional supplies of oil were finite and the world as a whole was approaching peak production. In 1969 Hubbert applied his model again and estimated total world supply to be about 2 trillion (2,000 billion) barrels of recoverable oil. From this he came up with two production curves (Figure 13.27). The more generous of the two curves has world peak production occurring in 2000, which is often referred to as **Hubbert's peak.** In an extensive 2000 study, the U.S. Geological Survey estimated total recoverable oil reserves to be 3 trillion barrels, which would place world peak production between 2025 and 2040. Many independent experts believe the estimated reserves by the government agency are far too optimistic, and calculate peak production around 2005. However, as will be described in the next section, the actual peak can not be defined until years after it occurs.

With respect to natural gas, it also is following similar discovery and production curves and is expected to reach peak production. Therefore, one can expect that in the future, supplies of natural gas will no longer be able to meet demand. In addition, because its density is much lower than crude oil, it is more difficult to transport large quantities of natural gas across oceans using tankers. The natural gas must first be cooled to a liquid state, called *liquefied natural gas* or LNG, and loaded onto specialized tankers

(Figure 13.28)—unloading also requires specialized shipping terminals. Because of the difficulty of overseas shipping, natural gas is transported primarily through land-based pipelines and is generally consumed on the same continent on which it is produced. This means that each continent has a unique gas production curve and a corresponding date where peak production is expected to occur. Although natural gas is an important energy source, we will focus on oil in the remainder of this section due to its critical role in transportation and in the petrochemical industry.

Past the Oil Peak

When the world reaches peak oil production it will likely occur over a period of time measured in months or years, creating more of a plateau in the production curve (Figure 13.27) as opposed to a sharp peak. This means that the actual date of Hubbert's peak will be known only when production moves beyond the plateau area and begins an overall downward trend. The world will then find itself permanently on the downward limb of the production curve. Of course our experience over the past 100 years has been on the rising limb of the curve, where production has always been able to exceed demand. Here oil has been relatively inexpensive and market prices have been stable. The cheap energy allowed industries to expand, whereas steady prices created a stable business environment for companies to grow. When the world enters the plateau period of global oil production, oil experts believe that it will be difficult for supply to satisfy demand, thereby causing higher prices and creating instabilities in the global oil market. This would result in dramatic fluctuations in price, in which high prices lead to lower demand, followed by short-lived increases in supply that bring prices down again. The cycle would then repeat as supply is again outstripped by rebounding demand.

  Many economists, however, do not agree with the peak oil theory. Instead they believe that higher prices will cause market forces to spur more exploration and create advanced technologies, thereby generating sufficient quantities of oil to meet the growing demand. As shown in Figure 13.29, the U.S. Department of Energy estimates that between 2003

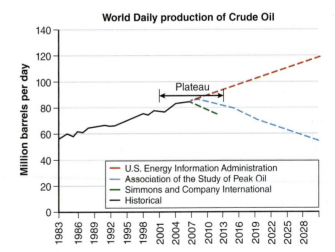

World Daily production of Crude Oil

Legend:
— U.S. Energy Information Administration
— Association of the Study of Peak Oil
— Simmons and Company International
— Historical

FIGURE 13.29 Graph showing the historic rate of world oil production along with projected trends based on different analyses. The U.S. Department of Energy projection assumes production will keep up with demand, whereas other projections indicate that production will peak around 2010.

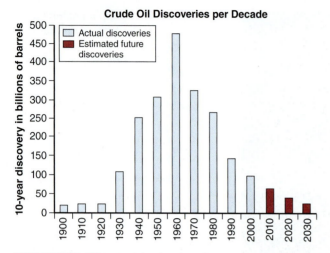

FIGURE 13.30   Worldwide plot of new oil reserve discoveries based on 10-year intervals. Discoveries peaked around 1960 and have been in decline ever since despite intense exploration efforts.

and 2030, world oil consumption will increase by nearly 50%. The obvious question is where will all the additional oil come from? Peak oil advocates point out that most of the reservoirs containing highly desirable light crude have already been located. In fact, a large portion of the world's oil supplies currently comes from giant fields, such as those in the Middle East and Alaska. Moreover, most of these giant fields were discovered prior to the 1970s during the major boom in new discoveries shown in Figure 13.30. These fields are all slowly being depleted, with many already past peak production and experiencing as much as 10% annual declines in production. This means that a progressively greater number of smaller fields must be found in order to make up for the lost production from the giants. Even more fields must be found to meet the projected increase in the demand for oil. Although there has been much hope for advanced technologies, the results so far have been more rapid depletion of existing reservoirs, with only modest increases in overall recovery. Therefore, it does not appear that technology will have much of an impact on the long-term supply of light crudes. Also, keep in mind that it requires energy to extract and process oil from a reservoir. Once the amount of energy needed to bring the oil to market equals the energy it can provide, production will stop regardless of how much remains in the ground. Price at this point becomes irrelevant.

What the world does have in abundance are deposits of *nonconventional oil* (Chapter 14), which are much more difficult and expensive to extract. These deposits generally consist of heavy crudes and hydrocarbons that are too viscous to be pumped from the ground, are tightly bound within shale, or exist in ice crystals on the seafloor. Higher prices have already made some of these resources economical to extract using existing technology, while others remain too costly. One of the concerns with relying on market forces to spur significant production of nonconventional oil is that prices could become so high as to have a crippling effect on demand, often referred to as *demand destruction*. In this scenario higher prices would force people to consume less gasoline as well as all the goods and services that require oil, creating an economic recession. Such a scenario occurred in 2008 when oil prices reached nearly $150 per barrel, then fell to a low of around $40. Although the price spike was due in part to market speculation that drove prices beyond normal supply and demand relationships, it caused demand destruction and a significant drop in consumption. This demand destruction, coupled with the global financial crisis of 2008, led to the collapse in oil prices. Experts believe that oil prices will eventually increase as the world economy rebounds, increasing the demand for oil.

Finally, as the world reaches Hubbert's peak and supplies remain tight, there is a danger that a sudden and sharp reduction in supply could send prices skyrocketing, plunging the global economy into a deep recession. For example, a serious and long-lasting oil shock would occur should there be a loss of production from Saudi Arabia, which produces about 13% of the world's crude. Such an event is a very real possibility given the region's history and previous attacks on the Saudi monarchy. Another worst-case scenario would be if Iran made good on its promise that, if attacked by Western nations, it would shut down oil shipments through the Straits of Hormuz in the Persian Gulf (Figure 13.31). If this were to happen, 20% of the world's crude oil supply could be shut off immediately, creating a crisis of unprecedented proportions.

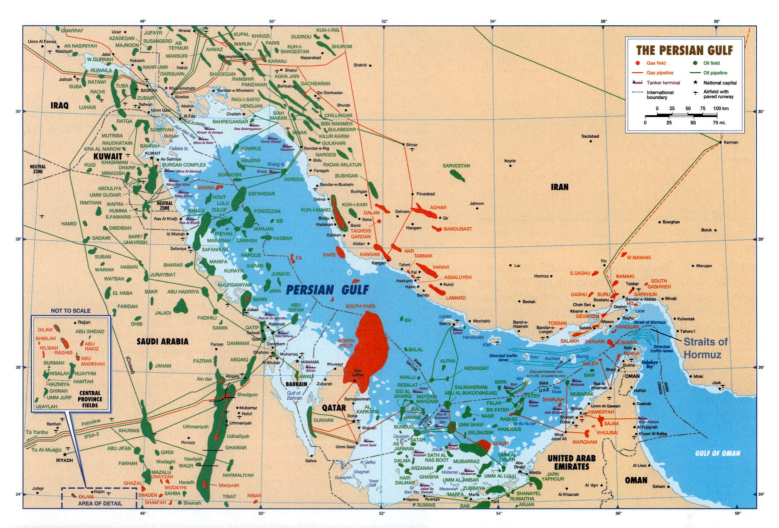

FIGURE 13.31   Nearly 20% of the world's crude oil supply is taken from the vast Persian Gulf oil fields and loaded onto supertankers. The huge ships must then pass through the Straits of Hormuz, a strategic choke-point that Iran has threatened to shut down should there be a military clash with Western nations.

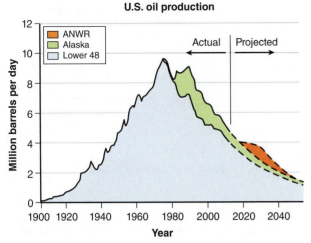

U.S. oil production

FIGURE 13.32 Graph showing total U.S. crude oil production, with production peaking around 1970. Note how the contribution of oil from Alaska delayed the overall production decline. Possible future production from the Arctic National Wildlife Refuge (ANWR) would have a similar effect.

Solving the Energy Crisis

To put it simply, we are in a bind. Continued population growth along with increasing per capita consumption of light crude oil is simply not sustainable. We will soon face the situation where increased demand will have outstripped our dwindling supplies of light crude. Although there are abundant deposits of nonconventional oil, they are much more difficult and expensive to extract. Therefore, we are not running out of energy, but rather running out of *cheap oil*, thereby putting our petroleum-based economic systems at risk. Like it or not, we are living during a watershed moment in human history where the world is moving beyond Hubbert's peak. The future will bring significant changes and plenty of challenges, forcing us to quickly develop nonconventional oil and renewable sources and expand our current use of coal and nuclear power. Further delay will likely produce consequences far more serious than anything we experienced during the oil crises of the 1970s.

We can use the United States' experience with peak oil to help understand the available options for solving the oil crisis on a worldwide basis. From the graph in Figure 13.32 one can see that the United States, once the world's leading oil producer, is now on the downward limb of its oil production curve. After 1970, when the United States moved past peak production, it was no longer energy independent and had to begin importing progressively more and more foreign oil. Notice how the additional production from the giant Prudhoe Bay field in Alaska did not reverse the overall production decline, but rather delayed the inevitable decline. Should the United States begin drilling in the Arctic National Wildlife Refuge (ANWR), or the much talked-about offshore deposits, the additional production will have a similar delaying effect. Due to the fact that the United States has already extracted most of its own oil reserves, domestic production will continue to decline regardless of whether it drills in ANWR or the continental shelf, or anywhere else for that matter. The country's domestic oil situation is analogous to being on a life raft that is leaking air at a faster rate than it can be pumped back up.

The fact that the United States will never again be energy independent with respect to oil does not mean that additional exploration will not be needed. Bringing new oil fields online will delay the overall production decline and buy the country more time to develop alternative energy sources, thereby lessening the impact of the crisis. Drilling by itself, however, will do nothing to avert the crisis. Success will depend on developing alternative sources combined with strong and immediate action to improve conservation and efficiency, particularly in the area of transportation. The world as a whole faces the same dilemma as the United States and the same set of options as it moves toward Hubbert's peak and beyond.

In this section we take a closer look at how the world can minimize the impact of the global energy crisis. One of the themes here will be quantity and scale. Because the world consumes such an enormous amount of oil, it will be necessary to scale-up production of alternative energy sources to generate the quantities we need. The other theme is *time and money*. Large sums of money will have to be invested to bring this energy online and also to develop more efficient transportation systems. Time is of the essence in order to keep conventional energy prices from rising too rapidly, creating market instabilities that wreak havoc on the global economy. We will briefly examine the nonconventional oil and alternative energy sources for which existing technology will allow us to quickly scale-up production; a more thorough description of nonconventional energy resources will be provided in Chapter 14.

Replacements for Conventional Oil

To get an appreciation for the size of the problem the world is facing, we can again use the United States as an example because it represents nearly a quarter of the world's entire energy consumption. The plot in Figure 13.33 illustrates the tremendous growth in U.S. energy consumption since World War II. This growth reflects the nation's surging economy, increasing population, and per capita consumption of energy after the war. Of particular interest is how fossil fuels make up the vast majority (84%) of the current U.S. energy needs and how comparatively little comes from nuclear and renewables (16%). Moreover, nearly 40% of all energy used in the United States comes from oil. Because the United States consumes such a large proportion of the world's energy, this 40% represents a very large volume of oil. The energy problem then boils down to one of scale. Developing other energy sources to replace even a fraction of what conventional oil now provides will be a daunting task.

The world faces a similar dilemma in terms of oil as does the United States, which brings up the question of whether there are existing energy sources that can replace conventional oil. Fortunately, significant deposits exist of heavy crude oil, including tar sands, which can be extracted and refined into fuels and petrochemicals. We also have the technology to make both oil and gas from our abundant coal reserves. We are currently producing crude oil from heavy oil and tar sands, but the amounts are relatively small compared to world demand. Although the production of oil from coal used to be quite common, production today is almost nonexistent largely because it is an inefficient process and generates considerable pollution. Because the technology already exists for producing crude from heavy oil, tar sands, and coal, the basic problem becomes one of scaling up production in time to prevent a serious shortage. The difficulty here is that special processing plants must be built, plus the resources themselves are generally more difficult to extract than conventional oil. Scaling up production then will require both time and money.

Another readily available replacement for conventional oil is natural gas. Here special processing plants are not needed because natural gas can be used directly in existing internal combustion engines, with relatively minor modifications. A drawback is the amount of space taken up by the larger fuel tanks. These tanks can store only relatively small amounts of natural gas, which means the driving range is much shorter compared to gasoline or diesel fuel. Another issue is that it will take years to convert a significant portion of our transportation fleet to run on natural gas and build the necessary filling stations. Perhaps the biggest drawback is that natural gas is already in high demand for use in heating buildings and generating electricity. Moreover, natural gas follows a similar production curve and will experience depletion just like oil, therefore it cannot serve over the long term as a replacement for oil in meeting our transportation needs.

Biofuels, particularly ethanol made from corn, are often touted as substitute fuels that can lead us to energy independence. The biofuels industry claims that growing corn to make ethanol is cost-effective and that it represents a 25% net gain in energy. However, scientists have estimated that by including the energy required to make fertilizers and pesticides, run irrigation and farm equipment, and operate the distillation process, then it takes 30% more energy to produce ethanol from corn than what it gives off. The scientists also claim that if current government subsidies were taken away, then large agribusinesses would stop making ethanol because it would no longer be profitable. As long as taxpayers are willing to support corn-ethanol subsidies, then ethanol can be used to help reduce the demand for oil.

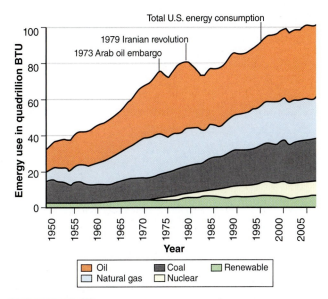

FIGURE 13.33 With the exception of the oil crises in the 1970s, total energy consumption in the United States has increased steadily. Oil presently makes up 40% of U.S. energy needs, representing a large amount of energy that will be difficult to replace.

The use of all-electric cars, plug-in hybrids, and those powered by hydrogen fuel cells are often promoted as a way of reducing the demand for oil. However, recall that electricity itself must be made from a primary energy source. The same goes for hydrogen, which requires energy to break down water (H_2O) or natural gas (CH_4) molecules in order to get free hydrogen (H_2). The hydrogen is then used to run a fuel cell that produces electricity for an electric engine. In either case, fossil fuels are currently the primary means for producing electricity or hydrogen for such vehicles. Because burning fossil fuels releases carbon dioxide and contributes to global warming, we would need to greatly increase electrical production through nuclear, wind, or solar energy for the large-scale use of electric or hydrogen-powered vehicles. While this is feasible, building the infrastructure for the additional electrical production and new generation of vehicles would take years, and thus be more of a long-term solution.

Although there are several suitable replacements for conventional oil, they all are going to require time to scale-up production. This leaves us in a very precarious position should there be a significant and sudden disruption in conventional supplies. Recent U.S. energy policy has focused on offering tax incentives designed to encourage conventional oil exploration. With world crude oil supplies already struggling to keep up with demand, the relatively small amount of additional U.S. production would do little to affect the price of oil during a crisis. This additional oil will also not reverse the overall U.S. production decline, nor make much of a difference when world production eventually begins to decline. There is an option, however, that would make an immediate and significant impact on oil supplies and buy us precious time for scaling-up the production of alternative resources. It is called *conservation*.

Increasing Supply by Reducing Demand

The world oil crisis is essentially one of decreasing supply and increasing demand. When prices get too high, demand destruction will occur. While many people like the idea of a totally free market system, letting the oil market correct itself via demand destruction is not desirable. In fact, proponents of the peak oil theory have long warned that a permanent and progressive decline in oil production would send the world economy into a major recession, or possibly cause it to collapse. To minimize this possibility, the most prudent approach would be to work both sides of the energy equation, namely to increase supply while at the same time taking steps to reduce demand.

Reducing demand can easily be accomplished through conservation efforts and increases in efficiency. The great thing about conservation is that it does not necessarily cost anything to implement, and it affects both supply and demand almost immediately. Because the United States consumes nearly 25% of the world's oil supply, serious conservation efforts in this country can translate into a surplus of oil on the global market, driving prices downward. A good example is the decline in crude oil consumption associated with the 2008 spike in oil prices and onset of an economic recession. In the United States alone, oil consumption declined nearly 6% from 2007 to 2008. Although part of the decline was due to the recession, equally important were the voluntary conservation efforts of citizens during the period of high gasoline prices. Another example is the dramatic decline in U.S. consumption that occurred in response to the oil crises of the 1970s (Figure 13.33). Again, the decline was due to an economic recession and conservation efforts, some of which were voluntary and others mandatory.

The twin Middle East oil crises of the 1970s also led to increases in energy efficiency in developed nations around the world. This caused a reduction in oil consumption with a lasting effect in many countries. Early in the first crisis, the U.S. government mandated efficiency standards for cars sold in the United States, resulting in a dramatic 40% increase in fuel efficiency by 1990 (Figure 13.34). Unfortunately, during the oil glut that followed the second crisis, further efforts at improving fuel efficiency standards failed to pass the U.S. Congress. Lawmakers instead offered tax breaks for low-mileage sport utility vehicles, or SUVs. In contrast, most of the other industrialized nations chose to impose stiff gasoline taxes, on the order of $3 to $4 per gallon, to encourage conservation and the purchasing of more fuel efficient vehicles. Some of the tax revenue was used to fund research and development of alternate energy resources. Denmark, for example, imported nearly 100% of its energy needs from the Middle East in 1973. Today it imports none. The United States, on the other hand, kept gasoline taxes around 40 cents per gallon, encouraging SUV ownership (Figure 13.34). Consequently, the United States began importing progressively greater amounts of oil to keep pace with rising consumption and to help offset the natural depletion of its own oil reserves.

Possible Strategy

As the world moves past Hubbert's peak, it should be clear that the United States needs to quickly take steps toward increasing its oil supply and reducing demand. First, the United States should begin serious conservation efforts and institute higher efficiency standards so as to reduce demand, thereby increasing supply. Next the country needs to take advantage of as many alternate resources as possible, both renewable and non-renewable. Below is a potential strategy that could buy the United States enough time to get past the immediate oil crisis, and then move toward a permanent solution involving alternate energy resources and new forms of transportation, including light and heavy rail.

1. **Reduce consumption by increasing efficiency.** This would both decrease demand and increase supply. Measures could include providing financial incentives for purchasing more fuel-efficient vehicles and encouraging development of mass transit, carpooling, and other energy-saving measures.
2. **Increase exploration and development of conventional crude oil.** Exploration would add to the current supply, but it will take years to bring new reserves to market.

FIGURE 13.34   After the oil crises of the 1970s the U.S. government mandated fuel efficiency standards for cars, resulting in a dramatic 40% increase in overall fuel efficiency. Efficiency slowed after 1990 largely due to Americans switching from cars to SUVs.

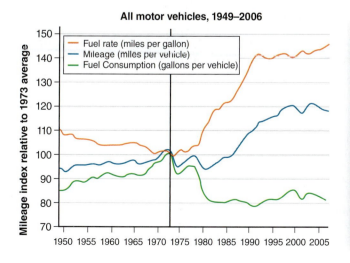

3. **Make gasoline and diesel fuel from coal and heavy oils.** This would also add to the oil supply, but it will take years to build processing plants and scale-up production to significant levels.

4. **Start converting the transportation sector to the use of natural gas.** This would reduce demand and create a surplus of oil. The advantage here is that older vehicles could quickly be converted to run on natural gas, thus providing an immediate impact. This measure would also require developing natural gas filling stations.

5. **Begin replacing natural gas used in electrical generation with other energy sources.** The idea is to use wind, solar, and nuclear power along with coal to free up natural gas supplies for meeting the higher demand created by vehicles running on natural gas.

6. **Sequester carbon dioxide generated by additional use of coal.** The additional use of coal will increase CO_2 emissions and exacerbate global warming. Sequestration involves capturing CO_2 from coal power plants and synthetic oil processing plants, then injecting the gas into subsurface reservoirs. Sequestration is currently being done on an experimental basis and will need to be developed further to make additional use of coal a viable option.

7. **Use biofuels where feasible.** Generating biofuels that are not a net energy loss should be developed to ease pressure on oil supplies. Ethanol from corn is a net energy loss, so money spent on subsidies might be better used on other objectives.

8. **Use remaining oil to generate aviation fuels and petrochemicals.** The idea here is to direct our limited oil supplies to those areas where suitable replacements do not exist.

It is clear that the United States has placed itself in a difficult position, particularly since China's rapid industrialization is putting additional pressure on the world's remaining oil supplies. However, the United States retains world-class research institutions and enormous technical capability. Many people believe that the U.S. government must develop a comprehensive energy policy, similar to what the country once had in the 1970s and early 1980s. The goal would be to take advantage of the nation's technical resources to move from an oil-based economy to one powered mainly by renewable energy resources. The United States could also follow the lead of other industrialized nations and use a gasoline tax to help pay for the research and new infrastructure all this will require. In fact, many people believe that the manufacturing and installation of wind towers and solar panels will help form the basis for a so-called *green economy* well into the twenty-first century. In the end, declining oil production and global warming are going to force us to switch to renewable energy resources whether we like it or not. So, why not get started on a green economy before it is too late?

SUMMARY POINTS

1. Energy is the capacity to do work, and it exists in many different forms (electrical, chemical, thermal, kinetic, nuclear, and solar). Humans make use of energy transformations to suit our needs, as in the conversion of fossil fuels into heat, electrical, or kinetic forms of energy.

2. A primary energy resource is a form of energy that humans can use directly, such as fossil fuels or wind—electricity is a secondary energy source as it has to be produced from a primary source.

3. Humans historically relied on wind, water, wood, and then coal to provide mechanical power. Petroleum became the dominant energy resource because it is a highly concentrated form of energy, and its liquid nature makes it convenient to use and transport.

4. Coal originates when organic-rich deposits in wetlands become buried. As temperature and pressure increases, water and volatile compounds are driven off, thereby transforming the organic matter into a carbon-rich combustible rock. Coal and mineral deposits are mined by similar methods.

5. Oil and gas originate from marine plankton found in organic-rich sedimentary rocks. Burial of these rocks causes temperature and pressure to reach the point where the organic matter is converted into hydrocarbon molecules, which then migrate upward into permeable rock layers.

6. Geologists locate traps where oil or gas can be stored in permeable reservoir rocks. If an exploration well finds an economical deposit,

a production well is installed to extract the oil and gas in a manner similar to that used for water wells.

7. Crude oil is refined into different products by a distillation process in which the oil is heated and the vapors are allowed to rise in a cooling tower. The densest hydrocarbon molecules condense near the bottom where temperatures are highest; the lighter, more volatile compounds condense at the top where it is cooler.

8. Coal and petroleum deposits are widely dispersed around the globe, but nearly 60% of all crude oil reserves are held by the Persian Gulf nations in the Middle East. Most industrialized nations do not have sufficient energy resources to meet their own needs.

9. The peak oil theory holds that production of conventional oil will follow a bell-shaped curve that will reach its peak as early as 2005 and as late as 2040. After the peak, rising demand is expected to outstrip supply, creating a permanent oil shortage.

10. Current replacements for conventional oil include synthetic oil produced from coal and deposits of tar sands and heavy oil. However, special processing plants must be built to scale-up production from these sources. Other types of nonconventional oil exist, but are not yet available for production.

11. Most solutions to the current energy crisis include increasing the supply of oil while at the same time decreasing the demand through conservation and increased efficiency. These efforts are needed to provide more time for scaling-up production of other energy resources in order to avoid a serious oil shortage.

KEY WORDS

cap rock   410
coal   405
crude oil   409
energy   399
energy efficiency   401
fossil fuels   398

gas window   409
Hubbert's peak   424
hydrocarbons   408
natural gas   408
oil window   409
peak oil theory   424

petroleum   408
petroleum reservoir   410
petroleum trap   410
work   399

APPLICATIONS

Student Activity

Either listen for the daily price of a barrel of oil or go to a website such as http://www.nymex.com/index.aspx. Track the price of oil for the next several weeks. Did it go up, down, or remain steady? If there was some movement on the price, was there a news item that triggered the price move (for example, a Japanese tanker was fired upon off the coast of Yemen, and the price of oil went up)? Are the news events really impacting the price of oil, or are they just causing speculation?

Critical Thinking Questions

1. What price per gallon of gasoline would cause you to take public transportation or carpool? If you are taking public transportation or carpooling, how much money a month are you saving?

2. How does the price of a barrel of oil influence exploration and recovery techniques?

3. Does the growing industrialization of India and China (with the world's largest populations) influence global energy production? If so, how?

Your Environment: YOU Decide

The Arctic National Wildlife Refuge (ANWR) is a vast area of tundra on the north shore of Alaska. It is a very important breeding area for migratory birds and many mammals. It also contains a vast oil field with approximately 7.7 billion gallons of oil. Should drilling be allowed to proceed, for the benefit of Americans who use oil, or should this land remain off limits to drilling?

Chapter **14**

Alternative Energy Resources

LEARNING OUTCOMES

After reading this chapter, you should be able to:

▶ Explain the difficulty in replacing existing crude oil supplies with oil produced from coal, heavy oils, and tar sands.

▶ Know why corn-based ethanol is a net energy loss and how it affects global food supply and climate change.

▶ Describe the basic reasons why the nuclear power industry in the United States stopped expanding during the 1990s.

▶ Describe the two ways solar radiation can be converted into more useful forms of energy and why solar energy is more cost-effective in some regions than in others.

▶ Know how wind and the tides are used to generate electricity, and describe the advantages and disadvantages of each method.

▶ Describe the basic way a geothermal heat pump operates and how it differs from using geothermal energy to generate electricity and to operate space heaters.

▶ Know how ocean thermal energy systems (OTEC) work and why these systems have geographic limitations.

▶ Describe the difficulty in moving from fossil fuels to renewable and non-carbon-based sources of energy.

The key to a sustainable future may lie in making use of available roof space for collecting solar energy, as in this solar housing development in Germany.

Introduction

In Chapter 13 you learned that as conventional crude oil production continues to decline, humans must look toward other energy sources, or else face higher oil prices that could cripple the world economy. For our purposes we will define **alternative energy sources** as those other than conventional coal, natural gas, and light crude oil that currently make up the bulk of the world's energy supply. Alternative sources will include renewable energy (e.g., wind and hydro power), nuclear, and unconventional fossil fuels, such as heavy oils and synthetic fuels made from coal and tar sands. Although alternative resources make up a significant portion of the energy supply in some nations, the United States and China continue to rely on fossil fuels for approximately 90% of their energy needs. Because these two countries are the world's biggest energy users, consuming 37% of the total supply, their energy usage has a major impact on worldwide supplies. Consequently, despite the success of many industrialized nations at reducing their use of fossil fuels, alternative energy sources still provide only 11% of the world's total energy supply. The continued reliance of the United States and China on fossil fuels not only threatens the global economy, but the release of carbon dioxide from the burning of these fuels is affecting Earth's climate system (Chapter 16). It is critical that the U.S. and China begin moving away from fossil fuels and toward alternative energy sources as quickly as possible.

The United States and China continue to rely on petroleum and coal primarily because these fuels are highly concentrated and convenient forms of energy, and they are comparatively cheap. In contrast, most alternative energy resources are intermittent in nature and are not as energy intensive. Scaling-up production so that alternative energy will make a significant impact will require building massive amounts of infrastructure (e.g., windmills and solar panels). Many Americans argue against moving toward alternatives based on the idea it will be too costly for the economy and result in a loss of jobs. What this argument ignores, however, is that the era of cheap oil is over anyway and that shortages can threaten the world economy (Chapter 13). Part of the reason petroleum and coal have remained inexpensive in the United States is because of the substantial subsidies the fossil fuel industry receives in the form of tax breaks. With the exception of corn ethanol, U.S. tax incentives for the renewable energy industry have been relatively minor and inconsistent, making it difficult for the new industry to develop. Environmentalists argue that if one adds the costs society must pay in terms of pollution and climate change caused by fossil fuels, then renewable energy is actually far cheaper.

The term *green economy* is commonly used today in reference to the inevitable conversion of our fossil fuel-based society to one that is sustainable and whose energy supply is dominated by renewable and non-carbon fuels. Environmentalists envision a green economy where people have jobs building and installing new infrastructure, ranging from solar panels and windmills to an expanded electrical grid to a vast rail system that moves both people and freight. In fact, the green revolution has already begun in Europe, where solar and wind industries are booming. Some strategists believe that China will be a future leader in green technology, in part because of its rapid industrialization and widespread air and water pollution. China may be compelled to develop a green economy because its citizens are literally choking on some of the world's worst air pollution, caused by its use of fossil fuels.

The goal of Chapter 14 is to provide an overview of alternative energy sources and to examine their future role in society. Although some of these resources are carbon-based, in the short term these will still be needed to help us get through the current oil crisis. Others do not contain carbon and are renewable, and therefore will be critical for combating global warming and building a more sustainable society. Many of the world's developed nations started transforming their economies after the oil crises of the 1970s, and have now built successful industries involving alternative energy. Meanwhile the United States has focused on expanding its supply of conventional oil and coal. Ultimately, the continued depletion of the world's oil reserves, combined with global warming, will force Americans to switch over to alternative energy sources. The United States has come to a crossroad. The current path continues toward reliance on fossil fuels for as long as possible, leaving the country exposed to supply disruptions and accelerated global warming. The other path leads to energy independence, reduced carbon dioxide emissions, and a more sustainable society.

Nonconventional Fossil Fuels

In this section we want to examine **nonconventional fossil fuels,** which we will define as all fossil fuels except for the traditional use of coal, natural gas, and light crude oils. Here we are interested in those fuels that might be used to replace declining production of conventional oil and natural gas, thereby satisfying the world's energy demand. Some nonconventional fossil fuels have already been used to produce synthetic oil, but at a higher cost and in relatively small amounts. Now with tighter crude oil supplies and higher prices, nonconventional fuels are becoming more competitive and profitable to produce.

Many people have taken note of the fact that nonconventional fossil fuels include deposits that are equivalent to very large reserves of crude oil, leading to the conclusion that oil will be abundant for the foreseeable future. However, extracting and then converting these deposits into actual crude oil is costly and time-consuming. It will be a very challenging task just to extract these resources in amounts that are sufficient to satisfy the world's enormous appetite for oil. Another problem is that nonconventional or synthetic crude requires special processing plants. Proponents of the peak oil theory warn that scaling-up production to avoid a serious oil shortage may be difficult simply because of the time required to build these processing plants. We will begin our discussion of nonconventional fossil fuels by taking a look at our abundant coal reserves.

Synthetic Fuels from Coal

Despite the long list of environmental problems associated with coal (air and water pollution, acid rain, mercury fallout, mining hazards, and global warming), coal remains the fuel of choice for generating electricity. Coal is very desirable as a fuel because it is a highly concentrated form of energy and is cheap, provided one does not factor in its environmental costs to society. Prior to the widespread availability of petroleum, coal was broken down chemically to produce gas, often referred to as *coal gas* or *town gas.* Coal gas was typically made at a local processing plant, and then piped throughout a city where it was used for lighting, cooking, and heating. As shown in Figure 14.1, gas lanterns made it possible for city streets to become fully lit at night for the first time in human history. Later when automobiles became popular, a similar process was developed for making gasoline and diesel fuel from coal. In fact, during World War II, Germany used coal as its primary source for gasoline and diesel fuel. Note that gas or liquid fuels produced from coal or heavy oil are commonly referred to as **synthetic fuels,** or simply *synfuels.*

With the exception of a single processing plant in South Africa, coal-based synthetic fuels have not been produced on a large scale in recent years because these fuels are not cost-competitive with those derived from conventional petroleum. Energy is required to transform coal into synthetic fuels,

FIGURE 14.1   Prior to large-scale electrical production, most towns and cities had their own processing plants where coal was converted into synthetic gas and used for lighting as well as for heating and cooking. As demonstrated by this photo (pre-1890) of London, England, gas lighting transformed night into day, allowing city life to thrive after sunset.

thereby creating an additional cost. However, as world crude oil prices continue to climb, more countries are developing plans for coal-liquefaction plants. Given a sufficient amount of time and investment, production could be ramped up to where significant quantities of synthetic gasoline and diesel are added to the global supply. This would reduce demand on conventional crude oil and ease the pressure on rising prices. China, a country with large coal reserves but limited crude oil, began operating a new $3.2 billion coal liquefaction plant in 2008. This plant produces approximately 1 ton of synthetic oil products for every 3.5 tons of coal, and by 2010 is expected to be producing 6 million tons of synthetic oil per year. Given that China alone consumed 345 million tons of oil in 2006, it becomes clear that large numbers of liquefaction plants would need to be built for this process to have a substantial impact on the world's oil supply.

In addition to the issue of scaling up production, there are some other drawbacks to coal-based synthetic fuels. While world coal reserves are large and expected to last several hundred years under current demand, mass production of synfuels could dramatically shorten the lifespan of the reserves. Coal therefore cannot be expected to serve as a long-term solution to declining crude oil production. Another issue is that the mining of coal would increase dramatically, magnifying the already serious environmental problems associated with mining and burning coal (Chapter 13). An even more serious concern is that the liquefaction of coal is less efficient, and produces considerably more carbon dioxide (CO_2) than does the refining of conventional crude oil. Large-scale coal liquefaction then would exacerbate the current problem of CO_2 emissions and global warming. Therefore, synthetic fuels produced from coal are not a viable option unless a practical means can be developed for pumping the CO_2 into suitable rock layers for long-term storage in the subsurface, a process called *carbon sequestration* (Chapter 16).

Heavy Oils and Tar Sands

Recall from Chapter 13 that crude oil consists of many different types of hydrocarbon molecules. Those oils with a higher proportion of lighter molecules and fewer sulfur atoms are referred to as **light crude oil,** and those with heavier molecules and more sulfur are called **heavy crude oil.** Heavy crudes are also referred to as "sour" since they contain more sulfur, whereas light crudes are called "sweet" due to the low sulfur content. This is important because heavy crudes command a lower price since the refining process produces smaller amounts of high-end fuels (gasoline and jet fuel). Refining costs are also higher given the need to remove the additional sulfur. Production costs are increased even further due to the fact heavy oils are more viscous, which makes them more difficult to extract and transport. Oil companies obviously prefer to find deposits of light crude as they are much more profitable. Consequently, after more than 100 years of intense exploration and production, most of the world's remaining oil reserves are on the heavier and more sulfur-rich end of the spectrum. Because of this preferential consumption of lighter crude, it is often stated that the world is running out of *cheap* oil.

There comes a point at which crude oil can be classified as nonconventional because it is so heavy and viscous that special techniques are required to extract it from a reservoir rock. For example, despite the additional cost and effort, heavy crude has been produced from several oil fields in California since the early 1900s. At these fields a technique has been developed whereby wells inject steam into the reservoir, heating the oil and reducing its viscosity so it can be pumped out. As illustrated in Figure 14.2, one approach is to inject steam into an individual well; then, after waiting for the oil to become thinner, the well is pumped. Pumping continues until the oil again becomes too thick,

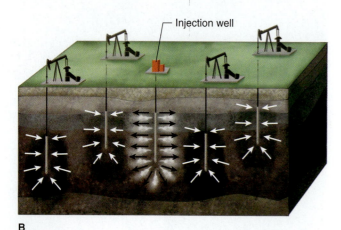

FIGURE 14.2 Some heavy oils are too viscous to be extracted directly by pumping. One approach is to inject steam into an individual well (A), which lowers the oil's viscosity. The thinner oil is then pumped until the resistance again becomes too great, at which point the process is repeated. Another method involves a dedicated injection well (B) surrounded by four pumping wells. Here steam is injected continuously while the pumping wells remain in full production.

at which point the process is repeated. A more elaborate and efficient method involves using separate wells for injection and extraction that operate continuously. With either method, once the oil is extracted it eventually cools down and becomes too thick to transport to a refinery. To make it transportable, the heavy oil's viscosity is reduced by mixing it with lighter crude. This, however, requires that lighter crude be available nearby.

A different form of heavy oil is sometimes found near the surface in what are called *tar pits*—a famous example is the La Brea Tar Pits in California, where prehistoric animal remains are preserved. The hydrocarbons in tar pits are in a form called *bitumen,* which creates a highly viscous fluid that is more like asphalt than normal crude oil. Bitumen is also found in sand deposits close to the surface which geologists call **oil sands** or **tar sands.** These deposits are generally believed to have formed when crude oil from petroleum reservoirs migrated upward and then collected in sand layers near the surface. Here the crude was broken down over time by bacteria, leaving behind bitumen. Two of the largest oil sand deposits in the world are located in Canada and Venezuela, and together are estimated to contain enough bitumen to produce the equivalent of several trillion barrels of conventional crude oil. However, only a fraction of this is considered recoverable. For comparison, about 1 trillion barrels of conventional crude remain available in the world today.

Presently the only significant production from oil sands is in Alberta, Canada (Figure 14.3). Canadian officials estimate recoverable reserves from these deposits to be 175 billion barrels of equivalent crude, which is second only to those of Saudi Arabia. The problem is that extracting the highly viscous bitumen is much more costly and labor intensive compared to

FIGURE 14.3   Synthetic crude oil is currently being produced from tar sand deposits (A) in Alberta, Canada. The hydrocarbons are in the form of bitumen, a highly viscous substance that is separated from the sand using steam. Photo (B) shows a bitumen sample whose viscosity has been lowered by heating. Approximately 60% of Canadian production involves strip-mining (C); the remainder is produced by steam injection and pumping wells.

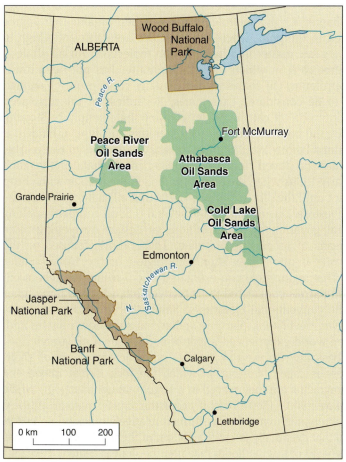

A

B

C   Athabasca oil sands, Muskeg River mine

conventional crude. Currently about 60% of all oil sand production in Canada is from deposits located relatively close to the surface, making it feasible to remove the material by strip mining (Chapter 12). Here the bitumen-saturated sand is loaded onto trucks and sent to an on-site processing plant where steam is used to separate the bitumen from the sand. Natural gas is then used in a process that converts the bitumen into synthetic light crude. For the Canadian deposits that are located too deep for strip mining, the bitumen is being extracted using steam-injection techniques, similar to those used for heavy oils.

Canada is currently producing over 1 million barrels of synthetic crude per day from its oil sands, and expects production to peak between 2.5 and 5 million barrels per day around 2020. While the Canadians have done an impressive job of scaling-up their production, the process is still basically a large mining operation with certain physical limits. For example, the processing of oil sands requires large volumes of water and natural gas, the supplies of which are not unlimited. Perhaps the biggest obstacle is natural gas, since it is already in high demand as a fuel for generating electricity and heating buildings. To make matters worse, natural gas production is expected to decline in the near future as more and more gas reservoirs undergo natural depletion as predicted by the peak oil theory (Chapter 13). It does not appear likely therefore that Canada's production of synthetic crude oil will go much beyond the projected 5 million barrels per day. When one considers that the world is presently consuming over 80 million barrels per day, it seems clear that oil sands by themselves are not going to solve the present oil crisis.

Oil Shale

Recall from Chapter 13 that petroleum forms when organic-rich source rocks are buried to the point where temperatures are sufficiently high to transform the organic matter into hydrocarbon molecules. An **oil shale** is basically a source rock in which the organic matter has not been transformed into petroleum. Oil shales do not actually contain oil, or enough clay minerals to be considered true shales. Regardless of the misplaced name, synthetic crude oil can be made from oil shale simply by heating the rock. In fact, the basic technology for generating this type of crude has been around since before 1900. Similar to what happened to the market for synthetic fuels produced from coal, interest in oil shale pretty much ended in the early 1900s when the oil boom flooded the market with cheap crude. It was not until the oil crises of the 1970s that oil shale again became appealing to investors. During this period several U.S. oil companies, along with the federal government, spent billions of dollars developing more economical means of processing oil shale. These efforts focused on the oil shale deposits in the Rocky Mountain region of the United States (Figure 14.4). This region holds an estimated 2 to 3 trillion barrels of oil, with 800 billion barrels considered to be recoverable, which is three times greater than the conventional reserves of Saudi Arabia. Commercial production, however, never took place as operations shut down in the mid-1980s when the world experienced a glut of conventional oil supplies.

There are basically two ways to generate synthetic crude from oil shale. The first process is called *retorting*, in which the rock is first mined and crushed. The fragments are then placed in a large rotating tank (kiln) where temperatures of over 900°F (500°C) literally cook the organic material. At these temperatures the organic matter is transformed relatively quickly into hydrocarbon droplets that eventually fall to the bottom of the tank and accumulate. One of the drawbacks of this method is that after the shale is heated, it expands to approximately 120% of its original volume. This creates a disposal problem

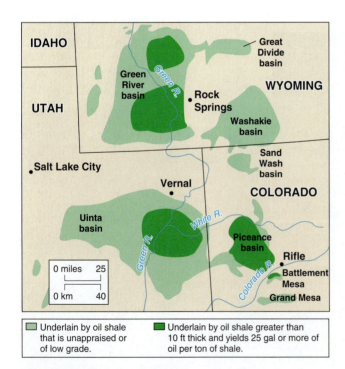

FIGURE 14.4 Over 60% of the world's reserves of oil shale are in the Rocky Mountain region of the United States. Although past U.S. efforts at commercial oil shale production have failed, promising new methods may soon prove viable. Important environmental issues remain and may limit production, such as the region's limited water supplies and concerns over increased carbon dioxide emissions.

since the original excavation is not large enough to accept the waste. In a large-scale operation this would require that the waste material be placed in nearby stream valleys, thereby disrupting the natural hydrology of the area and harming ecosystems. To avoid this problem, U.S. research efforts during the 1970s focused on cooking the rock underground—a process referred to as *in situ* (i.e., in place). Once the hydrocarbons formed in sufficient quantities, they could then be pumped to the surface using wells.

Although in-situ techniques eliminated the problem of large-scale waste disposal, commercial oil shale production never became viable because of high production costs. These costs are driven in part by the large amount of energy needed to keep the temperature at a high level during the cooking process. Production costs are also affected by the fact that oil shale produces relatively heavy oil. As with any heavy crude, the refining process requires natural gas to increase the output of highly desirable products such as gasoline and jet fuel. This leads to yet another issue—as long as fossil fuels are used in both the heating and refining processes, oil shale production will release far more carbon dioxide than conventional oil. Finally, there is the potential that some of the hydrocarbons generated during the in-situ process would be able to migrate from the site and end up contaminating nearby groundwater supplies.

Despite all the difficulties, the recent rise in crude oil prices has led to a renewed interest in oil shale. In 2004 Shell Corporation initiated a pilot program in Colorado that is based on an improved in-situ method. As illustrated in Figure 14.5, this design uses subsurface electric heaters where operators can cook the oil shale at a lower temperature and for a longer period of time—on the order of years rather than days or months. The design also involves pumping refrigerants into the subsurface via a series of small wells, creating a barrier of ice that surrounds the heated section of oil shale. Once the impermeable barrier of ice is in place and the electric heaters are turned on, wells are used to remove the groundwater from within the heated section. By removing the groundwater, operators can minimize the possibility that nearby water supplies will become contaminated. Another advantage of Shell's new approach is that cooking oil shale at a lower temperature generates lighter crudes, thereby eliminating the need for natural gas in the refining process.

In 2005 Shell Oil reported that their small-scale Colorado tests were successful, and announced plans to build a production test facility. Should this production test prove to be commercially viable, then large-scale production might finally be possible. This still leaves the question of whether or not production could be scaled-up in time to make much of an impact

FIGURE 14.5 Shell Oil's experimental oil shale project in Colorado uses electric heaters to slowly cook the oil shale for several years, producing lighter crude that requires fewer fossil fuel inputs. Groundwater contamination is minimized by using refrigerants to create an impermeable ice barrier around the heated section of oil shale.

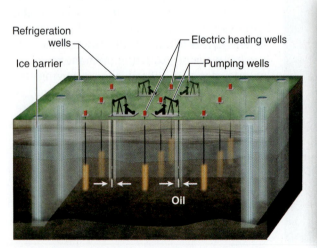

on the present oil crisis. Ironically, a limiting factor in all this may be water. While the in-situ process uses relatively little water compared to other techniques, water supplies in the Rocky Mountain region are already stretched thin. This brings into question the amount of water that could be made available to support a massive oil shale operation and all the associated growth and development that would go with it. Another serious obstacle will be supplying the vast amounts of energy required to produce oil shale. This energy would ultimately have to be produced in such a way that carbon dioxide is not released into the atmosphere.

Gas Hydrates

During surveys of the seafloor in the 1970s, researchers discovered deposits of **gas hydrates,** also called *methane hydrates,* an icelike substance composed of methane (CH_4) and water molecules bound together. More recent surveys have found thick accumulations in shallow polar waters and in lower latitudes on deeper parts of continental shelves. Deposits are also found buried in the permafrost on land throughout the Arctic. Scientists believe that gas hydrates originate in nutrient-rich waters when the remains of marine organisms fall to the seafloor and accumulate under oxygen-poor conditions. The organic matter slowly breaks down and forms methane gas, which can then bond with water to form gas hydrates. This process, however, occurs only under certain temperature and pressure conditions. Should gas hydrates experience a change in pressure or temperature, the icelike compounds can become unstable and begin releasing methane gas (Figure 14.6). Later in Chapter 16 you will learn that this process is important since methane is a major greenhouse gas. Because the effect of methane (CH_4) on atmospheric warming is about 20 times greater than carbon dioxide (CO_2), scientists are concerned that as Earth continues to get warmer, gas hydrates will begin releasing large volumes of methane, thereby accelerating the warming.

In addition to being a greenhouse gas, methane makes up the bulk of what we call *natural gas,* which means that gas hydrate deposits represent a potential energy resource. Based on worldwide surveys, scientists estimate that gas hydrates contain nearly twice as much carbon as does the world's current supply of fossil fuels. Of considerable interest is the fact that many of the deposits may be economical to extract due to their thickness and purity. One such deposit, located off the southeastern coast of the United States, is estimated to contain enough natural gas to meet current U.S. demand for nearly 30 years. Because gas hydrates represent a potentially large source of energy, researchers are now trying to develop ways of extracting methane from gas hydrate deposits. Physically removing the material through mining is not practical, so the best approach appears to be one that creates either a pressure decline within a deposit or a temperature increase. This causes the icelike substance to become unstable and begin releasing its methane gas. The methane could then be brought to the surface using wells similar to conventional gas wells. An international group of scientists has recently completed field and production experiments

FIGURE 14.6 Gas hydrates are icelike substances composed of methane gas and water molecules found in marine sediment and in permafrost on land. Should this icelike substance be exposed to either higher temperature or lower pressure, the material breaks down, releasing methane gas. These deposits represent a potential major source of energy.

at a site in Canada, demonstrating that large volumes of methane can indeed be produced by this technique.

Although gas hydrates hold much promise, major production is still years into the future due to technical issues and physical dangers. For example, destabilizing the gas hydrates in order to liberate methane also creates the risk that methane could be released in an uncontrolled manner. Such a release would pose an explosion and fire hazard to both the workers and the production facility. For offshore deposits, destabilizing the gas hydrates could trigger movements within the seafloor, which, in turn, could threaten production platforms whose supports rest on the seafloor. There is also a problem in that the economical deposits tend to be located in either deep marine environments or in harsh arctic climates. This will make it more difficult and expensive to produce the methane and also get it to distance markets. Even if the technical and safety issues can be overcome, it does not appear that production of gas hydrates could be scaled-up in time to have much of an impact on the current oil crisis.

Carbon-Free and Renewable Fuels

With the exception of biofuels, the alternative energy sources in this section do not involve the combustion of organic matter, and thus do not release carbon dioxide. Because of the serious consequences of global warming, these energy sources will be critical in helping us move from a fossil fuel based society to one that is largely carbon free. Many of the energy sources in this section are also considered to be renewable, and therefore will help us to build a more sustainable society. Note that hydrogen fuel cells are not included here because free hydrogen (H_2) is not naturally available in any meaningful quantities, which means it is not a primary energy resource. Similarly, as discussed in Chapter 13, electricity is not a primary resource because it has to be made from another energy source. Later you will learn that the long-term solution to the oil crisis and global warming will be to transform non-carbon-based energy resources into electricity and hydrogen.

Biofuels

Prior to the widespread use of coal and petroleum, humans had to rely on dried organic matter, such as wood, peat (organic soil), and animal manure (dung), as a fuel for heating and cooking. The term **biofuels** refers to any combustible material derived from modern (nonfossilized) organic matter. Liquid biofuels (e.g., ethanol) are being offered as a solution to the oil crisis since these can be used to run transportation systems. Such fuels are also being promoted as a means of combating global warming, because biofuels are considered to be *carbon-neutral;* that is, the carbon dioxide they emit is supposedly equal to what the original plant material removed from the atmosphere. In general, there are currently two types of liquid biofuels: *ethanol,* which is an alcohol made by fermenting plant material (e.g., corn and sugarcane), and *biodiesel,* which consists of fatty acids derived from vegetable oil (e.g., soybeans) and animal fats. We will focus on ethanol because it can be readily used in gasoline engines, and therefore is in greater demand.

Major biofuel production began in the United States during the energy crises of the late 1970s and early 1980s when a portion of the nation's surplus corn crop was turned into ethanol (Figure 14.7). The ethanol was blended with gasoline to create what was known as *gasohol* (10% ethanol and 90% gasoline). Although this effort required government subsidies to

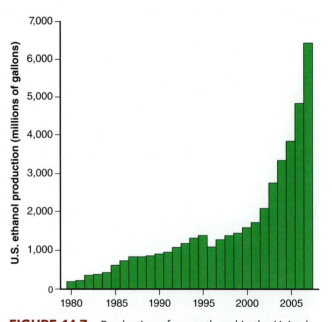

FIGURE 14.7 Production of corn ethanol in the United States as a transportation fuel began in 1980 in response to the energy crisis. Demand increased again after 1990 when ethanol started being used as an additive for making clean-burning oxygenated gasoline. After 2000, production further increased because of the growing oil crisis.

be profitable, it was considered reasonable during the oil crisis to use surplus corn to help extend the limited supplies of gasoline. The crisis eventually ended in 1986, but ethanol production and government subsidies continued. Then in 1990 the U.S. Clean Air Act of 1990 was passed and required pollution controls on vehicles and set new air-quality standards. To help meet the new standards, large metropolitan areas began requiring the sale of *oxygenated gasoline* since it produces less carbon monoxide (CO) and ozone (O_3). Because ethanol could be used to make oxygenated gasoline, the sudden demand for cleaner-burning gasoline created a new and important market for ethanol.

As concerns grew in the late 1990s over global warming and rising crude oil prices, industry groups promoted ethanol as renewable and carbon-neutral alternative to gasoline. The ethanol industry also began pushing to have auto manufacturers make so-called *flex-fuel* vehicles that can run on either standard gasoline blends or fuels composed of 85% ethanol, popularly known as *E85*. Proponents often point to Brazil's successful flex-fuel program where ethanol accounts for nearly half of the country's transportation needs. However, Brazil's ethanol is made from leftover waste material generated when sugarcane plants are processed into raw sugar, whereas the only parts of the corn plant that can be used are the actual ears of corn. Brazil's program therefore is naturally much more cost-effective because it produces two valuable commodities from a single crop. Since sugarcane does not grow well in North America, researchers in the United States are investigating different plants and waste products, with the aim of finding a more efficient means of producing ethanol.

Another problem with corn-based ethanol is the amount of energy that is needed to produce the fuel itself. Industry groups claim that corn ethanol produces 25% more energy than what is used in growing the corn and distilling it into ethanol. When independent scientists include the energy inputs for fertilizers, pesticides, irrigation pumps, and farm equipment, they calculate that corn ethanol consumes 30% more energy than what it can provide. Although the industry disputes there is a net loss in energy, critics point out that natural gas is used instead of ethanol as a heat source for the distillation process. If corn ethanol truly represents a net energy gain, then it would be more profitable to use ethanol over natural gas as a heat source. Critics also claim that large agribusinesses find corn ethanol to be profitable only because of continued government subsidies (i.e., tax dollars). In addition to production inefficiencies, ethanol as a transportation fuel provides 28% less energy compared to normal gasoline. This means that although you may pay the same price for ethanol-blended gasoline at the pump, your gas mileage will be less.

Finally, there is the issue of how biofuel production affects food supplies and global warming. Although nearly 15% of the U.S. corn crop was dedicated to producing ethanol in 2006, the volume that was generated met only 3.5% of the nation's demand for gasoline. Globally, biofuel production tripled between 2000 and 2007, but accounted for less than 3% of the supply of transportation fuels. The problem is that to supply even this limited volume of biofuels, a significant amount of farmland is being used that could have gone toward producing food. In 2008, the rapid rise in crude oil prices greatly increased the demand for biofuels, which, in turn, contributed to a steep rise in world food prices. The International Monetary Fund estimated that the high demand for biofuels in 2008 accounted for 70% of the increase in corn (maize) and 40% of the increase in soybean prices. The higher prices led to food riots in many countries, including Mexico where corn is the main food staple. In addition to its effect on food

supply, biofuel production may be contributing to global warming because it is encouraging the conversion of natural forests and grasslands into farmland. A recent scientific study has shown that natural landscapes and soils absorb significantly more carbon dioxide than agricultural lands. This means that in our quest to grow biofuels that are supposedly carbon-neutral, the creation of new farmland is actually adding carbon dioxide to the atmosphere and exacerbating the problem of global warming. Therefore, while biofuels may be useful in helping us address the immediate oil crisis, they are certainly not a long-term solution to our energy needs or to global warming.

Hydropower

In Chapter 13 you learned that humans have a long history of using mechanical energy from falling water, called **hydropower,** to perform a variety of work. Today we use falling water primarily to spin a turbine to make electricity, often referred to as *hydroelectricity.* The advantage of hydroelectricity is that it is both renewable and does not produce carbon dioxide. However, most hydroelectric plants involve damming a river and creating an elevated pool of water (reservoir). This allows managers to manipulate the release of water from the reservoir on a daily basis to meet the cyclic demand for electricity. However, by regulating the flow of a river, dams not only block the migration of fish, but the loss of natural variations in stream flow has a negative impact on entire ecosystems (see Chapters 8 and 11 for details). It is largely because of these environmental impacts that the construction of new dams today is commonly met with public and political resistance. Moreover, hydroelectric dams cannot be built just anywhere because the reservoir needs to be relatively deep, which requires a river valley that has steep sides and is somewhat narrow (Figure 14.8). Height is important because it gives the falling water more power to move the turbines, thereby making hydroelectric production more cost-effective.

Hydropower currently accounts for only 6% of the world's total energy consumption but represents nearly 20% of all electrical production, although just 6% of U.S. electricity comes from hydropower. Even though there remain enough suitable dam sites to double the current level of hydroelectric production, it is unlikely anything near this level could be achieved given the environmental and political obstacles. One interesting solution for generating additional hydroelectric

FIGURE 14.8   Hydroelectric power plants are most efficient in steeper terrain as it allows for a deeper reservoir. As height increases, so too does the power of the falling water as it moves through the turbines, which makes hydroelectric production more cost-effective. Shown here is Hoover Dam located on the Colorado River, Nevada.

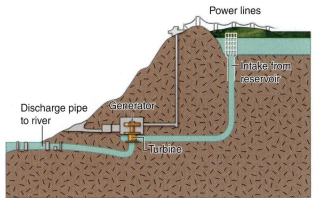

FIGURE 14.9 The Taum Sauk reservoir in Missouri is a pumped-storage system that generates electricity without creating the environmental problems associated with damming rivers. Water is pumped up into the reservoir during the nighttime when demand is low and there is a surplus of electricity. The stored water is then released during the daytime to create electricity during peak demand.

power is to use pumped-storage systems. As shown in Figure 14.9, rather than damming a river, a pumped-storage system uses an elevated reservoir built on a topographic high point. These systems take advantage of the fact that electrical demand peaks during the daytime and reaches a minimum at night. Because generators typically run continuously, this means there is a surplus capacity at night (more electricity is being produced than consumed). Engineers have found it cost-effective to pump water up into a storage reservoir at night using the surplus electricity, then release the water the next day to generate electricity during peak demand.

Nuclear Power

Recall from your chemistry study that the nucleus of an atom contains protons and neutrons, and that the energy that binds these particles together is called *nuclear energy*. Also, the various combinations of protons and neutrons of a given element are known as its *isotopes* (the number of protons is fixed, but neutrons can vary). Prior to World War II, physicists had determined that the nucleus of certain isotopes were *fissile*, meaning they were susceptible to being split apart. The term **nuclear fission** therefore refers to the process of splitting the nucleus of an atom, resulting in the release of neutrons and conversion of mass into energy. Research into developing the first atomic bombs focused on making use of the heat energy that would be released from the splitting of the fissile isotopes of the elements uranium and plutonium. One of the technical problems was that uranium's fissile isotope, U-235 (92 protons and 143 neutrons), makes up only 0.72% of the uranium found in the natural environment—99.28% is nonfissile U-238. Thus, U-235 atoms had to be separated from those of the more abundant U-238. Another option was to produce fissile plutonium in a nuclear reactor by bombarding U-238 atoms with neutrons. In 1945, American scientists succeeded in initiating the first uncontrolled nuclear chain reaction by splitting just a few pounds of uranium-235 (the first plutonium bomb was developed later that same year). Shortly afterward much more powerful atomic bombs, called hydrogen bombs, were developed in which hydrogen nuclei are forced to combine or fuse, a process called **nuclear fusion.**

Fission Reactors

In the years that followed World War II, physicists designed nuclear reactors that could be used for peaceful applications. Here the focus was on harnessing the heat from fission reactions to generate electrical power by steam-driven turbines and generators. In 1954 the first commercial nuclear power plant began generating electricity in the former Soviet Union (now Russia). Today most of the world's nuclear power plants use uranium dioxide pellets whose uranium-235 concentration has been increased (enriched) to somewhere between 3 and 8%. The uranium fuel is then bombarded with neutrons in what is called a *reactor core*. As illustrated in Figure 14.10, the neutrons split uranium-235 atoms, releasing heat and additional neutrons, which go on to split other U-235 atoms in what is called a *sustained nuclear reaction* (engineers use water or graphite to slow the velocity of the neutrons so that the reaction is self-sustaining). To keep the reactor core from overheating, circulating water is used to transfer heat away from the core. This hot water is then used to produce steam for driving the electrical generator. Figure 14.11 illustrates the two basic water-cooling designs used in modern reactors. Notice in both designs how the reactor core is housed in a heavily reinforced structure called a *containment building*, whose purpose is to keep radioactive materials from escaping in the event of an accident.

Nuclear power plants have the benefit of being able to generate electricity on a large scale, and at the same time not emit any carbon dioxide, sulfur dioxide, or particulate matter. Thus, nuclear power does not contribute to global warming or acid rain as do coal-fired plants (Chapters 12 and 15). However, fission reactors generate waste that is highly radioactive,

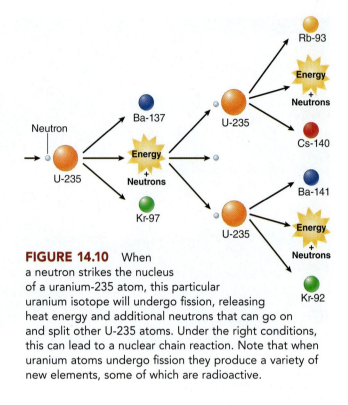

FIGURE 14.10   When a neutron strikes the nucleus of a uranium-235 atom, this particular uranium isotope will undergo fission, releasing heat energy and additional neutrons that can go on and split other U-235 atoms. Under the right conditions, this can lead to a nuclear chain reaction. Note that when uranium atoms undergo fission they produce a variety of new elements, some of which are radioactive.

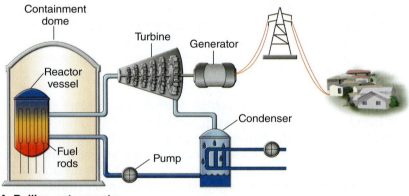

A Boiling water reactor

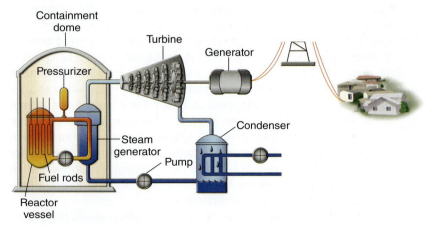

B Pressurized water reactor

FIGURE 14.11   Modern nuclear reactors use water for cooling the reactor core and for driving a steam turbine to generate electricity. A boiling water reactor design (A) uses the same water for cooling and generating electricity, whereas a pressurized water reactor (B) uses separate loops. Water around the reactor core in either design makes it possible for the nuclear reactions to be self-sustaining. Should there be a loss of cooling water, the nuclear reactions will stop.

Cooling towers

Containment dome

FIGURE 14.12   Aerial view of the Three Mile Island nuclear power plant on the Susquehanna River in Pennsylvania.

which presents issues of operational safety and long-term waste disposal (Chapter 15). Despite the safety and waste concerns, the construction of new plants boomed during the 1960s and 1970s. Many industrialized nations subsidized the development of nuclear power in an attempt to move away from fossil fuels as the primary means of generating electricity. Then in 1979 a cooling valve malfunctioned at the Three Mile Island nuclear plant in the United States (Figure 14.12), causing one of the plant's reactor cores to overheat and begin fusing together in what is known as a *meltdown*. Although the reactor was permanently destroyed, a complete meltdown was avoided, and only a small amount of radio-active steam managed to escape the containment building. Public attention, however, focused on the fact that had a complete meltdown occurred, the reactor core could have burned through the floor of the containment building, permanently contaminating the surrounding surface and groundwater supplies.

Then in 1985 a runaway nuclear reaction and fire at the Soviet Union's nuclear plant at Chernobyl sent massive amounts of radioactive gas and dust into the atmosphere. Radioactive particles eventually fell to the ground over parts of Eastern and Western Europe, with the worst contamination occurring closer to the plant in what are now Ukraine, Belarus, and Russia. A total of 50 workers and rescue personnel died by receiving lethal doses of radiation, and nine children died of thyroid cancer related to the radioactive fallout. In a 2005 United Nations report on the Chernobyl disaster, health experts estimated that 4,000 people will eventually die from radiation-induced cancers. It was also reported that with the exception of the highly contaminated 30-kilometer (19-mile) exclusion zone around the plant, radiation levels have returned to "acceptable levels" over most of the fallout area.

The Chernobyl accident, which occurred just seven years after Three Mile Island, only heightened concern in the United States over the safety of nuclear power plants. However, the design of nuclear reactors built in Western nations is fundamentally different from what the former Soviet Union used at Chernobyl. The Soviet reactor's nuclear fuel was surrounded with graphite, whose purpose is to lower the velocity of neutrons so that a sustained fission reaction can be achieved (Figure 14.10). Moreover, the Soviet design did not include a containment building. The accident at Chernobyl began when there was a loss of cooling water, whereas the graphite allowed the fission reactions to continue. The result was a buildup of heat. Eventually the reactor core exploded and the graphite caught fire, sending huge amounts of radioactive material into the atmosphere. In contrast, Western reactors achieve sustained fission by using water to slow the neutrons. This water also happens to be the same water used to cool the reactor. This means that if there is a loss of cooling water, the fission reactions will stop. The reactor may still overheat and begin to melt down, but there is no chance of a runaway reaction and massive graphite-fed fire as at Chernobyl. The worst-case scenario for a Western reactor would be a reactor core melting through the floor of the containment building, allowing radiation to escape into the environment.

The accidents at Three Mile Island and Chernobyl helped to cause a virtual halt in new construction permits for nuclear power plants in the

United States. In addition to concerns over reactor safety, there was the issue of the United States not having developed a long-term strategy for disposing of radioactive waste. This is a serious problem because some radioactive isotopes will remain dangerous for tens of thousands of years (Chapter 15). To make matters worse, at the time of the accidents utility companies were discovering that operating large nuclear power plants was more expensive than coal or natural gas-fired plants. The cost issue, combined with concerns over safety and waste disposal, caused U.S. utility companies to cancel plans for constructing additional nuclear power plants. The number of nuclear plants operating in the United States reached a peak of 112 in 1990 (Figure 14.13), falling to 104 by 2007. Some nations, however, continued to expand their nuclear power programs in an effort to become more energy independent. A striking example is France, which now generates 75% of its electrical needs from nuclear power, compared to just 19% for the United States.

After more than 30 years, utility companies in the United States have again starting applying for permits to build additional nuclear power plants. This renewed interest is due in part to the recent surge in energy prices, and the growing likelihood that some type of carbon tax will be imposed on fossil fuels to help combat global warming. However, before nuclear power can play an expanded and sustained role over the long term, the United States will have to address the issue of waste disposal. In particular the United States is going to have to open its long-planned waste-disposal site, which is specifically designed to permanently store high-level radioactive waste (Chapter 15). Note that this waste is currently being stored aboveground on a temporary basis at nuclear plants; some of this waste has been in storage for over 30 years.

Finally, there is another roadblock to the expansion of nuclear power, namely the limited supplies of uranium. Similar to other mineral resources, Earth contains a finite amount of uranium-bearing minerals that can be mined economically. Expanded mining will help, but it is not necessarily a long-term solution. One option is to construct special *breeder reactors* that are designed to maximize the production of plutonium from uranium. Recall that fissile uranium-235 atoms make up only 3 to 8% of the uranium fuel in a nuclear reactor. However, the much more abundant U-238 atoms within the fuel absorb neutrons and are transformed into fissile atoms of plutonium-239. Rather than continuing to discard the spent nuclear fuel as waste, the fissile plutonium can be extracted and used to fuel a reactor. Breeder reactors themselves therefore can serve as a long-term supply of nuclear fuel because they generate more fissionable atoms than they consume. In the past, however, the United States has discouraged the use of breeder reactors because they can serve as a ready source of plutonium for making nuclear weapons.

Fusion Reactors

Perhaps the most tantalizing energy source of all is *nuclear fusion* because its basic fuel supply is water, which is unlimited, and it does not produce long-lived radioactive waste products. Despite the obvious advantages, no successful fusion reactor has yet been built. Physicists are currently testing a variety of experimental approaches, mostly involving attempts to fuse two different hydrogen isotopes. One isotope called *deuterium* can be obtained from seawater, and the other is *tritium,* which can be produced from the element lithium. In order to create a self-sustaining fusion reaction, deuterium and tritium atoms must be brought up to a temperature of at least 100 million degrees centigrade. When this occurs, the gas molecules are stripped of their electrons and form what is called *plasma.* At the same

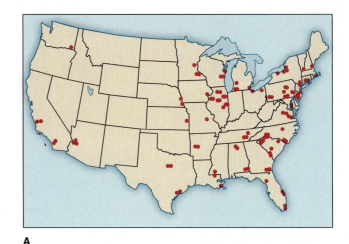

A

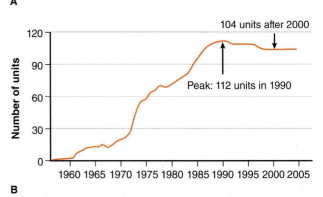

B

FIGURE 14.13   Map (A) showing the location of the 104 nuclear power plants operating in the United States as of 2007. Graph (B) showing how the construction of new nuclear plants came to a halt around 1990 due to cost-effectiveness and concerns over safety and waste disposal. In 2007, after more than 30 years, utility companies have again started applying for permits to build new plants.

time this is taking place, the superhot plasma must be confined under enormous pressure for a long enough period of time (seconds) to allow fusion reactions to begin. Most of the science and engineering problems revolve around trying to control the incredibly hot plasma, while simultaneously keeping it confined under terrific pressure. Note that fusion naturally takes place on the Sun, whose immense gravity provides the confining mechanism. Researchers here on Earth are trying to confine the plasma using either a magnetic field or a series of lasers.

Although physicists have yet to achieve a self-sustaining fusion reaction after 35 years of effort, they have made steady progress and are getting close to achieving the necessary temperature and pressure. Should they succeed, the next step would be to produce a commercially viable fusion power plant. The plant would generate electricity by using the immense heat from the fusion reactions to drive a steam turbine similar to that described above for fission plants. Fusion power plants, however, are not something we are going to see anytime soon because there are difficult scientific and engineering issues yet to be resolved.

Solar Power

Energy from the Sun is our most basic and critical source of energy. Sunlight warms our planet and makes it habitable, and it drives Earth's ecosystems and hydrologic cycle from which we obtain food and fresh water. Scientists often refer to solar energy as *solar radiation* because it is technically electromagnetic radiation that forms from nuclear fusion on the Sun. As described in Chapter 2, electromagnetic radiation is a continuous spectrum of wave energy that travels away from its source. Different portions of the spectrum have been given names based on the wavelength of the waves (e.g., X-ray, ultraviolet, visible, and infrared). When this traveling wave energy strikes the Earth it is transformed into different forms of energy. For example, plants convert energy from the visible part of the spectrum into cell tissue (biomass), which itself is a form of chemical energy; the fossil fuels we rely on represent the ancient transformation of sunlight into biomass. We as individuals experience another type of solar energy transformation every time we go outside on a sunny day. As the sunlight strikes our skin or clothing, the solar radiation is converted into heat energy and we begin to feel warmer. Sunlight can also be transformed directly into electricity, such as in solar-powered calculators. In this section we will focus on the two solar energy transformations that society can use directly as an energy source, namely solar to thermal and solar to electrical.

Solar Heating

The process called **solar heating** takes place when solar radiation strikes a solid object and is transformed into thermal or heat energy. People can make use of this free heat energy in one of two ways. The simplest approach is referred to as *passive* because it does not involve any mechanical effort (moving parts). Perhaps the best example is the use of glass windows that allow direct sunlight to enter a building. Solar radiation from the visible part of the spectrum passes through the transparent glass, and is then absorbed by physical objects within the building (walls, flooring, furniture, etc.). As the objects get warmer they release heat energy, which becomes trapped within the building and raises the indoor temperature (this same process is what makes a car become hot when the windows are up).

As illustrated in Figure 14.14, architects can create passive building designs that take advantage of free solar heating. The basic idea is to maxi-

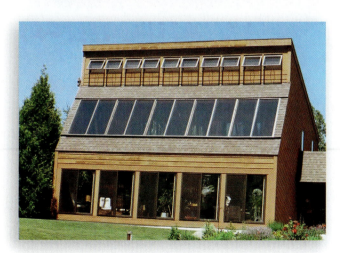

FIGURE 14.14 Passive solar homes are designed to let sunlight enter during the wintertime, and then make use of trees and roof overhangs to block the sunlight during the summer months. This solar-designed home in Michigan reportedly gets 85% of its heating needs from the sun.

mize the amount of sunlight that can enter a building by orienting the building and a bank of windows toward the sun (south-facing windows are used in the northern hemisphere, whereas north-facing would be required below the equator). In many climatic zones solar heating may be desirable in the winter, but is something to be avoided during the summer when the goal is to keep a building cool. This problem can be solved by using deciduous trees and roof overhangs that prevent sunlight from entering the windows when the sun is higher in the sky during the summer months. As wintertime approaches, the leaves will fall from the trees and the sun angle will become progressively lower, thereby allowing direct sunlight to stream through the windows again. In this way solar heating automatically takes place in the winter, and then is blocked during the summer when it is not needed.

The other form of solar heating is called *active* since it involves some type of mechanical system to help collect and distribute the heat. Figure 14.15 provides an example of an active solar heating system in which a solar collector (panel) is mounted on a roof and oriented to maximize the amount of direct sunlight striking the collector. During the daytime the collector transforms solar radiation into heat energy, which is then transferred to a circulating water system. Many homeowners choose to connect the thermal panel to their hot water supply, thereby reducing the energy demand on a standard electric or gas water heater. The water heater will only switch on when the temperature of the solar-heated water falls below a certain point, typically at night or on cloudy days. The savings can be significant since a conventional hot water heater accounts for about 15% of the average homeowner's utility bill. Another option is to pipe solar-heated water through radiators where the heat is then transferred to the airspace within the building.

Regardless of whether the system is active or passive, solar heating is highly desirable because once the energy savings pays for the cost of the system, the heating is essentially free. Although solar heating is renewable and inexpensive, it can never entirely replace the need for conventional forms of heating, particularly in areas where cloudy days are frequent. Nevertheless, if large numbers of homes and businesses were encouraged to make use of this free heat from the Sun, society could make significant reductions in its demand for carbon-based fuels.

FIGURE 14.15   Solar collectors mounted on a roof convert solar radiation into heat energy, which is then transferred to a circulating water system. The hot water from a solar system can be used to help reduce the energy load on a conventional hot-water heater, or it can be sent through a radiator to help warm the building's airspace.

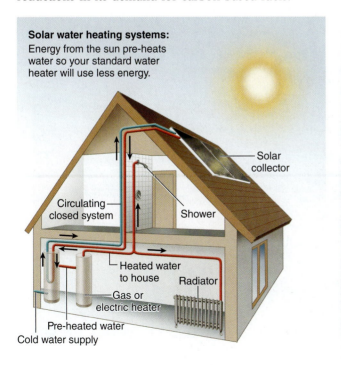

Solar water heating systems:
Energy from the sun pre-heats water so your standard water heater will use less energy.

Solar collector

Circulating closed system

Shower

Heated water to house

Radiator

Gas or electric heater

Pre-heated water

Cold water supply

Photovoltaic Cells

Due to society's great demand for electricity, solar power's greatest potential lies in the technology that allows us to directly transform sunlight into electricity. The key to this transformation lies in the fact that solar radiation contains tiny particles of energy called *photons*. As illustrated in Figure 14.16, when sunlight strikes a semiconductor, a thin slab of material composed of silicon alloys, some of the photons are absorbed, which dislodges electrons from the alloy. These electrons then start moving in a circular path between the semiconductor's surface and internal layers, creating an electrical current called a **photovoltaic cell,** sometimes called a *solar cell*. Solar panels consisting of individual photovoltaic cells are commonly placed on roofs and then wired to the electrical system of a building. Note that some home and business owners will install both thermal and photovoltaic panels in order to reduce the need for conventional heating and electricity.

One of the problems with photovoltaic cells is that the amount of electricity they produce depends on the intensity of the sunlight hitting the panels. Very little electricity can be generated when the skies are overcast or cloudy, and of course none is produced at night. Although photovoltaic systems can generate surplus electricity on bright sunny days, a major drawback is that we have a limited ability to store electricity (solar thermal systems can store energy by placing hot water in insulated tanks). Despite the fact electricity can be stored in banks of interconnected batteries, the problem is that batteries are expensive and cannot store large amounts of power. Rather than using batteries, many home and business owners with photovoltaic systems simply switch over to conventional utility power at night, and on cloudy days when their system cannot produce a sufficient amount of electricity. Regulations in the United States and some other countries require public utility companies to purchase electricity generated from small sources as well as major power plants. Because most homeowners are already connected to the utility grid, they can therefore sell surplus electricity from their photovoltaic system to a public utility company. Rather than making a direct payment, a utility company will typically issue a credit on the homeowner's electric bill. The ability to sell surplus electricity is important because it

FIGURE 14.16   Photovoltaic panels consist of individual cells composed of silicon alloys that transform solar radiation directly into electricity. Light particles called photons are absorbed by the uppermost layer, dislodging electrons which then move in a circulating manner. Electricity is obtained from this circulating flow of electrons. Photo showing both photovoltaic and thermal panels installed on a home.

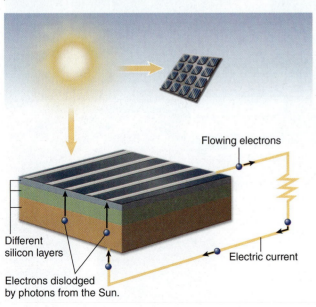

helps makes small-scale photovoltaic systems more cost-effective for individual homeowners.

The issue of cost-effectiveness is particularly important because the actual photovoltaic panels are still relatively expensive. Therefore, most people who purchase a photovoltaic system will want to save enough money on their electric utility bills to pay for their system over some reasonable amount of time. For example, if a system costs $20,000, but saves you only $500 a year, it would take 40 years just to recover your initial costs. From just an economic point of view, you may find it wise to invest your money elsewhere. One of the key factors that affects the so-called *pay back period* of the system is the amount of solar radiation your panels receive on an annual basis. Obviously, if the weather patterns in your area result in many days that are overcast or cloudy, then your system will have a longer payback period. Similarly, the farther north or south of the equator that you live, the lower the sun angle, which reduces the amount of radiation the solar panels can receive annually. You can see the effects of climate and latitude in Figure 14.17, which shows the average daily solar radiation across the United States. Obviously, photovoltaic systems are most cost-effective in the southwestern portion of the nation due to the region's lower latitude and less cloudy skies.

FIGURE 14.17 The primary factor determining the cost-effectiveness of solar panels is average annual sunshine, which, in turn, is controlled by latitude and weather patterns that determine the amount of cloud cover. From this map of average annual solar radiation, one can see that solar power is most cost-effective in the southwestern portion of the United States.

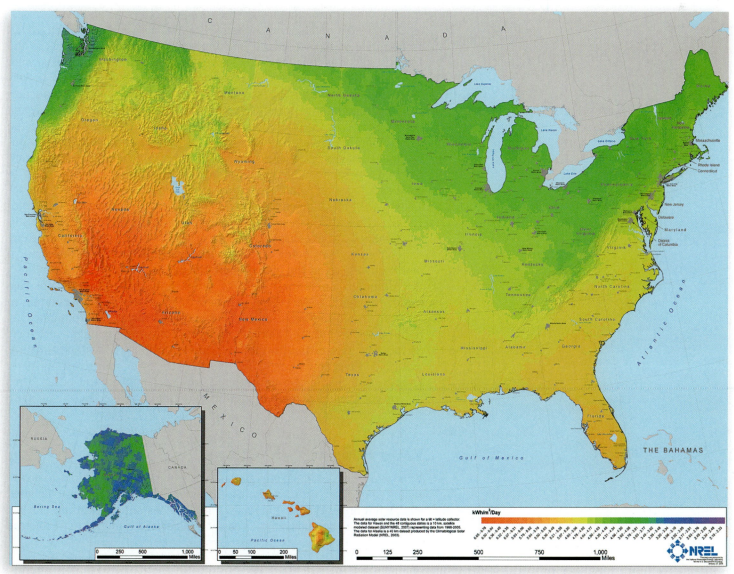

FIGURE 14.18 Thin-film photovoltaic panels are flexible and can provide modest amounts of electrical power at remote locations. As the price of thin-film panels continues to decline, this technology may revolutionize the solar energy industry.

The other major factor affecting the economics of photovoltaic systems is the cost of the systems themselves, which has largely been driven by the price of silicon used in making semiconductor cells. Back in the 1960s semiconductors were so expensive that photovoltaic cells were used only in applications where cost was not an issue, such as for powering satellites. Over time the increased demand has led to lower production costs—similar to how prices of computer chips have fallen over the years. Industry experts are predicting that prices may soon come down to the point where photovoltaic systems would be cost competitive with conventional means of generating electricity. In the meantime, recent advances in semiconductor technology have enabled manufacturers to paint thin layers of semiconductor material onto flexible films. These so-called *thin-film solar cells* are 50 times thinner than conventional photovoltaic cells, and are designed to be portable (Figure 14.18). Despite still being somewhat expensive, the cost of thin-film solar cells is offset by the convenience of being able to generate modest amounts of electrical power anywhere on the globe where the sun is shining.

Although generating electricity from sunlight is presently not cost competitive with fossil fuels, the use of solar power is almost certain to increase due to the need for non-carbon-based fuels that are renewable. Because the technology already exists for converting sunlight into electricity, now is the time to rapidly scale-up production, which, in turn, will help drive down costs and make solar power more cost-effective. Later in this chapter we will explore how solar power, and other forms of renewable energy, could be used to produce clean electricity in a more sustainable society.

Wind Power

Wind is a form of kinetic energy that has been used for centuries to power sailing ships and turn windmills. Today there is great interest in wind energy not only because it is renewable and does not emit carbon dioxide, but because its kinetic energy is easily transformed into electricity. Instead

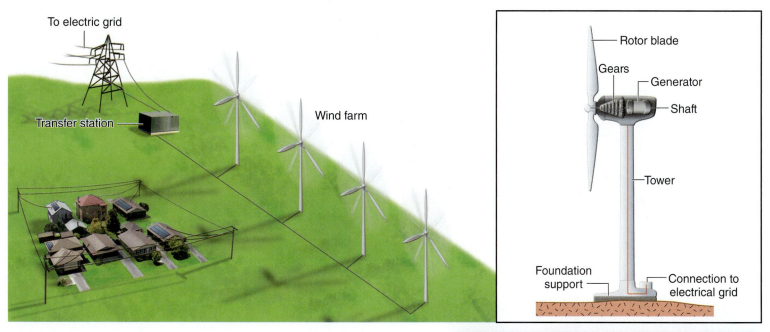

of spinning an electrical generator with hot steam or falling water, wind is used to turn a rotor similar to an airplane propeller, as shown in Figure 14.19. The amount of electricity a wind turbine can produce depends on how fast the rotor turns and how long it can be kept spinning. Clearly, the ideal location is someplace where the wind is both strong and steady. Mountainous areas often contain suitable sites because the mountains act as barriers, forcing the wind to pick up speed as it flows over ridges or funnels through gaps or passes (Figure 14.20A). Coastal areas and open plains also make excellent sites because of the tendency of the winds to be strong and steady in these areas. Vast amounts of electricity in the United States could be produced across the Great Plains and Midwest and at numerous localized areas in the mountains and around the Great Lakes (Figure 14.20B).

Similar to solar power, the cost-effectiveness of wind-generated electricity depends on both geographic location and the cost of the wind turbines themselves. This means that in most situations, the profits or savings from wind power must be enough to repay the initial investment over some reasonable amount of time. Because wind-generated electricity is presently much more profitable compared to photovoltaic systems, new firms are investing in wind energy projects and selling the power to electric utility companies. In order to maximize their profits, wind-energy firms are making use of the economies of scale and building multiple wind turbines in what are called **wind farms.** Part of the cost for a wind tower includes installing underground cables and building a transfer station so that the power can be fed into the regional electric grid (see Figure 14.19). By building multiple turbines relatively close together, the cabling and transfer station costs can be shared. In this way the costs go down as the scale or size of a wind farm increases.

Another way of using the economies of scale is to make the wind turbines themselves bigger, thereby increasing the amount of electricity an individual turbine can produce. Although the cost of a single tower becomes greater, the increased electrical production more than offsets the additional construction costs. There is, however, a limit to size due to physical

FIGURE 14.19   Wind is a form of kinetic energy that can be transformed into electrical energy by turning a set of rotor blades, which then spins an electrical generator. To take advantage of the economy of scale, multiple wind towers are usually constructed in what are referred to as wind farms.

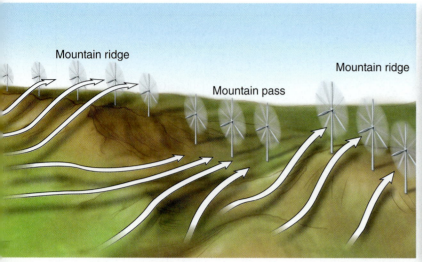

Near Palm Springs, California

A

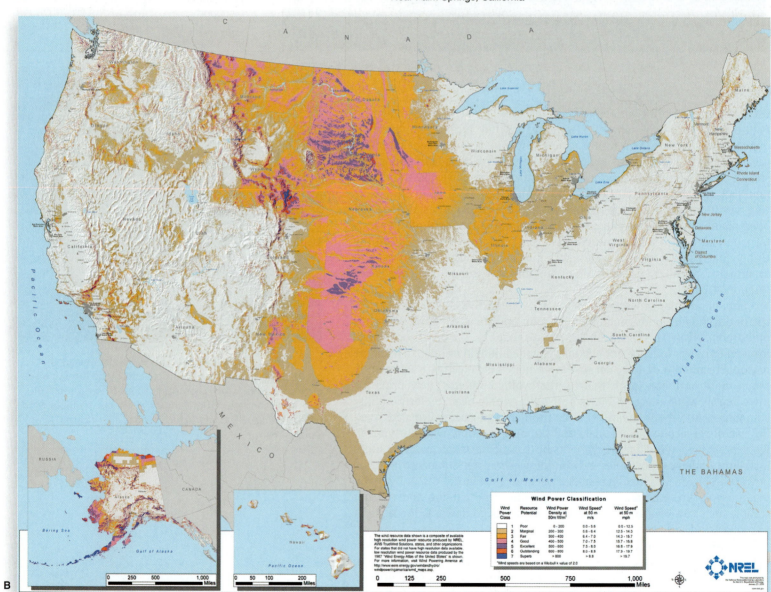

B

FIGURE 14.20   For wind farms to be cost-effective they need to be located where the winds are both strong and steady over much of the year. Such conditions are common on top of ridges and mountain passes (A) which act to funnel the wind. Other ideal locations include windswept open plains and coastlines. Map (B) showing areas in the United States that are favorable for developing cost-effective wind power.

constraints. In 2008 the world's largest wind turbine was built in Emden, Germany, on the North Sea coast. The rotor assembly is an impressive 413 feet (126 m) in diameter and is mounted to a tower that stands 453 feet (138 m) above the ground. This single wind turbine generates enough electricity to power 5,000 European households. Note that wind turbines can also be scaled down (Figure 14.21) and be cost-effective for an individual homeowner, depending upon the circumstances. Similar to solar systems, a homeowner can sell any surplus electricity to an electric utility company, thereby reducing the payback period for recovering the initial costs of the system.

In addition to being able to produce electricity more cheaply, wind turbines have an advantage over photovoltaic panels in that they will generate electricity on cloudy days and at night. Wind power therefore is likely to be a better choice than solar in climatic zones that are dominated by overcast skies. Keep in mind, however, that even in the most ideal locations there are times when the wind slows down or stops altogether. This means that both wind and solar are *intermittent* sources of energy, but they are usually intermittent at different times and in different places—we will examine the significance of this variation in a later section.

A   B

FIGURE 14.21   Large wind turbines (A) are more cost-effective because the economies of scale allows them to produce more power relative to their size. Residential turbines (B) come in a variety of sizes and can be cost-effective for individual homeowners.

Despite the fact wind power may be clean and renewable, wind farm projects are often met with opposition by local landowners. In some cases the projects are opposed because they alter the landscape in ways some people find unattractive. Other times the issue is noise. Residents who will be living in close proximity to a wind tower will have to endure the constant swooshing noise made by the rotating blades. There is also the issue of bird and bat mortality. One of the early wind farms in the United States happened to be located on a mountain pass near San Francisco that also serves as a migratory bird route. An estimated 5,000 birds were reportedly killed each year, raising serious concerns about the environmental impact of wind farms. The wind power industry has largely eliminated this problem by locating new wind farms away from migratory routes and by building turbines whose rotors turn more slowly. The result is that the number of birds being killed today by wind turbines is far lower than those killed flying into buildings and cars and trucks on the highways. The problem with bats, unfortunately, appears to be different. Researchers have found relatively high rates of bat mortality around nearly every wind facility surveyed in North America. The question of why bat are colliding with wind turbines remains unanswered, but the data do show that the mortality rate varies among different bat species. The highest rates are found associated with bats that prefer to roost in trees and tend to migrate long distances. This phenomenon is especially puzzling since bats rarely fly into buildings or moving cars. Researchers are focusing on whether the wind turbines were located in unknown bat migration pathways, or whether the bats are somehow attracted to the spinning blades. Groups representing

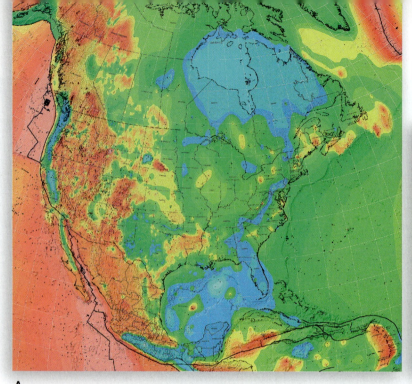

A

B

FIGURE 14.22 The rate of heat flow through Earth's crust (A) largely depends on rock type, crustal thickness, and tectonic history. Crustal rocks are warmer in areas of highest heat flow and can be used as a source of geothermal heat. The loss of heat through Earth's crust is analogous to heat escaping from a heated house (B) in the winter. Red represents the areas of highest heat flow, and blue represents the lowest.

FIGURE 14.23 Geothermal heat can be utilized by extracting water that comes into contact with hot rocks. A series of injection and pumping wells are used to circulate either natural groundwater or surface water pumped down into hot, dry rocks. The water can be used for space heating, and if hot enough, it can be turned into steam and used to make electricity.

the wind industry are actively supporting this new research and hope to develop ways to mitigate the problem.

Geothermal Power

Our planet receives energy from two inexhaustible and renewable sources: the Sun and Earth's interior. **Geothermal energy** refers to the heat contained within the Earth, most of which comes from the decay of radioactive elements. Recall from Chapter 4 that large convection cells within the mantle transport this heat toward the surface, where it eventually dissipates into space. As shown in Figure 14.22A, the outflow of heat through Earth's crust is not uniform. The higher rates of heat loss generally occur in areas of tectonic and volcanic activity (subduction zones, spreading centers, and hot spots), or where the crust is relatively thin. A useful analogy is how heat flows or escapes from a warm house on a winter night. A thermal camera (Figure 14.22B) reveals that heat is preferentially lost through windows and cracks, whereas very little heat escapes through the highly insulated walls.

Geothermal energy can be extracted from the subsurface in different ways, but perhaps the most common is to utilize the heat contained in water that has come into contact with unusually warm rocks. Areas with the greatest potential for geothermal energy (i.e., warmest rocks) are located where the rates of heat flow through Earth's crust are the highest. Volcanic regions, such as Yellowstone National Park (see Chapter 6), are ideal since shallow rocks are often hot enough to turn groundwater into steam. In the past, geothermal waters could only be accessed at *natural hot springs,* which are simply places where unusually warm groundwater is able to make its way to the surface (Chapter 11). Geothermal waters today can be extracted using wells drilled directly into aquifers that contain hot water, as shown in Figure 14.23. Moreover, because wells can extract water that is much closer to the heat source, geothermal wells typically obtain water that is warmer compared to that from natural hot springs. In fact, some geothermal well water is over 900°F (500°C), which is hot enough to melt lead. Figure 14.23 shows a situation where hot rocks at depth are not porous, thus do not contain groundwater. In this case explosives can be used to fracture the rocks deep within a well. Water is then injected into the fractured rocks so that the heat energy can be transferred to the water, which is then removed by a nearby pumping well.

Geothermal Electricity

The water temperature of a geothermal system plays a key role in determining how it can be used as an energy source. For example, if the temperature of the water being extracted is significantly higher than its boiling point of 212°F (100°C), it will quickly turn to steam when it is exposed to the low-pressure environment of the atmosphere. This steam can be used to turn a turbine and generator to create what is referred to as *geothermal electricity*. Such geothermal electricity is highly cost competitive because Earth's interior heat is being used to turn water into steam, as opposed to using coal, natural gas, or nuclear fuel. For this reason another advantage is that geothermal energy does not produce carbon dioxide and is generally renewable.

Several countries are currently producing geothermal electricity in volcanic regions where the rocks are sufficiently hot to produce steam. Many of the sites lie along the convergent boundary of the Pacific plate, often referred to as the Ring of Fire (Chapter 6). Geothermal electricity in the United States is presently being produced in several western states where the rocks remain hot from volcanic activity that occurred in the recent geologic past. The greatest production is in California, where geothermal power represents 5% of the state's electrical needs. Geothermal electricity is also being produced in Hawaii, where the heat is associated with a volcanic hot spot.

Geothermal Space Heating

The number of sites around the world where subsurface rocks are hot enough to generate steam is actually quite small. However, there are many places where groundwater temperatures are more than adequate to provide heat for buildings. *Geothermal space heating* involves piping hot water through radiators where the heat energy is then transferred to the air space inside a building (this method of indoor heating is the same as described earlier for solar thermal energy). Geothermal water can also be used to heat swimming pools. In fact, many resorts have to actually cool the heated groundwater by mixing it with water from another source in order to bring it under 104°F (40°C), which is about the maximum people can tolerate for any reasonable length of time.

Perhaps the best example of geothermal energy being utilized on a large scale is Iceland, a country with a population of only 300,000, located just below the Arctic Circle. Iceland itself is geologically unique in that it lies over a volcanic hot spot and also straddles a major spreading center (Figure 14.24). In fact, Iceland is the only place where the over 6,000-mile (10,000 km) long mid-Atlantic spreading center rises above sea level. Due to its abundance of both groundwater and geothermal energy, Iceland is able to heat 87% of all its households using geothermal space heating. In terms of electricity, geothermal plants produce 17% of the nation's electrical needs, with the remainder being generated by hydropower, which is abundant and more cost-effective. It is worth noting that Iceland presently uses just a small fraction of its geothermal energy resources, but plans to utilize some of its excess capacity to create a

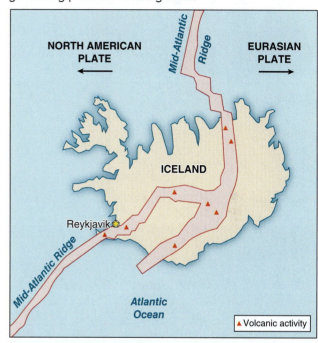

FIGURE 14.24 Iceland sits atop a hot spot on the mid-Atlantic spreading center (Chapter 4) where geothermal energy meets 45% of the country's total energy needs. Photo showing one of Iceland's year-round swimming facilities that is heated by the constant discharge of geothermal water from the electrical-generating plant in the background.

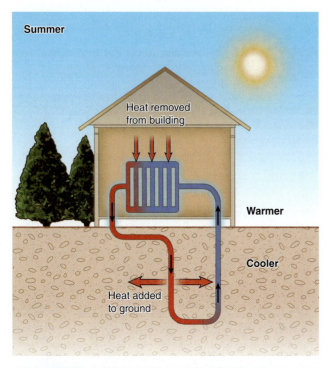

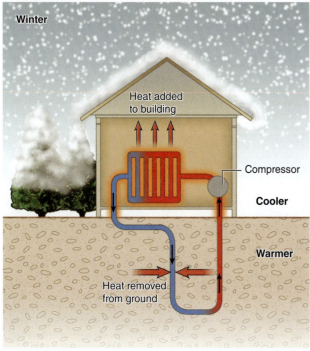

FIGURE 14.25 A geothermal heat pump uses a series of interconnected pipes to exchange heat between the ground and a building. During the summer, relatively cool water is piped into the building where it absorbs heat from the air. The warmed water then goes into the subsurface where the additional heat is transferred to the ground. In the winter, a compressor is used to concentrate heat energy from the ground, which is then released into the building.

transportation system based on hydrogen fuel cells. Such a program would make Iceland the first country to convert over to a hydrogen-based transportation system, and largely eliminate its need for imported oil.

Geothermal Heat Pumps

In most climatic zones the air temperature varies considerably from night to day, and on a seasonal basis. However, if you go about 10 feet (3 m) below the surface, the temperature of the rocks or soil no longer follows the atmospheric changes. Depending on latitude and elevation, the ground temperature will remain steady somewhere between 45° and 75°F (7–21°C), which is cooler than the average air temperature in the summer, yet warmer than the winter average. A **geothermal heat pump** is a mechanical system that supplies a building with warm or cool air by making use of the relative difference in temperature between the inside air and the ground. The basic principle involved is similar to how a cave's constant temperature will have a cooling effect on a person who enters during the summer, and then provides warmth during the winter.

As illustrated in Figure 14.25, a geothermal heat pump is actually a system of interconnected wells or pipes, where water circulates between the ground and a heat exchanger (metal coils) located inside a building. During the summer the heat exchanger receives water from the subsurface that is cooler than the air inside the building. As air from the building is blown through the heat exchanger, the water absorbs heat from the air. This reduces the temperature of the air, which is then returned back through the building. Meanwhile the water travels back through the subsurface loops where the ground absorbs the heat that had been removed from the building. In the winter the heat transfer occurs in the opposite direction, namely from the ground to the air. However, this requires an electric-powered compressor in order to extract enough heat to bring the air up to a comfortable temperature. Basically the system removes heat from the water coming up from the subsurface, then transfers this heat to a compressed gas (*Freon* is currently used, but will be phased out by 2010 and replaced by *Puron*). The gas is then allowed to decompress, at which point the concentrated heat in the gas is transferred to the air within the building. In this mode a geothermal heat pump is identical to how your refrigerator extracts heat from its contents (food and drinks), then transfers the heat to the air in your kitchen.

It is important to note that a conventional heat pump exchanges heat with the outside air, rather than with the subsurface as in geothermal systems. The advantage of a geothermal system is that it uses far less electricity to operate because the ground temperature remains constant and is comparatively close to the 70–75°F (21–24°C) range maintained in most buildings. The outside air temperature of course can vary widely, ranging from subzero to over 100°F (73°C). This means that the efficiency of a conventional heat pump is greatly reduced on days when the difference between the outdoor and indoor temperatures is large. The largest difference will naturally occur on days when the outside air temperature falls below freezing. A conventional heat pump, therefore, is a poor choice in cold climates since it will consume a great deal of electricity while providing a minimum amount of heat.

Future of Geothermal Energy

Geothermal energy is highly desirable because it emits little to no carbon dioxide, plus it can be very cost-effective because it makes use of free heat from the Earth's interior. Unfortunately, geothermal electricity and space heating are geographically limited to those relatively few areas where the heat flow through the crust is unusually high. Another problem is that rocks are poor heat conductors. This becomes a problem because some geothermal plants are able to remove heat at a faster rate than it can be replaced. In other situations the rate of water being withdrawn is greater than the recharge rate, which not only depletes the resource but can also lead to serious land-subsidence problems (Chapters 7 and 11). In the case of electrical production and space heating, geothermal energy therefore is not always renewable. Fortunately, the problems associated with land subsidence and withdrawing heat too quickly can be minimized by injecting the warm wastewater into the subsurface rather than discharging it on the land surface. Another useful technique is to extract geothermal heat on a cyclic rather than a continuous basis. This allows the thermal energy to build back up between production cycles. Nevertheless, even if mitigation steps are taken, geothermal electricity and space heating will always be restricted to relatively small geographic areas where suitable geothermal reservoirs are found.

In contrast, geothermal heat pumps can be utilized almost anywhere. This form of geothermal energy therefore has the potential of making a significant contribution to society's effort to reduce its use of fossil fuels. According to the U.S. Environmental Protection Agency (the EPA), geothermal heat pumps use 44% less electricity than air-based systems, and 72% less than electric heating and standard air conditioning systems. Despite the increased efficiency, builders and homeowners tend to avoid installing geothermal heat pumps because of the relatively high installation costs. However, these costs are usually recovered in two to ten years from the savings in reduced electrical consumption.

Ocean Thermal Energy Conversion

Every day the oceans absorb about 70% of the solar radiation that reaches the Earth, converting the sunlight into a vast amount of thermal energy. **Ocean thermal energy conversion** (OTEC) is a technique that makes use of a simple heat engine, which is driven by the heat energy stored within the oceans. An OTEC plant draws in warm water from the ocean surface along with cold water from great depths, and then uses the temperature difference of these two waters to drive a turbine and electrical generator. Experimental OTEC plants have been built by Japan and the United States and have shown the technology to be a cost-effective means of generating electricity. One OTEC design, shown in Figure 14.26, incorporates a closed-loop system that contains an ammonia-based fluid. When the warm seawater is pumped through a heat exchanger, the ammonia within the pipes begins to boil and turns into a gas. The ammonia vapor is then used to spin a turbine and electrical generator. From there the gas is piped through a second heat exchanger that is being flushed with cold seawater, causing the ammonia to condense back into a liquid. The other basic OTEC design takes the incoming warm seawater and places it under a vacuum, which causes it to boil. The resulting steam is then used to drive the electrical generator as opposed to vaporized ammonia. Engineers included this additional step of vaporizing seawater so that valuable freshwater could be produced as a by-product of generating electricity, thereby making the overall system more cost-effective.

Although OTEC plants produce a net gain of electricity (generate more power than they consume), a large-scale

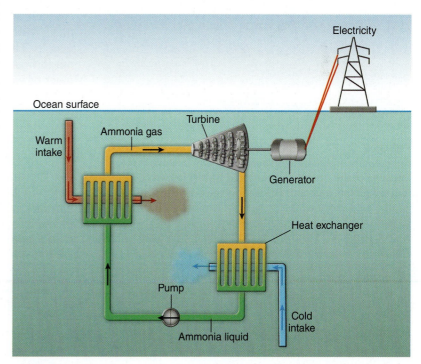

FIGURE 14.26 Ocean thermal energy conversion (OTEC) uses the heat energy stored in the oceans to make electricity. The simplest OTEC design pumps warm seawater through a heat exchanger in order to vaporize ammonia. The ammonia gas is then used to spin an electrical generator. Cold seawater is pumped through a second heat exchanger to turn the ammonia back into a liquid.

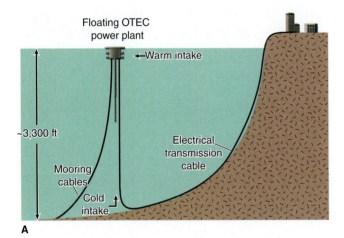

A

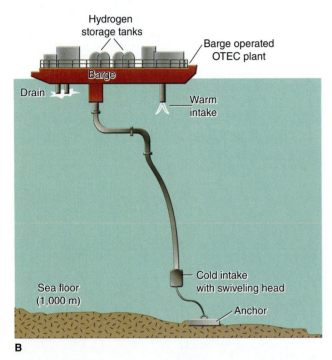

B

FIGURE 14.27 Potential land-based OTEC sites are restricted to tropical coastlines with a steep seafloor (A) because of the need to access cold water. To overcome this restriction, a proposed floating sea-based system (B) would have easier access to deep, cold water. This would require electricity to be sent to the mainland via a transmission cable, or be used to produce hydrogen, which could be shipped via tankers.

demonstration plant has yet to be built due to a lack of investors or government support. The primary problem is that conventional coal-fired power plants are still the cheapest way to produce electricity. Another problem is that for an OTEC plant to operate efficiently, it must be located where the temperature difference between the warm and cold seawater is at least 36°F (20°C), which restricts OTEC to tropical waters. In addition, to draw in water that is cool enough, the plant's cold intake pipe must reach a depth of around 3,000 feet (1,000 m). This means that OTEC plants can only be built on coastlines where the slope of the seafloor is quite steep. Otherwise, the distance from shore to where the water is sufficiently deep would require building and maintaining an excessively long intake pipe, thereby making the entire operation uneconomical. This presents yet another problem in that most of the suitable tropical sites that do exist are located far from major population centers and large electrical demands. One proposed solution (Figure 14.27A) is to build a floating OTEC plant that would send power to a city on the mainland via a transmission cable. Another solution is to have a floating plant use the electricity it produces to separate hydrogen from seawater (Figure 14.27B). The hydrogen would then be stored on the platform and eventually loaded onto tankers for transport to major population centers. From there the hydrogen could be used as a transportation fuel. While all this is feasible, any major contributions from OTEC to the world's energy supplies are at best many years away.

Tidal Power

Tidal power is different from the other alternate energy sources in that it is not ultimately derived from sunlight or Earth's internal heat, but from the cyclic lowering and raising of sea level caused by the gravitational interaction between the Earth, Moon, and Sun (Chapter 9). On a nearly continuous 12-hour cycle, coastal estuaries and inlets around the world are filled with seawater when an incoming tide laps up onto Earth's landmasses. This is followed by an outgoing low tide that drains the water back out toward the sea. Because Earth's oceans are so vast, this physical movement of water represents an enormous amount of kinetic energy. Similar to other forms of kinetic energy (e.g., wind and hydro), tidal power can be used to spin an electrical generator. As you will learn in this section, however, tidal power is cost-effective only in places were the shoreline is shaped in such a way that the tides are amplified or concentrated.

There are two basic ways tidal power can be used to generate electricity. The first involves constructing a low-lying barrier called a *tidal barrage* across an enclosed inlet or estuary as shown in Figure 14.28. The barrage is basically a dam with large gates that remain open while the incoming tide fills the estuary. At high tide the gates close, trapping the water before the tide starts to go back out to sea. With the reservoir behind the barrage full, water is then allowed to fall through turbines and generate electricity, similar to a hydroelectric dam. Electrical production naturally depends on the amount of kinetic energy the water has as it falls through the turbines, which, in turn, depends on the difference in height between the filled reservoir and low tide. It turns out that a tidal range of at least 16 feet (5 m) is required for a barrage system to function economically. Unfortunately there are relatively few places in the world with this type of tidal range.

Currently there are only two commercial barrage plants in operation, one built in France in 1967 and the other in Canada in 1984. Despite the fact these plants have been profitable and that other suitable sites exist around the world, no other major tidal plants have been built. The basic problem is that engineers did not take into consideration the environmental impact of barrage systems on coastal ecosystems. Recall from Chapter 9 that coastal estuaries and associated wetlands serve as the nursery grounds for nearly 70% of all commercial and sport fisheries. A tidal barrage, unfortunately, acts as a barrier for fish and other marine life attempting to enter an estuary to spawn. For those species that are able to enter, many end up being killed on the way back out as they pass through the turbine blades. Another problem is that a barrage alters the timing of the tides, and consequently impacts the entire food web which is highly sensitive to the natural ebb and flow of the tides.

Engineers have recently developed a new technique that harnesses the kinetic energy of tidal currents, which has less of an environmental impact compared to impounding water behind a barrage. This approach involves placing propeller-driven turbines in areas where the currents are naturally strong due to the tidal waters being funneled through narrow channels, similar to how mountain passes funnel the wind. In 2008 a British company installed the first set of commercial tidal turbines at its so-called SeaGen project, located at the entrance to a restricted bay in Northern Ireland (Figure 14.29). These turbines swivel back and forth 180 degrees so that electricity can be produced during both the incoming and outgoing tides (these turbines can also be raised for maintenance). Together the two turbines at this site are capable of producing 1.2 megawatts of power, which is enough to meet the electrical needs of over 1,000 homes. The company is also installing a much larger 10.5 megawatt tidal project off the Welsh coast and plans to begin operations by 2012.

Proponents of the British tidal power projects claim that the threat to marine life will be minimal because the propellers rotate at a relatively slow rate—10 to 15 revolutions per minute. Moreover, marine animals that are capable of swimming in areas with strong currents tend to be very agile and have great perceptive powers, which should enable them to avoid the rotating propellers. Although the risk to marine life is thought to be small,

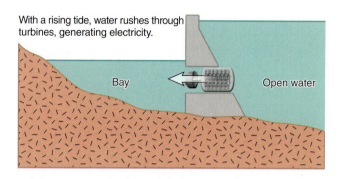

With a rising tide, water rushes through turbines, generating electricity.

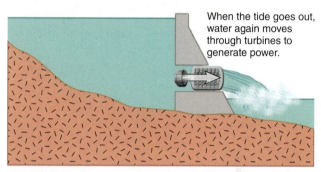

When the tide goes out, water again moves through turbines to generate power.

FIGURE 14.28   A tidal barrage is a barrier constructed across an inlet and is designed to trap the incoming high tide. Gates in the structure are closed at high tide, and then opened to allow water to fall through turbines as the tide goes out. Systems can also be set up to generate electricity with incoming tides as well.

A

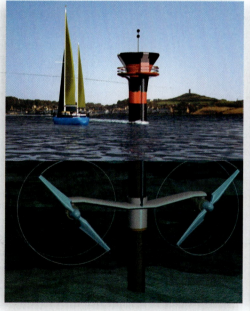

B

C

FIGURE 14.29   The first commercial tidal turbines were installed in 2008 at the entrance of Strangford Lough (A) in Northern Ireland. Here the incoming and outgoing tides are naturally forced to move through the restricted entrance to the bay, creating powerful currents that help make electrical production more cost-effective. An artist's illustration (B) of how the turbines are installed; an actual photo (C) showing the strength of the tidal current flowing past the turbine tower.

the company is funding an extensive environmental impact study led by university researchers. During the initial commissioning phase of the Sea-Gen project, the turbines will operate only during the daytime so that a full-time observer can monitor how marine mammals interact with the turbines. The environmental study also includes an underwater sonar system for monitoring the movement of seals.

Should the environmental impacts from the tidal turbines indeed prove to be minimal, it is likely that additional tidal power projects will soon follow. Because there are many sites around the world with sufficiently strong tidal currents, tidal-generated electricity has the potential for making a significant contribution to the world's energy supplies. One drawback, however, is that the production of electricity is uneven. Because the tide comes in and goes out twice each day, tidal currents reach a maximum four times in a given day. This causes the rate of electrical production to reach a peak four times a day, with production falling to zero between each peak as the tide switches from incoming to outgoing. Despite the uneven production of electricity, tidal power has an advantage over wind and solar in that the tides are consistent and highly predictable since they depend on neither the wind nor the Sun. Finally, it should be noted that researchers in Florida are testing the feasibility of underwater turbines located within the steady currents of the Gulf Stream. Although the installation of these turbines poses certain technical and logistical difficulties, the use of open-ocean currents is an attractive option because the production of electricity would be more continuous.

Conservation

An often overlooked source of energy is conservation. **Energy conservation** involves reducing energy consumption through decreased human activity, and by making more efficient use of what we do consume. An example of conservation through reduced human activity would be if you cut back on the number of trips you made driving to the grocery store. If you were to purchase a more fuel-efficient car, then you would be practicing conservation by becoming more efficient in your use of energy. Conservation is important because when a society embarks on a coordinated effort to reduce energy consumption, the result can be a dramatic increase in supply that can be used for other purposes. A classic example is the widespread energy conservation efforts that took place in response to the oil crises of the 1970s (Chapter 13). During this period people in developed nations reduced their own personal consumption by changing their driving habits and by keeping their homes at a lower temperature during the winter. Manufacturers produced more fuel-efficient vehicles and built appliances that consumed less electricity. Buildings were better insulated and engineers developed more efficient heating and cooling systems. All these efforts led to a sharp reduction in energy usage, particularly oil, which, in turn, helped alleviate the supply problem. Also important was the fact that by using less energy, individual homeowners and businesses were able to save money.

Although major gains in energy efficiency were realized after the oil crises of the 1970s, there is still much room for improvement. Consider that in 1975 the U.S. Congress passed legislation that set fuel efficiency standards for automobiles, resulting in a 40% increase in fuel efficiency by 1990. This also meant that car owners realized a 40% savings in terms of their annual fuel costs. Finally after more than 32 years, in 2007 Congress raised the fuel mileage standard from an average of 25 miles per gallon (mpg) to 35 mpg. Other efficiency improvements include compact fluorescent bulbs, which use approximately four times less electricity than standard incandescent light bulbs. Even greater amounts of energy could be saved if building codes were to require copper wiring that is only one size larger than the present standard. The thicker wire creates less resistance, which means less electricity will be lost by being converted into heat. Although the thicker wiring costs 25% more, the lower electric bills will result in a payback period of less than two years. Finally, note that all conservation measures have one thing in common: they save energy and money, both of which can be put toward more productive uses.

Post-Petroleum World

In Chapter 13 you learned that the current energy crisis is due to the natural depletion of conventional oil supplies combined with population growth and increasing per capita consumption of oil. Moreover, the growing oil shortage has the potential to cripple the global economy, with our transportation systems being the most vulnerable. For example, the transportation system in the United States is powered almost exclusively by gasoline, diesel and jet fuels, all of which are refined from conventional crude oil. While replacements for conventional oil do exist (tar sands, oil shale, coal), the amount of oil we consume is so large that it requires a rapid scale-up of production of alternative sources. Scaling-up production itself presents numerous financial and technological issues that must be addressed. Moreover, the burning of massive amounts of coal, natural gas, and oil for the past 150 years has greatly increased the carbon dioxide concentration in the atmosphere, and is now affecting Earth's climate system (Chapter 16).

Because of the critical nature of the oil and climate crises, societies around the world are moving away from the traditional use of fossil fuels and toward a new era of alternative and renewable energy resources. To successfully make it through this transition, society will need to tackle the following challenges: (1) develop a transportation system that emits a minimal amount of carbon dioxide; (2) generate electricity with non-carbon-based energy sources; and (3) scale-up the production of alternative sources to meet the current demand for energy. In the following section we will briefly examine how these goals might be achieved.

Transportation Systems

The transportation system in most developed nations generally consists of cars, trucks, trains, ships, and airplanes. Also included are heavy pieces of equipment, such as earth-movers, bulldozers, and backhoes. With the exception of jet aircraft, these machines are powered almost exclusively by internal combustion engines that burn either gasoline or diesel fuel. To transform our transportation system into one that produces a minimal amount of carbon dioxide will require some type of replacement for the internal combustion engine. At the present time the only feasible option for most applications is the electric motor—steam engines are not a viable option because they require a heat source for making steam. Although much of our transportation needs can be met with electric motors, certain applications, such as cargo ships and air travel, will need to be powered by conventional diesel and jet engines for the foreseeable future.

Electric motors are already being used in a wide variety of applications. They range in size from ones that run tiny fans on our computers to those that supply massive amounts of power to construction and mining equipment. The problem with using electric motors to power nonstationary machines is that they either must be attached to an electrical source via a transmission cable, or they must be run off a bank of batteries. The only other option is to have some means of producing electricity on the vehicle itself—examples include nuclear power plants on submarines and solar panels on satellites. Trains are perhaps the one segment of our transportation network that could most easily be converted over to electric motors. This would entail replacing diesel with electric-powered locomotives and installing transmission lines alongside existing rail beds. Because trains are one of the most efficient modes of transportation, they will likely play an increasingly important role.

Although trains will be critical in helping society meet its transportation needs, the fact that people in developed nations (particularly in the United States) tend to live in suburbs means that there will still be a need for automobiles. By making use of both a conventional engine and an electric motor, today's hybrid vehicles represent a significant improvement in fuel efficiency. However, hybrids are not a long-term solution because gasoline is still used as the primary source of power. Automobiles therefore will eventually need to be powered solely by electric motors. Electric power lines are not going to work with cars as they do for trains, so our only other option is batteries or some means of generating electricity onboard the vehicle itself. The problem with batteries has always been storing enough electricity to meet the power demands of conventional cars for any useful length of time. Because new battery technology has yielded only modest improvements over the years, the only way to increase the operating range of a vehicle has been to reduce its power demand. This has historically been accomplished by reducing the vehicle's weight. Electric cars therefore have never been popular because they must be small and still have a relatively short driving range. Despite the disadvantages, most commuters in the future will likely rely on electric vehicles for trips around town, then use trains for travel between cities. Note that batteries will not work for semi-tractor trucks and heavy equipment because their power demands are simply too great.

Another option would be for electric cars to be powered by electricity generated onboard using a *fuel cell*, thereby eliminating the battery storage issue altogether. The most promising is a **hydrogen fuel cell** that takes the chemical energy contained in hydrogen and oxygen atoms and converts it into electricity. As illustrated in Figure 14.30, hydrogen gas molecules (H_2) are stored in a tank and then broken down into hydrogen ions (H^+) along a plate called a *cathode* within the fuel cell. On the opposite side is an *anode* plate where oxygen (O_2) is drawn in from the atmosphere and broken down in oxygen ions (O^{2-}). The hydrogen ions are then allowed to pass through a membrane where they chemically combine with the oxygen ions, producing an electric current. Note that the only by-products of this reaction are electricity and water (H_2O)—fuel cells do not generate any carbon dioxide.

Nearly every major automaker in the world today is testing prototype electric vehicles powered by hydrogen fuel cells. Although the technology works, there are several technical and cost issues that must be overcome before these vehicles will ever be mass produced. One of the problems is because hydrogen molecules are so small, they tend to escape through just about any type of container. Moreover, hydrogen gas reacts with metals to produce hydrogen ions, which will penetrate the metal itself and make it brittle. An even bigger problem is that because of its low density, hydrogen gas contains far less energy than conventional gasoline. This means that for a vehicle to have a reasonable driving range, the hydrogen gas must either be highly compressed or cooled to a liquid. For example, it takes 26 gallons of hydrogen, compressed to 2,200 pounds per square inch (150 atmospheres), to generate the same amount of energy that is obtained from a single gallon of gasoline. The result is a hydrogen-powered car with a rather limited driving range. Cooling hydrogen into a liquid increases the driving range, but this process itself requires energy and therefore is less cost-effective. Similarly, the low density of hydrogen means that tanker

FIGURE 14.30   A hydrogen fuel cell (A) breaks down hydrogen (H_2) and oxygen (O_2) molecules into their respective ions (H^+ and O^{2-}), which are then combined chemically to produce electricity and water. The electricity can then be used to power a vehicle with an electric motor, thereby eliminating the need for banks of batteries. Hydrogen would be stored onboard in tanks, whereas oxygen would be drawn in from the outside air. Hydrogen fuel cell bus (B) used in a California demonstration project.

trucks can only transport relatively small volumes to future filling stations, thereby increasing the cost of the fuel.

In addition to cost and storage problems, hydrogen (H_2) gas is a secondary source of energy, which means it must be produced using energy from some other source. Most of the hydrogen today is generated by using steam to separate hydrogen from natural gas (CH_4). Unfortunately, natural gas supplies are not unlimited, plus this process generates carbon dioxide gas that must eventually be sequestered permanently, a technique which is still unproven (Chapter 16). The other method of producing hydrogen is through a process called *electrolysis,* whereby an electric current is used to split the water molecule—basically a fuel cell in reverse. This process therefore requires that electricity be generated to produce hydrogen, only to have the hydrogen converted back into electricity again, a grossly inefficient method that results in a net loss of energy.

For society to develop electricity-based transportation systems, vast amounts of electricity will need to be produced using renewable and non-carbon-based energy sources. In the next and final section we will take a brief look at how we may be able meet the future demand for electricity for all sectors of an economy.

Generating Electricity in the Future

One of the major difficulties in using electricity instead of fossil fossils is that most electricity is produced using coal and natural gas. In the United States, for example, 70% of all electricity is generated by coal and natural gas-fired power plants. Consequently, society will not only need to generate greater amounts of electrical power, but also produce progressively less of it with fossil fuels. Considering the vast amount of energy we consume, accomplishing this goal will be difficult. The solution is to make use of existing technology to ramp up electrical production on a massive scale, and from as many non-carbon sources as possible. At the same time we should reduce electrical consumption through the expanded use of solar heating and geothermal heat pumps. Equally important would be conservation and increased efficiency measures. Note that clean coal technology and carbon sequestration (Chapter 16) may make it feasible to continue generating electricity from coal, but it is not yet known whether the new technology will work on a large scale.

One of the common objections regarding the use of non-carbon and renewable methods of generating electricity is that the higher costs will harm the economy. From Table 14.1 one can see that coal and natural gas are generally the most economical means of producing electricity. However, if one includes the environmental costs associated with fossil fuels (e.g., pollution and global warming), these fuels become economically less attractive. Proposed carbon taxes and so-called cap-and-trade systems (Chapter 16) are designed to address the environmental impacts of fossil fuels and to make alternative sources more cost competitive. Also note in Table 14.1 that nuclear power is relatively expensive compared to coal and natural gas. As described earlier, the higher cost of nuclear power is one of the primary reasons U.S. utility companies stopped building new plants after 1990.

Another key aspect of electrical power is whether a particular energy source used to generate electricity is available continuously or intermittently. Because electricity cannot be stored in large quantities, utility companies are forced to generate enough electricity to meet whatever the demand is, twenty-four hours a day, seven days a week. As illustrated in Figure 14.31, electrical demand itself fluctuates, reaching a peak in the afternoon and then dropping to a minimum called *base load*. This base load always occurs at night when human activity is at its lowest. Power

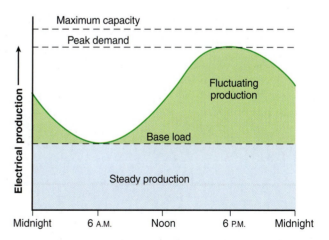

FIGURE 14.31 Electrical demand and production fluctuates on a daily cycle, which is related to human and industrial activity. Demand goes up during the day as human activity increases, then falls to a minimum or base level at night. Utility companies must provide a steady amount of power to meet base load requirements, yet maintain enough excess capacity to ensure that peak demand can be met throughout the year.

TABLE 14.1   Comparison of the different energy sources commonly used to produce electricity.

	Electrical Generation Method	Percent of World Power Demand	Cost (cents per kilowatt-hour)	Advantages and Disadvantages
Carbon-based and Nonrenewable	Natural gas	15%	3.9–4.4	Gas-fired plants are less expensive to build than coal or nuclear, but rising gas prices are expected to increase production costs. Gas burns cleaner than coal, but still produces carbon dioxide.
	Coal	38%	4.8–5.5	Coal reserves are plentiful, but coal produces far more carbon dioxide than other fossil fuels.
Non-carbon and Limited Fuel	Nuclear	24%	11.1–14.5	Risk of radioactive contamination and long-term waste disposal issues make nuclear power politically difficult in some countries. It produces large amounts of continuous power with no CO_2 emissions, and breeder reactors can create long-term fuel supplies.
Non-carbon and Renewable	Wind	1.4%	4.0–6.0	Presently wind is the most cost-effective renewable method that is readily available. However, wind is geographically restricted, intermittent, and meets some political resistance over noise and aesthetic changes to the landscape.
	Geothermal	0.23%	4.5–30	Geothermal energy is clean and renewable, but is geographically limited to areas of high heat flow through the Earth's crust. Some geothermal reservoirs become depleted if heat is removed at too fast of a rate.
	Hydro	14%	5.1–11.3	Hydro power currently contributes more energy to the global supply than all other renewable sources combined. Future potential is limited due to the limited number of dam sites and concerns over damage to aquatic ecosystems.
	Solar	0.8%	20–40	High cost of solar power has limited its use, but soon costs are expected to drop as the production of silicon increases. Although intermittent, solar power has the potential to revolutionize the generation of electricity and make distributed energy production possible.
	Tidal		2–5	New tidal turbines are cost-effective and the environmental impact appears to be low. Tidal power is geographically limited and cyclic in nature, but is highly predictable.
	Ocean thermal energy conversion		6–25	Experimental plants have shown the technology to be cost-effective, but potential commercial production is limited to the Tropics.

Source: Data from Cold Energy L.L.C. and the U.S. Dept. of Energy.

plants therefore must continuously produce enough power to meet base-load demand, and yet still have enough additional capacity to meet peak demand, each and every day. Electric utility companies generally prefer stored forms of energy, particularly fossil fuels and nuclear power, because they make it possible to adjust production to meet the constantly changing demand. In contrast, intermittent forms of energy, such as solar and wind, do not provide the same flexibility in adjusting electrical production to meet demand. Tidal power is highly reliable, but it is not continuous due to the cyclic nature of tides. Hydro and geothermal energy are stored forms of energy that can produce electricity continuously, but suitable sites are geographically limited, thus are not available everywhere. Finally, there is ocean thermal energy conversion (OTEC), which can generate electricity in a continuous manner. The problem is that OTEC is restricted to tropical waters, and large-scale commercial production has yet to occur.

Ultimately there are two basic issues that need to be resolved for the United States to replace fossil fuels as its primary means of generating electricity. The first is that the United States must finally open its disposal site for high-level radioactive waste from nuclear power plants (Chapter 15).

The waste disposal issue must be addressed before nuclear energy, which has a proven record of providing large amounts of continuous power, can be expanded significantly. The other issue is that intermittent forms of energy need to be integrated into the nation's electrical grid (collection of transmission lines) in such a way that utility companies can still meet daily peak demand. One solution would be to feed electrical power from solar, wind, and tidal sources into the grid intermittently, then use the flexibility of nuclear power plants to ensure that peak demand would always be met. Another option would be for power companies to produce excess electricity at night from wind and tidal, and then use it to produce hydrogen. The hydrogen could later be used in fuel cells to generate electricity and help meet peak demand, or to fill intermittent gaps in production due to cloudy skies or decreased winds. The hydrogen could also be sold to filling stations to power cars that run on hydrogen fuel cells.

Some energy experts believe that in the future, electricity will not be supplied solely from a relatively small number of large, centralized power plants as is the case today. As illustrated in Figure 14.32, what is envisioned instead is a more decentralized system in which electricity is produced from numerous sources distributed across the grid. Such a distributed system will likely include nuclear power plants to ensure that electrical demand is met during periods when various intermittent sources are not providing power. Although more complicated, this system is feasible because solar, wind, and tidal resources are usually intermittent at different times. For example, wind and tidal can still be generating power at night when solar is not available, whereas solar can provide energy during the day to help fill gaps in wind

FIGURE 14.32 Homes and businesses receive power from an electric grid system consisting of interconnected transmission lines. Most grids today receive power from a few centralized power plants, but as electrical production from alternative sources increases, grids will become more decentralized. Eventually rooftop photovoltaic panels (photo inset) will allow individual homes to feed excess electricity back into the grid, creating a more distributed system of power generation. Excess electricity could also be used to produce hydrogen for filling cars that run on hydrogen fuel cells.

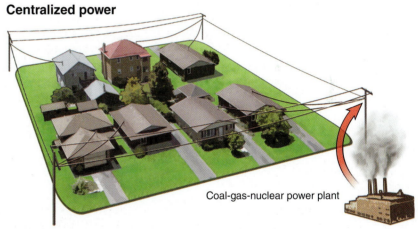

Centralized power

Coal-gas-nuclear power plant

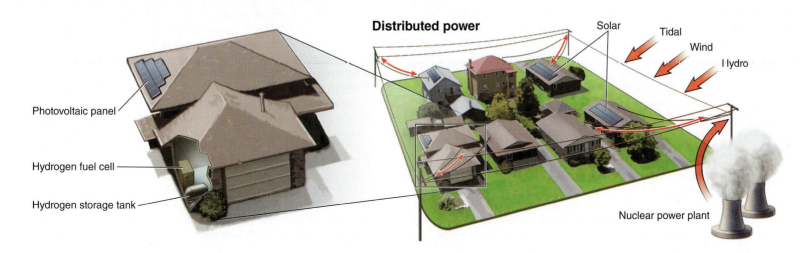

Distributed power

Solar
Tidal
Wind
Hydro

Photovoltaic panel

Hydrogen fuel cell

Hydrogen storage tank

Nuclear power plant

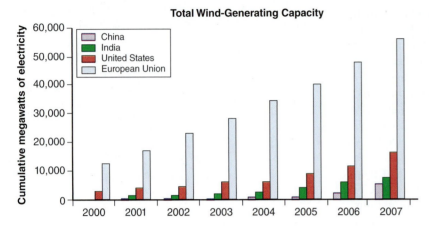

FIGURE 14.33 Graph compares the cumulative or total amount of wind power generating capacity that has been installed by the world's top energy consumers. The difference between the European and U.S. wind-generating capacity is even more dramatic when one considers that the United States consumes 25% more energy than all of Europe combined. Note that one megawatt of electrical production will meet the needs of 400 to 900 American homes, depending on climate.

and tidal production. Notice in Figure 14.32 that one of the key elements in a distributed power system is how individual home and businesses would generate electricity during the day using photovoltaic panels. Individuals therefore could generate power to meet their own electrical needs, and transfer any excess power to the electrical grid. The additional electricity could also be used to generate hydrogen gas from water, and then store it in tanks for use in a hydrogen car. The hydrogen could also run a household fuel cell to provide electricity at night.

A distributed or decentralized power system will create challenges for engineers who will need to manage the fluctuating power contributions of all the various intermittent sources. Moreover, in order for this system to be economically feasible, the cost of photovoltaic panels will need to come down. Continued advances in photovoltaic technology, combined with increased production, are expected to soon make solar-generated electricity more affordable. Ultimately though, the oil crisis and the need to make serious reductions in carbon dioxide emissions will likely force developed nations to meet more of their needs with non-carbon and renewable fuels. From the graph in Figure 14.33 one can see that European nations have already invested heavily and have taken the lead in the development of renewable energy, whereas the United States lags far behind. The basic difference has been that European governments have instituted policies and provided investment incentives that favor renewable energy. In contrast, U.S. government policies have tended to favor the use of conventional fossil fuels.

In the end, it will be necessary for both the United States and China, the world's two largest consumers of energy, to follow the lead of European nations in providing incentives for private business to invest in alternative energy sources. By committing themselves to the use of non-carbon and renewable fuels, these two countries would be making a major contribution to solving both the energy and global warming crises. In addition, such an effort would help the United States and China develop a green economy in which workers would be needed to help build and maintain new infrastructure, ranging from solar panels and wind turbines to an expanded electrical grid and rail system. Even more important, the move toward renewable fuels would be a critical step in creating a more sustainable global society.

SUMMARY POINTS

1. Tighter oil supplies and higher prices are making some nonconventional fossil fuels profitable to produce, such as synthetic fuels from coal, heavy oils, and tar sands. The problem lies in scaling-up production to the point where these fuels can make a significant impact on crude oil supplies.

2. Oil shale and gas hydrates are nonconventional fossil fuels with large potential reserves of equivalent petroleum, but are not yet being produced in significant quantities due to economic and technical reasons.

3. Biofuels from ethanol are considered renewable substitutes for gasoline, but provide only 72% of the energy as gasoline. While ethanol made from sugarcane has been shown to be cost-effective, corn-based ethanol is a net energy loss and requires government subsidies to be profitable.

4. Hydroelectric power represents a clean and renewable form of energy, but further development will likely be limited due to the restricted number of sites and the environmental impact of dams.

5. Nuclear fission is a proven means of generating large quantities of electricity in a way that does not produce carbon dioxide, but construction of new plants has slowed in some countries because of economic and safety issues and lack of long-term storage options for radioactive waste.

6. Passive solar heating involves allowing sunlight to enter a building where it naturally converts into thermal (heat) energy. Active solar heating utilizes outdoor panels to collect the heat and then transfer the energy to a fluid that circulates through the building.

7. Photovoltaic or solar cells consist of sheets of silicon alloys that convert sunlight into electricity. The use of photovoltaic panels has been limited because they are somewhat expensive and have a long payback period. This technology is also less cost-effective at higher latitudes and in areas with frequent clouds.

8. Electricity can be generated by wind using a rotor-driven generator. Wind power has an advantage over solar in that it can generate electricity at night and on cloudy days, plus it is more cost-effective and has a shorter payback period.

9. Geothermal energy can cause shallow crustal rocks to become hot enough for groundwater to reach its boiling point. When brought to the surface this water will turn to steam and can be used to drive an electrical generator. Nonboiling groundwater can be used for heating.

10. Geothermal heat pumps exchange heat energy between ground and the air space within a building. During the summer the constant temperature of the ground is used to cool a building, and in the winter it provides warmth.

11. Ocean thermal energy conversion (OTEC) is a technique that uses the difference in temperature of ocean water from different depths to operate a simple heat engine for producing electricity. Land-based OTEC systems are restricted to tropical shorelines where the seafloor drops off rapidly.

12. Tidal power can be used to generate electricity by trapping an incoming tide behind a damlike structure called a barrage. Barrage systems have fallen out of favor because they are highly disruptive to marine ecosystems. Tidal turbines have less of an environmental impact and are now being used to produce electricity on a commercial basis.

13. To address global warming and the oil crisis, society needs to: (a) develop new transportation systems that emit minimal amounts of carbon dioxide; (b) generate electricity with non-carbon fuels; and (c) scale-up the production of alternative sources to meet the current demand for energy.

KEY WORDS

alternative energy sources   436
biofuels   443
energy conservation   464
gas hydrates   442
geothermal energy   458
geothermal heat pump   460

heavy crude oil   438
hydrogen fuel cell   466
hydropower   445
light crude oil   438
nonconventional fossil
   fuels   437

nuclear fission   446
nuclear fusion   446
ocean thermal energy
   conversion   461
oil sands   439
oil shale   440

photovoltaic cell   452
solar heating   450
synthetic fuels   437
tar sands   439
tidal power   462
wind farms   455

APPLICATIONS

Student Activity

Do you think that you could go two hours without using electricity? How much electricity do you think you would save in two hours? Of course, this would be easiest to do during the day. Go through your house/apartment/dorm room and turn everything off. Computers, clocks, TV, cable, everything that consumes electricity. If there is a light still on the appliance after it is turned off, unplug it. This might take a while. It is impossible to turn off the refrigerator, so do not do it. If you have access to your electric meter, take a reading after you have unplugged everything. (Do not forget the garage door opener, lights, radio, etc.) Keep a log of how many times you try to use something electric. At the end of the two hours, take another reading of the meter if you can. How much electricity was still being used? How much do you think you saved? Could you do this every day for two hours? Was it difficult? What would happen if most of the United States did this on a regular basis?

Critical Thinking Questions

1. Is there an alternative energy resource that could be used in most of the United States? Which one would be the cheapest?

2. Currently, our economy is heavily reliant on fossil fuels. Is it a good idea to spend time and resources developing alternative energy resources and technologies instead of exploring for more oil and building more refineries? Explain your answer.

3. Canada is experiencing an oil boom in Alberta with the tar sand fields. What are some of the benefits and problems associated with this energy resource?

Your Environment: YOU Decide

NIMBY stands for "Not In My Back Yard." Would you move if a nuclear power plant was going to be built near your home? The plant would bring good jobs to the area and the additional tax revenue would boost the local school districts and services. It might even reduce individual property and local taxes. But, would the perceived risk outweigh the real benefits?

Chapter 15

Pollution and Waste Disposal

LEARNING OUTCOMES

After reading this chapter, you should be able to:

▶ Explain the difference between point source pollution and nonpoint source pollution.
▶ Describe why municipal waste is hazardous and how landfills can threaten our water supplies.
▶ Explain why most communities no longer use local landfills, but instead ship waste to distant sites.
▶ Know how hazardous waste is disposed of.
▶ Describe the basic way sewage is broken down in septic systems and wastewater treatment plants.
▶ Explain how humans can cause excessive algae growth and oxygen depletion in rivers and lakes.
▶ Describe why radioactive waste is a more difficult problem compared to other hazardous wastes.
▶ Know what gases cause acid rain and the types of human activity that generate these gases.
▶ Explain how humans are increasing mercury in the atmosphere and how it threatens human health.

Humans create enormous amounts of waste that must ultimately be disposed of in the natural environment. Our waste material oftentimes ends up polluting the very water and air that we depend on for survival. Shown here is the daily routine of trucks unloading solid waste at a large landfill located near Hong Kong, China.

Introduction

Pollution of the environment can be caused by either natural processes or human activity. Consider how volcanic eruptions and wildfires send pollutants directly into the atmosphere, and at the same time, generate loose sediment that can wash into nearby streams and cause sediment pollution. Although natural pollution can be significant, our focus will be on **anthropogenic pollution** that results from various human activities and society's consumption of natural resources. For example, our prolific use of petroleum periodically results in an accident where a tanker spills millions of gallons of crude oil, causing widespread damage to the marine environment. Consider how streams can be poisoned by runoff from the mining of coal and metallic minerals, which can be highly acidic and laden with heavy metals (Chapter 12). Anthropogenic pollution also occurs whenever we consume resources and then allow the waste by-products to enter the environment. A good example is the burning of fossil fuels in automobiles and power plants in which large quantities of unwanted gases and particulate matter are released into the air. By dumping these wastes into the atmosphere we not only make the air unhealthy to breathe, but are altering Earth's global climate system on which the entire biosphere depends.

Because anthropogenic pollution is a consequence of our human existence, it tends to increase in severity as our population continues to expand and increase its per capita consumption of resources. Exponential population growth and resource consumption results in the generation of more waste and the need to dispose of it. For example, if human waste (bodily waste or sewage) is not properly treated then we run the risk of serious waterborne diseases. In consumer societies such as the United States, people generate tremendous amounts of household trash, ranging from uneaten food to electronic equipment to a multitude of products that contain hazardous chemicals. These materials need to be properly disposed of in order to avoid contaminating our water supplies. Pollution also results from the agricultural chemicals we use to grow our food and from the hazardous waste that is generated during the production of consumer goods. Society is made up of individual consumers, therefore, each of us contributes to the problem of pollution. In Chapter 15 we will focus on how society can minimize pollution by encouraging citizens and industry to use resources more efficiently, and to properly dispose of their waste materials.

Historical Waste Disposal

For most of history Earth's human population was relatively small and the biosphere could easily process or absorb the amount of waste we generated. As people began living in densely packed towns and cities, however, the combined volume of human and animal waste became so large that it could no longer be processed by the surrounding biosphere. In most cases the waste was put out onto streets or in ditches, where it either slowly soaked into the ground or was washed into nearby rivers during heavy rains. The stench from the waste was quite powerful and represented one of the first forms of air pollution. Far worse was the fact that the waste would commonly infiltrate the same shallow aquifers that people were using as a water supply. This caused the spread of deadly diseases such as cholera. To remove both the stench and threat of disease, cities eventually began installing sewage systems consisting of underground pipes for collecting the human waste. The pipes would then transport the raw sewage to a nearby body of water where the waste was diluted, thereby reducing the risk of disease. Despite the fact that rivers historically have been used as a water supply, diluting human waste in rivers was the most common form of sewage treatment in the United States up until the 1920s and 1930s. Today the direct discharge of sewage into surface waters remains a common practice in many parts of world (Figure 15.1).

During the Industrial Revolution, factories routinely disposed of their liquid wastes in the same manner cities handled sewage, namely by direct

FIGURE 15.1 Human waste has historically been disposed of in open sewers that drain directly into nearby streams, where the only treatment consists of dilution. Open sewers such as the one shown here in Lucknow, India, are still common in many parts of the world.

discharge into bodies of water where the only treatment was dilution. Similarly, combustion by-products from industrial processes were discharged directly into the atmosphere. After World War II, the combination of direct-waste discharge and rapid industrialization in the United States resulted in air and water pollution that is almost unimaginable by today's standards. As indicated by the photos in Figure 15.2, pollution had become so bad that people could not help but take notice. During the 1960s scientists began measuring the amount of toxic chemicals in the environment and were able to show that toxins were moving up the food chain and into humans. The most influential of this early research was that of a biologist named Rachel Carson. In 1962 Carson's book, *Silent Spring*, documented how the indiscriminant use of commercial pesticides was impacting bird populations, and also warned of the potential health threat to humans. Her book revolutionized public opinion and initiated a grassroots movement in the United States. Citizens soon began demanding that the federal government begin controlling the discharge of chemicals into the environment.

Another defining moment came in 1969 when the Cuyahoga River in Cleveland, Ohio, caught fire after a spark ignited petrochemicals floating on the water (Figure 15.2B). Interestingly, this was not the first time the river caught fire nor was it the worst. Fires had been reported on the Cuyahoga in 1936, 1941, 1948, and 1952, plus rivers in other cities also had a history of catching fire. What made the 1969 Cuyahoga event so important was that it happened at a time when the public was demanding that its political leadership do something about pollution. The fact that America's rivers were catching fire was now a national embarrassment that most people would no longer tolerate. The Cuyahoga River then became a symbol of the nation's environmental neglect, and helped prompt the U.S. government into regulating the discharge of pollutants into the environment.

U.S. Environmental Laws

One of the initial steps the U.S. government took in regulating the discharge of waste into the environment was to pass the **National Environmental Policy Act (NEPA)** in 1969. Because NEPA put the government's commitment to protecting the environment into law, it is useful to read the stated purpose of the act:

> To declare a national policy which will encourage productive and enjoyable harmony between man and his environment; to promote efforts

FIGURE 15.2   Rapid industrialization in the United States resulted in dramatic increases in pollution as wastes were freely discharged into the nation's air and water. Pollution eventually became such a problem that citizens demanded action, prompting Congress to enact a series of environmental laws beginning in the 1970s. Photo (A) taken in 1947 showing unchecked industrial emissions in Tulsa, Oklahoma. In 1952 the Cuyahoga River (B) in Cleveland, Ohio, caught fire due to the industrial discharge of petrochemicals.

A

B

which will prevent or eliminate damage to the environment and biosphere and stimulate the health and welfare of man; to enrich the understanding of the ecological systems and natural resources important to the Nation; and to establish a Council on Environmental Quality.

Note that the emphasis is on protecting the environment for the benefit of *people* and the nation as a whole. The purpose of NEPA runs counter to the claim of anti-environmentalists that U.S. environmental regulations protect plants and animals at the expense of humans. Although NEPA is basically just a policy statement, it was crucial in establishing a national framework for protecting the environment. For example, it requires that all federal agencies consider environmental protection before undertaking any action that may significantly impact the environment. NEPA also created the Council on Environmental Quality, which eventually led to the establishment of the *Environmental Protection Agency (EPA)* by President Nixon in 1970. In addition to performing environmental research, the basic mission of the EPA is to set and enforce environmental standards.

After NEPA, the next major federal law resulting from the environmental movement was the **Clean Air Act** of 1970. This act was significant because it required the EPA to set minimum levels of air quality that cities must meet, and set emission standards for sources that emit hazardous substances, including automobiles. For the first time, the federal government was dictating how clean the nation's air should be and was setting limits on the emission of pollutants. The federal government soon began regulating water quality and the discharge of pollutants into surface waters with passage of the Water Pollution Control Act of 1972, better known as the **Clean Water Act.** Although this act did not include groundwater, it did make it illegal to discharge any pollutant from a discrete source into rivers and lakes. The Clean Water Act also provided funding assistance to communities for the construction of sewage treatment plants. Because of this funding, the discharge of raw sewage into the nation's waterways was largely eliminated.

Both the Clean Air and Clean Water acts have been amended over the years in order take into account changing needs as well as scientific and technological advances. Additional laws have been passed that are designed to deal with other pollution-related issues. For example, in 1980 Congress enacted the *Comprehensive Environmental Response, Compensation, and Liability Act (CERCLA)* for regulating the release, or potential release, of pollutants from hazardous waste sites. One of CERCLA's key provisions is that it holds those parties that created these sites to be liable for the cleanup costs. The problem was that many toxic waste sites around the country had been left abandoned, which meant the responsible parties were no longer in existence. In other cases the parties were still in business, but did not have the resources to pay for the cleanup. Rather than using taxpayer's money to remediate these so-called *orphaned sites,* CERCLA created a trust fund known as **Superfund,** whose primary source of revenue was an excise tax on the petroleum and chemical industries.

Although Superfund made it possible to clean up 966 of the nation's worst toxic waste sites as of 2005, nearly 600 sites still remain on the EPA list. Despite the fact that it has resulted in the successful cleanup of a large number of hazardous sites, Superfund remains highly controversial. Chemical and petroleum companies have long felt it was unfair for them to pay the cleanup costs for both their own hazardous sites and the orphan sites from other companies. In 1995 a sympathetic Congress refused to renew the Superfund tax, which allowed the reserve fund to slowly dwindle until it was finally exhausted in 2003. Today taxpayers incur 100% of the costs required to clean up orphaned sites for which there are no responsible parties. Many environmentalists now fear that most of the remaining orphaned sites will never be remediated due to federal budget reductions.

Another key federal regulation is the *Resource Conservation and Recovery Act (RCRA)* of 1976, which gave the EPA the authority to regulate hazardous materials throughout their entire life cycle, including generation, transportation, treatment, storage, and disposal. As a consequence, RCRA is often said to regulate hazardous waste from "cradle to grave." Note that CERCLA focuses on abandoned and historical waste sites, whereas RCRA regulates the handling of hazardous materials before they are discarded as waste.

A key result of these regulations is that the federal government exerted its authority over individual and states' rights in order to protect citizens and the nation as a whole. Although many people disagreed with this intrusion, the essential issue came down to whether people have the inherent right to clean air and water, or do individuals and businesses have the right to freely dump wastes into the environment. This issue of personal rights has led to a backlash against the environmental movement in recent years, led largely by well-funded industry and political action groups. During this debate people often forget that prior to these regulations, the states were unable to control pollution on their own and it took the power of the federal government to reign in polluters. Today the most visible and obvious forms of pollution have largely been eliminated, leaving more subtle but still dangerous forms of pollution. Because pollution in the United States is now largely invisible, citizens tend to believe it is no longer a problem.

Ironically, people in the rapidly industrializing countries of Asia are now experiencing levels of pollution (Figure 15.3) once found in the United States. It seems reasonable to expect that these people will eventually rise up and demand that pollution be brought under control just as Americans did 30 years ago.

FIGURE 15.3   Rapid industrialization throughout Asia is resulting in unrestrained pollution similar to what took place in the United States prior to the environmental movement. In time, citizens of these countries will likely demand that pollution be brought under control just as people did in Western nations. Shown here is pollution from industrial plants in Anshan, China.

Pollution and Contamination

Throughout most of history, humans have drunk freshwater directly from rivers and lakes, which naturally contain various types of compounds and dissolved ions from the weathering of rocks as well as bacteria. However, if the concentration of ions (salts) or bacteria becomes too high, it can have an adverse impact on human health. For example, *freshwater* is defined as containing less than 1,000 milligrams per liter (mg/l) of dissolved salts, whereas the EPA has set 500 mg/l as the safe limit for human consumption. Drinking ocean water can be fatal for humans because it averages around 35,000 mg/l dissolved salts, which is 70 times greater than the EPA's drinking-water limit.

In this chapter we will use the term **polluted** to describe those situations where a substance has been introduced into the air or water and has reached concentrations that are harmful to living organisms. In contrast, *contaminated* is when the concentration of a substance is above natural levels, but is not harmful. Although the two terms are technically different, they are often used synonymously because most people incorrectly assume anything that is contaminated must also be unsafe. Imagine you poured some milk in a glass of water. The water would technically be contaminated since milk is not naturally found in water, but yet the mixture would be perfectly safe to drink. It is also important to note that pollution is a relative term that varies from organism to organism. For example, the drinking limit of dissolved salts for humans is 500 mg/l, whereas most farm livestock can tolerate up to 5,000 mg/l. This means water with a salt content of 1,000 mg/l would be considered polluted for human usage, but yet safe for livestock. Moreover, some substances may be harmful to one species and not others. Elevated levels of copper and zinc, for example, are deadly to certain types of fish, but relatively harmless to humans.

One of the difficulties with respect to pollution is identifying those substances that are harmful to people. With the advent of the Industrial Revolution there has been a wide range of organic and inorganic substances making their way to our rivers and aquifers from factories and agricultural fields. Determining which of these chemicals is causing health problems in humans is complicated by the fact that over 75,000 chemicals are currently being manufactured in the United States alone, and over 1,000 new ones are being developed each year. Obviously, performing exhaustive and expensive laboratory studies on such a large number of chemicals is simply not feasible. Although drugs undergo considerable testing in the United States, government agencies have no health data on nearly 70% of all of chemicals entering the environment.

Because direct tests on humans involve ethical considerations and are very time consuming and expensive, scientists must extrapolate safe exposure levels for people based on animal testing. Mice or rats are commonly used because they are inexpensive and somewhat similar physiologically to humans (Figure 15.4). Different populations of mice are usually given unrealistically large doses of a substance, representing far more than what a human would likely ingest over a long time period at much lower concentrations. Based on the number of mice that develop a disease at different doses, scientists then extrapolate a concentration which people should be able to tolerate without developing the disease. A critical assumption is that a high dose over a short time period gives the same result as a low dose over a longer period. Moreover, mice and humans are assumed to have similar tolerances and respond in a similar manner to a given chemical. These assumptions obviously do not always hold, but given the limited amounts of time and money, such testing is the best method for setting "safe" exposure limits. Other types of studies examine unusually high rates of a specific disease among groups of people all exposed to the same substance, then make comparisons to rates found in the general population.

FIGURE 15.4   Mice are commonly used to extrapolate the health risk of chemical substances to humans. A major assumption is that mice and people respond similarly to the same substance. Although the extrapolated data may not accurately reflect long-term exposure risks to humans, such testing is the most practical and ethical means available for gathering quantitative information on chemicals with a high-risk potential for humans.

In the United States the EPA has been given the authority to determine which chemical substances to regulate, and then set standards for those deemed most likely to pose a risk to human health. In setting standards for a regulated contaminant, the EPA defines **Maximum Contaminant Levels** (MCLs) as the maximum allowable concentration for water supply systems providing drinking water to the public. The EPA has established MCLs for approximately 90 different chemical, microbial, and radiological contaminants. A representative sampling of these contaminants is provided in Table 15.1. Note in the table how contaminants come from a variety of sources and create specific types of health issues.

TABLE 15.1 The U.S. EPA sets maximum contaminant levels (MCLs) in drinking water for substances known to be detrimental to human health. Below is a small selection of contaminants regulated by the EPA.

	Contaminant	Treatment or MCL (mg/l)	Potential Health Effects from Ingestion of Water	Sources of Contaminant in Drinking Water
Micro-organisms	Cryptosporidium	Treatment only	Gastrointestinal illness (e.g., diarrhea, vomiting, cramps)	Human and fecal animal waste
	Viruses	Treatment only	Gastrointestinal illness (e.g., diarrhea, vomiting, cramps)	Human and fecal animal waste
Disinfection By-products	Chlorite	1.0	Anemia; infants and young children: nervous system effects	By-product of drinking-water disinfection
	Trihalomethanes	0.1	Liver, kidney or central nervous system problems; increased risk of cancer	By-product of drinking-water disinfection
Inorganic Chemicals	Arsenic	0.01	Skin damage or problems with circulatory systems, and may increase risk of cancer	Erosion of natural deposits; runoff from orchards, runoff from glass and electronics production wastes
	Cadmium	0.005	Kidney damage	Corrosion of galvanized pipes; erosion of natural deposits; discharge from metal refineries; runoff from waste batteries and paints
	Fluoride	4.0	Bone disease, children may get mottled teeth	Water additive which promotes strong teeth; erosion of natural deposits; discharge from fertilizer and aluminum factories
	Mercury	0.002	Kidney damage	Erosion of natural deposits; discharge from refineries and factories; runoff from landfills and croplands
	Nitrate-Nitrogen	10.0	Infants below the age of six months could become seriously ill and, if untreated, may die. Symptoms include shortness of breath and blue-baby syndrome.	Runoff from fertilizer use; leaching from septic tanks, sewage; erosion of natural deposits
Organic Chemicals	Atrazine	0.003	Cardiovascular system or reproductive problems	Runoff from herbicide used on row crops
	Benzene	0.005	Anemia; decrease in blood platelets; increased risk of cancer	Discharge from factories; leaching from gas storage tanks and landfills
	1,2-Dibromo-3-chloropropane	0.0002	Reproductive difficulties; increased risk of cancer	Runoff/leaching from soil fumigant used on soybeans, cotton, pineapples, and orchards
	Dichloromethane	0.005	Liver problems; increased risk of cancer	Discharge from drug and chemical factories
	Dioxin	0.00000003	Reproductive difficulties; increased risk of cancer	Emissions from waste incineration and other combustion; discharge from chemical factories
	Polychlorinated biphenyls (PCBs)	0.0005	Skin changes; thymus gland problems; immune deficiencies; reproductive or nervous system difficulties; increased risk of cancer	Runoff from landfills; discharge of waste chemicals
	Vinyl chloride	0.002	Increased risk of cancer	Leaching from PVC pipes; discharge from plastic factories

Source: U.S. EPA.

Also notice how some contaminants, such as fluoride, can be found at relatively high concentrations before posing a health risk, whereas dioxin is potentially deadly at exceedingly low concentrations.

Many of the contaminants regulated by the EPA are particularly worrisome because they impart no noticeable taste to the water and are potentially quite harmful at relatively low concentrations. Water that appears quite clean can actually be contaminated to the point where it is detrimental to human health. Moreover, scientists continue to be concerned because of the wide range of chemicals being introduced into the environment for which there is little health data. In the following sections we will examine the types of human activity from which various contaminants originate and how they enter our surface water and groundwater supplies.

Movement of Pollutants in Water

As illustrated in Figure 15.5, pollutants generally enter our water supplies in one of two ways. One is through a **point source,** which is a physically discrete point, such as a pipe discharging wastewater from a factory.

FIGURE 15.5 Pollutants entering our water supplies come from two types of sources. Point sources are discrete locations such as pipes discharging wastewater into streams from factories and sewage plants. Nonpoint sources are those where contaminants get picked up by water as it moves across the landscape and then enter a network of stream channels. Examples of nonpoint sources include chemicals and animal wastes from agricultural fields and residential lawns, and oil from parking lots and highways.

Because point sources are easy to identify, the EPA has been able to effectively reduce water pollution by requiring wastewater be treated prior to being discharged into the environment. **Nonpoint sources,** on the other hand, are those which release contaminants over a broad area and commonly consist of multiple sites. Examples include agricultural fields, parking lots, golf courses, and lawns. Here overland flow (Chapter 8) carries both natural and human-made contaminants off the landscape and into nearby stream tributaries. At the same time, infiltrating water transports the pollutants down into groundwater systems. Note that nonpoint source pollution has been more difficult to control because of the way the various sources are dispersed across the landscape, making it the leading cause of water-quality problems today.

When a pollutant enters an aquifer or open body of water it naturally begins to mix such that it disperses and becomes less concentrated, forming what is called a *contaminant plume* (Figure 15.6). In the case of a surface water body, concentration differences will cause a plume to grow by *molecular diffusion,* whereas the moving water molecules will cause it to spread by what is called *advection*. Both diffusion and advection also take place within an aquifer, but here the spreading is enhanced due to the fact water must flow around the individual grains making up the aquifer. In both surface and groundwater contamination, plumes are also affected by the way in which the pollutants are released. For example, a leaking underground storage tank may release pollutants in a continuous manner over a long period of time, whereas a tanker spill would release a large volume almost instantaneously.

Once water becomes contaminated, the natural flow of the water will eventually flush contaminants from the system. In rivers and lakes this natural cleansing action can take place on the order of days to years, whereas the sediment can remain contaminated for far longer. Because groundwater moves much more slowly than surface water (Chapter 10), it can take anywhere from decades to thousands of years for an aquifer to flush itself of contaminants. This means that once an aquifer is contaminated, it can no longer be used as a drinking-water source for the foreseeable future. Although there are ways of remediating or reducing the concentrations of pollutants in groundwater, the low flow rates and granular nature of most aquifers makes this exceedingly

FIGURE 15.6   A contaminant plume forms when polluted water enters uncontaminated surface water bodies like lakes and streams (A) or subsurface aquifers (B). In either case contaminants become diluted due to mixing when the plume grows in size, both laterally and in the direction of flow. Once a contaminant source has been eliminated, surface waters are generally able to flush themselves of pollutants in a relatively short time period. Groundwater moves much more slowly and can easily remain polluted longer than a human life span.

A Calumet River, near Chicago, Illinois

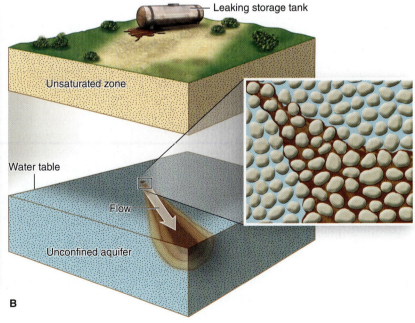

B

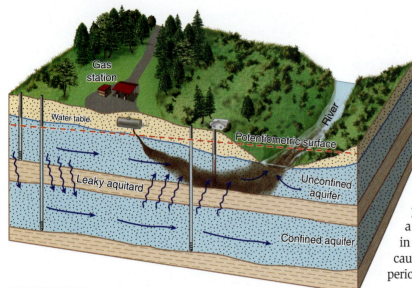

FIGURE 15.7   Groundwater pollution not only contaminates aquifers, but will eventually discharge into streams, polluting them as well. Unconfined aquifers are always highly susceptible to contamination from surface activities, whereas aquitards provide a critical layer of protection for confined systems. Note that in areas where vertical flow is upward, confined aquifers are largely immune from contamination by surface activities.

difficult—analogous to getting soap out of a sponge. It is far cheaper to prevent groundwater contamination than cleaning it up after the fact.

Finally, from Figure 15.7 we can see that unconfined aquifers are highly susceptible to contamination from surface activities, whereas aquitards (Chapter 10) provide a critical layer of protection for confined aquifers. Notice that in areas where vertical leakage across an aquitard is upward, confined systems are virtually immune from being contaminated by the overlying unconfined aquifer. Another key point here is that a contaminant plume not only can make water-supply wells unsafe, but will eventually discharge into a stream or lake, polluting it as well. Note that contaminated runoff normally enters a stream during overland flow events, whereas groundwater enters in a more continuous manner. Polluted groundwater therefore can cause serious water quality problems in streams, particularly during periods of low stream flow when dilution is minimal.

Solid Waste Disposal

If you took a survey asking people in the United States what kind of solid waste they generate, common responses would likely be things like cardboard and plastic containers, beverage cans and bottles, and of course, paper. Some may even remember to include discarded electronic items, tires, batteries, furniture, and so on. However, few people would think to include all the solid waste they are indirectly responsible for, such as the industrial waste that is created when making all those consumer products we buy. Take your car or truck, for example, where tremendous amounts of rock are removed from the earth in order to extract a relatively small amount of the necessary metals; the rest is discarded as waste. Each of us is also responsible for the waste that is generated to produce the food we eat, particularly meat. Here farmers grow crops to feed cows, chickens, and pigs, which then transform the feed into large volumes of manure. Although we generally do not see industrial and agricultural wastes, we nevertheless are responsible for creating it.

The more familiar household garbage or trash that is typically collected by local governments is referred to as **municipal solid waste**—this also includes trash from local businesses. The volume of municipal waste is surprisingly small, making up only 4% of all solid waste, whereas agricultural and mining wastes represent an astonishing 56% and 34%, respectively, and industrial waste around 6%. Interestingly, only municipal and industrial wastes are collected and disposed of in landfills designed to keep hazardous substances from seeping into the environment. This means the vast majority of the waste from mining and agricultural activity is simply placed on the land surface. The problem is that much of this waste contains heavy metals or nutrients that can make their way into streams and aquifers, polluting water supplies and damaging ecosystems. In the following section we will take a closer look at how society handles different forms of solid waste and how this impacts our water supplies. Because solid industrial waste is generally nonhazardous and ends up in regular landfills, it will be lumped in with municipal waste in our discussion.

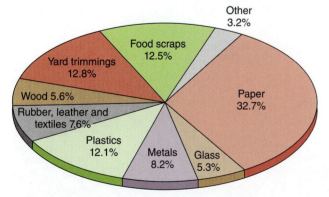

FIGURE 15.8   The stream of municipal solid waste generated in the United States can be broken down in several major categories, shown here as a weight percentage. Paper is by far the largest part of our waste stream. Note that due to recycling efforts much of this material is removed from the waste stream and is not buried in a landfill.

Municipal and Industrial Solid Waste

The term *solid waste stream* is often used to describe the steady flow of waste that leaves our homes and businesses, and which society has to dispose of. From Figure 15.8 we can see that in terms of total weight, paper

FIGURE 15.9 Prior to 1976, municipal waste in the United States was routinely disposed of in open dumps. Health concerns eventually led to regulations requiring the use of sanitary landfills. Shown here is an open dump in Alaska, date unknown.

products make up the largest portion of our solid waste stream. Of course not all of this material ends up in a landfill, as much of it is either recycled or composted. Many people today are unaware that the disposal of solid waste in the United States has dramatically changed since the start of the environmental movement and onset of EPA regulations. Prior to the regulations municipal governments commonly placed local garbage in open dumps (Figure 15.9), which were often located in nearby ravines, wetlands, and abandoned gravel pits. However, the open nature of these sites naturally attracted birds, rodents, and a variety of other pests; in addition, the dumps would periodically catch fire, sending noxious fumes into the surrounding air. Because open dumps posed a health threat to nearby residents, as early as the 1930s engineers developed what are known as **sanitary landfills,** where the trash is compacted by heavy equipment and then covered each day with a layer of dirt (Figure 15.10). Finally by 1983, open dumps were no longer allowed as the EPA began requiring that all municipal waste be placed in sanitary landfills. These landfills effectively isolate the trash from atmospheric oxygen (O_2), so decomposition rates are extremely low because of the lack of oxygen-dependent (aerobic) bacteria. Landfills therefore are more like big preservation pits than compost piles.

Sanitary landfills were a big improvement over open dumps because the trash could no longer attract animals, catch on fire, or foul the air. Figure 15.10 shows that water can infiltrate through the trash in sanitary landfills, leaving shallow aquifers highly susceptible to contamination. Rainwater percolating into the landfill is able to pick up bacteria, viruses, and various chemical compounds from the trash, forming a liquid referred to as **leachate**—similar to what would drain from a garbage can full of trash that had been allowed to soak in water for weeks. The contamination of shallow aquifers by leachate is made worse by the fact that municipal trash commonly contains *hazardous household wastes,* including items such as batteries, paint, solvents, and insecticides. In the presence of water, hazardous compounds are leached from these materials and carried into the groundwater system.

To prevent the escape of leachate, the EPA eventually developed new standards for what are called **containment landfills,** which are designed to minimize the formation of leachate and collect what does form. As

FIGURE 15.10 Sanitary landfills are excavated pits where heavy equipment is used to compact and cover the trash with dirt on a daily basis. Although an improvement over open dumps, such landfills do not prevent the escape of infiltrating water that interacts with the trash to form leachate. The escape of leachate from landfills has led to widespread groundwater pollution.

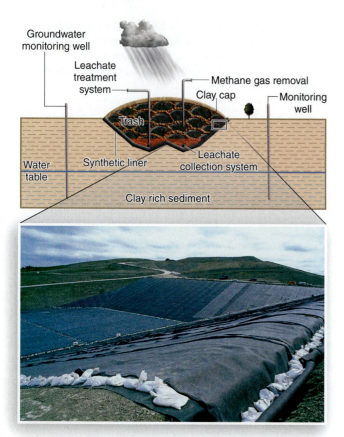

Groundwater monitoring well

Leachate treatment system

Methane gas removal

Clay cap

Monitoring well

Trash

Water table

Synthetic liner

Leachate collection system

Clay rich sediment

FIGURE 15.11 To prevent leachate from escaping, modern containment landfills use a compacted clay base overlain by an impermeable synthetic liner. For additional protection, landfills are typically located in areas of clay-rich sediment and surrounded by monitoring wells for detecting potential leaks. Clay caps are also used to minimize infiltration, whereas drainage pipes allow for the collection and treatment of leachate. Methane gas is also recovered and can be used for heating or generating electricity.

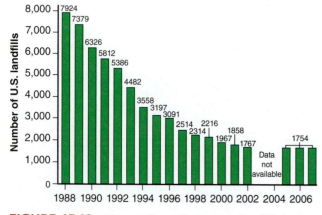

FIGURE 15.12 The number of operating landfills in the United States declined 78% after 1988 as new EPA design standards forced many landfills to shut down. Compounding the shortage of landfills has been citizen opposition to new landfills. This, in turn, greatly increased the cost of solid waste disposal as the remaining landfills are able to charge higher prices for their limited space.

shown in Figure 15.11, modern containment landfills are typically sited in areas of clay-rich material whose low permeability will limit the movement of leachate should the primary containment system fail. Here the primary containment system consists of a layer of compacted clay at the base of the excavation, and overlain with a synthetic liner. Once the landfill reaches its maximum capacity, it is covered with a clay cap and a final layer of topsoil. The clay cap reduces the infiltration of rainwater, thereby minimizing the formation of leachate—the vegetation helps reduce erosion so the cap stays in place. The leachate that does form will enter a system of buried pipes and then be brought to the surface. Here the leachate is either treated and discharged into the environment, or it is disposed of as hazardous waste. In addition to collecting leachate, the EPA also requires a series of groundwater monitoring wells surrounding the landfill, so officials can detect potential leaks. Because methane (natural) gas forms as the trash slowly decomposes, landfills are also required to have a system for collecting the gas. This prevents a dangerous buildup of the highly flammable gas, and provides an energy source for heating or generating electricity.

Although modern containment landfills have dramatically reduced the potential for groundwater contamination, thousands of older landfills still exist in the United States that were built prior to the newer EPA standards. Most of these older landfills were forced to shut down because they could not be upgraded to meet the new standards, and new landfills are more expensive to build and operate. This has resulted in a 78% reduction in the number of active landfills since 1988 (Figure 15.12). Compounding the problem of fewer landfills is the public opposition that invariably arises when new landfills are proposed. Understandably, people object over fears of groundwater contamination and lower property values. Although everyone wants to get rid of their trash, almost no one wants a landfill in their backyard—hence the expression NIMBY for "not in my backyard." This has resulted in a relatively small number of landfills that service a much broader area than in the past. Consequently, most municipalities have been forced to build *waste-transfer stations,* which is where local trash is pressed into bales and then loaded onto semi-tractor trailers for transport to distant landfills.

Another important consequence of the dramatic decline in the number of U.S. landfills is that waste disposal is now more expensive. Companies that own landfills now have something that is in short supply, namely available space; hence they can command a higher price. The solid waste business today is quite lucrative, which helps explain the seemingly strange practice of trucking garbage hundreds of miles just to bury it in a landfill. Because of the increased cost, many municipalities have taken steps to reduce the amount of trash leaving their waste-transfer stations. The basic approach utilizes the concept of a **waste management pyramid,** which is a strategy for minimizing the amount of solid waste being sent to a landfill. As illustrated in Figure 15.13, the most desirable method is *source reduction,* which involves efforts designed to keep material from entering the waste stream in the first place. For example, many municipalities no longer allow yard wastes to be mixed in with regular household trash, but instead have a separate curbside pickup for yard wastes. The next most desirable method in the pyramid is *recycling* and *reuse.* Recycling involves removing things from the waste stream that can be used again to make new products, such as paper, aluminum and glass containers, and various types of plastics and metals. Reuse entails using items like glass cups as opposed to disposable foam cups, and also repairing items or donating them to a charity rather than simply throwing them away.

The last option in the waste management pyramid before landfilling is *incineration,* where combustible materials are removed from the waste stream and then burned at a high temperature. In the past, various facilities across the United States, including hospitals, schools, and homeowners, routinely used incineration as a means of waste disposal. However, many types of materials in the solid waste stream create dangerous by-products during combustion. The combustion of bleached (white) paper is a particular problem as it is pervasive in trash and creates the extremely hazardous compound known as dioxin (Table 15.1). Because of the air pollution issue, the EPA now requires operators to install air pollution controls to meet certain emission standards. Consequently, the incineration of municipal waste today is largely performed in commercial operations in which recyclables are recovered before the remaining waste is shredded for incineration. Some operations are also equipped to capture the heat from incinerated waste to generate electricity, thus turning trash into valuable energy. Because burning trash generates carbon dioxide (CO_2), concern over global warming (Chapter 16) makes recycling and reuse far more desirable than incineration for reducing our volume of solid waste.

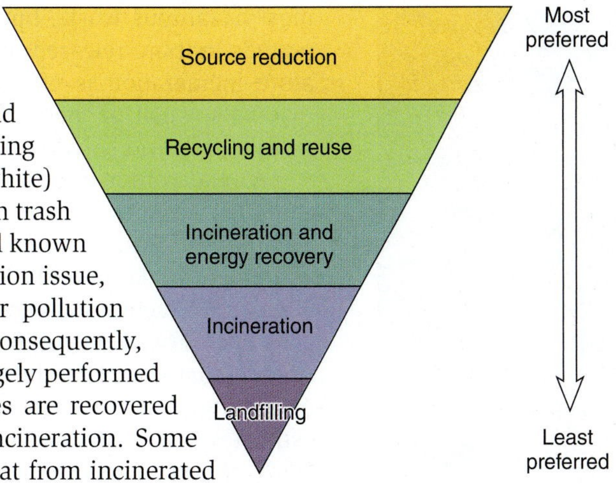

To give you some idea of the success of these programs, consider that the recycling rate in the United States went from 6.5% in 1960 to 33.4% in 2007. From Figure 15.14 we can see that the combined efforts of recycling, composting yard waste, and incineration have led to a reduction in the amount of solid waste being sent to landfills. Note that this reduction occurred despite the fact the total amount of waste being generated increased due to population growth and higher consumption rates. Although these efforts have been successful, the recycling rates shown in Figure 15.15 indicate that there is room for improvement as considerable amounts of recyclable materials are still making their way into landfills.

Solid Hazardous Waste

Earlier you learned that the EPA regulates a wide range of hazardous materials in a "cradle to grave" manner via RCRA. Industry is required to keep records on the handling of all hazardous materials, including documentation on how and where it was disposed. However, regular household trash is exempt from RCRA, which means citizens are allowed to place small amounts of hazardous household items into the municipal waste stream. Hazardous household material therefore ends up in a typical landfill. Industrial hazardous wastes, however, can only be disposed of in **secured landfills** that have special systems for collecting leachate and detecting leaks. Waste shipments require tracking permits and must be placed in separate containers and specific locations within the landfill. Keep in mind that despite the greater safeguards, even secure landfills will at some point begin to leak and allow leachate to escape into the environment.

In addition to secured landfills, solid hazardous waste can be disposed of in special high-temperature incinerators, commonly referred to as *hazardous-waste incinerators*. First the waste is shredded and sent into a primary combustion chamber, where the resulting ash and water are collected at the bottom. Gases and unburned particulate matter then enter a secondary chamber and are burned at an even higher temperature. The remaining gases pass through a *scrubber system,* which essentially removes everything but carbon dioxide (CO_2) gas and water vapor. However, scrubber systems are not completely efficient, so some hazardous material escapes into the atmosphere. In addition, the remaining ash is still a hazardous solid and must be disposed of in a secured landfill. Opponents of high-temperature incineration point out that incineration does not destroy

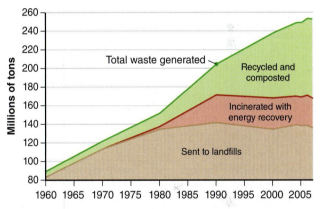

FIGURE 15.13   A waste management pyramid helps describe the strategy many communities use to minimize the amount of solid waste they must send off to a landfill. The most desirable option is to keep unnecessary waste from entering the waste stream in the first place. For the waste that is received, recycling, reuse, and incineration are all ways of reducing the volume needed to be landfilled.

FIGURE 15.14   Despite an overall increase in the amount of solid waste being generated, the combined efforts of recycling, composting, and incineration have resulted in a decline in the amount being sent to landfills since 1990.

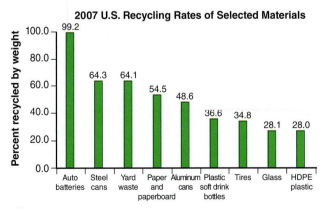

FIGURE 15.15   Percent recycling rates of selected materials in the United States in 2007. The recycling of these materials not only saves energy and valuable landfill space, but with the exception of yard wastes, also serves as a source of raw materials that can be used to make new products.

most hazardous waste, but converts it into fine ash and hazardous compounds that are released into the atmosphere. Nevertheless, high-temperature incineration is effective at destroying certain types of hazardous materials, such as pathogens in medical wastes and various types of petroleum-based solvents and oils.

A relatively new problem in solid waste management has been discarded electronic products, also called **electronic waste.** Although the EPA reports that TVs, computers, cell phones, and assorted audio and video equipment make up only about 2% of all municipal waste in the United States, the electronic components contain highly toxic substances such as lead, mercury, and cadmium. Moreover, the plastic cases and cables are usually coated with bromine-based flame retardants, which are suspected of causing a variety of health problems, including certain cancers. The EPA estimates that of all the televisions and computers sold between 1980 and 2007, approximately 235 million units have become obsolete and are in storage waiting for disposal. Figure 15.16 illustrates how only about 18% of the electronic waste is recycled in the United States. In order to reduce the influx of electronic waste going into landfills, the EPA and various state agencies have begun instituting electronic waste management programs similar to that described earlier for municipal waste.

The proliferation of hand-held electronics in recent years has also caused a sharp rise in the use of *dry-cell batteries,* which contain toxic mercury, lead, cadmium, and nickel metal. The EPA estimates that Americans purchase nearly 3 billion dry-cell batteries per year (Figure 15.17). The batteries used in automobiles and boats are referred to as *wet-cell batteries* because they contain lead plates surrounded by acid. Retailers freely exchange wet-cell batteries with the purchase of a new one, so recycling rates are now over 90% in the United States. Until recently, consumers have had few options for dry-cell battery disposal other than tossing them into the trash. Similar to electronic waste, special battery recycling programs at the state and local levels are being implemented across the United States.

A

U.S. Electronic Waste, 2006–2007

B

FIGURE 15.16 The volume of electronic waste has grown considerably in recent years (A), creating a disposal problem as many items contain toxic metals and plastics. To reduce the amount of electronic waste going to landfills, the EPA, state, and local agencies have set up reuse and recycling programs. As indicated in the plot (B), only about 18% of computers, cell phones, and televisions are currently being recycled.

FIGURE 15.17 This pile of depleted dry-cell batteries collected over one year by the author illustrates the size of the battery disposal problem. Because batteries contain heavy metals that are highly toxic, many state and local governments are instituting battery recycling programs to help remove batteries from the municipal waste stream.

Scrap Tires

There are literally hundreds of millions of cars around the world, all of which go through several sets of tires in their lifetime. In 2003 the EPA estimated that the United States alone generates nearly 290 million scrap tires each year, taking up landfill space and creating hazards in open dumps on the land surface (Figure 15.18). Due to the way tires hold rainwater, tire dumps make excellent breeding grounds for mosquito-borne diseases. These dumps also pose serious environmental hazards when the tires catch fire. The burning tires not only produce toxic smoke, but the intense heat also releases liquid petrochemicals that end up contaminating shallow aquifers and nearby streams. Today, all but two U.S. states have laws and recycling programs that deal specifically with the handling of scrap tires. Currently 35 states now require that tires be shredded prior to being placed in landfills, and 11 states ban even shredded tires in landfills. A key part of most of these programs has been developing markets for the shredded tires themselves. Today, scrap tires are used to make rubberized asphalt for running tracks, dock bumpers for boats, highway crash barriers, floor mats, soaker hoses, and more. According to the EPA, these marketing efforts have resulted in the percentage of scrap tires being recycled to increase from 25% in 1990 to over 80% by 2003. Individual state cleanup efforts have brought the number of tires in dumps down from approximately 800 million in 1994 to 275 million in 2004.

Liquid Waste Disposal

Recall from Chapter 11 that in terms of per capita water consumption, each of us shares responsibility for the water being used to manufacture products, irrigate crops, and generate electricity. What is important here is the fact that nearly all of our nondrinking applications cause the water to become contaminated to one degree or another. Scientists and engineers use the term **wastewater** to refer to water that has become contaminated during some human-related process. Perhaps the best example is how the use of flush toilets turns high-quality drinking water into wastewater, or *sewage*. Another example is the wastewater we generate when drinking water is used for processing animals that are the source of the meats we find in grocery stores.

In addition to wastewater, modern societies generate a variety of hazardous liquids, including acids, bases, and a variety of petroleum-based products, ranging from motor oil and gasoline to agricultural pesticides. Prior to the Clean Water Act, industries commonly disposed of both liquid wastes and wastewater by piping it directly into nearby streams. Because most petroleum-based solvents are *immiscible* and do not mix with water, those molecules that are denser than water will sink and the lighter ones will float. The result is that dense hydrocarbon compounds tend to accumulate in the sediment of streams and lakes, whereas the lighter compounds float on the water's surface. Today, the EPA regulates a wide variety of toxic substances, setting limits on the amounts a company can discharge along with their wastewater. In some cases the liquid waste is simply too toxic, or the volume too large, for it to be included in the wastewater. This requires that the waste either undergo a treatment process, or be disposed of as a hazardous material and permanently isolated from the environment. Compared to solid wastes, isolating liquid wastes is much more difficult because their fluid nature enables them to migrate more easily.

FIGURE 15.18   In the past, scrap tires have been either placed in landfills or stockpiled on the surface in large open dumps, such as the one shown here in Saskatchewan, Canada. Dumps create problems with mosquito-borne diseases and can produce toxic fumes should the tires catch fire.

FIGURE 15.19 Considerable amounts of liquid hazardous waste stored on-site have been left abandoned by bankrupt companies, leaving costly cleanup efforts to government agencies as shown in this photo from the State of Illinois.

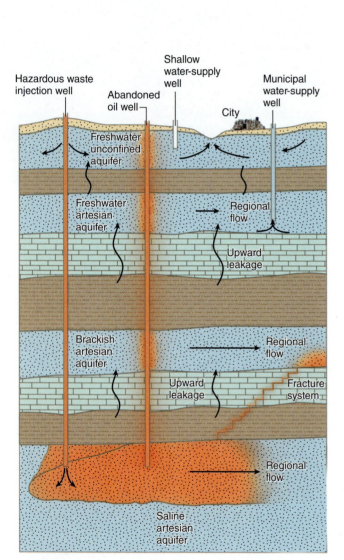

FIGURE 15.20 Deep-well injection involves disposing of liquid hazardous waste by pumping the material into deep, saline aquifers that are not used as a water supply. Shown here is a worst-case scenario where groundwater naturally flows upward, allowing the waste to contaminate freshwater aquifers via fractures and abandoned oil wells that have not been properly sealed. Contamination can also occur due to leakage along the casing of the injection well.

Liquid Hazardous Waste

Of all the industrial waste generated in the United States, the EPA estimates that 97% is in the form of wastewater. With passage of the Clean Water Act and RCRA regulations, pollutants could no longer be freely dumped into rivers and lakes. Industries were now forced to develop better management strategies for their liquid wastes. As with solid waste management, businesses learned that the most cost-effective approach is to reduce and recycle as much of their liquid waste as possible, thereby minimizing more expensive disposal options, particularly landfilling. The result has been more efficient manufacturing processes that generate far less hazardous waste. For the waste that is created, companies employ a variety of wastewater treatment processes in which harmful substances are chemically broken down or neutralized before being released into the environment. Unfortunately, wastewater treatment is not effective or practical for removing all hazardous compounds, leaving EPA-approved disposal methods as the last option. Note that companies can temporarily store wastes on-site while waiting for treatment or permanent disposal. In the past, however, this has led to leaks and the escape of hazardous liquids. Also, bankrupt companies have oftentimes simply abandoned the waste, leaving costly cleanup efforts to government agencies (Figure 15.19).

One common disposal technique is to take containers filled with liquid waste to a *secured landfill*. Another option is known as *deep-well injection* or *deep-well disposal,* whereby hazardous liquids are pumped down into a deep aquifer that is far removed from shallower aquifers that are being used as a water supply (Figure 15.20). This practice originated in Texas during the 1930s when oil companies began injecting wastewater from production wells back into petroleum reservoirs in order to maintain fluid pressure and increase oil production (Chapter 13). Eventually waste operators began using injection wells and old petroleum reservoirs to dispose of hazardous liquids from manufacturing industries. Because of its long history of oil and gas production, Texas today leads the United States in deep-well disposal.

Deep-well injection is rather controversial despite the fact abandoned petroleum reservoirs typically contain only saline groundwater, and are separated from water-supply aquifers by a thick sequence of sedimentary of rocks. The problem is that freshwater aquifers can become contaminated should the injected waste flow toward the surface along fractures, or around the steel casing of the injection well as shown in Figure 15.20. Of particular concern are abandoned oil and gas wells that have not been properly plugged and sealed. Such wells serve as open conduits and allow the waste to move quite rapidly. Proponents of deep-well injection, however, claim that such contamination is actually quite rare and that the practice is still far safer than placing hazardous liquids in a landfill at the surface.

FIGURE 15.21 In order to reduce the amount of hazardous household materials entering the municipal waste stream, some state and local governments now hold hazardous waste roundups where citizens can drop off household wastes for proper disposal by trained personnel. Photo from the State of Illinois.

The last option for disposing of hazardous liquid wastes is high-temperature incineration. Incineration is becoming a popular alternative to secured landfills and deep-well disposal since the waste material is "destroyed," thereby eliminating the potential risk of leaks associated with long-term storage. However, high-temperature incineration is controversial because it creates hazardous ash, plus the scrubber systems for removing hazardous air emissions are not 100% efficient. In the case of hazardous organic liquids derived from petroleum (Chapter 13), air emissions are less of an issue because if done properly, the only combustion by-products are carbon dioxide (CO_2) gas and water vapor.

As with solid waste, a typical household contains hazardous liquids, but these are exempt from RCRA regulations and therefore end up in the municipal waste stream and sent to a landfill. Examples of hazardous household liquids include used motor oil, insecticides, pesticides, oil-based paints, pool chemicals, and a wide variety of solvents and cleaning fluids. While the contribution from a single household is generally small, landfills are laced with hazardous liquids because nearly everyone at some point will discard unused or partially used liquid products. To help reduce the amount of hazardous liquids entering the municipal waste stream, the EPA and state agencies have developed public awareness campaigns that encourage people to buy hazardous liquids more sparingly. Also, some state and local governments now hold hazardous waste roundups in which citizens are encouraged to bring unwanted household hazardous wastes to a collection site on specific days (Figure 15.21). Trained personnel can then safely place the materials in drums for proper disposal. Note that some states and municipalities now have regulations that are more stringent than RCRA regarding hazardous household wastes. Therefore, depending on where you live, it may now be illegal to place certain household chemicals, batteries, electronics, and so forth, in the trash. Check your state or local government's website to find out where to send things in your area.

Human Waste

As late as the 1930s it was common for towns and cities in the United States to collect raw sewage (human waste mixed with water) from homes and businesses, and then discharge it directly into rivers and lakes. In rural areas people would dig a pit and place a wooden structure called an *outhouse* or *pit toilet* over the hole. Here human waste is partially broken down by *anaerobic bacteria,* which do not require oxygen, and is converted into compounds such as ammonia (NH_3). The wastewater within the pit then seeps into the ground and the remaining solids, called *sludge,* continue to build, eventually forcing the owner to dig a new pit and move the outhouse. Because of the threat of spreading serious diseases through

FIGURE 15.22 Photo showing a section of the Canoochee River in coastal Georgia that is completely covered with algae and is experiencing severe oxygen depletion. The algae here are being fed by excessive nutrients. As the algae die and fall to the bottom of the river, aerobic bacteria break down the material and remove oxygen from the water.

our water supplies, pit toilets and the discharging of raw municipal sewage are no longer acceptable practices in many countries.

Due to the type of things humans eat and drink, human waste largely consists of organic compounds rich in carbon, nitrogen, and phosphorous. However, the various types of bacteria and viruses within the waste threaten our water supplies and can lead to serious human illness and disease. In addition to the spread of disease, when *aerobic bacteria* break down organic matter in an oxygen-rich environment, the bacteria consume free oxygen (O_2) and produce carbon dioxide (CO_2). The problem is that the bacteria will keep consuming oxygen until either all the organic matter is decomposed, or the available oxygen is gone. Scientists use the term *oxygen depletion* to refer to the undesirable situation where aerobic bacteria remove enough of the dissolved oxygen from a lake or slow-moving stream that fish and other aquatic life begin to die. To help gauge the risk of oxygen depletion, biologists measure what is called *biochemical oxygen demand* or BOD, which is the amount of oxygen microorganisms would need to break down whatever organic matter is present within the water. When BOD is high, it is a sign that raw sewage may be entering a water body, creating both a risk of disease and oxygen depletion. Oxygen depletion, however, can also occur due to *excess nutrient loading,* which typically occurs in areas where agricultural fertilizers are used, followed by rains that wash nitrogen and phosphorus off into streams and lakes. The addition of these nutrients then promotes excessive algae growth (Figure 15.22), which eventually die and are broken down by aerobic bacteria. It is the excessive aerobic bacteria then that remove the oxygen from the water. Next, we will examine how dangerous pathogens and oxygen-demanding wastes are removed from wastewater before it is discharged into the environment.

Septic Systems

A septic system is a technique for treating wastewater from homes and small businesses that are not connected to municipal sewage systems. The EPA estimates that septic systems are used in 25% of all American homes. As illustrated in Figure 15.23, the system consists of two components: a *septic tank* and a *drain field.* Here wastewater is piped into the tank, which is made of concrete and sealed to prevent the waste from escaping. Most of the suspended organic matter settles out in the tank, forming a layer of sludge; the remainder either stays suspended or floats as a layer of scum. Because the tank is sealed, it contains very little oxygen, which means the organic wastes are only partially broken down by anaerobic bacteria—similar to a pit toilet. When additional wastewater enters the tank, some of the clarified wastewater will flow into the drain field where perforated pipes allow it to percolate through a bed of gravel. Because the gravel bed and underlying sediment are in the unsaturated zone, most of the remaining organic matter and harmful pathogens are either destroyed by aerobic bacteria, or filtered out as the wastewater percolates through the sediment.

Although a properly constructed and maintained septic system can last almost indefinitely, the EPA estimates that 10 to 20% of systems fail to operate properly. Perhaps the most common cause is that the drainage field is located in an area that is poorly drained. This allows the gravel bed and underlying sediment to become saturated, preventing the final aerobic breakdown of the waste. Poor drainage is usually the result of either the system being built too close to the water table, or the underlying sediment is not sufficiently permeable. A septic system can also fail when a homeowner does

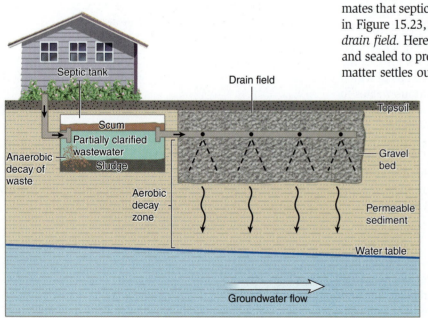

FIGURE 15.23 Septic systems are used for household wastewater in areas that are not connected to municipal sewage systems. Wastewater first goes into a holding tank where solids settle out and anaerobic bacteria break down some of the organic matter. Water and suspended waste then leave the tank and flow into a drain field and percolate through an unsaturated gravel bed where the remaining organic matter and pathogens are removed by aerobic bacteria.

not have a service company periodically pump the sludge from the tank. Over time the sludge will build up to the point where solids begin moving through the pipe that leads to the drain field (Figure 15.23). This eventually causes the drain field to become plugged with particulate matter, thereby limiting the amount of oxygen and inhibiting the aerobic breakdown of the waste.

While a properly functioning septic system is a very effective means of treating wastewater, environmental problems can still occur. Because septic systems remove very little of the nitrogen and phosphorus from the wastewater, these nutrients can end up in nearby water bodies, causing excessive algae growth and oxygen depletion. Excessive nutrient loading can be a real problem around recreational lakes, where large numbers of homes and cottages use septic systems. Another problem is that hazardous household chemicals are often incorporated in the wastewater, but are not broken down by either anaerobic or aerobic bacteria. These hazardous chemicals end up either settling out in the septic tank or percolating through the drain field, contaminating shallow aquifers and nearby streams.

Municipal Wastewater Treatment

Wastewater from homes and businesses in U.S. cities is collected by sanitary sewer lines and then piped to a centralized facility called a *wastewater treatment plant*—commonly known as a *sewage plant*. Here the wastewater typically undergoes a series of treatment processes before being discharged into a body of surface water. As shown in Figure 15.24, *primary treatment* involves using a large screen to first remove any coarse debris. The wastewater then flows into a series of chambers where solid particles are allowed to settle out, forming a layer of sludge. After this, the wastewater undergoes *secondary treatment* in aeration tanks, where fine bubbles of air move upward through the waste. With a nearly unlimited supply of oxygen, aerobic bacteria are able to flourish and convert much of the remaining organic matter into harmless solids and carbon dioxide gas. Some plants include a final step where the wastewater is disinfected with either chlorine or ozone in order to kill any remaining pathogens.

The combination of primary and secondary treatment results in approximately 99% of all harmful bacteria being removed from the wastewater along with 90% of the oxygen-demanding organic wastes. However, as with septic tanks, the treatment process removes only small amounts of

FIGURE 15.24 Modern wastewater treatment involves a series of processes to separate out solids and break down oxygen-demanding organic matter. Primary treatment involves the settling out of solids, followed by secondary treatment where aeration allows aerobic bacteria to break down most of the remaining waste. A final tertiary step involves a disinfection process to eliminate any harmful bacteria that may remain.

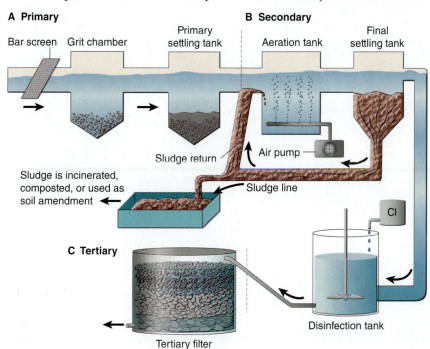

nitrogen and phosphorus. The discharge from a wastewater treatment plant therefore can lead to excessive nutrient loading and oxygen depletion. Note that primary and secondary treatment is also not very effective in removing heavy metals and many types of hazardous compounds. Consequently, some municipalities use *tertiary* or *advanced treatment* techniques that can remove nutrients and certain types of hazardous compounds.

Although wastewater treatment systems in the United States have been highly effective in eliminating waterborne disease, they also produce an enormous amount of sludge that must be disposed of. Sewage sludge has historically been applied to agricultural fields, as it contains many valuable nutrients necessary for plant growth. However, heavy metals and other toxic substances tend to accumulate in sludge during wastewater treatment. When sewage sludge is applied to the land surface, the oxygen-rich environment causes the heavy metals to become immobile such that they accumulate in the soil, potentially making soil toxic over time. Because of the heavy metal issue and the fact that sludge has a foul smell, people living in the area commonly oppose efforts to spread sewage sludge on the land surface. As a result, sludge disposal is now moving away from land applications and toward the more expensive options of landfilling and incineration.

Another common problem involving wastewater treatment goes back to the period when cities discharged raw sewage into surface-water bodies. Here municipalities used **combined sewer systems** consisting of a single set of underground pipes to collect both sewage and the storm water runoff from streets and parking lots (Figure 15.25). When the time came to build sewage treatment plants, municipalities were faced with the prospect of having to lay separate collection lines. Instead of laying the additional lines, many cities found it more cost-effective to treat both the storm water and sewage and use their existing collection system. The problem is that during large storms, the combined flow can become greater than what the treatment plant can handle. When this occurs, the excess flow is allowed to bypass the treatment plant and discharge directly into a body of surface water. This periodic release of raw sewage not only has a negative impact on aquatic ecosystems, it creates a potential human-health hazard. For municipalities that depend on rivers as

FIGURE 15.25   Many older cities have combined storm and sewage systems as opposed to separate collection systems. During heavy rains or snowmelt, the runoff is mixed with sewage and allowed to discharge directly into streams so as to not overwhelm the plant. The raw sewage can cause a host of problems, particularly for cities located downstream who use the river as their water supply.

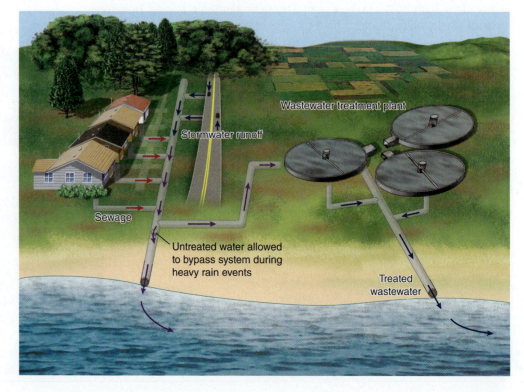

their source of drinking water, extra precaution must be taken to ensure that their water-supply plants can eliminate harmful bacteria coming from the occasional release of raw sewage from cities located upstream.

Agricultural and Urban Activity

There are many land-use activities in rural and urban settings that end up releasing toxic substances or excessive sediment and nutrients into our water supplies. The result is nonpoint source pollution that is both pervasive and widespread. For example, consider the chemicals that farmers and suburban homeowners apply to different types of plants. A significant portion of what they apply either infiltrates or is washed off the landscape during overland flow, polluting rivers, lakes, and aquifers. In this section we will explore some of the more important land-use activities that are related to nonpoint source pollution.

Agricultural Chemicals

Over the past 50 years, food production in the United States and other developed nations has increased dramatically due in part to modern machinery and irrigation techniques. Equally important has been the widespread use of petroleum-based synthetic pesticides for killing insects, herbicides for controlling weeds, and inorganic fertilizers for promoting plant growth (Figure 15.26). These same chemicals are also used in urban areas to maintain landscaped areas around both homes and commercial businesses. For an example of how much we rely on these chemicals, consider that for pesticides alone, there are over 20,000 products registered with the EPA. While the benefits of synthetic chemicals for increasing food production and beautifying urban areas is quite clear, they also cause water pollution that is undesirable.

Given the widespread use of synthetic chemicals, it is not surprising that they are being detected in our rivers and shallow aquifers. Pesticides are of particular concern with respect to human health because they are specifically designed to kill living organisms. From 1992 to 2001, the U.S. Geological Survey sampled streams in both rural and urban areas and detected pesticides in more than 90% of the streams that were tested and in over 50% of the shallow aquifers. However, of the streams, less than 10% had concentrations above human benchmark standards—keep in mind that just because pesticides can be detected does not necessarily mean that these trace concentrations are harmful to people. With respect to aquatic organisms, the study found that well over half the streams sampled had pesticide levels that are likely to have an adverse impact on aquatic life. Also of concern was that a large number of these streams had contaminated fish, whose pesticide concentrations were high enough to pose a risk to people who may eat the fish.

Although the concentration of agricultural chemicals in U.S. streams and aquifers is seldom at levels likely to affect humans, the risk varies seasonally and is much higher in areas of intensive agriculture. Based on sampling data,

FIGURE 15.26 Pesticides and herbicides are widely used in modern agriculture to control insects and weeds. These toxic substances end up in our streams and aquifers where they are found in relatively low concentrations. Studies have shown that these concentrations seldom exceed human health standards, but often exceed those for aquatic organisms.

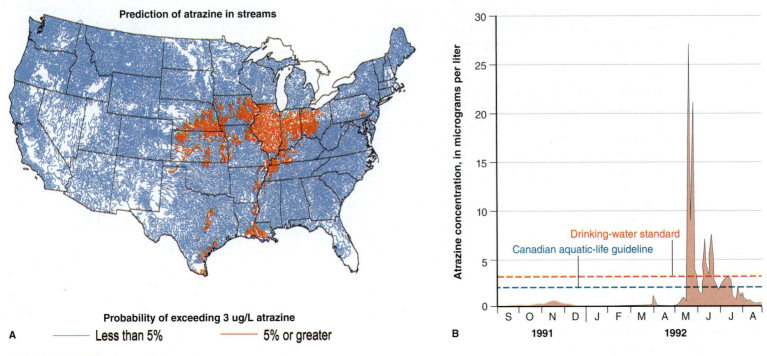

FIGURE 15.27   Map (A) showing areas of the U.S. Corn Belt where streams can be expected to contain elevated levels of the pesticide atrazine. Graph (B) from the Platte River in Nebraska showing how atrazine concentrations typically spike in the spring and early summer after the pesticide is applied to farm fields.

the U.S. Geological Survey predicts that the highest concentrations of the pesticide called atrazine will be found throughout the Corn Belt of the Midwest as shown in Figure 15.27. Note in the graph from the Platte River in Nebraska how atrazine concentrations typically spike in the month of May, reaching levels far above drinking-water and aquatic standards. Here spring rains wash some of the recently applied atrazine into nearby streams and shallow aquifers.

Developed nations also use synthetic fertilizers on agricultural fields and urban landscapes, some of which end up in streams and shallow aquifers. Here the resulting problem is largely one of oxygen depletion in streams and less of a human health issue. Unlike animal manure, synthetic fertilizers are made in chemical plants and consist of compounds containing nitrogen, phosphorus, and potassium, all of which are essential plant nutrients. When rainwater carries excess fertilizer off the landscape, these nutrients promote excessive algae growth, which, in turn, leads to oxygen depletion.

Animal Wastes

Farmers have historically collected livestock excretions called *manure,* and then apply it to their fields as a natural means of fertilizing their crops—this is often referred to as *organic farming.* Today, there are farms that specialize in raising large numbers of livestock; some are considered industrial operations where thousands of animals are confined in a relatively small area called *feedlots* (Figure 15.28). Here the volume of manure is so large that it presents a serious disposal problem. In order to reduce the volume, managers commonly put both the manure and urine into large pits or lagoons, where it then undergoes partial decomposition by anaerobic bacteria. From there the consolidated waste is typically spread out onto nearby fields so it can be taken up by plants or broken down by aerobic bacteria as it infiltrates the unsaturated zone.

Feedlot operators commonly find it difficult to acquire an adequate amount of land on which to spread the concentrated waste, creating the incentive to apply more waste than what the available land area can handle. Therefore, unbroken-down waste containing dangerous pathogens and substantial amounts of nitrogen and phosphorus can make its way into nearby streams and shallow aquifers. In addition to creating a biological hazard, the additional nutrients can easily lead to excessive algae growth and oxygen depletion. With respect to groundwater, the most common problem is that a plume of nitrate (NO_3^-) develops beneath the areas where the waste is being applied, thereby contaminating shallow drinking-water supplies.

Finally, the use of animal feedlots and synthetic fertilizers in the United States has led to a growing problem of oxygen depletion in the Gulf of Mexico. As shown in Figure 15.29, the Mississippi River drains a large portion of the agricultural land in the United States. The river then transports a tremendous amount of nutrients from the agricultural fields to the Gulf of Mexico. Here the nutrients fuel the familiar cycle of algae growth, followed by decay and oxygen depletion. In the Gulf of Mexico this has created what scientists refer to as a *hypoxic* or *dead zone*, where oxygen levels are too low for many types of marine organisms to survive. Although hypoxic zones are not uncommon, the dead zone near the mouth of the Mississippi is of concern because it has been growing in size since the 1980s, presenting a threat to the marine ecosystem in the Gulf of Mexico.

Sediment Pollution

Recall from Chapters 8 and 10 that *sediment pollution* occurs from agricultural and construction activity that allows excessive amounts of sediment to move off the landscape and into drainage systems. The filling of stream channels with sediment not only results in an increased risk of flooding, it is also highly destructive to aquatic ecosystems. Here the fine sediment can cover up coarse sands and gravels on the streambed that represents critical habitat for many types of species. Fine sediment also reduces visibility within the water as it stays suspended for relatively long periods of time. Reduced visibility, in turn, is highly detrimental to fish that require relatively clear water for seeing their prey. Because sediment pollution results from agricultural and construction activity that leaves soils exposed to erosion, the problem can be mitigated by using various techniques that minimize soil erosion. Examples include no-till farming and the use of silt fences around construction sites. See Chapters 8 and 10 for details.

FIGURE 15.28   Animal feedlots generate enormous amounts of manure and urine. To reduce the volume, the waste is put in holding lagoons where it is partially decomposed by anaerobic bacteria. The remaining waste is normally applied to fields. However, overapplication commonly leads to pollution and nutrient-loading problems in shallow aquifers and nearby streams.

FIGURE 15.29   The runoff of agricultural nutrients within the Mississippi drainage basin has led to a zone of oxygen-depleted water in the Gulf of Mexico. This hypoxic "dead zone" has been growing in recent years and is a threat to the marine ecosystem.

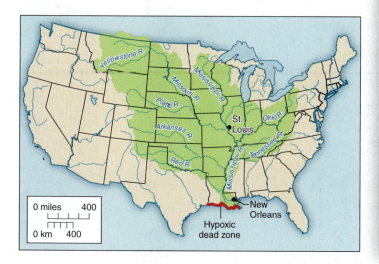

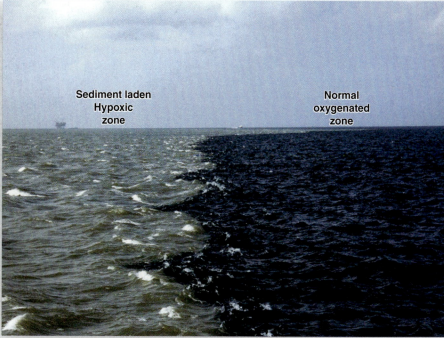

Radioactive Waste Disposal

Radioactive waste is different from chemical waste in that it emits dangerous forms of energy, collectively referred to as *radiation*. Recall that radiation originates from atomic nuclear reactions, and that the number of protons in the nucleus of a particular element is fixed, but the neutrons can vary. Scientists use the term *isotopes* when referring to the various combinations of protons and neutrons of different elements. What is important to our discussion here is that the nucleus of some isotopes is unstable. Unstable or so-called *radioactive isotopes* eventually reach a more stable configuration through the process of **radioactive decay,** where the nucleus undergoes a spontaneous nuclear reaction and releases energy (radiation).

Ever since the nuclear age began over 60 years ago, the number of applications involving radioactive isotopes has expanded from making atomic bombs to such things as generating electricity, industrial and scientific research, and medical treatments for various forms of cancer. All of these applications generate radioactive waste, and as their numbers have grown so too has the amount of waste being generated. The waste problem is compounded by the fact that anything that comes into contact with radioactive materials becomes contaminated, and therefore must be disposed of as radioactive waste. Unlike typical hazardous waste, radioactive waste cannot be treated or broken down by chemical processes. This is because nuclear decay is unaffected by chemical reactions that only involve orbiting electrons.

Radiation Hazard

Radioactive isotopes pose a human health risk because they emit radiation consisting of energetic particles and waves that can damage human tissue, which leads to development of cancer cells. As indicated in Figure 15.30, *alpha particles* are relatively large and slow-moving particles (2 neutrons and 2 protons) that rapidly lose energy as they encounter other forms of matter. Interestingly, alpha particles typically travel only a few inches through air, and do not penetrate even thin paper or the outer layer of human skin. On the other hand, *beta particles* are electrons that are ejected from the nucleus at high velocities. Although beta particles are much smaller than alpha particles, their higher velocity gives them enough energy to easily go through human skin and damage internal cells, but not enough to penetrate solid objects such as a door or wall. Another form of radiation is *gamma rays,*

FIGURE 15.30
When the nucleus of an unstable atom undergoes nuclear decay it emits radiation, consisting of alpha and beta particles and energy waves called gamma rays. Because gamma rays possess the most energy and are not particles, they have the greatest ability to penetrate objects and pose a threat to humans.

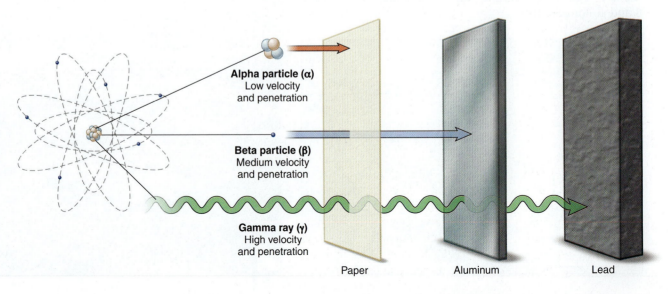

Alpha particle (α)
Low velocity and penetration

Beta particle (β)
Medium velocity and penetration

Gamma ray (γ)
High velocity and penetration

Paper   Aluminum   Lead

which are not particles, but just wave energy that is part of the electromagnetic spectrum (Chapter 2). Note that gamma rays are 10,000 times more powerful than visible light rays, making them the most energetic of all electromagnetic waves. Gamma rays are very powerful and do not consist of particles, which allows them to easily penetrate solid objects and thus pose the greatest risk to humans. People can be protected from gamma rays by thick layers of concrete or lead shields, similar to those used by dentists.

The radiation risk to humans largely depends on the type and intensity of radiation that is emitted by a particular isotope. Relatively few isotopes emit all three types of radiation (alpha and beta particles and gamma rays); most tend to emit two of the three types, and a few emit just one. Any isotope emitting high-energy gamma rays is a serious risk, whereas those that emit only alpha particles generally pose a threat only if the material is ingested or inhaled. Radioactive isotopes also have different decay rates, which means that the intensity and duration of the radiation varies among the different isotopes. Scientists use the term *half-life* to describe the exponential decay rate of a particular isotope, which is simply the time required for half of the radioactive atoms in a sample to decay into stable isotopes. Interestingly, the half-lives of natural and human-made isotopes range from fractions of seconds to hundreds of millions of years. Although short-lived isotopes decay more rapidly, they tend to emit more intense radiation, thus are more of a direct health risk. In contrast, long-lived isotopes remain radioactive for longer periods of time, so they present greater problems in terms of waste disposal.

The Disposal Problem

The U.S. Nuclear Regulatory Commission classifies radioactive waste into one of two categories. **High-level waste** consists of intensely radioactive by-products from nuclear power and weapons reactors, and the wastes generated from the reprocessing of spent reactor fuel. During nuclear fission reactions, uranium is converted into heat and highly radioactive by-products that consist of new isotopes and elements such as plutonium (Chapter 14). Eventually most of the original uranium is consumed, and the now highly radioactive fuel rods must either be discarded as waste or reprocessed into new fuel. During reprocessing, elements that can be used for fuel, mostly plutonium, are selectively removed and the remaining material is regarded as high-level waste. Until a national repository is operational (Case Study 15.1), high-level wastes in the United States are being stored on-site at commercial power plants and U.S. Department of Energy facilities. The spent nuclear fuel is either placed in special water-filled pools or put in dry storage. At federally operated plants, high-level liquid wastes are also being stored in large underground storage tanks made of stainless steel and concrete.

All radioactive waste in the United States that is not spent nuclear fuel is classified as **low-level waste**—tailings from uranium mines fall under mining regulations. This rather broad category includes contaminated materials that have come into contact with radioactive substances. Examples include contaminated reactor water residues, equipment and tools, protective clothing, wiping rags and mops, filters, medical needles and syringes, and laboratory animal remains. Consequently the amount of radiation emanating from low-level wastes can vary considerably, depending on the quantity of radioactive particles that have adhered to these materials.

Since 1962 commercial generators of low-level waste have been able to place the waste in sealed drums and ship the material to a federally licensed disposal site. Several disposal sites have been shut down due to subsidence and surface and groundwater contamination problems, leaving only three facilities in operation—one each in Washington, Utah, and South Carolina. In 1985 a federal law made each state responsible for the proper disposal of its

Long-Term Storage of Nuclear Waste in the United States

One of the reasons why there has been a lack of public support in the United States for building additional nuclear power plants is that the nation does not have a permanent means of storing high-level waste. The opening of a centralized repository has long been delayed in part because of the need to ensure that the chosen site can "permanently" isolate radioactive waste from the biosphere. This is necessary because the isotopes in high-level waste are long-lived and require thousands to tens of thousands of years before decaying to nonthreatening levels. Consequently, scientists as early as the 1950s recommended that such waste be buried in deep geologic formations. The ideal location would be one that is geologically stable and where severe earthquakes or volcanic eruptions are highly unlikely. Another important criterion is that the repository be located in a thick unsaturated zone, providing a buffer of dry material between the surface and the underlying groundwater system.

Although scientists found three suitable geologic sites in three different U.S. states, the final selection stalled because of public opposition. Citizens in these states were generally opposed to having theirs being the one to receive nuclear waste from across the country. Here the opposition focused on the potential risks and the fact the waste would need to be safely contained for approximately 10,000 years. Despite the fact that a great deal of scientific and engineering effort has gone into finding a suitable site and in developing physical containers to hold the waste, no one can guarantee the integrity of the repository for that length of time. In addition to potential changes in climatic and geologic conditions, people naturally asked what type of society, if any, would monitor the site for 10,000 years. Keep in mind that the Egyptian civilization ended only 3,000 years ago, and the U.S. government has been around for only a little over 200 years.

In 1982, Congress passed the Nuclear Waste Policy Act and designated that of the three initial sites, Yucca Mountain, Nevada, would be the sole candidate for the nation's centralized repository. Although the geology at all three sites was suitable, Yucca Mountain was chosen partly because it is located on a remote and highly restricted piece of federal property that has been used for conducting nuclear weapons tests since 1945 (Figure B15.1). The repository is to be built within a 1,200-foot-high, flat-topped ridge composed of compacted volcanic ash (tuff) that formed more than 13 million years ago. Because the site is in a desert climate that receives about 6 to 7 inches of precipitation a year, the unsaturated zone here is extremely thick. The repository itself would be located in the middle of the unsaturated zone and consist of tunnels mined within the volcanic rocks. Carefully packed containers of high-level waste would be placed in the tunnels and stored approximately 1,000 feet below the surface and 1,000 feet above the groundwater system.

Although extensive scientific and engineering studies have determined Yucca Mountain to be a suitable site, concerns remain. Much of the concern centers on the fact that the repository lies in an active earthquake zone. Here geologists have mapped numerous faults as well as a relatively recent volcanic cinder cone. Geologists have also found evidence of hydrothermal activity (Chapter 12) and numerous interconnecting fractures within the volcanic rock making up the repository. The fear is that should climate change bring increased precipitation and a higher water table, the fracture system could allow infiltrating water to carry radioactive waste away from the site much more rapidly. Nevertheless, in 2002 President G. W. Bush approved the Department of Energy's recommendation that Yucca Mountain be the national repository for high-level nuclear waste. Then in 2009 the Obama administration eliminated funding for opening the facility. Funds instead will be directed toward developing a new, comprehensive strategy that addresses both short and long-term storage and the reprocessing of spent fuel. It remains to be seen, therefore, if Yucca Mountain will someday be used to store nuclear waste.

FIGURE B15.1 Yucca Mountain in Nevada has long been viewed by the U.S. federal government as a good site for a central repository for high-level nuclear wastes. The waste would be stored in specially designed containers and placed in mined-out shafts within volcanic rocks approximately 1,000 feet below the surface and 1,000 feet above the water table. Although official scientific studies have found the site suitable, concerns remain over the possibility of seismic activity along nearby faults and a rising water table due to future climatic changes.

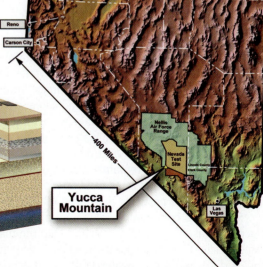

own low-level waste. Note that the law allows states to form compacts with other states in order to share construction and operating costs of new disposal facilities. These new facilities must meet federal guidelines requiring that low-level wastes be contained for as much as 500 years, depending on the type of waste. However, no new disposal sites have been built and the three existing facilities are now part of compact agreements with other states. A potential disposal problem may soon develop as the remaining facilities are beginning to restrict the amount of waste they will accept from non-compact states.

Air Pollutants and Fallout

Another major form of waste that society must discard is the gas and particulate matter generated during the combustion of fossil fuels by our cars, factories, and power plants. These combustion gases and particulates are simply released into the atmosphere (Figure 15.31), and are considered pollutants because they have an adverse impact on human health and natural ecosystems. Scientists use the term *anthropogenic* air pollution to describe that which occurs due to human activity, in contrast to natural forms of air pollution such as volcanic eruptions and wildfires.

There are many different types of anthropogenic air pollutants, but the most common are particulate (solid) matter, ozone (O_3), carbon monoxide (CO), nitrogen oxides (NO_x), and sulfur dioxide (SO_2). Like all combustion by-products, these pollutants have a limited residence time in the atmosphere.

FIGURE 15.31 Burning fossil fuels releases large volumes of combustion gases into the atmosphere. Carbon dioxide is affecting the global climate system, whereas sulfur dioxide and nitrogen oxides are causing acid rain. Another serious problem is the fallout of mercury associated with burning coal. Photo shows a tall smokestack releasing combustion gases from a coal plant—the smaller stack is a cooling tower that is emitting water vapor.

Eventually they react chemically with other gases in the atmosphere, fall back to earth with rain or snow, or fall directly as particulate matter. We will focus our attention on the fallout of acid rain and heavy metals and their impact on the geologic environment. Note that anthropogenic carbon dioxide (CO_2) can also be considered a pollutant because it affects the Earth's climate system, which, in turn, can have an adverse effect on human health. This topic will be covered extensively in Chapter 16.

Acid Rain

Natural precipitation is slightly acidic, a fact that has had a profound impact on the evolution of Earth's biosphere and weathering of rocks (Chapters 1 and 3). Recall that the pH of water is a measure of its hydrogen ion (H^+) concentration. The pH scale itself is logarithmic, where a pH of 1.0 (10^{-1} H^+ ions) is strongly *acidic* and 14.0 (10^{-14}) is strongly *alkaline*—a pH of 7 is *neutral*. Because carbon dioxide (CO_2), sulfur dioxide (SO_2), and nitrogen oxides (NO_x) gases readily dissolve in atmospheric water to form acids, the pH of natural precipitation averages around 5.6, and thus is slightly acidic. The term **acid rain** is commonly used to describe precipitation containing abnormally high levels of sulfuric and nitric acid related to the anthropogenic emission of SO_2 and NO_x. The EPA estimates that approximately 65% of all SO_2 and 25% of NO_x emissions in the United States come from coal and natural gas power plants that generate electricity. Other important sources of SO_2 include the refining of crude oil and smelting of sulfide minerals (Chapters 12 and 13).

During combustion, sulfur compounds in fossil fuels are transformed into SO_2, and atmospheric nitrogen is converted into NO_x. Table 15.2 shows that burning coal produces the highest SO_2 and NO_x emissions, whereas the emission from natural gas is relatively minor. The reason for this is that many coal deposits contain significant amounts of sulfide minerals, whereas oil and gas contain sulfur compounds in much smaller proportions. Since coal is used to meet 49% of the U.S. electrical demand, and natural gas only 21%, burning sulfur-rich coal is considered to be the leading cause of acid rain—refining of petroleum makes a significant, but smaller contribution. Figure 15.32 shows there is a strong correlation between the location of coal-burning power plants and the areas of acid rain in the United States. Because the smokestacks at these plants are quite tall (see Figure 15.31), the emission gases can be carried downwind far to the east before falling back to the earth in the form of acid rain.

One of the reasons acid rain is such an environmental problem is that Earth's biosphere has grown accustomed to rainfall with a pH around 5.6. Today, however, the average pH of rainfall in the eastern United States can be as low as 4.3. Because the pH scale is logarithmic, this means that the precipitation today is as much as 20 times more acidic than what existed prior to the Industrial Revolution. This is a dramatic

FIGURE 15.32 Map (A) showing the location of coal-fired power plants in the United States—those equipped with pollution controls for removing sulfur dioxide are shown in green. Map (B) shows the average pH of precipitation in 2007. Due to the prevailing winds, acid rain is most severe in the northeast as this region lies downwind of the major concentration of coal-fired power plants.

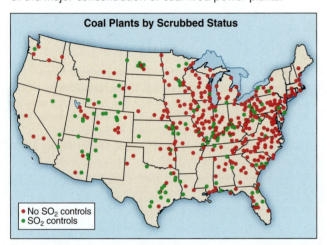

Coal Plants by Scrubbed Status

- No SO_2 controls
- SO_2 controls

A

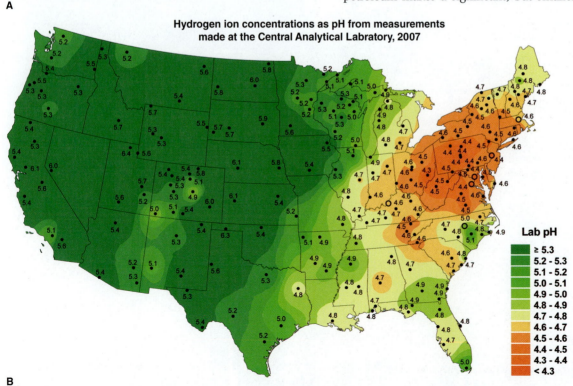

Hydrogen ion concentrations as pH from measurements made at the Central Analytical Labratory, 2007

Lab pH

- ≥ 5.3
- 5.2 - 5.3
- 5.1 - 5.2
- 5.0 - 5.1
- 4.9 - 5.0
- 4.8 - 4.9
- 4.7 - 4.8
- 4.6 - 4.7
- 4.5 - 4.6
- 4.4 - 4.5
- 4.3 - 4.4
- < 4.3

B

FIGURE 15.33   Photo from the Appalachian Mountains in North Carolina showing trees that were killed by acid rain, the source of which could have been hundreds of miles away upwind.

increase, particularly for the many aquatic organisms that are highly sensitive to pH changes. Keep in mind that acid rain impacts both aquatic organisms and land plants, such as forests and agricultural crops (Figure 15.33). Acid rain also accelerates the chemical weathering of concrete, limestone, and marble building materials. These materials all contain the mineral calcite, which readily dissolves in acidic solutions. The result has been significant damage to concrete highways and bridges and monuments made of limestone and marble (Figure 15.34).

Mitigating the Effects of Acid Rain

Several different methods are currently being used to reduce sulfur dioxide emissions from electrical power and industrial plants. Most methods employ a system called a **scrubber** that chemically removes SO_2 and other pollutants before the emission gases enter the smokestack (Figure 15.35). Here, finely ground limestone is injected into the emission gases as either a dry powder or mixed with water to form a slurry. Limestone particles

TABLE 15.2   Data showing the pounds of pollutants that are emitted for every billion BTUs of energy that are obtained from burning fossil fuels. On an equivalent energy basis, coal generates by far the most pollutants.

Pollutant	Natural Gas	Oil	Coal
Carbon dioxide (CO_2)	117,000	164,000	208,000
Carbon monoxide (CO)	40	33	208
Nitrogen oxides (NO_x)	92	448	457
Sulfur dioxide (SO_2)	1	1,112	2,591
Particulates	7	84	2,744
Mercury	0.0	0.007	0.016

Source: Energy Information Administration, Office of Oil and Gas; and Environmental Protection Agency.

FIGURE 15.34   Photo showing how acid rain has damaged the stonework and gargoyles of Winchester Cathedral in England. This church was built in 1079 and endured nearly 800 years of weathering by natural rainfall. Most of the damage shown here has occurred since the Industrial Revolution and the large-scale burning of fossil fuels.

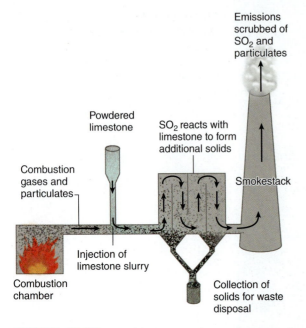

Emissions scrubbed of SO$_2$ and particulates

Powdered limestone

SO$_2$ reacts with limestone to form additional solids

Combustion gases and particulates

Smokestack

Injection of limestone slurry

Combustion chamber

Collection of solids for waste disposal

FIGURE 15.35 Scrubbers are a chemical means of removing pollutants from emission gases. In general, some form of powdered limestone is injected into the emission gases as they leave the combustion chamber of a coal-burning plant. The limestone then bonds with the SO$_2$ gas to form solids, which then fall to the bottom of a collection chamber where they are later removed and disposed of in a landfill.

then react with the SO$_2$ gas, forming a solid residue that falls to the bottom of a collection chamber. Modern scrubbers are capable of removing over 95% of the sulfur dioxide from emission gases. There are also scrubbing systems for nitrogen oxides, the most familiar being the catalytic converters required on the exhaust systems of U.S. autos and trucks since 1975.

Initial efforts at reducing acid rain began in 1970 when Congress passed the Clean Air Act, requiring power and industrial plants to start reducing their emission of SO$_2$ and other pollutants. Congress recognized that requiring industries to retrofit all the existing plants with scrubbing systems would be a costly and undue burden. Consequently, emission controls were mandated only on newly constructed facilities. The idea was that as old plants reached the end of their life-expectancy they would be replaced by new plants with scrubbers, allowing the additional costs to be phased in over time. By 1977 it was clear that new air-quality standards were not being met, in part because companies were upgrading old plants rather than building new ones with pollution controls. Congress then added a provision to the Clean Air Act stipulating that if a plant was modified beyond a certain dollar amount, it could no longer be considered "routine maintenance," but instead trigger the mandatory installation of a scrubber system.

By the 1990s the petroleum refining industry had largely complied with the updated regulations and had retrofitted most of their old refineries with scrubber systems. But, in the electric-utility industry many companies chose not to retrofit their older plants. Then in 2003 after a review by the Bush administration, the EPA increased the financial threshold for triggering mandatory scrubbers during plant upgrades from 0.75% to 20% of a plant's overall value. This meant that for a $1 billion plant, for example, the threshold for requiring pollution controls went from $7.5 million a year to $200 million. Utility companies therefore could now continue operating their old plants without any pollution controls. The end result of this long battle over the Clean Air Act is partly reflected in Figure 15.32A, which shows the high proportion of coal-power plants operating without pollution-control systems. Note that some of these plants have opted to utilize low-sulfur coal (Chapter 13) as a means of reducing SO$_2$ emissions rather than scrubbing systems.

Mercury Fallout

Similar to carbon and nitrogen, the element mercury (Hg) moves naturally in a cyclic manner between the hydrosphere, biosphere, atmosphere, and solid earth components of the Earth system. Mercury and other heavy metals (Chapters 12 and 13) tend to accumulate in the same environment in which fossil fuels originate; namely oxygen-poor swamps and wetlands. Heavy metals in fossil fuels are usually found in trace amounts where they bond with sulfur atoms. In addition to fossil fuels, heavy metals are found in various manufactured products, many of which eventually get placed into our municipal and medical waste streams. The problem is that when fossil fuels or these waste materials undergo combustion, some of the metals get transported into the atmosphere along with emission gases. Our focus here will be on mercury because of its extreme toxicity and the fact that it easily moves up through the food chain, posing a serious threat to human health. Notice in Table 15.2 that the combustion of coal releases more than twice as much mercury per BTU of energy than does oil—natural gas emits virtually none.

In studies of layered sediment, scientists have found mercury levels approximately two to five times higher than what existed prior to the Industrial Revolution. It is generally accepted that this increase is due to anthropogenic inputs of mercury into the atmosphere, which is estimated to range between

2,000 and 6,000 tons per year worldwide. Of this the United States accounts for 158 tons per year and China an estimated 1,000 tons. Of considerable interest to scientists and policymakers is how and where this anthropogenic mercury returns to the land surface. As could be expected, studies have found mercury concentrations in soils and aquatic systems to be significantly higher downwind from emission sources. However, scientist have also found elevated levels in polar regions, far removed from any emission source, indicating that mercury deposition is more complex than originally thought.

Recent research indicates that the key to mercury deposition lies in its physical and chemical properties, especially its low melting point that allows it to be a liquid at room temperature. As illustrated in Figure 15.36, when mercury compounds undergo combustion, the high temperature liberates elemental mercury atoms ($Hg°$)which form a gas. During this process some of the gaseous mercury becomes oxidized and forms ions (Hg^{2+}) that are chemically reactive and able to bond to particulate matter and dissolve in water. This oxidized mercury tends to fall out of the atmosphere downwind of emission sources during precipitation events—dry particulates can fall out on their own without precipitation.

Compared to its oxidized form, elemental mercury is nonreactive and may stay aloft in the atmosphere and circle the globe. Scientists have recently learned that the presence of certain oxidizing gases, like ozone (O_3) and those containing chlorine and bromine, act as triggers that cause the fallout of elemental mercury (Figure 15.36). The unusually high concentrations of mercury found in polar regions are now believed to be related to the accumulation of these oxidizing gases on snow and ice crystals during the long winter period of complete darkness. When the sun finally rises in the spring, these compounds are released and quickly convert the elemental mercury into oxidized mercury, which then falls from the sky. Researchers now understand that mercury fallout occurs regionally, downwind of emission sources, and also globally in a more complicated manner.

Mercury and Human Health

When mercury falls onto the landscape, the path that it takes is rather complex because of its ability to bond with a variety of organic and inorganic substances. Of particular concern is the form called **methylmercury** in which mercury bonds with carbon atoms. Methylmercury is easily ingested by aquatic lifeforms and then stored in the tissue of these organisms. Consequently, larger and older aquatic organisms tend to contain more mercury. This also means that as smaller organisms are eaten by larger ones, the concentration of methylmercury increases or magnifies as you move up the food chain. For example, the concentration of methylmercury in fish ranges from 1 million to 10 million times greater than the original concentration in the water. Because of the modern input of anthropogenic mercury, scientists predicted that concentrations of methylmercury in the food chain should now be higher than in the past. A recent study of modern polar bears found that they have mercury levels 11 to 14 times greater than bears that lived prior to global industrialization.

Because humans are at the top of the food chain, anthropogenic mercury is increasing the health risk of people who eat mercury-contaminated fish, shellfish, or mammals that eat seafood. During digestion, methylmercury is nearly completely absorbed by the blood and then distributed to tissue throughout the body. The primary problem is the mercury that accumulates in the brains of

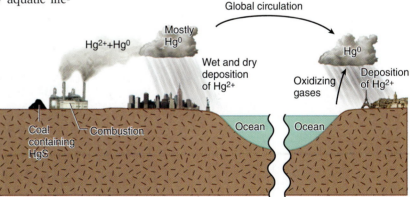

Hg^0 = Elemental mercury Hg^{2+} = Oxidized or reactive mercury

FIGURE 15.36   Anthropogenic mercury enters the atmosphere when materials containing mercury undergo combustion. During combustion, two different forms of gaseous mercury are liberated: elemental and oxidized. Oxidized mercury is highly reactive and tends to fall out relatively close to the emission source as dry particles and with precipitation. Elemental mercury is nonreactive and stays in the atmosphere, circling the globe until encountering oxidizing gases that convert it into oxidized mercury.

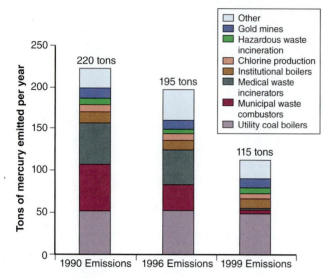

FIGURE 15.37 Anthropogenic mercury emissions in the United States have decreased significantly due to EPA regulations requiring scrubbing systems on medical and municipal waste incinerators. Today the single largest source of mercury released into the atmosphere are unregulated coal-fired electric power plants.

children whose nervous system is still developing. Here it acts as a powerful neurotoxin that is known to cause a reduction in cognitive ability (IQ). Pregnant women who eat contaminated fish easily pass methylmercury to their developing fetus, where it is known to cause serious birth defects and irreversible loss of cognitive ability. Because mercury concentrations greatly magnify in fish, the EPA and the U.S. Food and Drug Administration (FDA) recommend that young children and women of childbearing age restrict their fish consumption, particularly certain types of large fish.

Reducing Anthropogenic Mercury Inputs

Because of the health risks, Congress included mercury in the list of regulated air pollutants when the Clean Air Act was amended in 1990. The EPA then developed new rules governing the incineration of medical and municipal wastes, which resulted in a more than 90% reduction in mercury emissions from waste incineration. Figure 15.37 shows the large decline in U.S. emissions between 1990 and 1999 is due almost soley to the regulations on waste incineration. However, during this same period there was no reduction from coal-burning power plants, which by 1999 accounted for over 40% of the remaining mercury emissions. In 2000 the EPA issued a new rule under the Clean Air Act requiring U.S. power plants to begin installing control systems to scrub mercury from their emissions.

Before the EPA could implement the mercury regulations for coal-fired plants in 2000, the new Bush administration issued a revised rule exempting electric utility companies. It was argued that emission controls were an expensive and unnecessary burden, because mercury is a global problem and U.S. industries are contributing only 3% to the global mercury load. Independent scientists pointed out that while this is true with respect to elemental mercury, oxidized mercury is known to be deposited regionally. Nevertheless, the EPA later wrote a new rule in 2006 aimed at reducing mercury emissions using a cap-and-trade approach. Under this rule, operators could avoid installing mercury control systems on older coal plants by purchasing credits from plants whose emissions were below targeted levels. The EPA estimates that the cap-and-trade system will reduce mercury emissions from 48 tons to 15 tons per year by 2026. Under the original Clean Air Act amendment, emissions from coal-power plants were expected to be only 5 tons per year by 2008. Environmentalists have opposed the cap-and-trade system in part because the reduction targets were higher, and because old coal-fired plants could keep discharging untreated emissions, thereby exacerbating the existing problem of local and regional hot spots of mercury fallout.

Radon Gas

One of the most dangerous natural air pollutants is **radon gas,** which is an odorless radioactive gas that can cause lung cancer when it is allowed to accumulate in homes. According to the EPA, radon gas is responsible for 15,000 to 22,000 lung cancer deaths per year in the United States, making it the second leading cause of lung cancer after smoking. Radon (Rn) is one of several different elements that form from the radioactive decay of uranium-bearing minerals found in small amounts in igneous rocks. When igneous rocks undergo weathering, the uranium ends up accumulating in the resulting sediment. Consequently, radon gas is found in a variety of geologic materials, and in the groundwater that comes into contact with these materials.

As illustrated in Figure 15.38, radon becomes hazardous when it moves up through the soil zone and seeps into a building. Radon enters a dwelling through cracks and pipe joints, and from the use of groundwater. Once radon enters a building, it becomes trapped and can build up to dangerous levels.

When radon gas undergoes radioactive decay, it emits alpha radiation and decays into solid particles of the radioactive element polonium (Po). Radon gas is dangerous because it is easily inhaled, and then produces solid particles of polonium that become lodged in the lungs. Here the polonium emits alpha particles that cause cell damage within the lungs, which can eventually lead to cancer. Polonium particles can also form in the airspace of a house and become attached to dust particles that are then inhaled. Note that the most common radon isotope (Rn-222) has a half-life of 3.8 days, and the half-life of its polonium decay product ranges from fractions of a second to several minutes. This means that once radon gas is eliminated from the home, the radiation hazard will quickly fade away.

The EPA estimates that 1 out of every 15 homes in the United States has elevated levels of radon gas. People at highest risk are those living in well insulated homes and in areas with relatively high concentrations of uranium minerals in the underlying rock or sediment. Fortunately, homeowners can determine their radon levels through simple and inexpensive tests. If high concentrations are found, the radiation hazard can be minimized by (1) sealing cracks in the foundation and other possible entry points of radon; and (2) installing a simple ventilation system that prevents the gas from building up inside the home. For more information on radon in your area and how to have your home tested, see the list of state radon offices at the EPA's radon website—simply type *EPA* and *radon* into any Internet search engine.

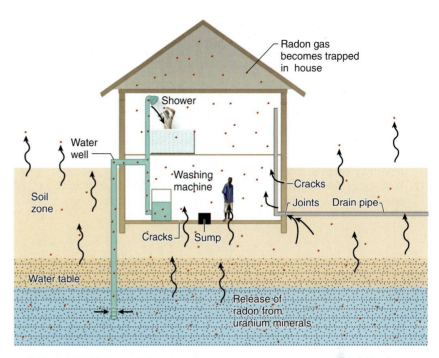

FIGURE 15.38 Uranium minerals release radioactive radon gas that can enter a house through cracks and joints and by well water. Once in the house, radon can become trapped and build up to dangerous levels. Radon is easily inhaled into the lungs, where it decays into radioactive particles of polonium that become trapped and damage lung tissue. The radon risk can be reduced by eliminating entry points into the home and installing a ventilation system that prevents the gas from building up.

SUMMARY POINTS

1. Modern societies generate a wide variety of solid, liquid, and gaseous wastes. Because many wastes are harmful to humans and other living organisms, they cause pollution if not disposed of properly.

2. Accelerated industrialization and population growth in the United States after World War II led to pollution levels that were unacceptable to the public. Congress and the president responded by creating the EPA and passing the Clean Air Act and Clean Water Act in the 1970s.

3. "Polluted" describes situations in which a substance's concentration in the environment has reached the point where it's harmful to living organisms. "Contaminated" is when the concentration becomes higher than natural levels, but is not necessarily harmful. What is considered harmful varies among different organisms and substances.

4. Pollutants often enter the environment from point sources, which consist of physically discrete locations such as smokestacks and wastewater discharge pipes. Pollutants also enter through nonpoint sources where contaminants are released over a broad area, as in the runoff from agricultural fields or parking lots.

5. Municipal and industrial solid wastes typically contain hazardous materials that react with water to form leachate, which can easily migrate and contaminate water supplies. Today the EPA requires

that municipal and industrial solid wastes be buried in containment-type landfills.

6. Tighter EPA standards have led to a dramatic decline in the number of U.S. landfills, resulting in local communities having to ship their waste to distant landfills. To lower costs, most municipalities have implemented waste management schemes that reduce the amount of waste being shipped. Efforts include recycling and rules that exclude yard wastes from municipal waste.

7. Hazardous waste can be buried in specially designed landfills or burned in hazardous-waste incinerators; liquid hazardous wastes can also be disposed of via deep-well injection. Some states still permit household hazardous wastes to be placed in the municipal waste stream, but others do not.

8. Human waste in the United States is required to undergo treatment before being released into the environment. Treatment options include individual septic systems or municipal sewage plants servicing an entire town or city. Both options utilize bacterial action to break down human waste into harmless compounds, but also generate sludge that is usually considered hazardous.

9. Animal wastes, fertilizers, overflow from combined sewer systems, and improperly operating septic systems can lead to algae blooms and oxygen depletion in lakes and streams.

10. Large amounts of pesticides, insecticides, and fertilizers are applied each year to agricultural fields, lawns, and gardens in order to increase food production and beautify urban landscapes. A significant portion of these chemicals either infiltrates or washes off the landscape, contaminating rivers, lakes, and aquifers.

11. Radioactive waste presents unique disposal issues due to the long half-life of some isotopes and the fact that radiation levels cannot be reduced using chemical methods. Low-level U.S. radioactive waste is currently being disposed of in one of three remaining facilities, whereas high-level wastes are being stored on-site until a national repository opens in 2017.

12. Emission of combustion gases, primarily from fossil fuels, causes air pollution and other environmental problems. Carbon dioxide con-tributes to global warming, whereas sulfur dioxide and nitrogen oxides result in acid rain. The fallout of mercury is a serious problem because it is a powerful neurotoxin that moves easily through the food chain and poses a threat to human health.

13. Both acid rain and mercury fallout can be greatly reduced by the installation of scrubbing systems that strip SO_2, NO_x, and mercury from the emission gases of industrial plants.

14. Radon gas is a natural form of air pollution that develops from the radioactive decay of uranium-bearing minerals. Radon is a serious hazard that can build up to dangerous levels in homes and cause lung cancer.

KEY WORDS

acid rain   500
anthropogenic pollution   474
Clean Air Act   476
Clean Water Act   476
combined sewer system   492
containment landfill   483
electronic waste   486
high-level waste   492
leachate   483

low-level waste   497
Maximum Contaminant Levels   479
methylmercury   503
municipal solid waste   482
National Environmental Policy Act (NEPA)   475
nonpoint source   481
point source   480
polluted   478

radioactive decay   496
radon gas   504
sanitary landfill   483
scrubber   501
secured landfill   485
Superfund   476
waste management pyramid   484
wastewater   487

APPLICATIONS

Student Activity   Few people wake up each day with the thought that they are adding pollution to our environment. Here you will do some research to find out how you might be contributing to pollution in your area.

1. Determine how much of the electrical power being supplied to your community is produced by coal, natural gas, or nuclear power plants. Describe the environmental problems that would be associated with each of these methods.

2. Contact your local government and ask for the location of the sanitary landfill that accepts your municipal trash. Has this landfill been cited for any pollution problems? Find out how long the landfill is expected to continue accepting trash before it becomes full. What does your community plan to do then? Are there plans to open a local landfill? If not, why?

3. Find out whether the wastewater in your house or apartment goes through a septic system, or a municipal sewage treatment plant. Describe the pollution issues that are associated with each of these treatment systems. Is any of your wastewater being reused for irrigation? If not, why?

Critical Thinking Questions

1. You read in the local news that someone in a neighboring county was caught illegally dumping hazardous waste near a stream. Think about the hydrologic cycle and describe the different ways that this illegal dumping could eventually affect you.

2. Explain why solid wastes (even biodegradable materials) in landfills do not break down very quickly. How does this relate to the long-term potential for groundwater pollution?

3. Describe the options for disposing of liquid hazardous wastes and list the advantages and disadvantages of each method.

4. What is the difference between high-level and low-level radioactive waste and how does this affect the method of waste storage?

Your Environment: YOU Decide

Much of the municipal and hazardous waste in the United States is sent to landfills located far from the populated areas where most of the waste is being generated. Even well designed landfills cannot be expected to safely contain the waste forever, and pollution of aquifers and nearby streams is almost inevitable. Moreover, the landfills are generally located in rural areas where people have little political power to prevent a landfill from being built. Is it fair that society forces these people to accept the waste and potential pollution problems, despite the fact they create very little of the waste themselves? What solutions would improve the situation?

Chapter 16

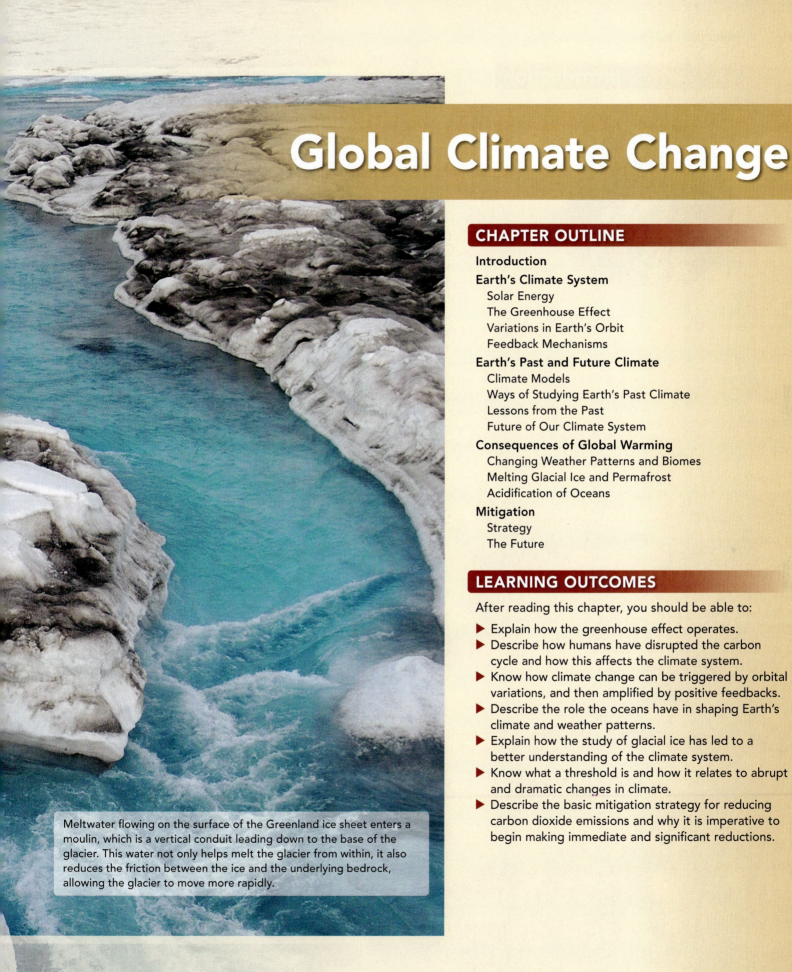

Global Climate Change

LEARNING OUTCOMES

After reading this chapter, you should be able to:

▶ Explain how the greenhouse effect operates.
▶ Describe how humans have disrupted the carbon cycle and how this affects the climate system.
▶ Know how climate change can be triggered by orbital variations, and then amplified by positive feedbacks.
▶ Describe the role the oceans have in shaping Earth's climate and weather patterns.
▶ Explain how the study of glacial ice has led to a better understanding of the climate system.
▶ Know what a threshold is and how it relates to abrupt and dramatic changes in climate.
▶ Describe the basic mitigation strategy for reducing carbon dioxide emissions and why it is imperative to begin making immediate and significant reductions.

Meltwater flowing on the surface of the Greenland ice sheet enters a moulin, which is a vertical conduit leading down to the base of the glacier. This water not only helps melt the glacier from within, it also reduces the friction between the ice and the underlying bedrock, allowing the glacier to move more rapidly.

Introduction

Climate change and global warming are potentially our most serious environmental issues today because human existence is so intimately tied to the planet's climate system. Earth's position within the solar system's habitable zone resulted in an atmosphere and a climate system that are ideal for complex plant and animal life because temperatures fall within the relatively narrow range where liquid water can exist at the surface. It is no coincidence, therefore, that most of Earth's human population resides in climatic zones where the combination of moderate temperatures and adequate water supplies allow for agricultural production. The problem today is that climate change and global warming are likely to cause a shift in temperatures and water supplies, thereby disrupting our existing agricultural and settlement patterns. Although for the past 200,000 years modern humans *(Homo sapiens)* have survived major global climatic changes, the difference is that we now have nearly 7 billion people living on the planet. Moreover, global warming is threatening to disrupt the Earth system, at the same time as the planet's limited resources are being pushed to their limits by our expanding population (Chapter 1).

Human history is full of examples where climatic change has had a profound effect on society. In some cases societies were able to flourish under the new climate, but others collapsed, particularly those already under environmental, political, or economic stress. Consider for example how the Viking settlements in Greenland came to an end in the 1400s when the climate became cooler in the Northern Hemisphere during what is known as the Little Ice Age. Although the settlers had survived for nearly

1859

500 years, the change in climate helped make it all but impossible for the Vikings to continue their way of life in the ecologically fragile environment. Another classic example is the collapse of the great Mayan civilization, located on the Yucatan Peninsula of present-day Mexico. This complex and highly advanced society had flourished for over a thousand years in the warm and humid climate, reaching a population estimated to be in the millions by 800 AD. However, the Mayans were experiencing a number of problems, including deforestation, soil loss, and limited water supplies. The land-use issues combined with limited water supplies meant that they could no longer expand their food production. Ultimately, a series of droughts is believed to have led to starvation, followed by warfare over the remaining resources. By the time Spanish explorers arrived in the early 1500s, the Mayan population is estimated to have fallen to around 30,000, a mere fraction of the millions who once lived there.

Evidence that climate change contributed to the collapse of ancient societies has taken on greater relevance in recent years as scientists continue gathering evidence that Earth's climate system is undergoing significant and rapid change. Climate change is resulting in a rise in the average global temperature, a phenomena commonly referred to as **global warming.** Perhaps the most visually striking evidence of global warming is the rapid melting of glacial ice on nearly every continent (Figure 16.1). The consequences of global warming, however, go far beyond just the loss of ice. Similar to the Vikings and Mayans, modern societies have adapted to the existing climatic conditions. Because global warming is

FIGURE 16.1   The drawing and photo show the dramatic loss of ice over a 142-year period from the Rhône glacier in Valais, Switzerland. Scientists have been gathering a variety of data around the world that provide strong evidence that the climate system is in a significant warming trend, commonly referred to as global warming.

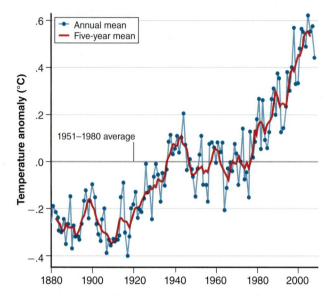

FIGURE 16.2 Graph showing the change in global average temperature relative to the 1951–1980 average. Data points and blue curve illustrate the variations from year to year, whereas the red curve represents a five-year average that smoothes out the yearly variations, making it easier to see the long-term trend. Data represent a combination of land temperatures from ground stations, and ocean temperatures from satellite and ship-based measurements.

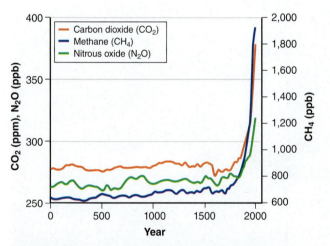

FIGURE 16.3 The burning of fossil fuels and other human activities since the Industrial Revolution have led to a sudden and dramatic increase in the concentration of greenhouse gases in the atmosphere. Because these gases cause the atmosphere to retain more of Earth's outgoing heat energy, their increased concentration is causing the atmosphere to warm. Note that CO_2 concentration is in parts per million (ppm), whereas CH_4 and N_2O are in parts per billion (ppb).

expected to cause climate zones to shift, it may become difficult for societies to maintain the systems that support our modern way of life. For example, most of the wheat in the United States is grown in the semiarid region known as the Great Plains (Chapter 11). Should climate change result in decreased precipitation for this region, then wheat production could fall dramatically. Another major consequence of global warming is accelerated sea-level rise. Although sea-level has changed throughout Earth's history, global warming is causing the oceans to rise at an accelerating rate, posing a threat to port cities and low-lying coastal areas around the world. Societies may soon be faced with the costly choice of defending coastal cities with seawalls and flood control structures, or moving to higher ground.

Because of the profound implications of global climatic change, numerous scientific investigations have been conducted in recent years with the goal of better understanding the present and past state of Earth's climate system. Included in this effort is the Intergovernmental Panel on Climate Change (IPCC) established by the United Nations. The IPCC is a scientific body that summarizes the research results on climate change from scientific institutions around the world. Starting in 1990, the IPCC has issued several major reports designed to provide the public and policymakers with objective scientific information. For example, Figure 16.2 shows land- and sea-based measurements from an IPCC report that clearly indicates that the planet is in a significant warming trend. The consensus among researchers is that global warming is largely a response of the Earth system to the sudden increase in the concentration of carbon dioxide (CO_2), nitrous oxide (N_2O), and methane (CH_4) gases in the atmosphere (Figure 16.3). Although these so-called *greenhouse gases* occur naturally, the dramatic increase is largely due to the burning of fossil fuels and other human activities since the Industrial Revolution.

Data from international scientific studies also show that Earth's climate has been remarkably stable for the past 10,000 years, a period that coincides with the rise of modern civilization. The long-term record, however, reveals that climate change does not necessarily occur gradually over hundreds or thousands of years as was once commonly assumed. These data clearly show that the climate system is capable of undergoing abrupt and sweeping changes in a matter of decades. There is concern among many climate researchers that the additional input of greenhouse gases into the atmosphere may destabilize the entire climate system, leading not to gradual warming, but a return to a system of abrupt and chaotic change. Although humans are highly adaptable, some question whether civilization could cope with wide climate fluctuations, particularly in light of what happened to the Vikings and Mayans who faced comparatively small changes.

Because global warming could present serious problems for entire societies, it has moved from being primarily a scientific issue to one that includes powerful economic and political interests. This, in turn, has caused the issue to become controversial and contentious, especially in the United States. Some interest groups attempt to portray global warming as an overly hyped threat pushed by alarmists and attention seekers; others frame the issue simply as a hoax. Arguments used to dismiss the threat commonly focus on the fact that Earth's climate varies naturally, and that climate researchers are uncertain as to how much of the warming is attributable to humans. There is also the familiar argument that we must choose between jobs and the environment, a position which ignores the fact that we humans and our economies are ultimately dependent upon the environment for our very survival.

In this chapter we will focus on what scientists know about Earth's climate system, both past and present, and why they believe human activity is playing a leading role in global warming. In particular we will examine Earth's climate record and explore what it might tell us about the future. Although predicting the future with much certainty is difficult given the exceedingly complex nature of the climate system, the potential consequences of global warming are so serious that it would be unwise to simply dismiss the threat. What we do know for certain is that the atmosphere is exceedingly thin compared to the solid Earth (Figure 16.4), and that the outpouring of greenhouse gases since the Industrial Revolution is causing the planet to retain more heat. Because Earth operates as a system, we can expect that this additional heat will have a ripple effect throughout the entire system. Recall that after the people of Easter Island (Chapter 1) degraded the island's environment, they ran out of food and had no means of escape. They simply had to suffer the consequences of their own actions. The situation today is analogous in that if human activity creates a climate where survival becomes more difficult, humanity has no place else to go. Earth is our only home.

FIGURE 16.4   Photo from the International Space Station showing how exceedingly thin the atmosphere is relative to the size of the solid Earth. The atmosphere has been absorbing large volumes of greenhouse gases released by human activity, causing the planet to retain more of the heat that radiates out into space. This process is disrupting Earth's heat balance and creating a ripple effect throughout the entire Earth system.

Atmospheric column

Earth's Climate System

The weather is undoubtedly one of the most common topics of conversation. Is it supposed to rain? How cold is it going to get tonight? For many of us the answer to these questions merely dictates the type of clothes we wear on a given day. However, for people such as farmers, construction workers, and ski resort operators whose livelihood depends on the weather, the answers take on greater consequences. This brings us to the often misunderstood distinction between weather and climate. Scientists define **weather** as the state of Earth's atmosphere at any given time and place, whereas **climate** represents the long-term average weather and its statistical variation for a given region. For example, the climate in tropical regions is characterized by having a comparatively large number of days when the weather is warm and rainy. Desert climates can be warm or cold, but rain or snow is a fairly rare event. Climate then is how weather behaves over extended periods of time, which is important because of its effects on people. In extreme cases, such as in Antarctica or the Sahara Desert, the climate is so harsh that practically no one lives there. Because a hospitable climate is necessary for humans to thrive, we will briefly explore those processes that regulate Earth's climate system.

Solar Energy

Recall from Chapter 2 that solar radiation (energy) is the primary driving force behind Earth's climate system. If the Sun's energy output were to change, our climate system would obviously have to change. Solar output has increased somewhat over geologic time, but Earth has remained within the solar system's *habitable zone*—where surface temperatures and atmospheric pressure allow liquid water to exist. This, in turn, has provided Earth with the ideal conditions and a sufficient amount of time for complex plant and animal life to evolve. Although the Sun has been generally stable, modern scientists have documented that its energy output oscillates back and forth in a rhythmic manner in conjunction with *sunspot* activity (i.e., relatively cool, dark areas on the Sun's surface). It has also been established that the number of sunspots during the past century varies on an 11-year cycle, with more sunspots corresponding to higher energy output. However, the variation in energy is relatively small compared to the effects caused by changes in greenhouse gas concentrations. Sunspot activity therefore is not considered to be the primary driver in the recent warming trend.

The Greenhouse Effect

When sunlight strikes an object, some of the light is transformed into thermal energy, a process called *solar heating* (Chapter 14). Solar heating naturally warms our planet whenever sunlight strikes gas molecules in the atmosphere, solid materials on land, or water molecules in the oceans. As these materials gain heat energy, they radiate heat back out into space—similar to how heat radiates off an asphalt parking lot. Like all planetary bodies, the Earth maintains what is referred to as a **heat balance,** where the amount of heat energy the planet radiates into space is equal to the energy it receives from the Sun. Should something disrupt this heat balance, Earth will either gain or lose heat until a new equilibrium can be established. For example, if the Earth were to suddenly gain heat due to a large asteroid impact, the rate at which heat escapes into space would increase. The planet would then begin to cool and eventually reach equilibrium again. Note that the term *climate forcing* refers to those processes that disrupt Earth's heat balance, forcing the climate to change.

In the 1820s a French mathematician named Joseph Fourier calculated the amount of heat energy Earth receives, and compared it to what is lost to space.

From this heat budget, Fourier's calculations showed that Earth's average surface temperature should be below freezing. Because the average temperature is actually well above freezing, he recognized that some phenomenon must be taking place that was not accounted for in his calculations. Fourier concluded that the atmosphere must be trapping part of the heat Earth radiates out into space, thereby keeping the planet relatively warm. Then in 1862, an Irish physicist named John Tyndall performed experiments proving that the atmosphere is largely transparent to incoming sunlight, but that carbon dioxide (CO_2) gas and water vapor absorb substantial amounts of Earth's radiant heat. Other researchers later showed that methane (CH_4), nitrous oxide (N_2O), and ozone (O_3) also absorb some of the heat Earth radiates into space.

The ability of certain atmospheric gases to trap heat and keep the planet relatively warm has become known as the **greenhouse effect.** Today scientists understand that the Sun produces a spectrum of electromagnetic waves whose peak energy lies in the *visible* portion of the spectrum (Chapter 2). As indicated in Figure 16.5, the atmosphere absorbs little of the incoming solar radiation. When this radiation strikes matter (solid, liquid, or gas), some of the light rays are reflected, whereas others are absorbed and transformed into thermal energy. Because the Earth is warm, the planet itself emits electromagnetic radiation. But notice in the figure that compared to the Sun's radiation, the peak radiation leaving the Earth is toward the longer wavelength portion of the spectrum. It is this longer wavelength, *infrared* radiation emitted by the

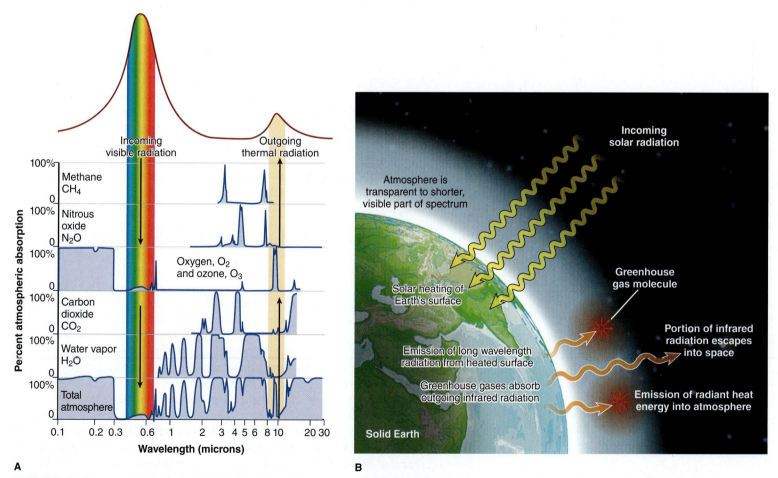

FIGURE 16.5   Graph (A) showing how incoming solar radiation and outgoing radiation emitted by the Earth lie in different portions of the electromagnetic spectrum. Greenhouse gases absorb wavelengths from different parts of the spectrum. Illustration (B) depicting how incoming visible radiation easily passes through the atmosphere, but some of the outgoing infrared rays are absorbed by greenhouse gas molecules. The molecules then radiate heat, warming the atmosphere. Note that the atmosphere is much thinner than shown here.

TABLE 16.1 Atmospheric gases and their percent volume in dry air—greenhouse gases shown in dark green. Although greenhouse gases make up a small fraction of Earth's atmosphere, their ability to trap heat has kept Earth's average surface temperature above freezing, allowing complex life to evolve.

Atmospheric Gas	Chemical Formula	Percent Volume in Atmosphere
Nitrogen	N_2	78.08
Oxygen	O_2	20.95
Water	H_2O	0 to 4
Argon	Ar	0.93
Carbon dioxide	CO_2	0.0360
Neon	Ne	0.0018
Helium	He	0.0005
Methane	CH_4	0.00017
Hydrogen	H_2	0.00005
Nitrous oxide	N_2O	0.00003
Ozone	O_3	0.000004

Source: M. Pidwirny, D. Budikova, and K. Vranes, 2007, "Atmospheric Composition" in Encyclopedia of Earth, http://eoearth.org/article/Atmospheric_composition

Earth that is absorbed by greenhouse gases in the atmosphere. As the greenhouse gas molecules gain energy, they radiate heat into the atmosphere, thereby producing what we call the greenhouse effect. Without this natural greenhouse effect, Earth's average surface temperature would be around 0°F (–18°C) as opposed to its present average of 57°F (14°C).

From Table 16.1 you can see that greenhouse gases make up only a small percentage of Earth's atmosphere. Although they are minor constituents, these gases have been critical in keeping Earth warm enough to allow liquid water to exist at the surface, thereby making it possible for life as we know it. Because natural processes such as decaying vegetation and volcanic eruptions release greenhouse gases, scientists use the term *anthropogenic greenhouse gases* to refer to those gases that result from human activities. Likewise, *anthropogenic climate change* is used to distinguish between human-induced and natural climate change.

The Greenhouse and Glacial Cycles

Prior to the discovery of the greenhouse effect, in 1836 a Swiss scientist by the name of Louis Agassiz began studying mountain glaciers in the Alps. There he found striations (scratches) carved into rocks, and large, isolated boulders at the foot of glaciers. Agassiz and other scientists began finding identical features all over northern Europe, far from any glaciers present at the time. Accounts were also coming in from North America of rocks with deep grooves and scratches. Even more intriguing were reports from Siberia, where people were finding well-preserved woolly mammoths. These extinct animals were found not in glacial ice, but rather in snow, which meant they had died relatively recently. Based on the evidence, Agassiz hypothesized that vast sheets of glacial ice had recently covered large portions of the continental landmasses in the Northern Hemisphere. Glacial geologists soon found the remains of forests buried beneath layers of sediment that had been deposited during glacial periods. Because the forests were composed of trees that grow only in more temperate zones, scientists recognized that sudden and dramatic climate change must have occurred in order for the forests to become buried by glacial sediment.

Scientists then began asking the obvious question of what could have caused Earth's climate to change so abruptly. Based on the earlier work of Fourier and Tyndall, a Swedish chemist named Svante Arrhenius hypothesized in 1895 that ice ages could be triggered by changes in carbon dioxide concentrations in the atmosphere. Arrhenius reasoned that lower CO_2 levels would reduce the greenhouse effect and lower atmospheric temperatures, plunging the world into an ice age. Conversely, an increase in CO_2 would raise temperatures and usher in what is called an *interglacial period*. Assuming lower CO_2 levels, Arrhenius performed thousands of laborious hand calculations and determined the heat balance for different parts of the globe. From this work he concluded that if atmospheric CO_2 concentrations were reduced between one-half to one-third of modern levels, then the average global temperature would drop by about 8°F. Such a temperature decline would have allowed glacial ice to advance and cover most of northern Europe and North America.

At the time of his work, Arrhenius had no way of knowing if carbon dioxide levels were indeed lower during the previous ice age. He simply proved that it was possible for the world to plunge into an ice age should the greenhouse effect be weakened by a substantial decline in CO_2. Arrhenius's ideas were finally confirmed in the 1960s when glaciologists were able to determine ancient CO_2 levels by measuring the composition of atmospheric gases trapped in glacial ice. They found that CO_2 concentrations were indeed about a third lower during the last ice age. Moreover, during his original work, Arrhenius had taken the time to calculate the increase in average global temperatures should the planet experience a doubling of CO_2 levels. Although his estimate of a 10°F increase is on the high side of modern forecasts using sophisticated computer models, his estimate is rather impressive considering the level of science at the time.

Arrhenius also recognized that the large-scale burning of fossil fuels, which was starting to take place at the time of his work, would alter the carbon dioxide concentration of the atmosphere and lead to global warming. Most scientists at the time scoffed at the idea of anthropogenic global warming, and believed that the oceans were more than capable of absorbing the additional CO_2. However, by the 1950s scientists began recording dramatic increases in the concentration of atmospheric carbon dioxide, vindicating Arrhenius's original hypothesis regarding a connection between CO_2 and global temperatures.

Variations in Earth's Orbit

Like Arrhenius and other scientists in the mid-to-late 1800s, the Scottish astronomer James Croll tried to find a mechanism that would explain Earth's ice ages. Although astronomers had previously suggested the ice ages were related to small changes in planetary orbits, Croll was the first to perform detailed calculations showing how orbital variations affect the distribution of solar radiation striking the Earth. He demonstrated that cyclical changes in the shape of Earth's orbit and timing of the seasons could affect the maximum and minimum amounts of solar heating. Then in the 1920s to the 1940s, a Serbian mathematician named Milankovitch took Croll's original work and developed a comprehensive mathematical model that computed changes in solar radiation. Results from Milankovitch's model showed that when certain orbital parameters were synchronized, the difference in solar heating between winter and summer would be large enough to initiate a glacial cycle. In honor of his work, the cyclical changes in solar heating are now called **Milankovitch cycles.** By the 1950s, the scientific community had largely accepted the idea that ice ages could be explained by a combination of astronomical and carbon dioxide (greenhouse) forcings.

The orbital parameters that Croll and Milankovitch incorporated into their models are illustrated in Figure 16.6. The first parameter is called *eccentricity*, which describes how Earth's elliptical orbit slowly becomes more circular over a 95,000-year period. As Earth's orbit changes shape, it affects the maximum and minimum distances to the Sun. This, in turn, affects the

FIGURE 16.6 Variations in Earth's orbit ultimately affect the distribution of solar radiation received by the planet in a given year. Eccentricity (A) controls the difference in solar heating between summer and winter, whereas the intensity of the seasons is governed by the amount of axial tilt (B). Precession of Earth's wobbling axis (C) affects where winter and summer will occur in Earth's orbit around the Sun. Note that drawings are not to scale.

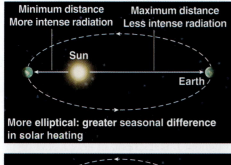

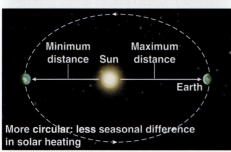

A Eccentricity: 95,000-year cycle

B Axial tilt: 41,000-year cycle

C Precession: 21,000-year cycle

intensity of the solar radiation and corresponding solar heating on the planet. When Earth's orbit is the most elliptical, the difference between the maximum and minimum radiation ranges from 20 to 30%, but then falls to around 6% when the orbit reaches its most circular shape.

The remaining two orbital parameters relate to how Earth rotates or spins about its axis. *Axial tilt* (Figure 16.6B) refers to how the inclination of Earth's axis varies between 22.1 and 24.5 degrees over a 41,000-year period. When this angle increases, the Northern Hemisphere tilts even farther away from the Sun during the winter, which produces less solar heating and cooler winter temperatures. However, during the summer, the Northern Hemisphere will tilt closer to the Sun, producing warmer temperatures. The net result is that the temperature difference between the two seasonal extremes (winter and summer) becomes greater as the tilt angle increases. The last orbital parameter is called *precession* (Figure 16.6C), which results from a wobbling motion in the axis that causes it to move in a somewhat circular path over a 21,000-year period. This movement in Earth's axis changes the date when a particular pole is closest to the Sun (summer) and then farthest away (winter). Precession is important because it slowly changes the timing of the seasons.

Milankovitch showed that each of the orbital parameters produces only minor differences in solar heating between summer and winter, and thus are not capable by themselves of causing an ice age. However, because each parameter repeats over a different time interval, all three will periodically coincide such that their effects are reinforced. As shown in Figure 16.7, Milankovitch cycles reinforce each other and produce more extreme summer and winter conditions when: (a) Earth's orbit is most elliptical; (b) precession alters the timing of the seasons to where winter and summer coincide with the maximum and minimum distances from the Sun; and (c) axial tilt is at its greatest. Using this model, Milankovitch demonstrated that ice ages are most likely to occur when summers are cooler, as opposed to cooler winters. Because glacial ice can accumulate even during relatively warm winters, it is the reduced melting when summers are cooler that tends to produce the greatest net increase in glacial ice.

The problem with using Milankovitch cycles to explain periodic glaciations is that the amount of potential cooling is too small to account for the growth of the vast ice sheets. Glaciologist therefore concluded that other factors must be involved. This led to the idea that cooler summers caused by orbital changes would trigger a decrease in atmospheric carbon dioxide. This, in turn, would reduce the intensity of the greenhouse effect and amplify the initial cooling. Climate scientists today accept the idea that carbon dioxide and orbital variations can alter the heat balance of the Earth, forcing the climate to change on a global scale. Next we will explore the carbon cycle and greenhouse effect in more detail along with important mechanisms that can amplify changes in climate.

Feedback Mechanisms

There have been many examples throughout this textbook of how a change in one part of the Earth system initiates changes in other parts. In the case of our global climate system, disrupting the heat balance sets in motion changes that ripple through the entire Earth system. The planet's overall heat balance is affected by changes in solar output, orbital characteristics, and concentration of greenhouse gases. In this section we will explore what scientists call **climate feedbacks,** which are processes within the Earth system that respond to a disruption in the heat balance, and then act to either increase or decrease the energy imbalance. A *positive climate feedback* is a process that increases or amplifies the system's response to a change in the heat balance, whereas a *negative feedback* works in the opposite sense and reduces or dampens the response. For

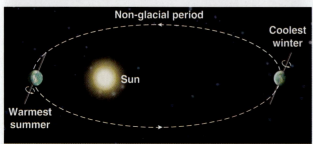

FIGURE 16.7 Milankovitch's model explains how glaciations correspond to periodic conditions where eccentricity, precession, and axial tilt coincide to produce cooler summers. Note that the shape of the Earth's orbit and the position of the Sun are exaggerated.

example, one positive feedback is the thawing of permanently frozen ground, called permafrost. As the ground thaws, it releases additional greenhouse gases (CO_2 and CH_4), which then amplify the initial warming. An example of a negative feedback is how elevated CO_2 levels cause plants to absorb CO_2 at a higher rate, which then reduces or dampens the effect of global warming.

A key point is that the intensity of global warming will depend on how much the heat balance is disrupted by the human input of greenhouse gases, and the degree to which positive and negative feedback mechanisms interact and compensate for one another. Although climatologists understand how the global climate system is responding, their quantitative forecasts have a significant amount of scientific uncertainty due largely to the complex and interactive nature of climate feedbacks. Compounding the uncertainty is the fact that some feedbacks are presently less well understood than others, particularly in regard to their rates of change. What is certain, however, is that climate feedbacks can take a relatively small disruption in Earth's heat balance and amplify it, producing a more dramatic change in the climate system. It is this ability of positive feedbacks to intensify Earth's current warming trend that makes global warming such a potentially dangerous problem. Next, we will explore some of the key feedback mechanisms and how they are expected to respond to global warming.

Reflection and Absorption of Solar Radiation

When sunlight strikes the Earth, a fraction of the incoming radiation is reflected, whereas the remainder is absorbed and transformed into thermal energy. It is this absorption of solar energy that heats the surface and ultimately warms the atmosphere through the greenhouse effect. Clearly, the more absorption that is allowed to take place the warmer the atmosphere will become. Scientists use the term **albedo** to describe the fraction of solar radiation that is reflected off a body. Light-colored objects, such as clouds and land covered with snow or ice, reflect more light and therefore have a high albedo. On the other hand, open bodies of water or forested landscapes have a low albedo because they are relatively dark, absorbing considerably more solar radiation than snow-covered surfaces. A good analogy is how a dark-colored shirt (low albedo) will absorb more solar radiation and feel much warmer than a white shirt.

When Earth's heat balance is disrupted and global temperatures begin to change, it naturally affects the distribution of vegetation and snow covering the planet. This, in turn, changes Earth's overall albedo and sets up a positive feedback mechanism. For example, the photos in Figure 16.8 illustrate how global warming is contributing to the loss of snow and ice, thereby

FIGURE 16.8 The loss of ice and snow cover over a 64-year period illustrates the dramatic changes in albedo that are taking place in arctic regions. Here reflective snow and ice are being replaced by darker rocks, vegetation, and open water. This produces greater amounts of solar heating, which, in turn, causes more melting. Photos showing Muir Glacier, Alaska.

1941

2005

exposing much darker bodies of water and rock. This allows for greater solar heating over an ever increasing area of the planet, generating even more heat and exacerbating the problem of global warming. Albedo changes also serve as a positive feedback when the planet is in a cooling mode. Cooler temperatures allow ice sheets to expand, thereby reducing the amount of solar heating. The result is even more cooling and continued expansion of ice-covered areas.

Greenhouse Gases and the Carbon Cycle

Atmospheric carbon dioxide (CO_2) and methane (CH_4) are part of what scientists call the **carbon cycle,** in which carbon atoms move in a cyclical manner through Earth's interior and near-surface environment. Although the element carbon makes up a very small percentage of the entire Earth, it becomes concentrated near the surface because of the planet's internal heat and tectonic cycle. For example, when rocks in the upper mantle begin to melt and form magma, carbon atoms are liberated and react with oxygen to form CO_2. Some of the CO_2 is eventually released into the atmosphere by volcanoes. Once in the atmosphere the CO_2 converts into other carbon compounds as it moves in a cyclical manner between the atmosphere, biosphere, hydrosphere, and solid earth. Although some of the carbon is returned to Earth's interior via the subduction process, over geologic time it tends to accumulate in the near-surface environment. In addition to plants and animals, carbon is concentrated in fossil fuels, carbonate rocks (e.g., limestone), and seawater.

Carbon is important because it is the basis for life on Earth, and is a key component of the greenhouse effect that makes our planet hospitable. However, if the Earth did not have mechanisms for removing CO_2 from the atmosphere, then the outgassing by volcanoes over time would have led to an ever-intensifying greenhouse, known as a *runaway greenhouse.* Here temperatures could have become high enough to drive most of Earth's water off into space—similar to what happened on Venus where surface temperatures today exceed 800°F (425°C). Scientists believe that Earth did not develop a runaway greenhouse in part because of the oceans' ability to absorb CO_2. But since the oceans would have long ago become saturated and stopped absorbing additional CO_2, it follows that a mechanism must exist which removes the gas from the oceans and stores it on a somewhat permanent basis. This process of removing carbon from the surface environment and placing it in storage is called **carbon sequestration.** The fact that Earth's climate has been relatively stable over long periods of geologic time tells us that the sequestration of CO_2 from the oceans has been keeping pace with volcanic outgassing.

The key process that sequesters carbon dioxide from the ocean-atmosphere system is the formation of carbonate sedimentary rocks (Chapter 3). When atmospheric CO_2 dissolves in water it combines with H_2O to form carbonic acid (H_2CO_3), making the water slightly acidic. This carbonic acid then breaks down to form carbonate ions (CO_3^{2-}), which bond with calcium ions to form the shells of marine organisms and limestone rock ($CaCO_3$). Over geologic time this process has produced thick sequences of limestone, thereby removing vast quantities of CO_2 from the ocean-atmosphere system. The sequestration of carbon dioxide by limestone deposition has been critical in regulating the greenhouse effect, keeping temperatures in a hospitable range for much of Earth's history.

While geologic processes in the carbon cycle operate on time scales covering millions of years, there are biological and chemical processes within the cycle that affect greenhouse gas concentrations on scales ranging from days to thousands of years. These more rapid processes are the key to understanding short-term changes in climate. In Figure 16.9 you can see that an estimated 762 billion tons of carbon is currently being held in the atmosphere,

2,260 billion tons in the biosphere, and an incredible 39,000 billion tons in the oceans (most of this remains deep and does not rapidly interact with the atmosphere). With respect to short-term climate change and global warming, what is important is not necessarily the size of the reservoirs, but the transfer rates of carbon between the reservoirs. Humans have been releasing CO_2 and CH_4 (methane) faster than the gases can be absorbed by the oceans and terrestrial biosphere, causing atmospheric carbon to increase by nearly a third over the past 250 years. In recent years this increase has been well documented by accurate CO_2 measurements on Mauna Loa, Hawaii (Figure 16.10).

Most climate scientists believe that the warming in modern times is largely being driven by anthropogenic greenhouse gases that have intensified the greenhouse effect. A major concern is that increasing temperatures may eventually shut down some of the negative feedbacks within the carbon cycle, which so far have been moderating the effects of global warming. For example, higher temperatures may begin to reduce the ability of the terrestrial biosphere and oceans to absorb atmospheric CO_2. Should the uptake of carbon dioxide be reduced, then the rate of warming would naturally accelerate. Higher temperatures may also set in motion positive feedbacks that amplify the warming. An example is the potential loss of forests due to higher temperatures and changing rainfall patterns. This would not only reduce the rate at which CO_2 is being removed from the atmosphere, but the decaying trees would soon transfer large quantities of CO_2 into the atmosphere.

Another positive feedback is related to frozen compounds of methane and water molecules called *gas* or *methane hydrates,* found in great quantities in permafrost soils and ocean sediment (Chapter 14). As temperatures continue to rise in polar regions, methane hydrates will begin to thaw, releasing large volumes of CH_4 into the atmosphere. Because methane traps 21 times more heat in the atmosphere than does carbon dioxide, the potential breakdown of methane hydrates represents a rather strong positive feedback. Interestingly, scientists have found evidence for a major global warming event 55 million years ago which many believe was caused by the rapid release of methane from the seafloor. Data show that this abrupt and extreme event caused global temperatures to increase as much as 13°F (7°C).

Water Vapor and Clouds

Recall that water vapor is a greenhouse gas, and therefore helps warm the atmosphere by trapping some of the radiation that Earth emits into space. This is why cloudy nights tend to be warmer than clear nights. However, clouds also have a cooling effect on the climate by reflecting incoming sunlight, which reduces the amount of solar heating at the surface. In terms of global warming, a warmer atmosphere is expected to hold more water vapor, but climatologists are not sure whether this will actually lead to the formation of more clouds. If more clouds do form, lower-level clouds are expected to have a net cooling effect, whereas upper-level clouds should have a warming effect. Because these complex interactions are not yet fully understood, the formation and role of clouds generates considerable scientific uncertainty in the computer models used to forecast future climatic conditions. Due to the uncertainty, the effect of water vapor and clouds is currently a major area of research among climatologists.

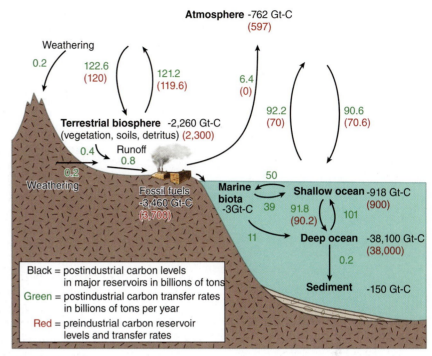

FIGURE 16.9   Earth's biochemical carbon cycle represents the movement of carbon between major carbon reservoirs. The estimated amount of carbon in reservoirs for the industrial period 1750–1994 is shown in black, and transfer rates in green. Red values represent preindustrial levels. Due mainly to the burning of fossil fuels, the atmosphere now holds nearly one-third more carbon compared to the preindustrial level of 597 billion tons.

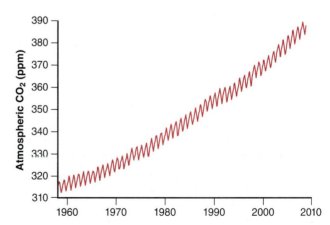

FIGURE 16.10   Graph showing the increase in atmospheric CO_2 from 1958 to 2009. Concentrations fluctuate on an annual basis due to the uptake of CO_2 by land plants in the Northern Hemisphere each summer. Data were measured on Mauna Loa, Hawaii, by Scripps Institution of Oceanography and NOAA, Earth System Research Laboratory.

Oceans

When the atmosphere warms during the day because of solar heating and the greenhouse effect, some of the heat energy is transferred to the oceans. Since water changes temperature more slowly than air, the oceans end up storing some of the heat that is removed from the atmosphere. This additional heat is later released back to the atmosphere whenever the air temperature becomes cooler than seawater. A familiar example is how a swimming pool becomes warmer during the summertime as it accumulates more heat during the day than it loses at night. With respect to the current global warming trend, Earth's oceans have been absorbing some of the additional heat being generated in the atmosphere, thereby slowing the overall increase in atmospheric temperatures. This ability of the oceans to moderate atmospheric temperatures helps explain the delayed and muted response of the climate system to the anthropogenic input of greenhouse gases.

The absorption and release of heat by the oceans also plays a key role in the *hydrologic cycle,* which itself is an integral part of Earth's climate system. Because incoming solar radiation is more intense near the equator, tropical regions undergo considerably more solar heating and evaporation than areas at higher latitudes. This causes a vast amount of water and heat to be transferred from the tropical oceans to the atmosphere. The resulting heat imbalance, combined with the rotation of the globe, sets in motion wind and ocean currents that transport both energy and water to higher latitudes. Here ocean circulation is driven by winds and by temperature and density differences of the water, with density being determined by temperature and salinity (salt water is denser than freshwater, and cold water is denser than warm). Oceanographers use the term **thermohaline circulation** to describe the collection of large-scale density and wind-driven currents that move in a convective manner through the ocean basins (Figure 16.11). Thermohaline circulation is an important part of Earth's climate system since it helps smooth out global temperature differences and distributes freshwater around the planet. For example, the warm surface current in the Atlantic, known as the Gulf Stream, keeps continental Europe significantly warmer than other land areas at similar latitude.

FIGURE 16.11
Thermohaline circulation is a collection of wind and density-driven currents that move in a convective manner through the ocean basins (red indicates warm, less dense water, blue represents cool, more dense water). This system of currents distributes both heat and freshwater around the planet and is a key component of Earth's climate system and hydrologic cycle.

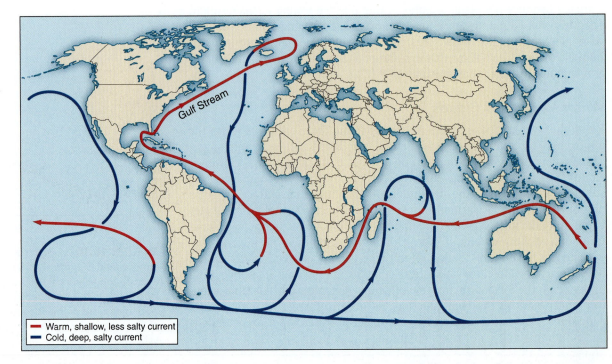

Gulf Stream

■ Warm, shallow, less salty current
■ Cold, deep, salty current

Another important ocean phenomenon that affects global weather patterns are periodic changes in water and air currents in the equatorial parts of the Pacific Ocean. As shown in Figure 16.12, under normal conditions, prevailing winds push equatorial waters westward across the Pacific, forming a pool of warm water that stands about 1.5 feet (a half meter) higher than waters off the coast of South America. Intense evaporation over the warm water results in cloud formation and a convective loop that brings rains over much of the western Pacific. On the opposite side of the ocean this normal pattern produces relatively cool and dry air along the coast of South America. This pattern is periodically disrupted when easterly trade winds slacken along the equator, allowing the elevated pool of warm water to migrate eastward toward South America, a phenomenon called **El Niño.** El Niño events are important since they alter weather patterns along the entire width of the equatorial Pacific, which, in turn, creates a ripple effect through Earth's global climate system. In the western Pacific (e.g., Australia and Indonesia) El Niños produce dry conditions, but commonly bring heavy rains and destructive flooding to parts of coastal North and South America. On the other side of the Andes Mountains in South America, precipitation over the Amazon rainforest typically decreases during El Niño events.

El Niño conditions generally occur every 3–5 years and last for 9–12 months, after which time the equatorial Pacific returns to its normal pattern of ocean and wind currents. When the system reverts to normal, the trade winds are sometimes stronger than usual as they push the warm water westward again. This enhanced movement of air and water allows cold water from higher latitudes to move down farther than usual into the equatorial

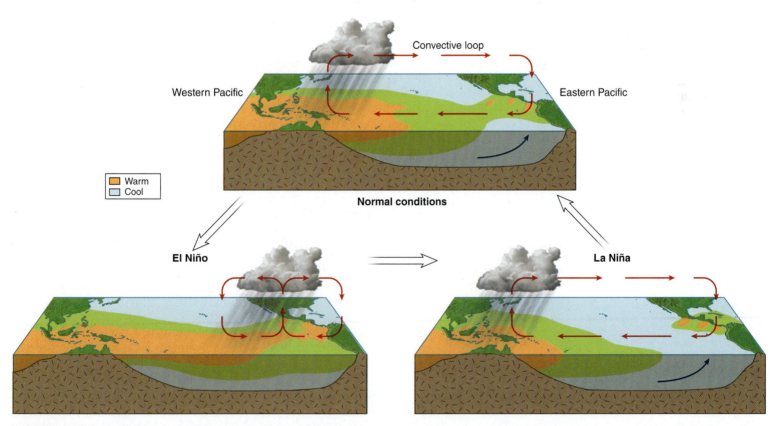

FIGURE 16.12   Under normal conditions, prevailing winds push warm surface waters into the western portion of the equatorial Pacific. This generates heavy rains in the western Pacific and cool, dry conditions along South America. El Niño occurs when trade winds slacken, allowing warm water to move eastward. This brings heavy rains to parts of North and South America and drought conditions to Australia and Indonesia. Before the system returns to normal, cooler waters sometimes push down along the North American coast in what is called *La Niña*, altering weather patterns across the United States.

Pacific, a condition referred to as **La Niña** (Figure 16.12). As the cold water pushes into the eastern Pacific it suppresses rainfall along the western United States and alters the course of the upper-level air currents (jet stream).

Together, El Niño and La Niña conditions represent an oscillating system that affects global weather patterns as vast amounts of heat energy move across the planet's largest ocean. Climate researchers have been able to find clear evidence of El Niño and La Niña conditions in the long-term climate record. For example, scientists have recently linked periodic droughts in the western United States to La Niña conditions. An important area of current research is whether global warming is causing El Niño and La Niña conditions to return more frequently, and persist for longer periods of time.

Earth's Past and Future Climate

There are two popular arguments against global warming and the need to curtail the emissions of anthropogenic greenhouse gases. The first is that scientists are simply too uncertain about the future state of the climate system. Here the focus is commonly on the *scientific uncertainty* associated with the computer models used to simulate Earth's global climate system. Other times it is the uncertainty of the temperature measurements themselves. In particular, the modern temperature rise is greatly exaggerated because cities become naturally warmer as they continue to expand, replacing vegetation with materials that absorb more solar radiation. This so-called *heat island effect* has been well documented by scientists, who have properly eliminated the bias from global temperature data. The other argument is that climate change is not uncommon and that the current warming trend is well within the range of natural variability. Implicit here is the assumption that the effect of anthropogenic greenhouse gases is small compared to natural factors that regulate Earth's long-term climate.

In this section we will take a brief look at the computer models used for studying Earth's climate system and the scientific uncertainty of the results. In addition to the models, we will examine Earth's climate record and how this record reveals a planet with a long history of sudden and dramatic climate change.

Climate Models

Atmospheric scientists today have come to rely on sophisticated computer models for studying different aspects of Earth's climate system, such as the paths of hurricanes. In its most basic form, a *model* is a simplified version of something in the real world. Examples of models include physical representations built to the correct proportion or scale, like a globe of the Earth or a toy airplane. A model can also be more abstract, as in the case of a flowchart illustrating the interaction of certain phenomena. A good example is the movement of carbon within the carbon cycle (see Figure 16.9). A *mathematical model* is a special type of abstract model in which physical processes are represented in terms of mathematical relationships. These models are particularly useful in that they allow scientists and engineers to quantify changes in whatever system or process is being modeled. Because a mathematical model typically involves complex equations and repetitive calculations, they are usually written in terms of a computer code or program so that a digital computer can perform the calculations.

Some of the most sophisticated mathematical models are those that simulate atmospheric processes as they must incorporate the many feedbacks between the atmosphere, oceans, and land. Some atmospheric models are regional in scale and are used to quantify the movement of heat (energy), air, and moisture on a time scale of hours to days. To simulate Earth's entire global climate system, scientists use what are called **general circulation models (GCMs),** where the simulation period is commonly on the order of years to decades. As illustrated in Figure 16.13, a GCM uses a three-dimensional grid to divide the atmosphere into a large number of cells. The grid consists of columns, where each column is divided into layers to produce individual *cells* (the lowest layer of cells represents the land or ocean surface). At the start of a simulation, every cell in every layer is assigned an initial set of conditions, such as temperature, wind, greenhouse gas concentration, and intensity of solar radiation. The model is then run forward in time, computing the rates at which heat, wind, and moisture move both vertically and horizontally between adjacent cells.

General circulation models (GCMs) are important because they are the most advanced scientific tools available for answering two key questions regarding global warming. The first is how much of the current warming trend is due to anthropogenic greenhouse gases versus natural climate forcings (solar output and orbital variations)? The other key question is what will the future state of Earth's climate system be should humans continue to release greenhouse gases? To try and answer these questions, research institutions around the world have performed detailed simulations using a number of different GCMs. Recent results reported by the Intergovernmental Panel on Climate Change (IPCC) are shown in Figure 16.14. From these graphs one can see that when both anthropogenic and natural forcings are considered, the overall average temperature from the models closely corresponds to the actual (observed) rise in surface temperatures since 1900. Note how the models accurately simulate the temporary cooling effects associated with major volcanic eruptions. Because of this ability to accurately portray the climatic system, scientists are confident in using GCMs to examine how

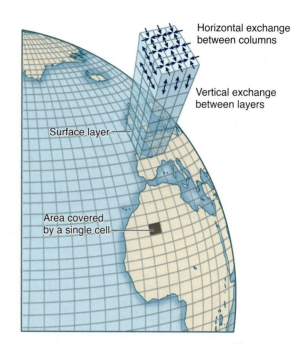

FIGURE 16.13 Atmospheric scientists use general circulation models (GCMs) to study Earth's global climate system. A GCM uses a three-dimensional grid to divide the atmosphere into a large number of individual cells. During a simulation the model computes the vertical and horizontal movement of heat, wind, and moisture between adjacent cells.

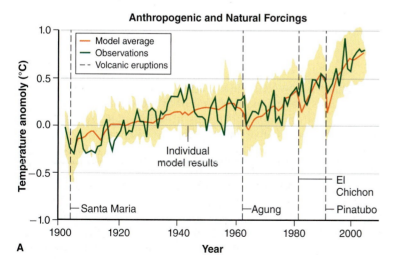

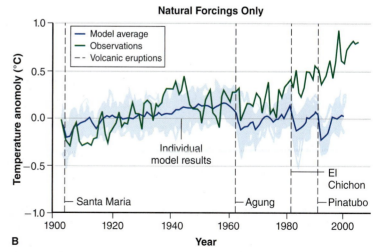

A   **B**

FIGURE 16.14 Graphs comparing actual surface temperatures (green line) with average results from GCM simulations (red and blue lines). When anthropogenic greenhouse gases and natural forcings are both considered (A), average GCM results (red) closely match the overall rise in temperature (green)—even the cooling effects of major volcanic eruptions are accurately portrayed by the models. When just natural forcings are considered (B), model results (blue) no longer match the rise in temperatures after the 1960s. This indicates that anthropogenic gases are largely responsible for the recent rise.

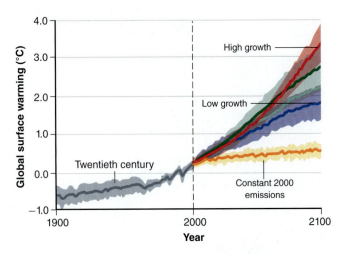

FIGURE 16.15   Plot showing measured changes in surface temperatures from 1900 to 2000, followed by GCM projections to 2100 under different scenarios of greenhouse gas emissions (temperature changes are relative to the 1980–1999 average). Scenarios range from maintaining 2000 conditions (orange), where equivalent CO_2 concentrations were 1.5 times greater than preindustrial levels, to levels as much as 3.5 times over preindustrial. The shaded portion of each plot represents ±1 standard deviation from annual averages.

the climate system might behave under a different set of conditions. For example, when GCMs are run under a scenario that does not include anthropogenic greenhouse gases, the results no longer correspond to the historical temperature record (Figure 16.14B). This type of analysis has led to the conclusion that while much of the warming prior to the 1960s can be attributed to natural forcings, the temperature rise over the last 30–40 years is largely due to the human input of greenhouse gases.

GCMs are also useful in forecasting how the climate system might respond under a variety of possible CO_2 scenarios. Figure 16.15 is from an IPCC report showing different GCM temperature projections out to the year 2100—temperature changes are relative to the 1980–1999 average. Here the models were run under scenarios ranging from greenhouse gas concentrations at 1.5 to 3.5 times greater than the preindustrial levels. It should be emphasized that GCMs do a good job of simulating temperature and precipitation patterns around the globe, but the results are statistical in nature and have a certain level of scientific uncertainty, as does all quantitative work in science (Chapter 1). Note that the uncertainty in each of the GCM projections (shaded areas) progressively increases over the simulation period—this is similar to how landfall projections for a hurricane become less certain the farther the storm is from shore. Despite the uncertainty about exactly how much the planet will warm, GCM results all indicate a significant rise in temperatures should humans continue to increase the concentration of greenhouse gases.

Ways of Studying Earth's Past Climate

Another useful tool for understanding Earth's climate system is to examine our planet's long history of climate change, a field of study called **paleoclimatology.** The key to any reconstruction of the climate record requires gathering temperature and precipitation data. Unfortunately, the oldest reliable instrument recordings go back only a few hundred years, and exist for just one or two sites in the world. To study global climate history requires data from different sites around the world, but here instrument data only go back several decades. Paleoclimatologists make use of *proxy data,* which are physical lines of evidence that are sensitive to changes in temperature or precipitation at the time they form. Proxy data can substitute for actual instrument recordings and allow scientists to extend the climate record farther back in time.

Sediment

Recall from Chapter 3 how geologists have been able to reconstruct environmental conditions on our planet going back more than a billion years based on the types of minerals and fossils in sedimentary rocks. Of considerable importance are pollen grains that are released by plants and become incorporated in the sedimentary record. Because different plants produce rather unique pollen grains, scientists are able to determine the types of plants that existed at the time the sediment was being deposited. Because certain types of plants grow in specific climatic zones, the study of pollen grains preserved in sediment has been a very useful tool for unraveling Earth's climate history. However, pollen data typically do not provide the necessary detail for resolving year-to-year variations in temperature or precipitation.

A particularly useful method for determining ancient temperature changes in the oceans is the analysis of stable oxygen isotopes found in carbonate (limestone) rocks and minerals, principally the mineral calcite ($CaCO_3$). Recall from Chapters 14 and 15 that the term *isotope* refers to the different combinations of protons and neutrons of a given element. In the case of oxygen, its two most common isotopes are *oxygen-16* (8 protons and 8 neutrons) followed by the heavier isotope called *oxygen-18* (8 protons and 10 neutrons). Because the isotopes have different masses, certain chemical and biological reactions have a physical preference for either a lighter or heavier isotope. When marine organisms precipitate calcite from calcium (Ca^{2+}) and carbonate (CO_3^{2-}) ions, the proportion (ratio) of oxygen-18 to oxygen-16 atoms entering the mineral structure depends on both the water temperature and the type of organism involved. Here the isotopic ratios are compared to those of a known standard and reported on a per mill (‰) basis as opposed to a percent (%). By analyzing the oxygen isotope ratios found in different layers of carbonate sediment, scientists can calculate temperature changes during the time layers were deposited.

Annual Tree Rings

One type of high-resolution climate data are annual growth rings in trees, an area of study known as *dendrochronology*. Annual tree rings (Figure 16.16) develop in climatic zones where seasonal differences result in rapid growth during the spring and summer, followed by minimal growth in the fall and winter. In climates where there is essentially an unlimited supply of water, tree growth is largely a function of temperature, which causes annual rings to be wider during warmer years and narrower in cooler years. In areas where water is a limiting factor for growth, rings are wider during relatively wet years and narrower in dry years. Thus, trees in a given area will exhibit the same annual growth-ring pattern since they all experience the same year-to-year variations in temperature or precipitation. By matching the overlapping tree-ring patterns from trees of different ages, scientists can obtain a reliable climate record that is as old as the oldest tree, which typically is hundreds or even thousands of years old. While tree-rings provide very useful data on year-to-year variations in climate, the records span only the most recent period of geologic time.

Glacial Ice

One of the most useful types of high-resolution climate data comes from glacial ice and the analysis of stable hydrogen and oxygen isotopes. Notice in Table 16.2 that the vast majority of hydrogen exists as simple hydrogen atoms that contain just a single proton in their nucleus. The other stable hydrogen isotope, called *deuterium,* is heavier as it has both a proton and a neutron—the isotope tritium is radioactive and is not used as a temperature proxy. With respect to the oxygen isotopes, *oxygen-16* atoms are by far the most common, followed by heavier *oxygen-18 atoms*. Although deuterium and oxygen-18 atoms are quite rare, they are found in a small but measurable number of water molecules (H_2O).

The fact that water contains different combinations of hydrogen and oxygen isotopes means that some individual water molecules are slightly heavier than others. This is important because when water evaporates, the lighter molecules tend to evaporate first. Later, when the water vapor condenses, the heavier molecules will generally condense first. Based on the fact that temperature ultimately controls evaporation and condensation rates, scientists have found a reliable and quantitative relationship between the isotopic composition of precipitation and the temperature at which it condenses from the atmosphere. Water that condenses at a cooler temperature is relatively light (lower deuterium/hydrogen and oxygen-18/oxygen-16 ratios) compared to water that condenses at a warmer temperature. By analyzing the isotopic composition of different layers within glacial ice, paleoclimatologists can reconstruct past temperature changes in great

FIGURE 16.16   Variations in temperature or precipitation create a distinctive pattern of annual growth rings in trees. By sampling trees of different ages and matching their overlapping growth-ring patterns, scientists can reconstruct past climatic conditions going back hundreds to thousands of years. The sample shown here is from a New Mexico pine tree. Note the narrow growth ring in 1407 indicating a drier year than normal.

TABLE 16.2   Common isotopes of hydrogen and oxygen and their natural abundances. Although deuterium and oxygen-18 atoms make up a tiny fraction of hydrogen and oxygen atoms, the fact that they are heavier causes them to condense from the atmosphere at different temperatures compared to their lighter and more abundant counterparts. By measuring the ratios of deuterium to hydrogen and oxygen-18 to oxygen-16 atoms, scientists can determine the temperature at which ancient snow and ice formed.

Hydrogen Isotopes	Percent of All Hydrogen Atoms
Hydrogen (^1H)	99.984
Deuterium (^2H)	0.016
Tritium (^3H)	1×10^{-17}

Oxygen Isotopes	Percent of All Oxygen Atoms
Oxygen-16 (^{16}O)	99.763
Oxygen-17 (^{17}O)	0.038
Oxygen-18 (^{18}O)	0.200

Source: Cornell University Stable Isotope Laboratory, http://www.cobsil.com/iso_main_stable_principles.php

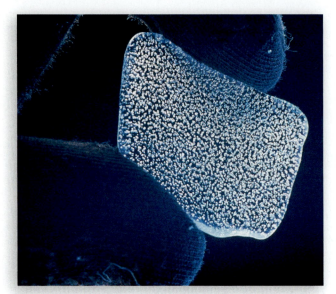

FIGURE 16.17 A thin slice of ancient glacial ice showing trapped air bubbles, which have preserved samples of the atmospheric gases that were present at the time the original snow turned into glacial ice.

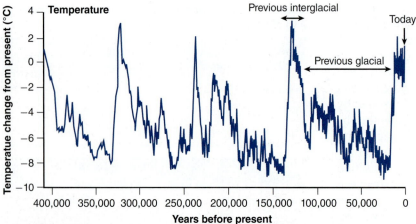

FIGURE 16.18 Graphs showing atmospheric temperature and carbon dioxide changes over the past 400,000 years as recorded in glacial ice at Vostok, Antarctica. During this period Earth's climate has fluctuated between glacial and interglacial periods, which were triggered by changes in orbital variations.

detail. Note that ancient ice deposits also have tiny air bubbles (Figure 16.17) that contain the atmospheric gases present at the time the original snow was transformed into glacial ice. Studies of glacial ice, therefore, provide a relatively long and detailed record of both atmospheric temperature and greenhouse gas concentrations.

Since the 1960s, climatologists have been drilling into glacial ice sheets in Antarctica and Greenland and various mountain glaciers around the world. Here researchers collect long cores of ice, then measure the isotopic ratios of hydrogen and oxygen and CO_2 concentrations over the length of the core. A good example is the data set shown in Figure 16.18 obtained from an 11,886-feet (3,623 m) hole, drilled into the antarctic ice. Data from this project have yielded a detailed temperature and CO_2 record going back over 400,000 years—a similar drilling project goes back 800,000 years. A striking feature of these ice records is the repetitive pattern of glacial and interglacial (warm) periods. Another key feature is the strong correlation between atmospheric temperature and carbon dioxide concentration.

Lessons from the Past

Glacial ice studies all show a strong correlation between greenhouse gas concentrations and atmospheric temperature. A key question, though, is what triggers the shift between glacial and interglacial periods and the transfer of carbon between the atmosphere, oceans, and biosphere? Recall that detailed studies have linked the cyclical glacial patterns with changes in solar heating as predicted by the Milankovitch cycles. Here the effects of orbital variations periodically reinforce each other, disrupting Earth's heat balance to the point where positive feedback mechanisms are activated, particularly carbon dioxide and ice albedo. Orbital forcings then act as a triggering mechanism that activates positive feedbacks, which then amplify the initial warming or cooling brought on by orbital changes. Climate scientists hypothesize that this triggering and feedback relationship has caused Earth's climate system to oscillate back and forth for the past 3 million years between two states of natural equilibrium, namely glacial and interglacial periods. Note that other glacial cycles have occurred throughout Earth history; the cycle beginning 3 million years ago is the most recent and well documented.

Skeptics of anthropogenic global warming often point to the ice core records (Figure 16.18) and note that when the climate system goes into a cooling (glacial) period, temperatures can begin to decrease long before carbon dioxide levels start to fall. Because CO_2 sometimes lags behind temperature, they argue that the release of CO_2 by human activity must not be the primary driving force behind the current warming trend. However, just because greenhouse gases amplify changes initiated by orbital forcings does not mean that these gases cannot take the lead in triggering climate change. For example, it is well accepted that a rapid release of methane (CH_4) from the seafloor 55 million years ago caused global temperatures to increase by as much as 13°F (7°C). Other evidence that greenhouse gases are capable of initiating climate change includes results from general circulation models (GCMs). Moreover, GCM simulations have shown that solar forcings alone cannot account for the current rate of warming, leaving elevated greenhouse gas concentrations as the primary driving force.

Another common argument against human-induced global warming is that the current warming trend is well within the natural range of past temperature fluctuations. For example, the 2,000-year Greenland record in Figure 16.19 clearly shows that temperatures in the Northern Hemisphere have been higher than today, such as during the so-called Medieval Warm Period. It was during this warm interval that the Vikings succeeded in colonizing Greenland. This interval, however, was followed by a cooling episode known as the Little Ice Age, ultimately causing the Viking colonies to collapse.

The problem with basing conclusions about global warming on the 2,000-year climate record is that it represents a rather brief period of Earth history. A much different picture of climate variability emerges when one examines the longer record shown in Figure 16.20. This graph displays 50,000 years of temperature changes in Greenland, taking us back through the entire interglacial period and into the previous ice age. One of the most striking features of this record is the abrupt and dramatic changes in climate that took place during the last glacial period. Of particular interest is how the climate began to slowly warm from about 20,000 to 15,000 years ago, but then suddenly went into a several-thousand-year cold snap known as the *Younger Dryas*. This abrupt and dramatic cooling event occurred despite the fact that orbital forces were pushing the planet into a warm interglacial period. This and other dramatic shifts in the record illustrate how Earth's complex climate system is highly sensitive to change.

Another important lesson we can learn from the Greenland record is how remarkably stable the climate system has been for the past 10,000 years, a period in which human civilization began to flourish. It is obvious from Figure 16.20 that the climate fluctuations during the past 2,000 years are minor compared to the sweeping changes that took place during the previous ice age. Of concern to many scientists is the possibility that anthropogenic global warming could push the climate system beyond some unknown threshold, activating powerful feedback mechanisms that destabilize the planet's sensitive climate system. If this were to occur, civilization could face climatic swings far more dramatic than anything we have seen in the past 10,000 years. Consider that as the planet began to enter the interglacial period after the Younger Dryas, temperatures in Greenland rose more than 18°F (10°C) in a single decade. For comparison, the temperature swing between the Medieval Warm Period and the Little Ice Age was only about 3°F (1.7°C). This relatively minor temperature swing was enough to drive the Vikings from Greenland and bring famine to civilized Europe.

Future of Our Climate System

In terms of geologic time, the release of greenhouse gases over the past 250 years by human activity has been a rather sudden event. There is little question among climatologists that global warming is being driven in large part by a more intense greenhouse effect. This, in turn, is disrupting Earth's heat balance and driving the climate system toward a higher energy state. Climatologists have been predicting that this shift to a higher energy level would lead to certain changes, such as more intense storms and shifting rainfall patterns. However, because of the statistical nature of weather and climate, it has been difficult to show scientifically that these changes are

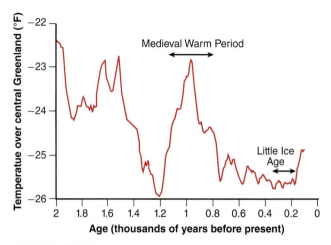

FIGURE 16.19   Graph showing temperature variations over Greenland during the past 2,000 years as determined from isotopic analysis of glacial ice. The Medieval Warm Period coincided with the expansion of human settlements in Europe and Greenland, whereas the cooling period known as the Little Ice Age brought famine and disease.

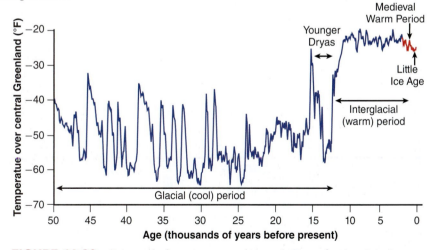

FIGURE 16.20   A record of temperature change in Greenland, extending back 50,000 years. Note the abrupt and dramatic climate changes during the previous ice age. These wide swings in temperature are unlike anything seen during the past 2,000 years (in red) of the current interglacial period. In fact, Earth's climate over the past 10,000 years has been remarkably stable, allowing human civilization to flourish.

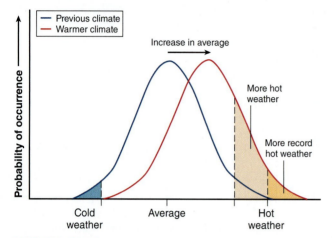

FIGURE 16.21 The climate of a region is actually the statistical variation in daily weather events, where the number of days with cool weather roughly equals the days with warm weather. Global warming is causing a shift in climates worldwide, resulting in more hot weather events and record highs.

taking place. Remember that *weather* is the state of the atmosphere at any given time and place, whereas *climate* represents the long-term average weather and its statistical variation.

Suppose that you recorded the average daily temperature (i.e., weather) at your house for a year, and then counted the number of days a given temperature occurred. If you plotted the data it would likely follow a bell-shaped curve similar to the one in Figure 16.21. Such a curve represents the overall climate and its statistical variation. Notice that the number of days with cooler than average weather is roughly equal to the number of days of relatively warm weather. A shift to a warmer climate simply means that there will be a greater number of warm days compared to the previous climate. There will also be an increase in the number of record highs, and a decrease in record lows. One of the convincing arguments for global warming is the graph in Figure 16.22, which shows a significant increase in the number of years where global temperatures have been above the recent long-term average. Also note how record highs have been occurring more frequently. In fact, 8 of the 10 hottest years on record have taken place since 1998. Keep in mind that winters will still occur in a warmer world, but exceptionally cold winters will be less frequent.

Depending on the rate at which humans continue to emit greenhouse gases, the Intergovernmental Panel on Climate Change (IPCC) reports that by 2100, global average temperatures will likely increase somewhere between 3.2° and 7.1°F (1.8–4.0°C). Although the shift to a warmer climate appears to be taking place gradually, it is quite possible that future changes will be abrupt and drastic, similar to those found in the Greenland record. However, the IPCC projections are based on global climate models, which are good at simulating long-term trends, but currently lack the sensitivity to accurately forecast abrupt changes. Policymakers therefore generally assume that global warming will continue to be a gradual process relative to human time scales.

Because climate change is such a serious issue, scientists are keenly interested in learning what makes the climate system undergo abrupt and dramatic changes. The first clue came back in the 1990s when paleoclimatologists discovered the rapid temperature swings in the Greenland record. After gathering additional data from different sites around the world, some researchers hypothesized that the sudden climate shifts were linked to disruptions in *thermohaline circulation* within the North Atlantic. It is thought that a warming trend could send large volumes of glacial meltwater flowing into the North Atlantic, reducing the salinity of the warm surface current known as the Gulf Stream. If the Gulf Stream were disrupted, then Greenland and much of Europe could experience a period of sudden and rapid cooling.

This meltwater hypothesis led to the concept of a **climate threshold,** in which some component of the climate system is pushed beyond a critical point, causing the entire system to suddenly change. Richard Alley, an expert on Greenland's climate, uses the example of a canoe to illustrate how climate thresholds operate. A person can safely rock a canoe back and forth, but only to a certain point. If they go past this critical point or threshold, the canoe will suddenly overturn. Similarly, the climate system is thought to have multiple tipping points or thresholds. Should a critical threshold be crossed, the resulting change would then become amplified by positive feedbacks, creating a ripple effect through the entire system. Eventually the cli-

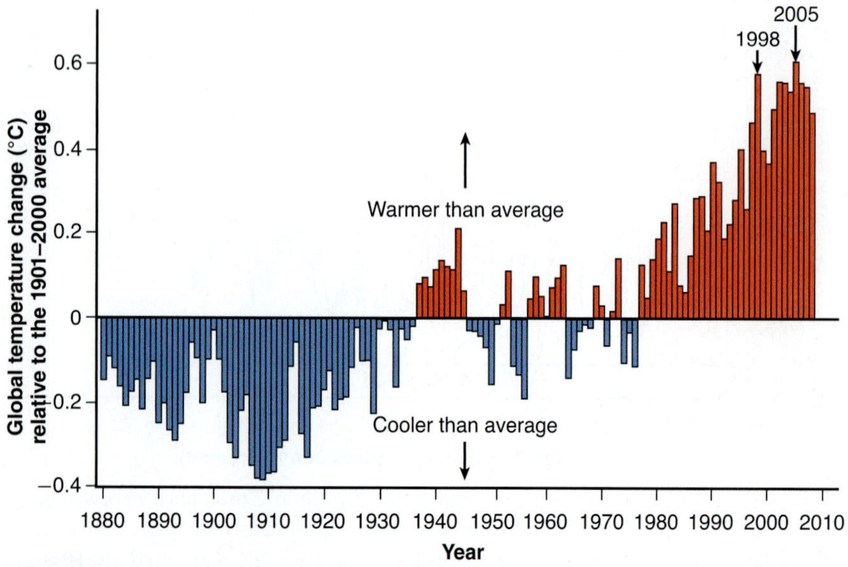

FIGURE 16.22 A historical record showing yearly changes in average global temperature relative to the long-term 1900–2000 average. Note that 8 of the 10 hottest years have occurred since 1998.

mate system would reorganize into a new and stable configuration, where it may remain for centuries or millennia before being pushed past another threshold.

In addition to thermohaline circulation, climatologists have identified another possible climate threshold related to the distribution of heat in the tropical oceans. Recall that the oscillating manner in which heat is transported away from the equatorial parts of the Pacific generates El Niño and La Niña conditions, affecting rainfall patterns around the globe. The concern is that a prolonged El Niño would cause an extended drought in the Amazon rainforest. Should enough trees die, it could result in a permanent loss of soil moisture that pushes the Amazon ecosystem past a critical threshold, beyond which it could not recover. The collapse of the rainforest ecosystem would not only affect regional climates, but represent a significant loss in the ability of the biosphere to remove carbon dioxide from the atmosphere. To make matters worse, the decaying trees would release substantial volumes of greenhouse gases. Global warming would then accelerate as the rainforest switches from operating as a negative feedback that dampens the effect of greenhouse gas emissions, to a positive feedback that amplifies the warming.

To get a better sense of how human activity might be pushing the climate system toward critical thresholds, it is useful to examine the 800,000-year climate record from Antarctica shown in Figure 16.23. Notice how carbon dioxide concentrations have consistently ranged between 180 and 280 parts per million (ppm) as the climate system oscillated back and forth between two operating modes, namely glacial and interglacial. Likewise, methane levels have ranged between 350 and 700 parts per billion (ppb). Again, this pattern is due to orbital variations that first trigger a change in Earth's heat balance, followed by positive feedbacks that reinforce and amplify the initial change, and negative feedbacks that keep the system in check. This has produced an oscillating and self-regulating system with relatively firm limits or stops, such as the historic carbon dioxide concentrations of 180 and 280 ppm. However, in just 250 years, humans have taken 240 billion tons of carbon, once sequestered in sedimentary rocks as fossil fuels, and transferred it to the atmosphere. The result is that by 2008, CO₂ levels had climbed to 385 ppm, a level that is far beyond the climate system's 180 to 280 ppm range for the past 800,000 years. Likewise, methane concentrations are now far greater than at anytime during the past 800,000 years.

Scientists do not know for certain how the climate system is going to respond to this recent input of greenhouse gases, but it is becoming abundantly clear that the system is now undergoing significant change as it gains energy.

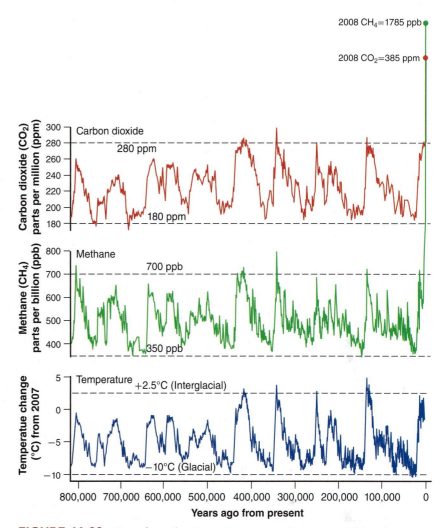

FIGURE 16.23   Data from the Dome ice core project in Antarctica showing variations in temperature, methane, and carbon dioxide for the past 800,000 years. The data indicate that the climate system has been operating within a certain range of temperature and greenhouse gas concentrations. Note that current CO₂ and CH₄ concentrations are now far beyond the normal operating range. The abnormally high greenhouse gas concentrations could push the climate system beyond critical thresholds, causing abrupt rather than gradual climate change.

Consequences of Global Warming

Based on our scientific understanding of how Earth operates as a system, we can expect that global warming will impact nearly every aspect of the Earth system. Because humanity has adapted itself to environmental conditions that have been relatively stable for thousands of years, climatic changes brought on by global warming could have potentially severe consequences. Of particular concern is the possibility that the warming may push the system past critical thresholds, initiating a period of abrupt and dramatic climate change. No one knows whether the climate system will undergo gradual or abrupt change, so our focus in this section will be on the expected changes themselves. In general, the consequences for society will revolve around water and food supplies, shoreline and ecosystem changes, and human health. The degree to which global warming poses challenges for humanity will depend in large part on how much society can reduce its emission of greenhouse gases.

Changing Weather Patterns and Biomes

Many of the problems with global warming are related to how higher temperatures are affecting the mechanisms that redistribute heat around the planet, namely ocean and air currents and the hydrologic cycle. For example, a warmer world has resulted in an increase in the frequency of certain types of extreme weather events, particularly heavy rains, droughts, and heat waves. Worldwide temperature and precipitation patterns are also changing, with some areas of the globe experiencing greater change than others. Because the location of Earth's *biomes* (groupings of similar types of plants, animals, insects, and bacteria) is largely a function of temperature and precipitation patterns, plant and animal communities are now beginning to migrate.

Climatic change and shifting biomes have occurred repeatedly throughout our planet's history, and are therefore a natural part of the Earth system. Humans, of course, are part of the biosphere and have settled into those biomes we find most hospitable. Should a significant shift in biomes occur because of global warming, we would naturally migrate and follow our preferred biomes. Being quite adaptable, humans could probably cope with slowly shifting biomes. However, if climate change becomes abrupt, it would be very difficult to move existing populations quickly enough to regions with adequate water supplies and conditions suitable for agriculture.

Extreme Weather

As Earth's climate system gains additional heat energy, evaporation rates will naturally increase. This and the fact that a warmer atmosphere is capable of holding more water vapor will result in a greater potential for unusually heavy rain and snow events. In a warmer world we should also expect to see a greater number of record highs, which will lead to increased drought conditions. Keep in mind that climate is statistical in nature, which means that an individual weather event cannot be linked directly to global warming. Scientists typically analyze historical weather records and determine whether or not a statistically significant relationship exists between global warming and the frequency of certain types of extreme events.

As the amount of data has increased in recent years, climatologists have been able to show statistically that atmospheric temperatures and heat waves have increased worldwide. Weather data since 1950 also show that while daytime temperatures have risen, nighttime temperatures have increased at a significantly faster rate. This means that on a global scale, less cooling is taking place at night because of the more intense greenhouse effect. A direct consequence of this is that occasional heat waves can be more severe with respect to human health, particularly in areas where air conditioning is less common due to normally mild temperatures. People not only experience greater levels of heat stress during the day, but find less relief at night. Take for example the 2003 heat wave in Europe, where 35,000 people, many of whom were elderly and living in homes without air conditioning, died of heat-related causes.

Climatologists have also documented that as temperatures have increased there has been a corresponding increase in heavy precipitation events, elevating the risk of flooding. More frequent flooding clearly poses problems for society as it puts human lives at risk and leads to higher property and agricultural losses (Chapter 8). Although heavy rain events may become more frequent, higher temperatures will accelerate the evaporative loss of water from soils in between rain events, thereby increasing the risk of drought conditions and agricultural losses.

Another important research question is whether global warming is causing tropical storms and hurricanes (Chapter 9) to occur more frequently and with greater intensity. A key factor in the formation of tropical storms is that surface ocean temperatures need to be at least 80°F (26.5°C). As ocean temperatures continue to rise, scientists have hypothesized that there will be an increase in both the number and intensity of hurricanes. Although hurricane frequency has increased in the Atlantic, the data do not show this to be true worldwide, leading some experts to believe the changes in frequency are simply part of natural oscillations within the climate system. Others believe that global warming may be affecting upper-level winds such that it inhibits the formation of these complex storms. While the question of increased frequency remains open, new studies have shown a measurable increase in maximum wind speed and duration of hurricanes worldwide. In addition, studies have found a worldwide increase in the number of more powerful category 4 and 5 storms, as opposed to tropical storms of all sizes. This increase in the power of hurricanes suggests a link to global warming.

Shifting Precipitation Patterns and Climatic Zones

In addition to the intensity and frequency of weather events, global warming is influencing the timing of the spring rains, and is suspected of causing climatic zones to shift toward Earth's poles. Here winters are becoming milder, and in certain areas snow is beginning to accumulate later in the fall and starting to melt earlier in the spring. Moreover, spring rains in some areas are occurring earlier, which means soils are exposed to dry, summer conditions for longer periods of time. Soil moisture depletion therefore is more likely to occur during the summer, increasing the potential for drought conditions and reduced agricultural production. Warmer temperatures can then cause natural biomes and agriculturally productive areas to shift toward higher latitudes where soil moisture depletion is less severe.

By decreasing the period in which snow is able to accumulate, global warming is also causing snowpacks to be thinner in mountainous areas. Snowpacks then may no longer be thick enough to last through the summer, creating serious water-supply problems for cities and farmers that rely on meltwater streams. In the United States, officials have been worried about declining snowpacks in the Sierra Nevada Mountains of California and throughout the Rocky Mountains. Both of these mountainous areas are the source of critical water supplies for extensive agricultural operations and millions of people throughout this arid region.

Scientists are also detecting changes in the large-scale circulation patterns that control global weather patterns. As illustrated in Figure 16.24, the climate system contains three large circulation cells that lie in belts parallel to the equator in each hemisphere. These cells circulate vast amounts of heat energy and water vapor between the oceans, land, and atmosphere. Note how precipitation rates are highest in regions where evaporating water is drawn up into the atmosphere by rising air currents. Then in areas where the depleted air masses descend, they generate dry, arid conditions at the surface. In this way the circulation cells create alternating bands of humid and arid zones that generally lie parallel to the

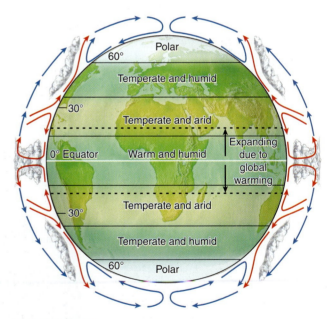

FIGURE 16.24   Earth's climate system has three sets of circulating cells on either side of the equator. Humid zones form in belts where rising humid air masses generate high rates of precipitation, whereas arid zones are generally found where dry air descends. Global warming is enlarging the mid-latitude cells, threatening to push arid zones poleward, disrupting existing agricultural production and water supplies in temperate and humid zones. This shift is also allowing tropical diseases to move northward.

equator. The size and shape of these zones are strongly influenced by the position of the continents and major mountain ranges.

As the ocean-atmosphere system continues to gain heat energy, climatologists have found measurable changes in these global circulation patterns. Higher evaporation rates have led to increased rainfall over the Tropics, but in the arid belts where dry air masses descend, there has been an overall decline in precipitation. In temperate and humid zones located at higher latitudes, precipitation associated with the rising air currents has generally increased. In essence then, the areas that are normally humid tend to receive more precipitation, whereas arid regions receive even less. This presents a serious problem in drier climates where water supplies are already limited and agriculture activity is marginal.

Global warming may also be responsible for the climatic zones making a measurable shift away from the equator and toward Earth's polar regions. Researchers have found that from 1979 to 2005, the set of circulating cells located on both sides of the equator (Figure 16.24) expanded about 70 miles (110 km) toward their respective poles. This, in turn, has caused the major arid belts to move toward the poles. From the image in Figure 16.25, you can see that if the arid belt in Africa and the Middle East continues to shift northward, desertlike conditions could expand into the heavily populated areas of Europe and Asia. Moreover, as indicted by the desertification vulnerability map in Figure 16.26, shifting weather patterns around the world have the potential to convert many semiarid areas into deserts. This is of particular concern because much of the current grain-producing regions of North America and Asia lie in semiarid climates. While warmer temperatures may open up parts of Canada and Siberia to grain production, farming operations would likely collapse in many existing areas.

In addition to food and water supplies, global warming will present society with a host of problems related to changes in Earth's *biomes.* As circulation cells along the equator expand, warm tropical conditions will

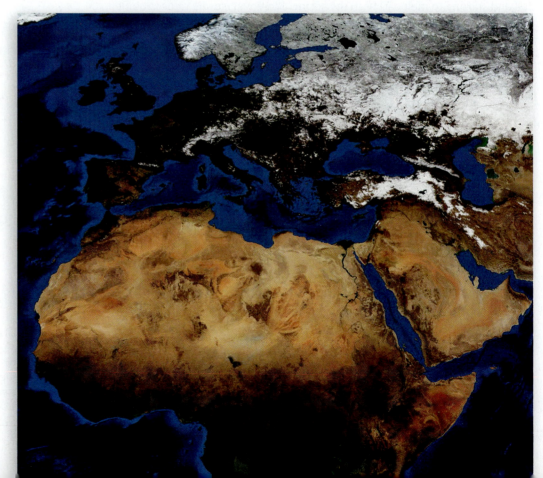

FIGURE 16.25 A satellite image showing the present arid belt running through Africa and the Middle East. As this arid zone continues to move poleward, desertlike conditions will expand into the heavily populated, grain-producing regions of Europe and Asia. Note the areas blanketed by snow in this wintertime image.

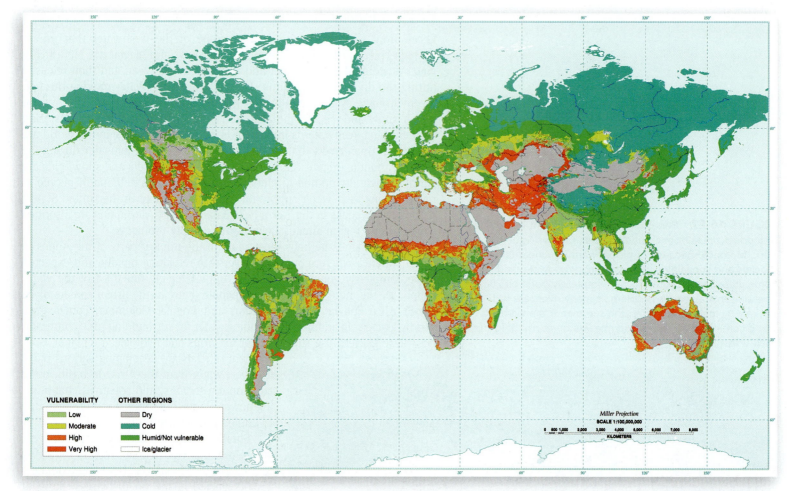

FIGURE 16.26   A map showing the current location of major deserts (gray) and the vulnerability of different areas to undergo desertification. As global temperatures increase, the existing deserts are expected to become larger.

move toward higher latitudes, bringing tropical diseases like malaria and dengue fever to more heavily populated areas. To make matters worse, people currently living at higher latitudes generally have never been exposed to tropical diseases, hence have no natural resistance. For example, in 2007, Italy experienced a local epidemic of dengue fever, representing the first modern outbreak of a tropical disease in Europe. Shifting climatic zones can also bring new organisms into contact with humans living in polar zones. During the summer of 2006, people in Alaska's Prince William Sound began contracting food poising by eating oysters contaminated with a bacterium never before found in waters at such high latitude.

Melting Glacial Ice and Permafrost

Although Earth is presently in a warm interglacial period, massive ice sheets still cover Antarctica and Greenland, and mountain glaciers still exist, even in the Tropics. In addition to glacial ice, permafrost conditions remain over vast tracks of Asia and North America. We will explore some of the expected consequences should global warming accelerate the melting of glacial ice and thawing permafrost. As with changing climatic zone and biomes, there will be benefits to these changes, but overall, the potential problems associated with disruptions to established population patterns and economies will likely far outweigh any benefit. Moreover, should the climate system pass critical thresholds and enter a mode of abrupt change, then humans would probably find it even more difficult to cope with changes in glacial ice and permafrost.

Water Supplies

Recall from Chapters 8 and 10 that some of the freshwater that moves through rivers comes from the melting of snow and glacial ice. Here freshwater is stored during the wet and cool part of the year, and then released during the dry summer months. The volume of meltwater moving through the hydrologic cycle naturally depends on Earth's climate and the rate at which snow and ice accumulates, and then melts. This process of storing and releasing freshwater is important because many people in the world are dependent upon rivers that are fed almost entirely by meltwater during the summer months (Figure 16.27). Today mountain glaciers and snowfields are shrinking worldwide due to global warming, putting millions of people at risk of losing their water supply for much of the year. In the United States, reduced snowpacks in the Sierra Nevada and Rocky Mountains are a serious concern for a region already facing water-supply problems.

The problem is perhaps most acute in the tropical parts of the Andes in South America where mountain glaciers are retreating rapidly, and expected to be completely gone by 2050. Here approximately one-third of the water supply for the more than 11 million people living in the urban centers of La Paz-El Alto, Bolivia; Quito, Ecuador; and Bogotá, Colombia, comes from nearby glaciers. In addition to providing water for surrounding rural communities, these glaciers also supply water for crops and hydroelectric plants. Government officials are developing various plans to divert water from the Amazon basin, build additional reservoirs, and construct coal-burning power plants, but funding is a problem due to limited financial resources.

Similar problems are expected in Asia where mountain glaciers are rapidly retreating throughout the Himalayas. Here approximately 15,000 glaciers make up the largest store of freshwater on the planet after Antarctica and Greenland. Meltwater from these glaciers contributes as much as 70% of the summer flow in the Ganges, Indus, and Brahmaputra Rivers, which, in turn, provide water for nearly 500 million people living in Pakistan, India, and Bangladesh. Researchers are increasingly worried that the rapid retreat of these glaciers could soon limit the amount of water available for irrigating crops and generating hydroelectric power. Governments in the region, like those in the Andes, have very limited resources for constructing water-supply projects that could ease future shortages.

Rising Sea Level

Currently, 2.2% of Earth's water is being stored on land as glacial ice. During glacial periods sea level routinely falls as additional water is removed from the oceans and stored as glacial ice. Sea level then rises when the climate warms and the ice begins to melt. As shown in Figure 16.28, around 20,000 years ago during the depths of the last ice age, sea level was approximately 460 feet (140 m) lower than today. Such changes are not

FIGURE 16.27 Photos showing a significant loss of ice in modern times from Grinnell glacier, Glacier National Park, Montana. Because of global warming, mountain glaciers and snowfields are retreating worldwide, and in some areas are expected to be completely gone by 2050. This presents a serious problem as many rivers are fed entirely by meltwater during the summer. If the warming trend continues, millions of people will be at risk of losing their water supply.

1940

2006

unusual, which means there is nothing special about the current level that modern societies have grown accustomed to. Although the climate has been quite stable for the past 10,000 years, sea level has slowly risen as the Earth attempts to reach equilibrium with respect to its heat balance—from 1900 to 2000, sea level has risen only 0.6 feet (0.2 m). However, if all of the glacial ice remaining today were to melt, sea level would rise an additional 260 feet (80 m). While no one is predicting such a scenario, how much of the current ice will melt and how fast it will melt are major concerns in regard to global warming.

Based on temperature projections from climate models run under different greenhouse gas scenarios, the IPCC reported in 2007 that sea level will likely rise between 0.6 and 1.9 feet (0.2–0.64 m) by 2100. The lower estimate being equivalent to the rise over the past century, which was relatively minor. However, even at today's comparatively low rate, people around the world are beginning to feel the effects of sea-level rise. For example, low-lying Pacific island nations are at increasingly higher risk during storms, with many of the islands likely to be abandoned within the next 50 years. Also consider Bangladesh, one of the poorest nations on Earth, where many of its 150 million residents live on a low-lying river delta. According to the 2007 IPCC assessment report, nearly 17 million citizens in Bangladesh are expected to be displaced by 2100. Where will these people go and who will feed them?

Although the IPCC's upper limit of a 1.9-foot (0.64 m) rise in sea level over the next century is based on good science, there is geologic evidence that it could be much greater. Here it is important to understand some of the assumptions and limitations of the modeling studies of the IPCC. Because the massive ice sheets covering Antarctica and Greenland are very complex, and good data are limited, the models had to assume that the ice sheets will melt somewhat similar to blocks of solid ice. However, glacial ice sheets are actually slow-moving bodies with numerous fractures called *crevasses*. As shown in Figure 16.29A, lakes can form on the surface of ice sheets during the summer, which increases solar heating and accelerates the melting. This meltwater also moves downward through crevasses and large vertical

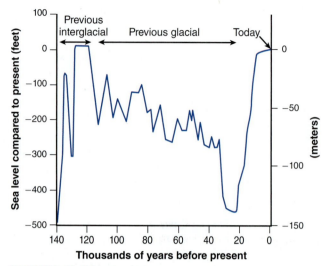

FIGURE 16.28   Graph showing dramatic changes in sea level during the previous glacial and interglacial periods. About 20,000 years ago enough freshwater was stored in glacial ice to lower sea level more than 400 feet compared to today. Most of this water has since been returned to the oceans, and continued melting is slowly raising sea level even further.

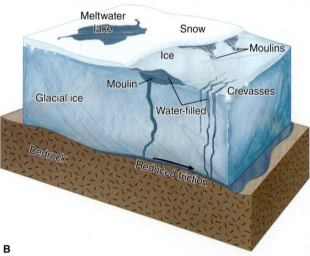

A   B

FIGURE 16.29   Photo (A) showing a lake and numerous crevasses on the surface of the Greenland ice sheet. The presence of lakes increases solar heating and accelerates melting. Meltwater also infiltrates through crevasses and moulins (B), causing the ice to melt from within. Water that reaches the base reduces friction with the underlying bedrock and allows the ice to flow more rapidly.

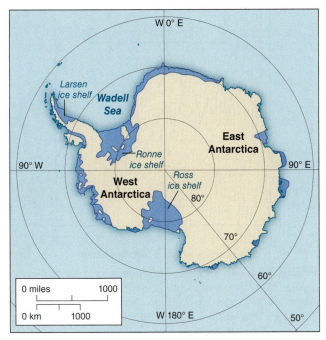

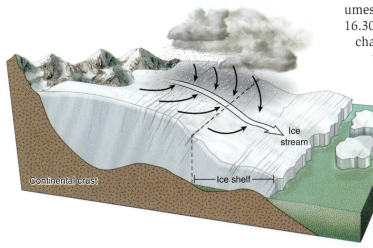

FIGURE 16.30 Glacial ice moving from the interior of West Antarctica funnels through areas called ice streams and moves out into the sea, forming extensive ice shelves. Several ice streams are now beginning to accelerate, creating the possibility that large sections of the West Antarctic ice sheet will destabilize and flow into the sea, causing a rapid sea-level rise.

conduits called *moulins,* causing the ice to melt from within (Figure 16.29B). Some of this water will build up and increase the pore pressure (Chapter 7) at the base of the ice, thereby reducing the friction between the glacier and underlying bedrock. The reduced friction allows the ice to flow more rapidly, and in some cases, triggers a sudden and rapid increase in flow called a **glacial surge.** The combined effects of internal melting and glacial surging act as a positive feedback, which can cause large ice sheets to break up much more rapidly compared to current model calculations.

Glaciologists are now concerned that positive feedbacks could cause the Greenland and Antarctic ice sheets to reach a critical threshold such that large sections of the sheets begin to disintegrate. Were this to occur, sea level would rise much higher and faster than predicted by the IPCC. In Greenland, much of the ice is forced to flow through narrow glacial valleys, called *fjords,* before entering the sea. Warmer ocean temperatures and the buildup of water along the base of the ice are believed to be responsible for the recent increase in glacial flow through Greenland's fjords. Since these narrow passageways act as buttresses that help support the overlying ice sheet, glaciologists believe that if the flow becomes too great, parts of the ice sheet could destabilize, sending large volumes of ice into the sea.

At the other end of the globe in Antarctica, there is a possibility that the West Antarctic ice sheet could destabilize and send even larger volumes of ice into the sea compared to Greenland. As illustrated in Figure 16.30, ice in Antarctica flows away from the continental interior and is channeled into rapidly moving zones called *ice streams.* Upon reaching the coast, some ice streams are able to extend a great distance over the sea and form what is known as an *ice shelf.* Today there is mounting evidence that several of the major ice streams in West Antarctica are beginning to accelerate. Here it is suspected that rising sea level and warmer ocean temperatures are playing a role in destabilizing the ice streams where they extend offshore. The concern is that further destabilization could allow large volumes of ice from the main ice sheet to surge into the sea, resulting in rapid sea-level rise. Although Antarctica is one of the few places in the world where glaciers are still growing, destabilization of the West Antarctic ice sheet could still occur.

Scientists estimate that if both the Greenland and West Antarctic ice sheets were to destabilize, sea level would rise 33 feet (10 m) over the course of a few centuries, not millennia as projected by the IPCC. This worst–case scenario would have catastrophic consequences for nations around the globe. To help appreciate the severity of a 33-foot (10 m) rise, consider the inundation map of southern Florida in Figure 16.31. Although such a rise may take centuries, the consequences would be severe considering that nearly 25% of the current U.S. population lives within the inundation zone. Flooding would progressively destroy vast amounts of the nation's existing infrastructure and displace tens of millions of people, all of which would likely lead to serious economic and social upheaval.

Melting Sea Ice and Release of Stored Methane

In addition to glaciers, important changes are taking place in areas of the planet covered by sea ice and permafrost. Unlike the South Pole, there is no landmass situated over the North Pole. Here sea ice covers much of the Arctic Ocean, expanding in size during the winter, only to shrink again each summer. By using satellites to monitor the maximum and minimum extent of Arctic ice cover since 1979, scientists found a dramatic reduction in sea ice in 2007, followed by a slight increase in 2008 (Figure 16.32). Although

the ice cover had been on a downward trend for 29 years, the magnitude of the 2007 decline was so great that it appears that the Arctic Ocean may now be in an accelerated warming mode. Note that because sea ice floats on the ocean, this melting will not lead to sea-level rise (similar to how melting ice cubes do not cause a cup of water to overflow).

While much media attention has been given to the extinction of polar bears as Arctic sea ice is lost, a more pressing problem is how the loss of ice may trigger a positive feedback that amplifies global warming. As the ice continues to shrink, there is a reduction in albedo (less reflected light), which increases solar heating and causes the Arctic Ocean to become warmer. This sets up a positive feedback, where warmer ocean temperatures increase the rate of ice loss. The concern is that the loss of sea ice may soon pass a critical threshold, beyond which the ice would not return for the foreseeable future. Moreover, exposing the Arctic Ocean would represent a significant change in Earth's overall albedo

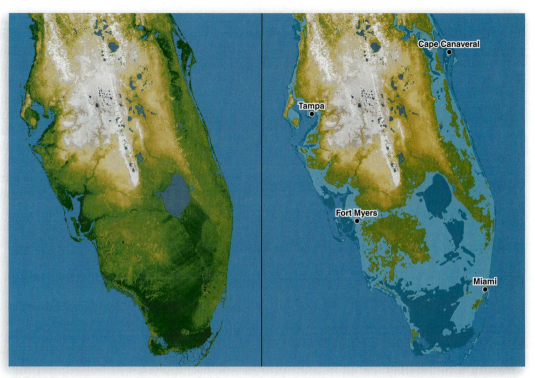

FIGURE 16.31   A shaded relief map of southern Florida showing the areas that would be inundated by the worst-case scenario of a 33-foot (10 m) rise in sea level.

that would tend to amplify global warming. Climatologists also express concern that a newly exposed Arctic Ocean would alter existing global weather patterns and ocean currents in ways not yet understood. However, there are actually two benefits of an ice-free Arctic. First, it would open up the famed Northwest Passage (Figure 16.32), greatly reducing the shipping distance between North America and Europe with Asian countries. Another is that vast new areas of continental shelf will be accessible for oil and gas exploration—burning additional fossil fuels, however, will only exacerbate global warming.

A warmer Arctic may trigger yet another powerful feedback, namely the release of methane gas. Recall from Chapter 14 that *gas hydrates* are frozen compounds containing methane (CH_4) gas. The Arctic contains enormous quantities of gas hydrates currently being sequestered in marine sediment

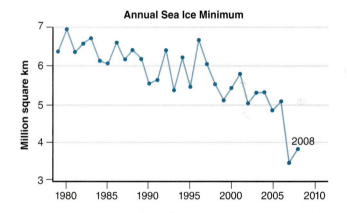

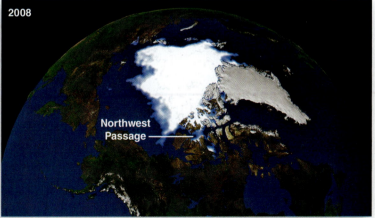
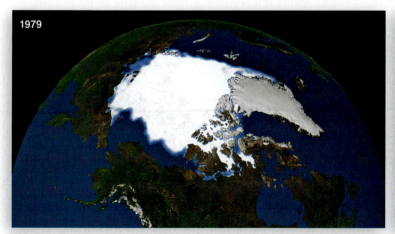

FIGURE 16.32   Minimum summer coverage of Arctic sea ice had been on a 29-year decline, but then abruptly decreased after 2006. The resulting reduction in albedo may cause the Arctic Ocean to undergo accelerated warming, which would then affect the global climate system and lead to an even greater loss of ice. Note how the Northwest Passage is now nearly ice-free during the summer.

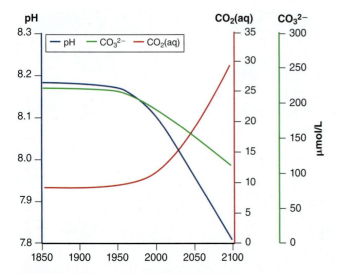

FIGURE 16.33 Graph showing how the increase in dissolved CO_2 is causing the pH of the oceans to lower, or become more acidic. With increased acidity there are fewer carbonate ions (CO_3^{2-}) available for marine organisms to make protective shells. If present CO_2 emission trends continue, by 2100 ocean acidity would increase 80% over 2007 conditions.

and permafrost. Another source of methane is bogs found throughout the Arctic. Here thick layers of saturated organic material, known as *peat,* have been sequestering carbon dioxide in the cold climate. When bogs begin to warm, CO_2 is converted into CH_4. As the Arctic continues to warm, scientists expect that increasing quantities of methane gas will be released from both gas hydrates and from bogs. Because methane has over 20 times the warming effect of carbon dioxide, a large release of methane from the Arctic is expected to cause a sharp spike in atmospheric temperatures.

Acidification of Oceans

As part of the carbon cycle, the oceans have been absorbing roughly 30% of the carbon dioxide humans have released into the atmosphere. While this has helped reduce the rate of global warming, it also has caused the oceans to become more acidic, a process called **ocean acidification.** Recall from Chapter 15 how the pH scale is a measure of the number of hydrogen ions (H^+) in a solution, which is referred to as *acidity.* On this logarithmic scale, a pH of 1.0 (10^{-1} H^+ ions) is strongly *acidic,* and on the other end, a pH of 14.0 (10^{-14}) is strongly *alkaline,* or basic. In historical times the average pH of seawater has been 8.16, meaning the oceans are slightly alkaline (pH of 7.0 is neutral). However, as the oceans are absorbing some of the additional CO_2 released by humans, the dissolved CO_2 is reacting with water molecules and freeing up more hydrogen ions (H^+), making the oceans slightly more acidic. In Figure 16.33 notice that after dissolved CO_2 began to increase, by 2007 the average pH of seawater had declined from 8.16 to 8.06. A decline of 0.10 pH units may not sound like much, but the pH scale is logarithmic, so this decline means that the acidity of the oceans has increased by a factor of 1.25, or 25%.

Ocean acidification is a problem because many marine organisms are sensitive to pH changes. Of particular concern are tiny marine organisms, such as plankton, that form the base of the marine food web. Similar to many marine organisms, these tiny creatures make protective shells from carbonate minerals such as calcite ($CaCO_3$). As the oceans become progressively more acidic, there are fewer carbonate ions (CO_3^{2-}) available to make the protective shells. Consequently, the release of anthropogenic CO_2 into the atmosphere may ultimately affect the entire marine food web. If carbon dioxide emissions continue to increase at their present rate, scientists expect that the pH of seawater will fall to 7.8 (Figure 16.33), representing an 80% increase in acidity over the record 2007 conditions. Should this occur, it would represent a level of ocean acidity not seen for at least the past 20 million years.

Mitigation

Based on what you learned in this chapter, there should be little question that human activity is altering the carbon cycle and impacting the climate system. Moreover, if carbon dioxide emissions continue to increase at the present rate, we could possibly push the system beyond critical thresholds. The result could be abrupt and dramatic climate change and a more rapid rise in sea level. Although the human species has proven to be extremely adaptable, the speed and degree of these changes might make life very difficult for people in many parts of the world that are currently considered habitable. Simply put, the changes could be catastrophic for civilization as we know it.

The fact that global warming is largely due to human activity also means that we have the ability to reduce its impact. Fortunately, the technology currently exists that will enable us to begin making substantial reductions in greenhouse gas emissions. The basic problem lies in generating the political will necessary to make significant reductions in emissions, and to do so quickly enough to avert more serious consequences. Many policymakers are working under the assumption that the warming will be gradual, as in the IPCC model projections. This has made some hesitant to take steps toward reducing emissions that they perceive will impact economic growth. However, as the climate record shows, there is a very real possibility that climate change may be sudden and severe. This in itself would have economic consequences far more dire than those associated with reducing emissions.

In 1997, the international treaty known as the *Kyoto Protocol* was signed by 170 countries, with the aim of reducing greenhouse gas emissions. Under the Kyoto agreement, developing nations were exempt from having to meet specific emission reductions. The rationale behind the exemption was that most of the excess CO_2 in the atmosphere today was released over many decades by the existing industrialized nations. Developing nations would therefore be able to grow their economies, and then later reduce their emissions. However, in 1998 the U.S. Senate failed to ratify the Kyoto Protocol largely because China was exempt from emission reductions due to its developing nation status. Notice in Figure 16.34 that China and the United States are by far the largest producers of CO_2. Although China's per capita CO_2 emissions are still relatively low, continued economic growth combined with its large population means that China's total CO_2 emissions could be much higher in the future.

While the U.S. objection to the Kyoto Protocol seemed reasonable to some, others felt that it should have ratified the agreement, setting an example that would have pressured China to follow suit. Because of the impasse, the world's two largest producers of greenhouse gases have actually increased their emissions since 1998. During this period other industrialized nations began meeting their reduction targets under Kyoto. In 2009, representatives from nations around the world met in Copenhagen, Denmark, to try and develop a new treaty to lower CO_2 emissions. Leading climate experts believe that without the United States and China making substantial CO_2 reductions soon, there is no chance that the world will be able to lessen the impact of global warming. Some believe there may be as little as a decade before the climate system goes past a critical threshold. While no one knows for certain whether or not this will happen, the potential consequences are so great that it seems reasonable for the world to take immediate action.

For industrialized nations whose economies rely on cheap fossil fuels, making rapid and substantial reductions in greenhouse gas emissions will not be an easy task. Switching to alternative, non-carbon-based energy sources will take time, primarily because of the sheer scale of our energy needs (Chapters 13 and 14). Since no single alterative energy resource or conservation measure will lead to an overall reduction in CO_2 emissions, the collective contribution of multiple reduction strategies is needed to avoid continued global warming. In the United States, researchers at Princeton University developed a useful concept called a *stabilization triangle*. As depicted in Figure 16.35 a **stabilization triangle** represents the amount that CO_2 emissions must be reduced to keep future emissions at a steady or stable level. Individual *stabilization wedges* represent the contribution of different reduction strategies, such as the use of clean wind and solar power, improved energy efficiency, and conservation. When taken collectively, the individual wedges or strategies will make it possible for us to stabilize carbon emissions.

It is important to note that human activity has been releasing CO_2 into the atmosphere at a faster rate than it can be removed by natural processes. This means that even if we stopped all emissions today, atmospheric temperatures will continue to increase as the climate system attempts to

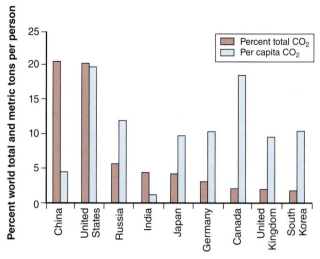

FIGURE 16.34 Percentage of world total and per capita carbon dioxide emissions by country.

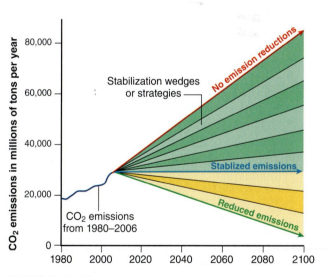

FIGURE 16.35 A stabilization triangle represents the amount that CO_2 emissions need to be reduced to keep future emissions at some stable level. This can be accomplished through the collective contribution of individual wedges, or CO_2 reduction strategies. To prevent continued warming by existing gases, climate experts recommend additional wedges that will lead to an overall reduction in emissions. Note that the emission levels shown for 2100 are for illustrative purposes only and do not represent actual reduction targets.

come to a new equilibrium. To minimize the possibility that the climate system will move beyond a critical threshold, most climate experts believe that future emissions must actually decrease, not just stabilize.

Strategy

To minimize the chance of the climate system moving past a critical threshold, it will be necessary to concentrate our efforts on stabilization wedges that produce the greatest reduction in emissions in the shortest period of time, and at the least cost. For example, conservation is an approach that can have an immediate impact and involves little or no cost to society. Building nuclear power plants, on the other hand, is expensive, and they can take 10 years or more to build. Another key element is to focus our initial efforts on reducing methane emissions, since the warming effect of methane is 21 times greater than carbon dioxide. After concentrating on those wedges that are easy to implement and have the greatest impact, we can turn our attention to more costly and difficult measures. Keep in mind that as technology improves, additional wedges will become feasible and more cost-effective.

Because it is beyond the scope of this text to provide a thorough examination of possible stabilization wedges or strategies, the following measures are intended to give the reader a general idea of what can be done. Ultimately, the best solution is to minimize the use of fossil fuels and begin switching over to non-carbon-based fuels. For details on energy usage, efficiency, and conservation, refer to Chapters 13 and 14.

Conservation

Without question, conservation is the most effective means of making immediate reductions in greenhouse gas emissions. Here businesses and individual citizens only need to cut back on energy use wherever possible. Individual efforts can include such things as eliminating unnecessary trips, carpooling, raising or lowering the thermostat, and turning off the lights when leaving a room. Businesses can also find ways of reducing energy consumption that do not add to their overall operating costs. In fact, one of the advantages of conservation efforts is that users end up saving money because of reduced energy consumption.

Increased Efficiency

In contrast to conservation, increasing efficiency typically requires some physical change in a mechanical system that uses energy. This, in turn, usually involves some cost for purchasing new equipment or making modifications to existing equipment. Similar to conservation efforts, increasing efficiency can be purely voluntary or government mandated. A good example is the higher fuel efficiency standards passed by Congress in 1975, which resulted in a 40% increase in fuel efficiency by 1990 (car owners also realized 40% savings in terms of their annual fuel costs). In addition to automobiles, manufacturers began producing a wide range of more energy-efficient appliances, and builders began constructing well-insulated homes and businesses, all of which led to a significant reduction in energy consumption.

Although great progress was made after the 1970s, there are still many areas in which efficiency gains can be realized. One key area is transportation. Not only can automakers develop even more efficient vehicles, but policymakers can encourage the development and use of mass-transit systems. Getting people out of their cars and into buses and light-rail systems will reduce emissions and ease traffic congestion. Another big efficiency gain with respect to transportation could come from a national effort to rebuild the vast network of railroads that once existed in the United States. Although trucks have proven to be highly flexible and convenient in terms of delivering goods, trains are two to four times more fuel efficient on a ton-per-distance-traveled basis. Here trains would be far better suited for long-distance transport, leaving trucks for more short-distance deliveries. Likewise, passenger rail service could shuttle people between regional destinations far more efficiently than airplanes or automobiles.

Incentives

Because fossil fuels remain the cheapest form of energy, most policymakers agree that some form of incentive is needed for businesses and industries to switch to non-carbon-based energy sources. One option is a **carbon tax** levied on fossil fuels, that is phased in over a period of time to soften the impact. In this approach, the tax rate is based on the amount of CO_2 released by a particular type of fuel, including that which is released during the fuel's extraction and processing. For example, the rate for coal would be higher than for natural gas because more CO_2 is released during the mining and burning coal, per unit of energy. The total amount of tax levied would then depend on total usage and the type of fossil fuel being used. This would produce a strong economic incentive to use the cleanest burning fuel. Moreover, the tax revenues could be used to provide tax credits to encourage the development of non-carbon energy sources such as wind and solar power.

The other type of incentive being proposed is a **cap-and-trade system,** where government regulators set a limit or cap on the amount of emissions for a particular type of user, such as power plants. Here the emission cap represents a total that is lower than what all the plants combined had previously been releasing. The cap is then divided up into small permits and allocated to individual plants based on their previous emission totals. After an initial phase-in period, companies are no longer allowed to exceed their total number of permits. Because the cap-and-trade system allows plants to buy and sell individual permits, a financial incentive exists for operators to develop more-efficient systems, or use a cleaner fuel. In this way, more-efficient plants can sell unused permits to other plants that are unable to meet their emission limits. Some policy experts prefer a carbon tax over a cap-and-trade system because it would lead to more immediate reductions in CO_2 emissions.

Decarbonization of Energy Use

The most effective means of reducing greenhouse gas emissions is for society to begin switching from fossil fuels to carbon-free and low-carbon energy sources. Because of our dependence on fossil fuels and the large amount of energy we consume, switching over to clean energy sources in significant quantities will take time. In many instances the switch will require the development of new technology. For example, through the use of scrubber systems (Chapter 15) to remove CO_2 from combustion gases, the process called **carbon capture and storage** may someday turn coal and other fossil fuels into low-carbon energy resources. As illustrated in Figure 16.36, the CO_2 that is removed from a power plant would be compressed, and then pumped into a permeable rock formation, such as a depleted petroleum reservoir or deep aquifer. However, carbon capture and storage is still largely experimental, and it presents a host of issues, not the least of which are high installation and operating costs. On the technical side, it has yet to be shown whether the CO_2 will remain permanently sequestered or escape into the atmosphere. There is also the issue of finding a suitable geologic formation near each of the plants in the world that

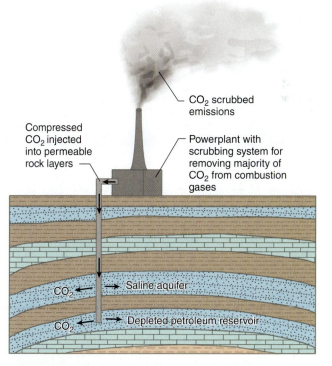

FIGURE 16.36 Carbon dioxide can be captured or scrubbed from the combustion gases at plants that burn or process fossil fuels. The CO_2 is then compressed and injected into geologic formations that are capable of sequestering CO_2 from the atmosphere for an indefinite period of time. This carbon capture and storage process is still experimental, and it has not yet been shown that carbon can be sequestered effectively on a large scale.

burn or process fossil fuels. Storage in the deep ocean is possible, but doing so risks exacerbating the problem of *ocean acidification.*

Because of the pressing need to reduce greenhouse gas emissions as quickly as possible, our focus should be on clean energy sources that are available using existing technology. The most promising, of course, are solar and wind power. Recall from Chapter 14 that wind power is already a less expensive means of producing electricity than is nuclear power. The cost of wind-generated electricity is even approaching that of coal-fired power plants, particularly if the environmental costs of using coal are included. Although solar-generated electricity is still not yet cost-competitive, the technology is improving rapidly, leaving many to predict it will soon be able to compete effectively.

The Future

Mitigating the effects of global warming does not necessarily mean that people in developed nations will no longer be able to enjoy the benefits of modern society. However, it most definitely will require that we make dramatic and fundamental changes in the way we use energy and the types of energy we use (Figure 16.37). If humans hope to keep the climate system from going past a critical threshold, then these changes cannot take place incrementally over the next 50 years or so, but rather they must begin now and on a large scale. Humans have been slow to recognize the signs of climate change, but the danger is now clear.

Although we have enjoyed tremendous growth and prosperity from fossil fuels for powering our societies, we are now entering a period of consequences. Our release of greenhouse gases into the atmosphere has upset a highly sensitive part of the Earth system, namely the climate. Similar to all systems, the Earth is now responding to this imbalance, leaving us with no choice but to live with the consequences of our own actions. There is still time to take corrective action, and hopefully avert disaster. We just need

FIGURE 16.37 In order to avoid pushing the climate system past a critical threshold, humans need to make fundamental changes in the way we use energy, from using more efficient modes of transportation (A) to finding clean sources of electricity (B) to power our cities.

the political leadership and collective action of all citizens to make it happen. Ultimately, this crisis will force us to face the fact that Earth is our only home; one that provides everything we need: clean air and water, food, and a hospitable climate. Remember, if we alter the Earth system so that it is no longer hospitable for humans, we have nowhere else to go.

SUMMARY POINTS

1. Weather is the state of Earth's atmosphere at any given time and place, whereas climate represents the long-term average of weather and its statistical variation.

2. Earth's atmosphere is warmed by two processes: solar heating and the greenhouse effect in which carbon dioxide, methane, nitrous oxide, and water vapor absorb outgoing infrared radiation.

3. Earth's climate varies naturally because of cyclic changes in solar output and in Earth's orbital characteristics that affect solar heating of the planet. Changes in solar heating are then amplified by positive climate feedbacks and dampened by negative feedbacks.

4. Due to the release of greenhouse gases by humans, the atmosphere is retaining more heat, which is now acting as the primary driving force in the current global warming trend.

5. Thermohaline circulation and cyclic El Niño and La Niña conditions in the Pacific greatly affect the distribution of heat and freshwater around the planet, and, in turn, play a key role in Earth's climate system and hydrologic cycle.

6. Using general circulation models for studying the global climate system, scientists have shown that anthropogenic greenhouse gases are largely responsible for the current warming trend. Model projections indicate gradual warming between 3.2° and 7.1°F over the next century.

7. From isotopic analysis and gases trapped in glacial ice, scientists have found clear evidence for glacial and interglacial cycles for at least the past 800,000 years. These data also show a strong correlation between atmospheric CO_2 and temperature.

8. Ice studies have proven that the climate system commonly goes through cycles of abrupt and dramatic change. Over the past 10,000 years the climate system has been remarkably stable, allowing human civilization to flourish.

9. Scientists are concerned that anthropogenic warming is pushing the climate toward a critical threshold, beyond which the system may begin to undergo abrupt and dramatic change.

10. Regardless of whether climate change is gradual or abrupt, most of the major impacts on society will revolve around water and food supplies, more severe storms, shoreline and ecosystem changes, and human health.

11. To avoid pushing the climate system past a critical threshold, experts believe that society needs to immediately begin reducing greenhouse gas emissions.

12. Because no single course of action will bring emissions down fast enough to avoid making global warming worse, the best mitigation strategy is to concentrate on specific options that provide the greatest reduction in emissions in the shortest period of time and for the least amount of money.

13. Most experts suggest that any mitigation strategy involve incentives for reducing carbon emissions, conservation efforts, increased efficiency, and development of non-carbon energy sources.

KEY WORDS

albedo   519
cap-and-trade system   543
carbon capture and storage   543
carbon cycle   520
carbon sequestration   520
carbon tax   543

climate   514
climate feedbacks   518
climate threshold   530
El Niño   523
general circulation models
   (GCMs)   525

glacial surge   538
global warming   511
greenhouse effect   515
heat balance   514
La Niña   524
Milankovitch cycles   517

ocean acidification   540
paleoclimatology   526
stabilization triangle   541
thermohaline circulation   522
weather   514

APPLICATIONS

Student Activity   The phrase "global warming" has become commonplace in our language. When did this phrase enter into the popular culture? Use your library's search portals to find the earliest reference in pop culture. Then see if you can use the science portals to find the earliest reference to this phrase in a scientific paper. How many years did it take to cross over? How did it cross over?

Critical Thinking Questions

1. How does the greenhouse effect work? Is carbon dioxide the only greenhouse gas? If not, what is another one?

2. Describe the carbon cycle. Is this a natural cycle? How have humans interrupted the cycle?

3. Does the Earth go through normal cycles of heating and cooling?

4. Can the Earth experience dramatic and relatively fast climate changes? If so, how?

Your Environment: YOU Decide

Movies like *The Day after Tomorrow* and *The Core* have tried to take some science out of the scientific literature and into the popular culture. Is this a good way to publicize science? Do you think that all of the science in those movies is real? What do your friends think? Do you think that science should be in the public eye, or should it remain behind closed doors?

Appendix

Answers to End-of-Chapter Questions

Chapter 1

Critical Thinking

1. We all think in a scientific way; thinking scientifically is how we decide where to park, where the best buy is, and so on.
2. Answers will vary, but we all must interact with the Earth.
3. Humans tend to view time on a much finer scale than the scale over which geologic processes operate. To an individual person, decades or even hundreds of years may seem like a long time, but in terms of geology, it commonly takes millions of years for significant changes to take place.
4. Answers will vary, but they must touch upon food and energy resources.

Your Environment: YOU Decide

Answers will vary, but they should explain that although the developed world has more money and lower birth rates than the developing world, it consumes the vast majority of Earth's limited resources. This means that sustainability will also be difficult to achieve in developed countries.

Chapter 2

Critical Thinking

1. Collision and coalescence should be included in the answer describing the formation of the solar system.
2. Extremophile bacteria can live in extreme conditions that are very hot, very cold, or under high pressures, for example. They are important to life because they may have been the first, earliest life on Earth.
3. Mass extinction is an event where a significant number of species become extinct over a relatively short period of geologic time.
4. Comets tend to be composed of ice and travel in an elliptical orbit around the sun, whereas asteroids are rocky and may or may not have stable orbits around the sun.

Your Environment: YOU Decide

Answers should include a discussion about humans finding solutions to our own environmental problems here on Earth, and how the difficulty of space travel makes populating another planet nearly impossible.

Chapter 3

Critical Thinking

1. The rock cycle begins and ends with magma. The answer should include all three types of rocks and how they relate to each other.
2. Yes, some minerals are composed of just a single element. For example, diamond and graphite are composed of just carbon (C) atoms, and a gold nugget is made of almost all gold (Au) atoms.
3. Yes, rocks can be made up of only one mineral: some sandstones have only quartz in them, and both marble and limestone are made up of calcite. There are other examples.
4. There are three rock types. Igneous rocks form from cooling magma, sedimentary by erosion and deposition of sediment, and metamorphic by altering older rocks through heat and pressure.

Your Environment: YOU Decide

Yes, a rock can be both sedimentary and igneous; for example, volcanic tuffs are made from volcanic ash and are compacted and cemented together via the sedimentary process. An example of a rock that is both igneous and metamorphic is migmatite, half of which melted and the other half metamorphosed.

Chapter 4

Critical Thinking

1. The forces that deform rocks are compressive stress, tensional stress, and shear stress.
2. Scientists determined the interior structure of the Earth through the study of seismic waves.
3. The major types of plate boundaries are divergent, convergent, and transform boundaries. Ridges, mountains, and volcanoes are features that develop at these boundaries.
4. The theory of continental drift lacked the mechanism for movement of the plates. Plate tectonics explained the plate movements.

Your Environment: YOU Decide

No, California will never fall into the Pacific Ocean because there is a transform plate boundary there. Los Angeles and San Francisco will slide past each other; Los Angeles is basically moving north, while San Francisco is moving south.

Chapter 5

Critical Thinking

1. Answers will vary. This question is to get students thinking about where they live. Answers should include any or all material from the section on Reducing Earthquake Risks. Again, this is to get students thinking about the dangers/payoffs of living in an earthquake-prone area.
2. Seismographs are not located at the tops of buildings because seismic waves do not travel through the air, and there is building sway which causes background noise. A seismograph must be attached in the ground to accurately record earthquake activity.
3. If the geology is right and a relatively minor earthquake hit an area that has a high density of people and cheaply made buildings, such as in the slums of Mumbai or Mexico City, then there would be a lot of death and destruction (high on the Mercalli scale) without a lot of seismic energy being released (low on the Richter scale). If a major earthquake (high on the Richter scale) hit a relatively uninhabited area, such as Siberia, there would be little death and destruction (low on the Mercalli scale).

Your Environment: YOU Decide

Answers will vary; hopefully the students will see the value in both sides. If an area is already prone to earthquakes, then the money should be spent on building improvements. If an area has a hit-or-miss record with earthquakes, then funding to research earthquake prediction would be the way to go.

Chapter 6

Critical Thinking

1. The Ring of Fire is mainly a subduction zone, where the descending plate causes melting in the asthenosphere, which generates magma that commonly comes to the surface via volcanoes.
2. Basaltic eruptions are more fluid (lower viscosity) because they have higher temperatures and lower SiO_2 content than rhyolitic eruptions.
3. The sides of a shield volcano are less steep than those of a composite cone because the lava that formed them was more fluid.
4. A hot spot is where a rising plume of mantle material causes the overlying lithosphere to partially melt, generating fairly constant volcanic eruptions. Yellowstone National Park and the islands of Hawaii are two examples of hot spots.

Your Environment: YOU Decide

Answers will vary, but lava from shield volcanoes can at times be diverted, and they tend not to have devastating eruptions.

Chapter 7

Critical Thinking

1. The main factors that affect slope stability are steepness of the slope, type of material, presence of water, existence of planes of weakness, and amount of vegetation.
2. Water destabilizes a slope by adding weight and reducing friction through increases in pore pressure.
3. Yes, snow avalanches are a form of mass wasting, except that snow moves downslope instead of rock and debris.
4. The risk of mass wasting can be minimized by reducing the angle of the slope, reducing the amount of water, planting more vegetation, and using rock bolts or mesh to shore up weakness planes.

Your Environment: YOU Decide

Answers should include a discussion about the money that logging companies would have to spend to reduce mass wasting, versus the societal costs associated with the environmental consequences, such as sediment pollution and increased flooding.

Chapter 8

Critical Thinking

1. The description of the hydrologic cycle should include a discussion of precipitation, evaporation, transpiration, infiltration, and overland flow, and how these processes relate to each other.
2. Some factors that affect flooding are the intensity and duration of the precipitation events, the size of the area where precipitation falls, ground conditions, and type of vegetation cover.
3. The higher the velocity, the greater the amount and the larger the size of sediment that can be transported. As the velocity decreases, lesser amounts and only smaller size sediment can be carried.

Your Environment: YOU Decide

Answers will vary, but should include some discussion on how flood controls and removal of wetlands has allowed increased development in floodplains, which ultimately has led to increased flooding and greater property losses.

Chapter 9

Critical Thinking

1. Tides are the result of the gravitational forces between the Earth, Moon, and Sun, and the spinning motion of the Earth.

2. Most waves are caused by the transfer of energy from the wind to a body of water.
3. If you are caught in a rip current, do not panic and fight it, but instead start swimming parallel to the shore. When you no longer feel the rip current, you should then swim back to shore.
4. In the United States, more people are living in coastal areas, which has resulted in more development and more property to be damaged. Better technology has led to better prediction of where a hurricane will land, resulting in greater early warning time and fewer storm-related deaths.

Your Environment: YOU Decide

Answers will vary. You could threaten to sue, move, install a groin of your own, or put in a breakwater to give yourself more beach. If you install a groin or a breakwater, this will cause your neighbor's beach down current to erode.

Chapter 10

Critical Thinking

1. Generally, the more black or dark brown the soil, the higher the organic content.
2. Plants and animals are important to soil formation because they introduce and incorporate organic material into the soil, and they break it up so that water and oxygen can circulate.
3. Soil is considered an important resource because without soil, there would be basically no plants on Earth's landmasses. Soils also can be an important mineral resource, such as bauxite for aluminum, and kaolinite for clay.
4. The development of the soil horizons begins with the weathering of the parent rock to form loose sediment, and then organic material is worked into it. As water leaches through the sediment, various types of soil horizons can form.

Your Environment: YOU Decide

Answers will vary. The mineral material of the soil is not living, but for a soil to be productive, there must be microbes and other living things within it.

Chapter 11

Critical Thinking

1. Unconfined aquifers are open to the atmosphere and are close to the ground surface. Confined aquifers are sealed off from the atmosphere by an overlying aquitard, creating a pressurized system.
2. A confined aquifer would be better suited for drinking-water, because its water is not in direct contact with contaminants found in the surface environment.
3. Groundwater mining is the process of removing water faster from an aquifer than it is being replenished by natural recharge, thereby depleting the supply.
4. Off-stream water use is taking water from one source and returning it to a different one. Consumptive water use is when water is lost and not returned to a supply source, such as with irrigation or making concrete. Flushing of toilets and washing dishes is not consumptive use because the water is not lost, but is eventually discharged into another supply source, commonly a river.

Your Environment: YOU Decide

Answers will vary, but there should be mention of the fact that humans can survive without oil as an energy source, but cannot live without water for drinking and growing food.

Chapter 12

Critical Thinking

1. A mineral resource is a mineral deposit that is feasible to mine using existing technology. This definition simply means that we have the

technical ability to extract the minerals, but it says nothing about profitability. A mineral reserve, on the other hand, is a deposit that is currently economical to extract.

2. Some of the U.S. strategic minerals are chromium, cobalt, manganese, and platinum. These minerals are critically important to the defense industry, yet must be imported in significant quantities.

3. Underground mining, placer mining, and surface mining are important mining methods. Each has their own set of environmental problems: tailings, ore-processing contaminants, and alteration of the original landscape are just a few.

4. Up to the modern age, salt was hard to mine and very difficult to "make" by evaporation. Now, the technology is there to mine it anywhere and to evaporate it in many locations.

Your Environment: YOU Decide

Answers will vary; students should see the value in purchasing only conflict-free, registered diamonds. Some students should question the registration process as a marketing tool. Others should argue the value of a bargain.

Chapter 13

Critical Thinking

1. Answers will vary, but students should be thinking about how they live. Some may say that they will always drive themselves, while others have taken public transportation all of their lives.

2. As the price of a barrel of oil goes up, it becomes much more profitable to use new exploration and recovery techniques and technology. As the price stays high, more money will be put into research and development. If the price falls, then only the most profitable technology and techniques will be used.

3. Yes. As populations increase their ability to manufacture products, there has to be a way to market or export these products. There is also more money for the workers who first trade-up to motorcycles and cars. This growing industrialization leads to an overall higher demand for energy, thus driving up energy prices.

Your Environment: YOU Decide

Answers will vary; but students should see the value in both sides of the debate. The United States needs to increase domestic production, but developing ANWR risks disrupting a sensitive ecosystem. Also, the potential oil from ANWR is limited, which still leaves us with the development of alternate energy sources as the best long-term solution.

Chapter 14

Critical Thinking

1. Solar, geothermal, and wind are alternative energy resources that could be used in most of the United States. Passive solar energy to supplement hot water heating is probably the cheapest form of alternate energy because of its low infrastructure and operational costs.

2. Answers will vary, but there should be a balanced argument for either side. Students should consider the costs, not just in money but also to the environment. Another point would be the investment into changing our views on alternative energy, a shift which has been happening over the last few years.

3. Extracting oil from tar sands is very difficult. It is a mining operation, not just drilling, so the footprint is much bigger. However, this resource was not previously utilized because of the cost. As the price per barrel increases, the profitability of the sands increases.

Your Environment: YOU Decide

Answers will vary, but students should see the value in both sides. Although there is a small chance of a meltdown and radioactive contamination of water supplies, Western nations have benefited from nuclear power by safely generating large amounts of electricity. The perceived risk of a catastrophic Chernobyl-type accident is actually nonexistent due to the design of nuclear power plants used in the United States. The real issues involve the cost and safety concerns related to the disposal of spent fuel.

Chapter 15

Critical Thinking

1. The illegal waste can pollute both groundwater and surface water bodies, which, in turn, may be your source of drinking water. The pollution can also severely impact the ecology of streams and lakes, and thereby have a negative effect on fishing and other types of recreational activity.

2. The lack of oxygen in landfills greatly slows down the breakdown of waste. This means that groundwater contamination can occur long after the expected life span of the landfill's containment system.

3. Liquid hazardous waste can be stored in a secured landfill, incinerated, or injected into deep aquifers. The disadvantage of landfills and deep-well injection is that there is the potential for groundwater contamination, whereas incineration creates the possibility of dangerous air pollution.

4. High-level waste consists of intensely radioactive by-products from nuclear reactors. This waste contains long-lived isotopes that must be safely isolated from the environment for thousands of years. Low-level waste is a broad category that includes all radioactive waste that is not spent nuclear fuel. Because the amount of radiation emanating from low-level waste varies considerably, storage facilities in the United States only have to isolate the waste for 500 years.

Your Environment: YOU Decide

Answers will vary, but students should understand that everyone in society generates waste, and that the waste must be stored where it can cause the least amount of harm. Student solutions to the social inequities of situating landfills are likely to include ways of compensating those who must live near landfills, and finding ways of minimizing the potential pollution problems.

Chapter 16

Critical Thinking

1. In the greenhouse effect, gases absorb some of the outgoing radiation Earth emits into space, increasing the temperature of the atmosphere. Besides carbon dioxide, nitrous oxide, methane, and water vapor are also greenhouse gases.

2. The carbon cycle refers to the cyclical movement of carbon atoms through Earth's interior and surface environment, and has been a natural part of the Earth system for most of the planet's history. Humans have disrupted the carbon cycle by adding large volumes of greenhouse gases to the atmosphere through the burning of fossil fuels and other human activities.

3. Yes, the Earth goes through warming and cooling cycles as seen with the glacial cycles.

4. Yes, volcanic eruptions and the impacts of asteroids and comets can all cause fast and severe climate change. Relatively rapid climate change can also occur when critical climate thresholds are passed, such that changes are then greatly amplified by positive feedback mechanisms.

Your Environment: YOU Decide

Answers will vary; students should see the difference between fact and fiction.

A

absolute age The actual age of rocks in terms of years. This age is normally determined by *radiometric dating,* a technique that is based on the known decay rates of different radioactive elements.

accretion Process whereby planets eventually form when gravitational attraction causes individual particles to clump together into larger masses.

acid mine drainage Acidic leachate, normally rich in heavy metals, that drains from tailings and underground mines, and which forms from the chemical reaction between water and sulfide minerals commonly found in ore deposits and coal.

acid rain Term used to describe precipitation containing abnormally high levels of sulfuric and nitric acid, which results from the release of sulfur dioxide and nitrous oxide gases by human activities.

albedo Term used to describe the fraction of solar radiation that is reflected off a solid body. Light-colored objects, such as clouds and land covered with snow or ice, reflect more light and therefore have a high albedo.

alternative energy sources Can be defined as energy sources other than the conventional coal, natural gas, and light crude oil that currently make up the bulk of the world's energy supply.

anthropogenic pollution The pollution results from various human activities and society's consumption of natural resources, in contrast to pollution that occurs solely due to natural processes.

aquifer A permeable geologic material (rock or sediment) that readily transmits water. Often used as a supply source for groundwater.

aquifer storage and recovery A water management technique that involves storing surplus surface water in an aquifer, which is then removed at a later time when surface water supplies are more limited. Sometimes referred to as *water banking.*

aquitards A geologic material (rock or sediment) with low permeability that inhibits the flow of groundwater. Also acts as confining layers for *artesian* (confined) aquifer systems.

artificial levees Earthen mounds or concrete panels constructed along river banks by humans for the purpose of keeping a river from overflowing its banks and inundating its floodplain. Those made of concrete are often called *floodwalls.*

asteroids Small bodies orbiting the Sun that are composed primarily of rocky and metallic materials. Most asteroids lie in what is known as the main asteroid belt between Mars and Jupiter.

asthenosphere The weak zone within the Earth that lies near the top of the mantle where silicate minerals are near their melting points. Tectonic plates ride over this weak layer.

B

back swamps Those areas of the floodplain that are poorly drained and can remain wet for extended periods of time. Drainage is inhibited by natural levees, high water tables, and fine-grained sediment beneath the swamps.

barrier islands Elongate sediment deposits that parallel a shoreline and are separated from the mainland by open water, lagoons, tidal mudflats, or saltwater marshes.

basalt A common type of extrusive igneous rock found in oceanic crust that is rich in plagioclase feldspar and ferromagnesian minerals.

base level The lowest level to which a stream can erode its channel. Sea level is often referred to as *ultimate base level* because the oceans represent the end or low point of most rivers.

bauxite A sedimentary deposit of highly enriched, aluminum minerals that form from the chemical weathering of silicate rocks over a long period of time in climatic zones with considerable rainfall.

beach nourishment The process of artificially adding sand to a beach in order to reduce erosion, and to enhance its recreational value.

big bang theory The scientific explanation for the origin of the universe in which all matter at one time existed at a single point, but then about 14 billion years ago began expanding outward in all directions.

biofuels Any combustible material derived from modern (nonfossilized) organic matter. Examples include ethanol and biodiesel fuels.

body waves Seismic waves that travel through Earth's interior, which includes *primary (P) waves* and *secondary (S) waves.*

breakwaters An engineering structure consisting of large rocks placed parallel to shore and designed to keep waves from breaking onto land, thereby creating a protected area.

C

calcite A calcium carbonate ($CaCO_3$) mineral that is a major constituent in the group of rocks called *limestone*, and used as the raw material for making cement and concrete.

caldera A circular depression that forms after a volcanic eruption when rocks begin collapsing or subsiding into the now empty magma chamber. Some geologists also refer to large *craters* that form during explosive eruptions as calderas.

cap rock A low-permeability rock layer, typically composed of clay or evaporite minerals, that overlies a petroleum reservoir and greatly limits the ability of oil and gas to escape from the reservoir.

cap-and-trade system A strategy for reducing carbon emissions where regulators provide permits and set emissions limits for a particular type of user. Because users are allowed to buy and sell individual permits, a financial incentive exists for operators to develop more efficient systems, or switch to a cleaner fuel.

carbon capture and storage A technique under development for removing carbon dioxide from the emission gases of industries burning fossil fuels, then compressing the CO_2 and injecting it into permeable rock formations for permanent storage. Also see *carbon sequestration.*

carbon cycle Refers to the cyclical and natural movement of carbon atoms through Earth's solid interior and surface environment (biosphere, hydrosphere, and atmosphere).

carbon sequestration The process of removing carbon from the surface environment and placing it in storage for long periods of time. Can occur naturally (formation of limestone rock and fossil fuels) or artificially by humans (subsurface injection of carbon dioxide). Also see *carbon capture and storage.*

carbon tax A strategy for reducing carbon emissions by levying a tax on fossil fuels, thereby providing an incentive for businesses and industries to switch to non-carbon-based energy sources.

channelization A flood control technique that involves straightening and deepening a stream channel so that its *discharge capacity* is increased, thereby reducing the probability that water will overflow the banks.

chemical weathering Any chemical process that causes minerals within rocks to decompose into simpler compounds and individual ions.

cinder cones Relatively small volcanic features that form when lava is ejected into the air and cools into *cinders,* which then fall and accumulate around the vent.

clay minerals A group of aluminum rich, silicate minerals that typically form by the chemical weathering of other silicate minerals.

Clean Air Act Passed by the U.S. Congress in 1970 requiring the EPA to set minimum levels of air quality for cities, and emission standards for sources that emit hazardous substances, including automobiles.

Clean Water Act Passed by the U.S. Congress in 1972 giving the EPA the authority to regulate water quality and the discharge of pollutants into surface waters—technically known as the *Water Pollution Control Act of 1972.*

climate The long-term average weather and its statistical variation for a given region: contrasts with *weather* which refers to the state of Earth's atmosphere at any given time and place.

climate feedbacks Processes within the Earth system that respond to a disruption in the planet's heat balance and act to further increase or decrease the energy imbalance.

climate threshold The concept in which some component of Earth's climate system is pushed beyond a critical point, causing the entire system to suddenly change. Positive feedback mechanisms would then amplify the change.

coal A combustible sedimentary rock that forms when accumulated plant material becomes deeply buried under new sediment. The corresponding higher levels of heat and pressure drive off water and other volatile compounds, leaving behind a solid that is highly concentrated in carbon.

combined sewer system A wastewater collection system consisting of a single set of underground pipes for collecting both sewage and storm water runoff from city streets and parking lots.

comets Relatively small bodies, 0.6 to 6 miles in diameter, composed of small rocky fragments embedded in a mass of ice and frozen gases. Most have highly elliptical orbits around the Sun.

composite cone A cone-shaped volcano with steep slopes that consists of alternating layers of pyroclastic material and lava flows; also called a *stratovolcano*.

compressibility A property that describes the ability of a material to compact and reduce its volume when placed under a force or load.

compression A force that pushes on a rock body from opposite directions, causing it to become shorter.

cone of depression A cone-shaped depression in the water table or potentiometric surface of an aquifer, which forms when the water level *(hydraulic head)* within a well is lowered by pumping.

confined aquifer An aquifer that is overlain by an aquitard such that it is sealed off from the atmosphere and surface environment, creating pressurized conditions within the aquifer. Also known as an *artesian* aquifer.

consumptive water use Refers to those human activities where water is removed from its source and is lost or consumed. Examples include the irrigation of crops and production of concrete.

containment landfill A type of sanitary landfill designed to minimize the formation of leachate, and to collect what does form. These landfills are also lined with a synthetic fabric in order to prevent the escape of any leachate.

continental arc A volcanic mountain range associated with a subduction zone in which magma rises up through a continental tectonic plate to form a string of volcanoes.

convection cells The circular motion of heat and matter within the Earth that is driven by temperature-induced changes in the density of material.

convergent boundary A tectonic plate boundary that is dominated by compressive forces such that the two plates move toward one another.

crater A circular depression around a volcanic vent that forms during an eruption as pyroclastic material is ejected into the air.

creep An exceptionally slow type of mass wasting process where repeated expansion and contraction causes unconsolidated materials to move downslope.

crude oil The liquid phase of petroleum consisting of a mixture of heavier hydrocarbon molecules, which are separated during a distillation process (refining) to produce a variety of products, including gasoline, lubricating oil, and agricultural chemicals.

crust Earth's outermost layer consisting primarily of silicate-rich rocks whose density is lower than those in the underlying mantle.

D

deposition The process where transported earth materials begin to accumulate. Solid sediment will accumulate in low-lying areas of the terrain, whereas dissolved ions accumulate in either a body of surface water or a groundwater system.

desalinization The process of removing the dissolved ions from saline water, by either distillation or reverse osmosis, in order to generate freshwater that is suitable for human consumption.

disseminated deposits Low-grade hydrothermal deposits where the ore minerals are widely dispersed in a zone surrounding an igneous intrusion, and within the intrusion itself. Also see *hydrothermal deposits*.

dissolution A type of chemical weathering reaction in which minerals completely dissolve or disassociate in water, leaving only individual ions in the solution.

divergent boundary A tectonic plate boundary that is dominated by tension forces such that the two plates move away from one another.

downstream flood A large volume flood that tends to rise slowly and stay above flood stage for extended periods of time. Commonly occurs in the lower portion of drainage systems where streams have wide channels and natural floodplains.

drainage basin The land area that collects water for an individual stream or river; also referred to as a *watershed*.

drainage divide A topographic line that follows the crests in the landscape and marks the point where surface water is forced to flow into different drainage networks.

E

earth resources Any natural resource that comes from the solid earth. Examples include water, soil, mineral, and energy resources.

Earth systems science The concept where the Earth is viewed as a dynamic, constantly changing system composed of four major components: the *atmosphere, hydrosphere, biosphere,* and *solid earth.*

earthquake Ground shaking associated with the vibrational wave energy that results when a rock body suddenly fails and releases its accumulated strain.

earthquake precursors Various physical phenomena that can occur just prior to the release of energy associated with a main earthquake shock.

ecological footprint Refers to the amount of *biologically productive* land/sea area needed to extract the resources needed by humans, and to absorb the waste they generate.

El Niño The periodic change in water and air currents in the equatorial parts of the Pacific Ocean, which bring dry conditions to the western Pacific and heavy rains to parts of coastal North and South America.

elastic limit The maximum amount of strain that a rock body can accumulate before either fracturing or undergoing plastic deformation.

elastic rebound theory Explains how earthquakes originate when a rock body deforms and accumulates strain such that it reaches its elastic limit, at which point the rock suddenly fails and releases its stored energy.

electromagnetic radiation A type of energy that travels in a continuous series of waves in which individual waves vary in terms of their wavelength and amount of energy they contain.

electronic waste Discarded electronic products, including TVs, computers, cell phones, and assorted audio and video equipment. Typically contains highly toxic substances such as lead, mercury, and cadmium.

energy The capacity to perform work or transfer heat. See *work*.

energy conservation The process of reducing energy consumption through decreased human activity, and through the more efficient use of the energy itself.

energy efficiency A measure used to describe the amount of unwanted heat that is lost during energy transformations, such as during the conversion of electrical energy to radiant light.

enrichment factor The degree to which geologic processes have concentrated a mineral resource above its average concentration in the crust. A deposit may be deemed economical to mine only if it reaches a certain concentration factor.

environmental geology A branch of geology that examines the interaction between humans and the geologic environment. Common issues include resources, hazards, and pollution.

environmental risk The chance that some natural process or event will produce negative consequences for an individual, or society as a whole. Risk is characterized in terms of probability and consequences.

epicenter The point on the surface that lies directly above the focus, which is the place where an earthquake originates in the subsurface.

erosion The process whereby rock or sediment is removed from a given area through chemical reactions or by being physically picked up or worn down by abrasion.

essential nutrients The most critical elements necessary for plant growth, which include nitrogen, phosphorous, potassium, calcium, magnesium, and sulfur.

evaporites A class of economic sedimentary minerals that form when water bodies undergo evaporation, causing salts to precipitate. Deposits can be either marine or freshwater in origin.

expanding clays Those clay minerals that are capable of incorporating large numbers of water molecules within their structure, thereby producing significant volume changes. Synonym: *swelling clays*.

exponential growth When the amount added over successive time increments keeps increasing such that it plots as a curve as opposed to a straight line.

extremophile bacteria Bacteria that thrive under extreme conditions, such as ancient Antarctic ice, superhot vents on the seafloor and rocks located deep underground. Earth's complex plant and animal life is generally believed to have evolved from extremophile bacteria.

F

fall A type of mass wasting that involves the rapid movement of earth materials falling through air.

fault A fracture plane along which slippage or movement has occurred along opposite sides of the fracture.

feldspars A group of rock-forming silicate minerals that are rich in aluminum (Al), and are commonly transformed into clay minerals by chemical weathering.

ferromagnesian minerals Rock-forming minerals that contain relatively high proportions of iron (Fe) and magnesium (Mg).

fertilizers Natural or synthetic materials containing essential nutrients that are applied to soil in order to increase plant growth and agricultural productivity.

flash floods A flood in which a stream rises and falls rapidly. Also known as *upstream floods* since small streams in the upper reaches of a drainage system tend to quickly overflow their banks during heavy rain events.

flood stage The height at which a river begins to overflow its banks. Stage height changes along a river because the channel continually decreases in elevation as it moves downstream.

flow A type of mass wasting involving loose material that accumulates enough water so that internal friction is reduced, allowing it to behave like a fluid and start flowing downslope.

focus The point within a rock body where accumulated strain is suddenly released, causing an earthquake.

foliated texture The parallel realignment of minerals within a rock caused by the increased pressure associated with regional metamorphism.

fossil fuels A collective term used to describe a combustible form of energy that has been derived from the remains of ancient plants and animals. Major types include coal, oil, and natural gas.

freshwater Low-salinity water (contains few dissolved ions) that originates as precipitation within the hydrologic cycle.

G

galaxies Groupings or clusters of stars within the universe, some of which form a planar, rotating disk of stars.

gamma-ray burst A short-lived burst of very high energy waves that can destroy ozone molecules in the upper atmosphere, which shield the biosphere from dangerous ultraviolet radiation.

gas giants The four outermost planets that are largely composed of hydrogen and helium gas and have surfaces marked by clouds of swirling gases.

gas hydrates An ice-like substance composed of methane and water molecules found in thick accumulations in shallow polar waters, deeper parts of continental shelves at lower latitudes, and in permafrost throughout the Arctic. Represents a potentially large source of new energy. Also called *methane hydrates*.

gas window The depth range where subsurface temperatures are high enough to transform organic compounds into natural gas. Beyond the window, higher temperatures cause the hydrocarbons to be transformed into the mineral graphite.

general circulation models (GCMs) Sophisticated mathematical models that use a three-dimensional grid to simulate atmospheric processes on a global scale, including the many feedbacks between the atmosphere, oceans, and land.

General Mining Act A law passed by the U.S. Congress in 1872 that still governs the mining of precious metals on public lands. The act allows prospectors to purchase public land for no more than $5 per acre, provided that they extract the existing mineral resource.

geologic hazard A geologic condition, natural or artificial, that creates a potential risk to human life or property. Examples include earthquakes, volcanic eruptions, floods, and pollution.

geologic time A term used to imply extremely long periods of time over which geologic processes take place, typically measured in intervals of millions or billions of years. In contrast, human time is often measured in intervals ranging from seconds to decades.

geologic time scale The worldwide rock record classified according to the relative or chronological age of individual rocks. This time scale uses various names to subdivide Earth's rock record into progressively smaller time intervals.

geology The study of the solid earth, which includes the materials it is composed of and the various processes that shape the planet.

geothermal energy The heat energy contained within the Earth that can be used to drive a steam turbine for producing electricity, or serve as a heat source and sink for a geothermal heat pump.

geothermal gradient The rate of temperature increase in the Earth with increasing depth.

geothermal heat pump A mechanical system that supplies a building with warm or cool air by making use of the relative difference in temperature between the inside air and the ground.

glacial surge A sudden and rapid increase in the flow of glacial ice, believed to be triggered by the buildup of meltwater at the base of glacier, which reduces the friction between the glacier and underlying bedrock.

global warming An overall rise in the average global temperature of Earth's atmosphere. The term is commonly used to refer to the current warming trend.

granite A common type of intrusive igneous rock found in continental crust that contains relatively few ferromagnesian minerals, but is rich in quartz and potassium feldspar.

greenhouse effect The natural warming of Earth's atmosphere due to the presence of certain gases, which absorb outgoing radiation and release the corresponding heat energy into the atmosphere.

groin An engineering structure built perpendicular to the beach consisting of rocks too large to be moved by wave action. The purpose is to reduce shoreline retreat by widening the beach through the trapping of sand moving with the longshore current.

ground amplification A phenomenon that occurs when seismic waves encounter weaker materials and begin to slow down, causing an increase in wave amplitude and ground shaking.

ground fissures Large open cracks that form over a wide area of the landscape during an earthquake.

groundwater Fresh or saline water that resides within the void or pore space of subsurface materials.

groundwater baseflow Water that moves through subsurface materials and then flows into the surface environment, such as a stream, lake, wetland, or ocean.

groundwater mining The undesirable situation where the human withdrawal of groundwater is greater than the rate of natural recharge, causing water levels in an aquifer to get progressively lower over time.

H

habitable zone That relatively narrow zone around a star where the surface temperature of orbiting planets would be such that liquid water could exist, creating a greater potential for the development of life.

hardpan A soil layer whose physical characteristics limit the ability of either roots or water to penetrate the soil. Commonly consists of dense accumulations of clay minerals or soil particles that have been cemented together by minerals.

headwaters The upper portion of a drainage network where there are numerous small channels, which eventually merge to form progressively larger streams.

heat balance The dynamic relationship where Earth naturally attempts to balance the amount of heat energy it radiates into space with the amount of energy it receives from the Sun. When the heat balance is disrupted, Earth either gains or loses heat until a new equilibrium can be established.

heavy crude oil Oils with a higher proportion of heavier hydrocarbon molecules and more sulfur atoms.

high-level waste Intensely radioactive by-products from nuclear power and weapons reactors, and also the wastes generated from the reprocessing of spent reactor fuel.

historical geology A branch of geology that examines Earth's past by unraveling the information held in rocks.

hot spots Rising plumes of mantle material that cause partial melting in the overlying lithospheric plate, creating magma that moves upward through weak zones within the plate.

Hubbert's peak The projected peak in world oil production as predicted by the peak oil theory; named in honor of M. King Hubbert who developed the theory.

hurricane A large, rotating low-pressure storm system that originates in tropical oceans, and has sustained winds of over 74 miles per hour. Also called *typhoons* or *cyclones* depending on where they form in the Tropics.

hydraulic conductivity A measure of the ability of a material to transmit a fluid, and is dependent on both the property of the material (permeability) and properties of the fluid itself (density and viscosity).

hydraulic gradient The slope or steepness of a water table or potentiometric surface, and is calculated based on the difference in hydraulic head and distance between any two points in an aquifer system.

hydraulic head A measure of the potential energy within an aquifer system, and is represented by the height of the water table or potentiometric surface.

hydraulic sorting The process where flowing water separates sediment grains based on their size, shape, and density.

hydrocarbons Organic molecules composed chiefly of hydrogen and carbon atoms that are the basis of petroleum (oil and gas), a combustible substance that forms when organic-rich shales become deeply buried and exposed to progressively higher levels of heat and pressure.

hydrogen fuel cell A device that produces electricity by allowing hydrogen and oxygen atoms to chemically combine to form water molecules.

hydrologic cycle The cyclic movement of water within the Earth system, driven by solar radiation that causes water to evaporate from the oceans and land surface.

hydrolysis A type of chemical weathering reaction in which water molecules directly take part in the breakdown of minerals, releasing both dissolved ions and producing new minerals called *weathering products*.

hydropower Mechanical energy obtained from falling water, commonly used today for generating electricity.

hydrothermal deposits Mineral deposits associated with igneous intrusions, which result from the chemical interaction of hot, mineral-rich fluids with the surrounding rock. Commonly occur as sulfur-rich deposits in veins and disseminated grains in a zone around the intrusion.

hypothesis A scientific explanation of data or facts. Hypotheses must be testable such that it is possible to show them to be false or incorrect. Supernatural explanations are not considered scientific because they are not testable and cannot be shown to be false.

I

igneous rocks A major class of rocks that form when minerals crystallize from cooling magma.

inertia The physical tendency of objects at rest to stay at rest, which plays a key role in causing damage during earthquakes when structural foundations are suddenly forced to move.

infiltration capacity The ability of the land surface to absorb water. Water that is unable to infiltrate is generally forced to move downslope as overland flow.

inner core The innermost part of the Earth that consists of a solid metallic sphere.

intraplate earthquakes Earthquakes that occur far from a plate boundary or active mountain belt.

ion exchange The process where dissolved ions attach themselves to soil particles, and are then removed in a selective manner by growing plants and by water moving through the soil zone.

ions Individual atoms that have either gained or lost electrons, thereby acquiring either a positive or negative electrical charge.

island arc A landmass surrounded by ocean which forms when magma rises up through a buckled tectonic plate along a subduction zone, producing a string of volcanic islands.

J

jetties An engineering structure built perpendicular to shore at the mouth of inlets, whose purpose is to keep inlets open for navigation by blocking the longshore movement of sand.

L

La Niña A pattern of water and air currents in the Pacific Ocean that sometimes follow a periodic El Niño event, suppressing rainfall along the western United States. Also see *El Niño.*

lag time The time difference between a rain event and the resulting peak discharge in a stream.

land subsidence The lowering of the land surface due to the closing of void spaces within subsurface materials; commonly triggered by the withdrawal of subsurface fluids (water or oil) or by the collapse of natural cavities or mining voids.

lava domes A steep-sided mound of cooling lava that is built from more viscous magma that does not flow very readily.

lava flow A body of lava that flows out onto the land surface and eventually cools and solidifies into an igneous rock.

law A scientific term that describes some phenomena in which the relationship between different data occurs regularly and with little deviation. Such a relationship often can be described in terms of mathematics.

layered intrusions A type of igneous intrusion in which ore minerals had become separated from the magma during the crystallization process, forming a concentrated layer of valuable minerals.

leachate The liquid that forms when rainwater percolates into a landfill and interacts with the trash, incorporating bacteria, viruses, and various chemical compounds in the water. Another form of leachate forms during the chemical process where fluids are used for extracting metals from crushed ore. See *leaching.*

leaching A technique of processing metallic ores in which a solution is allowed to permeate through crushed ore, with the resulting chemical reactions then liberating the desired element(s). See *leachate.*

leakage A term used to describe the process where groundwater moves between two different aquifers by flowing across an aquitard (i.e., a confining layer).

light crude oil Oil with a higher proportion of lighter hydrocarbon molecules and fewer sulfur atoms.

limestone A type of clastic or chemical sedimentary rock that is composed chiefly of the mineral calcite ($CaCO_3$).

linear growth When the amount added over successive time increments remains the same so that it plots as a straight line as opposed to a curve.

liquefaction An earthquake phenomenon that occurs when the ground shaking causes sand-rich layers of sediment to behave as fluid.

lithosphere Rocks from both the crust and upper mantle, forming a brittle layer that is broken up into individual tectonic plates that move over the relatively weak asthenosphere.

longshore current The parallel movement of water and sand to land, which occurs in the surf zone when waves approach a shoreline at an angle.

low-level waste A broad category of radioactive waste which incorporates any radioactive material that is not spent nuclear reactor fuel. Low-level waste includes contaminated materials that have come into contact with radioactive substances, such as protective clothing and equipment.

M

magma Molten rock material that forms within the Earth and can cool to form igneous rock. *Lava* is used to describe magma that cools on Earth's surface.

magma chamber A zone or reservoir of molten material that forms within the lithosphere.

magmatic earthquakes Earthquakes that result from the strain that accumulates as rising magma forces its way through crustal rocks.

mantle The rocky shell surrounding Earth's metallic center that is composed of iron-rich silicate minerals.

mass extinction When large numbers of Earth's species go extinct in a relatively short period of time.

mass wasting The general process of earth materials moving downslope due to gravity; the terms *landslide* and *avalanche* are often used synonymously, but technically involve specific types of movement.

massive sulfide deposit A thick, hydrothermal mineral deposit associated with mid-oceanic ridges where active volcanism takes place along divergent plate boundaries. See *hydrothermal deposits*.

Maximum Contaminant Levels Standards set by the EPA regarding the maximum allowable concentration of specific contaminants in a water-supply system that provides drinking water to the public.

Mercalli intensity scale A qualitative means of ranking the intensity of earthquakes based on first-hand human observations, particularly the amount of structural damage.

metamorphic rocks A major class of rocks that form when preexisting rocks are altered by some combination of heat, pressure, and fluids.

methylmercury A highly toxic compound in which mercury bonds with carbon atoms. This form of mercury is easily ingested by aquatic life-forms, where it then gets passed on through the food chain, posing a human health risk.

mid-oceanic ridges The chain of submarine mountains, circling nearly the entire globe, where rising magma and seafloor spreading produce new oceanic crust.

Milankovitch cycles Refers to the cyclical changes in solar heating of the Earth caused by periodic changes in the way the planet orbits the Sun. When the orbital parameters (eccentricity, axial tilt, and precession) reinforce each other, the resulting change in heat balance can trigger an overall cooling or warming trend.

mine tailings Noneconomical rock and processed ore from mining operations, which is considered as waste material and is placed on the land surface.

mineral A naturally occurring inorganic solid composed of one or more elements in fixed proportions, and where the individual atoms have an orderly arrangement called a *crystalline structure*.

mineral reserve A mineral deposit that is economical to extract using existing technology.

mineral resource Any rock, mineral, or element that has some physical or chemical property humans find useful. This term does not necessarily mean a particular mineral deposit is economical to extract. See *mineral reserve*.

moment magnitude scale The modern earthquake magnitude scale that provides a more accurate measure of the amount of ground motion.

mouth The lowest point in a drainage system where a river discharges into an ocean, lake, or another river.

multiple working hypotheses Common in the early stages of an investigation where researchers develop more than one plausible hypothesis for a given set of data. The number of hypotheses normally decreases over time as new data show one or more hypotheses to be false.

municipal solid waste Trash or garbage from households and local businesses that is typically collected by local governments, and then disposed of in a landfill or by incineration.

N

National Environmental Policy Act Passed by the U.S. Congress in 1969 stating the country's commitment to protecting the environment for the benefit of people, and the nation as a whole.

natural floodplain The flat portion of a river valley underlain by sediment, which has been deposited over time as the river periodically overflows its banks.

natural gas The gaseous phase of petroleum consisting of a mixture of lighter hydrocarbon molecules, including methane and propane.

natural levees A pair of sand ridges that run parallel to stream banks, and form over time as a river overflows its banks and experiences an immediate decrease in velocity, causing sediment to fall from suspension.

natural vibration frequency The frequency at which a building will naturally vibrate when the ground shakes during an earthquake.

nebular hypothesis Describes how all solar system objects originally formed from a rotating cloud of dust and gas called a *nebula*.

nonconventional fossil fuel Can be defined as all fossil fuels except for the traditional use of coal, natural gas, and light crude oils. Includes heavy crude and synthetic liquid fuels from coal and tar sands.

nonpoint source A pollution source that releases pollutants into the environment over a broad area, commonly consisting of multiple input sites. Examples include agricultural fields, parking lots, golf courses, and lawns.

nuclear fission The process of splitting the nucleus of an atom, resulting in the release of neutrons and conversion of mass into energy. Used in nuclear power plants to drive steam turbines for generating electricity.

nuclear fusion A natural process that takes place in the Sun in which the nuclei of hydrogen atoms combine or fuse, releasing large amounts of energy. Humans have used this process to create nuclear weapons, and may someday harness the energy to generate electricity.

O

ocean acidification The process where the acidity of the oceans increases over time due to higher carbon dioxide levels in the atmosphere. As the oceans absorb additional carbon dioxide, the dissolved gas is converted into carbonic acid, making the oceans more acidic.

ocean currents The physical movement of ocean water from one location to another. Localized *tidal currents* form in coastal areas when tidal forces funnel water through inlets and river channels. In the open ocean large-scale *surface currents* are driven mainly by winds, whereas *density currents* move throughout the ocean basins in response to differences in temperature and salinity.

ocean thermal energy conversion A technique for producing electricity that makes use of the temperature differences within the deep ocean to power a simple heat engine, which then drives a turbine and electrical generator.

ocean tides The periodic rise and fall of sea level along coastlines, caused by the combination of gravitational forces between the Earth, Moon, and Sun and the spinning motion of the Earth on its axis.

ocean trenches Narrow, steep-sided depressions associated with subduction zones that run parallel to adjoining landmasses.

off-stream use Refers to those human activities where water is removed from one supply source, but then returned to a different source after being used. An example would be the withdrawal of groundwater that is later discharged into a stream.

oil sands A near-surface sand deposit that contains a highly viscous form of hydrocarbons called bitumen, which can be extracted to produce synthetic crude oil. Also called *tar sands*.

oil shale A fine-grained sedimentary rock containing abundant organic matter, which can be used to produce synthetic crude oil by simple heating of the rock.

oil window The depth range where subsurface temperatures are high enough to transform organic compounds into oil and natural gas. Beyond the window, higher temperatures cause all oil molecules to be transformed into natural gas.

open-pit mine A type of surface mining operation that commonly involves excavating large volumes of a low-grade deposit, such as disseminated ores.

ore deposit A body of rock or sediment containing some mineral that has value to society and whose concentration is sufficiently high that the deposit is economical to extract. The term is often used to describe metallic deposits.

outer core The liquid metallic shell within the Earth that surrounds the solid metallic inner core.

overburden Undesirable rock or sediment that must be removed during a mining operation in order to gain access to a valuable mineral deposit.

overland flow The process where water moves downslope in thin sheets over the land surface.

oxidation/reduction A type of chemical weathering reaction in which electrons that are gained or lost take part in the breakdown of minerals, releasing both dissolved ions and producing new minerals called *weathering products*.

ozone depletion The thinning of the atmosphere's outermost layer, composed of ozone (O_3) molecules, that shields the biosphere from dangerous ultraviolet radiation.

P

paleoclimatology The field of study that examines Earth's history of climate change using a variety of tools, including sediment, hydrogen and oxygen isotopes, tree rings, and glacial ice.

paleosol A soil sequence that has been buried during some geologic event, representing a distinct time marker that can be useful in scientific investigations.

parent material The original weathering product or organic material from which soil horizons develop.

peak oil theory A statistical analysis, based on historical exploration and production data, which predicts that oil production will reach a peak, then follow a permanent production decline.

permafrost A subsurface horizon that remains frozen throughout most or all of the year.

permeability A property that describes the ease with which a fluid is able to flow through a porous material.

petroleum A general term geologists use to describe the various types of organic compounds that make up both crude oil and natural gas.

petroleum reservoir A permeable rock in the subsurface where hydrocarbons (oil and gas) have accumulated, and are being stored in the rock's pore spaces.

petroleum trap Any configuration of rocks in the subsurface that allows hydrocarbons (oil and gas) to accumulate over time. Common traps include dome structures and faults.

photovoltaic cell A thin slab of material composed of silicon alloys, which produces electricity when sunlight is allowed to strike the slab.

physical geology The study of the solid earth and the processes that shape and modify the planet.

physical weathering Any process that causes rocks to disintegrate into smaller pieces or particles by some mechanical means.

placer deposit A concentrated sedimentary deposit consisting of dense, chemically resistant minerals that have been hydraulically sorted from the rest of the sediment load.

planets Solar system bodies that are large enough that their gravity is able to dominate their orbital zones, sweeping it clear of debris.

plasticity A property that describes the ability of a material to deform without breaking when a force is applied.

point source A pollution source that releases pollutants into the environment at a physically discrete point, such as a pipe discharging wastewater from a factory.

polluted Can be defined as those situations where a substance has been introduced into the air or water and has reached concentrations that are harmful to living organisms.

pore pressure The pressure from fluids (water, oil, or gas) within the void spaces of rocks that acts outward in all directions.

porosity A measure that describes the amount of void (air) space within rock or sediment, usually expressed as a percentage of the material's overall volume.

potentiometric surface Represents the height that water in a well will rise above a confined aquifer, and is a measure of the amount of potential energy within the aquifer.

primary (P) waves Seismic waves that cause solid particles to vibrate in same direction the wave is traveling such that the rocks alternately compress and decompress.

pyroclastic flow A dry avalanche consisting of hot rock fragments, ash, and super-heated gas that rush down the side of a volcano at great speed.

pyroclastic material Particles of pulverized rock and lava that are ejected into the surface environment during explosive volcanic eruptions.

Q

quartz A mineral composed entirely of silicate ions (SiO_4^{4-}) and commonly found in continental crustal rocks along with feldspars. Because it is resistant to chemical weathering, quartz is also abundant in soil and sediment.

R

radioactive decay The process where the nucleus of an unstable isotope undergoes a spontaneous nuclear reaction, releasing radiant energy (radiation) along with *beta* particles (electrons) and *alpha* particles (neutrons and protons).

radiometric dating A general term applied to absolute dating techniques involving any type of radioactive element and its decay product. Since different radioactive elements decay at different rates, scientists can obtain reliable dates for events ranging anywhere from thousands to billions of years old.

radon gas An odorless, radioactive gas that forms naturally from the decay of uranium-bearing minerals found in igneous rocks and sediment. Radon can cause lung cancer in humans when it is allowed to accumulate in poorly ventilated homes.

rare Earth hypothesis The scientific idea that life is probably common throughout the universe, but complex animal life similar to Earth's is likely to be exceedingly rare.

recurrence interval A statistical calculation representing the frequency at which a particular value of stream discharge can be expected to repeat itself.

relative age The geologic age of a rock, fossil, or event in relative terms as compared to the age of another rock, fossil, or event. The relative age of sedimentary layers is based on the principle that the bottom layers were deposited first, thus are the oldest.

resonance A phenomenon that occurs when the natural vibration frequency of a building matches that of the seismic waves, causing the building to shake more violently.

retaining wall An engineering structure designed to strengthen an oversteepened slope; commonly used when a flat surface is needed in sloping terrain for a roadway, building, or parking lot.

retention basins An engineering structure designed to reduce flooding by temporarily storing excess surface water before it can reach a stream channel.

Richter magnitude scale The original earthquake magnitude scale, developed by Charles Richter, used to quantify the amount of ground motion.

rift valley A linear valley, also called a *graben*, that forms when tension forces cause the land to down-drop in a stair-step fashion along parallel faults.

rip current A dangerous current that flows away from a beach, created when the backwash from breaking waves funnels through a breech in an underwater sand bar; sometimes inappropriately referred to as a *rip tide*.

rock An aggregate or assemblage of one or more types of minerals.

rock bolts An engineering technique that utilizes a steel rod and anchoring system to prevent fractured blocks of rock from falling onto highways and rail lines. Also used for stabilizing walls and ceilings in tunnels and underground mines.

rock cycle A geologic concept that describes the recycling of rocks from one rock type to another by various geologic processes.

rock-forming minerals The relatively small set of minerals that make up most of the rocks in Earth's outermost layer called the *crust*.

S

salinization A process in which the salinity of soil water increases to the point that plant growth is reduced. Common in arid areas where agricultural irrigation causes mineral salts in the soil to dissolve.

saltwater intrusion The undesirable situation where the human withdrawal of groundwater allows saline water to migrate into the freshwater portion of an aquifer that is being used as a water supply.

sandstone A type of clastic sedimentary rock that is dominated by sand-sized rock and mineral fragments.

sanitary landfill An excavation used for disposing of municipal solid waste in which the trash is covered with dirt on a daily basis.

scientific method The process by which the physical world is examined in a logical manner. Data or facts are gathered via observations or experiments, which are then explained through hypotheses, theories, and laws.

scrubber A system which is installed at industrial boilers and power plants that chemically removes sulfur dioxide and other pollutants prior to the emission gases being released into the atmosphere.

seafloor spreading A hypothesis that describes how new oceanic crust forms as mid-oceanic ridges spread or open up over time, and are then filled by erupting magma.

seawall An engineering structure constructed of concrete, steel, or large rocks that is placed parallel to shore at the top of a beach. The purpose is to prevent shoreline retreat by creating a barrier that absorbs the impact of breaking waves.

secondary (S) waves Seismic waves that cause solid particles to vibrate perpendicular to the wave path, which creates a shearing (side-to-side) motion.

secured landfill A highly regulated type of landfill designed to handle industrial hazardous wastes. Uses special systems for collecting leachate and detecting leaks, and requires tracking permits for individual waste shipments. Waste must also be placed in separate containers within the landfill.

sediment Fragments of rock and minerals grains that are produced when rocks are broken down by undergoing physical weathering.

sediment pollution The movement of excessive sediment off the landscape and into drainage systems. Channels can become filled with sediment, destroying the ecology of streams and increasing the frequency and severity of flooding.

sedimentary rocks A major class of rocks that form when weathered rock fragments, or mineral grains that chemically precipitate from dissolved ions, are reassembled to form a layered rock sequence.

seismic gap A tool for predicting the likelihood of an earthquake, based on sections of an active fault where the strain has not been released for an extended period of time.

seismic waves Vibrational waves that travel through solid earth materials that are caused by earthquakes, sudden impacts, or explosions.

seismographs Instruments that are used to measure (quantify) the amount of ground motion during an earthquake.

sensitivity An engineering term used to describe how easily a soil will lose its strength when it is disturbed.

shale A type of clastic sedimentary rock that is dominated by fine, clay-sized particles.

shear A force that pushes on a rock body in an uneven manner, causing it to become skewed such that different sides of the body move in opposite directions.

shield volcanoes Large volcanic landforms that are composed primarily of a series of basaltic lava flows.

shoreline retreat The landward migration of a shoreline caused by the erosion that results from the interaction between waves and a landmass.

silt fences A technique for reducing sediment pollution and flooding. Consists of a barrier, placed downslope of construction sites, made of a synthetic fabric that is fine enough to trap sediment, but yet allows some water to pass.

sinkholes Circular depressions created by the collapse of solution cavities in limestone rock. In areas with large numbers of sinkholes, the landscape takes on a pitted or cratered appearance referred to as *karst* terrain.

slate A fine-grained and highly foliated type of metamorphic rock that forms when clay minerals in a shale are transformed into platy minerals of the mica family.

slide A type of mass wasting involving masses of rock, earth, or debris (mixtures of rock and earth) that moves in a sliding manner along a zone of weakness (bedding planes, faults, fractures, and foliation planes).

slump A complex form of mass wasting involving unconsolidated material where sliding takes places near the top of the slump, transitioning to a flowing mechanism toward the bottom.

smelting A technique of processing metallic ores in which the minerals are heated to the point where their chemical bonds are broken, producing liquid metal that is then allowed to cool.

snow avalanche The term used to describe mass wasting events involving snow as opposed to rock or sediment.

soil A natural mixture of mineral and organic material that is capable of supporting plant life.

soil erosion The movement or transport of soil particles away from their place of origin.

soil fertility The ability of a soil to supply the elements necessary for plant growth.

soil horizons Horizontal layers within the soil zone that form as the result of chemical and biological processes and the physical transport of material vertically within the soil.

soil loss The net loss of soil that occurs when soil erosion exceeds the rate of natural soil formation.

solar heating The process that takes place when solar radiation strikes a solid object and is transformed into thermal or heat energy.

spring A place where groundwater discharges into the surface environment in a concentrated manner.

stabilization triangle A concept illustrating the amount that carbon dioxide emissions would need to be reduced in order to keep future emissions at a steady level. Emission reductions are achieved through *stabilization wedges*, which represent individual techniques, such as increased use of wind power and higher fuel mileage.

storm surge The rapid rise in sea level during a hurricane that inundates areas above the normal high-tide line. Results from a dome of water that follows a hurricane ashore, which itself is created by the storm's high winds and low air pressure.

strategic minerals Minerals that a particular country considers critical to its civilian and defense industries, but must be imported in significant quantities.

stream discharge The volume of water moving through a channel over a given time interval, commonly measured in units such as cubic feet per second (ft^3/s).

stream gradient The rate of elevation change or steepness of a stream channel. Water velocity increases in areas where the gradient is higher.

strength A property that describes the ability of a material to resist being deformed.

strip mine A type of surface-mining operation in which the excavation follows a mineral deposit that lies parallel to the surface; commonly associated with sedimentary deposits.

subduction The process of one lithospheric plate descending beneath another, where it then undergoes melting and becomes incorporated into the mantle.

subduction zone earthquakes Earthquakes that occur when an oceanic plate is overridden by another plate, generating some of the more powerful earthquakes on record.

Superfund A trust fund created by the U.S. Congress in 1982, whose purpose was to pay the cleanup costs for toxic waste sites that had been abandoned, or for which the responsible parties did not have sufficient resources. The fund's primary source of revenue was an excise tax on the petroleum and chemical industries.

surface waves Seismic waves that travel along Earth's surface; examples include Rayleigh and Love waves.

sustainability The ability to maintain a system or process for an indefinite period of time. A *sustainable society* is one that lives within the Earth system's capacity to provide resources such that they remain available for future generations.

swelling clays Those clay minerals that are capable of incorporating large numbers of water molecules within their structure, thereby producing significant volume changes. Synonym: *expanding clays*.

synthetic fuels Gas or liquid fuels produced from coal or heavy oil; also called *synfuels*.

T

talus pile A cone-shaped deposit of rocks that accumulates at the base of exposed rock bodies due to mass wasting processes.

tar sands A near-surface sand deposit that contains a highly viscous form of hydrocarbons called bitumen, which can be extracted to produce synthetic crude oil. Also called *oil sands*.

tectonic plates Individual slabs of the brittle lithosphere (crust and upper mantle) that move over the relatively weak asthenosphere.

tension A force that pulls on a rock body from opposite directions, causing it to become stretched or lengthened.

terracing A technique where a series of flat surfaces are cut into a hillside, with retaining walls commonly used to support oversteepened portions of the slope. The flat areas (terraces) can be used for growing food, constructing buildings, or reducing the chance of rocks tumbling onto highways.

terrestrial planets The four planets closest to the Sun that have outer shells composed of rocky, earth-like materials.

theory A scientific term used to describe the relationship between several different and well-accepted hypotheses, providing a more comprehensive or unified explanation of how the world operates.

theory of plate tectonics A major theory in geology that describes how Earth's lithosphere is broken up into rigid slabs that are in motion due to forces associated with the planet's interior heat.

thermohaline circulation A term used to describe the collection of large-scale density and wind-driven currents that move in a convective manner through the ocean basins. This circulation plays a key role in Earth's climate by transferring heat energy around the globe.

tidal power Energy that is obtained from the cyclical lowering and raising of sea level of the tides, and used to turn a generator to produce electricity.

tragedy of the commons An environmental concept in which the self-interest of individuals results in the destruction of a common or shared resource.

transform boundary A tectonic plate boundary that is dominated by shear forces such that the two plates slide past one another.

transportation The natural process of moving earth materials from one location to another through some combination of gravity, running water, glacial ice, and wind.

triggering mechanisms Processes or events that lead to a mass wasting event by reducing the frictional forces on a slope and/or increasing the effect of gravity.

tsunami A series of ocean waves that form when energy is suddenly transferred to the water by an earthquake, volcanic eruption, landslide, or asteroid impact.

U

unconfined aquifer A type of aquifer that is open to the surface environment, and includes a water table that marks the boundary between the saturated and unsaturated zones.

V

viscosity The ability of a fluid to resist flow that is caused by internal friction within the fluid. In magma, greater SiO_2 (silica) content and cooler temperatures result in greater internal friction, hence increased viscosity.

volcanic ash Fine pyroclastic fragments that are ejected from a volcano, then fall from the sky downwind of the volcano, sometimes traveling hundreds or even thousands of miles.

volcanic landslide The rapid downslope movement of rocks, snow, and ice that can occur when the steep flanks of a volcano become unstable and then fail; also called a *debris avalanche*.

volcanic mudflow A mixture of ash, rock, and considerable amounts of water that tends to rush down the stream valleys that lead away from a volcano; also called a *lahar* or *debris flow*.

volcano An accumulation of extrusive materials around a vent through which lava, gas, or pyroclastics are ejected into the surface environment.

W

waste management pyramid A management strategy designed to minimize the amount of solid waste being sent to a landfill. Common elements include source reduction, recycling, reuse, and incineration.

wastewater Water that has become contaminated during some human-related process. Examples include water used in industrial processing and for flushing toilets.

water table Marks the boundary between the unsaturated and saturated zones, below which all the pore spaces are filled with water.

water waves The rhythmic rise and fall in the surface of a water body that normally results from the transfer of energy from wind to water. As the energy moves through a water body, water molecules move in a circular manner such that the surface rises and falls.

wave attenuation The steady decrease in seismic wave energy that occurs as the waves travel away from their point of origin at the focus.

wave base The water depth at which water molecules are no longer affected by passing waves. The depth of the wave base is directly proportional to the amount of wave energy.

wave refraction The bending of wave fronts caused by a progressive decrease in velocity that occurs when wave energy encounters the seafloor, thereby generating greater frictional resistance.

weather The state of Earth's atmosphere at any given time and place; contrasts with *climate* which represents the long-term average weather and its statistical variation for a given region.

wind farms A collection of individual wind turbines used to generate electricity from the wind.

work Involves moving an object (mass) some distance against a given force (work = mass × distance).

X

xeriscaping A water conservation technique in which landscaping is done with native plants that require little to no irrigation water.

Credits

Photographs

Front Matter

P. x: © Getty RF.

Chapter 1

Opener: © Arthus-Bertrand/Altitude; 1.1(left): © Vol. 19/PhotoDisc/Getty RF; 1.1(right): © Corbis RF; 1.2A: © Dr. Parvinder Sethi; 1.2B: © Vol. 88/PhotoDisc/Getty RF; 1.2C: © Vol. 1/PhotoDisc/Getty RF; 1.3A: © Elisa Haberer/Corbis; 1.3B: USGS, photo by Edwin L. Harp; 1.4: Courtesy of Drs. Barrie McKelvey and N.C.N. Stephenson, Dept of Geophysics, Univ. of New England; 1.7: USGS, photo by J.D. Griggs; 1.8(left): © Getty RF; 1.8(right): © Vol. 44/PhotoDisc/Getty RF; 1.9(top): © Corbis RF; 1.9(bottom): © Vol. 7/PhotoDisc/Getty RF; 1.12: © Vol. 89/PhotoDisc/Getty RF; 1.14A,B: © Digital Globe/Getty Images; 1.16B: US Bureau of Reclamation; 1.17(all): USGS/NASA; 1.18: © Frans Lanting Photography; 1.21A: USGS/NASA; 1.21B: © Carlos Oscar Ruiz; 1.24: USDA and NRCS; 1.26(right): © Brand X/PunchStock RF; 1.26(left): © Vol. 1/PhotoDisc/Getty RF; Box 1.1B: © Brand X/PunchStock RF.

Chapter 2

Opener: © Brand X/PunchStock RF; 2.1: NASA; 2.2A: NASA and StSci; 2.2B: National Radio Observatory; 2.3: X-ray (NASA/CXC/M. Karovska et al.); Radio 21-cm image (NRAO/VLA/J.Van Gorkom/Schminovich et al.); Radio continuum image (NRAO/VLA/J. Condon et al.); Optical (Digitized Sky Survey U.K. Schmidt Image/STScI); 2.5: Lunar and Planetary Lab NASA; 2.7A: © Brand X/PunchStock RF; 2.8: NASA and Hubble Heritage Team; 2.9: NASA; 2.10: Alessandro Dimai/Astronomical Assoc. of Cortina; 2.11: NASA/The Galileo Project; 2.13: © David Hardy/www.astroart.org/STFC; 2.15A,B: NASA; 2.16: © Brand X/PunchStock RF; 2.17: European Southern Observatory; 2.18(right): Chris Angelos and Richard Nolthenius; 2.18(left): Courtesy of STSCI; 2.20B: © Brand X/PunchStock RF; 2.21(left): Galileo Project/NASA; 2.21(right): NASA; 2.22: NASA; Box 2.1A: NASA; Box 2.1B(bottom): NASA/JPL/Arizona State University; Box 2.1B(top): NASA; Box 2.2: NASA; 2.23, 2.24, 2.26: NASA; 2.27: USGS; 2.28A: NASA; 2.28B: Reto Stockli, NASA Earth Observatory; 2.29B: Virgil L. Sharpton, University of Alaska, Fairbanks; 2.30A: NASA, ESA, and H. Weaver and E. Smith (STScI); 2.30B: NASA and STSci.

Chapter 3

Opener: © Doug Sherman/Geofile; 3.1A,B: © Jim Reichard; 3.6A,B: © J. Geisler; 3.7: © Doug Sherman/Geofile; 3.9A-D: © J. Geisler; 3.10: USGS; 3.11A-C, 3.15A: © Jim Reichard; 3.15B: © Andrew McConnell/Getty Images; 3.15C: © Doug Sherman/Geofile; 3.17A: © Norris Jones; 3.17B(left): © Charlie Jones; 3.17B(right): Courtesy of Dr. Stan Celestian; 3.18: © Charlie Jones; 3.19A: NOAA; 3.20: © John Karpovich/University of Virginia; 3.21A,B: © Getty RF; 3.22(right): © Dr. Marli Bryant Miller; 3.22(left): © Doug Sherman/Geofile; 3.24A: C.J. Northup/Boise State University; 3.25A-C: © J. Geisler; 3.27: © Doug Sherman/Geofile; 3.28A,B, 3.29: © Jim Reichard; 3.30: NASA/JPL/Malin Space Science Systems.

Chapter 4

Opener: Amante, C. and B. W. Eakins, ETOPO1 1 Arc-Minute Global Relief Model: Procedures, Data Sources and Analysis, National Geophysical Data Center, NESDIS, NOAA, U.S. Department of Commerce, Boulder, CO, August 2008; 4.2(top): Courtesy of Phil Dombrowski; 4.2B(Bottom): © Jim Reichard; 4.5: National Geophysical Data Center/NOAA; 4.8: NOAA; 4.10, 4.12: NOAA/NGDC; 4.18: © Jim Reichard; 4.21(left): NASA; 4.23A,B, 4.24: National Geophysical Data Center/NOAA; 4.25B(left):NASA; 4.26B: National Geophysical Data Center/NOAA; 4.28: USGS, Menlo Park, CA.

Chapter 5

Opener: © Chien-min Chung/Getty Images; 5.1: Photo by M. Celebi, Courtesy of USGS; 5.2B(both): © Digital Globe/Getty Images; 5.4(top): USGS; 5.4(bottom): Photo by Univ. of Colorado; Courtesy of National Geophysical Data Center, Boulder, CO; 5.6B: © Anne Rippy/Getty Images; 5.15: USGS; 5.17(right): © AFP/Getty Images; 5.17(left): © Jiang Yi/Xinhua Press/Corbis Images; 5.18A: D. Perkins/USGS; 5.18B: M. Mehrain, Dames and Moore/NOAA; 5.19A: NOAA; 5.19B: C.L. Langer/USGS; 5.20: Reinsurance Company, Munich, Germany/NOAA; 5.21: NOAA/NGDC; 5.22B: C. Arnold, Building Systems Development, Inc./NOAA; 5.24: Edward H. Field, 2001/U.S. Geological Survey Open-File Report 01-164; 5.26B: NOAA/NGDC; 5.27: © Dr. Ross W. Boulanger/University of California at Davis; 5.28: Karl Steinbrugge/Earthquake Engineering Research Center, University of CA, Berkeley; 5.29A: Library of Congress Prints and Photographs Division; 5.29B: USGS/Photo by M. Rymer; Box 5.1, 5.34(both), 5.35(both), 5.36A: Earthquake Engineering Research Center, University of CA, Berkeley.

Chapter 6

Opener: © Joel W. Rogers/Corbis; 6.1A: Photograph by R. Janda, USGS, Cascade Volcano Observatory/USGS; 6.4: USGS/Photo by J.D. Griggs; 6.6A,B, 6.7A: USGS; 6.7B: Michael Clynne, USGS and USDA; 6.7C: USGS/Long Valley Observatory/Photo by R. Bailey; 6.9A: USGS/Photo by G.E. Ulrich; 6.9B: USGS; 6.10B: © Steve Reynolds; 6.11B: © Image Plan/Corbis RF; 6.13B: © Doug Sherman/Geofile; 6.14A,B: USGS/Photo by J.D. Griggs; 6.15, 6.16: USGS/Lyn Topinka; Box 6.3(Both): USGS; Box 6.4A: USGS; Box 6.4B: USGS/Mike Doukas; 6.19: R. Hoblitt, USGS; 6.20: Photo by Underwood & Underwood, Courtesy of the Library of Congress; 6.21A: NASA; 6.22: Photograph by M. Mangan, U.S. Geological Survey, 1997; 6.24, 6.25: USGS/Steven Brantley; 6.26: USGS/Photos by J. Vigil; 6.27A: 2002 MBARI (Monterey Bay Aquarium Research Institute); 6.27B: © J.B. Paduan; 6.28A: NASA; 6.28B: © Jonathan Blair/Corbis; 6.30: © Jim Reichard.

Chapter 7

Opener: © Bruce Chambers/Corbis; 7.1: British Columbia Ministry of Forests and Range/Kevin Turner; 7.2: USGS; 7.3(both): Montana Dept of Transportation; 7.5, 7.6: © Jim Reichard; 7.8(top): Courtesy of Myron McKee/River of Life Farm; 7.8(bottom): Canadian Landscapes: http://gsc.nrcan.gc.ca/landscapes/details_e.php/photoID = 899, July 2001. Reproduced with the permission of Natural Resources Canada 2009, courtesy of the Geological Survey of Canada (Photo 2002-585 by Réjean Couture); 7.11: © Jim Reichard; 7.13: © Rick Guthrie; 7.15: Photo courtesy of Clatskanie, Oregon, PUD employee Kerry Kallunki; Box 7.1(main); Photo by Robert A. Larson; Box 7.1(top): USGS/Photo by R.L. Schuster; 7.17A: © Mark Rikkers; 7.17B: © Toby Weed; 7.18: Gary Greene 2000 MBARI (Monterey Bay Aquarium Research Institute); 7.20: USGS; 7.23A: California Department of Transportation; 7.23B: © Jim Reichard; 7.24: California Department of Transportation; 7.25: © Doug Sherman/Geofile; 7.26: © King's Material, Inc.; 7.27: © Jim Reichard; 7.28A,B: Profile Products; 7.31: © Doug Sherman/Geofile; 7.32: © Jim Reichard.

Chapter 8

Opener: © Doug Sherman/Geofile; 8.1: Larry Robinson USGS; 8.9A,B: © Jim Reichard; 8.10A, 8.12A,B: © The McGraw-Hill Companies, Inc./John A. Karachewski, photographer; 8.13B: Image courtesy Liam Gumley, Space Science and Engineering Center, University of Wisconsin-Madison and the MODIS science team; 8.14B: © Doug Sherman/Geofile; 8.20: Carrie Olheiser, The National Operational Hydrologic Remote Sensing Center (NOHRSC)/NOAA; 8.21A: National Geophysical Data Center, NESDIS, NOAA, Big Thompson Drainage from USGS NED, J. Varner CIRES; 8.21B1, B2: USGS; 8.22: National Park Service/Johnstown Are Heritage Assoc; 8.24: NASA; 8.25B: Paul Ankcorn USGS; 8.27: © Jim Reichard; 8.31B: Jocelyn Augustino/FEMA; Box 8.1: Courtesy of the University Libraries, The University of Texas at Austin; Box 8.2: US Army Corps of Engineers; Box 8.5A: NASA; 8.5B: Jocelyn Augustino/FEMA; 8.32: © Jim Reichard; 8.33: Robert Criss, Washington Univ., St. Louis; 8.34: © Skip Metheny; 8.35: © Jim Reichard; 8.37: The National Operational Hydrologic Remote Sensing Center (NOHRSC), NOAA.

Chapter 9

Opener: Program for the Study of Developed Shorelines, Western Carolina University; 9.1B: NASA; 9.2(left): NOAA/Photo by Rear Admiral Harley D. Nygren, Corps (ret.); 9.2(right): NOAA/Photo by Ralph F. Kresge; 9.8(left): © Michael J. Walsh; 9.8(right): States of Alderney Photo Library/Ilona Soane-Sands, photographer; 9.10: USGS; 9.12B: NOAA and Cooperative Institute for Meteorological Satellite Studies/University of Wisconsin-Madison; 9.13B, 9.14A,B: NOAA; 9.17B: USGS; 9.18: © AP Wide World Photos; 9.19: NOAA; 9.22: Dave Gatley/FEMA; 9.24A: U.S. Navy photo by Photographer's Mate 3rd Class Tyler J. Clements (RELEASED); 9.24B: USGS/Guy Gelfenbaum; 9.25(both): © Digital Globe; 9.26B: Wendy Carey, Delaware Sea Grant; 9.27: USGS; Box 9.2A: Getty RF; 9.2B: USGS; 9.28: Weiss and Overpeck, The University of Arizona; 9.29: Courtesy of Olsen Associates, Inc.; 9.30B(both): © Jim Reichard; 9.31: © Doug Sherman/Geofile; 9.32: © Wallace C Smith Jr.; 9.34: Courtesy of Olsen Associates, Inc.

Chapter 10

Opener: USDA-NRCS; 10.1: USDA-NRCS; 10.2: Howard Woodward/Plant Science Department, South Dakota State University; 10.3A,B: © Jim Reichard; 10.5A,B: Photo by Jim Fortner, USDA-NRCS; 10.12B: Jukka Käyhkö University of Turku; 10.15C: © Goodshoot/Fotosearch; 10.16: Paul McDaniel /University of Idaho; 10.17B(top): Colorado Geological Survey/Photo by Dave Noe; 10.17B(bottom): P. Camp, USDA-NRCS; 10.20B: USDA-NRCS; 10.22: Photo courtesy of Cynthia Stiles, USDA-NRCS; 10.23: USDA-NRCS; 10.25A: Lynn Betts, USDA-NRCS; 10.25B, 10.26, 10.27, 10.28A,B: USDA-NRCS; 10.29: © Doug Sherman/Geofile; 10.30A,B: USDA-NRCS; 10.31: © PhotoLink/Getty RF; 10.32: USDA; 10.33A: © Jim Reichard; 10.34A: USDA-NRCS; 10.36: Joe Moore USDA-NRCS.

Chapter 11

Opener: Jeff Vanuga, USDA NRCS; 11.3: PRISM Climate Group, Oregon State University; 11.6A: Photo Deutsches Museum; 11.7: © Jim Wark/Airphoto; 11.11: A. Rivera, Geological Survey of Canada; 11.14A: © The McGraw-Hill Companies, Inc./John A. Karachewski, photographer; 11.14B,C, 11.15: © Jim Reichard; 11.19A: © Richard T. Nowitz/Corbis; Box 11.1: Ground Water Atlas of Colorado (2003): Colorado Geological Survey Special Publication 53, modified after U.S. Geological Survey Open File Report 99-267; Box 11.2A(bottom): © Corbis RF; Box 11.2A(top): © Getty RF; Box 11.2B: NASA; 11.20A(main): P.W. Bush and R.H. Johnson (1988)/U.S. Geological Survey. Professional Paper 1403-C; 11.20B,C: Krause, R.E., and R.B. Randolph (1989), U.S. Geological Survey Professional Paper 1403-D; 11.22B: Alan V. Morgan, Earth and Environmental Sciences, University of Waterloo; 11.24: IDE Technologies; 11.25B: Rainwater Services, LLC; 11.26A: © David Winger Landscape Photography; 11.26B: USDA/Photo by Lynn Betts; 11.27A: © Vol. 39/Getty RF; 11.27B: © Corbis RF; 11.28: USDA/NRCS, photo by Lynn Betts.

Chapter 12

Opener: © Lester Lefkowitz/Getty; 12.1(main): Ohio Department of Natural Resources; 12.1(inset): Salt Institute; 12.2B: U.S. Navy photo by Journalist 1st Class Mark H. Overstreet (RELEASED); 12.5B: © David Waters, Oxford University; 12.7: © Fernando Tornos, Instituto Geológico y Minero de España; 12.8B: NOAA; 12.9A: USGS/Photo by P.D. Rowley; 12.9B: © 2005 Alluvium Construction; Box 12.1: USGS Microbeam Laboratory; Box 12.2: © Hulton-Deutsch/Corbis; Box 12.3: © Donald Johnson/Corbis; 12.10B, 12.13A: © Jim Reichard; 12.13B: © Jim Wark/Airphoto; 12.16(left): R.D. Ramsey, Utah State Univ., Remote Sensing and GIS Lab; 12.16(right): Bureau of Land Management; 12.17: © Jim Reichard; 12.18(left): R.D. Ramsey, Utah State Univ.; 12.18(right): © Vol. 39/Getty RF; 12.19A: USGS; 12.19B: West Virginia Office of Miner's Health, Safety, and Training; 12.20: © Digital Vision/Punchstock RF; 12.21: California Geological Survey Library; 12.22B: Southern Peru Earthquake Geo-Engineering Extreme Event Reconnaissance (GEER) team; 12.22A(top): Courtesy of Barrick Gold Corporation; 12.22A(left): © Doug Sherman.Geofile; 12.26(left): © Jim Wark/Airphoto; 12.26(right): EARTHWORKS; 12.27A: Philip L. Verplanck, U.S. Geological Survey; 12.27B: © Jim Wark/Airphoto; 12.29: IntraSearch Inc.; 12.30: USGS/Photo by C.R. Dunrud; 12.31: Doug Paddock.

Chapter 13

Opener: © Vol. 7/Getty RF; 13.2(right): © Photolink/Getty RF; 13.2(left): © Vol. 51/Getty RF; 13.4: Texas Energy Museum, Beaumont, TX; 13.5: Fred Rich, Georgia Southern Univ.; 13.8: USGS/Photo by C.R. Dunrud; 13.10(both): Don Alexander/Ohio Valley Environmental Coalition; 13.11: © iStock; 13.13: IntraSearch; 13.16A, 13.17B: BP Oil; 13.18: NOAA; 13.19: Exxon Valdez Oil Spill Trustee Council; Box 13.1(top): ANWR base map/U.S. Fish and Wildlife; Box 13.1(bottom): U.S. Fish and Wildlife; 13.28: BP Oil; 13.31(top): By A.F. Alhajii, reprinted with permission World Oil, September 2007; 13.31(bottom): BP Oil; 13.34: © Richard Michael Pruitt/Dallas Morning News/Corbis Sygma.

Chapter 14

Opener: www.wagner-solar.com; 14.1: © Corbis; 14.3B: Image courtesy of Syncrude Canada Ltd; 14.3C, 14.5: Shell Oil, Inc.; 14.6(both): MARUM, Bremen University, Germany; 14.8: © ThinkStock/Superstock RF; 14.9: © Jim Wark/Airphoto; 14.12: © Doug Sherman/Geofile; 14.14: Urban and Northern Options Sustainable Education Handbook; 14.15: Solahart/ESTIF; 14.16: Conergy AG; 14.17: NREL, National Renewable Energy Lab of the U.S. Dept. of Energy;

14.18(left): © Misty Tosh; 14.18(right): Courtesy of Brunton/Fiskars Outdoor - Americas; 14.19: © Sarah Leen/National Geographic Image Collection; 14.20(right): © Jim Reichard; 14.20(bottom): U.S. Dept. of Energy National Renewable Energy Lab; 14.21A: © Gary Smith/Age Fotostock; 14.21B: Shawn Lessord, Rochester Solar Technologies; 14.22A: © 2004 AAPG, reprinted by permission of the AAPG whose permission is required for further use; 14.22B: Courtesy of Chris Moll/Radio 101 Thermography; 14.24: © Javier Larrea/Age Fotostock; 14.29B,C: Marine Current Turbines LTD; 14.30B: Filmsight Productions, LLC; 14.32: © TopFoto/The Image Works.

Chapter 15

Opener: © Jason Hawkes/The Image Bank/Getty Images; 15.1: © Howard Davies/Corbis; 15.2A: Tulsa Historical Society; 15.2B: The Cleveland Press Collection/Cleveland State University Library; 15.3: © Colin Garratt/Corbis; 15.4: © Digital Vision/Getty Images; 15.5(left): © Digital Vision/PunchStock; 15.5(right): © Jim Reichard; 15.6A: United States EPA; 15.9: U.S. Fish and Wildlife Service /Photo by Luther Goldman; 15.10, 15.11: © Doug Sherman/Geofile; 15.16: © Digital Vision/PunchStock RF; 15.17: © Jim Reichard; 15.18: Government of Saskatchewan; 15.19, 15.21: Illinois Environmental Protection Agency; 15.22: © Sylvia Smith-Reynolds; 15.24: © Ceatas/PunchStock RF; 15.26: © Digital Vision/PunchStock RF; 15.27A: National Water Quality Assessment Program, USGS, USGS Fact Sheet 2006-302B, By Robert J. Gilliom and Pixie A. Hamilton; 15.28: USDA; 15.29: Dr. Nancy Rabalais, Louisiana Universities Marine Consortium; Box 15.1A,B: U.S. Dept. of Energy; 15.31: © Argus/Peter Arnold, Inc; 15.32B: National Atmospheric Deposition Program (NRSP-3). 2009. NADP Program Office, Illinois State Water Survey; 15.33: © Dr. Parvinder Sethi; 15.34: © Sally A. Morgan/Corbis.

Chapter 16

Opener: © James Balog/Getty Images; 16.1A: Rhone 1859 etching from photograph courtesy Stephan Wagner; 16.1B: Rhone 2001 photograph © Gary Braasch; 16.4: Image Science and Analysis Laboratory, NASA-Johnson Space Center; 16.8A: National Snow and Ice Data Center, W.O. Field, B. F.; 16.8B: National Snow and Ice Data Center, B.F. Molnia; 16.16: © Henri D. Grissino-Mayer, Department of Geography, The University of Tennessee; 16.17: Science Images/CSIRO Publishing; 16.25: Reto Stockli, NASA Earth Observatory; 16.26: USDA-NRCS, Soil Survey Division; 16.27A: Glacial National Park Archive; 16.27B: Karen Holzer/USGS; 16.29A: © James Balog/Corbis; 16.31: NASA/JPL/NGA; 16.32A(both): NASA/Goddard Space Flight Center Scientific Visualization Studio; 16.37A: © Getty RF; 16.37B: © Corbis RF.

Text and Line Art Credits

Chapter 1

B1.1: Modified after University of Texas Library, http://www.lib.utexas.edu/maps/australia.html; 1.16A: Data from the World Commission on Dams; 1.20: Data from US Census Bureau, http://www.census.gov/ipc/www/idb/worldpopinfo.php; 1.22, 1.23: Data from Population Reference Bureau, http://www.prb.org/

Chapter 2

2.12: Modified after Lutgens and Tarbuck, 2006; 2.19: Modified after NASA and Space Telescope Science Institute, http://hubblesite.org/newscenter/archive/releases/2003/18/image/c; 2.20A: Modified after NASA, http://www.nasa.gov/centers/goddard/news/topstory/2004/0801frozenworlds.html; 2.25: After Raup and Sepkoski, 1982, and Wilson, 1992; 2.28: Data from Earth Impact Data Base, http://www.unb.ca/passc/ImpactDatabase/; 2.31: Modified after B. Walsh, University of Arizona, http://nitro.biosci.arizona.edu/courses/EEB105/lectures/impact/impact.html

Chapter 3

3.3A: Data from US Geological Survey, http://pubs.usgs.gov/gip/dynamic/inside.html; 3.4: Modified after Tarbuck and Lutgens, 2002; 3.5: Modified after TutorVista, http://www.tutorvista.com/; 3.13: Modified after Blatt, 1997

Chapter 4

4.4: Data from US Geological Survey, http://pubs.usgs.gov/gip/dynamic/inside.html; 4.7A,B: After US Geological Survey, http://pubs.usgs.gov/gip/dynamic/historical.html; 4.11: After US Geological Survey, http://pubs.usgs.gov/gip/dynamic/developing.html; 4.13: Data from NASA, http://denali.gsfc.nasa.gov/dtam/seismic/; 4.14: Modified after Hamblin and Christiansen, 1998; 4.15: After US Geological Survey, http://pubs.usgs.gov/gip/dynamic/slabs.html; 4.16: Modified after US Geological Survey, http://pubs.usgs.gov/publications/text/Vigil.html; 4.19: Modified after NASA, http://denali.gsfc.nasa.gov/dtam/gtam/; 4.26: Modified after Tarbuck and Lutgens, 2002; 4.27: Modified after US Geological Survey, http://pubs.usgs.gov/gip/dynamic/understanding.html

Chapter 5

5.2A: Modified after Australian Government Department of Defense, http://www.defence.gov.au/defencemagazine/editions/20050201/features/specialfeatures1b.htm; 5.4A-D: Modified after Blatt, 1997; 5.6B: Modified after UPSeis, http://www.geo.mtu.edu/UPSeis/waves.html; 5.9: Data from US Geological Survey, http://quake.usgs.gov/research/deformation/modeling/papers/lake_elsman.html; 5.10: Data from Natural Resources Canada, http://earthquakescanada.nrcan.gc.ca/histor/20th-eme/1925/19250301-eng.php; 5.12: Data from US Geological Survey, http://earthquake.usgs.gov/regional/states/seismicity/; 5.13: Data from California Department of Conservation, http://www.consrv.ca.gov/cgs/rghm/psha/ofr9608/index.htm; 5.14: Modified after US Geological Survey, http://www2.nature.nps.gov/geology/usgsnps/province/cascade2.html; 5.15A: Data from US Geological Survey, http://earthquake.usgs.gov/regional/states/events/1886_09_01.php; 5.16: Data from US Geological Survey, Fact Sheet 0007-02, http://pubs.usgs.gov/fs/fs-0007-02/; 5.18: Modified after Pipkin, Trent, and Hazlett, 2005; 5.25: Modified after T. Pratt, US Geological Survey; 5.26A: Modified after USGS Fact Sheet 131-02, http://pubs.usgs.gov/fs/fs-131-02-p3.html; 5.30: After US Geological Survey Fact Sheet 150-00, http://pubs.usgs.gov/fs/fs150-00/; 5.31: Data from US Geological Survey Open-File Report 99-36, http://earthquake.usgs.gov/research/hazmaps/products_data/Alaska/aks.php; 5.32: Data from US Geological Survey Open-File Report 2008-1128, http://earthquake.usgs.gov/research/hazmaps/index.php; 5.33A-B: Data from US Geological Survey Fact Sheet 152-99, http://geopubs.wr.usgs.gov/open-file/of99-517/; 5.34: Modified after NASA, http://www.nasa.gov/worldbook/earthquake_worldbook.html; 5.36B: Modified after Seismic Science at the Epicenter, http://www.exploratorium.edu/faultline/activezone/slides/retrofit-slide.html

Chapter 6

6.1: After US Geological Survey Circular 1073, http://vulcan.wr.usgs.gov/Vhp/C1073/hazard_maps_risk.html; 6.2: Data from US Geological Survey, http://vulcan.wr.usgs.gov/Graphics/framework2.html; 6.3: Data from US Geological Survey, http://volcanoes.usgs.gov/images/pglossary/VolRocks.php; 6.8: After US Geological Survey, http://vulcan.wr.usgs.gov/Volcanoes/PacificNW/Maps/pacificNW_volcanics.html; 6.12: Data from US Geological Survey, http://vulcan.wr.usgs.gov/Imgs/Gif/VolcanoTypes/shield_vs_composite.gif; B6.1: After NOAA and US Geological Survey; B6.2A: Data from Pierce and Morgan, 1992, and GSA Memoir 179; B6.2B: Modified after US Geological Survey Fact Sheet 100-03, 2004, http://pubs.usgs.gov/fs/fs100-03/; 6.13: After H. Williams, 1962, *The Ancient Volcanoes of Oregon;* 6.17: Data from US Geological Survey Fact Sheet 165-97, http://pubs.usgs.gov/fs/1997/fs165-97/; 6.18: Modified after US Geological Survey Fact Sheet 058-00, http://pubs.usgs.gov/fs/2000/fs058-00/; 6.21B: Modified after US Geological Survey Fact Sheet 030-97, http://pubs.usgs.gov/fs/fs030-97/; 6.23: Data from US Geological Survey, http://erg.usgs.gov/isb/pubs/teachers-packets/volcanoes/lesson3/lesson3.html; 6.25: After US Geological Survey Open File Report 95-642, http://vulcan.wr.usgs.gov/Imgs/Gif/Rainier/OFR95-642/rainier_mudflows.gif

Chapter 7

7.10, 7.14: Modified after US Geological Survey Fact Sheet 2004-3072, http://pubs.usgs.gov/fs/2004/3072/fs-2004-3072.html; 7.12, 7.15: Modified after Government of British Columbia, http://www.em.gov.bc.ca/Mining/Geolsurv/Surficial/landslid/ls2.htm; B7.2: Data from US Geological Survey Open File Report 2005-1067, http://pubs.usgs.gov/of/2005/1067/508of05-1067.html#conchitafig04; 7.22: Data from US Geological Survey Fact Sheet 110-02, http://pubs.usgs.gov/fs/fs-110-02/

Chapter 8

B8.3: Data from US Army Corps of Engineers; 8.6: After National Atlas of the United States/DOI, http://www.nationalatlas.gov/articles/geology/a_continentalDiv.html; 8.7: After US Army Corps of Engineers, http://www.mvn.usace.army.mil/pao/bro/misstrib.htm; 8.19: Data from US Geological Survey, http://waterdata.usgs.gov/nc/nwis/rt; 8.28(right): Data from US Geological Survey Fact Sheet 076-03, http://pubs.usgs.gov/fs/fs07603/; 8.30B: Modified after Thompson and Turk, 2003; 8.36: Modified from City of Tulsa Public Works, Flood Control, Stormwater Management Program; 8.38: Data from US Geological Survey Fact Sheet 024-00

Chapter 9

B9.1: Data from The Johns Hopkins University Applied Physics Laboratory, http://meto.umd.edu/~stevenb/hurr/05/katrina/index.html; 9.1, 9.19, 9.20A, 9.21, 9.23: Data from NOAA; 9.3: Data from US Geological Survey Open File Report 96-000; 9.10(right): Modified after NOAA, http://www3.csc.noaa.gov/beachnourishment/html/geo/barrier.htm and Blatt et al., 1980, *Petrology of Sedimentary Rocks;* 9.12A: Modified after University Corporation for Atmospheric Research, http://meted.ucar.edu/hurrican/strike/text/htc_t5htm; 9.13A: Modified after NOAA, http://www.srh.noaa.gov/srh/jetstream/tropics/tc_structure.htm; 9.17A: Modified after NOAA, http://www.nhc.noaa.gov/HAW2/english/storm_surge.shtml; 9.20B: Data from US Geological Survey; 9.25A: Modified after Australian Government Department of Defense, http://www.defence.gov.au/defencemagazine/editions/20050201/features/specialfeatures1b.htm; 9.29A: Modified after NOAA Nautical Chart Map 11513 (1997); 9.30A: After Goudie, 2000

Chapter 10

10.6: After US Department of Agriculture, http://soils.usda.gov/technical/manual/print_version/chapter3.html; 10.13: Data from US Department of Agriculture, http://soils.usda.gov/use/worldsoils/mapindex/order.html; 10.17B: After US Geological Survey, 1989, Swelling Clays Map of the Conterminous United States; 10.20A: After Schreiner and Brown, 1938; 10.24: Data from Wolman, 1967

Chapter 11

11.2: Data from US Geological Survey; 11.4: Data from US Geological Survey Circular 1268, http://pubs.usgs.gov/circ/2004/circ1268/index.html#toc; 11.5: Data from American Water Works Association, http://www.drinktap.org/consumerdnn/Default.aspx?tabid=85; 11.6: Modified after Central Basin Municipal Water District, http://www.centralbasin.org/waterSupplySystem.html; 11.9: After US Geological

Survey Water Supply Paper 2220, 1983; 11.18: Data from Georgia Southern University well field; 11.19: Water Resources Research Center, University of Arizona, http://ag.arizona.edu/AZWATER/publications/sustainability/report_html/cover.html

Chapter 12

12.12: Data from H. Winchell, American Mineralogist, v 30, 1945, http://www.minsocam.org/MSA/collectors_corner/arc/knoop.htm; 12.3: After US Geological Survey Circular 831, 1980; 12.4: Data from BRC DiamondCore, http://www.brc-diamondcore.com/s/Technical.asp; 12.8: Data from NOAA, http://www.pmel.noaa.gov/vents/chemistry/fluid.html; 12.12: Modified after Berry and Mason, 1959; 12.14: After US Geological Survey Professional Paper 820, 1973; 12.15: After The Salt Institute, http://www.saltinstitute.org/News-events-media; 12.20: After Robertson GeoConsultants, http://www.robertsongeoconsultants.com/index.php; 12.23: Data from Natural Resources Canada, http://gdr.nrcan.gc.ca/minres/index_e.php; 12.24: Data from US Geological Survey Mineral Commodity Summaries, 2008; 12.25: Data from US Geological Survey Circular 1221, 2002, http://pubs.usgs.gov/circ/2002/c1221/; 12.29: After US Geological Survey Open File Report 95-23, http://pubs.usgs.gov/of/1995/ofr-95-0023/summit.htm

Chapter 13

13.1: Modified after Blatt, 1997; 13.3: Data from US Department of Energy, http://www.eia.doe.gov/emeu/aer/eh/frame.html; 13.6: Data from US Geological Survey Professional Paper 820, 1973; 13.6(inset): After Kentucky Geological Survey, http://www.uky.edu/KGS/coal/coalkinds.htm; 13.7: After American Coal Foundation, http://www.teachcoal.org/aboutcoal/articles/coalreserves.html; 13.10: After *The Washington Post,* 2008, http://www.washingtonpost.com/wp-dyn/content/graphic/2008/04/19/GR2008041900169.html; 13.11: After Chemistryland, http://www.chemistryland.com/ElementarySchool/BuildingBlocks/BuildingOrganic.htm; 13.12: After Boreham and Powell, 1993; 13.15: After Baker Hughes, http://www.bakerhughesdirect.com; 13.19: Data from State of Alaska, http://www.evostc.state.ak.us/facts/spillmap.cfm; 13.20: Data from British Petroleum, 2008 Statistical Review of World Energy, and Central Intelligence Agency, 2008 World Factbook; 13.21, 13.22, 13.33, 13.34: US Department of Energy, 2007 Annual Energy Review; 13.23, 13.24: Data from British Petroleum, 2008 Statistical Review of World Energy; 13.25: Data from InflationData, http://www.inflationdata.com/inflation/Inflation_Rate/Historical_Oil_Prices_Table.asp; 13.26A: After W. Youngquist, *Geotimes,* July 1998; 13.26B: Data from US Department of Energy, Strategic Significance of America's Oil Shale Resource, 2004; 13.27: After M. K. Hubbert, 1962; 13.29: Data from G. Tverbergy, Association for the Study of Peak Oil, 2008; 13.30: Data from K. Aleklett, Association for the Study of Peak Oil, 2005; 13.32: Data from S. Shelton, Strategic Energy Institute, Georgia Institute of Technology, and C. Campbell, Association for the Study of Peak Oil

Chapter 14

14.3: After Alberta Geological Survey, http://www.ags.gov.ab.ca/energy/oilsands/alberta_oil_sands.html; 14.4: US Department of Energy, OST-05-04, http://www.netl.doe.gov/technologies/oil-gas/Petroleum/projects/EP/ImprovedRec/OST-05-04.html; 14.7: Data from Renewable Fuels Association, http://www.ethanolrfa.org/industry/statistics/; 14.9: After Tennessee Valley Authority, http://www.tva.gov/heritage/mountaintop/index.htm; 14.11: Modified after US Nuclear Regulatory Commission, http://www.nrc.gov/reading-rm/basic-ref/students/reactors.html; 14.13A: After World Nuclear Association, http://www.world-nuclear.org/info/inf41.html; 14.13B: Data from US Department of Energy, http://www.eia.doe.gov/emeu/aer/eh/frame.html; 14.15: Modified after European Solar Thermal Industry Federation, http://www.estif.org/126.0.html; 14.16: Modified after Rowan University

Clean Energy Program, http://www.rowan.edu/colleges/engineering/clinics/cleanenergy/case_study_summaries/solar.htm; 14.24: After US Geological Survey, http://pubs.usgs.gov/gip/dynamic/understanding.html; 14.27A: After Carbon Zero Planet, http://www.carbonzeroplanet.org/renewables/wave-tidal-oceanthermal.php; 14.27B: After Institute of Ocean Energy, Saga University, http://www.ioes.saga-u.ac.jp/english/news/index.html; 14.29A: After Marine Current Turbines, http://www.marineturbines.com/18/projects/19/seagen/; 14.30A: Modified after Australian Government, http://www.environment.gov.au/index.html; 14.33: Global Wind Energy Council 2007 Report, http://www.gwec.net/index.php?id=90

Chapter 15

15.8, 15.15: Data from US Environmental Protection Agency, http://www.epa.gov/epawaste/basic-solid.htm; 15.12, 15.14: Data from US Environmental Protection Agency, http://www.epa.gov/osw/nonhaz/municipal/msw99.htm; 15.13: After US Environmental Protection Agency, http://www.epa.gov/epawaste/basic-solid.htm; 15.16: Data from US Environmental Protection Agency, http://www.epa.gov/epawaste/conserve/materials/ecycling/manage.htm; 15.27B: After US Geological Survey Circular 1225, http://pubs.usgs.gov/circ/circ1225/html/intensively.html; 15.29: After US Environmental Protection Agency, http://www.epa.gov/msbasin/marb.htm; 15.30: Modified after Blatt, 1997; 15.32A: Data from Platts, http://www.platts.com/ProductServicesHome.aspx; 15.35: Modified after US Environmental Protection Agency, http://www.epa.gov/apti/bces/module6/sulfur/control/control.htm; 15.37: Modified after US Environmental Protection Agency, http://www.epa.gov/mercury/control_emissions/emissions.htm

Chapter 16

16.2: Data from J. Hansen, M. Sato, R. Ruedy, K. Lo, D. Lea, and M. Medina-Elizade, NASA Goddard Institute for Space Studies; 16.3, 16.9, 16.14, 16.15, 16.21: Data from Working Group I, Fourth Assessment Report of the United Nations IPCC, 2007; 16.5: Modified from Howard, 1959, and R. Goody and G. Robinson, 1951; 16.6: After NASA, http://earthobservatory.nasa.gov/Features/Milankovitch/milankovitch_2.php; 16.10: Data from Scripps Institution of Oceanography, and NOAA, Earth System Research Laboratory; 16.12: Modified after NOAA, http://www.pmel.noaa.gov/tao/elnino/el-nino-story.html; 16.13: Modified after United Nations IPCC, http://www.ipcc-data.org/ddc_gcm_guide.html; 16.18: Carbon dioxide data from Barnola, Raynaud, Lorius, and Barkov, 2003, http://cdiac.ornl.gov/trends/co2/; temperature data from Petit, Raynaud, Lorius, Delaygue, Jouzel, Barkov, and Kotlyakov, 2000, http://cdiac.ornl.gov/trends/temp/vostok/jouz_tem.htm; 16.19, 16.20: Data from R. Alley, Pennsylvania State University, http://www.ncdc.noaa.gov/paleo/pubs/alley2000/alley2000.html; 16.22: Data from NOAA, National Climatic Data Center, http://www.ncdc.noaa.gov/oa/climate/research/anomalies/index.html; 16.23: Carbon dioxide data from D. Lüthi et al., 2008; methane data from L. Loulergue et al., 2008; and temperature data from J. Jouzel et al., 2007, http://www.ncdc.noaa.gov/paleo/icecore/antarctica/domec/domec_epica_data.html; 16.24: Modified after NASA, http://rst.gsfc.nasa.gov/Sect14/Sect14_1c.html; 16.28: Data from K. Lambeck and J. Chappell, 2002, and K. Lambeck, Y. Yokoyama, and A. Purcell, 2002; 16.29B: Modified after Global Warming, http://www.global-greenhouse-warming.com/index.html; 16.30(top): Modified after US Geological Survey Fact Sheet 2005-3055, http://pubs.usgs.gov/fs/2005/3055/; 16.30(bottom): Modified after NASA, http://www.agu.org/sci_soc/articles/bind.html; 16.32: Data from NASA, http://svs.gsfc.nasa.gov/vis/a000000/a003500/a003563/; 16.33: Data from Woods Hole Oceanographic Institution, http://www.whoi.edu/oceanus/viewArticle.do?id=17726; 16.34: US Department of Energy, 2006 Annual Energy Review; 16.35: Modified after Carbon Mitigation Initiative, Princeton University, http://cmi.princeton.edu/wedges/intro.php

Index